T0297204

Seaweed Ecology and Physiology

Second Edition

In coastal seas, from the tropics to the poles, seaweeds supply the energy required to support diverse coastal marine life and provide habitat for invertebrates and fish. Retaining the highly successful approach and structure of the first edition, this is a synthesis of the role of seaweeds in underpinning the functioning of coastal ecosystems worldwide. It has been fully updated to cover the major developments of the past 20 years, including current research on the endosymbiotic origin of algae, molecular biology including "omics", chemical ecology, invasive seaweeds, photobiology, and stress physiology.

In addition to exploring the processes by which seaweeds, as individuals and communities, interact with their biotic and abiotic environment, the book presents exciting new research on how seaweeds respond to local and global environmental change. It remains an invaluable resource for students and provides an entry into the scientific literature on a wide range of topics.

Catriona L. Hurd is an Associate Professor in the Institute for Marine and Antarctic Studies, University of Tasmania, Australia. She is known for her work on water motion, photosynthetic and nutrient physiology along environmental gradients in the intertidal and subtidal zones, and most recently on ocean acidification.

Paul J. Harrison is Professor Emeritus in the Department of Earth and Ocean Sciences at the University of British Columbia, Vancouver, Canada. He is an expert in nutrient ecophysiology of seaweeds and phytoplankton, with over 40 years of research experience.

Kai Bischof is Head of the Department of Marine Botany at the University of Bremen, Germany. His research covers all aspects of seaweed ecophysiology, with a focus on the field of light acclimation and oxidative stress management.

Christopher S. Lobban is Professor of Biology in the Division of Natural Sciences at the University of Guam, Mangilao, USA. He has over 40 years of experience working with marine algae, including 25 years on coral reefs in Guam. He is currently investigating biodiversity of coral reef diatoms.

Seaweed Ecology and Physiology

Second Edition

Catriona L. Hurd

Institute for Marine and Antarctic Studies, University of Tasmania, Australia

Paul J. Harrison

Department of Earth and Ocean Sciences, University of British Columbia, Vancouver, Canada

Kai Bischof

Department of Marine Botany, University of Bremen, Germany

Christopher S. Lobban

Division of Natural Sciences, University of Guam, Mangilao, USA

CAMBRIDGE
UNIVERSITY PRESS

Shaftesbury Road, Cambridge CB2 8EA, United Kingdom

One Liberty Plaza, 20th Floor, New York, NY 10006, USA

477 Williamstown Road, Port Melbourne, VIC 3207, Australia

314–321, 3rd Floor, Plot 3, Splendor Forum, Jasola District Centre, New Delhi – 110025, India

103 Penang Road, #05–06/07, Visioncrest Commercial, Singapore 238467

Cambridge University Press is part of Cambridge University Press & Assessment, a department of the University of Cambridge.

We share the University's mission to contribute to society through the pursuit of education, learning and research at the highest international levels of excellence.

www.cambridge.org
Information on this title: www.cambridge.org/9780521145954

First edition © Cambridge University Press & Assessment 1994
Second edition © C. L. Hurd, P. J. Harrison, K. Bischof and C. S. Lobban 2014

First published 1994
Second Edition 2014
4th printing 2020

A catalogue record for this publication is available from the British Library

Library of Congress Cataloging-in-Publication data
Hurd, Catriona L.
Seaweed ecology and physiology / Catriona L. Hurd, Institute for Marine and Antarctic Studies, University of Tasmania, Australia, Paul J. Harrison, Department of Earth and Ocean Sciences, University of British Columbia, Vancouver, Canada, Kai Bischof, Department of Marine Botany, University of Bremen, Germany, Christopher S. Lobban, Division of Natural Sciences, University of Guam, Mangilao, USA. – Second edition.
       pages cm
Originally published: Seaweed ecology and physiology / Christopher S. Lobban, Paul J. Harrison. Cambridge ; New York : Cambridge University Press, 1994.
Includes bibliographical references.
ISBN 978-0-521-14595-4 (Paperback)
1. Marine algae–Ecophysiology.   I. Harrison, Paul J. (Paul James), 1941–   II. Bischof, Kai.   III. Lobban, Christopher S.
IV. Title.
QK570.2.L64 2014
579.8´177–dc23   2013038001

ISBN   978-0-521-14595-4   Paperback

To our families and friends for their encouragement and inspiration, especially Philip and Phoebe, Victoria, Sandra and Finn, and Maria.

Contents

List of contributors   *page* xi
Preface   xiii

1   Seaweed thalli and cells   1
    1.1 Introduction: the algae and their
        environments   1
        1.1.1 Seaweeds   1
        1.1.2 Environmental-factor interactions   5
        1.1.3 Laboratory culture versus field
              experiments   8
    1.2 Seaweed morphology and anatomy   9
        1.2.1 Thallus construction   9
        1.2.2 The Littler functional-form model   11
        1.2.3 Unitary, clonal, and coalescing
              seaweeds, and modular
              construction   15
    1.3 Seaweed cells   17
        1.3.1 Cell walls   17
        1.3.2 Cytoplasmic organelles   21
        1.3.3 Cytoskeleton and flagella apparatus   25
        1.3.4 Cell growth   32
        1.3.5 Cell division   32
    1.4 Molecular biology and genetics   35
        1.4.1 Advances in seaweed molecular
              biology   35
        1.4.2 Seaweed genetics   42
        1.4.3 Nucleocytoplasmic interactions   45
    1.5 Synopsis   47

2   Life histories, reproduction, and
    morphogenesis   48
    2.1 Introduction   48
    2.2 Theme and variations   48

2.3 Environmental factors in life histories   54
   2.3.1 Seasonal anticipators and
       responders   55
   2.3.2 Temperature   56
   2.3.3 Light: photoperiod and wavelength   58
   2.3.4 Other factors   65
2.4 Fertilization biology   68
2.5 Settlement, attachment, and establishment   74
   2.5.1 Settlement   74
   2.5.2 Attachment   78
   2.5.3 Establishment   81
2.6 Thallus morphogenesis   83
   2.6.1 Cell differentiation   85
   2.6.2 Development of the adult form   88
   2.6.3 Seaweed growth substances   91
   2.6.4 Wound healing and regeneration   94
2.7 Synopsis   98

3 Seaweed communities   100
3.1 Intertidal zonation patterns   100
   3.1.1. Tides   100
   3.1.2 Vertical zonation on intertidal
       rocky shores   101
   3.1.3 Factors controlling vertical
       zonation   104
3.2 Subtidal zonation on rocky shores   111
3.3 Seaweed communities   113
   3.3.1 Tropical   113
   3.3.2 Temperate   116
   3.3.3 Polar   117
   3.3.4 Tide pools   118
   3.3.5 Estuaries and salt marshes   119
   3.3.6 Deep-water seaweeds   120
   3.3.7 Floating seaweeds   120
   3.3.8 Other seaweed habitats and
       communities   121
3.4 Invasive seaweeds   121
3.5 Community analysis   124
   3.5.1 Vegetation analysis   124
   3.5.2 Population dynamics   128
3.6 Synopsis   133

4 Biotic interactions   136
4.1 Foundation species and facilitation   136
4.2 Competition   138
   4.2.1 Interference competition   138

4.2.2 Epibionts and allelopathy   140
   4.2.3 Exploitative competition   145
4.3 Grazing   148
   4.3.1 Impact of grazing on community
       structure and zonation   148
   4.3.2 Seaweed-herbivore interactions   156
4.4 Chemical ecology of seaweed-herbivore
   interactions   164
   4.4.1 Bioactive chemicals   164
   4.4.2 Chemical defenses against grazers   166
4.5 Symbiosis   170
   4.5.1 Mutualistic relationships   171
   4.5.2 Seaweed endophytes   172
   4.5.3 Kleptoplasty   173
4.6 Synopsis   174

5 Light and photosynthesis   176
5.1 An overview of photosynthesis   176
5.2 Irradiance   179
   5.2.1 Measuring irradiance   179
   5.2.2 Light in the oceans   182
5.3 Light harvesting   188
   5.3.1 Plastids, pigments, and pigment-
       protein complexes   188
   5.3.2 Functional form in light trapping   194
   5.3.3 Photosynthesis at a range of
       irradiances   197
   5.3.4 Action spectra and testing the
       theory of complementary
       chromatic adaptation   200
5.4 Carbon fixation: the "dark reactions"
   of photosynthesis   202
   5.4.1 Inorganic carbon sources and
       uptake   202
   5.4.2 Photosynthetic pathways in
       seaweeds   206
   5.4.3 Light-independent carbon fixation   207
5.5 Seaweed polysaccharides   210
   5.5.1 Storage polymers   210
   5.5.2 Wall matrix polysaccharides   211
   5.5.3 Polysaccharide synthesis   215
5.6 Carbon translocation   215
5.7 Photosynthetic rates and primary
   production   218
   5.7.1 Measurement of photosynthesis
       and respiration   218

5.7.2 Intrinsic variation in
          photosynthesis   222
5.7.3 Carbon losses   225
5.7.4 Autecological models of
          productivity and carbon budgets   227
5.7.5 Ecological impact of seaweed
          productivity   229
5.8 Synopsis   235

6 Nutrients   238
6.1 Nutrient requirements   238
    6.1.1 Essential elements   238
    6.1.2 Essential organics: vitamins   239
    6.1.3 Limiting nutrients   240
6.2 Nutrient availability in seawater   240
6.3 Pathways and barriers to ion entry   242
    6.3.1 Membrane structure and ion
              movement   242
    6.3.2 Movement to and through the
              membrane   243
    6.3.3 Passive transport   243
    6.3.4 Facilitated diffusion   243
    6.3.5 Active transport   244
6.4 Nutrient-uptake kinetics   244
    6.4.1 Measurement of nutrient-uptake
              rates   246
    6.4.2 Factors affecting nutrient-uptake
              rates   249
6.5 Uptake, assimilation, incorporation,
      and metabolic roles   255
    6.5.1 Nitrogen   256
    6.5.2 Phosphorus   264
    6.5.3 Calcium and magnesium   267
    6.5.4 Sulfur   274
    6.5.5 Iron   274
    6.5.6 Trace elements   274
6.6 Long-distance transport (translocation)   275
6.7 Growth kinetics   276
    6.7.1 Measurement of growth kinetics   276
    6.7.2 Growth kinetic parameters and
              tissue nutrients   276
6.8 Effects of nutrient supply   279
    6.8.1 Surface-area:volume ratio and
              morphology   279
    6.8.2 Chemical composition and
              nutrient limitation   280

6.8.3 Nutrient storage and nutrient
          availability   282
6.8.4 Growth rate and distribution   284
6.8.5 Effects of nutrient enrichment
          on community interactions   286
6.9 Synopsis   290

7 Physico-chemical factors as environmental
    stressors in seaweed biology   294
7.1 What is stress?   294
7.2 Natural ranges of temperature and
      salinity   295
    7.2.1 Open coastal waters   295
    7.2.2 Estuaries and bays   296
    7.2.3 Intertidal zone   297
7.3 Temperature effects   300
    7.3.1 Chemical reaction rates   300
    7.3.2 Metabolic rates   303
    7.3.3 Growth optima   304
    7.3.4 Temperature tolerance   308
    7.3.5 Physiological adaptation to
              changes in temperature   312
    7.3.6 Temperature tolerance in polar
              seaweeds   315
    7.3.7 El Niño   319
    7.3.8 Temperature and geographic
              distribution   321
7.4 Biochemical and physiological effects
      of salinity   325
    7.4.1 Water potential   325
    7.4.2 Cell volume and osmotic
              control   326
    7.4.3 Effects of salinity changes on
              photosynthesis and growth   328
    7.4.4 Tolerance and acclimation to
              salinity   330
7.5 Further stresses related to water
      potential: desiccation and freezing   333
    7.5.1 Desiccation   333
    7.5.2 Freezing   336
7.6 Exposure to ultraviolet radiation   337
7.7 Variation in seawater pH and
      community-based impacts of ocean
      acidification   340
7.8 Interaction of stressors, oxidative
      stress, and cross adaptation   343

7.9   Physiological stress indicators   346
7.10 Synopsis   347

8   Water motion   349
  8.1 Water flow   349
      8.1.1 Currents   349
      8.1.2 Physical nature of waves   350
      8.1.3 Laminar and turbulent flow over
            surfaces   352
      8.1.4 Methods for measuring seawater
            flow and wave forces   355
  8.2 Water motion and biological processes   356
      8.2.1 Function and form in relation to
            resource acquisition   356
      8.2.2 Synchronization of gamete and
            spore release   359
  8.3 Wave-swept shores   360
      8.3.1 Biomechanical properties of
            seaweeds   360
      8.3.2 Wave action and other physical
            disturbances to populations   366
  8.4 Synopsis   372

9   Pollution   374
  9.1 Introduction   374
  9.2 General aspects of pollution   374
  9.3 Metals   378
      9.3.1 Sources and forms   378
      9.3.2 Adsorption, uptake, accumulation,
            and biomonitors   380
      9.3.3 Mechanisms involving tolerance
            to toxicity   383
      9.3.4 Effects of metals on algal metabolism 384
      9.3.5 Factors affecting metal toxicity   388
      9.3.6 Ecological aspects   389
  9.4 Oil   390
      9.4.1 Fate of oil in the ocean   392
      9.4.2 Effects of oil on algal metabolism,
            life cycles, and communities   394
      9.4.3 Ecological aspects   396
  9.5 Synthetic organic chemicals   398
  9.6 Eutrophication   400
      9.6.1 Sewage effluent and impacts of
            nutrient enrichment on algal
            communities   400

      9.6.2 Sewage effluent and toxicity   403
      9.6.3 Other anthropogenic nutrient
            sources   407
  9.7 Radioactivity   408
  9.8 Thermal pollution   409
  9.9 Synopsis   410

10 Seaweed mariculture   413
  10.1 Introduction   413
  10.2 *Pyropia/Porphyra* mariculture   415
        10.2.1 Biology   416
        10.2.2 Cultivation   417
        10.2.3 Problems in *Pyropia* culture   419
        10.2.4 Future trends   420
  10.3 *Saccharina/Laminaria* for food and
        alginates   421
        10.3.1 Cultivation   422
        10.3.2 Utilization and future
               prospects   423
  10.4 *Undaria* for food   424
        10.4.1 Cultivation   425
        10.4.2 Food products and future
               trends   425
  10.5 *Kappaphycus* and *Eucheuma* for
        carrageenans   426
        10.5.1 Biology   426
        10.5.2 Cultivation   426
        10.5.3 Production, uses, and future
               prospects   427
  10.6 *Gelidium* and *Gracilaria* for agar   428
        10.6.1 *Gelidium* production and
               products   428
        10.6.2 *Gracilaria* production and
               products   429
  10.7 Tank cultivation   430
  10.8 Offshore/open-ocean cultivation   431
  10.9 Integrated Multi-Trophic Aquaculture
        (IMTA) and biomitigation   431
  10.10 Other uses of seaweeds   434
  10.11 Seaweed biotechnology: current
        status and future prospects   435
  10.12 Synopsis   437

References   440
Subject Index   536

Contributors

Jonas Collén

UPMC Paris 06
UMR 7139 Marine Plants and Biomolecules
Roscoff
France

Glenn A. Hyndes

Centre for Ecosystems Management
School of Natural Sciences
Edith Cowan University
Joondalup, Western Australia
Australia

Bruce A. Menge

Oregon State University
Department of Zoology
Corvallis, Oregon, USA

John A. Raven

Division of Plant Sciences
University of Dundee at the James Hutton Institute
Invergowrie, Dundee, UK

Gary W. Saunders

Department of Biology
University of New Brunswick
Fredericton, New Brunswick
Canada

Britta Schaffelke

Australian Institute of Marine Science
Townsville, Queensland
Australia

Preface

There have been very significant advances in many areas of phycology since the last edition nearly 20 years ago. In particular, the advances in our understanding of the endosymbiotic origin of algal plastids, and molecular aspects and genetics, stand out. The wealth of new literature alone in all the areas has warranted adding two new co-authors, Catriona Hurd, who focuses on water motion and seaweed physiological ecology, especially in the southern hemisphere, and Kai Bischof, who is well known for his research on photobiology and stress physiology. Hence, the previous edition's chapter on "Temperature and salinity" has been expanded to include other environmental stressors such as UV radiation, ocean acidification, oxidative stress responses and the interactions between stressors.

Seaweed Ecology and Physiology is a textbook for senior undergraduates and a reference book for researchers. The rapid growth of knowledge in this field is both exciting and daunting. Our goal was to select papers that help put together a coherent story on a wide variety of ecological and physiological aspects. This book provides an entry to the literature, not a systematic literature review. With two of our co-authors having experience in the tropics and the temperate southern hemisphere, we have tried to avoid the typical temperate northern hemisphere bias.

The previous large introductory chapter on morphology, life histories, and morphogenesis has been divided into two chapters because of the many advances in these areas. We have included an encapsulation of algal structure and life histories, but we still expect that students using this book will have learned these subjects in more detail or will be learning about them concurrently in a general phycology course. The chapter on mariculture has been expanded because of the continuing increased interest in aquaculture, multi-trophic aquaculture and biomitigation, and algal biotechnology.

Finally, we have invited six renowned phycologists to give their personal perspectives on currently active areas of research. These essays are included as six boxes in the most appropriate chapters. We hope that this book has been greatly enhanced by the personal stories of how these essayists became interested in their research topics and how their career path sometimes took unexpected/unplanned turns.

Many colleagues contributed to this book in a variety of ways. Various chapters of the book have been critically read by Ricardo Scrosati, Mike Hawkes, Richard Taylor, Dave Schiel, Christine Maggs, Svenja Heesch, Christopher Hepburn, Giselle Walker, Patrick Martone, Chuck Amsler, Ivan Gómez, Dieter Hanelt, Christian Wiencke, Alwyn Rees, Morten Pedersen, Matt Bracken, John Berges, Britta Eklund, Murray Brown, Charles Yarish, Wendy Nelson and Thierry Chopin. We thank Mike Guiry for checking the taxonomy using AlgaeBase.

CLH especially thanks Rochelle Dewdney for her invaluable help in compiling references, checking species names, and acquiring copyrights.

Seaweed thalli and cells

1.1 Introduction: the algae and their environments

1.1.1 Seaweeds

The term "seaweed" traditionally includes only macroscopic, multicellular marine red, green, and brown algae. However, each of these groups has microscopic, if not unicellular, representatives. All seaweeds at some stage in their life cycles are unicellular, as spores or gametes and zygotes, and may be temporarily planktonic (Amsler and Searles 1980; Maximova and Sazhin 2010). Some remain small, forming sparse but productive turfs on coral reefs (Hackney et al. 1989) while others, such as the "kelps" of temperate reefs, can form extensive underwater forests (Graham et al. 2007a). Siphonous algae such as *Codium*, *Caulerpa* and *Bryopsis* that form large thalli are, in fact, acellular. The prokaryotic Cyanobacteria have occasionally been acknowledged in "seaweed" floras (e.g. Setchell and Gardner 1919; Littler and Littler 2011a). They are widespread on temperate rocky and sandy shores (Whitton and Potts 1982) and are particularly important in the tropics, where large macroscopic tufts of Oscillatoriaceae and smaller but abundant nitrogen-fixing Nostocaceae are major components of the reef flora (Littler and Littler 2011a, b; Charpy et al. 2012). Benthic diatoms also form large and sometimes abundant tube-dwelling colonies that resemble seaweeds (Lobban 1989). An ancient lineage of (mostly) deep-water green algae, the Palmophyllales, that includes *Verdigellas* and *Palmophyllum*, have a palmelloid organization with complex thalli built from an amorphous matrix

with a nearly uniform distribution of spherical cells (Womersley 1971; Zechman et al. 2010). On a smaller scale are the colonial filaments of some simple red algae, such as *Stylonema* (previously *Goniotrichum*). A "seaweed" is therefore problematic to precisely define: here "seaweed" refers to algae from the red, green, and brown lineages that, at some stage of their life cycle, form multicellular or siphonous macrothalli. In this book we shall consider macroscopic and microscopic marine benthic environments and how seaweeds respond to those environments.

The algae are evolutionarily diverse, but are related to one another through the endosymbiotic events that gave rise to plastids. The traditional classification of seaweeds as "red", "green", and "brown" is still fitting, but our understanding of how these groupings arose and their relatedness to each other and other eukaryotes has been transformed over the past 20 years as our understanding of endosymbiosis has grown (e.g. Walker et al. 2011). The evolutionary origin of the algae continues to be the subject of considerable research effort and debate (e.g. Brodie and Lewis 2007; Archibald 2009; Keeling 2010; Yoon et al. 2010; Burki et al. 2012; Collén et al. 2013). The taxonomic position of a species can be viewed as a "working hypothesis", and as such is subject to change as new information arises (Cocquyt et al. 2010). Unraveling algal evolution is complex because, in addition to the multiple endosymbiotic events, there are other complicating events such as horizontal (lateral) gene transfer (HGT; see Brodie and Lewis 2007). Knowledge of the relatedness of the different seaweed groups, and their relations to

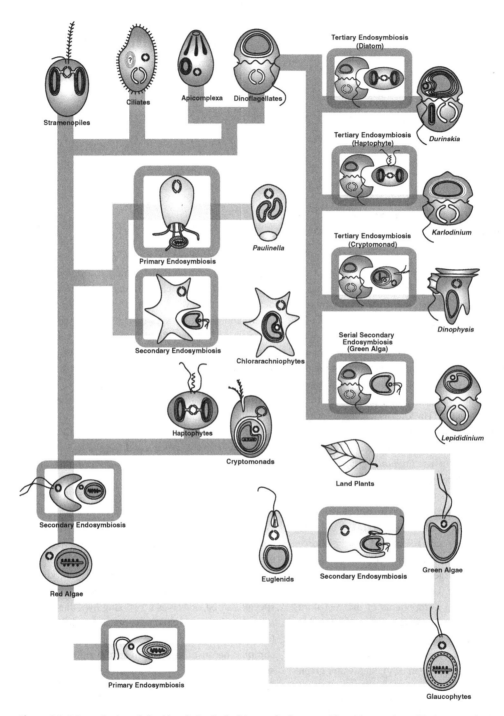

Figure 1.1 Schematic view of plastid evolution in the history of eukaryotes. The various endosymbiotic events that gave rise to the current diversity and distribution of plastids involve divergences and reticulations whose complexity has come to resemble an

other eukaryotes, is helpful in predicting aspects of their physiology and ecology.

There are several hypotheses on how algal plastids have arisen. A leading hypothesis is that a primary endosymbiotic event (~1.5 billion years ago), in which a free-living cyanobacterium was engulfed and incorporated within a heterotrophic eukaryote, gave rise to three major lineages: (1) the glaucophytes, (2) the green lineage in which the green algae are ancestral to the terrestrial plants, and (3) the red lineage which includes the red seaweeds (Yoon *et al.* 2004; Keeling 2010; Fig. 1.1). But there may have been more than one primary event, and the glaucophytes could have arisen separately from the green and red lineages (see Graham *et al.* 2009). At least three secondary endosymbiotic events (eukaryote + eukaryote) have occurred. It is fairly certain that two separate secondary events involving unicellular green algae gave rise to the euglenoids and chlorarachniophytes (reviewed by Keeling 2009, 2010; Fig. 1.1). Less clear, however, are the secondary endosymbiotic event(s) involving unicellular red algae (Burki *et al.* 2012). The chromalveolate hypothesis proposed by T. Cavalier-Smith (1999) suggests that a single secondary endosymbiotic event involving a red alga gave rise to six lineages (Fig. 1.1): ciliates, dinoflagellates, apicomplexa, haptophytes, cryptomonads, and the stramenopiles (heterokonts), with the first three belonging to the Alveolata. The chromalveolate hypothesis is "highly contentious" but considered by Keeling (2009) and others as the "hypothesis to beat". At the time of writing (2013),

the consensus is that the stramenopiles and Alveolata group with Rhizaria, forming the "SAR" clade; the haptophytes form a closely related sister group to the SAR clade, and the position of the cryptomonads is equivocal (Walker *et al.* 2011; Burki *et al.* 2012). Within the stramenopiles, the unicellular diatoms share a common ancestor with the multicellular brown seaweeds (Phaeophyceae; Patterson 1989a; Andersen 2004). However, phylogenies based on carbon storage and cell wall polysaccharides suggest that the stramenopiles arose separately from the Alveolates, and that a related, but distinct, red algal plastid was incorporated into an ancestral stramenopile in a second endosymbiotic event (Michel *et al.* 2010a, b). The dinoflagellates arose from tertiary or serial secondary endosymbioses (Fig. 1.1). As new information arises, and new molecular and bioinformatic techniques are added to the existing repertoire, hypotheses on eukaryotic evolution and speciation will continue to develop.

Ocean vegetation is dominated by the algae. No mosses, ferns, or gymnosperms are found in the oceans, and only a few angiosperms (the seagrasses) occur in marine habitats. That there are relatively few marine angiosperms may reflect the problems of adaption to the sea, including ion regulation and pollination (Ackerman 1998). The water column is chiefly the domain of the phytoplankton, but populations of floating seaweeds that have been detached from the substratum are common and provide an important mechanism of dispersal (sec. 3.3.7). Intertidal rocky shores are abundantly covered with a macrovegetation

Caption for Figure 1.1 (*cont.*) electronic circuit diagram. Endosymbiosis events are boxed, and the lines are shaded to distinguish lineages with no plastid (dark gray), plastids from the green algal lineage (light gray) or the red algal lineage (mid-gray). At the bottom is the single primary endosymbiosis leading to three lineages (glaucophytes, red algae, and green algae). On the lower right, a discrete secondary endosymbiotic event within the euglenids led to their plastid. On the lower left, a red alga was taken up in the ancestor of chromalveolates. From this ancestor, haptophytes and cryptomonads (as well as their non-photosynthetic relatives such as katablepharids and telonemids) first diverged. After the divergence of the rhizarian lineage, the plastid appears to have been lost, but in two subgroups of Rhizaria, photosynthesis was regained: in the chlorarachniophytes by secondary endosymbiosis with a green alga, and in *Paulinella* by taking up a cyanobacterium (many other rhizarian lineages remain non-photosynthetic). At the top left, the stramenopiles diverged from alveolates, where plastids were lost in ciliates and predominantly became non-photosynthetic in the apicomplexan lineage. At the top right, four different events of plastid replacement are shown in dinoflagellates, involving a diatom, haptophyte, cryptomonad (three cases of tertiary endosymbiosis) and green alga (a serial secondary endosymbiosis). Most of the lineages shown have many members or relatives that are non-photosynthetic, but these have not all been shown for the sake of clarity. (From Keeling, 2010, reproduced with permission.)

that is almost exclusively seaweeds, although in western North America surf grass *(Phyllospadix* spp.) is an exception. Seaweed surfaces themselves are colonized by benthic microalgae and bacteria, with which they may have intimate ecological relationships, and seaweed microstages grow on and within larger seaweeds. Muddy and sandy areas have fewer seaweeds, because most species cannot anchor there, though some siphonous greens (e.g. some species of *Halimeda, Caulerpa,* and *Udotea)* produce penetrating, root-like holdfasts that also serve in nutrient uptake (Littler *et al.* 1988). In such areas, seagrasses become the dominant vegetation, particularly in tropical and subtropical areas (Larkum *et al.* 2006). There is also a paucity of freshwater macroalgae. Freshwater red and brown algae are represented by relatively few genera and species, and Ulvophyceae are also scarce with only a few genera (e.g. *Cladophora)* having penetrated fresh waters (Wehr and Sheath 2003).

Most seaweeds are multicellular most of the time. What does this imply for physiological ecology? Multicellularity confers the advantage of allowing extensive development in the third dimension of the water column. Such development can be achieved in other ways, however. Siphonous green algae form large multinucleate thalli that are at least technically single cells (acellular rather than unicellular), supported by turgor pressure *(Valonia),* ingrowths of the rhizome wall (trabeculae) in *Caulerpa,* or interweaving of numerous narrow siphons *(Codium, Avrainvillea)* (Fig. 1.2). Colonial diatoms, both tube-dwelling and chain-forming, also build three-dimensional structures, as do zooxanthellae (dinoflagellates) in association with corals. Multicellular algae often grow vertically away from the substratum; this habit brings them closer to the light, enables them to grow large without extreme competition for space, and allows them to harvest nutrients from a greater volume of

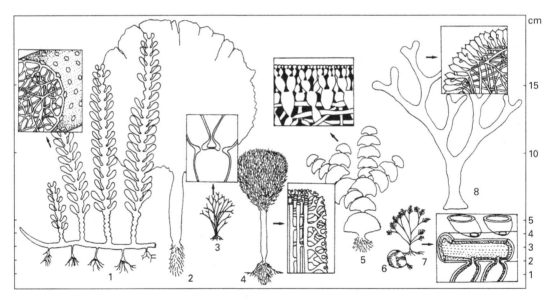

Figure 1.2 Thallus morphology and construction in siphonous green algae. Thalli drawn to scale; insets (not to scale) show principles of construction: (1) *Caulerpa cactoides*: network of trabeculae. (2) *Avrainvillea gardineri*: tightly woven felt of filaments. (3) *Chlorodesmis* sp.: bush of dichotomously branched siphons, constricted at the bases of the branches (inset). (4) *Penicillus capitus*: calcified siphons form a multiaxial pseudotissue in the stem (inset), but separate to form bushy head. (5) *Halimeda tuna*: segmented, calcified thallus of woven medulla and cortical utricles (inset). (6) *Halicystis* stage of *Derbesia*, a single ovoid cell (shown at gametogenesis). (7) *Bryopsis plumosa* gametophyte: pinnately branched free siphons. (8) *Codium fragile*: interwoven uncalcified siphons form multiaxial branches. (From Menzel 1988, with permission of Springer-Verlag, Berlin.)

water. On the other hand, there are creeping filamentous algae, even endophytic and endolithic filaments (e.g. *Entocladia*), as well as crustose algae such as *Ralfsia*, and *Porolithon*, that do not grow up into the water column. Support tissue usually is not necessary for this upward growth, because most small seaweeds are slightly buoyant, and the water provides support. Support tissue is metabolically expensive, however strength and resilience are required to withstand water motion. Some of the larger seaweeds (e.g. *Pterygophora*) have stiff, massive stipes, but others (e.g. *Hormosira*) employ flotation to keep them upright. Many of the kelps and fucoids have special gas-filled structures, pneumatocysts (Dromgoole 1990; Raven 1996), whereas in other seaweeds (e.g. erect species of *Codium*) gas trapped among the filaments achieves the same effect (Dromgoole 1982).

A second important feature of multicellularity is that it allows division of labor between tissues; such division is developed to various degrees in seaweeds. Nutrient (and water) uptake and photosynthesis take place over virtually the entire surface of the seaweed thallus, in contrast to vascular land plants. Differentiation and specialization among the vegetative cells of algal thalli range from virtually nil (as in *Ulothrix*, where all cells except the rhizoids serve both vegetative and reproductive functions), through to *Porphyra* [many species of this genus are now treated in other genera, with most being in *Pyropia*; Sutherland *et al.* 2011; see sec. 10.2] and *Ulva* whose blades are morphologically simple but are differentiated into regions with distinct physiologies (e.g. Hong *et al.* 1995; Han *et al.* 2003), to the highly differentiated photosynthetic, storage, and translocation tissues in a variety of organs, including stipe, blades, and pneumatocysts, that occur in fucoids and kelps (Graham *et al.* 2009). Of course, no seaweed shows the degrees of differentiation seen in vascular plants. Even in vascular plants, the cells are biochemically more general than animal cells: the organs of vascular plants (stems, leaves, roots, flowers) all contain much the same mix of cells, whereas animal organs each contain only a few specialized cell types. The low diversity of cells in a seaweed thallus means that each cell is physiologically and biochemically even more general than vascular plant cells.

The evolution of multicellularity entails the co-ordinated growth of cells, which, in turn, requires cell-to-cell communication. The detection of genes coding for receptor kinases (signaling molecules that are found in all multicellular eukaryotes) in *Ectocarpus*, and their absence in the related unicellular diatoms, suggests that these molecules were a pre-requisite for multicellularity. Another pre-requisite is cell–cell adhesion via a sticky extracellular matrix. Integrin-related proteins that have a key role in cell adhesion in animals are also present in *Ectocarpus*, but not in diatoms (Cock *et al.* 2010a). In the red seaweeds, pit plugs are considered a vital step in the evolution of multicellularity, by providing structural integrity within the otherwise loosely packaged cells of pseudo-parenchymatous construction (Graham *et al.* 2009, p. 319; Gantt *et al.* 2010).

1.1.2 Environmental-factor interactions

Benthic algae interact with other marine organisms, and all interact with their physico-chemical environment. As a rule, they live attached to the seabed between the top of the intertidal zone and the maximum depth to which adequate light for growth can penetrate. Among the major environmental (abiotic) factors affecting seaweeds are light, temperature, salinity, water motion, and nutrient availability. Among the biological (biotic) interactions are relations between seaweeds and their epiphytic bacteria, fungi, algae, and sessile animals; interactions between herbivores and seaweeds (both macroalgae and epiphytes); and the impact of predators, including humans. Each propagule contains the genetic information that will allow the maturing seaweed to develop a phenotype that is suited to its environment: in fact, there can be a high degree of phenotypic plasticity even within a genetically uniform population grown under the same environmental conditions (Fig. 1.3). Individual patterns of growth, morphology, and reproduction are overall effects of all these factors combined.

An organism's physico-chemical environment, consisting of all the external abiotic factors that influence the organism, is very complex and constantly varying. In order for us to discuss or study it, we need to reduce it

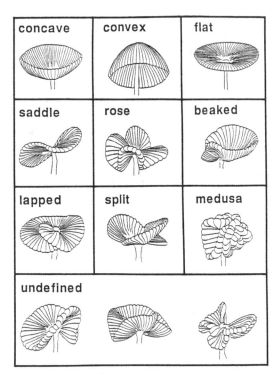

concave	convex	flat
saddle	rose	beaked
lapped	split	medusa
undefined		

Figure 1.3 Variation in cap morphologies of *Acetabuaria acetabulum*, the progeny of which were raised in the same experimental conditions. "Concave" included a minor variant ("concave-bell") in which the rim of a concave cap was flattened. "Convex" are mirror images of concave. "Flat" caps are usually perpendicular to the stalk. "Saddles" have two opposing quadrants of the cap curved up and the other two down. In "beaked", one or both halves of the cap are adpressed and parallel to the stalk. "Split" caps have rays that are not all fused so that the cap is divided into halves, quarters or sixths. In "lapped", the rays adjacent to the two that had not fused overlap each other. "Rose" and "medusas" are the most convoluted cap shapes. "Undefined" are caps which combine two or more of the above morphologies. (From Nishimura and Mandoli 1992, reproduced with permission.)

to smaller parts, to think about one variable at a time. And yet, each of the environmental "factors" that we might consider – temperature, salinity, light, and so forth – is really a composite of many variables, and they tend to interact. Most importantly, the organization of life is now best understood as constitutive hierarchies (Mayr 1982, p. 65), in which at each new level or system there are emergent properties that are not predictable from study of the component parts. This is most evident in comparing the properties of individuals (say, humans) to the properties of the next level of components (organ systems, e.g. nervous system), but it also works upward from the individual through populations, communities, ecosystems, and the biosphere. It has major implications when we attempt to predict community or ecosystem properties from studies at the species (population) level, as we usually must. The following paragraphs are intended to paint the big picture, before we go on to study it pixel by pixel.

Factor interactions can be grouped into four categories: (1) multifaceted factors, (2) interactions between environmental variables, (3) interactions between environmental variables and biological factors, and (4) sequential effects.

1. Many environmental factors have several components that do not necessarily change together (i.e. multifaceted factors). Light quality and quantity, which are important in photosynthetic responses and metabolic patterns, both change with depth, but the changes depend on turbidity and the nature of the particles. In submarine caves, light quantity diminishes with little change in quality. Natural light has the additional important component of day length, which influences reproductive states. Salinity is another complex factor, of which the two chief components are the osmotic potential of the water and the ionic composition. Osmotic potential affects water flow in and out of the cell, turgor pressure, and growth, while the concentrations of Ca^{2+} and HCO_3^- affect membrane integrity and photosynthesis, respectively. Hydrodynamic forces affect thallus survival and spore settlement on wave-swept shores, and water motion also has important effects on the boundary layers over seaweed surfaces and thus on nutrient uptake and gas exchange. Nutrients must be considered not simply in their absolute concentrations but also in the amounts present in biologically available forms; concentrations of trace metals may create toxicity problems, particularly in polluted areas. Pollution, as a factor, may include not only the toxic effects of component chemicals but also an increase in

turbidity, hence a reduction in irradiance. Emersion often involves desiccation, heating, or chilling, removal of most nutrients (except CO_2), and, frequently, changes in the salinity of the water in the surface film on the seaweeds and in the free space between cells.

2. Interactions among environmental variables are the rule rather than the exception. Bright light is often associated with increased heating, particularly of seaweeds exposed at low tide. Light, especially blue light, regulates the activities of many enzymes, including some involved in carbon fixation and nitrogen metabolism. Temperature and salinity affect the density of seawater, hence the mixing of nutrient-rich bottom water with nutrient-depleted surface water. Thermoclines can affect plankton movements, including migration of the larvae of epiphytic animals. Temperature also affects cellular pH and hence some enzyme activities. The seawater carbonate system and especially the concentration of free CO_2 are greatly affected by pH, salinity, and temperature, while the availability of ammonium is pH-dependent, because at high pH the ion escapes as free ammonia. Water motion can affect turbidity and siltation as well as nutrient availability. These are examples of one environmental variable affecting another. There are also examples of two environmental variables acting synergistically on seaweed; for instance, the combination of low salinity and high temperature can be harmful at levels where each alone would be tolerable. In some seaweeds, the combined effects of temperature and photoperiod regulate development and reproduction.

3. Interactions between physico-chemical and biological factors are also the rule rather than the exception. The environment of a given seaweed includes other organisms, as we have seen, with which the seaweed interacts through intraspecific and interspecific competition, predator–prey relationships, associations with parasites and pathogens, and basiphyte–epiphyte relationships. These other organisms are also affected by the environment, as are their effects on other organisms. Moreover, other organisms may greatly modify the physico-chemical environment of a given individual. Protection from strong irradiance and desiccation by canopy seaweeds is important to the survival of understory algae, including germlings of the larger species. Organisms shade each other (and sometimes themselves) and have large effects on nutrient concentrations and water flow. Other interactions stem from the way the biological parameters, such as age, phenotype, and genotype, affect a seaweed's response to the abiotic environment, as well as the effects that organisms have on the environment. The chief biological parameters that condition a given seaweed's response to its environment are age, reproductive condition, nutrient status (including stores of N, P, and C), and past history. By "past history" is meant the effects of past environmental conditions on seaweed development. Genetic differentiation within populations leads to different responses in seaweeds from different parts of a population. The seasons can also affect certain physiological responses, aside from those involved in life-history changes; these responses include acclimation of temperature optima and tolerance limits.

4. Finally, there are factor interactions through sequential effects. Nitrogen limitation may cause red algae to catabolize some of their phycobiliproteins, which will in turn reduce their light-harvesting ability. In general, any factor that alters the growth, form, or reproductive or physiological condition is apt to change the responses of the seaweed to other factors both currently and in the future. A good example of a sequential effect, and also biotic–abiotic interaction, was seen by Littler and Littler (1987) following an unusual flash flood in southern California. Intertidal urchins (*Strongylocentrotus purpuratus*) were almost completely wiped out, but the persistent macroalgae suffered little damage from the freshwater. Subsequently, however, there was a great increase in ephemeral algae (*Ulva*, Ectocarpaceae) because of the reduction in grazing pressure. The complexity of the interactions of variables in nature often confounds interpretation of the effects even of "major" events, such as El Niño warm-water periods (Paine 1986; sec. 7.3.7).

Testing the effects of the various factor interactions described above requires a multifaceted approach that includes quantitative field observations, field manipulations and targeted laboratory experiments; for each approach a rigorous experimental design is essential so that the appropriate statistical analyses can be applied to detect differences among experimental treatments. In laboratory experiments, usually one variable is tested at a time, and all other factors are held constant, or at least equal in all treatments. Experiments in which two (or occasionally three) factors are varied are possible but the number of culture vessels required for independent replication of treatments can be technically difficult to achieve especially with large seaweeds: it is important to avoid pseudoreplication in both laboratory and field experiments (Hurlbert 1984). It is also important to understand how field manipulations can confound results, and to include appropriate controls. For example, Underwood (1980) criticized some field experiments designed to determine the effects of grazer exclusion because the fences and cages used to keep out grazers also affected the water motion over the rock surface and provided some shade. Furthermore, field studies that use correlation analyses to elucidate whether an environmental factor causes a specific biological pattern (e.g. growth, onset of reproduction) can be misleading because the key environmental factors that regulate seaweed biological processes are themselves tightly correlated, for example light, temperature, and nitrate concentration. As Schiel and Foster (1986, p. 273) explain "The existence of patterns and abundance of species constitutes evidence that these physical factors and biological interactions may affect the structure of these communities. They do not at the same time, however, demonstrate the importance or unimportance of these factors in producing observed patterns."

1.1.3 Laboratory culture versus field experiments

Several considerations confound the interpretation of field reality via laboratory studies. First, while laboratory studies provide much more controlled conditions than are found in nature, they are limited in some important ways and contain some implicit assumptions, such as the following: (1) High nutrient levels common in lab experiments do not alter the seaweeds' responses to the factor under study. (2) The reactions of seaweeds to uniform conditions (including the factor under study) are not different from their responses to the factor(s) under fluctuating conditions. To a certain extent these assumptions are valid. Culture media can be very rich in nutrients, to compensate for lack of water movement and exchange, but it is unlikely that this substitution can give precisely the same results. Other culture conditions are also generally optimal, except for the variable under study, and the results may not elucidate the behavior of seaweeds in the field, which are subject to competition and often suboptimal conditions (Neushul 1981). Another important difference between laboratory and field is that in culture, species usually are tested in isolation, away from interspecific competition and grazing. Furthermore, culture conditions are uniform (at least on a large scale), whereas in nature there often are large and unpredictable fluctuations in the environment (e.g. Gorospec and Karl 2011). Microscale heterogeneity in culture conditions should not be overlooked (Allen 1977; Norton and Fetter 1981). In the culture flask, one cell may shade another, and cells form nutrient-depleted zones around them, creating a mosaic of nutrient concentrations through which cells pass. In the field, scale also needs to be considered at the large end – for instance, the amount of space needed for a patch of a given alga to establish itself (Schiel and Foster 1986). In essence, for both field and laboratory experiments, informed decisions must be made on the experimental conditions that are provided, and it is important to be aware that these conditions will affect the outcome and interpretation of the results.

Second, the timescale over which an experiment is conducted affects the interpretation of the data (Raven and Geider 2003). In short-term physiological experiments (seconds to minutes), a single factor can be varied (e.g. different levels of UV-B radiation) and a response (e.g. the production of reactive oxygen species) is measured. This physiological response is at the level of *regulation* i.e. the up- or down-regulation of pre-existing enzymes, and reveals the physiological potential of that organism to respond to an immediate environmental change. In medium-term experiments (hours to days) *acclimation* to new environmental conditions may occur. Acclimation involves gene expression, and the synthesis of new proteins such as enzymes. *Adaptation* to particular environmental factors occurs over a longer timescale (up to millennia) and is a mechanism for speciation (sec. 7.1).

Third, when a single species occurs in widely different latitudes or longitudes, its physiology and ecology may be quite different. For many topics, only one study or a few studies have been done, and a phenomenon demonstrated in a particular alga under certain conditions will not necessarily turn out to be the same in other algae or under other conditions. In Australia, for example, the kelp *Ecklonia radiata* dominates across 3000 km of coastline, from the southeast to southwest. However, the morphology and ecology of *Ecklonia* on the east coast is very different to *Ecklonia* on the west and south coasts, with the result that different coastal management plans are required for these different regions (Connell and Irving 2009). Equally, very few natural populations or communities have been studied often enough to assess how much variability is present from place to place (ecotypic variation). The kelp beds of southern California are exceptional in that they have been repeatedly analyzed by different people along the coast since the 1960s (Steneck *et al.* 2002; Graham *et al.* 2007a). For *Macrocystis*, there is no typical kelp bed; environmental parameters differ from one kelp bed to another, and parameters such as specific growth rate versus nitrogen supply vary among populations (Kopczak *et al.* 1991).

In this first chapter we shall review the foundations of seaweed construction, cell biology, molecular biology and genetics on which any understanding of seaweed physiological ecology must rest. In Chapter 2, we continue this review by tracing the development of seaweed thalli from gametes and spores to reproductive individuals. In both these chapters, we build upon the fundamental information on seaweed anatomy and development that is described in algal text books, particularly van den Hoek *et al.* (1995) and Graham *et al.* (2009).

1.2 Seaweed morphology and anatomy

1.2.1 Thallus construction

Diversity of thallus construction in algae contrasts strongly with uniformity in vascular plants. In the latter, parenchymatous meristems (e.g. at the shoot and root apices) produce tissue that differentiates in a wide variety of shapes. For seaweeds, parenchymatous construction is prevalent only in the brown orders. For example, in kelps, fucoids, and Dictyotales, this mode of construction has given rise to internal and morphological complexity (Fig. 1.4). The larger seaweeds, especially Laminariales and Fucales, have several different tissue and cell types, including photosynthetic epidermis, cortex, medulla, sieve tubes, and mucilage ducts (Graham *et al.* 2009). The ontogeny of the parenchyma in the Dictyotales (Fig. 1.4d–m) has been followed in detail by Gaillard and L'Hardy-Halos (1990), who cite many sources, and by Katsaros and Galatis (1988). However, the great majority of seaweeds either are filamentous or are built up of united or corticated filaments. Large and complex structures can be built up this way, for example *Codium amplivesiculatum* (previously *C. magnum)* can reach several meters long (Dawson 1950). Cell division may take place throughout the alga, or the meristematic region may be localized. If localized, it is most commonly at the apex, but may be at the base or somewhere in between (intercalary).

A simple filament consists of an unbranched chain of cells attached by their end walls and results from cell division only in the plane perpendicular to the axis of the filament. Unbranched filaments are uncommon among seaweeds; examples are *Ulothrix* and *Chaetomorpha*. Usually, some cell division takes place parallel to the filament axis to produce branches *(Cladophora, Ectocarpus, Antithamnion;* see Fig. 1.17). Filaments consisting of a single row of cells (branched or not) are called uniseriate. Pluriseriate filaments, i.e. two or more rows of cells, are seen in genera such as *Blidingia, Bangia,* and *Sphacelaria* (Fig. 1.4a; Graham *et al.* 2009). Branches need not grow out free, but may creep down the main filament, forming cortication, as seen in *Ceramium* (Fig. 1.5a) and *Ballia.* In some of the

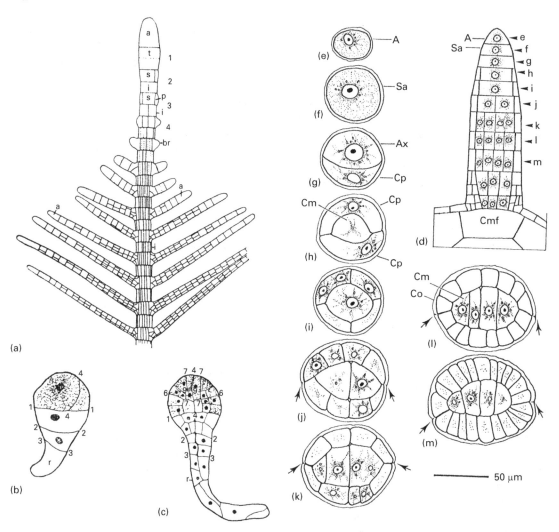

Figure 1.4 Parenchymatous development in seaweeds. (a) *Sphacelaria plumula* apex showing first transverse division (t), followed by pairs of cells (i, s), of which s forms branches, but i does not. (b, c) *Fucus vesiculosus* germination showing successive cell divisions (numbered) (divisions 5 and 8 in the plane of the page). (d-m) *Dictyota*: development of parenchyma; (d) long section through adventive branch, showing locations of cross sections at each level (diagrammatic); (e-m) serial cross sections to show sequence of periclinal divisions. Arrows indicate junction between original two pericentral cells (first shown in h). For the sake of clarity, the proportions of the cells were changed; the adventive branch is actually half as long and twice as wide as shown. A, apical cell; Sa, subapical cell; Ax, axial cell; Cp, pericentral cell, Cm, medullary cell; Co, cortical cell. (Parts a-c from Fritsch 1945, based on classical literature; d-m from Gaillard and L'Hardy-Halos 1990, with permission of Blackwell Scientific Publications.)

larger Rhodomelaceae, such as *Laurencia* and *Acanthophora*, the cortication becomes so extensive that the origin of the structure is obscured. Pseudoparenchymatous construction is when neighboring filaments adhere to one another and form a structure that looks very much like parenchymatous bodies (Graham *et al.* 2009). A detailed study by Kling and Bodard (1986) of axis development in *Gracilariopsis longissima* (previously *Gracilaria verrucosa*) (uniaxial) showed how complex, and difficult to interpret, pseudoparenchymatous growth patterns can be (compare Fig. 1.5g–n with the parenchymatous construction of *Dictyota* Fig. 1.4d–m).

Many of the larger seaweed thalli are multiaxial, produced by the adhesion of several filaments. This is particularly common among the red algae (Fig. 1.5d–f) (van den Hoek *et al.* 1995; Graham *et al.* 2009). Multiaxial construction is most readily seen in the less tightly compacted thalli of *Nemalion* or *Liagora*. The contrast between multiaxial and uniaxial growth can be seen within thalli of *Dumontia contorta* (previously *Dumontia incrassata*) (Fig. 1.5d–f), in which bases are multiaxial, but upper branches are uniaxial (Wilce and Davis 1984). The adhesion of filaments can also produce a pseudoparenchymatous crust (*Peyssonnelia*, *Neoralfsia*) or blade (*Anadyomene*; Fig. 1.5b,c). Many siphonous green algae, including *Halimeda* and *Codium*, are formed by the interweaving of numerous filaments (Fig. 1.2). In the Corallinaceae, multiaxial apical growth forms the hypothallus (in crusts) or central medulla (in erect forms), while intercalary meristems on the lateral branches form the epithallus and perithallus (cortex in erect axes) (Cabioch 1988).

Cell division in two planes can result in a monostromatic sheet of cells, as in *Monostroma*. *Ulva* spp. are distromatic and develop from a uniseriate filament that becomes a pluriseriate filament, which in turn can become either a hollow tube (e.g. *Ulva intestinalis*, previously *Entermorpha intestinalis*) or a two-layered blade (e.g. *Ulva lactuca*). Interestingly, for both *Monostroma* and *Ulva* the development of a thallus depends on the presence of epiphytic bacteria (Matsuo *et al.* 2005; sec. 2.6.2).

Plasmodesmata are a feature shared by land plants and parenchymatous green and brown seaweeds, and they connect neighboring cells allowing cellular communication (Raven 1997a). The red seaweeds, however, do not exhibit parenchymatous construction, nor do they have plasmodesmata. Characteristic of florideophycean red seaweeds are pit connections with pit plugs (Peuschul 1989). Primary pit plugs are granular protein masses that literally "plug the hole" that is left following incomplete cell division. Secondary pit plugs can form between cells of different filaments within a pseudoparenchymatous structure; they can also form between individual germlings as part of the coalescence process that gives rise to the chimeric organization that is common in red seaweed (Santelices *et al.* 1999 – see below). Although less common, pit connections and plugs do occur in the Bangiophyceae. For example in *Pyropia yezoensis* (previously *Porphyra yezoensis*), they are present in the filamentous sporophyte phase (conchocelis) but absent in the bladed gametophyte (Ueki *et al.* 2008).

1.2.2   The Littler functional-form model

The construction of the thallus has importance for developmental physiology. Similar morphologies can be constructed in different ways; the overall morphology is important to ecological physiology. Among different algal classes, certain morphologies are repeated, which, as noted by Littler *et al.* (1983a), indicates convergent adaptations to critical environmental factors. On the other hand, species face divergent selection pressures: those favoring more productive, reproductive, and competitive thalli, versus those favoring longevity and environmental resistance (Littler and Kauker 1984; Russell 1986; Norton 1991). Many seaweeds show a variety of morphologies within one life history (see Chapter 2). Heterotrichous seaweed with crustose bases and erect fronds within one generation (e.g. *Corallina*) and heteromorphic seaweeds with crustose/filamentous and frondose generations (e.g. *Scytosiphon*) (Fig. 2.2) are both common. How can we assess the significance of morphology when we are faced with convergence between classes on the one hand and diversification within species on the other hand?

The functional-form model was advanced by Littler and Littler (1980) and subsequently tested extensively by both themselves and others. The model has also

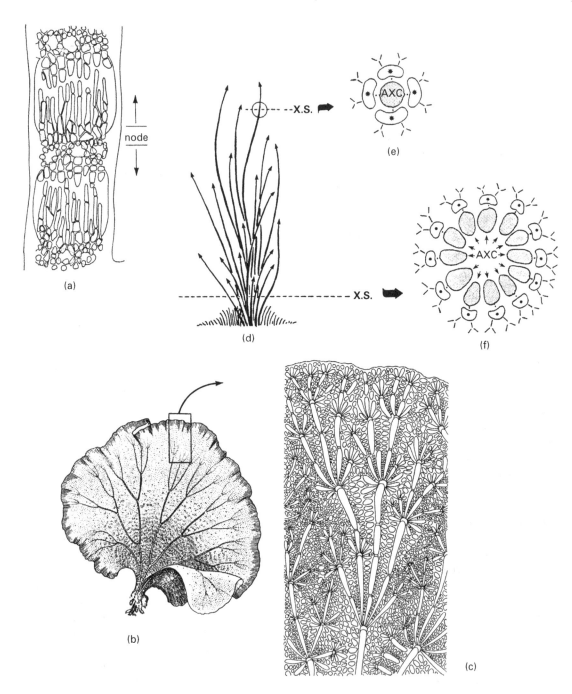

Figure 1.5 Filamentous thallus construction. (a) Small portion of a *Ceramium* axis with cortication growing upward and downward from a node between axial cells. (b, c) Formation of bladelike thallus from filaments in *Anadyomene stellata* (b, × 1.82; c, × 13.65). (d–f) Growth of *Dumontia contorta* (previously *Dumontia incrassata*) showing schematically the axial filaments and apical cells (arrows); cross section in the uniaxial part of the thallus near the tip (e) shows a single axial cell (AXC) surrounded by four pericentral cells (*) that have in turn produced cortical cells; (f) cross section through base shows multiaxial construction with a core of axial cells, each with one pericentral cell.

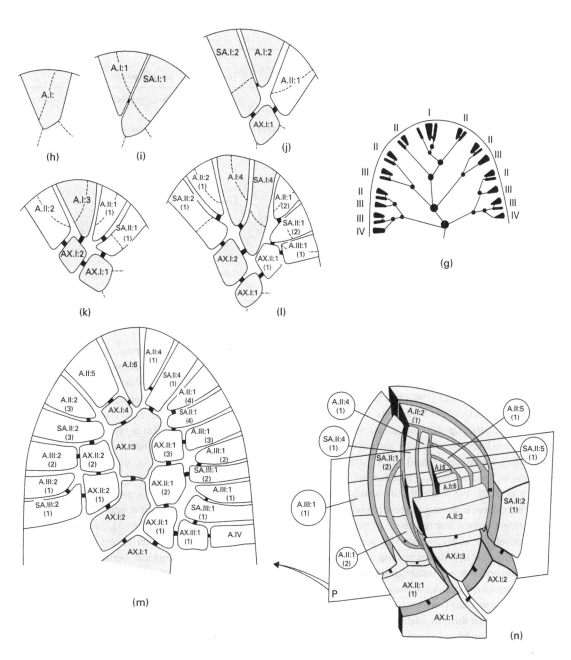

Caption for Figure 1.5 (*cont.*) (g–n) Apical growth of *Gracilariopsis longissima* (previously *Gracilaria verrucosa*). (g) A primary apical cell (I) occurs at the tip of the main axis, and secondary apical cells (II, III, etc.) occur at the tips of lateral filaments. (h–m) Division of the apical cell (A.I), shown by dotted line in (h), gives rise to a subapical cell (SA.I: 1) and a new apical cell (A.I:1)(i). In (i–j), the subapical cell is shown dividing to form an axial cell (AX.I:1) and a secondary apical cell (A.II:1), while the new apical cell (A.I:1) cuts off another subapical cell (SA.I:2) and becomes A.I:2. The lineages can be traced further with the help of the pit connections (represented as dark bars between cells). (n) The three-dimensional arrangement is complex because the apical cell divides on three faces. P is the plane of the vertical section in (m). (Part a from Taylor 1957; band c from Taylor 1960, with permission of University of Michigan Press; d–f from Wilce and Davis 1984, with permission of *Journal of Phycology*; g–n from Kling and Bodard 1986, with permission of *Cryptogamie: Algologie*.)

Table 1.1 Functional-form groups of macroalgae

Functional-form group	External morphology	Internal anatomy	Texture	Sample genera
Sheet group	Thin, tubular, and sheetlike (foliose)	Uncorticated, one to several cells thick	Soft	*Ulva, Pyropia, Dictyota*
Filamentous group	Delicately branched (filamentous)	Uniseriate, multiseriate, or lightly corticated	Soft	*Centroceras, Polysiphonia, Chaetomorpha, Ectocarpus*
Coarsely branched group	Coarsely branched, upright	Corticated	Fleshy-wiry	*Laurencia, Chordaria, Caulerpa, Penicillus, Gracilaria*
Thick, leathery group	Thick blades and branches	Differentiated, heavily corticated, thick-walled	Leather, rubbery	*Laminaria, Fucus, Udotea, Chondrus*
Jointed calcareous group	Articulated, calcareous, upright	Calcified genicula, flexible intergenicula with parallel cell rows	Stony	*Corallina, Halimeda, Galaxaura*
Crustose group	Prostrate, encrusting	Calcified or uncalcified parallel rows of cells	Stony or tough	*Lithothamnion, Ralfsia, Hildenbrandia*

Source: Littler *et al.* (1983b), with permission of *Journal of Phycology.*

been modified by Steneck and Dethier (1994), and by Balata *et al.* (2011) who propose a system that has 35 functional groupings, compared to Littler and Littler's 6. The Littler and Littler (1980) model holds that the functional characteristics of seaweeds, such as photosynthesis, nutrient uptake, and grazer susceptibility, are related to form characteristics, such as morphology and surface-area:volume (SA:V) ratios (Table 1.1). One can thus set up predictions of function from an examination of form. For example, the sheet group are predicted to have high rates of growth, photosynthesis and nutrient uptake, low resistance to herbivory, and low competitive ability. Functional groupings have been used to test hypotheses relating to algal primary production and nutrient uptake, resistance to herbivory, tolerance to physiological stress, and successional stage of communities (reviewed by Padilla and Allen 2000). Functional form in relation to light harvesting and nutrient uptake are discussed in secs. 5.3.2 and 5.7.2, respectively.

Functional-form models have proven valuable in predicting physiological rates, because nutrient and inorganic carbon uptake are strongly related to surface-area:volume ratio (e.g. Taylor *et al.* 1999). There is a trend of declining physiological rates and

specific growth rate from group 1 to group 6 (e.g. Fig. 5.25). However, power-scaling approaches can be equally useful as predictors of net photosynthesis, respiration, and growth (Enríquez *et al.* 1996; de los Santos *et al.* 2009). On a tropical reef, the productivity of the unicellular and filamentous components of a turf-forming community could be determined accurately without knowledge of the individual species (Williams and Carpenter 1990). However, functional groups are less successful in predicting the susceptibility of seaweeds to herbivores, and successional stage (Padilla and Allen 2000, Table 1). Also, categorizing specific morphologies is not always simple because there are no sharp boundaries between some groups. For example, 15 and 20% of species could not be allocated to a functional group in the studies of Phillips *et al.* (1997) and de los Santos *et al.* (2009), respectively.

The allocation of species to particular functional groups requires little taxonomic expertise, and as such it can be an attractive method of examining ecosystem biodiversity and detecting long-term changes (Collado-Vides *et al.* 2005; Balata *et al.* 2011). Phillips *et al.* (1997) compared functional groups and full taxonomic classification as methods of detecting shifts in seaweed communities along a wave-exposure gradient. The functional grouping method was less able to detect differences between communities, and resulted in a substantial loss of biodiversity information. In the Florida Keys, USA, four genera of calcareous green seaweeds (*Halimeda, Udotea, Penicillus*, and *Rhipocephalus*) fall into the same functional group (jointed, calcareous), but a 7-year study revealed that each species had very different seasonal patterns in abundance. Once again, grouping the different genera into one functional group lead to a loss of information (Collado-Vides *et al.* 2005). However, using their expanded functional-form model, Balata *et al.* (2011) were able to detect differences between Mediterranean seaweed assemblages that were exposed to different environmental stressors. In summary, functional groups have proven useful in assessing seaweed metabolic processes, but further testing is required if they are to be rigorously applied to other aspects of ecology and biodiversity (Padilla and Allen 2000; for further discussion on functional form and grazing see sec. 4.3.2).

1.2.3 Unitary, clonal, and coalescing seaweeds, and modular construction

In the 1970s, terrestrial plant ecologists distinguished between "unitary plants" (also termed "aclonal" and "non-clonal") which have leaves and roots connected to a main axis and grow predominantly in the vertical direction and "clonal plants" which can spread laterally and vegetatively over the soil surface; these distinctions equally apply to seaweeds (reviewed by Santelices 2004a; Scrosati 2005). Unitary seaweeds originate from unicellular propagules (haploid or diploid), have just one axis that grows vertically from the holdfast, tend to have morphological and physiological differentiation, and do not produce ramets (defined below). Examples include canopy-forming seaweeds such as *Durvillaea antartica, Saccharina latissima* (previously *Laminaria saccharina*), and *Lessonia nigrescens*, and also smaller seaweeds such as *Fucus* species, *Ulva taeniata* and *Colpomenia tuberculata* (Santelices 2004a; Scrosati 2005). Clonal seaweeds are defined by Scrosati (2005) as those for which the "holdfast produces a number of fronds vegetatively, each

frond having the potential capacity for autonomous life if it becomes physically isolated from the rest while remaining attached to the substrate by an original portion of holdfast". *Genet* is a "genetic individual" defined as "the free-living individual that develops from one original zygote, parthenogenetic gamete or spore and produces ramets vegetatively during growth" (Scrosati 2002a). A *ramet* is the smallest potentially physiologically independent unit of the genet or "any algal fragment with the ability to reattach to the substratum and develop as a new individual" (Collado-Vides 2002a). Seaweeds exhibiting clonal growth include *Mazzaella parksii* (previously *M. cornucopiae* and *Iridaea cornucopiae*), *Caulerpa*, and *Ascophyllum nodosum*. Clonal seaweeds can be further categorized as either coalescing clonal (e.g. *Mazzaella parksii*) or non-coalescing clonal e.g. *Pterocladiella capillacea* (previously *Pterocladia capillacea*) (Scrosati 2005). The genets of some clonal seaweeds establish cellular connections and coalesce to form chimeras[1] (Fig. 1.6a). Coalescence is widespread in the red seaweeds, but rare in the other seaweed phyla (Santelices *et al.* 1996, 1999; Santelices 2004a).

Both unitary and clonal seaweed groupings contain examples of modular construction. The term "module" can refer to any part of an organism that is a reiterated unit. *Fucus* species, for example, are unitary modular because each branch and associated apical cell is repeated as a result of growth, whereas *Laminaria* species are unitary non-modular because each individual has only one meristem and this pattern is not repeated within an individual. *Gelidium* species are examples of clonal-modular seaweeds, in which the branched ramets are the repeated units, whereas *Mazzaella* is clonal non-modular because each ramet is unbranched (Ricardo Scrosati, pers. comm.).

Until the early 2000s, most physiological and ecological research has considered seaweeds as unitary

---

[1] The terms "chimera" and "genetic mosaic" are often used interchangeably but this usage is incorrect. Both are genetically heterogeneous (i.e. not homogenous), but mosaics are more common as they arise from intrinsic genetic variations caused by for example somatic mutations whereas chimeras are the result of genetic mixing by genetically distinct individuals (Santelices 2004b).

organisms, including in the functional-form models of sec. 1.2.2. Santelices (2004a) suggests that the way in which unitary, clonal, and coalescing organisms interact and respond physiologically and morphologically to their abiotic and biotic environment will be different. For example, the number of erect axes formed from basal crusts of *Gracilaria chilensis* and *Mazzaella laminarioides* increases with the number of spores forming the coalescence, and within the first 60 days of life, the chimeric "individuals" grow faster than the unitary ones (Fig. 1.6b): such differences are likely to influence resource acquisition in the field (Santelices *et al.* 2010).

(a)

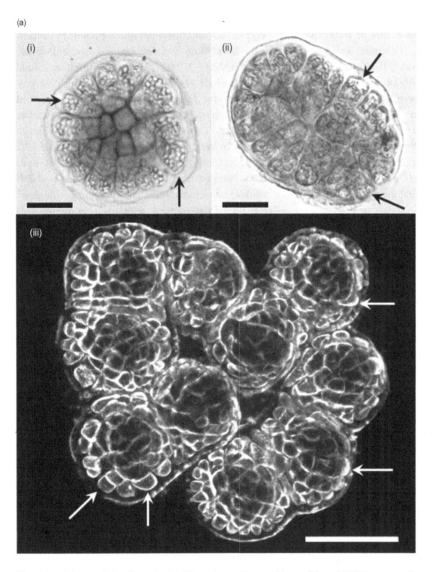

Figure 1.6 (a) Sporelings of *Gracilaria chilensis* formed by one (i), two (ii), and 10 (iii) spores. Observe the proportional reduction in free marginal cells with increasing number of fusions. Arrows indicate free marginal cells. Scales are 50 μm in (i) and

For species diversity, many indices are based on the numbers of individuals but an individual clonal-modular organism can be very large and cover a considerable surface area of substrate (Santelices 1999; see sec. 3.5.1). The ability of seaweeds to coalesce also raises the question of what constitutes an individual (Santelices 1999). Furthermore, self-thinning rules also do not apply equally to clonal versus unitary seaweed (Scrosati 2005; sec. 4.2.3). There is clearly a need for a more holistic model that combines the traditional functional-form groupings of Littler and Littler (1980) with developing theories on modularity and coalescence (e.g. Santalices 2004a; Scrosati 2005).

1.3   Seaweed cells

Although there is interaction between the morphology of the whole seaweed and the environment, the physiological responses to the environment, as well as the mechanisms by which the overall morphology

(b)

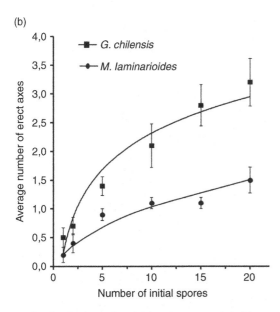

Caption for Figure 1.6 (*cont.*) (ii) and 100 μm in (iii). (b) Average number of erect axes differentiated by 30-d sporelings of *Gracilaria chilensis* and *Mazzaella laminarioides* as a function of the number of initial spores forming the sporeling. (From Santelices *et al.* 2010, reproduced with permission.)

is generated, occur within the individual cells (Niklas 2009). Cells are protected by walls and membranes, and are compartmentalized with membrane-bound organelles, and it is through these membranes and walls that contact with the environment must take place. The structure and composition of cell components thus provide a necessary background to the study of physiological ecology.

Certain components and functions of algal cells are similar to (though not necessarily identical to) the systems worked out in other organisms (e.g. rats or bacteria). Mitochondrial structure and function, genetic material and its translation into proteins, and membrane structure are fundamental features of eukaryotic cells. Other cell components are distinctive in the algae; these include cell wall composition and structure, flagellar apparatus, the cytoskeleton, and the thylakoid photosystem structure. See Pueschel (1990), Van den Hoek *et al.* (1995), Larkum and Vesk (2003), Katsaros *et al.* (2006) and Graham *et al.* (2009) for reviews of algal cytology; see Buchanan *et al.* (2000) and Beck (2010) for reviews of higher plant cell biology.

Algal cells also contain unique structures, many of which contain bioactive secondary metabolites. Brown algal cells characteristically contain physodes (Fig. 1.7), phlorotannin-containing vesicles that fulfill a wide range of roles at the cellular and organismal level including cell wall formation, wound healing (sec. 2.6.4), adhesion of propagules to the substrate (sec. 2.5.2), protection from UV radiation (sec. 7.6), herbivore deterrents (sec. 4.4) and detoxifying metals (9.3.3) (Schoenwaelder 2002). The *corps en cerise* (cherry bodies), specific to *Laurencia* species, are storage vesicles for halogenated compounds which are trafficked to the cell surface where the released contents act as herbivore deterrents and anti-foulants (Salgado *et al.* 2008). "Gland cells" common in the red algae also contain secondary metabolites which defend against bacteria (Paul *et al.* 2006a; sec. 4.2.2).

1.3.1   Cell walls

Cell walls do not merely provide rigidity. They are essential to cell growth and developmental processes, such as axis formation in zygotes and branching in

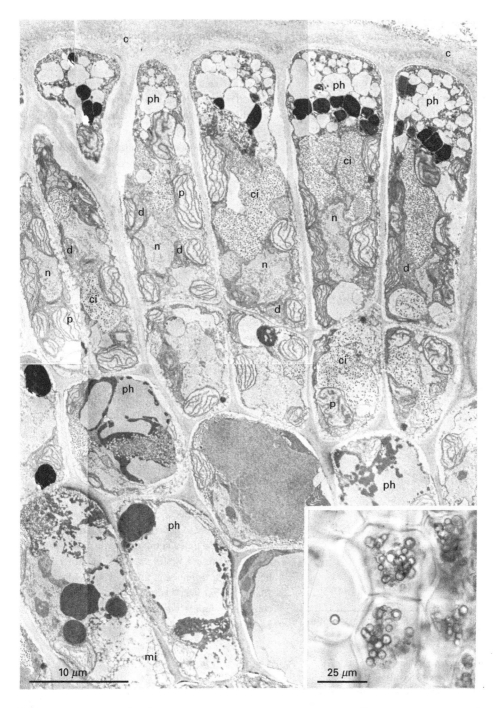

Figure 1.7 Cross section of the fucoid *Cystoseira amentacea* var. *stricta* Montagne (previously *Cystoseira stricta*) showing differentiation of tissues. The cells at the top of the view are the outer, meristodermal cells; those at the bottom are promeristematic. Inset shows fresh section stained with caffeine to reveal physodes; c, cuticle; ci, iridescent body; d, Golgi body; mi, mitochondrion; n, nucleus; p, chloroplast; ph, physode. (From Pellegrini 1980, with permission of The Company of Biologists.)

growing seaweed. Walls are crucial in mating, in the release and adhesion of reproductive cells, and as the outermost surface of many algae they are the first line of defense against pathogens and grazers (secs. 4.2.2 and 7.8). The abundance of matrix material relative to fibrillar components, the extensive sulfation, and the extensive intercellular matrix are characteristics of seaweeds that suggest environmental adaptations (e.g. to wave force and desiccation) (Kloareg and Quatrano 1988; see secs. 5.5 and 8.3.1). Cell walls also contain structural proteins, which are well studied in terrestrial plants and unicellular green algae (e.g. review by Cassab 1998) but have proven difficult to extract and characterize in red seaweeds (Deniaud *et al.* 2003). For *Palmaria palmata* the composition of structural proteins differed between blades that were soft versus rigid, indicating a role in "cell development and specialization" (Deniaud *et al.* 2003). So important are cell walls that Szymanski and Cosgrove (2009) consider it "more useful to think of the wall as another cell organelle ... and regulated by cytoplasmic and membrane processes that control pH, ion activities, reactive oxygen species, the concentration of metabolites, enzyme content and structural components".

Since the early days of electron microscopy, plant cell walls have been viewed as a meshwork of cellulose microfibrils in an amorphous matrix (Mackie and Preston 1974). There is a bewildering array of matrix polysaccharides in algae and considerable research effort has been exerted to identify and catalogue these, largely because of their potential commercial importance (Vreeland and Kloareg 2000; see sec. 10.3, 10.5 and 10.6). The biosynthetic pathways of polysaccharides are not fully understood, especially for brown seaweeds (Charrier *et al.* 2008), although putative pathways have been identified for *Ectocarpus* based on gene content; the next challenge is to rigorously test gene function (Michel *et al.* 2010a; sec. 5.5.2). The fibrillar components of algal walls are made of cellulose, β-(1,4) D-manan, β-(1,3) or (1,4)-D-xylan and, while they constitute just a small component (5–15%) of the wall dry weight, they are essential in providing tensile strength to the cells (Tsekos 1999; Lechat *et al.* 2000). The hypothetical model of algal wall structure

advanced by Kloareg and Quatrano (1988, Fig. 1.8a) has changed little (Michel *et al.* 2010a), and a similar but more detailed model has been proposed for green genus *Ulva* (Lahaye and Robic 2007; see sec. 5.5.2). Nevertheless, since the mid-1990s, the application of freeze–fracture electron micrograph and molecular biological techniques have resulted in substantial progress in understanding the sophisticated cellular machinery responsible for the synthesis of cellulose microfibrils and their assembly within plant and algal cell walls (reviewed by Tsekos 1999; Doblin *et al.* 2002; Saxena and Brown 2005).

Cellulose microfibrils are created by terminal complexes (TCs), comprised of cellulose synthases, which move through the cell membrane manufacturing microfibrils in a two-step process. First UDP-glucose is polymerized into β-1,4-linked glucan chains and second the chains are crystalized together into microfibrils (Tsekos 1999; Saxena and Brown 2005; Roberts and Roberts 2009). In higher plants, TCs are "rosettes" of six subunits but in the red, green, and brown seaweeds the TCs are linear. For example, the TCs of the brown seaweed *Pelvetia* are organized in a single line with 10–100 subunits, whereas those of the red *Pyropia yezoensis* are 2–3 rows deep. The structure of the TC shapes the dimensions and morphology of the cellulose microfibrils (see both Table 1 and Fig. 7 in Tsekos 1999). For instance, there are two forms of red algal microfibrils, either a "squarish" rectangular parallelepided or "flat and ribbon-like" orthogonal structure. Cellulose synthase genes have been sequenced for *P. yezoensis* and *Ectocarpus siliculosus* and they are being used to unravel the evolutionary origins of cellulose synthesis (Roberts and Roberts 2009; Michel *et al.* 2010a).

Cellulosic cell walls are made of layers of parallel cellulose microfibrils. The organization of these microfibrils is determined by the route that the TCs take as they move through the cell membrane. In terrestrial plants, this route is guided by the cortical microtubules, but much less is known for seaweeds except that in *Fucus* zygotes F-actin provides the roadmap (Bisgrove and Kropf 2001). In some genera, such as *Chaetomorpha* and *Siphonocladus*, the microfibrils in successive layers are oriented at steep angles to each

(a)   (b)

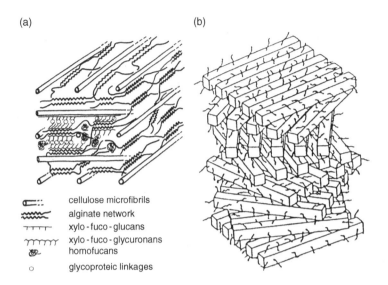

⊂⃗ᵏ	cellulose microfibrils
⌇⌇⌇	alginate network
⊤⊤⊤	xylo - fuco - glucans
⋎⋎⋎⋎	xylo - fuco - glycuronans
☞⌇	homofucans
○	glycoproteic linkages

Figure 1.8 Algal cell wall construction. (a) Brown algal wall showing fibrillar and matrix components. (b) Cell wall with helicoidal stack of hemicellulose molecules, as found in some green algae. The backbone of each molecule is represented by a rod, and the flexible side chains by squiggles. (Part a from Kloareg *et al.* 1986, with permission of Butterworth and Co.; b from Neville 1988, with permission of Academic Press, Inc.)

other (90º = orthogonal). In other algae, or in certain walls, including aplanospores of *Boergesenia forbesii*, eggs of *Silvetia compressa* (formerly *Pelvetia fastigiata*), zygotes of *Fucus serratus*, and vegetative walls of *Spongomorpha arcta* and *Boodlea coacta*, the angle changes much more slowly, giving a helicoidal arrangement typical of higher plants (Fig. 1.8b). In many algal walls, however, the microfibrils in each layer have no preferential orientation (Kloareg and Quatrano 1988; Tsekos 1999). Most red algae, for instance, have microfibrils randomly distributed within each layer although there are exceptions in which they are parallel (*Spermothamnion johannis*, *Polysiphonia denudata* (previously *P. variegata*), and *Herposiphonia secunda* f. *tenella* (previously *Herposiphonia tenella*) (Tsekos 1999).

Cellulose is the fibrillar material throughout the brown algae and most of the reds, but it is not the only fibrillar structural polysaccharide in algal walls as xylans also form microfibrils, and mannans form short rods. Xylans and mannans are particularly common in unicellular and coenocytic members of the Chlorophyta but, compared to cellulose, they have been little studied (Dunn *et al.* 2007; Fernández *et al.* 2010). Some seaweeds feature a biochemical alternation of generations in which different ploidy levels have different fibrillar or matrix polysaccharides. For instance, the diploid thallus of *Acetabularia* and *Codium* has mannans, and yet the walls of reproductive phases are mostly cellulose (Kloareg and Quatrano 1988). *Pyropia yezoensis* blades produce xylan while the filamentous sporophyte produces cellulose (Tsekos and Reiss 1994). No reason for these biochemical differences between generations has been deduced.

Most red seaweeds have an outer, multilayered, proteinacious cuticle covering their surface (Craigie *et al.* 1992) that may confer protection against herbivore grazing, desiccation, and bacterial degradation (Hanic and Craigie 1969; Gerwick and Lang 1977; Estevez and Cáceres 2003). The iridescence typical of some species including *Chondrus crispus* gametophytes and *Mazzaella* is the result of a thick multilaminated cuticle, in which many thin layers of alternating higher and lower refractive indices produce interference, as in a soap bubble. *C. crispus* sporophytes have fewer laminae (3–7) compared to gametophytes (6–14) and they are irregularly arranged, explaining why sporophytes of this species are not iridescent (Craigie *et al.* 1992). The utricles of the green seaweed *Codium vermilara* also have a cuticle, but its structure has yet to be detailed (Fernández *et al.* 2010). Other algae are well known for impregnating their walls with calcium carbonate, and these seaweeds may be vulnerable to ocean acidification (see secs. 6.5.3, 7.7 and Essay 4, Chapter 7). Martone *et al.* (2009) were the first to discover lignin and secondary cell walls in the calcifying red seaweed *Calliarthron cheilosporioides*, both characteristics of terrestrial plants, previously unknown in seaweeds.

The complexity and molecular specialization of wall surfaces are being revealed by the use of monoclonal antibodies and related techniques (Vreeland *et al.* 1987; Eardley *et al.* 1990; Vreeland *et al.* 1992; see Jelinek and Kolusheva 2004 for a review of methods). Different parts of a thallus have different wall structures. The high proportion of polyguluronic acid in adhesive alginate is well known (Craigie *et al.* 1984; Vreeland and Laetsch 1989; Vreeland *et al.* 1998; and see sec. 5.5.2). The difference between rhizoidal and thallus poles has been detected even in germinating zygotes and regenerating protoplasts, again using antibodies to different carbohydrate fractions (Boyen *et al.* 1988). In a detailed study of *Fucus serratus* sperm, Jones *et al.* (1988) were able to distinguish several regions, including the tip of the anterior flagellum (crucial in egg recognition; sec. 2.4), the mastigonemes on the anterior flagellum, and the sperm body. Localization of certain wall components also occurs during zygote germination, when carbohydrates are directed from their Golgi body to the appropriate piece of wall. The actin/Arp2/3 network is involved in this process in *Fucus* (see sec. 2.5.3).

1.3.2 Cytoplasmic organelles

Plastids and mitochondria are cellular organelles that originated via endosymbiosis of once free-living cyanobacteria (Fig. 1.1) and alpha-proteobacteria, respectively. As they became assimilated, most of their genes (90–95% for plastids) were transferred to the host nucleus, making them reliant on the host for essential gene products although some essential proteins are still made by the plastid (see below). These products are coded by nuclear DNA, synthesized in the cytoplasm and then imported into the organelles. This process is facilitated by transit peptides (TOC and TIC for plastids, and TIM and TOM for mitochondria), which are terminal peptides that attach to the nuclear-encoded pre-proteins and act as an address label that is recognized by membrane component(s) of the target cellular organelle. Once they have crossed organellar membranes, the transit peptides are cleaved (Reyes-Prieto *et al.* 2007; Graham *et al.* 2009; Weber and Osteryoung 2010; Delage *et al.* 2011). Thus, the once free-living prokaryotic cells became semi-autonomous organelles, but they retain many characteristics of their free-living ancestors. For example, both organelles divide by binary division but the genes for organelle division are now encoded in the nucleus which regulates division, explaining why plastids are unable to replicate in cell-free suspensions (Grant and Borowitzka 1984; Miyagishima and Nakanishi 2010).

Although most of the DNA associated with plastids and mitochondria was lost to the host's nucleus, the cpDNA and mtDNA that remains (within the nucleoids and ribosomes of the respective organelles) codes for essential core metabolic functions of the cells. The nucleus controls gene expression in organelles, termed antegrade signaling, and in return the organelles exert some control on nuclear gene expression by sending "retrograde" signals (Nott *et al.* 2006). Mitochondrial genome sizes of seaweeds are 25 836 bp for *Chondrus crispus*, 36 753 bp for *Porphyra purpurea* and 36 392 bp for *Fucus vesiculosus* (see Table 1 in Barbrook *et al.* 2010). The mitochondrial genomes include "core" protein-coding genes that are involved in oxidative phosophorylation and translation, and also RNA genes coding for large (LSU) and small (SSU) subunits of ribosomal RNA (rRNA). The cpDNA of almost all photosynthetic organisms contains a "core of genes" that are responsible for photosynthesis, including genes for photosystem I and II, cytochrome b6f, ATP synthase, RuBisCO (ribulose-1,5-bisphosphate carboxylase/oxygenase) and components of LSU and SSU (see Table 3 in Barbrook *et al.* 2010). Plastid genomes of several seaweeds have been fully sequenced, including the floridiophyte red seaweeds *Gracilaria tenuistipitata* (183 883 bp) (Hagopian *et al.* 2004), *Calliarthron tuberculosum* (178 981 bp), *Chondrus crispus* (180 086 bp), and *Grateloupia lanceola* (188 384 bp) (Janouškovec *et al.* 2013), the bangialean red seaweeds *Porphyra purpurea* (191 028 bp; Reith and Munholland 1995) and *Pyropia yezoensis* (191 954 bp; Smith *et al.* 2012), the brown seaweeds (*Ectocarpus siliculosus* (139 954 bp) and *Fucus vesiculosus* (124 986 bp) (Le Corguillé *et al.* 2009), and the greens *Bryopsis hypnoides* (153 429 bp) (Lü *et al.* 2011), *Acetabularia* which has the largest cpDNA of all photosynthetic organisms (1 500 000 bp; Mandoli 1998a), and *Codium fragile* which has the smallest at 89 000 bp (reviewed by Simpson and Stern 2002).

The mode of plastid and mitochondrion inheritance from parent to offspring varies between the seaweed phyla, and with the mode of reproduction (oogomy, anisogamy, and isogamy). For brown seaweeds that are oogamous, including *Fucus vesiculosus*, *Saccharina angustata* (formerly *Laminaria angustata*), and *Alaria esculenta*, mitochondria and plastids are inherited maternally (Motomura 1990; Kraan and Guiry 2000; reviewed by Motomura *et al.* 2010). Sperm plastids in *Saccharina angustata* zygotes remain small and do not divide, although they do survive, whereas mitochondria are enclosed in endoplasmic reticulum and digested in lysosomes (Fig. 1.9a) (Motomura 1990). For the isogamous brown seaweeds *Ectocarpus siliculosus* and *Scytosiphon lomentaria*, plastids are inherited from either

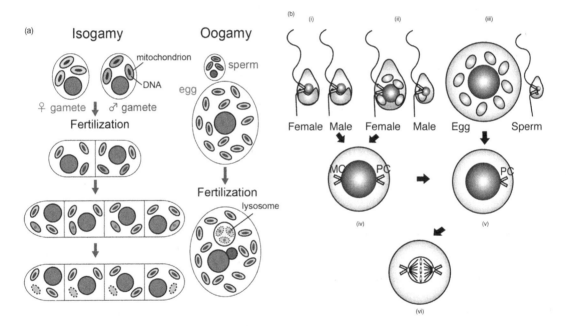

Figure 1.9 (a) Diagram of the cytoplasmic inheritance of mitochondria in isogamy (*Scytosiphon lomentaria*) and oogamy (*Saccharina angustata* – previously *Laminaria angustata*) in brown algae. In isogamy, mitochondrial DNA (or mitochondria) is selectively eliminated from the four-celled sporophyte after fertilization. In oogamy, sperm mitochondria are digested in the lysosome soon after fertilization. (b) Schematic representation of paternal inheritance of centrioles in fertilization of brown algae. (i) Isogamy, (ii) anisogamy, (iii) oogamy. In isogamy and anisogamy, the female gamete attracts the male gamete via sexual pheromones after settling. (iv) Immediately after fertilization, the zygote has two pair of centrioles (= flagellar basal bodies) derived from male and female gametes. (v) Subsequently, the maternal centrioles selectively disappear.(vi) Just before mitosis, the paternally derived centrioles duplicate and each pair of centrioles locates to the opposite pole of the spindle. MC, maternal centrioles; PC, paternal centrioles. (Part a from T. Motomura *et al.* 2010, Figure 1; b from Nagasato 2005, Figure 1, reproduced with permission.)

parent (biparentally) whereas mitochondria are maternally inherited. However, the timing of male mtDNA degradation in isogamous browns differs from that of oogamous species: mtDNA from male and female gametes survives in the offspring until the four-cell stage, at which time the male mtDNA is selectively broken down – the mechanism for this selective process is unknown (Peters *et al.* 2004a; Kimura *et al.* 2010). In all brown seaweeds, centrioles are inherited from the male gamete although, again, the timing of female centriole degradation differs between reproductive modes (Fig. 1.9b): for oogamous browns, the female centrioles disappear during oogenesis and the male centrioles are subsequently introduced as flagella basal bodies while in anisogamous and isogamous reproduction, centrioles from both parents are present in the zygote, but the maternal pair are subsequently degraded (Nagasato 2005).

Maternal inheritance of both plastids and mitochondria is common for green algae. For *Bryopsis maxima* and *Derbesia tenuissima* the male cpDNA and mtDNA degenerate during sperm gametogenesis (Lee *et al.* 2002). For *Acetabularia caliculus* and *Dictyosphaeria cavernosa* degeneration occurs in the zygote, following fertilization (reviewed in Lee *et al.* 2002). There are exceptions, however, and for *Ulva compressa*, the cpDNA is from mt+ (mt = mating type), whereas for some crosses of different genetic lines, mtDNA can be inherited from mt+, mt-, or both (Kagami *et al.* 2008; Miyamura 2010).

Plastids

Plastid diversity in eukaryotes is remarkable and reflects the numerous gains, losses, and replacements via endosymbiosis (Howe *et al.* 2008; sec. 1.1), and the wide diversity of plastids and their various functions are reviewed by Wise (2007). The term "chloroplast" has historically been used to describe plastids from all algal lineages and this usage is still common. However, it is also used to refer specifically to the plastids of higher plants and green algae that contain chlorophylls *a* and *b* (Purton 2002; Howe *et al.* 2008), and other terms exist, e.g. rhodoplast for red algal plastids (Wise 2007). Here, we follow Graham *et al.* (2009) and use

"plastid" as a general term to encompass the light-harvesting plastids of the red, green, and brown algal lineages. The plastids of green and red algae have two outer membranes as a result of primary endosymbiosis, while secondary endosymbiosis resulted in the 3–4 membranes typical of brown algae (Fig. 1.10; Larkum and Vesk 2000; Archibald 2009; see sec. 5.3.1 and Fig. 5.8). Some siphonous green algae (*Caulerpa, Halimeda, Udotea,* and *Avrainvillea*) have colorless amyloplasts in addition to plastids, which are used for starch storage (van den Hoek *et al.* 1995). In terrestrial plants, amyloplasts are involved in gravitropism (Palmieri and Kiss 2007), but such a role in algae has not been reported except for their role in orientation of regenerating rhizoids in *Caulerpa* (sec. 2.6.4).

Photosynthetic algal cells contain one or more plastids (some Acetabularia species may have 10^7–10^8 per giant cell). In thick thalli, medullary cells that are shaded from light and blocked from rapid gas exchange by overlying cortical cells usually lack plastids or have vestigial plastids. Plastids have characteristic shapes that are useful for taxonomy; they may be discoidal, stellate, band-shaped, or cup-shaped (Larkum and Barrett 1983; van den Hoek *et al.* 1995; Graham *et al.* 2009). All have photosynthetic pigments in thylakoids (and red algae also have phycobiliproteins that occur on thylakoids), and the arrangements of thylakoids are taxonomically significant (Larkum and Vesk 2003; Su *et al.* 2010; see sec. 5.3.1 and Fig. 5.8). Red algal thylakoids are single while brown algae typically have three per lamella, and in green algae they range from two to many. Some plastids in the more advanced Florideophycidae have a peripheral thylakoid just inside the plastid envelope (Fig. 1.10a), and brown algal plastids have endoplasmic reticulum tightly associated on the outside, termed periplastidal endoplasmic reticulum (PER) (Fig. 1.10b). Some plastids have a pyrenoid (again there are characteristic shapes) comprising chiefly RuBisCO; in others, this key Calvin cycle enzyme is dispersed in the matrix (Tanaka *et al.* 2007). Some pyrenoids (e.g. those of *Bryopsis maxima)* are also the sites of nitrate reductase (Okabe and Okada 1990).

The differences among the shapes and arrangements of plastids are used as key characters to assess phylogenetic relations (although in some cases

(a)

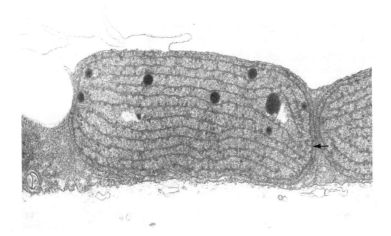

(b)

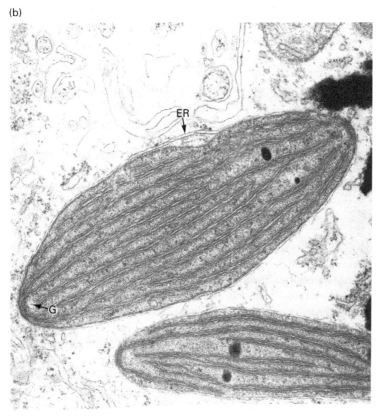

Figure 1.10 Algal plastids. (a) Plastid of the red alga *Osmundea spectabilis* (previously *Laurencia spectabilis*) showing parallel single thylakoids and one thylakoid (arrow) surrounding the others, just inside the plastid membrane. (b) Plastid of a brown alga (*Fucus* sp.) showing characteristic triple thylakoids, the genome (G), and endoplasmic reticulum (ER) surrounding the organelle. Scale: 1 μm (Courtesy of Dr T. Bisalputra.)

similarities may represent convergent evolution). For example, a specific organization of stellate plastids of brown seaweeds are a key character that, in combination with molecular phylogenetics, led Peters and Clayton (1998) to establish the brown algal order Scytothamnales. Although the significance to physiology is not entirely clear, differences in plastid shape and sizes may reflect different evolutionary response to the reduction in light absorbance by "packaged" pigments, as compared to pigments in uniform solution (the "package effect") (Osborne and Raven 1986; Dring 1990; sec. 5.3.2)

The plastids of siphonous green algae are unusual in that they have much greater autonomy than those of other algae and higher plants (Lü *et al.* 2011). For example, if the protoplasm of *Bryopsis* is experimentally squeezed out of the cell (or sucked out by herbivores), the plastids can aggregate and form a special integument around themselves. This extra membrane encloses a small amount of cytoplasm. Isolated plastids from *Codium* and *Caulerpa* do not swell or burst in distilled water. The integument may prevent the plastids of these species from being digested when they are eaten by saccoglossan mollusks (sea slugs), thus allowing the plastids to continue photosynthesis in a type of symbiotic relationship (termed kleptoplasty) with the animal (Grant and Borowitzka 1984; sec. 4.5.3).

Plastids may migrate within a cell. Dramatic diel migration of plastids takes place in *Halimeda* (Drew and Abel 1990). More than 100 plastids from each surface utricle pass along cytoplasmic strands through narrow constrictions into medullary filaments. They end up below the carbonate exoskeleton (Fig. 1.11), leaving the plant looking bleached. Inward migration is triggered by the onset of darkness (at any time of the day). Outward migration begins before dawn, apparently on an endogenous rhythm. Endogenously controlled plastid movements are also evident in the formation of new *Halimeda* segments. A proto-segment of colorless filaments is first formed and then, at night, plastids stream into the new segment, aided by microtubules and microfilaments, which becomes fully green within 3–5 h (Larkum *et al.* 2011). In the intertidal species *Dictyota*, plastid movements are a mechanism of photoprotection, with plastids moving away from the

light during midday (Hanelt and Nultsch 1990, 1991). In *Caulerpa*, amyloplasts are even more mobile than plastids and are transported on microtubules, whereas plastids are moved by the actin-myosin system (Menzel and Elsner-Menzel 1989).

1.3.3 Cytoskeleton and flagella apparatus

The cytoskeleton in algal cells plays fundamental roles in mitosis, cytokinesis, karyokinesis, polarity, and morphogenesis in zygotes and vegetative cells, organelle trafficking (including plastids and physodes) and cytoplasmic streaming, cell growth, flagella apparatus, and wound healing (e.g. Menzel 1994; Fowler and Quatrano 1997; Schoenwaelder and Clayton 1999; Katsaros *et al.* 2006; Bisgrove 2007). In algae, the cytoskeleton is composed of microtubules (MTs, ~25 nm diameter) and filamentous actin (F-actin) microfibrils (~5–7 nm), which are assembled and disassembled from their component protein subunits of tubulin and actin respectively (e.g. Hable *et al.* 2003; Taiz and Zieger 2010). Fucoid zygotes have been studied extensively as a model system for fertilization, polarization, and cell division because of their large size, accessibility and the apolar nature of the egg (sec. 2.5.3). The cytoskeleton of green seaweeds is also well studied, especially *Acetabularia*, but the picture is less complete for red seaweeds although the application of immunolabelling and confocal microscopy to *Griffithsia japonica*, *Aglaothamnion oosumiense*, and protoplasts of *Palmaria palmata* has revealed details of its organization (Garbary and McDonald 1996; Kim *et al.* 2001a; Le Gall *et al.* 2004), and actin genes and their expression have been reported for *Pyropia yezoensis* (Kitade *et al.* 2008).

An example of the way the cytoskeleton shapes cells is seen in the development of cysts in *Acetabularia* (Menzel 1994; Mandoli 1998b; Mine *et al.* 2008). During vegetative growth (Fig. 1.12), bundles of actin microfibrils are arranged along the axis of the cell (Fig. 1.13). After the cap has formed, the diploid primary nucleus undergoes one round of meiosis then divides mitotically into several thousand haploid "secondary" nuclei. The nuclei migrate along the actin microfibrils to the rays of the cap, where cyst formation

(a)

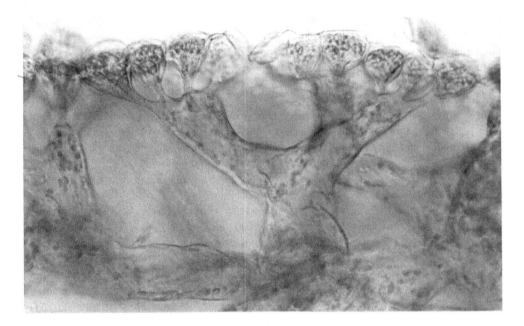

(b)

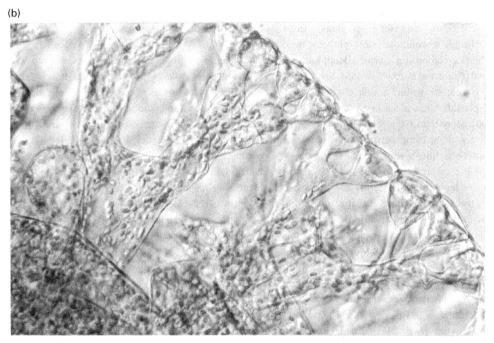

Figure 1.11 Migration of plastids in *Halimeda*. (a) Daytime cross section shows surface (primary) utricles packed with plastids. (b) Nighttime section shows that the chloroplasts have migrated below the calcified layer into the secondary utricles and medullary filaments. (From Drew and Abel 1990, with permission of Walter de Gruyter and Co.)

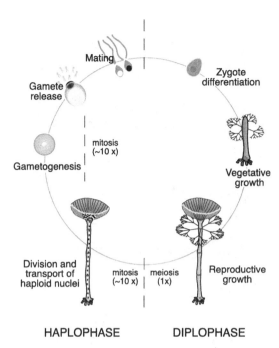

HAPLOPHASE   DIPLOPHASE

Figure 1.12 The basics of the haploid and diploid portions of the life cycle of *Acetabularia acetabulum*. Major portions of the life cycle that compromise a unit of development, i.e. all of gametogenesis, are united by an arc subtending the relevant label. For the sake of clarity, cartoons of the organism are not to scale. (From Mandoli 1998a, reproduced with permission.)

occurs. Then the actin network, which served as an organellar transport system in the vegetative phase, begins to disintegrate; once this is complete, the position of the secondary nuclei becomes fixed and cyst formation ensues (Fig. 1.13). The entire surface of each secondary nucleus acts as a microtubule organizing structure. The microtubules radiate, then draw organelles, including a single plastid, towards the nucleus and finally the nucleus, organelles and some cytoplasm become enclosed in a cyst. The shape of the developing cyst is determined by cytokinetic actin rings. In the final stage, the rings contract and cyst walls form (Fig. 1.13).

Brown and green seaweeds have flagella-bearing motile cells at some stage of their life cycle whereas red seaweeds do not. This difference between algal phyla is linked to the presence of centrioles in green

and brown seaweeds (and animals), but in red algae along with angiosperms and higher fungi, centrioles have been lost (Azimzadeh and Marshall 2010). Centioles act as microtubule organizing centers during mitosis and are essential in flagella synthesis (Azimzadeh and Marshall 2010; Kitagawa *et al.* 2011). In flagella the centrioles are known as "basal bodies", and their role is to synthesize the flagella at the cell membrane surface. Basal bodies comprise of a central cartwheel that has a hub from which nine spokes radiate, and each spoke joins via a pinhead to one of the nine triplet microtubules (Fig. 1.14a). There is a transition zone between the basal body and flagellum. The flagella themselves consist of a $9 + 2$ arrangement of microtubules, collectively called the axoneme (Fig. 1.14b), a structure that is highly conserved among flagella-possessing eukaryotes (Ginger *et al.* 2008; Marande and Kohl 2011). The nine outer tubules are doublets (A- and B-tubules), connected by the protein nexin, and flagella movement is brought about by the actions of dynein motor proteins that are attached to the A-tubule (see Chapter 18 of Lodish *et al.* 2008).

The basal bodies of flagella are joined by striated fibers and are anchored into the cytoskeleton by four microtubular rootlets (one pair with two tubules, and one pair with three to five tubules). One of these rootlets anchors the "eyespot" in position (in those cells that have them). The flagella also have striated "system II" roots reaching back around the nucleus (van den Hoek *et al.* 1995). The two sets of striated fibers are involved in Ca^{2+}-dependent contractions and are made of centrin, an acidic phosphoprotein with a molecular weight of about 20 000. The centrin protein family is one of ~350 ubiquitous "eukaryotic signature proteins" that are highly conserved and critical for eukaryotic cell function (Salisbury 2007).

The mechanism of flagella synthesis, the intraflagella transport (IFT) system, was first identified in the unicellular green microalga *Chamydomonas reinhardtii*, which is a model organism for flagella/cilia structure and function (reviewed by Cole 2003; Vincensini *et al.* 2011). Flagella are manufactured from more than 500 component proteins (representing over 3% of the *Chlamydomonas* genome) that are synthesized within the cell cytoplasm, and then enter the

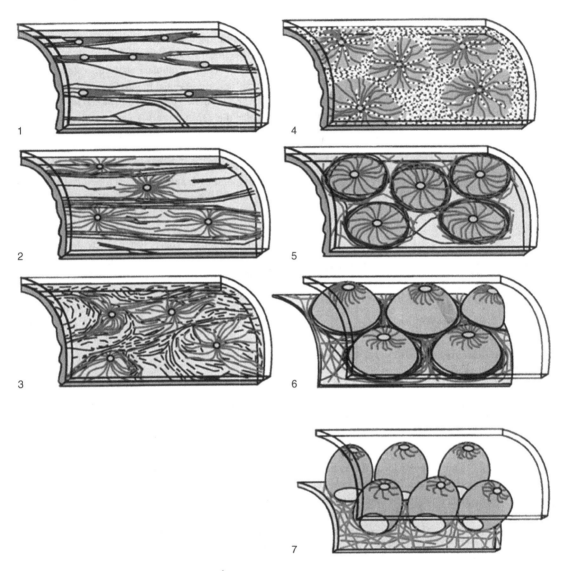

Figure 1.13 Schematic representation of cyst morphogenesis in *Acetabularia* in seven stages. Microtubules = dark gray lines; actin filament bundles = black lines; nuclei = light gray circles. (1) Migration of secondary nuclei along actin cables in the cap rays. (2) Beginning of immobilization of the nuclei and the extension of radial perinuclear microtubule systems. (3) Breakdown of actin cables causing irregular contractile events in the cytoplasm. Nuclear positions become rearranged. (4) Actin breakdown completed. Maximal radial expansion of perinuclear microtubules, gathering of chloroplasts and other organelles in disks around each nucleus. (5) Microtubules have become fragmented at their distal ends and the fragments gave rise to a second peripherally located microtubule-system, cytokinetic actin rings have formed around each domain. (6) Cyst domains begin to bulge out, actin rings contract. (7) Advanced state of contraction of the actin rings. Cyst protoplasts are being shaped. Note counter clockwise bending of perinuclear microtubules. This configuration eventually gives rise to microtubule band of the lid-forming apparatus. (From Menzel 1994, reproduced with permission.)

(a)

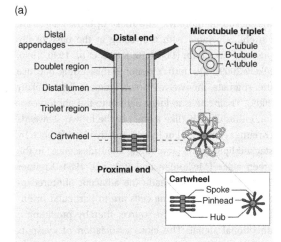

(b)

(i) Centrosome

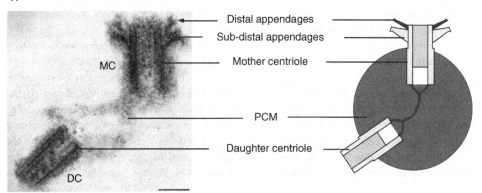

(ii) Cilia/flagella

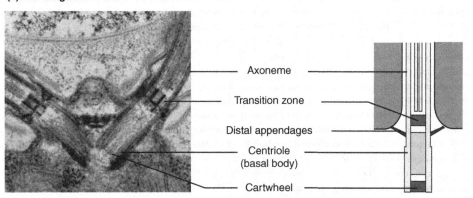

Figure 1.14 (a) Centriole structure. Centrioles are microtubule arrays composed of nine triplets of microtubules organized around a cartwheel structure. The triplets are connected to the cartwheel through the A-tubule, the first to assemble during

(*cont. over*)

flagella through the basal body; these proteins can be regarded as the "cargo" of the IFT system (Fig. 1.15). At the basal body, the cargo proteins are "loaded" onto ITF particles which are then conveyed by motor proteins (kinesins) along the outer doublet microtubule to the construction site at the flagella tip (anterograde IFT). The cargo is unloaded and used to extend the flagella. Another group of motor proteins, dyneins, moves the ITF particles back to the flagella base (retrograde IFT), where they are re-loaded (Fig. 1.15; Marande and Kohl 2011). The IFT system also regulates flagellum length (see Vincensini et al. 2011).

Flagella of seaweed microstages have two key roles: locomotive and sensory. Locomotion is important for getting gametes together and helping propagules swim to the seabed (sec. 2.5.1). For brown seaweeds, the anterior flagellum bears hairs (mastigonemes), and pulls the gametes through the water, while the smooth, posterior flagellum acts as a steering rudder (Jékely 2009; Fig. 1.16). Flagella also act as specialized recognition and adhesion organelles during mating (sec. 2.4) and selection of a suitable substratum for settlement (sec. 2.5.1).

Most lineages of algae contain species with pigmented eyespots (stigmata) on their motile cells, but their structure, position, and function differs between groups (reviewed by Hegemann 2008; Jékely 2009). The eyespots of motile seaweed cells are patches of lipid droplets, orange or red because of associated carotenoid pigments. Those of green algae are located on the outermost region of the plastid, directly under the plastid membranes, and those of brown algae are closely associated with a swelling at the base of the posterior flagellum (Fig. 1.16; Kawai et al. 1990, 1996; Jékely 2009). The term eyespot is misleading because the stigmata themselves do not detect light (Jékely 2009). Their role is to focus light onto the photoreceptor, either directly (like a lens) in the brown seaweeds (Kreimer et al. 1991) or by constructive interference by stacked lipid layers (something like iridescence) in the green algae (Melkonian and Robenek 1984; Kreimer 2001). Eyespots also shade the adjacent photoreceptors when the swimming cells are in particular orientation relative to a light source, thereby providing a directional signal. The close association of eyespots with the microtubular rootlets of the flagella, and their placement relative to the flagella, are critical for co-ordinating phototactic swimming behavior of motile cells of green and brown seaweeds (Hegemann 2008; Miyamura et al. 2010).

Two photoreceptor proteins (channelrhodopsin 1 and 2) on the outer surface of the plastid of Chlamydomonas perceive light and a "photoreceptor current" is generated. This current is mostly carried by Ca^{2+} but also H^+ and K^+. Flagella currents are subsequently generated when the photoreceptor current reaches a critical level, and this results in an adjustment in the plane, pattern, and frequency of the flagella beating (Hegemann 2008). For the motile gametes of brown

Caption for Figure 1.14 (cont.) centriole assembly and the only complete microtubule in a triplet. The B- and C-tubules are incomplete microtubules. In vertebrates and in Chlamydomonas, the C-tubule is shorter than the A- and B-tubules and the distal end of the centriole is thus formed by doublet microtubules. The cartwheel is formed by a central hub from which emanate spokes terminated by a pinhead structure that binds the A-tubule of the microtubule triplet. The very distal end of the centriole is decorated by nine-fold symmetric distal appendages (or transition fibers) required for anchoring the centrioles at the plasma membrane when they act as basal bodies. (b) (i) In animal cells, centrioles form the core structure of the centrosome, the main microtubule-organizing center. Quiescent cells (GØ) or proliferating cells in the G1 phase of the cell cycle contain a single centrosome. The centrosome is formed by one mature centriole, the mother centriole (MC), and one non-mature centriole, the daughter centriole (DC), linked together and surrounded by a protein matrix called the pericentriolar material (PCM). In vertebrates, the mother centriole is decorated by two sets of ninefold symmetrical appendages: the distal and sub-distal appendages, required for ciliogenesis and for the stable anchoring of microtubules at the centrosome, respectively. The distal appendages are observed throughout eukaryotes, whereas the sub-distal appendages are only found in animal centrosomes. (ii) In animals as well as in most other eukaryotes, centrioles are also required for the assembly of cilia/flagella. Centrioles, often referred to as basal bodies in this case, dock to the plasma membrane through their distal appendages and template the assembly of the nine outer microtubule doublets of the axoneme, the cytoskeletal core of cilia/flagella. A distinct structure called the transition zone separates the basal body from the axoneme. Shown are electron micrographs of the Chlamydomonas flagellar apparatus. (From Azimzadeh and Marshall 2010, reproduced with permission.)

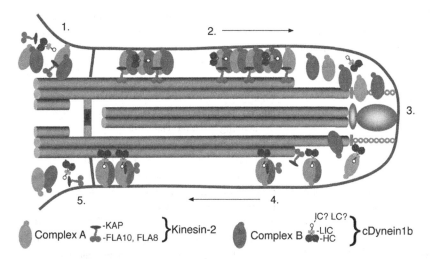

Figure 1.15 Intraflagellar transport. (1) Gathering of IFT particles and motors in the peribasal body region. (2) Kinesin-2-mediated anterograde transport of IFT complexes and inactive c Dynein1b. (3) Dissociation of IFT complexes. (4) Active cDynein1b transports everything back into the cell body. (5) IFT components are recycled to the cell body. cDynein1b: Cytoplasmic dynein1b; IC: Intermediate chain; IFT: intraflagellar transport; LC: Light chain. (From Marande and Kohl 2011, whose figure was modified from Pedersen *et al.* 2006, reproduced with permission.)

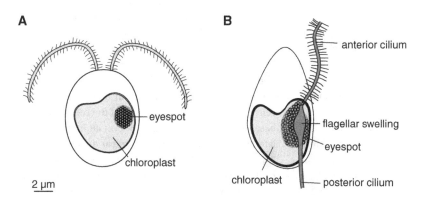

Figure 1.16 Location of eyespot in relation to flagella and chloroplasts in (a) a green alga and (b) a heterokont zoospore. Scale bar = 2 μm. (From Jékely 2009, reproduced with permission.)

seaweeds, the flagella swelling, upon which the stigma focuses light, is involved in photoreception, and also the posterior flagellum which contains at least two fluorescent compounds (a flavin and a pterin) that cause it to autofluoresce (Kawai *et al.* 1996; Fujita *et al.* 2005). Swimming *Ectocarpus* gametes, and other brown and green algal motile cells, roll as they swim, and when they are moving at a sufficient angle to the light, the photoreceptor receives flashes of light as the cell rolls. This stimulation is thought to cause the posterior flagellum to beat, acting as a rudder. When the cell is swimming parallel to the light, the photoreceptor is continually shaded by the cell (Kawai *et al.* 1990). The action spectrum of *Ectocarpus* phototaxis has peaks in the blue region (Kawai *et al.* 1990, 1996). Algal photoreceptors are discussed further in sec. 2.3.3.

Although red algal spores have no flagella, some are motile. The spores of 21 species of red seaweeds have "amoeboid gliding or shuffling movements" and for archeospores of *Pyropia pulchella* (formerly *Porphyra pulchella*) the movements are driven by the actin–myosin motility system (Pickett-Heaps *et al.* 2001; Ackland *et al.* 2007).

1.3.4   Cell growth

Cell growth is driven by water influx and is restricted by the cell wall. Plant and seaweed cells are normally turgid, because water tends to flow into them by osmosis (sec. 7.4). The layers of fibrils in the wall (sec. 1.3.1) resist swelling and stop net water influx. In terrestrial plants, cell growth is achieved by locally controlled loosening (yielding) of the cell wall in unison with water influx (reviewed by Szymanski and Cosgrove 2009); the role of expansins and auxins in cell growth is reviewed by Choi *et al.* (2008) and Perrot-Rechenmann (2010), respectively. Compared to terrestrial plants, there has been very little work on the physiological and molecular mechanisms underpinning algal cell expansion and growth but the finding, using bioinformatics, that *Ectocarpus* appears to lack known families of enzymes that are thought to be involved in cell wall expansion (cellulases, expansins, and alginate lyases) may stimulate new research in this field (Michel *et al.* 2010a).

Garbury and Belliveau (1990) list four modes of cell growth: (1) uniform throughout the wall which is typical of green plants; (2) localized in the tip of the cell with the remainder of the cell remaining rigid, for example the apical dome of the apical cell of *Pyropia yezoensis* sporophytes, and fucoid zygotes (Tsekos 1999); (3) band deposition typical of many red seaweeds, and (4) diffuse wall deposition, characteristic of the red algal orders Arcochaetiales and Ceramiales. Some of the mechanisms involved in the expansion of the apical cell have been elucidated for the filamentous sporophytes of *P. yezoensis* (reviewed by Tsekos 1999). The linear terminal complexes (TCs) are more abundant in the tips and, along with the Golgi apparatus, are responsible for synthesizing the new cell wall. Wall expansion is a dynamic process, with Golgi vesicles trafficking synthetic cell wall materials while also

delivering lytic enzymes that loosen the wall, allowing it to stretch; however, the wall thickness remains stable as new wall material is laid down by the TCs and Golgi apparatus.

Species that feature localized growth are useful as experimental material (e.g. Garbary *et al.* 1988; Fig. 1.17). The location of cell growth can be followed by labeling existing cell wall polysaccharides with a fluorescent stain such as Calcofluor White M2R (a brightener at one time used in laundry detergents) (Waaland 1980; Belliveau *et al.* 1990). If cell growth occurs by extension of existing wall material, the dye will be uniformly diluted. If, on the other hand, cell growth occurs by localized synthesis of new wall, dark bands will appear on the cells when seen under ultraviolet (UV) light, because the new wall will not be stained.

Intercalary cell extension in some Ceramiales, studied by Waaland and Waaland (1975), Garbary *et al.* (1988), and others, takes place through localized additions of wall material at each end of the cell (Fig. 1.17). The number and locations of the bands are characteristic of a species. In the *Antithamnion* illustrated, there is a strong basal growth band and a small apical band in axial cells, and only a basal band in determinate laterals. The location of band growth in this species is under apical control: if, for instance, the apex of a main axis is removed, the main growth band in those axial cells will switch to the other end of the cell, remaining basal relative to the nearest apex on an indeterminate lateral. This is an example of apical dominance (sec. 2.6.1). Cell growth may follow or be followed by cell division (sec. 1.3.5). Meristematic cells divide and grow repeatedly; other cells may stop growth and enter a stage of differentiation.

1.3.5   Cell division

Cell replication consists of two processes that do not necessarily happen together: nuclear division (karyokinesis) and cell division (cytokinesis). The brown algae, some green algae, and the Bangiophycidae have uninucleate cells, but coenocytic algae have many nuclei in cells, and thus karyokinesis and cytokinesis may be separated. In an unusual case in *Ascophyllum nodosum*, cell division can occur without mitosis,

(a)   (b)

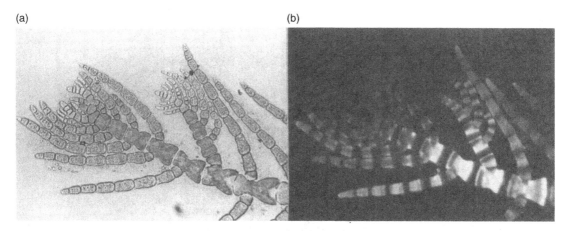

Figure 1.17 Cell growth in *Antithamnion defectum* visualized with Calcofluor White, as seen in bright field (a) and under UV light (b). A main axis with apical cell bears one indeterminate lateral and several determinate laterals. Under UV, dark bands of new, unstained wall are visible. The main axial and indeterminate lateral cells have two growth bands, the determinate laterals only one. Notice also the pit connections in the main axis. (From Garbary *et al.* 1988, reproduced by permission of the National Research Council of Canada from the *Canadian Journal of Botany*, vol. 66.)

perhaps a form of programed cell death (Garbary *et al.* 2009).

The cytological details of cell division have been studied particularly in the green algae, where the eight types of cell division that are recognized can be used as a taxonomic tool (van den Hoek *et al.* 1988, 1995). For the green seaweeds, the Ulvophyceae are characterized by having a persistent nuclear membrane (closed mitosis) and persistent telophase spindle microtubules. In coenocytic taxa (Dasycladales, Bryopsidales, Cladophorales, as defined by van den Hoek *et al.* 1988), mitosis is not immediately followed by cytokinesis. In uninucleate taxa (Ulvales, Codiolales), a cleavage furrow forms across the cell, and Golgi-derived vesicles are added to create the new cell wall. In the division of the apical cell of *Acrosiphonia*, more of the nuclei are partitioned to the apical cell than to the subapical cell; the apical cell remains meristematic, whereas the other cell rarely divides again (Kornmann 1970).

The application of new techniques for preparing samples for electron microscopy has allowed more accurate pictures of cytokinesis in brown seaweeds. Here, the mitotic spindle is more similar to that of animals than terrestrial plants and during mitosis one centrosome is at each mitotic pole (Fig. 1.18a). Spindle microtubules spread out from the centrosome

(Motomura and Nagasato 2004), and a small polar fenestration (pore) forms in the nuclear membrane, which otherwise remains intact until anaphase (Graham *et al.* 2009). The cytokinetic plane is set by the centrosomal position. The centrosomes act as microtubule organizing centers (MTOC), but there are no cortical microtubules in brown algae (Fig. 1.18a). For most, cytokinesis involves an outgrowth of the cell partition membrane, but *Sphacelaria* is an exception in which the plasma membrane becomes furrowed (note that previously this mechanism was considered the norm) (Katsaros *et al.* 2009; Motomura *et al.* 2010; Nagasato *et al.* 2010). For *Silvetia babingtonii*, the new cell partition membrane is formed by the flat plate cisternae (FC, unique to brown algae) together with Golgi vesicles (GVs), both of which accumulate at the future cytokinetic plane (Fig. 1.18b). They then fuse, forming an extended flat cisterna (EFC), and additional GVs supply fucoidin to the EFC which forms a membranous network (MN). The MN in turn develops into a membranous sac (MS), alginate deposition within the sacs begins, then the gaps between the sacs disappear and a continuous cell partition membrane is formed. Finally, cell wall materials including cellulose are deposited within the cell partition membrane and a new cell wall is formed (Nagasato *et al.* 2010).

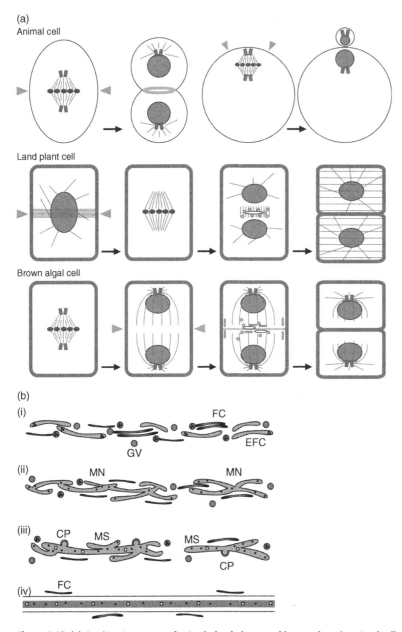

Figure 1.18 (a) Cytokinetic patterns of animals, land plants, and brown algae. In animal cells, the cytokinetic plane is determined by the position of the spindle (arrowheads), and is adapted during polar body formation. Cytokinesis proceeds via a contractile ring of actin (gray band). In land plant cells, there is no centrosome, and cortical microtubules (MTs) are well developed. The cytokinetic plane is determined by the microtubular preprophase band (arrowheads). Cytokinesis proceeds by outgrowing of a cell plate, which is mediated by the phragmoplast. Golgi vesicles participate in cell plate formation. In brown algal cells, a centrosome exists as a definite microtubule organizing center (MTOC), and no cortical MTs are observed. The cytokinetic plane is determined by the position of two centrosomes after mitosis (arrowheads). Cytokinesis proceeds by outgrowing of

Red algae are notable for the extensive evagination of the nuclear envelope that occurs during mitosis and for their nuclear associated organelles (NAOs, previously termed "polar rings"), that substitute for centrioles as microtubule organizing centers. As for the browns, they exhibit closed mitosis except for polar fenestrations, and cytokinesis begins with characteristic centripetal furrowing of the plasma membrane (Graham *et al.* 2009; Ueki *et al.* 2009).

Mitosis frequently occurs on a diurnal rhythm, with most cell division taking place at night (e.g. Austin and Pringle 1969; Kapraun and Boone 1987; Makarov *et al.* 1995; Kuwano *et al.* 2008). Two mechanisms controlling diurnal patterns of cell division have been identified in seaweed. The first is an endogenous circadian clock, in which the diurnal cycle of cell division continues even when the light/dark cue is removed, i.e. seaweeds are grown under constant light or darkness (see sec. 2.3.3 and 5.7.2 for other examples of endogenous rhythms). This has been reported for some brown, e.g. *Laminaria* and *Pterygophora* (Lüning 1994; Makarov *et al.* 1995), and red, e.g. *Porphyra umbilicalis* (Lüning *et al.* 1997), seaweeds. The molecular mechanisms underlying circadian clocks have been well studied in terrestrial plants, the unicellular green algae *Chlamydomonas*, and cyanobacteria (e.g. Harmer 2009; Johnson 2010; Schulze *et al.* 2010), but not yet in seaweeds.

The second mechanism, "circadian gating", was first identified in a seaweed by Kuwano *et al.* (2008) for *Ulva compressa* (formerly *Enteromorpha compressa*). Here, the light–dark cycle drives cell-cycle progression. The gate for cell division is situated in G1 and is opened only following a specific time in the dark; during darkness a "dark-induced substance" accumulates within a

cell until a critical level is reached, at which time the "gate opener" is triggered and cells can enter S-phase and undergo mitotic division. Cells must also be at a critical size to enter mitosis; if they are too small they will not undergo mitosis even if the gate is open, but if a cell grows fast enough during the dark phase, they can undergo a second round of cell division. No endogenous clock is involved in this mechanism, because in continuous light or dark, cell division ceases immediately. Moreover, if the timing of the light–dark cycle is changed, cell division is immediately re-synchronized, which is too fast for regulation by an endogenous clock. The question of how widespread gating control is within the three seaweed phyla requires further testing (Kuwano *et al.* 2008).

1.4 Molecular biology and genetics

1.4.1 Advances in seaweed molecular biology

The "molecular revolution" has had a profound impact on most aspects of seaweed research (Essay 1, Fig. 1), initially for taxonomy, phylogenetics, and biogeography, and increasingly for ecology and physiology. Early applications to seaweeds included the sequence data for the small subunit of cytoplasmic ribosomal RNA for *Costaria costata* by Bhattacharya and Druehl (1988) to evaluate its relatedness to other organisms. Electrophoretic patterns of plastid DNA were used to assess populations and species over geographic areas (Goff and Coleman 1988) and to address kelp phylogeny (Fain *et al.* 1988), and plastid genome sequences were generated for *Griffithsia pacifica* and

Caption for Figure 1.18 (*cont.*) cell partition membranes, which is mediated by the actin plate (gray band). Golgi vesicles and flat cisternae participate in cell partition membrane formation. (b) Schematic diagram of transitional membrane configuration during cytokinesis in brown algae. (i) Fusion of GVs to FCs transforms FCs into EFCs. GVs put fucoidan into FCs. *Dots* show the accumulation of fucoidan. (ii) Fusion of EFCs and supply of GVs produce MN. (iii) MN grows into MS with disappearance of gaps in the MN. MSs appear in patches. Clathrin-coated pits (CP) are detected on the MS. Alginate indicated by *open squares* begin accumulating. (iv) MSs become a continuous new cell partition membrane. Crystalline cell wall material deposits within it. EFC, expanded flat cisterna; FC, flat cisterna, GV, Golgi-derived vesicle; MN, membranous network, MS, membranous sac. (Part a from Motomura *et al.* 2010; b from Nagasato *et al.* 2010, reproduced with permission.)

Essay 1 Molecular techniques and their profound impact on contemporary phycological studies

Gary W. Saunders

I expect that few grow up anticipating a career as a systematist, let alone the seaweed variety. It certainly wasn't the case for me. Growing up along the coast of Nova Scotia, I developed a deep and enduring love of marine biology – that part of the path was obvious. During undergraduate studies, I was required to take a botany course, phycology, as it was widely referred to at the time, and was thus the logical choice. Much of the course was spent scuba diving and identifying algae – I was hooked! Moving to graduate studies under the supervision of Dr Jack McLachlan, I was tasked with resolving the life history of *Rhodophysema georgei* in light of exciting discoveries for another species of this genus (DeCew and West 1982). Exploring the literature surrounding my research, I started speculating on the relationships among the taxa under study and without conscious effort morphed into the realm of systematics. A pivotal moment, while pontificating my views on acrochaete evolution, occurred when Dr Christine Maggs (completing postdoctoral studies in the lab), asked "have you read Kylin?" – "ch who"... He is now, of course, iconic to me.

Leading to the early 1980s, Kylin's system of classification had worked so well for the majority of red algae that it, regrettably, attained universal acceptance, impeding efforts for reform. In this light, the words of Papenfuss (1958), in his review of Kylin's (1956) exceptional volume on red algal systematics, ring ironic – "As a former student of Professor Kylin, I know that the highest reward that he would have liked for his labors would be, for this, his last work, soon to become obsolete as a result of the intensive studies certain to be inspired by it."

It was during this time of universal acceptance that I endeavored to transfer certain algae from the "primitive and ancestral" Acrochaetiales to the "more derived" Palmariales. Needless to say, my manuscript was repeatedly rejected. Combining naivety and the arrogance-of-youth, I took the outcome personally – the old guard blocking the work of a young up-start. Turns out this was not the case, it was the state of red algal systematics at the time.

A few had succeeded in rendering change. In proposing the family Palmariaceae, Guiry (1974) departed from the axiomatic features of female reproductive anatomy and post-fertilization to emphasize tetrasporangial development. Guiry (1978) later argued that there was little save cruciate tetrasporangia, a state reported for species in all of Kylin's orders, to ally the Palmariaceae to the Rhodymeniales, and the Palmariales was proposed. Shortly thereafter Van der Meer and Todd (1980) published on a new life-history type and Pueschel and Cole (1982) provided ultrastructural observations supporting Guiry's proposals (Saunders and Kraft 1997). Pueschel and Cole further recognized a number of segregate orders thus implementing the first major revisions to Kylin's system. What frustrated me in reading these works is why Guiry had taken over 4 years to establish the order Palmariales and Pueschel had limited phylogenetic speculation in his manuscript to a short paragraph in which a very putative association between what were then thought to be divergent orders was outlined. I was fortunate through conferences and other communications to acquire from both of these colleagues explanations for these perceived shortcomings. Indeed it was not personal, for these two exceptional scientists were also not able to publish fully, or in a timely manner, their ideas on red algal evolution. Something new was needed, a tool so strong that critics would have to provide justification for rejecting new ideas that reached deeper than "it's simply wrong".

Near the end of my MSc studies I attended a seminar by Dr Linda Goff. She talked about comparative genomics, that by looking at the DNA of organisms and comparing it, we could understand evolutionary relationships. The remainder of my path was clear – I had to learn and apply these tools to my systematic hypotheses.

I packed my cultures and moved to the lab of Dr Louis Druehl, the only place in Canada at that time applying molecular tools to macroalgae, and outlined my exciting research agenda, explained the excellent culture resources that I had amassed – this was to be our finest moment! Apparently not sharing my enthusiasm, the response was "we work on kelp here". And so it was. I collected kelp, and set about learning the tools of the trade. As it turns out, kelp are very interesting.

Techniques were primitive in the early days and 2 years (a day's task now) were dedicated to generating eight small subunit ribosomal DNA (SSU) sequences for kelp in an effort to confirm relatedness in the face of what was also a widely established system of classification. However, the SSU was too conservative to resolve relationships among most genera. All was lost, or was it? At this point another valuable aspect of molecular data was presented – they can be used to estimate past divergence dates. Using molecular clock analyses we predicted that the derived kelps shared a common ancestor as recently as 16, and at most 30, million years ago in contrast to the 200 million years postulated for the group at that time (Saunders and Druehl 1992). This result was exciting, and of course controversial, but it did match with paleontological (Estes and Steinberg 1988) and paleooceanographic (Lüning and tom Dieck 1990) data. Most importantly, publication could not be blocked by the "it's simply wrong" argument (although some tried). Molecular clock analyses have matured greatly since those fledgling efforts and have provided numerous insights into the timing of past events for which detailed fossil records are lacking, which is the case for most algal lineages (e.g. Silberfeld *et al.* 2010).

Although the previous was exciting, the task of confirming that the system of classification for the kelps was natural had not been accomplished. One of the strengths of molecular tools is that an entire genome worth of characters is available for exploration. If the first gene tested is too conservative, simply try something more variable. A second common marker in use at the time was the internal transcribed spacer region (ITS) of the ribosomal cistron. We had progressed to PCR by that point, and were even experimenting with direct sequencing of the amplicons (i.e. foregoing the tedious and time-consuming cloning steps), which allowed me to generate these data in just under a year. Although a bit too variable across all of the advanced kelp, the ITS did allow for "state-of-the-art" (methods that are now largely considered unacceptable) phylogenetic analyses of kelp evolutionary relatedness (Saunders and Druehl 1993). Something apparently went horribly wrong. Why was *Lessoniopsis*, a paradigm representative of the Lessoniaceae resolving deeply in the Alariaceae? Was the molecular systematics dream just a farce as advocated by critics at the time? On the exposed coast of British Columbia, Louis elucidated the anomaly for me – *Lessoniopsis*, it turns out, has the splitting of the Lessoniaceae, and the paired sporophylls of the Alariaceae. In this light, it was hard to comprehend how kelp classification had become so firmly accepted. Indeed the taxonomists who established the system, Setchell and Gardner (1925), commented "the tribe of the Lessoniopseae might perhaps be placed with equal propriety either under Lessoniaceae or under Alariaceae, since the sole genus, monotypic, has the characters of each of these families". The shortcomings fully acknowledged, in essence a challenge put forth to resolve the conundrum, overlooked as the classification rooted. Kelp systematics remain a passion and I still have students working on various aspects, notably a recent multi-gene phylogenetic study (Lane *et al.* 2006) that, although more robust and taxon-rich, supported the conclusions of our rudimentary study.

Molecular tools have had an impact on many questions outside the realm of phylogenetics. Owing to the observation in culture, less commonly from nature, of putative intergeneric hybrids of different kelps (see Druehl *et al.* 2005b, Table 1 for a summary), a myth was perpetuating that kelp could interbreed freely, and that traditional species concepts may not apply. This concept never sat well with my way of thinking and, indeed, other possible hypotheses to explain these morphologically anomalous individuals (putative hybrids), by direct development from female (parthenogenesis) and male (apogamy) gametophytes, were largely ignored (with notable exceptions; see Druehl *et al.* 2005b). We designed species-specific primers to test the parentage of putative culture-reared hybrid individuals (Fig. 1a) with the result that only one individual was truly a hybrid. In this case the hybrid was between the species *Saccharina angustata* and *S. japonica* (as *Laminaria* spp.), which are very closely allied. As discussed in Druehl *et al.* (2005b), promiscuity among kelp genera, although exciting speculation, is more myth than reality and failed the molecular test.

It is not news that identifying algal collections to known species (even for experienced systematists) is a frustrating task. Difficulties arise from key commonalities among algal species, viz. simple morphologies, rampant convergence, phenotypic plasticity in response to environmental conditions, incompletely resolved life histories (alternations of heteromorphic generations), and abundant cryptic species (Saunders 2005, 2008; Lindstrom 2008; Le Gall and Saunders 2010). We have thus come to rely increasingly on molecular tools for the resolution of species (e.g. Saunders and Lehmkuhl 2005) and for assigning cryptic field specimens to known species (e.g. Lane and Saunders 2005; Fox and Swanson 2007). Under the label DNA barcoding (Hebert *et al.* 2003) a substantial database of COI-5P sequences (as well as other markers) is being established to facilitate the rapid identification of any biological specimen (see www.boldsystems.org). Ultimately this endeavor will provide scientists and managers worldwide with a powerful ally in the important task of species identification. An obvious practical outcome will be the ability to identify rapidly and accurately introduced species to an area. A key initiative of my research group is to complete a contemporary floristic account of the marine macroalgae of Canada using the DNA barcode as a preliminary screening tool of species diversity and distribution (e.g. Saunders 2008, 2009; Lane *et al.* 2007; Kucera and Saunders 2008; McDevit and Saunders 2009). With over 9800 COI-5P sequences generated for various red and brown algal collections, we have uncovered well over 100 new records or species in the Canadian flora. This minimally represents an increase of *c.* 10%, which means that if you walk on a beach in Canada and pick up 10 different algal species, on average, one of them is not currently known to science, or at least

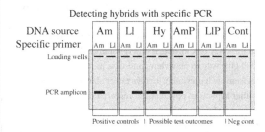

Detecting hybrids with specific PCR

Fig. 1(a) ITS sequences vary among species facilitating the design of specific PCR primers. In the example here, primers for *Alaria marginata* (Am) and *Lessoniopsis littoralis* (Ll) are reciprocally tested (positive controls) to confirm specificity. For progeny of hybrid cross experiments (or field-collected putative hybrids) there are three possible outcomes: Hy – both markers give positive amplification indicating hybridization; AmP – parthenogen of *Alaria*; or LlP – parthenogen of *Lessoniopsis*. Neg cont – is the negative control (no DNA added to the PCR reaction) and should have no amplification. (Original Saunders figure.)

Essay 1 (cont.)

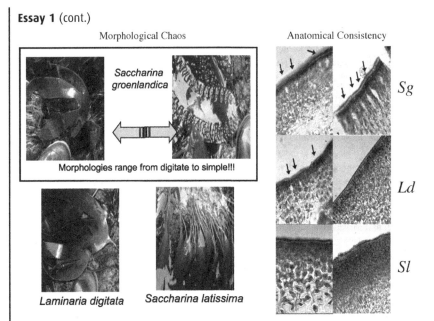

Fig. 1(b) DNA barcoding has uncovered an additional kelp species, the Pacific *Saccharina groenlandica* (*Sg*), in Eastern Canada, which masquerades as either *L. digitata* (*Ld*) or *S. latissima* (*Sl*) depending on the local environment. In agreement with the molecular data, the presence or absence of mucilage ducts (arrows) in the blade and/or stipe distinguish these three species. (Modified from McDevit and Saunders 2010.)

to the Canadian flora. In short, molecular data are completely changing perspectives of algal biodiversity and biogeography. These discoveries are not limited to small fuzzy reds and browns, even kelp species have gone undetected. McDevit and Saunders (2010) uncovered *Saccharina groenlandica*, currently considered a Pacific species, widely throughout the Atlantic Provinces, as well as in Hudson Bay. Whereas this seaweed can take on the gross morphology of *Laminaria digitata* or *Saccharina latissima* depending on the local environment, there are key anatomical differences in support of the molecular results (Fig. 1b).

   And what of those cultures for species that putatively belonged to the Palmariales rather than the Acrochaetiales? Following graduate school, I moved to Melbourne in search of mentorship with a significant figure in traditional red algal systematics, Dr Gerry Kraft. We embarked on a substantial phylogenetic investigation of many red algal lineages, generally emphasizing the Gigartinales sensu lato (see Saunders and Kraft 1997; Saunders and Hommersand 2004; Le Gall and Saunders 2007). During those critically formative years, I had the pleasure of accomplishing many objectives with regards to red algal systematics, explore side projects dealing with a variety of chromophytic lineages (in collaboration with Dr Robert Andersen), and, yes, generate data and complete analyses with regards to my MSc work. In 1995, we published a phylogenetic study in which the molecular data supported my view of red algal evolution (Saunders *et al.* 1995). Many lessons were learned along the way, adventures explored and discoveries made. To the graduate students reading this essay, do not subscribe to dogma, challenge the paradigms, and work hard to test your hypotheses. Remember, research is not about vindication, or about being right; the objective is to advance your chosen field and make contributions to the body of knowledge that is science. I have no idea how long my own contributions will stand, only that eventually some will give way to new views and data as should be expected, indeed encouraged, in science. I am, however, certain that future revisions will inevitably involve molecular data, the profound impacts of these powerful tools sure to dominate systematics and other aspects of phycological research into the future.

* Gary Saunders is a Professor and Research Chair at the Centre For Environmental and Molecular Algal Research at the University of New Brunswick, Fredericton, NB, Canada. In recent years he has expanded his research from largely systematics based questions to encompass issues of algal biogeography and tempo and modes of speciation. His current work focuses on isolated island floras and the Canadian Arctic exploring species richness, but also the origins of that diversity in these recently colonized habitats and the connectivity to adjacent floristic regions.

Pyropia yezoensis (Li and Cattolico 1987, Shivji 1991): today complete plastid genome maps have been assembled for various seaweeds (e.g. Fig. 1.19). Molecular methods have proven invaluable in clarifying the identity of species with extreme morphological variation such as the Laminariales (Essay 1) and Fucales (e.g. *Durvillaea*, Fraser *et al.* 2009), and species within genera such as *Pyropia* and *Porphyra*

that have few obvious morphological features to distinguish them from one another (Sutherland *et al.* 2011). In biogeography, molecular methods have helped to resolve the origin of genera such as *Fucus* which was thought to be Atlantic because of the high species diversity there, but Coyer *et al.* (2006a) show a North Pacific origin, with *Fucus distichus* as the ancestral form. "DNA barcoding" is an approach to

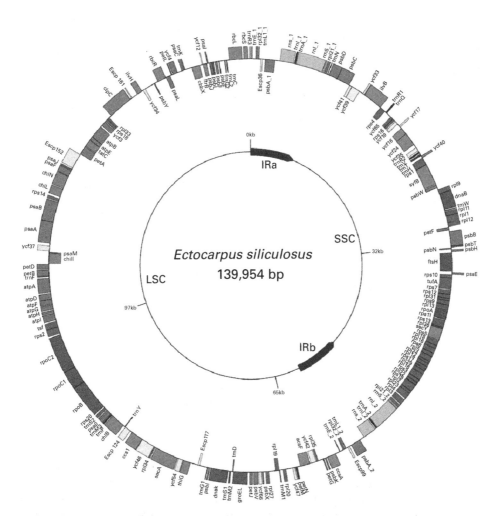

Figure 1.19 Plastid genome maps of *Ectocarpus siliculosus* and *Fucus vesiculosus*. Genes on the outside of the circles are transcribed clockwise, whereas those on the inside counter clockwise. Annotated genes are shaded according to the functional categories shown in the legend and the tRNA genes are indicated by the single-letter code of the corresponding amino acid. Abbreviations: IR, inverted repeats; SSC, small single-copy region; LSC, large single-copy region. (From Le Corguillé *et al.* 2009, reproduced with permission.)

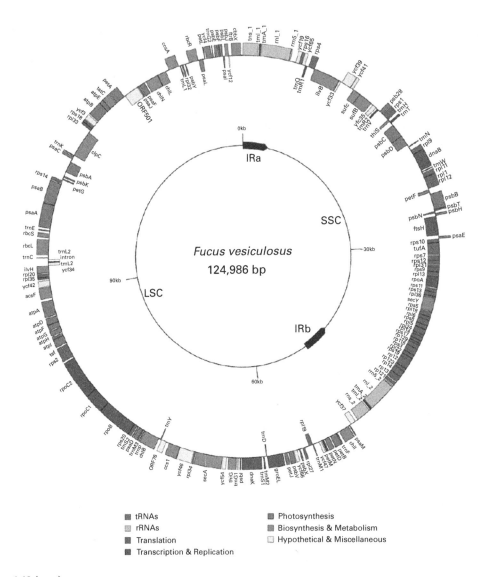

Figure 1.19 (*cont.*)

identifying species, based on gene sequences, and provides a rapid way of cataloging species; the mitochondrial gene cytochrome oxidase (Cox 1) used for animals has been applied to red algae (Saunders 2005). Combining molecular markers from the nuclear, plastid, and mitochondrial genomes has improved the resolution of phylogenies of problematic genera such as *Gracilaria* (Pareek *et al.* 2010), and

almost complete genome sequences are available for several seaweeds (Cock *et al.* 2010a; Collén *et al.* 2013), and genetic maps have been assembled (Heesch *et al.* 2010). The application of various molecular biological tools, including potential pitfalls, are discussed in Graham *et al.* (2009 Chapter 5) and Cock *et al.* (2010b).

We start this section by reviewing some recent advances in seaweed molecular genetics (the "omics") that are particularly relevant to seaweed physiological ecology, and the basis for the selection of model seaweeds. We then move on to classical genetic studies that used color mutants and cross-breeding to elucidate inheritance patterns, and then focus on non-Mendelian transmission of genetic material and hybridization.

Genomics is the study of an organism's entire genome and is a powerful tool that can tell us which genes are potentially available for use by an organism. Cells perceive external stimuli and respond by switching the appropriate genes on and off; gene expression is mostly regulated at the level of transcription (*transcriptomics*) and controlled by transcription factors (TFs) which function as "switches" (Rayko *et al.* 2010). For example, Collén *et al.* (2006) used Expressed Sequence Tags (ESTs) to compare the genes expressed by *Chondrus crispus* exposed to desiccation stress with non-desiccated controls and thereby identify putative genes involved in desiccation tolerance (Essay 5, Chapter 7). In their transcriptomic study of *Ectocarpus siliculosus*, Dittami *et al.* (2009) have begun to unravel the molecular mechanisms that underpin regulation and acclimation to physiological stress. Gene expression is not static, however (Clark *et al.* 2010). For instance, in an EST study of *Sargassum aquifolium* (formerly *Sargassum binderi*), genes coding for alginate synthesis were not detected: these genes must be present in *Sargassum* but it seems that at the time of collection alginate synthesis was not taking place due to slow rates of growth (Wong *et al.* 2007). The absence of an expressed gene should therefore be interpreted with due caution (Clark *et al.* 2010).

Genes code information for protein synthesis, and *proteomics* studies which proteins are available and their cellular functions. This is an emerging area of research for seaweeds, with *Gracilaria* being one of the first seaweeds in which proteome annotation has been attempted (Wong *et al.* 2006). Proteins in turn synthesize metabolites and the study of metabolite production is termed *metabolomics*. Metabolic profiling in combination with gene expression analysis is a powerful technique that can be used to link physiological processes to the underlying genetic control. Gravot *et al.* (2010) used these techniques for *Ectocarpus siliculosus* to examine the effect of CO_2 and O_2

concentrations over a diurnal cycle on key metabolites such as citrate, glutamine, mannitol, and the expression of genes for carbonic anhydrase.

Several model seaweeds have been selected for genomic studies; this approach allows resources to be focused on understanding in great depth a particular organism, and the tools developed can be subsequently applied to other species. Peters *et al.* (2004b) proposed *Ectocarpus siliculosous* as the model brown seaweed, and its genome was the first seaweed to be fully sequenced (Cock *et al.* 2010). The criteria for *Ectocarpus* as model seaweed were: short life cycle (2 months), small and easy to grow in the laboratory, reproductive traits and life cycles are well known from classical studies (especially Dieter Müller's work), genetic crosses can be easily carried out, and it has a relatively small genome size (214 Mbp) so that sequencing was faster than for other candidate brown algae (~650 Mbp for the kelp *Laminaria digitata* and 1095 Mbp for *Fucus serratus*). The red seaweeds selected for genome projects are *Porphyra umbilicalis* (~270 Mbp) (Gantt *et al.* 2010; Chan *et al.* 2012a), and *Chondrus crispus* (105 Mbp) which represent the two major evolutionary lines of red seaweed (Florideophyceae and Bangiophyceae). *C. crispus* is the first red seaweed genome to be fully sequenced (Collén *et al.* 2013). Pearson *et al.* (2010) propose *Fucus* as a model seaweed for ecological genomics. *Ulva* is an obvious candidate for a model green seaweed (Waaland *et al.* 2004), as is *Acetabularia*, which is already a model system for nuclear–cytoplasmic interactions (Mandoli 1998a; sec. 1.4.3).

The complete sequencing of the *Ectocarpus* and *Chondus* genomes has led to some exciting discoveries (see *New Phytologist*, Volume 188(1), 2010; Collén *et al.* 2013). Despite its simple morphology, isomorphic life history and small genome size, *Ectocarpus* is an "advanced" brown alga, most closely related to the Laminariales, and has 16 256 protein-coding genes (Cock *et al.* 2010a). Genes coding for 23 enzymes were discovered that may enable *Ectocarpus* to grow epiphytically on kelp by protecting it from kelp defensive systems. It has a large family of genes coding for reactive oxygen species, thought to be adaptive to the extreme environmental fluctuations typical of the intertidal zone (see sec. 7.1). *Chondrus* has a rich diversity of genes, with 52% of those discovered being previously unknown. There were surprisingly few genes (12) responsible for starch biosynthesis, and some of the cellulose synthases are ancient, having been acquired before the primary endosymbiotic event (Collén *et al.* 2013). Genomic studies thus offer the opportunity for studying eukaryotic evolution, the molecular basis for adaptation to stressful environments, and novel metabolic pathways (Gantt *et al.* 2010; Kamiya and West 2010; Collén *et al.* 2013).

1.4.2   Seaweed genetics

Seaweed genetics has lagged behind that of unicellular algae and terrestrial plants, but the discovery of color mutants in the red algae in the 1970s substantially advanced our understanding of mating systems. However, such breeding experiments fell out of favor for some ~20 years, largely because of the time-consuming nature of crossing experiments, their limited applications in answering some genetic questions, and a strong interest in developing molecular techniques that could be applied to seaweed genetics for example, sex determination of *Gracilaria* (Martinez *et al.* 1999) and heterosis (hybrid vigour) in *Gelidium* (Patwary and van der Meer 1994). The field has come full circle, with scientists now using an integrative approach that combines classical genetic and molecular tools (e.g. Yan and Huang 2010).

Breeding experiments by J. P. van der Meer and co-workers using color mutants of red seaweeds, particularly *Gracilaria tikvahiae*, shed much light on mechanisms of genetic inheritance in seaweeds (reviewed by Kain and Destombe 1995). These studies began with the discovery of two spontaneous green mutants in gametophyte populations raised from spores (van der Meer and Bird 1977), which allowed a study of Mendelian inheritance. The two green mutants, with reduced phycoerythrin, were stable, different from each other, and recessive. Besides a rainbow of color mutants, van der Meer also accumulated morphological and reproductive mutants (van der Meer 1986a, 1990). Some of the mutations are located

in the plastid DNA and show non-Mendelian inheritance: Tetrasporophytes have the phenology of the maternal gametophyte (Fig. 1.20) (van der Meer 1978). Color mutants have been used to study the genetics of other *Gracilaria* species (see Plastino *et al.* 2003), other red seaweeds including *Champia parvula* and *Chondrus crispus* (Steele *et al.* 1986), *Porphyra purpurea* (Mitman and van der Meer 1994), *Pyropia yezoensis* (Niwa *et al.* 2009), and life histories (e.g. van der Meer and Todd 1980; Maggs and Pueschul 1989) including spore coalescence (Santelices *et al.* 1996).

Color mutants of *Gracilaria* sp. were also used to show the existence of mitotic recombination. Crossing-over of chromosomes normally occurs in meiosis but can also take place during mitosis, with the result that one daughter cell in a heterozygous diploid gets both copies of one gene (wild type, +, in the example illustrated), while the other cell gets both copies of the mutant gene (*grn*) (Fig. 1.21; van der Meer and Todd 1977). The sex-determining gene (mt^m/mt^f) is also involved in the recombination, so that the color patches become diploid male and female gametophyte tissue and produce diploid gametes.

In addition to the primary sex-determining locus (*mt*), there is a second sex-determining gene regulating the dioecious condition in *Gracilaria tikvahiae*. A spontaneous bisexual mutant (*bi*) produced strange results in crosses, except with normal haploid females, and even then the F1 had females, males, and bisexual individuals in a 2:1:1 ratio, suggesting that the mutation is expressed only in males (van der Meer *et al.* 1984). The bisexual allele cannot substitute for the female allele mt^f, and subsequent analysis (van der Meer 1986b) has suggested that the *bi* + allele actually represses expression of female-specific genes. The significance of the *bi* mutation is that it allows the production, by selfing, of homozygous diploids. Bisexuality (mixed-phase reproduction) is also common in haploid generations of other red seaweeds, but here mitotic recombination is not responsible because it can occur only in diploids. In *Dasysiphonia chejuensis* (Ceramiales), for example, haploid individuals can form tetrasporophyte, male and female reproductive structures. Here sexuality is

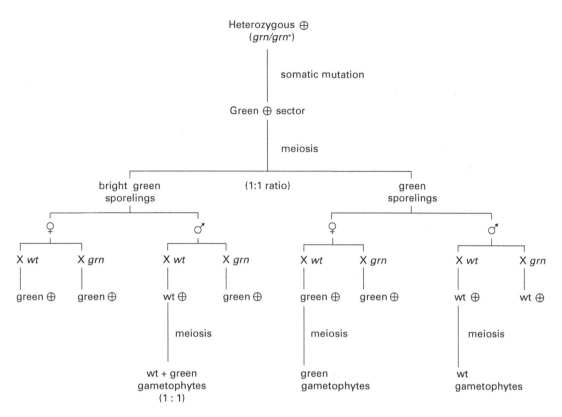

Figure 1.20 Non-Mendelian inheritance of a green somatic mutation in *Gracilaria tikvahiae*: wt, wild-type color (phenotype), grn, grn[+], the mutation for green color and its normal allele; ⊕, tetrasporophyte. (From van der Meer 1978, with permission of *Phycologia*.)

thought to be "controlled by multiple alleles at complex mating-type loci", compared to the two alleles (*mt* and *bi*) for *Gracilaria* (Choi and Lee 1996).

Unstable mutants of *Gracilaria tikvahiae* also occur (van der Meer and Zhang 1988). These may be the result of transposable (genetic) elements (transposons) during genome rearrangement. Insertion of a transposon disrupts the gene function, and removal restores it. The temporary change may be visible as an unstable mutation. Some transposons are autonomous, that is, they control their own insertion and excision. Transposons and/or retrotransposons have since been discovered in *Ectocarpus siliculosus* (Cock *et al.* 2010a; Dittami *et al.* 2011) and *Pyropia yezoensis* (Peddigari *et al.* 2008). In *Ectocarpus*, transposons were among the most variable

gene sequences detected, and they may be important in enabling genetic adaptation of seaweeds to environmental stress (Dittami *et al.* 2011). The genome size of *Chondrus crispus* appears to have increased rapidly within the last 300 000 years due to the action of transposable elements (Collén *et al.* 2013).

In addition to the classical (Darwinian) theories on the vertical transfer of genetic information, that is from parents to progeny, there is increasing awareness that other processes have also shaped the speciation of organisms. Horizontal gene transfer (HGT) is the movement of genes between species, via endosymbiosis in eukaryotes, and in prokaryotes via the viral transduction of genes between organisms. For example, genes coding for trehalose synthesis by *Ectocarpus* were

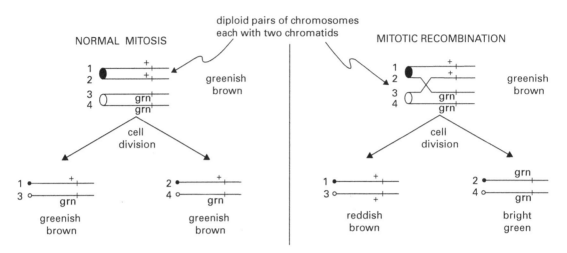

Figure 1.21 Mitotic recombination (right) compared to normal mitosis in heterozygous, diploid tetrasporophytes of *Gracilaria tikvahiae*. Diploid pairs of chromosomes are shown, each with two bivalents; the chromatids are numbered 1–4; +, wild-type color gene; grn, green mutant gene. (From van der Meer and Todd 1977, reproduced by permission of the National Research Council of Canada from the *Canadian Journal of Botany*, vol. 55.)

vertically inherited from an ancestral red algal symbiont but genes coding for D-mannitol metabolism arose via HGT from a species of Actinobacteria, and those responsible for sulfated-fucans typical of brown algal cell walls may have come from the eukaryotic host involved in the original primary endosymbiotic event that separated plants from animals (Cock *et al.* 2010a).

Furthermore, sexual reproduction and Darwinian theories on natural selection are not the only mechanism of generating genetic variation (reviewed in Monro and Poore 2009a). Modular organisms lack germ–soma segregation, and do not undergo meiosis but instead pass on genetic information via mitosis (somatic embryogenesis) to form ramets. Asexually produced *Asparagopsis armata*, for example, has a high level of phenotypic plasticity, illustrating that sexual reproduction is not a pre-requisite for morphological variability and, consequently, adaptation. The costs and benefits of sexual recombination versus clonal transmission of genetic material are discussed by Monro and Poore (2009a). Epigenetic phenomena, which affect gene transcription via changes in histone proteins and DNA methylation, but do not change the DNA sequence, are also likely to be important in generating genetic variation (Maumus *et al.* 2011).

The question of whether closely related seaweeds can hybridize has also benefited from molecular biological techniques (see Essay 1, Fig. 1). Before ~2000, laboratory crosses were used to demonstrate putative hybridization (see Table 1 in Kamiya and West 2010; Table 2 in Bartsch *et al.* 2008), but the application of molecular tools has allowed such studies to be extended to assess the level of hybridization in field populations. One of the most studied genera is *Fucus*. Species within this genus are closely related and have radiated relatively recently (Serrão *et al.* 1999a). Their thalli display substantial morphological variation and putative hybrids have been described for over 100 years (Coyer *et al.* 2002, 2006b). Coyer *et al.* (2002) verified in Eastern Norway the existence of hybrids between native *Fucus serratus* and introduced *Fucus evanescens*. Introgression (backcrossing) and paternal leakage are other determinants of *Fucus* population genetics. For example, during its northward postglacial range extension, estuarine *Fucus ceranoides* carried the introgressed cytoplasm of *Fucus vesiculosus* with it, the first marine example of "alien cytoplasm surfing the wave of range expansion" (Neiva *et al.* 2010). Paternal leakage, whereby paternal mtDNA is not degraded but instead transmitted into the F1 generation (heteroplasmy), from *F. serratus* to *F. evanescens* has been detected in their hybrid zone in Denmark (Hoarau *et al.* 2009). Hybridization can also occur between *F. vesiculosus* and *F. spiralis*, but this is considered a relatively rare event because they have different mating strategies: *F. vesiculosus* is dioecious and out-crossing while *F. spiralis* is predominantly a selfing hermaphrodite (Coleman and Brawley 2005; Engel *et al.* 2005; Perrin *et al.* 2007; Billard *et al.* 2010). The result is that *F. vesiculosus* has greater within-population genetic diversity, but *F. spiralis* has greater between-population diversity, because mutations are retained within the selfing population.

Self-fertilization has also been detected in the Laminariales. For *Macrocystis pyrifera* in California, selfing is thought to negatively affect local populations by increasing inbreeding depression (Raimondi *et al.* 2004). On the other hand, Barner *et al.* (2010) detected no negative effects of selfing for *Postelsia palmeformis*, and suggest that selfing is an important mechanism enhancing fertilization success in the extremely wave-exposed habitats in which it thrives.

Nuclear genome size varies widely among algae, and if the red, green, and brown lineages of seaweeds are taken together, there is a 1300-fold variation (Gregory 2005; Kapraun 2005). The size of nuclear genomes of plants and animals is typically reported as a C-value (pg) which refers to the amount of nuclear DNA in a haploid nucleus (e.g. sperm for animals). However, if diploid cells are studied (e.g. blood of animals) then the value reported is 2C. For an accurate assessment of the C-value, the ploidy level of the plant/alga is needed. For a polyploid, DNA content may be reported as 4C, 8 C or 16 C. Also, the C-value does not necessarily relate to the number of chromosomes. For terrestrial plants and seaweed, because cells are continually dividing, and neighboring cells may be at a different stage of cell division, it can be difficult to know precisely the ploidy level (Gregory 2005). Kapraun (2005) conducted the first wide-scale analysis of 2C nuclear DNA content in seaweeds. He found a minimum 2C nuclear genome size of 0.2 pg for all groups, with the Chlorophyta having the greatest range of up to 6.1 pg, the Rhodophyta up to 2.8 pg, and the Phaeophyceae 1.8 pg. This study raised some interesting questions: within the brown seaweeds, for

example, orders with oogamous reproduction tended to have more nDNA than isogamous or anisogamous modes. There was also a trend of cold-water Fucales (*Ascophyllum* and *Fucus*) having larger nuclear genomes than warm-water genera (*Sargassum* and *Turbinaria*), a finding that was supported by the nuclear DNA content of 19 Fucalean seaweeds from Spain, although whether a larger genome size causes enhanced cold tolerance requires testing (Garreta *et al.* 2010).

In addition to nuclear, plastid, and mitochondrial genomes, some seaweed contain plasmids – small loops of DNA. Plasmids have been identified in red seaweeds including *Gracilaria* spp., *Gracilariopsis* spp., and *Porphyra pulchra* (e.g. Goff and Coleman 1990; Moon and Goff 1997) and the coenocytic green seaweeds *Ernodesmis verticillata* and *Valonia ventricosa* (previously *Ventricaria ventricosa*) (La Clair II and Wang 2000). Plasmid size ranges from 1.6–8 kbp, compared with ~110–190 kbp for plastid genomes in a range of red algae (Goff and Coleman 1988). Their functions in seaweeds are still unclear (Moon and Goff 1997). The plasmid complement appears to be a stable species character, not the result of infection (e.g. by viruses or parasites) (Goff and Coleman 1990).

1.4.3 Nucleocytoplasmic interactions

Eukaryotes have three main compartments for gene expression: the nucleus/cytosol, with the perforated nuclear membrane partially separating them, the plastids and mitochondria (Nott *et al.* 2006; sec. 1.3.2). Nucleocytoplasmic interactions include the full range of interactions between these compartments. In a giant, uninucleate cell like *Acetabularia*, with thousands of plastids and mitochondria, the organellar DNAs are significant components (Mandoli 1998a). *Acetabularia* has a very high cpDNA content compared to other algae and plants; much RNA synthesis takes place in the plastids but the extent to which cpDNA may be involved in cell morphogenesis, and the interactions between plastid and nuclear genomes, has yet to be resolved (Mandoli 1998a). Also, despite its large cpDNA size, the DNA is repeated and so the actual number of *Acetabluaria* genes may be fairly

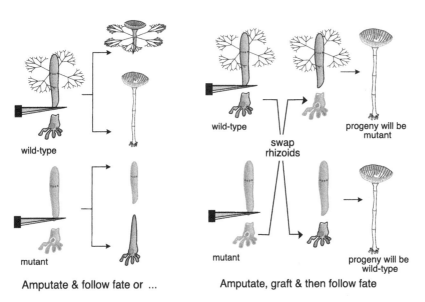

Amputate & follow fate or ...   Amputate, graft & then follow fate

Figure 1.22 Responses of *Acetabularia acetabulum* to amputation and grafting. The vertically aligned cartoons are to scale with each other, but horizontally aligned cartoons are not necessarily at the same scale. (Left) Typical fates of the enucleate apex and the nucleate rhizoid to amputation are shown for the wild type and for a Class I type defect which is arrested in development, e.g. *kurkku* (Mandoli and Hunt 1996). (Right) The fates of graft chimeras of a reciprocal graft in which rhizoids of a wild-type and mutant are exchanged. Progeny of these graft chimeras are shown. The depiction may also be valid for some other species and some combinations of interspecific graft chimeras. (From Mandoli 1998a, reproduced with permission.)

similar to other seaweeds (Simpson and Stern 2002). *Acetabularia* has been a model system for studying nucleocytoplasmic interactions since Hämmerling's classic studies in the 1930s. The main advantage of *Acetabularia* is that interspecific grafts can be made: both nuclei and cytoplasm can be transferred between species (Fig. 1.22; reviewed by Menzel 1994; Mandoli 1998a,b; Mine *et al.* 2008).

Hämmerling concluded, long before messenger RNA (mRNA) was known, that "morphogenetic substances" were released from the nucleus into the cytoplasm, where they could be stored for some time, but were gradually used up. *Acetabularia* cells can still form a cap if, after reaching about one-third of their final length, the nucleus is removed. There are apico-basal and baso-apical gradients of morphogenetic substances. That these substances come from the nucleus has been shown by transplanting nuclei into opposite ends of enucleated cells (Fig. 1.22). The type of cap formed by an enucleated stalk is characteristic of the species, but if

a nucleus from another species or mutant is inserted, either as an isolated nucleus or by grafting on a basal fragment (Fig. 1.22 – right-hand panel), the caps formed are first intermediate and then have the characteristics of the nucleus donor species. This, and the fact that enucleated cells can form a cap only once, provides evidence that the morphogenetic substances are used up in cap formation.

Hämmerling's morphogenetic substances are now known to be mRNA, and mRNAs specific to morphological and reproductive developmental processes in *Acetabularia acetabulum* have been identified. Homeobox genes are the "master control genes" of eukaryotes with important roles in morphological development (Buchanan *et al.* 2000). For example, homeobox gene *Aaknox1* is evenly expressed along the stalk of mature, vegetative *A. acetabulum* but at the onset of reproduction mRNA it is localized at the base, close to the primary nucleus, indicating post-transcriptional control of gene expression (Serikawa and Mandoli 1999). Mine *et al.*

(2005) found two mRNAs (as polyadenylated RNA) for *A. peniculus*. "Poly(A)$^+$ RNA striations", derived from the primary nucleus, are distributed evenly within the stalk cytoplasm and associated with actin bundles and filaments. These mRNA striations are thought to be a transport form of mRNA and are involved in vegetative developmental processes such as stalk elongation and whorl development because they disappear during the early stages of cyst formation. The second type of mRNA, the "perinuclear poly(A)$^+$ RNA mass", creates a mass around each secondary nucleus and is closely associated with ribosomes, indicating active translation of these genes during cyst formation. This mRNA is also associated with microtubules, rather than actin filaments. Indeed, there are multiple mRNA gradients along the stalk of mature, vegetative *Acetabularia acetabulum* and despite being a unicellular algae there is considerable regional differentiation (Serikawa *et al.* 2001; Vogel *et al.* 2002).

This completes our review of seaweed cells and thalli. In Chapter 2, we focus on seaweed life cycles, reproduction, and morphogenesis, and how these processes are affected by both intrinsic and extrinsic factors.

1.5 Synopsis

Benthic ocean vegetation is dominated by macroscopic multicellular and unicellular members of the red, green, and brown seaweeds. The term "seaweeds" represents an ecological grouping of disparate taxa, which are related by the endosymbiotic events that gave rise to plastids. Seaweeds have a microscopic phase to their life history, as gametes, spores, or zygotes, and for some as a free-living alternate life-history stage. The most common forms of seaweed construction are filamentous and pseudoparenchymatous, with parenchymatous construction being prevalent only in the Phaeophyceae. Seaweeds may be unitary or clonal, and some clonal species coalescence to form chimeras. Modularity is common in all the seaweed phyla. Seaweed cells differ from the cells of "higher" plants in general by their broader range of metabolic functions. Some special features of seaweed cells include vesicles that may store bioactive secondary metabolites, a high proportion and diversity of cell wall matrix polysaccharides, a wide variety of plastid structures and pigmentation, the different arrangement of flagella in motile cells, and the details of cell division. The unique characteristics of some seaweeds has led to their use as model organisms, e.g. *Acetabularia* for nucleocytoplasmic interactions and *Fucus* for cell polarization. Classical seaweed genetics, especially using color mutants in some red algae, have shown Mendelian and non-Mendelian inheritance. For asexually reproducing seaweed populations, genetic variation of clones arises via somatic embryogenesis. The application of molecular biology techniques has had a profound impact on phylogenetics, biogeography, and population biology. *Ectocarpus siliculosus* was the first seaweed to have its genome fully sequenced, and applications of other "omic" techniques are providing novel insights into seaweed physiological ecology.

Life histories, reproduction, and morphogenesis

2.1 Introduction

The basic patterns of alternation of sporophyte and gametophyte must be regarded as a theme on which many variations are played (Fig. 2.1). Each generation may reproduce itself asexually, and sexual reproduction should be taken to include meiosporogenesis as well as gametogenesis and mating (Clayton 1988). Asexual reproduction allows an economical population increase but no genetic mixing, whereas sexual reproduction allows genetic mixing but is more costly because of the waste of gametes that fail to mate (Clayton 1981; Russell 1986; Santelices 1990). Most seaweeds use both means of reproduction, and, as Russell (1986) has noted, where there are isogametes, these can function equally as asexual swarmers. Vegetative reproduction by "multicellular propagules", defined by Cecere *et al.* (2011) as "a vegetative, multicellular structure which detaches from the parent thallus and gives rise to a new individual", is also common, for example *Halimeda* (Walters *et al.* 2002). However, their roles in species' dispersal, and forming overwintering and resting "organs" that allow the survival of unfavorable environmental conditions, is unknown (Russell 1986; Cecere *et al.* 2011). Clonal seaweeds may spread by stolons and/or rhizomes, giving a significant competitive edge in the space race (sec. 1.2.3, 4.2.3). Some floating algal populations depend entirely on vegetative reproduction by fragmentation (sec. 3.3.7).

Culture studies are critical in establishing the range of possible life histories that can occur. Sufficient variations have been discovered in the basic pattern – between and within species – that today's generalizations must be viewed only as working hypotheses. New variations in what are considered to be well-known life cycles are regularly uncovered. For example, male gametophytes of *Laminaria digitata* can reproduce themselves via fragmentation (Destombe *et al.* 2011). Although a basic alternation of a sporophyte (typically diploid) and a gametophyte (typically haploid) is common among seaweeds various extras and shortcuts are known (Fig. 2.1).[1] Indeed, a better generalization may be that almost any alternation is possible, and even no alternation at all. Moreover, the term "alternation" is a misnomer, in that it implies only two phases and a regular progression from one to the other; clearly that is not always the case (e.g. *Scytosiphon*, Fig. 2.2). Maggs (1988, p. 488) concluded that "life-history patterns seem to be more labile than morphological features, and the role of life-history variability in speciation, and in ecological success, should not be underestimated".

2.2 Theme and variations

Three basic types of algal life histories are recognized (e.g. Dring 1982; Bold and Wynne 1985; Graham *et al.* 2009). An alternation of two phases is called haplodiplontic (Fig. 2.1). Genera such as *Ulva* and *Chondrus* are examples of an isomorphic alternation of generations, having sporophytes and gametophytes that are vegetatively indistinguishable (not counting the

---

[1] Life-history diagrams herein are as follows: *Acetabularia* (Fig. 1.12), *Scytosiphon* (Fig. 2.2), *Halimeda* (Fig. 2.3), *Ectocarpus* (Fig. 2.4), *Laminaria* (Figs. 2.5 and 2.8), *Nereocystis* and *Pyropia* (Fig. 2.10). See also Fig. 2.7.

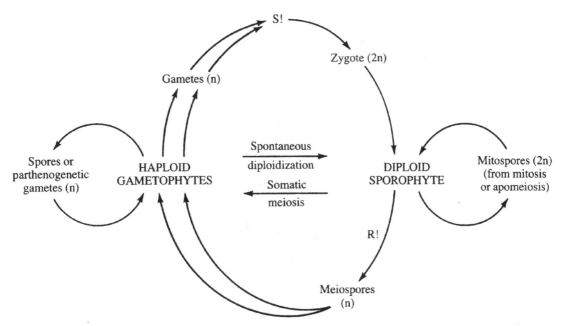

Figure 2.1 The basic pattern of an alternation between haploid (*n*) gametophyte and diploid (2*n*) sporophyte generations (the central cycle in the figure) is a theme upon which there are many variations. Some possible seaweed life-history progressions are indicated by the smaller cycles to the right and left of the figure. Most species use only a small part of this range. R!, meiosis (reduction division); S!, syngamy.

carposporophytes of red algae, which are not free living). Sometimes chemical differences occur between isomorphic phases, as in *Chondrus* in which different forms of carrageenan are found in the walls, even of the carpospores and tetraspores (Bellgrove *et al.* 2009; sec. 5.5.2). At reproduction, the two phases may become distinguishable by the reproductive structures. Heteromorphic generations often fall into two different functional-form groups, such as erect fronds versus creeping filaments or crusts (Figs. 2.2, 2.7, 2.8, and 2.10). A classic example is the well-known story of Drew's (1949) linking of *Conchocelis* (filamentous) and *Porphyra unbilicalis* (a blade) and its impact on the Japanese nori industry (sec. 10.2.1). Similarly, the crustose red *Erythrodermis allenii* is part of the *Phyllophora traillii* life history (Maggs 1989), and the unicellular green seaweed "*Codiolum*" is a life-history phase of various green seaweeds (see Graham *et al.* 2009). Some seaweeds exist in only one phase (Figs. 1.12 and 2.3). Most common are diplontic life cycles, for example the Fucales and *Codium* in which

the vegetative phase is diploid (2*n*). Here, the gametes are formed by meiosis and the gametes are the only haploid phase of the life cycle. In haplontic life cycles, the vegetative phase is haploid with the zygote being the only diploid phase. Here, meiosis occurs in the zygote, thereby restoring the haploid phase. *Chlamydomonas* and dinoflagellates are examples of algae with a haplontic life cycle, but in seaweeds this condition is quite rare.

Florideophycidae were included in the foregoing life-history generalizations in spite of the interpolation of a "carposporophyte", a diploid tissue that forms on the female gametophyte following fertilization. This structure is often regarded as an additional diminutive diploid phase, that is hemi-parasitic because it obtains some nutritive support from the female gametophyte (Maggs *et al.* 2011). (The term "triphasic" is often used to describe such life cycles, which consist of free-living gametophytes and tetrasporophytes, plus the carposporophyte). The carposporophyte is responsible for zygote amplification; from one fertilization event

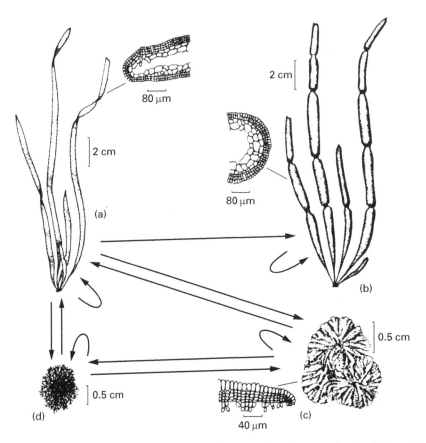

Figure 2.2 Life history and anatomical features of the *Scytosiphon lomentaria* (previously *Scytosiphon simplicissimus*) complex: (a) complanate form; (b) cylindrical form; (c) crustose form; (d) filamentous plethysmothalli. (From Littler and Littler 1983 (partly after Clayton), with permission of *Journal of Phycology*.)

thousands of carpospores (produced by mitosis) may be released and germinate into the free-living tetrasporophyte generation (see Graham *et al.* 2009). Some Bangiophycidae, including *Porphyra*, also multiply the zygote by forming "zygotosporangia", packets of diploid cells which are released upon breakdown of the cell wall, and germinate, giving rise to the conchocelis phase (Guiry 1990; Nelson *et al.* 1999). The zygote in *Palmaria palmata* develops into a large diploid phase, morphologically like the male gametophyte, that overgrows the tiny female gametophyte and produces spores by meiosis (van der Meer and Todd 1980). Replication of the zygote is one of several ways to amplify the results of sexual

reproduction (see sec. 2.4; Hawkes 1990; Graham *et al.* 2009; Maggs *et al.* 2011).

Ploidy levels within a particular life-cycle phase are often assumed, but studies have sometimes demonstrated the unexpected. For example, most *Codium* species studied in the western Atlantic are diploid and reproduce via haploid gametes. However, *C. fragile* ssp. *fragile* (formerly known as ssp. *tomentosoides*) is haploid, reproducing parthenogenetically (Kapraun and Martin 1987; Prince and Trowbridge 2004). Furthermore, various morphological forms may be expressed at a given ploidy level, and the same morphology may be formed at different ploidy levels: There is no necessary connection between ploidy and

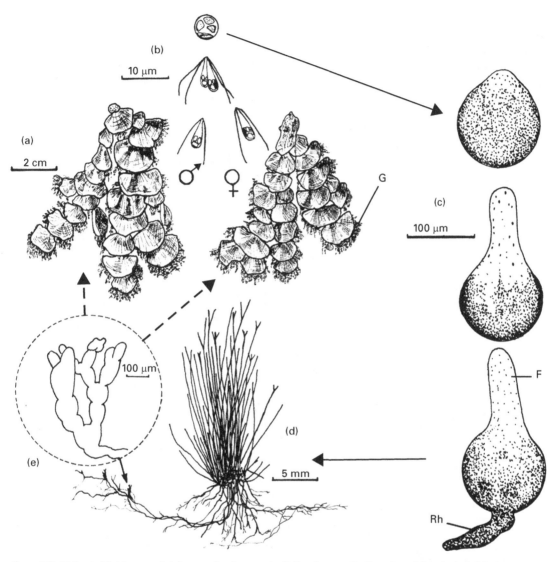

Figure 2.3 Diplontic life history and siphonous development in *Halimeda tuna*. Fertile male and female thalli (a), shown hanging downward, release biflagellate gametes from external gametangia (G) to form a zytoge (b). (c, d) Bipolar germination of the zygote leads to rhizoids (Rh) and erect free filaments (F). Subsequently (e), buds form on horizontal filaments and grow into the calcified, segmented fronts. (From Meinesz 1980; *Phycologia*, with permission of Blackwell Scientific Publications.)

form, even though, in general, gametophytes are haploid and sporophytes are diploid. Several seaweeds are known in which an apparent alternation of generations occurs with no ploidy change (e.g. *Petalonia* and *Scytosiphon*, Kapraun and Boone 1987). *Desmotrichum* can form filamentous microthalli or parenchymatous macrothalli, depending on temperature and day length. Also in response to temperature and day length, *Elachista stellaris* alternates between a diploid macrothallus, which produces meiotic spores, and a microthallus that can reproduce the macrothallus by spontaneous diploidization. Diploidization has also been reported in *Boergesenia forbesii*, which has an isomorphic life history (Beutlich *et al.* 1990); in this case it results in a preponderance of diploids in the population. Polyploidy can be developed in *Gracilaria tikvahiae* mutants; polyploid tetrasporophytes (e.g. 3*n*, 4*n*) were found to be robust, but polyploid gametophytes (again, 3*n*, 4*n*) were stunted (Zhang and van der Meer 1988; sec. 1.4.2). Endopolyploidy (chromosome division without nuclear division) is apparent for

Saccharina latissima (formerly *Laminaria saccharina*) and *Alaria esculenta*, and for vegetative sporophytes of both species the nuclear DNA content ranged from 2C to 16C (Garbary and Clarke 2002).

Variations in these basic patterns include several reproductive shortcuts (Fig. 2.1). In sporangia that are expected to be meiotic, such as red algal tetrasporangia and brown algal unilocular sporangia, spores may be formed by mitosis instead, giving rise to more seaweeds of the same ploidy level; this is called apomeiosis (e.g. *Ectocarpus*, Fig. 2.4). For instance, *Dasya*

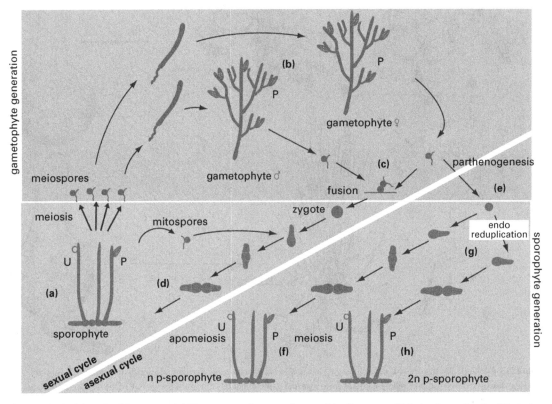

Figure 2.4 The life cycle of *Ectocarpus*. (a) Mature sporophytes produce upright filaments, which develop unilocular (U, single-compartment) sporangia. The first cell division in unilocular sporangia is meiotic and followed by a number of mitoses to give around 100 male and female meiospores. No septa are laid down between these daughter cells. (b) Meiospores grow into the dioecious gametophyte generation which has only one reproductive structure, the plurilocular (P, many compartments) gametangium, in which male or female gametes are produced through mitosis. (c) Gametes can fuse to form a zygote that grows into (d) a heterozygous sporophyte or, alternatively, (e) can develop parthenogenetically to give rise to haploid parthenosporophytes. (g) Endoreduplication may occur in a proportion of parthenosporophytes, giving rise to (h) diploid parthenosporophytes. Sporophytes and parthenosporophytes may also produce plurilocular sporangia, in which spores are formed by mitosis (= mito-spores) and grow as clones of the parent. (From Bothwell *et al.* 2010, reproduced with permission.)

ocellata has a sexual life history in Southern Portugal with tetrasporophyte and gametophyte generations, whereas in Northern Ireland the populations are apomeiotic tetrasporophytes (Maggs 1998). Five types of complications can arise in red algal life histories (Maggs 1988; see also Hawkes 1990): (1) formation of monosporangia, bisporangia, polysporangia, parasporangia, or vegetative propagules in a species that also forms tetraspores; (2) simultaneous occurrence of gametangia and tetrasporangia (mixed-phase reproduction); (3) bisexuality in a normally unisexual species; (4) direct development of tetrasporophytes from tetraspores (exclusively or mixed with gametophytes); (5) direct development of gametophytes from carposporophytes (exclusively or mixed with tetrasporophytes).

Parthenogenesis is the development of a gamete without fertilization. For isogamous species, parthenogenesis can occur in both male and female gametes, in anisogamous reproduction it is typically the female gamete, and in oogamous reproduction it is the female egg that develops parthenogenetically (Oppliger *et al.* 2007). Parthenogenesis is rare in red algae (Kamiya and West 2010) but widely reported in laboratory studies of brown and green seaweeds: within the orders Laminariales (Gall *et al.* 1996; Oppliger *et al.* 2007), Fucales (Maier 1997; Clayton *et al.* 1998), and Ectocarpales (Bothwell *et al.* 2010), and for green seaweeds, *Ulva* (Stratmann *et al.* 1996) and *C. fragile* ssp. *tomentosoides* (Kapraun and Martin 1987; Prince and Trowbridge 2004). However, its occurrence in natural populations and ecological significance have yet to be determined (Oppliger *et al.* 2007). Apospory is a process whereby diploid gametophytes are produced directly by sporophyte cells (i.e. without spores). Thus apospory differs cytologically from somatic meiosis in that no ploidy change occurs, but the morphological effect is the same. Apogamy is the production of haploid sporophytes directly from gametophyte cells, and it differs from spontaneous diploidization in having no ploidy change. Apospory and apogamy are detectable only in heteromorphic life histories, such as those of *Alaria crassifolia* (Nakahara and Nakamura 1973), *Desmarestia* species (Ramirez *et al.* 1986) and a number of red seaweeds (Murray and Dixon 1992).

Ectocarpus siliculosus, including mutant strains, is being used as a model alga to elucidate life-cycle events and gain a genetic understanding of the variable relation between ploidy level and morphology, a line of inquiry that will lead ultimately to the roles and origins of alternating generations. The basic life cycle of *E. siliculosus* is that of an alternation between sporophyte and gametophyte, but there are many developmental alternatives (see Charrier *et al.* 2008 who review D. Müller's extensive contributions; Peters *et al.* 2008; Bothwell *et al.* 2010). Sporophytes can form either plurilocular sporangia that mitotically produce spores (mitospores) that give rise to genetically identical sporophytes, or they can produce unilocular sporangia within which meiosis gives rise to $1n$ meiospores which germinate to give the male and female gametophyte generation (Fig. 2.4). The gametophytes produce plurilocular gametangia that release gametes, which typically fertilize to become a heterozygous diploid sporophyte. However, for any un-fused gametes there are two possible developmental pathways: (1) they can form haploid sporophytes parthenogenetically (*n*-parthenosporophytes) which themselves can form either plurilocular sporangia and mitospores, or unilocular sporangia by apomeiosis; or (2) they undergo endoreduplication (i.e. nucleus divides without cell division) which results in $2n$-parthenosporophytes (Fig. 2.4). Diploid sporophytes can also be produced from tetraploid sporophytes (Coelho *et al.* 2007). This "extreme developmental plasticity" of *Ectocarpus* is controlled genetically and not by the ploidy level of a particular stage in the life cycle (Bothwell *et al.* 2010). Using mutants, a regulatory locus "Immediate upright" (*IMM*) that controls aspects of the sporophyte developmental program was identified (Peters *et al.* 2008), and Coelho *et al.* (2011) found the "master regulator" *OUROBOROS* which controls the transition from gametophyte to sporophyte, probably by repressing the gametophyte developmental program.

Haplodiplontic life cycles are thought to have evolved repeatedly within each seaweed lineage and therefore the life cycles should represent adaptations to particular environments (Bessho and Iwasa 2010). A question that has received much attention is why both generations persist, and various genetic and ecological theories have been forwarded (e.g. Bell 1997; Hughes and Otto 1999; Bessho and Isawa 2009, 2010). Genetic models include: (1) DNA damage can be repaired only in diploids favoring their retention in the life cycle; (2) Deleterious mutations accumulate in the diploid phase but because they are often recessive their effect is masked; in the haploid generation such deleterious alleles are selectively eliminated; (3) Diploids will favor the accumulation of advantageous mutations; (4) Parasites may prefer diploid hosts. Some ecological models that have been proposed are: (1) Haploids are often smaller and therefore have lower nutrient and energetic requirements, including lower energetic costs of DNA replication; (2) For heteromorphic alternations, having both haplontic and diplontic generations means two niches can be inhabited at once, thereby exploiting differences in the environment (e.g. light, temperature), or escaping herbivory (sec. 4.3.2). Bessho and Iwasa (2009) suggest that a heteromorphic alternation of generations is better adapted to environments with marked seasonality than isomorphic life cycles. Results from "The *Porphyra* Genome" project indicate that cytoplasmic ribosomal proteins (RPs) are differently expressed in the haploid conchocelis phase compared to the diploid bladed phase; it is hoped that this line of inquiry will provide further insights into both the developmental regulation of life-history phases, and the biochemical mechanisms underpinning niche adaptation of different life-history phases (Chan *et al.* 2012a).

Another interesting question is how in some isomorphic seaweeds one phase is numerically dominant over the other (e.g. Van der Strate *et al.* 2002b; Scrosati and Mudge 2004). Fierst *et al.* (2005) reported 34 species of red algae for which one phase was dominant within the population compared to the alternate phase (see their Table 2). There are also differences between orders, with gametophytes dominating the Gigartinales while tetrasporophytes prevail in the Gracilariales and Ceramiales (Thornber 2006). Reasons for these disparities require testing but candidate theories include differential reproductive output of the different phases and differential susceptibility to various biotic and abiotic factors (Thornber and Gaines 2004; Fierst *et al.* 2005). Two studies support the latter hypothesis.

Thornber *et al.* (2006) found that the snail *Chlorostoma funebralis* (previously *Tegula funebralis*) much preferred gametophytes of *Mazzaella flaccida* (previously *Iridaea flaccida*) compared to sporophytes, because the carposporphytes on the gametophytes are large and protrude further out from the blade surface, making them easier to graze. Verges *et al.* (2008) found that the sea hare *Aplysia parvula* preferentially grazes male gametophytes of *Asparagopsis armata*, whereas the female cystocarps are strongly chemically defended and consumed much less, explaining why the sex ratio of *Asparagopsis* gametophytes is 1:1 early in its growth season but biased towards females (70%) later in the year.

2.3   Environmental factors in life histories

The life history of a species is a continuous interaction between the organism and its biotic and abiotic environments (Fig. 2.5). A seaweed begins life as a single undifferentiated cell, with the potential to produce the whole organism through the expression of its genetic information. The genotype interacts with the environment to produce the phenotype. The environment of a cell consists of the physical and chemical influences of the other cells in the seaweed, plus the environment of the seaweed itself. The environmental history of a seaweed, because it affects growth and form, in a sense becomes recorded in its body (e.g. Waaland and Cleland 1972; Niklas 2009). Thus individual seaweeds of the same genotype that are grown under different environmental conditions will grow into phenotypically distinct individuals.

The switch from vegetative growth to reproduction (which in most seaweeds involves very little growth) often depends on environmental factors such as temperature and light (Lüning and tom Dieck 1989; Lüning 1990; Santelices 1990). Kelp gametophytes, for example, may reproduce when they are only a few cells in size, or they may grow vegetatively almost indefinitely, depending on light quality and quantity (reviewed by Bartsch *et al.* 2008). They have been used for studies of minimum irradiance requirements for growth and reproduction because of their extreme

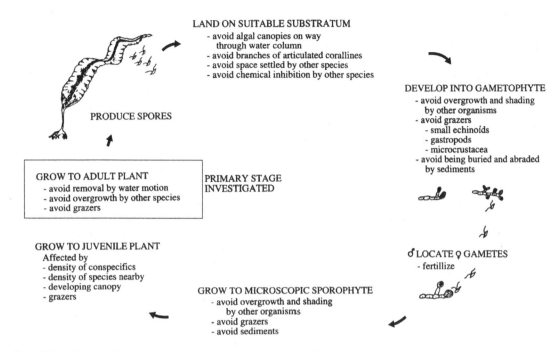

LAND ON SUITABLE SUBSTRATUM
- avoid algal canopies on way
  through water column
- avoid branches of articulated corallines
- avoid space settled by other species
- avoid chemical inhibition by other species

DEVELOP INTO GAMETOPHYTE
- avoid overgrowth and shading
  by other organisms
- avoid grazers
  - small echinoíds
  - gastropods
  - microcrustacea
- avoid being buried and abraded
  by sediments

PRODUCE SPORES

GROW TO ADULT PLANT
- avoid removal by water motion
- avoid overgrowth by other species
- avoid grazers

PRIMARY STAGE
INVESTIGATED

GROW TO JUVENILE PLANT
Affected by
- density of conspecifics
- density of species nearby
- developing canopy
- grazers

♂ LOCATE ♀ GAMETES
- fertillize

GROW TO MICROSCOPIC SPOROPHYTE
- avoid overgrowth and shading
  by other organisms
- avoid grazers
- avoid sediments

Figure 2.5 Life history of a laminarian alga, showing some of the major (chiefly biotic) environmental hazards that must be overcome at each stage. In addition, success will be affected by abiotic factors such as light, temperature, water motion. (From Schiel and Foster 1986, *Oceanogr. Mar. Biol. Rev.*, with permission of Aberdeen University Press, Farmers Hall, Aberdeen, AB9 2XT, UK.)

shade environment and ease of culture. A pre-requisite for growth, of course, is that the energy trapped and carbon fixed must exceed the totals used in respiration. Chapman and Burrows (1970) showed that development of *Desmarestia aculeata* gametophytes depends on the mean daily irradiance (i.e. (irradiance × photoperiod)/24). At the lowest irradiances tested, gametophytes did not mature, though they survived and were able to develop later when irradiance was increased. More detailed studies by Lüning and Neushul (1978) showed that various kelp gametophytes were saturated for vegetative growth at 4 Wm^{-2} (about 20 μmoles photons m^{-2} s^{-1}; see sec. 5.2.1 for explanation of units) but required two to three times that irradiance for reproduction. Blue light, alone or as part of white light, is required for kelp gametogenesis; in red light, gametophytes grow only vegetatively. The ability of these algae to grow

vegetatively in extremely dim light and reproduce only when irradiance increases provides a mechanism for populations to retain space after the canopy of parent sporophytes is lost. The effects of light wavelength on photosynthesis is discussed further in Chapter 5, and the importance of environmental factors in seaweed aquaculture are discussed in Chapter 10.

2.3.1 Seasonal anticipators and responders

One of the most important ways in which algae (and all organisms) respond to their environment is in the timing of reproduction, because in reproduction lies the key to the survival of the species (Santelices 1990; Pearson and Serrão 2006). Reproductive responses to the environment are particularly evident in algae with strongly heteromorphic generations, such as kelps and *Pyropia*, where different growth forms are adapted to different environments. When conditions are suitable for the growth of one form, vegetative growth or asexual propagation is likely to occur, whereas conditions poor for that form are likely to prompt a reproductive switch to the alternative morphology. However, because of the lead time sometimes needed for reproduction, some seaweeds may need to anticipate the changes in seasons, using an appropriate cue. A useful framework for categorizing the ability of a seaweed to modify its reproductive status in response to an environmental cue is that of "anticipators" and "responders", terms coined by Kain (1989); Lüning and co-workers (e.g. Lüning 1991; tom Dieck 1991) had the same idea and termed them Types II and I, respectively (Kain 1989).

For anticipators, seasonal patterns of growth and reproduction are controlled by a free-running endogenous clock. In the absence of environmental cues, the free-running rhythm is slightly different from that observed in nature (e.g. 9–10 months for *Laminaria hypoborea*), but is entrained by environmental cues, termed "Zeitgebers". In seaweeds, light is an important Zeitgeber (Lüning 1994; reviewed by Bartsch *et al.* 2008), and for animals and other algae e.g. *Euglena* and the dinoflagellate *Gonyaulax*, temperature or nutrients are also Zeitgebers (e.g. Roenneberg and Mittag 1996; Rensing and Ruoff 2002). The maximum growth rate of anticipators may occur when environmental conditions do not seem optimal. For the red seaweed *Delesseria sanguinea*, this is in late winter when light levels are low, and the growth rate declines in summer (Kain 1989). Another example is that of the perennial brown seaweeds of Antarctica, *Desmarestia anceps*, *D. menziesii*, and *Himantothallus grandifolius*, for which reproduction and sporophyte development occurs in winter (Wiencke 1990a; Wiencke *et al.* 2009). Seasonal "responders" on the other hand sense and respond directly to the prevailing environment, and seasonal patterns of growth and reproduction are not governed by a circannual endogenous clock. Examples of responders include growth of the giant kelp *Macrocystis pyrifera* and the red seaweed *Plocamium cartilagineum* (Kain 1989; Reed *et al.* 1997).

A major criterion used to demonstrate the presence of an endogenous rhythm is "self-sustainment for more than one cycle in conditions giving no seasonal information" (Schaffelke and Lüning 1994), and such patterns have been observed in various kelp including *Pterygophora californica*, *Laminaria setchellii*, and *Laminaria hyperborea* (Lüning 1991, tom Dieck 1991; Schaffelke and Lüning 1994). Sporophytes of *L. hyperborea* were grown for 2 years under constant conditions of light, photoperiod (12:12), nutrients and temperature and yet they still continued the seasonal patterns of blade growth observed in the field, with periods of fast growth followed by periods of zero growth. The Zeitgeber that entrained the endogenous clock was photoperiod, and this was clearly shown when the experimental year was shortened from 12 months to 6 or 3 months: the same seasonal pattern occurred, but instead of one growth period per year there were 2 or 4, respectively. An endogenous pattern of reproduction has been shown for *Dictyota dichotoma* (Müller 1963) and *Ulva pseudocurvata* (Lüning *et al.* 2008). The biochemical and molecular mechanisms underlying circannual endogenous clocks are not currently known for seaweeds (see Bartsch *et al.* 2008, p. 36).

For both seasonal responders and anticipators, environmental factors modulate growth and the onset of reproduction, and can trigger changes from one life-history phase to another. Modulating factors include temperature, light (photon dose, photoperiod, wavelength), nutrients, lunar and tidal cycles, desiccation, salinity, water motion, and biological factors including grazing and bacteria (Santelices 1990; Pearson and Serrão 2006). These factors do not act in isolation and Bartsch *et al.* (2008) sum up the effects of environmental factors on reproduction in *Laminaria*: "tissue location, temperature, irradiance as well as competition and life strategy modify reproductive output, for no simple parameter is the sole decisive trigger".

2.3.2 Temperature

Although temperature seems an obvious seasonal cue in middle to high latitudes, it should be remembered that changes in seawater temperature are directly related to the amount of light reaching the sea (which causes warming), and there is often an inverse relationship between inorganic nitrate and temperature:

thus culture studies are required to confirm if correlations with temperature are directly triggering a response, or acting indirectly. Kain's (1989) analysis of seasonal patterns of reproduction in subtidal seaweeds revealed few direct responses of temperature on seasonality of growth and reproduction, but she recognized that temperature is a key factor controlling biogeography, and the ability of seaweeds at their geographic limit to reproduce (sec. 7.3). Temperature can affect reproduction through its effects on metabolic rates. For example, in *Fucus* it may influence the timing of reproduction by speeding up gamete development; Ladah *et al.* (2008) suggest that at lower temperatures the oogonial sheath remains for longer and this results in a reduced dispersal of eggs. Similarly the time taken for sorus induction of *Laminaria* is temperature dependent, and it has been suggested temperature and photoperiod together could allow kelps to distinguish between spring and fall (Lüning 1988, see Bartsch *et al.* 2008). A further example is that of the Antarctic red seaweed, *Iridaea cordata*, for which the gametophytes take 21 months to form reproductive structures at typical Antarctic seawater temperatures of 0°C, whereas in sub-Antarctic waters of 5°C it takes just 12 months (Wiencke 1990b).

Nevertheless, some differences in the kinds of reproduction at different temperatures have been noted in seaweeds. Müller (1963) found that *Ectocarpus siliculosus* produced unilocular sporangia at 13°C and plurilocular sporangia at 20°C, although for other strains this is not temperature dependent (Charrier *et al.* 2008). Changing the temperature in which *Myriotrichia clavaeformis* sporophytes are grown determines whether sexual or asexual organs are produced, whereas for gametophytes the effect is triggered by photoperiod (Peters *et al.* 2004b). The formation of erect thalli in some isolates of *Scytosiphon lomentaria* var. *complanatus* (currently *Scytosiphon complanatus*) studied by Correa *et al.* (1986) was dependent on temperature and independent of photoperiod (Table 2.1), in contrast to the better known photoperiodism of the typical variety, as discussed later. (In other isolates of *S. lomentaria*, this morphogenetic switch apparently is not responsive to either temperature or photoperiod.) In some species, different steps

Table 2.1 Influences of temperature and day length on formation of upright fronds by *Scytosiphon complanatus* (previously *Scytosiphon lomentaria* var. *complanatus*) in Nova Scotia.

Temperature (°C)	Day length (h)	Crusts with uprights (%)
0	14	100
5	14	100
10	8	100
10	12	100
10	16	100
15	12	3.8
15	16	0.3
20	12	0
20	16	0

Source: Correa *et al.* (1986), with permission of Blackwell Scientific Publications.

in reproduction have different temperature optima. In the conchocelis stage of *Pyropia tenera* (previously *Porphya tenera*) in Japan, the temperature optimum for monosporangium formation is 21–27°C, whereas for monospore release it is 18–21°C (Kurogi and Hirano 1956; see also Dring 1974). Chen *et al.* (1970) found that conchosporangia of *Wildemania miniata* (previously *Porphyra miniata*) from Nova Scotia were formed at higher temperatures (13–15°C in this case), but conchospores were released only with low temperatures (3–7°C) and short days. Such interactive effects of light and temperature on seaweed reproduction appear common, for example: *Desmarestia firma* (Anderson and Bolton 1989), *Helminthocladia stackhousei* (previously *Helminthora stackhousei*; Cunningham *et al.* 1993), *Liagora californica* (Hall and Murray 1998), *Hydropuntia cornea* (previously *Gracilaria cornea*; Orduña-Orjas and Robledo 1999), *Lessonia variegata* (Nelson 2005). In some cases the responses are quantitative (e.g. higher fertility at lower temperatures), in others qualitative (i.e. fertile vs. nonfertile) (Maggs and Guiry 1987). For instance, the initiation of growth of the macrothalli of *Dumontia contorta* is strictly controlled by day length, but the initials do not grow out unless the temperature is less than 16°C (Rietema 1982).

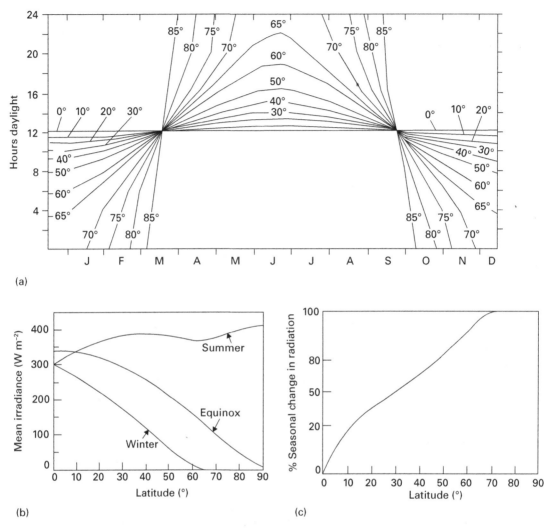

Figure 2.6 Variations in day length and irradiance at different latitudes. (a) Daylight hours in the northern hemisphere (southern hemisphere values may be obtained by 6-month transposition of the abscissa scale). (b) Mean energy flux with a cloudless sky for the months containing the equinoxes and solstices (mean values for both hemispheres). (c) Percentage seasonal change in energy flux with a cloudless sky (recalculated from b). (Part a from Drew 1983, with permission of Clarendon (Oxford University) Press; b and c from Kain 1989, with permission from British Phycological Society.)

2.3.3   Light: photoperiod and wavelength

Although temperatures are involved in reproductive cues, temperatures may undergo seasonal cycles, and such cues are erratic. A more dependable seasonal cue is day length (photoperiod) (Fig. 2.6). As one progresses to farther northern and southern latitudes, day length changes becomes increasingly pronounced, and it has been known since the 1930s that flowering plants respond to photoperiod. Some plants flower when days are long (LD plants) or short (SD plants), others require a specific sequence of long days followed by short days (long-short-day plant), or short days followed by long days (short-long-day plant), while others are insensitive to day length i.e. day-neutral (e.g. Taiz and Zeiger 2010). Notwithstanding these names, plants actually measure the length of uninterrupted night, not day length (Buchanan *et al.* 2000). Photoperiodic responses were expected in temperate seaweeds, but not until 1967 was a true photoperiodic response demonstrated, for the conchocelis phase of *Porphyra* (Dring 1967). For terrestrial plants, the biochemistry and molecular biology of photoperiodic responses are well understood (Buchanan *et al.* 2000) but we know little of these processes in seaweed.

Most seaweeds that show photoperiodism are SD. Dring (1988) suggested that this bias could be due to inadequate controls and that there was no reason to suppose that algae would respond more often to SD than LD. Nevertheless, only a few studies demonstrate LD responses (see Pang and Lüning 2004). Higher plants, green algae, cyanobacteria, and fungi, have a family of photoreceptors, the phytochromes, which detect light in the red region of the spectrum and are involved in their systems for measuring and responding to light/dark cycles (Mathews 2006; Sharrock 2008). The presence of phytochromes has not been confirmed for red and brown seaweeds, but phytochrome-like proteins have been detected in *Corallina* and *Gelidium* and responses to red/far-red light suggest its presence in *Porphyra* (López -Figueroa *et al.* 1989; Figueroa *et al.* 1994; see Kain 2006). Red-light effects could occur through red light absorbed by phycobiliproteins (in red algae), which have structures very similar to that of phytochrome (reviewed by Rüdiger and López–Figueroa 1992).

Photoperiodic effects have to be distinguished from the effects of the total irradiance received, which also changes seasonally (Fig. 2.6b,c) and can have strong effects on seaweed growth and development. For example, some kelp and *Desmarestia* gametophytes, have minimum requirements for accumulated total daily irradiance or a certain irradiance intensity in order to reproduce (Chapman and Burrows 1970; Lüning and Neushul 1978; Wiencke 1990a). The classic means of demonstrating a true photoperiodic effect

in terrestrial plants is the nightbreak experiment, in which a long night (e.g. 16 h) is broken by a short exposure to weak light; night breaks show that the alga is counting time rather than an accumulation of photosynthate over 24 h (Lüning *et al.* 2008). For flowering in terrestrial plants, the length of night break and the timing during the dark period both influence the effectiveness of the break on initiating flowering. The effect on SD plants is to spoil the inductive effect; the plant measures two short nights and a LD response is triggered. In LD plants, a night break is usually inductive. Converse experiments, either (1) with a long day (16 h) broken by a short dark period and a regular night, or (2) with the cycle extended (e.g. to 32 h) to give a long night and a long day, are inductive in SD plants, just as is a regular light:dark 16:8-h day. Phytochrome has two alternate forms which absorb red light (Pr) and far-red light (Pfr), respectively, and when these wavelengths are given in sequence in a night break, the last one determines the effect: red light spoils the long night, but far-red light does not (and even counters the effect of red light). The night-break technique remains a widely used tool for screening plants and seaweeds for photoperiodism (Hwang and Dring 2002; Taiz and Zeiger 2010). However, some seaweeds that have SD photoperiodic responses are insensitive to a night break and other pigment systems may be involved (Rüdiger and López-Figueroa 1992). Therefore, rigorous application of the night break as a diagnostic tool for photoperiodism might obscure photoperiodic responses that are mediated by physiological mechanisms other than phytochrome (Dring 1988; Hwang and Dring 2002).

There are also many records of blue-light effects, such as the formation of uprights in *Scytosiphon* (Fig. 2.7) (Dring and Lüning 1975; see reviews by Dring 1984a, 1988; Rüdiger and López-Figueroa 1992). Tetrasporangium formation in *Rhodochorton purpureum* takes place during short days and is inhibited by a night break of red light but not far-red light, and yet the red-light inhibition is not reversed by subsequent exposure to far-red light (in contrast to the case with flowering plants). Moreover, a night break by blue light is also inhibitory (Dring and West 1983). Eggs are released from oogonia of *Dictyota* when blue light is

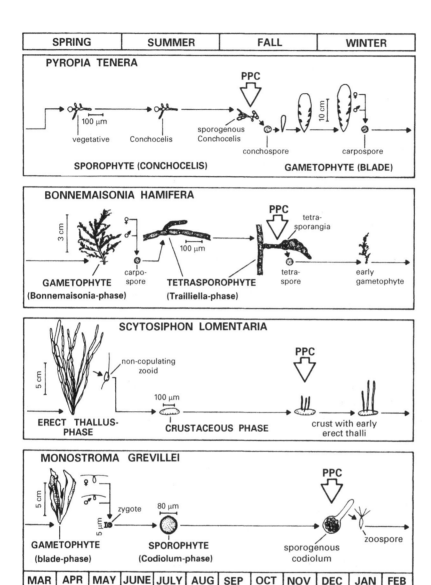

Figure 2.7 Annual cycles of four short-day algae. PPC, short-day signal. The responses are as follows: *Pyropia tenera* (previously *Porphyra tenera*) forms conchospores; *Bonnemaisonia hamifera* forms tetrasporangia; *Scytosiphon lomentaria* (in Europe) forms new erect thalli from the crust; *Monostroma grevillei* (codiolum phase) forms zoospores. (From Dring and Lüning 1983, *Encyclopaedia of Plant Physiology*, new series, vol 16B, with permission of Springer-Verlag, Berlin.)

switched on (*Dictyota*), whereas egg release by *Laminaria* is inhibited by blue light (Dring 1984a; Fig. 2.8). Blue-light effects (see Table 2.3) are typically attributed to "cryptochromes", which are a sub-group of the large cryptochrome/photolyase (CPF) family of blue-light sensitive flavoproteins, and found in animals, cyanobacteria, the unicellular red algae *Cyanidioschyzon merolae*, unicellular green algae, terrestrial plants,

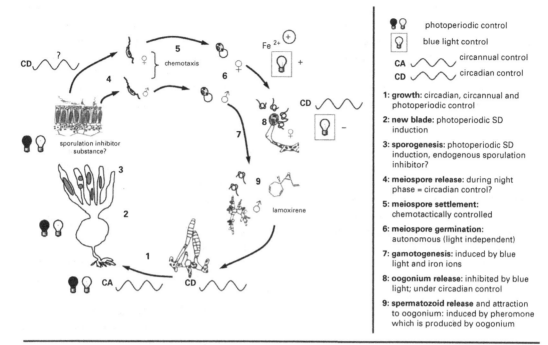

Legend content from figure:

- 🔌💡 photoperiodic control
- 💡 blue light control
- CA ～～ circannual control
- CD ～～ circadian control

1: **growth:** circadian, circannual and photoperiodic control

2: **new blade:** photoperiodic SD induction

3: **sporogenesis:** photoperiodic SD induction, endogenous sporulation inhibitor?

4: **meiospore release:** during night phase = circadian control?

5: **meiospore settlement:** chemotactically controlled

6: **meiospore germination:** autonomous (light independent)

7: **gamotogenesis:** induced by blue light and iron ions

8: **oogonium release:** inhibited by blue light; under circadian control

9: **spermatozoid release** and attraction to oogonium: induced by pheromone which is produced by oogonium

Growth of sporophytes
light saturation: 20-100 ∝mol m^{-2} s^{-1}
minimal annual light requirement: 40-96 mol photons m^{-2} y^{-1}
optimum temperatures: 5–15 °C
nutrients modulate growth, but are not triggers

Fertility:
gametophytes (optimum): 5-18 °C, 4-90 ∝mol photons m^{-2} s^{-1}
sporophytes: 1 to18 °C, 5-200 ∝mol photons m^{-2} s^{-1}

Figure 2.8 Schematic representation of life-cycle control in *Laminaria sensu lato* by abiotic and endogenous factors assuming that regulation processes are similar within the genus. (From Bartsch *et al.* 2008, reproduced with permission.)

and diatoms (e.g. Cashmore 2005; Asimgil and Kavakli 2012). Plant cryptochromes (plant CRY) regulate circadian rhythms and growth. The question of how CPFs evolved to perform a very wide range of physiological roles in prokaryotic and eukaryotic lineages is receiving much attention. However, although CPFs have been found in the red and brown lineages, and cryptochromes have been implicated in blue (and green) light detection in red seaweeds (Kain 2006), experimental validation of their existence in these seaweeds is required.

Other red- and blue-light photoreceptors have been found in seaweeds (Hegemann 2008; see sec. 1.3.3). The filamentous green seaweed *Mougoutia* uses neochrome to detect the red/far-red and UV-A/blue regions of the spectra (e.g. Suetsugu *et al.* 2005). The blue-light receptor aureochrome has been found in *Fucus, Vaucheria*, and the diatom *Thalassiosira* suggesting that it is common to the photosynthetic stramenopiles (Takahashi *et al.* 2007; Ishikawa *et al.* 2009). Dring (1988, p. 169) concluded that "photoperiodic responses in algae may be controlled by a variety of pigment systems analogous to the variety of pigments involved in algal photosynthesis". The application of molecular tools to identify candidate genes involved in photoreception, and of methodologies developed for photoperception and photocontrol in higher plants, would help advance this field for seaweeds.

Many seaweeds exhibiting photoperiodic control over reproduction have heteromorphic life histories, in which the algae use the cue to switch to a different phase, assumed a priori to be better adapted to the conditions in the next season. Best studied are species of Laminariales, which are strongly heteromorphic (Bartsch *et al.* 2008; Figs 2.5 and 2.8). *Laminaria* sporophytes grow vegetatively under long days under the control of circadian and circannual rhythms and photoperiod; new blade production is triggered by SD. The onset of sporogenesis is cued by SD and sori form on the older, distal tissue; this process may also involve a "sporulation-inhibiting substance" released by meristematic tissue but not distal tissue (also see *Ulva* later). The microstages of the life cycle are not under photoperiodic control. Meiospore germination is dependent on light dose, whereas gametophytogenesis is triggered by a specific dose of blue light, which means that when the sporophyte canopy is removed, increased blue light triggers gametogenesis. Iron (Fe^{2+}) also induces gametogenesis, while other nutrients (nitrogen, phosphorus) modulate the life cycle but do not act as a cue (Fig. 2.8).

Sporophyte development by *Undaria pinnatifida* is a rare example of a long-day photoperiodic response (Pang and Lüning 2004). This is a "facultative" LD response whereby LD triggers sporophylls to develop, but in the absence of a LD signal, sporophylls will eventually develop under a SD regime. *Undaria* are winter annuals with sporophyll production occurring in spring (i.e. longer days), and so the LD response clearly matches this seasonal cycle. The formation of hairs on the blade surface are also triggered by LD and these probably enhance nutrient acquisition during spring (see secs. 2.6.2 and 6.4.2).

Early field studies on *Ulva* on the Pacific coast of the United States showed a periodicity of propagule release, with gametes being released at the beginning of the spring tide series and spores 2–5 days later (Smith 1947). Over 30 years later, the endogenous, environmental and biochemical control of these cycles are being elucidated (Lüning *et al.* 2008). In a field experiment that lasted 2 years, *Ulva pseudocurvata* was shown to release gametes for a 1–5-day period every 14 days in fall and winter, and every 7 days in summer. In the laboratory, *U. pseudocurvata* has a free-running "sloppy" cycle of gamete release about every 7 days, that can be synchronized to a 1-month cycle by providing artificial moonlight. The difference between the summer (7-day) and winter (14-day) cycles may be due to insufficient light in winter to supply the energy required for reproduction. Gamete release occurs only after a minimum of 1 h darkness (quantified as < 0.001 μmole photons m^{-2} s^{-1}), followed by 5–9 minutes of light, explaining the sunrise release of gametes observed in the field. Red and blue wavelengths were equal in triggering gamete release, and gametophytes were more sensitive to these wavelengths (PDF of 0.01 μmole photons m^{-2} s^{-1}) than green light (0.1 μmole photons m^{-2} s^{-1}); the ability to detect such low PFDs suggests highly sensitive photoreceptors (phytochromes, cryptochromes, and rhodopsins). Gametes are only released from the marginal tissue, whereas tissue close to the holdfast is purely vegetative: this is explained by a "swarming inhibitor" (see below, Stratmann *et al.* 1996). Another finding of this study is that *Ulva* gametophytes become progressively smaller as the seasons progress from fall to winter, as they "sporulate away their thallus toward winter in a controlled way governed by the annual course of day length" (Lüning *et al.* 2008, p. 871). This is a photoperiodic SD response, because a night break spoils the effect. In winter, a perennial holdfast remains and therefore *Ulva pseudocurvata* is not an annual, it is a pseudo-perennial, with an overwintering holdfast that grows a new blade again in spring.

In *Ulva mutabilis* three regulatory factors have been identified that control the life-cycle progression from vegetative growth to gametogenesis (Stratmann *et al.* 1996), confirming the suggestion of Nilsen and Nordby (1975) that vegetative thalli release substances that inhibit sporulation, and Jónsson *et al.* (1985), who found that complex glycoproteins inhibit gametogenesis. Sporulation Inhibitor-1 (SI-1) is a very high molecular mass glycoprotein associated with the extracellular matrix (ECM) proteins that is released into the surrounding medium, and acts to maintain a vegetative thallus. Concentrations of SI-1 decrease as the thallus ages and when concentrations are below a critical level, gametogenesis occurs. A second compound, SI-2, with a low molecular mass, was isolated

Ulva mutabilis

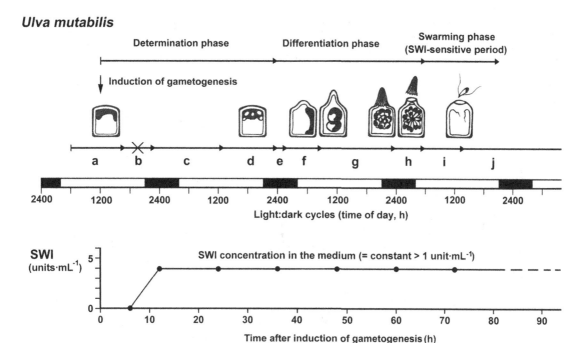

Figure 2.9 Time course (h) of induced gametogenesis and discharge of gametangia from *Ulva mutabilis* in relation to swarming inhibitor (SWI) –synthesis and action (summary of new and previous results). Top panels: Gametogenesis was induced in *U. mutabilis* [mutant slender sl-G (mt+)]. The "determination phase" and "differentiation phase" are defined as described by Stratmann *et al.* (1996). The "swarming phase" is the time period when gamete release can be induced by light or a medium change, or later may occur spontaneously. (a) Regular vegetative G1 cell-cycle phase in which gametogenesis can be induced by removal of sporulation inhibitors SI-1 and SI-2 from the medium. (b) Regular vegetative S cell-cycle phase in which the genome is replicated, normally, or, after induction of gametogenesis, the period of SWI synthesis and excretion. (c) Next G1-phase after induction of gametogenesis. (d) Next S-phase after induction of gametogenesis and accumulation of starch granules. (e) Time of irreversible commitment to gametangium differentiation. (f) Period of progamete formation, chloroplast reorientation, and papilla initiation. (g) Period of progamete multiplication to 16 cells, and papilla maturation. (h) Period of gamete and pore cap maturation. (i) Period when the exit pores are open and when gamete release can be induced by light and (or) depletion of SWI in the medium. (j) Period when the gametangia become insensitive to SWI, and gamete release may occur spontaneously and asynchronously. Bottom panel: The time course of SWI accumulation in the medium was measured in parallel to gametogenesis by the SWI assay at the times indicated in the figure in a sample containing 10 mg *Ulva* fragments · mL^{-1}. (From Wichard and Oertel 2010, reproduced with permission.)

from the space between the two cell wall layers, and is thought to act as positive regulator through interactions with SI-1 because concentrations do not change as the life cycle progresses. The conversion of vegetative cells to gametangia occurs only when SI-1 is removed: when this happens the cells enter a "determination phase" during which they can return to vegetative growth if SI-1 is added back (Fig. 2.9). The length

of this phase is 23–46 h, and depends on the time of day at which induction of the cell-cycle progression was begun. Cells then enter a "differentiation phase" that is 28 h long; they are now committed to becoming gametes and are no longer susceptible to SI-1. The flexible timing of the determination phase means that irrespective of the time of day (i.e. time of cell cycle) when the life-cycle progression is induced, gamete

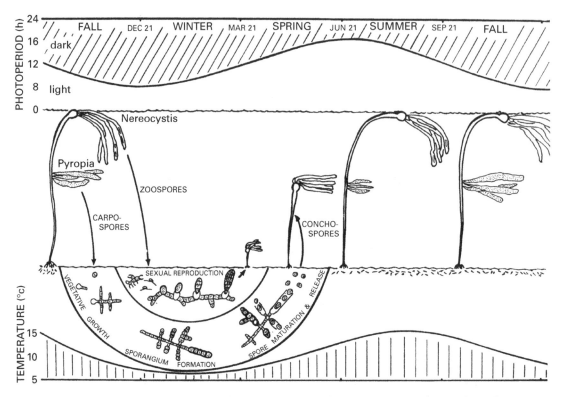

Figure 2.10 Life history and seasonal occurrence of the annual, epiphytic alga *Pyropia nereocystis* (previously *Porphyra nereocystis)* and its annual host seaweed, *Nereocystis luetkeana*. The top part of the diagram shows seasonal photoperiod variation at the Puget Sound study site; the lowest part of the diagram traces water temperatures. Carpospores from *Pyropia* blades form the shell-boring conchocelis stage, which releases conchospores in response to long days after short days, as the new annual crop of *Nereocystis* sporophytes elongates. Zoospores from *Nereocystis* form microscopic male and female gametophytes, and sexual reproduction (not photoperiodic) results in sporophytes. (From Dickson and Waaland 1985, with permission of Springer-Verlag, Berlin.)

release always occurs between 0500 and 0800 h. A third compound, a low molecular weight "Swarming Inhibitor" (SW-1) is released during the determination phase, triggering the release of motile gametes (Stratmann *et al.* 1996). SW-1 is also present in *Ulva lactuca*, although there are species-specific differences in timing of gametogenesis and its relationship to SW-1 concentrations (Wichard and Oertel 2010).

The more critical the timing of reproduction, the more complex the environmental cues need to be. Short days, for instance, occur in both fall and spring, as well as through the winter. *Pyropia nereocystis*

(previously *Porphyra nereocystis*) grows on the stipes of the annual kelp *Nereocystis luetkeana*. The bladed phase of the *P. nereocystis* appears when the host stipe has completed elongation but before the stipe becomes covered in other algae. Moreover, the stipes are high in the water column, whereas the sporophyte of *P. nereocystis* is on the bottom, in old shells. To time its spore release for spring, *P. nereocystis* responds to a dual photoperiod: prolonged short days followed by prolonged long days (Fig. 2.10) (Dickson and Waaland 1985). (Tests were run at 8:16 and 16:8 light:dark photoperiods, and critical photoperiods were not determined.) The response was also better in cooler water, typical of spring, than in warmer water, typical of fall. The conchospores are released in slime strands that may produce a "bola" effect for increasing the chances of snagging and sticking to the slippery young kelp stipes. Related bladed species have less critical photoperiodic control. For example, in *Pyropia torta* (previously *Porphyra torta)* from the same region (Puget Sound), conchospores can form in any photoperiod, but they mature and are released only when there are short days (Waaland *et al.* 1987); this species is a winter annual on rocky intertidal substrates.

"Short day" and "long day" obviously are relative terms, and for a seaweed with a wide latitudinal range, what is a short day in higher latitudes may be a long day in lower latitudes. Compare, in Figure 2.11, for instance, the effects of 11-h days on *Scytosiphon* from Tjörnes (66°N) and from Punta Banda (32°N) (Lüning 1980). However, intraspecific differences in critical day length do not always correlate with latitude, as Rietema and Breeman (1982) found in *Dumontia contorta*. Moreover, photoperiod responses sometimes are altered by temperature (Fig. 2.11 and Table 2.2) and may not be exhibited in the presence of high nitrogen levels (such as are created in standard culture media).

We still know very little about reproductive phenology in the tropics. There are seasonal changes in the environment, albeit more subtle than those in midlatitudes and there are strong seasonal variations in growth and reproduction of the flora, but the cues are unknown (Price 1989). A dramatic example of reproductive synchronicity in the tropics is the mass spawning of 17 species of Bryopsidales on Caribbean coral reefs (Clifton 1997; Clifton and Clifton 1999). These algae become fertile overnight, then release their entire cellular contents as gametes (termed holocarpy) in the morning, forming a dense cloud in the water. Up to nine species release gametes on the same morning, but do so at slightly different times, which could reduce hybridization. The entire event from onset of fertility to death of the parent material lasts just 36 h. All that is left of the adults are white cell walls that mostly disappear within 24–48 h, either via water motion or grazing, most likely freeing up space on the

reef for the propagules to settle. While the precise cues that trigger the onset of fertility are unknown, they may include a combination of light and temperate because on cloudy mornings, gamete release was delayed by around 20 minutes and a laboratory study indicated a trend of spawning being delayed by 8 minutes per 1°C decline in temperature.

The more equable conditions in the tropics are perhaps reflected in the apparently low numbers of tropical algae with heteromorphic life histories. Nevertheless, since photoperiodic effects have been shown in temperate isomorphic species, and there are day length changes except very close to the equator (e.g. the range in Guam, at 13° N, is 11–13 h) (Fig. 2.6a), it would be interesting to look for latitudinal effects in widely distributed heteromorphic species such as *Asparagopsis taxiformis* or *Tricleocarpa fragilis* (previously *Tricleocarpa oblongata*). If tropical algae near their northern or southern limits (e.g. in Bermuda, Hawaii, or southern Queensland) show photoperiodic responses, what do they do near the equator? And if they are day-neutral, to what environmental cues do they respond?

2.3.4   Other factors

How do the results of laboratory experiments on the temperature and photoperiod requirements for reproduction relate to conditions in the "real world", especially of the intertidal zone where other environmental factors (secs. 3.1 and 7.1) come into play? Breeman and Guiry (1989) described how tides alter reproductive timing in *Bonnemaisonia hamifera* sporophytes. These are SD algae, requiring a narrow range of warm temperatures (Table 2.2). Lüning (1980a) had predicted reproduction only during a short time in early fall, when the days become short enough but the sea is still warm (Fig. 2.12 and Table 2.2). Whereas, in general, phenology *in situ* bore out the predictions (Breeman *et al.* 1988), two factors confounded the predictions: (1) High spring tides at the beginning or end of the day shortened the effective day length, allowing reproduction to start earlier than predicted. (2) Low water of spring tides in the middle of the day exposed seaweeds to warm air temperatures,

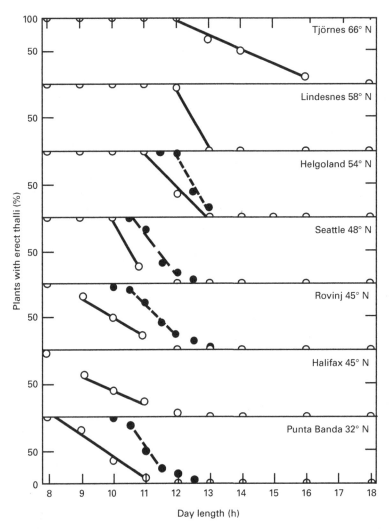

Figure 2.11 Effects of day length on erect thallus formation by different geographic isolates of *Scytosiphon lomentaria* at 10°C (open circles) and 15°C (filled circles). Each value is based on a count of 250 individuals. (From Lüning 1980a, with permission of The Systematics Association.)

when water temperatures were below the threshold, and allowed reproduction to resume. Brief exposures to a suitable combination of conditions sufficed for induction, and the reproductive period stretched from September into December. [At their study sites in Ireland, the times of high and low spring tides are always the same; that is not the case everywhere (sec. 3.1.1)].

In another example (Breeman *et al.* 1984), light was so reduced at high tide, because of turbidity and the fucoid canopy, that SD conditions prevailed all year for mid-intertidal populations of *Rhodochorton purpureum*.

Lunar and tidal cycles have also been found to trigger reproduction. Gametogenesis in *Dictyota diemensis*

Table 2.2 Effects of photoperiod and temperature on tetrasporangium formation in the trailliella phase of *Bonnemaisonia hamifera*[a]

Parameter	Response to day length (at 15°C)										
Hours light per day	8	9	10	10.5	11	12	12.5	13	14	15	16
Percentage fertile	93	92	48	16	6	0	0	0	0	0	0
Parameter	Response to water temperature (at 8 h light per day)					
Temperature (°C)	10	12	15	17	20	23
Percent fertile	0	0	97	73	0	0

[a] 150 plants in each experiment were grown in enriched seawater (containing less than 20 μM NO_3^-).
Source: Lüning (1981b), with permission of Gustav Fischer Verlag.

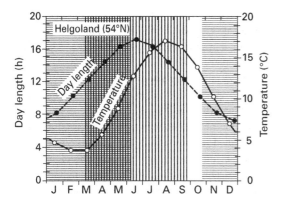

Figure 2.12 Predicted tetrasporogenesis in the trailliella phase of *Bonnemaisonia hamifera* in Helgoland, approximately the same latitude as the site in Ireland where Breeman and Guiry tested the prediction. The "window" for reproduction in September–October occurs between too warm seas (vertical hatching) and too short photoperiod (horizontal hatching). (From Lüning 1981b, *Ber. Deutsch. Bot. Ges.*, vol. 94, with permission.)

begins the day after a full moon and is completed with gamete release 10 days later (Phillips *et al.* 1990). For *Monostroma*, tides, rather than endogenous clocks, were considered responsible for synchronizing gamete release (Togashi and Cox 2001; compare to *Ulva* above). Some fucoid algae have fortnightly periodicity of gamete release, and the timing differs for the same species growing at different geographic locations, suggesting that the timing is entrained by tidal and diurnal cues, rather than lunar cycles

(reviewed by Pearson and Serrão 2006). Mid-intertidal *Fucus vesiculosus* tends to release eggs late in the day and at a low tide compared to high-intertidal *F. spiralis* which releases eggs throughout the day, at both low and high tides. The suggestion is that as a selfing-hermaphrodite, *F. spiralis* requires less synchrony of egg release compared to dioecious *F. vesiculosus* (Ladah *et al.* 2008).

Seaweeds that do not experience strong seasons of temperature and photoperiod may still require environmental cues for reproduction. The deep-water brown alga *Syringoderma floridana* has macroscopic sporophytes and microscopic gametophytes. Most of the two-celled gametophytes develop right on the sporophyte, because the zoospores have very limited motility (Henry 1988). In culture, sporogenesis was induced by low-temperature shock or by transfer to a nutrient-rich medium. [Sudden changes in the surrounding medium can induce reproduction in some seaweeds (Chapman 1973; DeBoer 1981).] Gametophytes matured and released gametes predictably 2 days after settlement of zoospores, at 20°C. Henry (1988) suggested that the arrival of a water mass high in nutrients (probably also relatively cool) or a low-temperature water mass followed by a warm one would induce simultaneous sporogenesis throughout a local population. Synchrony evidently is vital to seaweeds with such small and short-lived gametophytes, and here temperature acts as a non-seasonal cue. The effects of UV radiation and water motion on reproduction are discussed in secs. 5.2.2 and 8.2.2, respectively.

2.4 Fertilization biology

After reproduction has been initiated in response to environmental cues, sporogenesis or gametogenesis takes place, and finally spores or gametes are released. Three types of sexual reproduction are traditionally recognized: isogamy, anisogamy, and oogamy (e.g. Graham *et al.* 2009). Oogamy involves a non-motile female gamete or egg, for example the brown orders Fucales and Laminariales, and all red seaweeds, that is fertilized by a smaller male gamete. Among brown algae, many so-called isogamous and anisogamous species actually behave oogamously, with the female gamete settling before fertilization, and Motomura and Sakai (1988) show that *Saccharina angustata* (previously *Laminaria angustata*) eggs have vestigial flagella that are shed when the egg is released from the oogonium.

Santelices (2002) provides a framework for the fertilization ecology of seaweeds, based on the "brooders" versus "broadcasters" scheme used for marine invertebrates. "Broadcasters" (external fertilization) include species that liberate gametes into the water column, and for these to ensure fertilization success, there must be tightly controlled synchronicity, for example the mass spawning events on coral reefs (Clifton 1997) or synchronous release of gametes in the Fucales (Pearson and Serrão 2006). Most green and brown seaweeds are placed in the broadcaster category, and this mechanism is also common in marine invertebrates (e.g. sea urchins). Brooders (internal fertilization), in contrast, require efficient sperm collection mechanisms, a mechanism analogous to pollen collection in terrestrial plants. Many red seaweeds fall into this category, with their trichogyne acting as the "sperm-collector" (see Figs. 2.16 and 2.17). The trichogyne, consisting of a tube along which sperm travel to mate with the large egg at the end, resembles that of a flowering plant's pollen tube. Kelp gametophytes in which the egg remains on the female could also be classified as brooders. Santelices (2002) acknowledges that his framework may be an oversimplification given the diversities of algal life cycles, but nevertheless it provides a useful structure for testing hypotheses relating to the fertilization ecology of seaweeds.

In species with free-living male and female gametophytes, triggering of gamete release and attraction of one gamete to the other increase the chances of successful syngamy. In the Laminariales and Desmarestiales male antheridia do not release their sperm until they detect the pheromone from mature female gametophytes; the same compound (lamoxirene, Fig. 2.13) acts as antheridium releaser and sperm attractant (Müller *et al.* 1985; Müller 1989). This can result in a "mass release" of sperm, and the time from chemical signal to response is just 8–12 seconds for *Laminaria digitata* (Pohnert and Boland 2002). In the field, male and female kelp gametophytes need to be sufficiently close to one another that the male can detect the pheromone released by the female, and for *Macrocystis pyrifera* and *Pterygophora californica*, spore density must be greater than 1 spore mm^{-2} for successful recruitment (Reed 1990a).

The role of pheromones in sexual reproduction is best studied in brown algae for which 12 have been identified, and these each have structural isomers, leading to a high diversity of these signaling molecules (Fig. 2.13; Pohnert and Boland 2002). However, not all brown algae use pheromones. *Sargassum muticum* is monoecious, and fertilization takes place under a blanket of mucilage, while the oogonia are still strapped to the conceptacle by their mesochiton; presumably, self-fertilization can occur. *Himanthalia* eggs also apparently do not chemically attract sperm. Moreover, *Dictyopteris* and *Feldmannia mitchelliae* (previously *Hincksia* [*Giffordia*] *mitchelliae*) secrete compounds similar to those in Figure 2.13, from both gametophytic and sporophytic tissues, but these substances do not act as pheromones in these genera (Kajiwara *et al.* 1989; Müller 1989).

Although pheromones have been detected in a large number of brown algae, there is surprisingly little variation in their structure and they are not distributed along taxonomic lines (Müller 1989; Pohnert and Boland 2002). Female eggs typically release a blend of several pheromones, although usually one of these compounds is the most biologically active. All the

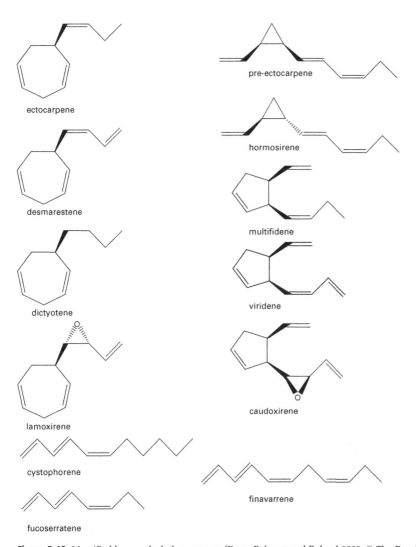

ectocarpene

pre-ectocarpene

hormosirene

desmarestene

multifidene

dictyotene

viridene

lamoxirene

caudoxirene

cystophorene

finavarrene

fucoserratene

Figure 2.13 Identified brown algal pheromones. (From Pohnert and Boland 2002, © The Royal Society of Chemistry 2002.)

compounds are simple, non-polar, volatile hydrocarbons, either open-chain or cyclic olefinic hydrocarbons. The structural similarity between the 12 brown algal pheromones indicates a common biosynthetic pathway from fatty acids, and a few tailoring enzymes are required for their activation (Pohnert and Boland 2002; Rui and Boland 2010). Their insolubility and volatility prevent their concentrations building up in the water and enable the female gametes to maintain steep concentration gradients at their surface. The range of attraction probably is no more than 0.5 mm (Müller 1981). The pheromones are highly active, and so only minute quantities of attractant are released: 5 million *Macrocystis* eggs yielded 2.9 μg of lamoxirene (Müller *et al.* 1985).

The behavior of male gametes in the presence of an attractant varies from one species to another (Fig. 2.14). *Laminaria* sperm head straight for the

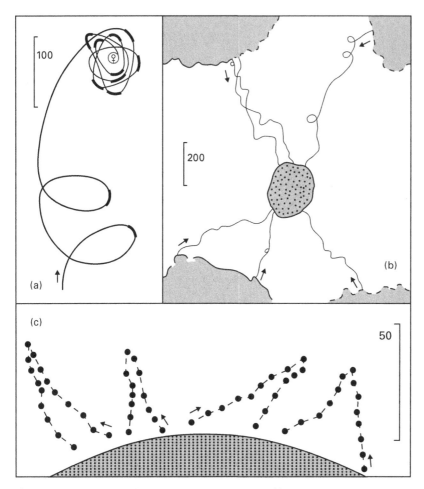

Figure 2.14 Different types of gamete approach in brown algae. (a) Chemo-thigmo-klinokinesis in *Ectocarpus siliculosus;* emphasized parts of male track indicate periods of hind-flagellum beat. (b) *Laminaria digitata:* impregnated silica particle as pheromone source in center, with tracks of individual sperm. (c) *Fucus spiralis:* return responses of individual sperm near a fluorocarbon droplet containing fucoserratene. Scales in micrometers. (From Müller 1989, with permission of John Wiley/Alan R. Liss Inc.)

egg. *Ectocarpus* males have a more complex pattern. In the water column they swim in straight lines, periodically changing direction abruptly. When they encounter a surface, they change to a wide, looping path along the surface. In the presence of attractant from the female gamete, the male changes to a circular path, the diameter of which decreases as the hormone concentration increases.

Gamete recognition is a critical stage in sexual reproduction. Because female gametes release a blend

of pheromones, they may attract the sperm of different species. In *Fucus*, the same attractant works for at least three species. Similarly for the kelps, (1'R, 2S, 3R)-lamoxirene was the bioactive component of pheromones released by *Laminaria, Alaria, Undaria,* and *Macrocystis* and no species specificity of the pheromone was detected (Maier *et al.* 2001). Rather than gamete recognition by a species-specific pheromone, gamete selectivity arises from the recognition of complementary surface carbohydrates, and syngamy is

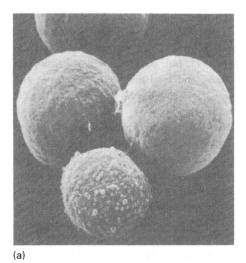
(a)

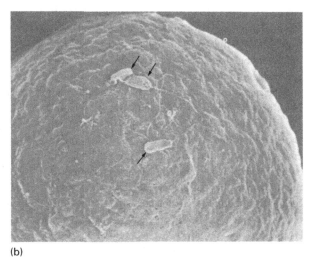
(b)

Figure 2.15 Scanning-electron-microscope views of eggs, sperm, and zygotes of *Fucus serratus*. (a) Group of cells 10 min after mixing eggs and sperm. Smooth cells have been fertilized and have formed a fertilization membrane; the rough cell in the foreground is an unfertilized egg (×450). (b) Detail of fertilized egg with three sperm (arrows); the tip of the anterior flagellum of the middle sperm is embedded in secreted cell wall material (×1600). (From Callow *et al.* 1978, with permission of The Company of Biologists.)

thus prevented (Schmid 1993; Schmid *et al.* 1994). Glycoconjugates are carbohydrates that are covalently linked to another chemical (e.g. glycoprotein = carbohydrate + protein), and are involved in a wide range of cell–cell and cell–substrate interactions (Lodish *et al.* 2008). The processes of recognition and fusion have been studied in some Fucales and Ectocarpales, taking advantage of the fact that surface receptors are not obscured by cell walls (Evans *et al.* 1982). The egg membrane initially appears lumpy because of protrusion of cytoplasmic vesicles (Fig. 2.15a; Callow *et al.* 1978). The spermatozoid probes the surface of the egg with the tip of its anterior flagellum, apparently seeking specific binding sites (Friedmann 1961; Callow *et al.* 1978). Attachment takes place first by the flagellum tip, and later also by the body of the cell (Fig. 2.15b). Egg membrane surfaces carry special glycoproteins with fucose and mannose units in particular patterns that fit into carbohydrate-binding sites (ligands) on sperm membrane proteins, analogous to a lock-and-key mechanism (Bolwell *et al.* 1979, 1980; Wright *et al.* 1995a, b). Some of the *Fucus* sperm surface domains that have been

distinguished by monoclonal antibodies probably are specific for egg recognition. The sperm protein has been characterized; it partially activates the egg when it locks to its surface (Jones *et al.* 1988; Wright *et al.* 1995a, b).

For Florideophyte red seaweeds, fertilization is initiated by the cell–cell recognition between the liberated sperm and the trichogyne. For *Aglaothamnion sparsum*, recognition was mediated by fucose and mannose units, similar to the fucoid example above (Kim *et al.* 1996). For *Aglaothamnion oosumiense*, however, male gametes use a "double-docking" process for recognizing the appropriate female trichogyne, utilizing at least two carbohydrate moieties and the complementary receptors which are lectin-like molecules (Kim and Kim 1999). They suggest that the carbohydrate-based system observed in some red algae may not be species specific, but rather a generic mechanism, and that other currently unidentified mechanisms will exist for specific species.

When one sperm has entered the egg, no more are needed. Indeed, polyspermy is lethal; fucoid germlings develop abnormally and die after a few days (Brawley

1987, 1991, 1992a; Brawley and Johnson 1992). Mechanisms have evolved to minimize the percentage of eggs that are fertilized by more than one sperm, even though in monoecious species fertilization often takes place when oogonia and antheridia are newly released and the sperm concentration is likely to be high. A fast block to polyspermy has been shown in *Pelvetia* and *Fucus* (Brawley 1987); this block is Na$^+$ mediated and is replaced within about 5 min by a "slow block" corresponding to the formation of the cell wall, and an intermediate block in which the sperm receptors on the surface of the egg are degraded (Brawley 1991; reviewed by Pohnert and Boland 2002). Polyspermy blocks do not always work perfectly and for *Fucus ceranoides* 1–9% of eggs shed were polyspermic (Brawley 1992a). Polygamy was also evident in *Bryopsis* but it is not known if it is lethal in this genus (Speransky *et al.* 2000). In red algae, there is indirect evidence of mechanisms to prevent polyspermy, but this has not been confirmed (Santelices 2002).

Red algal sperm do not have flagella and therefore cannot swim towards the egg located on the female gametophyte. Correspondingly, sexual attractant pheromones have not been detected in red seaweeds. The lack of sperm motility has led to the traditional view that red seaweeds are ill-equipped for fertilization and dispersal in the marine environment, and that they compensate for this through zygote amplification (sec. 2.2; Searles 1980). However, this theory has not been born out in quantitative field studies, which show that fertilization rates of red seaweeds are similar to those of brown and green seaweeds (Kaczmarska and Dowe 1997; Engel *et al.* 1999). This led Santelices (2002) to call for a new explanation for the "triphasic" life cycle of the red seaweeds. Brawley and Johnson (1992) suggest that immobilization of the female gametes (i.e. brooders) might be an adaptive advantage to enhance fertilization: with only one gamete moving in the water column the chances of missing the other might be reduced. A range of mechanisms can enhance fertilization success despite non-motile flagella. The ability of spermatia to reach a trichogyne is improved when they are released in slime strands, as in *Tiffaniella snyderae* (Fetter and Neushul 1981). The fibrillar mucilage is elastic; it stretches out in water

flow, and when it attaches to the female alga it tends to sweep the surface and deposit spermatia on extended trichogynes. There are cone-shaped appendages on spermatia from *Aglaothamnion neglectum* that are not sticky and bind only with trichogynes and hairs, though the binding is not species specific (Magruder 1984).

A popular theory in fertilization biology is that of "sperm limitation". The idea is that for broadcast-spawning marine organisms, fertilization success would be limited due to dilution of sperm in a turbulent water column. However, for fucoid seaweeds many species have fertilization levels of close to 100% indicating mechanisms that ensure fertilization success, such as timing gamete release to coincide with periods of calm water (Serrão *et al.* 1996; Berndt *et al.* 2002; sec. 8.2.2). For red seaweeds too, there is little evidence of sperm limitation (Maggs *et al.* 2011). Engel *et al.* (1999) conducted an elegant field study in which they mapped and sampled each of the 64 tetrasporophytes, 37 male and 26 female gametophytes of *Gracilaria gracilis* in a rock pool, at low tide. Using two microsatellite loci as genetic markers, they identified the exact males within the population that were responsible for fertilizing 72% of the 350 cystocarps sampled. Only 11% of males were from populations outside the rock pool. While they found no evidence of sperm limitation, there were strong differences in the male's ability to fertilize the females, indicating either male–male competition or female choice.

Sexual reproduction in red algae has been extensively studied because of its importance in systematics (reviewed by Hommersand and Fredriq 1990). Post-fertilization events in red seaweeds are complex. O'Kelly and Baca (1984) were able to observe the timing of reproductive stages in *Aglaothamnion cordatum*, which in culture produced one new axial cell per day. Like all Ceramiales, this species has a four-celled carpogonial branch (Fig. 2.16), and auxiliary cells are produced only after fertilization, in this case from the support cell and an additional auxiliary mother cell. Gamete fusion (including spermatium attachment, plasmogamy, transfer of the male nucleus down the trichogyne, and karyogamy) took 5–10 h. Carpogonia divided to form two daughter cells. Auxiliary cells

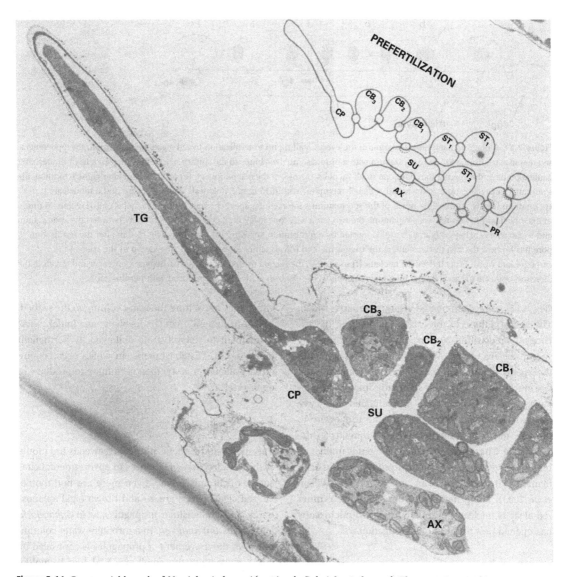

Figure 2.16 Carpogonial branch of *Neosiphonia harveyi* (previously *Polysiphonia harveyi*). Electron-micrographic section and diagram of prefertilization appearance. AX, auxiliary cell; $CB_{1...3}$, carpogonial branch cells; CP, carpogonium; PR, pericarp; $ST_{1,2}$ sterile cells; SU, support cell; TG, trichogyne. (From Broadwater and Scott 1982; with permission of *Journal of Phycology*.)

formed after about 40 h, and at around 72 h were diploidized; that is, the original haploid nucleus was partitioned off into a foot cell, and a diploid nucleus from a carpogonium daughter cell was transferred via a connecting cell. Some key aspects of this process, such as spermatial release and attachment to the trichogyne, have been elucidated for a few species (Pickett-Heaps and West 1998; Santelices 2002; Wilson *et al.* 2003). The events following a mass release of spermatia from *Bostrichya* leading to fertilization

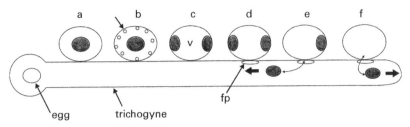

Figure 2.17 A diagrammatic representation of the events leading up to fertilization in red algae. Each spermatium represents a sequential stage from a to f. (a) Spermatium with a prophase nucleus binds to the surface of a trichogyne; (b) a few minutes after gamete binding, the spermatium develops small vacuoles (arrow) at the cell periphery; (c) at 20–30 min after gamete binding, the spermatial nucleus resumes mitosis and is usually complete within 15 min. Cytokinesis does not occur and a binucleate spermatium results. At this time much of the spermatium's intercellular space is occupied by a vacuole (v); (d) within 45 min of male nuclear division the walls separating the two gametes erode away until cytoplasmic continuity between the spermatium and trichogyne is achieved (gametes are connected by a fertilization pore, fp); (e) approximately 15 min after the fertilization pore has formed, the first nucleus enters the trichogyne and travels either towards the trichogyne tip or the base of the carpogonium which bears the female nucleus; (f) soon after, the second nucleus enters the trichogyne cytoplasm and travels in the direction opposite to the first migrating male nucleus. (From Wilson *et al.* 2003, reproduced with permission.)

(Fig 2.17) were documented using time-lapse video. The sperm adhere to the fragile filaments of the trichogyne, and within 30 min a mitotic division of the sperm nucleus occurs. At 50–80 min after attachment, the sperm membrane seals with the trichogyne and two nuclei move from the spermatia into the trichogyne. These two nuclei are differentiated and have different fates: one migrates down the trichogyne to the egg, the other stops moving and remains in the trichogyne. The mechanism of movement is unknown, but for *Aglaothamnion* and *Murrayella periclados* actin filaments are involved (Kim and Kim 1999; Wilson *et al.* 2003). The generality of these findings to other red algae is unclear, but time-lapse video microscopy has opened new avenues of investigation.

2.5   Settlement, attachment, and establishment

Once the reproductive cells have been released from the parent generation they must get to a surface and stick to it, prior to germination. The following scheme was developed by Fletcher and Callow (1992) to track the fate and development of spores that are released into the water column: (1) "Settlement" or "encountering the substrate". Settlement is the mechanism by

which they move from the water column to the seabed. (2) "Attachment" which involves an initial weak attachment to the substrate, followed by permanent adhesion. (3) "Establishment" in which cells acquire a cell wall and polarity, then germinate to produce an apex and rhizoid.

2.5.1   Settlement

The spores and gametes of many seaweeds are motile and show directional responses to environmental cues such as light and gravity. Even more are non-motile, i.e. red algal spores, green- and brown algal aplanospores, and multicellular propagules, as in *Sphacelaria* and *Sargassum muticum*. In a turbulent water column, however, the movement of propagules is controlled by water movement; the small size and low Reynold's number (see Chapter 8) of motile propagules means that their powers of motility are limited, similar to flagellated phytoplankton. However, water velocity and turbulence decreases towards the seabed, and at the seabed surface there are slow-moving (1–10 mm s^{-1}) and non-moving layers of water that form around all submerged objects (Denny 1988; Amsler *et al.* 1992; sec. 8.1.3). Estimates of the thickness of the viscous region (viscous sub-layer) of the velocity boundary layer are 5–150 μm for various surfaces, whereas red algal spore sizes, for instance, are 15–120 μm (Coon *et al.* 1972; Neushul 1972). In order to get into the "safe zone" of the boundary layer where they have time to attach, cells must travel through moving water.

Non-motile cells get to the seabed by strictly physical forces (Coon *et al.* 1972; Amsler *et al.* 1992; Stevens *et al.* 2008). Gravity tends to pull cells downward at ever-increasing speeds, but drag also increases with speed, so that a maximum (terminal) velocity is reached. This terminal velocity, V_t, depends partly on the density and radius of the spore (i.e. Stokes law). Coon *et al.* (1972) measured V_t for several species of red algal spores using time-exposure photomicrographs. *Sarcodiotheca gaudichaudii* carpospores were fastest, sinking at 116 μm s^{-1}, but that is much less than typical water-current velocities. Neushul (1972) estimated that it would take a *Cryptopleura* carpospore 10 min to fall through perfectly still water from the cystocarp on the adult to the seabed. Taylor *et al.* (2010) examined the effect of turbulence on settlement and attachment rates of eggs and zygotes of five species of fucoid algae. In still water, unfertilized eggs sank at a rate that was correlated to their size: the smallest eggs were of *Durvillaea antarctica* (29 μm diameter) and these had the slowest sinking rate (0.029 cm s^{-1}), the largest eggs were produced by *Cystophora torulosa* and *Pelvetiopsis limitata* (~100 μm) and sank at ~0.062 cm s^{-1}, whereas those of *Hormosira banksii* and *Fucus distichus* (previously *Fucus gardneri*) were intermediate in size and sinking rates. However, egg density was not related to sinking rate, with the eggs of *Durvillaea* having the greatest density, and they concluded that the stickiness and mucus characteristics of eggs are important in explaining differences amongst species. For fucoids, egg buoyancy appears to be the most important factor controlling sinking rates, and this is dependent on zygote properties such as the density of the mucus coat and how fast this dissolves (Stevens *et al.* 2008). The newly liberated spores of the red seaweed *Laurencia* are surrounded by an acid polysaccharide, which upon hydration, may increase the weight of the spores and their rate of deposition (Bouzon and Ouriques 2007). Unequal distributions of heavy (e.g. plastids and nuclei) and light (e.g. lipid bodies) organelles

within the cells might result in specific cell orientation in the water column, and "also influence propagule behaviour" (Amsler *et al.* 1992).

Swimming speeds of motile cells are "puny in relation to the [current and wave] forces that beset them" (Norton 1992), and motility probably has little influence on the ability of cells to move from the mainstream seawater to the boundary layers at the seabed. Suto (1950) reported speeds of 0.13–0.30 mm s^{-1} for various zooids, and similarly *Ulva* zoospores swim at 0.2 mm s^{-1} (Granhag *et al.* 2007) and *Laminaria* at 0.16 mm s^{-1} (Fukuhara *et al.* 2002). The energy required to sustain swimming is derived from photosynthesis for some species, for example, the motile spores of *Macrocystis pyrifera* and *Pterygophora californica* can swim for up to 120 h in the light, and only 72 h in the dark (Reed *et al.* 1992). This ability differs among species, however, and for *Saccharina japonica* (previously *Laminaria japonica*) swimming was greater in the dark and there was no effect of photosynthesis on swimming. In this case the energy reserves may come from energy stores, such as the lipid stores observed for *Macrocystis pyrifera* and *Pterygophora californica* (Brzezinski *et al.* 1993; Reed *et al.* 1999). While in the water column, spores and gametes are part of the phytoplankton, and Graham (1999) reported between 1360 to 18 868 spores L^{-1} depending on season. The length of time spent in the plankton varies between species and this is determined by environmental factors but also their life-history stage and strategy (Fletcher and Callow 1992).

Several seaweeds have evolved interesting means of improving the chances of spores reaching substratum (Fletcher and Callow 1992). *Nereocystis* blades float far above the seabed, but the entire sorus, which sinks readily, is shed (reviewed by Springer *et al.* 2010). Sorus shedding takes place for a few hours around dawn, giving the spores the best chance for photosynthesis and survival. Spore release begins before sorus abscission, continues as the sorus sinks, and is completed within about 4 h (Amsler and Neushul 1989a, b). *Postelsia*, which grows in very high-energy intertidal habitats, releases its spores when the first waves of the incoming tide splash over the seaweed; water and spores flow down channels in the drooping blades and drip onto the rock and parent seaweed's holdfast (Dayton 1973). *Fucus* releases its eggs still held together in the oogonium; the mass of eight eggs sinks faster than would a single egg. The invasive seaweed, *Sargassum muticum*, has a very effective settling mechanism in which the eggs released from the conceptacles remain attached to the outside of the receptacle, where they are fertilized and develop into small germlings, usually without rhizoids, before they drop to the seabed. As a result of their relatively large size (mean 156 μm), these propagules sink at an average rate of 530 μm s^{-1} in still water (Deysher and Norton 1982), some 5–10-times faster than unicellular spores. Once rhizoids start to grow out, they increase the drag and slow the sinking rate (Norton and Fetter 1981).

Significant numbers of propagules can reach the seabed in the fecal pellets of grazers (Santelices and Paya 1989). Herbivores ingest vegetative and reproductive tissues, sometimes preferring the latter (Santelices *et al.* 1983), and spores and tissue fragments often survive passage through their guts. Such fragments can form swarmers or protoplasts that will give rise to new individuals, especially in opportunistic algae like *Ulva*. Cells in fecal pellets have several advantages: The pellets are heavy, sinking 8–22-times faster than *Sargassum* propagules and 40–100-times faster than algal spores; the stickiness of pellets greatly improves attachment; the pellets provide protection against desiccation in the intertidal zone, giving sensitive germlings a chance to establish; and nutrient availability may be higher in the pellets.

Although swimming ability is ineffective compared to the waves and currents of mainstream flow, mobility is important in keeping cells at the water surface, or locating the seabed. A range of tactic responses have been observed for motile cells. Laminarian zoospores, for example, swim randomly, changing direction frequently. These zoospores do not have an eyespot and chemotactic responses to nitrate and phosphate guide them to the substratum (Amsler and Neushul 1989b, 1990; Fukuhara *et al.* 2002). Some motile cells can orient with respect to light, indicating the presence of a photoreceptor (sec. 1.3.3); some of these are

negatively phototactic and swim toward the seabed but others are positively phototactic and continuously swim upwards which may aid dispersal (Amsler and Searles 1980; Hoffman and Camus 1989; Clayton 1992). Swarmers of *Scytosiphon* exhibit either phototactic or thigmotactic swimming: of 34 swarmers released, 22 had a spiral phototactic swimming pattern, 10 swam in a circular path near to the coverslip surface, characteristic of thigmotactic swimming, and two were initially thigmotactic but became phototactic (Matsunaga *et al.* 2010). The zoospores of *Feldmannia irregularis* (previously *Hincksia irregularis*) from North Carolina, USA, were positively phototactic, whereas those from the Florida panhandle were negatively phototactic; Greer and Amsler (2004) caution that these could be different cryptic species. For *Monostroma grevillei*, and also in *Ulva lactuca*, gametes are initially positively phototactic which may enhance fertilization success (see below); they become negative upon pairing facilitating movement towards the seabed (Kornmann and Sahling 1977). Zoospores, which must move back to the intertidal, are positively phototactic.

For green seaweeds, there may be a link between the type of sexual reproduction, the possession of a photoreceptor and the location within the water column in which fertilization occurs (Togashi *et al.* 2006). For isogamous or slightly anisogamous green seaweeds such as *Ulva*, male and female gametes each have two flagella and a single eyespot. They are positively phototactic, and migrate to the water surface where fertilization takes place. This migration is thought to be advantageous particularly in shallow water because the water surface is essentially a two-dimensional plane in which the chances of gametes of the opposite sign meeting is enhanced (compared to a three-dimensional water column). The resulting planozygote has four flagella and two eyespots (one pair of flagella and one eyespot from each gamete), and is negatively phototactic, allowing the zygote to move to the seabed where it will settle. Green seaweeds, such as *Bryopsis*, *Caulerpa*, and *Halimeda*, possess "markedly anisogamous" gametes and in their case only the female gametes have an eyespot and exhibit phototaxis – the male has none and its movements are random (Togashi *et al.* 2006). The male does not, therefore, migrate to the water surface, and the motile female is attracted to the male via pheromonal cues. The planozygote inherits the single eyespot from the female gamete and, again, is negatively phototactic (Miyamura *et al.* 2010). In other markedly anisogamous seaweeds (e.g. *Udotea*, *Derbesia*) neither gamete has an eyespot, and in this case pheromonal attraction and fertilization takes place near the substratum so that the planozygotes are ideally placed to follow chemical signals to the seabed (Togashi *et al.* 2006).

Motile cells can "choose" their settlement site (Callow and Callow 2006). The behavior of *Ulva* spores changes from random movements to a "searching pattern" as spores near the seabed. In pre-settlement behavior, spores of *Ulva* "probe" the substrate with their flagella, and make repeated contacts with the apical papilla upon which they "spin like tops" (at 240 rpm) as the spore senses the chemical, physical, topographic, and biological characteristics of the surface (Callow and Callow 2006; Michael 2009). Spores are attracted to chemical signals released from surface biofilms (Joint *et al.* 2000) and the bacterial quorum sensing signal molecules *N*-acylhomoserine lactones (AHLs) are involved in this response (Joint *et al.* 2002; Fig. 2.18). For *Ulva flexuosa* and *U. lactuca* (previously *U. fasciata*), glycoconjugates on the surface of the zoospore flagella, apical dome, and anterior surface show molecular compatibility with biofilms that are suitable for settlement (Michael 2009).

Self-Assembled-Monolayers (SAMs) have been used to illustrate how *Ulva* zoospores choose their settlement site. These are synthetic surfaces which can be manufactured to have the same physical (e.g. topographical) and chemical properties, but differ in their wettability, also termed "surface energy" or "surface tension". High-energy surfaces are hydrophilic (wettable), and low-energy surfaces are hydrophobic. *Ulva* spores can sense the hydrophobic regions of the SAMs upon which they preferentially settle, and the signaling molecule nitric oxide (NO) triggers spore settlement (Thompson *et al.* 2010). They also choose surfaces with the most complex topographies on a

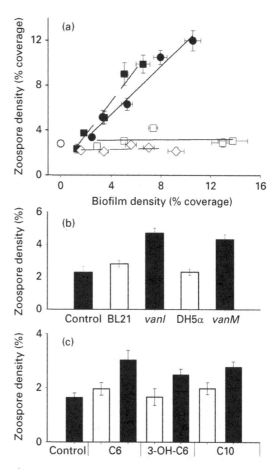

Figure 2.18 (a) Settlement of *Ulva* (previously *Enteromorpha*) zoospores on wild-type *Vibrio anguillarum* and mutants, expressed as percentage of surface covered. ○, control surface (clean cover glass); ●, wild type; □, *vanM* mutant; ■, *vanI* mutant; ◇, *vanIM* mutant. (b) Attachment of zoospores to *Escherichia coli* strains BL21 and DH5α with and without the insertion of plasmids expressing *vanI* – producing 30, C_{10}-HSL – and *vanM* – producing both C_6-HSL and 30H, C_6-HSL, respectively. There was no enhancement of attachment to either strain containing the vector plasmids alone. (c) Settlement of zoospores is not enhanced in the presence of three AHLs that have been treated to open the lactone ring structure (white bars) but is restored when the ring is closed (black bars). All error bars indicate ±2 SE. (From Joint *et al.* 2002, reproduced with permission.)

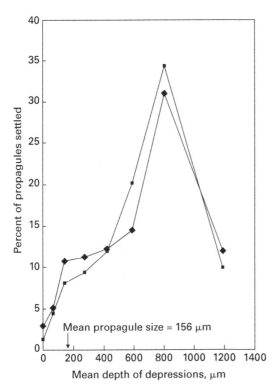

Figure 2.19 Effect of substratum roughness on settlement of *Sargassum muticum* propagules. The substratum consisted of sand-coated microscope slides on a surface with a jet of water flowing over it. Two independent experiments were run; within each experiment, several water velocities were used, and the results were pooled. (Drawn from data in Norton and Fetter 1981, with permission of Cambridge University Press.)

microscopic scale, which provide a greater surface area for the adhesive to stick to than a smooth surface; spores will chose depressions and crevices, which have a higher energy surface. Furthermore, they settle gregariously, responding to yet unidentified cues from each other, which may be beneficial in terms of space and post-germination factors like protection from desiccation and UV-R (Callow *et al.* 1997).

Surface roughness on the slightly larger scale of millimeters to centimeters is also important in affecting the passive settlement of motile and non-motile propagules. Clean glass slides (a favorite experimental surface in the past) are unnatural surfaces, to which macroalgal cells do not adhere well (see

Fig. 2.18). Natural surfaces, in contrast, usually are rough. Evidence from a number of experiments shows that surface roughness is an important factor in settlement (Vadas *et al.* 1992). Essentially, cells are deposited by eddies, in the same way that sand grains are deposited on the lee side of a sand dune. Norton and Fetter (1981) built a "waterbroom" to study the effect of surface roughness on settling of *Sargassum muticum* propagules in moving water. They found that settlement of *Sargassum* propagules was best on a surface with a mean depression depth of ~800 μm (Fig. 2.19), no matter what the water speed (range of 0.22–0.55 m s^{-1}). The propagules attached not because of sinking but because of turbulent deposition. There was a low rate of attachment in the depression of largest size because it was sufficiently large to be swept clean by water flow, rather than creating depositional eddies. Algal turfs also provide a rough surface, and they facilitate the settlement of reproductive cells, for example the zygotes of *Silvetia compressa* (previously *Pelvetia fastigiata*; Johnson and Brawley 1998). Of course, where cells can settle, so can sediment which can have a negative effect on recruitment (see sec. 8.3.2).

2.5.2 Attachment

Once a suitable substratum has been selected, the cells must stick to it. The ability of cells (i.e. their mucilage) to stick to a surface depends on the surface energy, which in turn depends on the nature of the substratum, including any coatings. Any material submerged in the ocean will quickly be coated by a biofilm of bacteria and their associated mucilage, which increases the surface energy and makes the surface much more suitable for macroalgal settlement (Fletcher *et al.* 1985; Dillon *et al.* 1989). In contrast, treatment of surfaces with hydrophobic coatings, such as silicone elastomers, in conjunction with engineered microtopographies, reduces algal settlement and is an effective antifouling technique (e.g. Fletcher *et al.* 1985; Callow and Callow 2006; Schumacher *et al.* 2007).

Zoospore or gamete ultrastructure and settling have been studied in a few species (Fletcher and Callow

1992), including *Ulva* species (= *Enteromorpha intestinalis*) (e.g. Evans and Christie 1970; see Callow and Callow 2006), *Scytosiphon* (Clayton 1984), a variety of Laminariales (Henry and Cole 1982), and several red seaweeds (Bouzon *et al.* 2006; Bouzon and Ouriques 2007). All these cells initially lack a cell wall and have among their organelles numerous cytoplasmic vesicles that contain adhesive material (Fig. 2.20a). Attachment first takes place by the tip of the anterior flagellum in kelps and *Ulva*. For *Ulva*, an initial, elastic bond is made with the substrate, but this bond is weak and the algae are easily removed, leaving the "blob" behind. For fucoid zygotes and red seaweed spores, neither of which have flagella, the initial attachment is via sticky mucilage (e.g. Ouriques *et al.* 2012). Once cells commit to settlement, vesicles containing adhesive material are trafficked to the cell surface and enormous quantities of adhesive are released (Fig. 2.21). This process is regulated by Ca^{2+} signaling in *Ulva* and *Fucus* (Roberts *et al.* 1994; Thompson *et al.* 2007). Following exocytosis, the cell membrane of *Ulva* is "dynamically recycled" to prevent it from stretching and expanding (Thompson *et al.* 2007). Newly settled spores of *Ulva* and *Cladophora surera* undergo rapid morphological changes including the adsorption of the flagellar axonemes and the shape changing from pear-shaped to round, and cell wall formation begins (Callow *et al.* 1997; Cáceres and Parodi 1998; Fig. 2.20b).

Fucoid eggs are initially covered by a mucilaginous layer of alginates and fucoidan that attaches them to the oogonium wall. The eggs are expelled from the conceptacle still enclosed in this layer, which is called the mesochiton. In *Fucus* and *Himanthalia* the mesochiton soon breaks down, and the zygotes attach to the substrate by the sticky zygote wall. In *Pelvetia canaliculata* the mesochiton persists, probably to protect zygotes from drying out in the very high shore habitat of this species (Moss 1974; Hardy and Moss 1979). The mesochiton, rather than the zygote wall, attaches the pairs of *Pelvetia* zygotes to the substratum. Within 24 h of settling, the *Pelvetia* zygote develops a firm alginate wall inside the mesochiton. Each zygote divides once or twice and then pushes out a group of up to four rhizoids, each from a single cell. These rhizoids grow

down into the substratum, entering minute crevices, if these are available, and the mesochiton splits open. The time between fertilization and the formation of rhizoids in this species is about 1 week. For various seaweeds, the surface energy of the substratum affects the morphology of the germlings, especially their rhizoids. Many species (but not all), when on the preferred high-energy surfaces, form compact, well-attached basal filaments or rhizoids, whereas on low-energy surfaces the filaments or rhizoids spread widely and are poorly attached (Fletcher *et al.* 1985).

The composition of the adhesive substances is thought to vary widely between species. The precise chemical composition of adhesives and the nature of the bonds made with the substratum is incompletely understood, but an area of active research (see Vreeland *et al.* 1998; Callow and Callow 2006). Glycoproteins and/or sulfated polysaccharides are involved in initial attachment of red, brown, and green algal spores/zygotes (e.g. Ouriques *et al.* 2012). After spores have initially stuck to a surface, they begin to improve their adhesion by hardening the adhesive. Hardening (curing) of the attachment mucilage in various seaweeds involves the formation of crosslinks between polymer molecules, especially Ca^{2+} bridges between alginate chains or between sulfate ester groups of fucoidan. The rhizoid wall and the rest of the zygote wall have different alginates, as shown by antibody labeling (Boyen *et al.* 1988). *Fucus* embryos grown in sulfate-free seawater form normal rhizoids, but crosslinking cannot occur, and the rhizoids cannot adhere to the substratum (Crayton *et al.* 1974). Moreover, sulfation is necessary for intracellular transport of fucan (sec. 2.5.3). Attachment of single cells mechanically released from *Prasiola stipitata* thalli also requires sulfation of a cell wall polysaccharide; inhibitors of sulfation (such as molybdate) and of protein synthesis prevent attachment (Bingham and Schiff 1979). The protein may be complexed with the polysaccharide or may be an enzyme involved in the sulfation process. In brown seaweeds, phlorotannins are released at the same time as the initial adhesion and Vreeland *et al.* (1998) suggest that three adhesion precursors (a sulfated carbohydrate, a phenolic compound and a haloperoxidase), released together, form

(a)

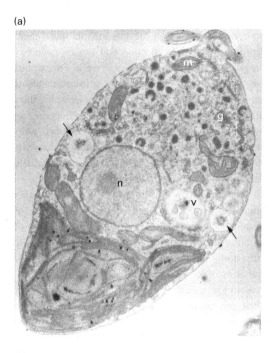

(b)

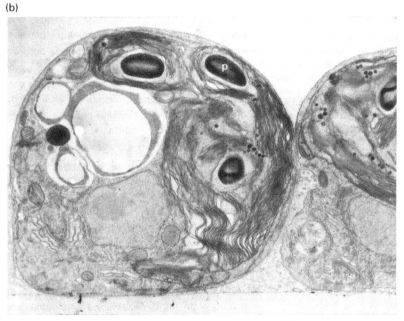

Figure 2.20 Ultrastructure of swimming (a) and newly settled (b) zoospores of *Ulva intestinalis*. In the anterior of the swimming cell can be seen numerous vesicles filled with adhesive (arrows). Also visible are part of the nucleus (n) and flagella bases, Golgi body (g), vacuole (v), and mitochondrion (m). (b) A mass of secreted adhesive lies in the triangle between the two cells, and there are virtually no vesicles remaining in the cell. (The attachment surface is parallel to the bottom of the photograph.) p, pyrenoid. Scales: (a) × 9000; (b) × 10 000. (From Evans and Christie 1970, with permission of the Annals of Botany Company.)

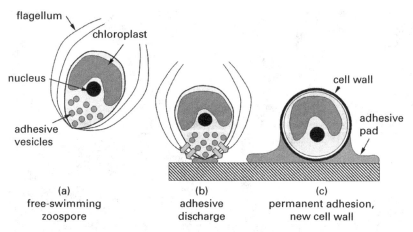

flagellum
chloroplast
nucleus
adhesive
vesicles
cell wall
adhesive
pad

(a)
free-swimming
zoospore

(b)
adhesive
discharge

(c)
permanent adhesion,
new cell wall

Figure 2.21 Cartoon depicting the course of events involved in the settlement and adhesion of *Ulva* spores. (From Callow and Callow 2006, © Springer-Verlag Berlin Heidelberg 2006.)

the glue. The phlroratannins stick to the substrate by "hydrogen bonding, metal complexation, and hydrophobic interactions". They suggest that this mechanism is common across the three seaweed lineages, although the form of carbohydrate, phenolic compound and haloperoxidase will vary between groups.

Attachment and hardening take time, and thus experiments designed to dislodge settled cells have shown that with a given water pressure, the numbers of cells that are washed away decrease the longer the cells have been allowed to settle (Christie and Shaw 1968). Other experiments have shown that the hydrodynamic force that a settled cell can withstand increases with time. *Ascophyllum* zygotes apparently cannot settle unless there is an adequate period of very calm water (Vadas *et al.* 1990). It takes 8 h for *Ulva* spores to be fully attached, at which time it is extremely difficult to remove them using hydrodynamic pressure. Also, spores that settle in groups are more resistant to hydrodynamic pressure (Callow and Callow 2006).

2.5.3 Establishment

After attachment, a cell wall is deposited around the spore (Fletcher and Callow 1992; Ouriques and Bouzon 2003; Fig. 2.22). Prior to germination, cells acquire polarity. In non-polar cells, for example an unfertilized egg, cellular components such as organelles and actin filaments are evenly distributed (Hable and Kropf 2000). A polarized cell (zygote or vegetative cell) has an asymmetrical arrangement of organelles, proteins, or cytoskeleton (Varvarigos *et al.* 2004). Fertilized eggs (zygotes) of fucoid algae are used as a model system for early developmental processes because of several advantages over terrestrial plants: (1) The eggs are free living and thus have no maternal influence on their polarity unlike angiosperm eggs which have maternally induced polarity prior to fertilization. (2) The eggs are easily accessible while those of angiosperms are contained within the ovule and are inaccessible. (3) The zygotes are relatively large (75–100 μm) facilitating micromanipulation, can be collected in quantity, aseptically, and they develop in synchrony in response to environmental cues. Their patterns of polar development and embryogenesis are morphologically similar to those of many algae (some exceptions will be discussed later) and angiosperms (Bisgrove and Kropf 2007; Bisgrove 2007)

Polarity is a key process that sets developmental patterning in algae, allowing the cell to respond to appropriate environmental cues that will ensure the correct orientation of the germling, juvenile, and adult.

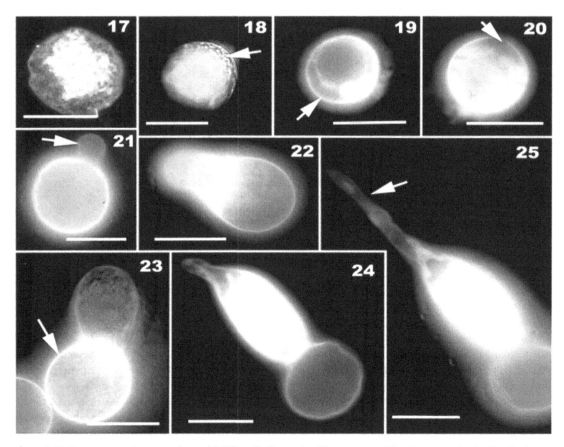

Figure 2.22 Tetraspores and tetrasporelings of *Gelidium floridanum* in different stages of development stained with calcofluor. (17) Non-germinated tetraspore without cell wall. (18) Beginning of cell deposition on a pole (arrow). (19) Tetraspore surrounded by the cell wall. (20) Polar disorganization of the cell wall signaling the start of the germination process (arrow). (21–23) Germ tube in varying stages of development covered by a thin cell wall. (24–25) Tetrasporelings showing a very thin cell wall on the rhizoids (arrow). Scale bar = 20 μm. (From Bouzon *et al.* 2006, reproduced with permission.)

For members of the Fucales, polarization begins upon fertilization of the egg (Fig. 2.23). The entry of the sperm into the egg triggers an actin/Arp2/3 network at the entry site. In the absence of any subsequent environmental stimuli such as a light gradient, blue light, temperature and pH, the site of sperm entry is the rhizoid pole. However, the polarity of the zygote remains labile for 8–14 h after fertilization and environmental cues perceived by the zygote trigger the disassembly of the sperm-induced actin/Arp2/3 network and its reassembly at a new location (Hable and Kropf 2000). Following fertilization, the zygote begins cell wall synthesis and secretes mucilage around itself, which facilitates attachment to the substratum. Once attached, each side of the cell will receive different environmental stimuli, for example the attached side will be shaded while the upper side will receive light. Such a differential light stimulus triggers the relocation of the rhizoid pole to the shaded side of the zygote. The mechanisms for sensing the environmental stimuli are unknown but it is likely that light is sensed by a rhodopsin-like protein (Gualtieri and Robinson 2002).

About 4 h after fertilization, axis amplification occurs. Ca^{2+} and H^+ ion transporters accumulate at the actin/Arp2/3 network that is located at the rhizoid pole, and an ionic concentration gradient forms across the cystol (Pu and Robinson 2003). The Ca^{2+} gradient is dependent on a Reactive Oxygen Species (ROS; see secs. 7.1 and 7.8) gradient which forms simultaneously and may act to modulate the intracellular Ca^{2+} signal (Coelho *et al.* 2008). Axis fixation occurs about 10 h after fertilization, and the polarity is no longer labile (Hable *et al.* 2003). Germination quickly follows, with the rhizoid tip emerging from the now pear-shaped cell. Soon thereafter, the first cell division occurs, resulting in the apical cell that will become the thallus and reproductive structures, and rhizoid cells that will develop into rhizoids and holdfasts. Numerous extensions of the nuclear membrane project toward the rhizoidal pole, and there is an accumulation of vesicles at the pole (Fletcher and Callow 1992). The vesicles, derived from the Golgi apparatus, are filled with a highly sulfated fucoidan (F2) that is deposited in the cell wall at the pole and serves to anchor the cell to the substratum: the apical cell does not produce F2 nor the associated binding proteins (Fowler and Quatrano 1997). Negative charges on the sulfate ester groups or perhaps on the vesicle surfaces may be needed to draw the material toward the positively charged pole. Sulfation of the fucan requires new enzyme synthesis; if synthesis is prevented by cycloheximide, or if SO_4^{2-} is lacking, there is no movement of fucan to the rhizoidal pole.

The fucoid algae have served as a model system to such an extent that germination in other seaweeds has received little attention, except for morphological studies. At the morphological level there is great variety. Even within the Fucales, *Himanthalia* shows a different pattern, which Ramon (1973) suggested was not oriented by light. Not all germinating spores or zygotes first divide parallel to the substratum (or perpendicular to the light gradient). Horizontal germination is common, in which a single filament (germ tube) or basal crust is formed. In some species, such as *Coelocladia arctica* (Dictyosiphonales), the protoplast migrates into the germ tube, leaving the spore wall empty (Pedersen 1981). Other brown algal zoospores

push out several lobes that are then cut off as cells; this stellate kind of germination leads to a monostromatic crust. Five patterns of germination have been described for the Rhodophyta although many do not fit into these general groups (reviewed by Murray and Dixon 1992). Recent studies have visualized aspects of red algal germination in *Gelidium floridanum* (Bouzon *et al.* 2005, 2006; Fig. 2.22) and *Laurencia arbuscula* (Bouzon and Ouriques 2007). In *Porphyra spiralis* var. *amplifolia* germination starts with the differentiation of a vacuole, which then pushes the cellular organelles into the developing germ tube (Ouriques *et al.* 2012). Spores of Corallinaceae divide in patterns characteristic for particular species (tetraspores and carpospores show the same pattern) (Chamberlain 1984). For carpospores of *Bangia fuscopurpurea* unipolar germination occurred when the day length was greater than 12 h, whereas germination was bipolar when it was less than 12 h (Dixon and Richardson 1970).

2.6 Thallus morphogenesis

The process by which unspecialized cells differentiate into cells with a specific role within a seaweed body and then multiply to acquire the familiar architectural form of the adult, has been studied classically using microscopic examination to follow each cell division (e.g. Figs. 1.4 and 1.5). More recently, the cellular developmental patterns of many (20–50) replicate individuals have been followed so that the robustness, or plasticity, of a particular developmental sequence can be determined statistically (Le Bail *et al.* 2008). Such data are also used in computer model simulations in which strict set of developmental "rules" are applied, that dictate the number and plane of cell divisions or the effect of a new cell on the development of a neighboring cell. These simulations "build" adult seaweed with an architecture that is characteristic of a particular species (e.g. Corbit and Garbary 1993; Lück *et al.* 1999; Billoud *et al.* 2008).

Bisgrove and Kropf (2007) ask the question "what factors determine the developmental fate of newly formed cells?" and suggest three mechanisms (that

(a) Polarization

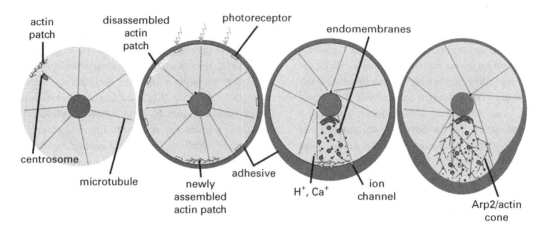

(b) Centrosomal alignment

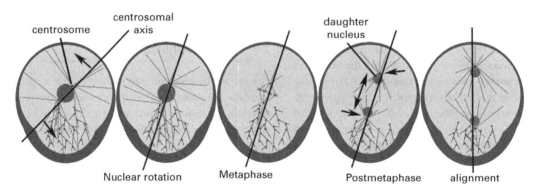

(c) Cytokinesis

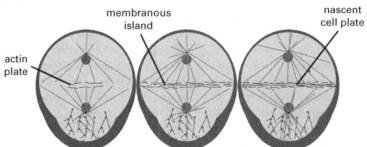

Figure 2.23 Mechanism of asymmetric cell division in zygotes of fucoid algae. (a) Fertilization induces formation of a cortical actin patch that marks the rhizoid pole of a default axis. Photopolarization causes disassembly of the sperm-induced patch and

are not mutually exclusive) by which cells acquire their own specific identities and roles within a plant or seaweed body: (1) "Intrinsic" or "cell-autonomous" development occurs when each new cell receives a different set of "cytoplasmic instructions". (2) "Extrinsic" or "non-cell-autonomous" development describes the influence of the cell's environment on its development. The "environment" here can include signaling from neighboring cells (e.g. by phytohormones; sec. 2.6.3), or external environmental signals (e.g. light, sec. 2.3.3). (3) Size and/or shape differences between the daughter cells can affect their developmental pathway. For the fucoid algae, all three of these processes affect developmental patterning as follows: Extrinsic signals, including sperm entry and light, set the initial polarity of the zygote (sec. 2.5.3). The first asymmetric cell division gives rise to daughter cells that are morphologically and cytologically distinct, and these give rise to the rhizoid and apical cell. During this first division, there is also an asymmetric allocation of mRNA to the daughter cells, indicating intrinsic signaling.

In Chapter 1 we reviewed the fundamental information on thallus construction and cell growth, and further details are found in van den Hoek *et al.* (1995) and Graham *et al.* (2009). In the following sections we review some classical and recent studies on cell differentiation, then consider how the development of the adult form is affected by abiotic (e.g. light, nutrients, temperature) and biotic (e.g. grazing), and how key developmental and physiological processes are regulated by seaweed growth substances (phytohormones).

2.6.1 Cell differentiation

The classical study of Ducreux (1984) follows the influence of the apical cell on the developmental fate of neighboring cells of the morphologically simple *Sphacelaria*. Here the apices have large cells with clearly defined functions. Organelles in the apical cell are concentrated near the tip, and the nucleus is also in the distal half of the cell. In many species the apical cell undergoes regular mitosis to form a symmetrical subapical cell. This cell, in turn, divides to produce two cells with different morphogenetic potentials: The upper one of the pair (the nodal cell) will branch, but the other (internodal) cell will not (Figs. 1.4a and 2.24). The asymmetry of the apical cell apparently is essential to its role as an apical cell and is also dependent on contact with older cells, as shown by regeneration experiments. If the apical cell is cut off, the subapical cell will become polarized and take over as a new apical cell (before or after dividing) (Fig. 2.24 b, c). An isolated subapical cell will form a new axis (Fig. 2.24 d), whereas an isolated apical cell will retain its polarity and continue to divide as before (Fig. 2.24 e). The polarity and shape of *Sphacelaria* apical cells is controlled by the cortical actin filaments of the cytoskeleton (Rusig *et al.* 1994; Karyophyllis *et al.* 2000).

Ectocarpus is "architecturally plastic" in response to environmental stimuli and there is also substantial morphological variation within populations grown in the same culture conditions. Le Bail *et al.* (2008) followed 20–50 *Ectocarpus* individuals, from two geographically isolated regions, from germination to the

Caption for Figure 2.23 (*cont.*) assembly of a new patch at the shaded pole. Endomembrane cycling then becomes focused to the rhizoid pole as the nascent axis is amplified, and cytosolic ion gradients are generated. At germination, the actin array is remodeled into a cone nucleated by the Arp2/3 complex. During early development the paternally inherited centrosomes migrate to opposite sides of the nuclear envelope and acquire microtubule nucleation activity, but microtubules play only an indirect role in polarization. (b) Centrosomal alignment begins with a premitotic rotation of the nucleus that partially aligns the centrosomal axis (defined by a line drawn through the two centrosomes) with the rhizoid/thallus axis. When the metaphase spindle forms it is partially aligned with the rhizoid/thallus axis. Postmetaphase alignment brings the telophase nuclei into almost perfect register with the rhizoid/thallus axis. *Arrows* indicate directions of nuclear movements.(c) Cytokinesis is positioned between the two daughter nuclei. A plate of actin assembles in the midzone between the nuclei, then membranous islands are deposited in the cytokinetic plane. The islands consolidate and cell plate materials are deposited in the division plane. All of these structures mature centrifugally, beginning in the middle of the zygote and progressing outward to the cell cortex. (From Bisgrove and Kropf 2007, © Springer-Verlag Berlin Heidelberg 2007.)

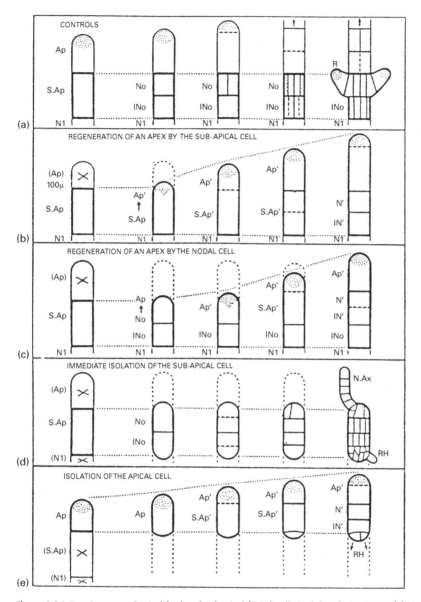

Figure 2.24 Development of apical (Ap) and subapical (S.Ap) cells in *Sphacelaria cirrosa*. (a) Normal ontogenesis of the subapical cell on control axes. (b, c) Regeneration of an apex after removal of the apical cell. If the subapical cell has formed recently (b), it transforms itself into an apical cell. If it is older (c), it undergoes a first division, and the nodal cell regenerates the apical cell. (d) Development of the subapical cell when isolated immediately after formation: modified development sequence leads to formation of a complete new axis. (e) An isolated apical cell continues normal sequence of cell divisions, except for the development of rhizoid initials (RH). N.Ax., newly formed axis; No, nodal cell resulting from transverse partitioning of the subapical cell; INo, corresponding internodal cell; NI, IN 1, . . ., successive nodal and internodal segments; R, branch. Abbreviations in parentheses indicate cells removed. (From Ducreux 1984, with permission of *Journal of Phycology*.)

100 cell stage, and assessed the robustness of the developmental pathways. Two distinct cell types, round and elongate, were observed, with the elongate cells being the "default" type that differentiated into round cells as filament growth progressed. A picture emerges of a tightly regulated pattern of branching which "ensured a stereotyped architecture", in spite of the inherent morphological variability. In a continuation of this work, Le Bail *et al.* (2011) created a morphological mutant, *étoile* (etl), that was "hyper-branched". They found branching to be controlled by a single, recessive locus *ETL*, the expression of which causes increased production of the round, branch-initiating cells and relatively fewer elongate cells.

When thalli or cells are injured, cells may dedifferentiate or redifferentiate. Cells isolated from the parent individual may behave as zygotes or spores and may form a whole new organism; the ability of a cell to do this is called totipotency. Single cells from simple thalli such as *Prasiola* (Bingham and Schiff 1979) readily regenerate the whole thallus, but cells from complex algae, such as cortical cells from *Laminaria* (Saga *et al.* 1978), can also regenerate the thallus under appropriate culture conditions. There probably are few cells in algal thalli that are irreversibly differentiated; anucleate sieve elements of *Macrocystis* provides an obvious example. Yet cells released from constraints imposed by neighboring cells do not always grow into a seaweed of the same generation. Examples are seen in the phenomenon of apospory: The diploid sporophytes of kelp tissue raised in stagnant culture for 3–4 months became bleached, leaving only isolated epidermal cells alive (Nakahara and Nakamura 1973). A few of these epidermal cells germinated, giving rise to gametophytes, which were shown to be diploid. These early studies on totipotency gave rise to an important area of applied seaweed research: cellular biotechnology and the search for methods of propogating seaweeds with superior characteristics (e.g. agar quality, growth rate) for aquaculture. Early work focused on producing viable protoplasts, in which the cell walls are digested and cells regenerated: Polne-Fuller and Gibor (1984) working with *Porphyra*, Ducreux and Kloareg (1988) with *Sphacelaria*, Fujimura *et al.* (1989) with *Ulva*, and Butler *et al.* (1989) with *Laminaria*. Protoplasts have

been an important tool in understanding cell polarization (e.g. Varvarigos *et al.* 2004) and cellular biotechnology (e.g. Reddy *et al.* 2008a; see sec. 10.11).

Although some cells from seaweeds, if isolated, are totipotent and can regenerate into a whole seaweed, others cannot. Blade cells of *Ulva mutabilis* are unable to form rhizoidal cells, but form vesicular thalli one cell thick (Fjeld and Løvlie 1976). However, isolated rhizoidal cells of this species can form the whole seaweed. A repressor is present in the thallus cells, as shown by Fjeld's study of a mutant, *bubble* (*bu*), that behaves like isolated blade cells (Fjeld 1972; also see Fjeld and Løvlie 1976). The mutant gene is recessive and chromosomal. Curiously, *bu* spores from meiotic sporangia on heterozygous algae *(bu +/bu)* develop a partly or completely wild-type phenotype in the first generation. When these are propagated asexually, subsequent generations are completely the mutant type. The explanation appears to be that there are repressor or rhizoid-forming genes in normal blade cells and *bubble* in the cytoplasm of *bu* spores, as well as in wild-type mutant cells and that this repressor is removed during sporogenesis, so that spores can form rhizoids when they settle. The *bu+* wild-type gene is thus responsible for removal of the repressor, and its transcription takes place before meiosis, so that the de-repressor is present in the cytoplasm of *bu* spores. (This substance is not diluted through many cell generations, because the number of rhizoidal cells is small.) The rhizoid-forming genes of both types are re-repressed early in development, but the mutant gene cannot de-repress them when it forms spores.

Although coenocytes such as *Caulerpa* and *Bryopsis* are technically single cells, there are differences between regions of their cytoplasm, allowing the same kinds of differentiation that occur in multicellular thalli. *Caulerpa mexicana*, for example, has four morphologically distinct thallus regions: a creeping stipe, rhizoids, erect blades, and petiole-like stalks (Fagerberg *et al.* 2010). The structure of the cytoplasm differs among the peripheral, central, and trabecular regions of the thallus, and there are at least three types of trabeculae (Dawes and Barilotti 1969; Fagerberg *et al.* 2010). The trabeculae are thought to have a key role in providing mechanical support and controlling developmental processes including cell shape.

2.6.2 Development of the adult form

The morphological form developed by a seaweed is strongly affected by their abiotic and biotic environment. Seaweeds exhibit considerable morphological plasticity which can vary over small spatial scales (e.g. *Ascophyllum*, Stengel and Dring 1997) and seasonally (e.g. *Ecklonia radiata*, Wernberg and Vanderklift 2010). Morphological plasticity is thought to be a mechanism by which seaweeds (and terrestrial plants) optimize resource acquisition and allocation to growth and/or reproduction in a spatially and temporally heterogeneous environment. Understanding how organisms interact with their environment is one of the "grand challenges" in organismal biology (Schwenk *et al.* 2009), and the effect of phenotypic plasticity on ecological organization is an emerging research field (Miner *et al.* 2005). The degree of phenotypic versus genotypic response of seaweeds to the environment, and evidence for incipient speciation, can be assessed by reciprocal transplant experiments (e.g. *Ecklonia radiata* Fowler-Walker *et al.* 2006; *Rissoella verrucosa*, Benedetti-Checchi *et al.* 2006), common garden experiments (e.g. *Eisenia arborea*, Roberson and Coyer 2004), and molecular studies (e.g. *Pelagophycus porra*, Miller *et al.* 2000). Morphological variation within a species can be so great that some specimens have been considered separate species, for example the kelp *Ecklonia brevipes* was found to be an extreme morphological variant of *Ecklonia radiata* (Wing *et al.* 2007; and see Essay 1, Fig. 1). Similarly, *Macrocystis* exhibits considerable morphological and reproductive plasticity across is wide geographic range, but evidence suggests it is just one species (Demes *et al.* 2009a). The propensity for phenotypic versus genotypic influence in response to environmental change seems to vary largely depending on species (Stengal and Dring 1997), and the ability to adapt and evolve to an environment may depend on if it is unitary or modular (Monro and Poore 2009a).

Branching is a characteristic developmental step in many algae and is important in establishing the final morphology of a seaweed (e.g. Coomans and Hommersand 1990; Waaland 1990). Branching patterns are consistent enough in many species (e.g. among Ceramiales) to be used as taxonomic criteria. In other species branching patterns are variable, and under apical control. For instance, when *Pterocladiella capillacea* loses its apical meristem due to wave action, the branching pattern changes from 1–3 orders to 2–5, that is, the seaweed becomes bushier (Scrosati 2002b). The effect of water motion on seaweed morphology is explored in Chapter 8.

The formation of erect uniaxial or multiaxial thalli from crustose germlings or microthalli requires that one or a number of erect filaments have their tips converted into meristems (reviewed by Murray and Dixon 1992). Formation of the meristems (macrothallus initials) and the outgrowth of the erect thalli are separate events, and in *Dumontia contorta (D. incrassata)* these events are controlled by different environmental cues (Rietema 1982, 1984). Production of the initials in this species depends solely on photoperiod: Short days (long nights) are required. Outgrowth of the initials also requires short days, but in addition the temperature must be 16°C or lower.

The amount of light received can have marked effects on branching and elongation patterns. Two modular seaweeds, *Asparagopsis* (Rhodophyta) and *Caulerpa* (Chlorophyta), have very similar morphological responses to light quantity. In low-light environments, they are sparsely branched with long stolons, and in high light they form short, densely branched stolons and highly branched ramets (Peterson 1972; Collado-Vides 2002b; Monro and Poore 2007, 2009a, b). This response to light environment is also typical of modular terrestrial plants, and is considered a phenotypic adaptation that allows plants to "optimally forage" for resources (i.e. maximize resource acquisition) or "escape" competition. The sparsely branched "guerrilla" types allow individuals to invade new habitats, while the dense "phalanx" form is highly competitive and typical of densely populated environments. Another modular seaweed, *Codium fragile*, also possesses two distinct morphologies, "filamentous" and "spongy" (Nanba *et al.* 2005). In laboratory culture, vegetative fragments of filamentous thalli can give rise to spongy thalli and vice versa, and the type of thallus that is formed depends

Table 2.3 Non-photosynthetic effects of blue light on marine macroalgae

Description of response	Genus
1. Photo-orientation responses:	
Induction of polarity in germinating zygotes	*Fucus*
Negative phototropism of haptera	*Alaria*
Negative phototropism of rhizoids	*Griffithsia*
Plastid displacement	*Dictyota, Alaria*
2. Effects on carbon metabolism and growth:	
Stimulation of protein synthesis and mobilization of reserves	*Acetabularia, Dictyota*
Stimulation of dark respiration	*Codium*
Stimulation of uridine diphosphate glucose phosphorylase	*Acetabularia*
3. Effects on vegetative morphology:	
Induction of two-dimensional growth	*Scytosiphon*
Induction of hair formation	*Scytosiphon, Dictyota, Acetabularia*
4. Effects on reproductive development:	
Stimulation of cap formation	*Acetabularia*
Induction of egg formation	*Laminaria, Macrocystis*
Stimulation of egg release	*Dictyota*
Inhibition of egg release	*Laminaria*
5. Photoperiodic effects:	
Blue light alone effective as night break	*Scytosiphon*
Blue and red light effective as night break	*Ascophyllum, Rhodochorton*
Blue light effective as day extension	*Acrosymphyton*

Source: (From Dring 1984b, with permission of Springer-Verlag, Berlin.)

on the interactive effects of irradiance and water velocity. Spongy thalli were formed from filamentous thalli under high irradiances (50 and 100 µmol quanta m^{-2} s^{-1}) and at flows of 10 cm s^{-1} whereas filamentous thalli arose from spongy thalli at all irradiances tested (10, 50, 100 µmoles quanta m^{-2} s^{-1}) but only under no or slow flows (0 and 3 cm s^{-1}). This example also fits with the guerrilla (spongy, dense thalli in high light) and phalanx (diffuse thallus, low light) framework for optimal foraging.

Specific wavelengths of light also can affect morphogenesis. Blue light (BL) is a very important environmental signal in the marine environment because it is transmitted to the greatest depths through the water column, compared to red light which is absorbed in the top 4 m (see sec. 5.2.2). Many photomorphogenetic effects are due to blue light (BL) (Table 2.3; Lüning 1981b; Dring 1984b; Schmid 1984). Algae are equipped with a range of photoreceptors although studies on the biochemical and molecular mechanisms underlying BL-mediated responses are largely unknown (Hegemann 2008; sec. 2.3.3). In terrestrial plants, phototropin senses the blue region of the spectrum (390–500 nm) and mediates a range of responses including plastid movement and solar tracking that facilitate optimal light harvesting (Ishikawa *et al.* 2009). This protein has two "light-oxygen-voltage" (LOV) domains at the N-terminus that sense blue light (BL) and an effecter at the C-terminus, and has been identified in the unicellular green alga *Chlamydomonas*. The filamentous, mud-flat dwelling *Vaucheria* (Xanthophyceae) shows a range of BL-mediated responses including a morphogenetic response of preferential branching at the BL irradiated side of the thallus, in this case the receptor is aureochrome (Takahshi *et al.* 2007).

Red and blue light have differing photomorphogenetic effects. For *Saccharina japonica*, red-light-grown sporophytes had larger holdfasts and longer stipes than those grown in blue or white light (Mizuta *et al.* 2007). This supports earlier work on red or far-red light effects on the growth of kelp stipes *(Nereocystis leutkeana* and *Saccharina japonica;* Duncan and Foreman 1980; Lüning 1981b). Red light causes specimens of the red alga *Calosiphonia vermicularis* to grow shorter and bushier than they do under white or blue light (Mayhoub *et al.* 1976).

A role of nutrition in morphogenesis has been shown in *Petalonia fascia* (Hsiao 1969) and *Scytosiphon lomentaria* (Roberts and Ring 1972). In *Petalonia* from Newfoundland, Canada, Hsiao (1969) found that zoospores from plurilocular sporangia on the blade could form protonemata (sparsely branched uniseriate filaments), plethysmothalli (profusely branched filaments), or *Ralfsia*-like crusts, any one of which could reproduce itself via zoospores or give rise directly to the blade. Protonemata and plethysmothalli survived in iodine-free medium, but formation of *Ralfsia-like* thalli or blades required iodine (5.1 mg L^{-1} and 0.51 mg L^{-1}, respectively). Plethysmothalli formed blades in progressively shorter times as the iodine concentration increased. Roberts and Ring (1972) found that changes in the proportions of filamentous and crustose microthalli correlated with nitrogen and phosphorus levels.

Some seaweeds develop hairs in response to a nutrient shortage, including *Acetabularia acetabulum, Ceramium virgatum* (previously *Ceramium rubrum*), *Fucus* sp. *Undaria pinnatifida* and *Codium fragile* (DeBoer 1981; Norton *et al.* 1981; Benson *et al.* 1983; Hurd *et al.* 1993; Pang and Lüning 2004). These hairs can enhance nutrient acquisition (secs. 6.8.1 and 8.2.1). For *Acetabularia acetabulum* blue light and red light directly affect hair formation (Schmid *et al.* 1990). If this species is grown in red light, no hairs form, and growth gradually slows. If a pulse of blue light is given and then growth in red light is continued, hair whorls are produced. Blue light induces the response; the red light is used solely in photosynthesis, and there is no evidence for a red/far-red receptor.

Temperature affects the morphological complexity of *Chondrus crispus* (Kübler and Dudgeon 1996). When *Chondrus* was grown at 20°C, a highly branched thallus formed compared to the more sparsely branched thalli produced by specimens grown at 5°C. The growth rate was also higher in the higher temperature treatment. The greater propensity for branching at

20°C may be an adaptation to increase the surface area of thallus that is available for light harvesting and nutrient uptake, enabling greater procurement of light and nutrients to fuel the higher metabolic demands at higher temperatures. The higher branching may also increase evaporative cooling when exposed to the atmosphere (Bell 1995).

Growth of kelp haptera is oriented by negative phototropism, not geotropism, with blue light being the most strongly orienting part of the spectrum (Buggeln 1974). Thigmotropism takes over when the elongating hapteron touches the substratum (Lobban 1978a). Many phototropic responses, both positive and negative, have been recorded (reviewed by Buggeln 1981; Rico and Guiry 1996). Orientation of unicellular rhizoids is more rapid and easier to interpret than orientation of multicellular haptera. Unilateral irradiance is detected by some pigment, possibly phototropin (see above). The information can be stored for several hours, with the response exhibited in subsequent darkness. However, gravity may be the stimulus for rhizoid orientation in *Caulerpa prolifera*. When rhizomes are inverted, rhizoid initiation is preceded by a movement (sinking) of amyloplasts toward the lower side, and rhizoid initials contain numerous amyloplasts (Matilsky and Jacobs 1983).

Morphogenesis depends partly on attachment, if only because orientation depends on consistency in the direction of environmental cues. Unattached seaweeds may remain in place as loose-lying individuals, with little change in morphology, if there is little water movement. If there is extreme water motion, seaweeds will be tossed ashore. Moderate water motion, if it tumbles the thalli, can lead to growth in all directions to form balls, a habit technically called aegagropilous (Norton and Mathieson 1983). Such algae are often morphologically distinct from their attached counterparts (see sec. 3.3.5 on salt-marsh *Fucus*). Aegagropilous forms of coralline algae ("rhodoliths") are harvested as maerl in Europe (Nelson 2009). Species such as *Chondrus crispus* and *Gracilaria tikvahiae* in cultivation tanks also form balls. Many filamentous algae form hemispherical tufts that are restricted by the substratum from growing downward; when free, these will easily form balls. As balls develop from

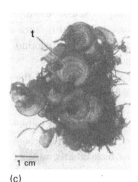

(a)   (b)   (c)

Figure 2.25 Morphological plasticity in *Padina jamaicensis*. (a) In heavily grazed areas, a prostrate, branching thallus with single apical cells forms a dense turf. (b) After grazing is reduced, the typical fan-shaped, calcified thallus begins to form; a row of apical cells develops along the tip of the thallus. (c) Foliose form on older turf after 8 weeks of reduced herbivory. The fan-shaped blades have produced concentric rings of tetrasporangia (t) on their upper surfaces. (From Lewis *et al*. 1987, with permission of the Ecological Society of America.)

fragments, abrasion and grazing damage will tend to increase their compactness by promoting regeneration and proliferation (Norton and Mathieson 1983).

A biotic factor, the presence of epiphytic bacteria, controls the development of the characteristic adult form of some thallose green seaweeds, including *Ulva lactuca*, *Ulva intestinalis*, and *Gayralia* spp. (Provasoli and Pintner 1980; Matsuo *et al*. 2005; Marshall *et al*. 2006). In the normal course of development of *Gayralia oxysperma* (previously *Monostroma oxyspermum*), the biflagellate swarmer produces a filament that divides in three planes to give a little sac, which subsequently ruptures, yielding a flat, monostromatic sheet (Tatewaki 1970). However, if placed in axenic culture, the germinating swarmer will form only a 2-cell thallus consisting of an apical cell (which will slough off cells during subsequent divisions) and a basal, rhizoidal cell. Normal morphology can be restored by addition of exudates of axenically cultured brown and red seaweeds, by growing *Gayralia* in bialgal axenic culture with a red or brown seaweed, and by extracts of seven marine bacteria (out of over 200 isolates tested) in the genera *Caulobacter*, *Cytophaga*, *Flavobacterium*, and *Pseudomonas* (Tatewaki *et al*. 1983). The morphogenetic inducer was isolated by Matsuo *et al*. (2005), who named it "Thallusin". Thallusin (a pyridine) concentrations in natural seawater are very low, and

epiphytic bacteria growing on the seaweeds provide a continuous supply to the thallus.

Another example of biotic control of morphology is the morphogenetic switch induced by fish grazing that has been found in *Padina sanctae-crucis* (previously *Padina jamaicensis*; Lewis *et al*. 1987). When grazing is intense, they grow as uncalcified, straplike, creeping branches formed from a single apical cell. In the absence of herbivory, a marginal row of apical cells forms, and the typical erect, calcified, fan-shaped thallus develops (Fig. 2.25). Interestingly, this morphological switch also occurs in *P. boergesenii*, but the effect is dependent on season, indicating an interactive effect between herbivory and abiotic factors (Diaz-Pulido *et al*. 2007a).

2.6.3 Seaweed growth substances

Growth is an oriented process: polarities in cells and thalli are established from the start and are maintained throughout development. For the "accurate execution of developmental programs", communication between different cells and tissues of an organism is essential (Pils and Heyl 2009). Hormones are messenger molecules responsible for communicating between adjacent cells (paracine hormones), or for long-distance communication between different tissues (endocrine hormones) of animals, terrestrial plants, and seaweed (Buchanan *et al.* 2000; Taiz and Zeiger 2010). Tarakhovskaya *et al.* (2007) list 10 hormones that are found in terrestrial plants which have been identified in seaweeds (see their Table 2). Note that some of the plant hormones listed by Tarakhovskaya *et al.* (2007) (polypeptides and jasmonic acid) are considered by others to be "signaling" or "elicitor" molecules rather than hormones (Buchanan *et al.* 2000; Taiz and Zeiger 2010). For terrestrial plants, the biosynthesis, regulation and action of plant hormones (phytohormones) and other growth substances is well understood, but there are few studies on algae (Tarakhovskaya *et al.* 2007; Stirk *et al.* 2009). The presence of phytohormones in algae suggests some metabolic roles similar to those in terrestrial plants, and recent studies on cytokinins and polyamines support this view. Nevertheless, there is much work to do before a complete picture is built of the biosynthetic pathways and integration of phytohormones in regulation of seaweed growth, reproduction, and physiology.

Early research on algal hormones involved the exogenous application of terrestrial plant hormones to algae, and examining any developmental or physiological response (reviewed by Bradley 1991); more recently such studies have been conducted in conjunction with other environmental variables such as day length or temperature (e.g. Lin and Stekoll 2007). The best-studied phytohormones in seaweeds are cytokinins, which are involved in signal transduction in terrestrial plants; cytokinin-based signal transduction evolved first in the ancestral green algae and is considered a key step in the colonization of terrestrial environments by plants (Pils and Heyl 2009). There are two major groups, "isoprenoid" cytokinins and "aromatic" cytokinins. Stirk *et al.* (2003) identified 19 types of cytokinins in 31 intertidal species from the green, brown, and red lineages. Interestingly, the cytokinin profiles of all 31 were very similar to each other, despite the different evolutionary lines and vertical positions on the shore. They also found little similarity between the cytokinin profiles of seaweeds and terrestrial plants, concluding that "different pathways for regulating cytokinin concentrations operate in macroalgae than in higher plants".

Seasonal patterns of phytohormones, and patterns relating to the zonal position on the shore, have been observed for seaweeds. iPRMP is a ribotide that higher plants use to synthesize cytokinins and for intertidal *Dictyota* and *Ulva*, a seasonal pattern was observed indicating a physiological function relating to higher growth rates in summer (Stirk *et al.* 2009). These findings support earlier work on *Macrocystis* in which increased "cytokinin-like" activity corresponded to higher seasonal growth rates (DeNys *et al.* 1990, 1991). Cytokinin activity was higher for *Ulva* from the high intertidal zone than low-shore *Dictyota*, which may indicate differences between species or a role in signaling environmental stress (Stirk *et al.* 2009). Abscisic acid (ABA) signals environmental stress in terrestrial plants and is a growth inhibitor, and appears to have similar roles in algae. ABA levels in high shore rock-pool *Ulva* were higher than those in low-shore *Dictyota* (Stirk *et al.* 2009). ABA has been detected in sporophytes of *Laminaria digitata*, *Saccharina japonica* and *L. hypoborea* (Schaffelke 1995a). For *L. hypoborea*, endogenous levels of ABA were inversely correlated to seasonal patterns of growth rate, and the external application of ABA inhibited growth under short-day but not long-day conditions. There is also a correlation between mannitol, laminarin, and ABA levels (Schaffelke 1995b).

The auxin indole-3-acetic acid (IAA) has been identified in a range of seaweeds at concentrations similar to those of terrestrial plants. Auxins affect thallus branching, rhizoid development, and the establishment of cell polarity in *Fucus distichus* and *Ectocarpus siliculosus* (Basu *et al.* 2002; Le Bail *et al.* 2010), growth rate of *Pyropia* sporophytes (Lin and Steckoll 2007), growth and sorus formation in *Saccharina japonica*, *Undaria pinnatifida*, and *Alaria crassifolia* (Kai *et al.* 2006; Li *et al.* 2007), and the induction of cell division and reproductive structures (see Tarakhovskaya *et al.* 2007). The metabolic pathways and candidate genes that regulate IAA production, and its role in communicating cell–cell positional information and regulating developmental patterning, have been elucidated for *Ectocarpus* (Le Bail *et al.* 2010; Fig. 2.26). The red seaweed *Grateloupia americana* (previously *Prionitis lanceolata*) has a symbiotic relationship with a

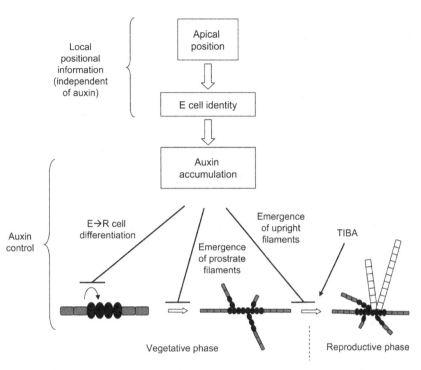

Figure 2.26 Model for the role of auxin in the development of the *Ectocarpus siliculosus* sporophyte. Cells positioned at the apices of the filament acquire the E identity. A higher concentration of auxin is present in these cells, which prevents them from differentiating into R cells and/or inducing branching. As the filament grows, subapical E cells get localized farther from the apex and perceive lower auxin concentrations, which progressively induce their differentiation into R cells as well as branching. Later, auxin maintains its control on the progession of the life cycle by negatively controlling the emergence of the upright filament and thereby the shift to the reproductive phase. Auxin control would then depend on active transport, allowing the apices to maintain control on distant tissues. (From Le Bail *et al.* 2010 ©American Society of Plant Biologists.)

gall-inducing bacteria, and within the galls levels of IAA were on average three-times greater than the surrounding tissue. Whether these high IAA levels in the galls are due to the symbiotic bacteria or host tissue is unclear (Ashen *et al.* 1999).

Polyamines (PAs) (predominantly putrescine, spermidine, and spermine) are low molecular weight aliphatic amines that modulate a wide variety of physiological processes from membrane stabilizations to senescence in all organisms (see Sacramento *et al.* 2004). In seaweeds, they are found in concentrations similar to those in terrestrial plants (Marián *et al.* 2000) and are synthesized from L-arginine via the L-ornithine decarboxylase (ODC) and arginine decarboxlayse (ADC) pathways (Sacramento *et al.* 2004).

They are involved in cystocarp development and sporulation in *Grataloupia imbricata* with levels of PAs and ODC decreasing during the transition from infertile to fertile thalli, and the expression of the gene encoding for ODC (GiODC) follows the same pattern (Sacramento *et al.* 2004, 2007; García-Jiménez *et al.* 2009). Polyamines are also involved in stress responses to hyposalinity in *Grateloupia* and seven intertidal green seaweeds (Lee 1998; García-Jiménez *et al.* 2007). Internal levels of polyamine can be adjusted by *de novo* synthesis or by the mobilization of bound endogenous PAs. In *Grateloupia*, moderate hyposaline conditions (18 PSU) caused decreased activity of the enzyme TGase (responsible for binding free PAs), which triggered an increase in free PAs. When PAs were applied exogenously, the maximum photosynthetic rates of the hyposaline treatment increased relative to the untreated control, indicating that PAs were involved in physiological adaptation to low salinity (García-Jimenez *et al.* 2007). PAs that are bound to the thylakoid membranes of plants and unicellular green algae protect the photosynthetic apparatus from UV-B damage, and this may also be the case for *Pyropia cinnamomea* (previously *Porphyra cinnamomea*) for which PA synthesis was upregulated upon exposure to UV-B radiation (Schweikert *et al.* 2011).

Ethylene is a gaseous hormone, found in the atmosphere in trace levels and is involved in the production and destruction of the ozone layer. In terrestrial plants, it has a well-known role in fruit ripening, but given its gaseous nature it was not considered a likely candidate for seaweed growth regulation. However, Plettner *et al.* (2005) demonstrate ethylene production by *Ulva intestinalis*, levels of which increased when low-light-grown samples were placed under stressful high-light conditions.

Several more phytohormones have been detected in seaweeds but their precise roles are not known (Tarakhovskaya *et al.* 2007). The oxylipins, jasmonic acid (JA) and methyl jasmonate (MJ), that regulate biosynthetic pathways of secondary metabolites in plants have been found in red, green, and brown algae (see Tarakhovskaya *et al.* 2007). Arnold *et al.* (2001) provide evidence that MJ triggers phlorotannin production in *Fucus vesiculosus* suggestive of a role in secondary metabolite induction. However, Wiesemeier *et al.* (2008) could not detect JA or MJ in *F. vesiculosus* nor in six other brown seaweeds, and the exogenous application of the hormones had no effect on the secondary metabolites studied: Any role of JA and MJ in brown seaweeds requires further clarification. Gibberellins are present in *Caulerpa* (Jacobs 1993) and *Fucus vesiculosus* (Table 2 in Tarakhovskaya *et al.* 2007), and have a growth-promoting effect on *Pyropia* sporophytes (Lin and Steckoll 2007). Rhodomorphin was found in *Griffithsia* (Waaland and Cleland 1972, see sec. 2.6.2), and lunularic acid, which signals environmental stress in

liverworts, is found in *Enteromorpha* (Table 2 in Tarakhowskaya *et al.* 2007).

2.6.4 Wound healing and regeneration

Thallus damage is a fact of life for seaweeds and the major sources of injury are herbivores, parasites, epiphytes, sand abrasion, and wave forces. Seaweeds must be able to heal the injury and a different sequence of events takes place for wound healing in multicellular seaweeds compared to coenocytic seaweeds. In multicellular seaweeds there is no need for the cut cells to recover; instead, the wound is sealed, a process that involves changes in the underlying cells. Siphonous algae, however, risk lethal cellular hemorrhaging if a wound cannot be sealed immediately; wounding triggers a biochemical cascade that leads to the formation, within 2 min, of a gelatinous "wound plug" of cellular material that seals the cell, and under this plug a new cell wall can be assembled (Welling *et al.* 2009). Menzel (1988) presented a generic six-step sequence of events applicable to most siphonous green algae (Fig. 2.27): (1) repair of the cell membrane; (2) contraction of the cut edge of cytoplasm; (3) extrusion of the plug precursor material from vacuoles in which they are stored; (4) restoration of turgor pressure; (5) formation of a wound plug; and (6) formation of a new cell wall.

There are two general biochemical processes for wound plug assembly (step 5) in siphonous green algae: lectin-carbohydrate interactions (*Dasycladus vermicularis*, *Bryopsis*, *Microdictyon umbilicatum*, *Chaetomorpha*, and *Codium*) and protein cross-linking (*Caulerpa taxifolia*) (Welling *et al.* 2009). Lectins are proteins that bind reversibly with carbohydrates, and algal lectins differ from those of higher plants by having a lower molecular weight and high affinity for oligosaccharides and glycoproteins (rather than monosaccharides of higher plants). In 2005, Kim *et al.* isolated a novel lectin from *Bryopsis plumosa*, which they called bryohealin, which is more closely related to the fucolectins of some invertebrates than the lectins of higher plants (Yoon *et al.* 2008). When *B. plumosa* is injured, the cellular

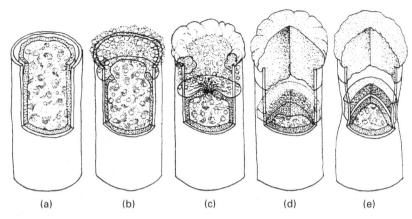

(a)   (b)   (c)   (d)   (e)

Figure 2.27 Wound healing in *Bryopsis*. Cutaway diagrams show changes in cell contents in the hour following a wound. (a) Undamaged siphon has a peripheral layer of cytoplasm and a large central vacuole filled with plug precursor material. (b) At 15–30 s after the siphon is cut, the cytoplasm begins to retract and form a concentric closure. Plug precursor is expelled; it swells and adheres to the edge of the cut wall. (c) After about 1 min, the cytoplasmic contraction is almost complete; the plug precursor coagulates, and the wound plug begins to form. (d) At 5–10 min after wounding, the wound plug begins to develop internal and external layers. (e) Within an hour, new cell wall has formed under the internal plug and begins expanding. (From Menzel 1988, with permission of Springer-Verlag, Berlin.)

contents spill out but the organelles then aggregate, and generate a new cell membrane around themselves, forming a protoplast which can eventually grow into a new individual (Kim *et al.* 2001b; Grossman 2005; Fig. 2.28). Bryohealin not only controls organelle aggregation, it also confers protection of the protoplast from bacterial contamination in a manner similar to the anti-pathogenic role of fucolectins in invertebrates (Pak *et al.* 1991; Kim *et al.* 2005; Yoon *et al.* 2008). In another example, the one-celled protoplasts formed following wounding of *Microdictyon umbilicatum* have two fates: 30% become mature *Microdictyon*, but the rest become quadriflagellate reproductive swarmers, and this could be a mechanism for dispersal (Kim *et al.* 2002). Similarly, protoplasts of *Chaeotomorpha aerea* develop into aplanospores or biflagellate swarmers (Klotchkova *et al.* 2003).

The biochemical signaling pathways for cell repair are being eludicated. When *Dasycladus vermicularis* is wounded, the cellular contents are extruded and within 1–2 min they form a sticky, protective gel. This solidifies at 10 min post-injury, and is fully "hardened" by 35–45 min (Ross *et al.* 2005a, b). The hardening

process is preceded by a nitric oxide (NO) emission at 25 min post-injury, and followed by a burst of reactive oxygen species (ROS; oxidative burst; sec. 7.1 and Essay 5) at 45 min that causes the final cross-linking of plug constituents (Ross *et al.* 2006). NO is a common signaling molecule in higher plants and animals, but this was the first record of it in seaweed. In another first, Torres *et al.* (2008) show that for *D. vermicularis* and *Acetabularia acetabulum*, extracellular ATP (eATP) released upon wounding signals the production of ROS and NO.

Caulerpa taxifolia employs protein cross-linking to seal damaged cells. In this case, the secondary metabolite caulerpeyne (which is a mild anti-feedant) is de-acetylated to oxytocin 2 (a 1,4-dialdehyde), 30 s after wounding. Oxytocin 2 is an aggressive protein cross-linker and proteins are recruited to it within a few seconds of wounding. The resulting polymer of proteins and secondary metabolites not only forms the foundation of the wound plug in *Caulerpa*, it also reduces palatability to herbivores possibly by reducing the food quality (Jung and Pohnert 2001; Weissflog *et al.* 2008).

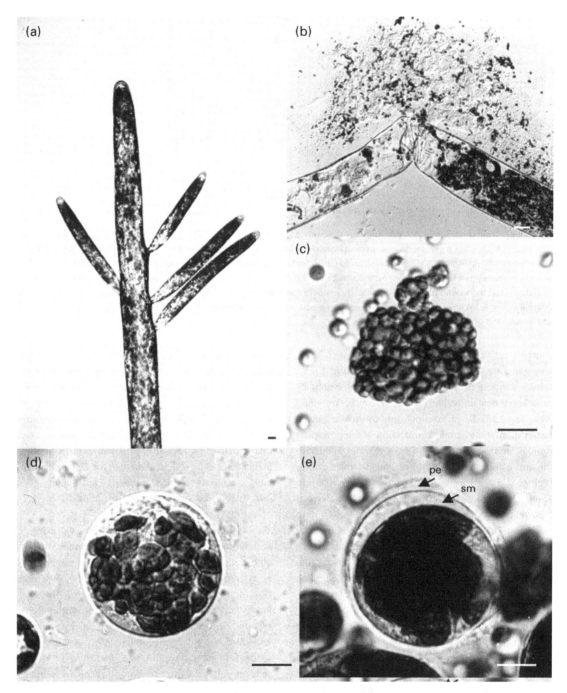

Figure 2.28 The sequential process of sub-protoplast regeneration from disintegrated cells of *Bryopsis plumosa*. (a) Vegetative plant with distichous branches. (b) Protoplasm comes out from the wounded cell and spreads in seawater. (c) Aggregation of the extruded cell organelles in seawater. (d) Regenerated sub-protoplast with a primary envelope 20 minutes after wounding. (e) The secondary lipid-based membrane inside the primary envelope 12 h after wounding (pe, primary envelope; sm, secondary membrane). Scale bars = 10 μm. (From Kim *et al.* 2001a © The Company of Biologists Ltd.)

A different process is seen in Siphonocladales, an order that is characterized by unique segregative cell division (e.g. *Ernodesmis, Boergesenia, Valonia)* (La Claire 1982a, b). In most cases, no plug is formed; rather, the cytoplasm retracts from the wound, again the work of actin microfibrils (La Claire 1989, reviewed by Shepherd *et al.* 2004), and then closes around the central vacuole in one or a few pieces, or breaks up into many protoplasts. The latter process looks much like segregative cell division and the production of gametangia.

Wound healing in multicellular algae has been most thoroughly studied in *Fucus vesiculosus* (Moss 1964; Fulcher and McCully 1969, 1971), *Sargassum filipendula* (Fagerberg and Dawes 1977), *Kappaphycus alvarezii* (previously *Eucheuma alvarezii*; Azanza-Corrales and Dawes 1989), and *Ecklonia radiata* (Lüder and Clayton 2004), and the processes for each of these seaweeds are similar. In the fucoids, the thin, perforated cross-walls of the medullary filaments are plugged after about 6 h with newly synthesized sulfated polysaccharide (presumably fucoidan). Later there is general accumulation of polysaccharide at the wound surface. Medullary cells adjacent to the damaged cells round off and become pigmented. After about a week they give rise to lateral filaments, which elongate and push through to the wound surface, where they branch repeatedly to form a protective layer. Cortical cells undergo longitudinal division (parallel to the wound surface), and the outer cells assume the cytological and functional characteristics of epidermal cells (e.g. they become pigmented). Cells of the medulla may also contribute to the formation of new epidermis. For *Ecklonia*, phlorotannins help seal cells by precipitating proteins, and also protect against microbial attack (Lüder and Clayton 2004). For *Kappaphycus*, cut cells lose their contents, while proteinaceous and phenolic substances accumulate at the pits of cortical and medullary cells just below the cut. After a few days, cellular extensions begin to grow from underlying cells, proliferate, and form a layer of new, pigmented cortical cells below the wound (Azanza-Corrales and Dawes 1989).

Wound healing is commonly followed by either regeneration or proliferation. The simplest kinds of

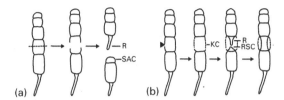

Figure 2.29 Cell regeneration (a) versus cell repair by cell fusion (b) in *Griffithsia*. When the filament is severed, a rhizoidal cell (R) and a new-shoot apical cell (SAC) form, and two separate filaments develop. If an axial cell is killed (KC), the rhizoidal cell fuses with a repair-shoot cell (RSC) and makes a new living link in the filament. (From Waaland 1989, with permission of John Wiley/Alan R. Liss Inc.)

regeneration involve uniseriate (branched or unbranched) filaments having apical growth. After the wound has healed, growth continues, as in the example of *Sphacelaria* (Fig. 2.24). There are well-documented cases of wound healing and regeneration in *Anotrichium tenue* (*Griffithsia tenuis*) and *Griffithsia pacifica* (Waaland and Cleland 1974; Waaland 1989, 1990). If filaments are severed, a rhizoid is produced from the base of the apical portion, and a new apical cell is regenerated on the basal portion (Fig. 2.29a). If, instead, an axial cell is killed, but the wall remains intact, the filament repairs itself (Fig. 2.29b). A regenerating rhizoid is produced by the apical fragment, and a special repair-shoot cell, not an apical cell, is produced by the basal fragment. This repair-shoot cell is induced by species-specific rhodomorphins, which diffuse out of the regenerating rhizoid. The repair-shoot cell grows toward and fuses with the regenerating rhizoid (Watson and Waaland 1983, 1986). Some seaweeds produce proliferations from cut surfaces; these are lateral outgrowths of cortical filaments, as in the red algae *Gigartina* (Perrone and Felicini 1976) and the brown *Dictyota* (Gaillard and L'Hardy-Halos 1990). The type of tissue produced in the reds – rhizoidal or bladelike – depends on the position of the wound with respect to the apex or base of the thallus. In other words, there is a correlation with an internal thallus polarity.

Regeneration in *Caulerpa* is also correlated with cell polarity. Excised blade "leaves" regenerate rhizomes and rhizoids from the basal end and new leafy shoots

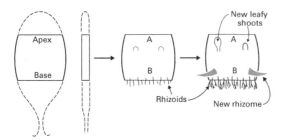

Figure 2.30 Regeneration from a portion of a "leaf" of *Caulerpa prolifera*. Leafy shoots form at the original apical end, and rhidoids form at the basal end. (From Jacobs 1970, with permission of the New York Academy of Sciences.)

from the apical end (Fig. 2.30). Rhizomes regenerate first rhizoids from the apical end, and later rhizoids from the basal end plus rhizome and leafy shoots from the apical end (Jacobs 1970, 1994). In Jacobs' experiments, "leaf" segments 30 mm long formed only rhizoids; if 40 mm long, half the specimens also formed a rhizome and a new leaf; if 50 mm long, all regenerated completely. However, leafy shoot production and rhizoid production from the rhizome of *Caulerpa* also respond to gravity, as shown by Jacobs and Olson's (1980) experiments, in which uninjured thalli were turned upside down. Rhizoids were produced from the new lower side, and leafy shoots from the new upper side (the rhizome did not twist, so polarity had been reoriented). Following wound healing, fragments of the tropical, calcifying Caulerpalean seaweed *Halimeda discoidea* also produce rhizoids; tiny pieces of 15 mm^2, that were cut on three of four edges, were able to produce rhizoids in 3 days, a mechanism of vegetative reproduction (Walters and Smith 1994).

The process of regeneration in *Fucus* is unusual in that distinct embryos, rather than lateral branches, are formed, although damaged holdfasts can regenerate adventitious shoots (McCook and Chapman 1992). During the process of wound healing in this genus, epidermal cells in certain regions of the wound begin to divide perpendicular to the wound surface, forming groups of branch initials (visible macroscopically after 4–6 weeks in culture), which develop directly into adventive embryos (Fulcher and McCully 1969, 1971). The midrib region of the thallus regenerates much

more rapidly than the wings (Moss 1964), correlating with the abundance in the midrib region of medullary filaments, which are primarily responsible for formation of new epidermis. Regeneration from vegetative branches always gives rise to vegetative shoots. Regeneration of strips cut from the discolored frond beneath spent receptacles of the deciduous species *F. vesiculosus*, although extremely slow, results in branches with small receptacles at their tips. Branches regenerated from strips cut from male thalli bear male receptacles, and those from female thalli bear female receptacles (Moss 1964).

This concludes our survey of the seaweed life histories, modes of reproduction, and morphological development. In the following chapters we shall examine the communities and habitats in which seaweeds live, the biotic factors they face, and the ways in which they are affected by abiotic factors.

2.7 Synopsis

Seaweed life histories can follow several patterns, depending on the species and the environment. An alternation between two free-living stages – one a haploid gametophyte, the other a diploid sporophyte – is common, but many variations exist. Some seaweeds have dissimilar sporophytes and gametophytes; others have only one free-living stage. There is no direct relation between ploidy level and morphology and so many variations of life cycles are possible, including changes between microthalli and macrothalli of the same chromosome number.

The life of a seaweed is a complex sequence of interactions between its genetic information and its abiotic and biotic environment. Development, from the initial polarization of the spore or zygote to the production and release of reproductive cells, is a highly co-ordinated process. Light (quality and quantity), photoperiod (usually the length of uninterrupted darkness), and temperature are the principal environmental cues. In some seaweeds, the onset of reproduction is triggered as a "response" to the prevailing environment conditions. Others "anticipate" the seasons using an endogenous clock which is entrained by environmental triggers (Zeitgebers; typically photo-period). Most seaweeds exhibit short-day responses to photoperiod. Seaweeds also respond to different wave-lengths of light but the nature of the receptor pigments and their mode of action are largely unknown. Several glycoproteins have been identified as regulatory factors that control the life-cycle progression of *Ulva*.

Sexual reproduction may be isogamous, anisoga-mous, or oogamous. "Broadcasters" liberate gametes into the water column and have external fertilization, whereas "brooders" have internal fertilization as the egg is retained on the female. Syngamy is regulated by cell recognition mechanisms on cell/flagella surfaces. Motile gametes in brown algae may be attracted to each other or to a stationary egg by volatile phero-mones. In red algae, sexual reproduction often involves complex post-fertilization development of a carposporophyte for zygote amplification.

Settlement of spores or other reproductive struc-tures depends a great deal on water motion (turbu-lence and eddies), notwithstanding the limited capacity of some cells for oriented swimming. Spores of *Ulva* can select their settlement site by responding to chemical cues from bacteria in the biofilm, and to surface energy. Cells then attach to the surface with adhesive substances. At first they are susceptible to being resuspended, but the chemical bond to the sur-face becomes stronger over time. Following attach-ment, cells acquire polarity and then germinate into rhizoidal cells and thallus-forming cells.

Erect thalli characteristically have an apico-basal polarity, which is expressed in the position and kind of regenerative outgrowths on wounded thalli and sometimes in apical dominance. Morphological devel-opment and the adult form depends on intrinsic factors such as phytohormones, extrinsic factors including light, temperature, and nutrients as well as grazing and epiphytic bacteria, and size/shape inequities between daughter cells. Some cells are totipotent and regenerate an entire thallus, whereas others cannot be regenerated. The morphological development of many seaweeds is plastic, and the genes responsible for regu-lating such development are being identified.

Wound healing is an important function in seaweeds, which are continually subjected to damage by grazers and abrasion. In siphonous algae, rapid plugging of the wound takes place to prevent cytoplasm loss. In multi-cellular algae, cut cells usually die, and wound healing is accomplished by the underlying cells. Regeneration commonly takes place from cut surfaces, with either frondlike or rhizoidlike tissue produced as a function of the distance from the dominant apex.

Seaweed communities

Seaweeds exist as individuals, but they also live together in communities with other seaweeds and animals – communities that affect and are affected by the environment. Ecologists and physiologists alike are drawn to coastal marine ecosystems because of the easy access to strong environmental gradients over short spatial scales. Marine organisms grow in often distinctive vertical or horizontal "zones" or "bands" along these gradients, thereby providing "natural laboratories" in which to study environmental (abiotic) and biological processes shaping the communities. Zones of vegetation are also found in terrestrial habitats, but here the spatial scales are typically much greater. On a mountain, for example, vegetation is zoned with altitude, but the vertical distance over which changes occur can be in the order of 1000 m rather than several meters in the intertidal zone (Raffaelli and Hawkins 1996). Vertical gradients in the intertidal are easily observed at low tide, but also extend underwater where the surface irradiance can be reduced to 1% at 15 m depth in many coastal waters (Lüning and Dring 1979; sec. 5.2.2). Horizontal gradients include the salinity gradients of estuaries and salt marshes, and wave exposure (Raffaelli and Hawkins 1996).

In Chapters 1 and 2, we reviewed the morphologies, life histories, and developmental processes of seaweeds as species. In this chapter we consider the patterns and processes in marine benthic communities as a starting point for later factor-by-factor dissection of the environment. We open with an overview of zonation patterns seen in the intertidal and subtidal environments.

3.1 Intertidal zonation patterns

3.1.1 Tides

Almost all marine shores experience tides, although tidal amplitudes vary greatly from place to place. The pattern of high and low waters also varies from place to place, depending on the interaction between the tide waves, and the standing waves caused by water slopping back and forth in the ocean basins; for details, see texts such as those by Gross (1996), Denny and Wethey (2001), and Trujillo and Thurman (2010). More important to seaweed ecology and physiology are the changes that occur at one place. Besides the progression from neap to spring tides twice a month, the times of high- and low-water change during the lunar month, and often from season to season (see Raffaelli and Hawkings 1996); this is important because desiccation and UV radiation (UV-R) stress in the intertidal zone and water-temperature stress in tide pools and reef flats are increased when summer low tides occur during the day. Sea levels in the tropical Pacific can also change because of El Niño/southern oscillation (ENSO) events (sec. 7.3.7).

Tidal cycles are of three types (Fig. 3.1). Diurnal tides have one high and one low per day; this is an unusual type, occurring in parts of the Gulf of Mexico and Do-San, Vietnam (Raffaelli and Hawkins 1996). Semi-diurnal tides rise and fall twice a day, with successive highs and lows more or less equal in height; this type is common along open coasts of the Atlantic Ocean. Mixed tides are variations on the first two themes and are named depending on whether they

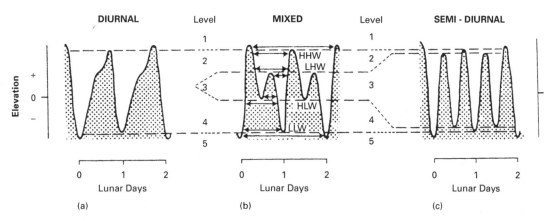

Figure 3.1 Physical division of the seashore. Daily (first-order) critical tide levels (CTLs) in diurnal, mixed, and semi-diurnal tide regimes during a neap tidal cycle. Stippling indicates submergence; arrows indicate duration of continuous exposure or submergence immediately above and below CTLs. CTLs divide the intertidal region into four or five levels. (From Swinbanks 1982, reprinted with permission of Elsevier Biomedical Press, Amsterdam.)

are more similar to a diurnal or semi-diurnal type: mixed dominant semi-diurnal tides occur twice a day but have clearly unequal highs and lows (Fig. 3.1b), whereas mixed dominant diurnal types have either one or two tides per day depending on the lunar cycle (Raffaelli and Hawkins 1996). Mixed tides are characteristic of Pacific and Indian Ocean coasts, as well as in smaller basins such as the Caribbean Sea and the Gulf of St. Lawrence (Gross 1996). In addition, there are storm tides, with irregular periods, usually of several days, caused by barometric-pressure changes and winds. Where the tide range is < 1 m, such as on the Swedish west coast or much of the Mediterranean Sea, atmospheric-pressure changes and onshore and offshore winds may combine to produce very irregular and unpredictable changes in water level (Fig. 3.2). Such shores are called atidal (Johannesson 1989).

3.1.2 Vertical zonation on intertidal rocky shores

The intertidal (littoral) zones are often described as the harshest habitats on Earth because of the rapid fluctuations in environmental conditions: when the tide is in, seaweeds are submerged, fully hydrated, with attenuated light and access to inorganic nutrients, but when the tide recedes, they are subject to

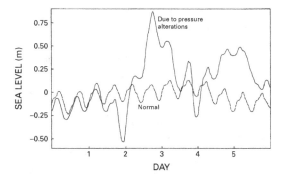

Figure 3.2 Sea-level changes on an atidal shore (Tjärnö, Sweden). One period features more regular, tide-dominated variations; the other is characterized by dramatic changes due mainly to atmospheric-pressure alterations. For example, during day 2, the water level changed nearly 1.5 m (0 m = mean sea level). (From Johannesson 1989, with permission of Munksgaard International Publishers.)

terrestrial conditions (Norton 1991). Temperature, salinity, light (quality, quantity, and UV-R), dissolved inorganic carbon sources for photosynthesis (CO_2 and HCO_3^-), and inorganic nutrient supply (e.g. nitrogen and phosphorus) all change simultaneously on time-scales of seconds to minutes as the tide rises and falls (see secs. 5.2.2, 6.4.2, 7.2.3). Seaweeds and sessile invertebrates grow in distinct vertical "bands" on the shore, and these banding patterns are a distinctive feature of intertidal zones worldwide (see Bertness *et al.* 2001; Connell and Gillanders 2007; Kaiser *et al.* 2011). Bands are particularly conspicuous at low tide on rocky shores, where we now focus our attention.

Despite the extent of global shorelines, there are general patterns of intertidal zonation upon which local variations are superimposed. Shores can be divided according to tidal levels, based on a known chart datum (high, middle, and low intertidal), or by using a biological scheme. A vertical strip of the shore may be divided into zones determined by the organisms present: a *Fucus* zone, a barnacle zone, kelp zone, and so forth. The Stephensons' (1949) universal zonation scheme is based on such biological features and has proven a very useful scheme that can be applied in a general way to rocky intertidal shores worldwide (Fig. 3.3a). These biological zones are also influenced by wave action, which decreases the duration of atmospheric exposure of a spot on the shore, thereby increasing the zonal band-widths (Fig. 3.3b; Lewis 1964; Raffaelli and Hawkins 1996). Many examples of intertidal zonation can be found in the books by Lewis (1964) and Stephenson and Stephenson (1972), and Raffaelli and Hawkins (1996) list key publications that describe intertidal zonation patterns from around the world (see their Table 2.2 and Fig. 2.2).

There are difficulties inherent in defining zones on the basis of the organisms found in them, however. First, there is variation in space. On the small scale, most shorelines consist of irregular rocks and boulders and are likely to present very confusing patterns of organisms, with zones breaking down into patches. On the large scale, flora and fauna change geographically. The scale at which a study is conducted can affect the outcome of the experiment (Benedetti-Cecchi 2001, see sec. 3.5.1 and Essay 2). Most early experimental work focused on single rocky shores, and provided precise information on local processes controlling community structure, whereas more recently, macroecology examines patterns and processes in controlling large-scale (hundreds to thousands of kilometers) phenomena (Connell 2007a; Santelices *et al.* 2009; Witman and Roy 2009). "The problem confronting contemporary ecologists is not whether one should test

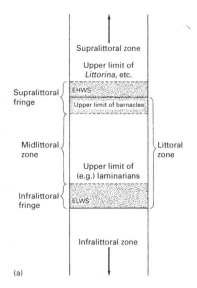

(a)

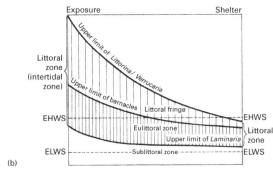

(b)

Figure 3.3 Biological division of the seashore. (a) The Stephensons' universal scheme for intertidal zonation. (b) Lewis' scheme for intertidal zonation, illustrating the effects of wave exposure in broadening and raising the zones (toward the left of the diagram). EHWS, extreme high water of spring tides; ELWS, extreme low water of spring tides. (Part a from Stephenson and Stephenson 1949, with permission of Blackwell Scientific Publications; b from Lewis 1964, with permission of the author.)

for the existence of general phenomena (macroecology) or specific phenomena, but what balance should be sought between the two" (Connell and Irving 2009).

Variability in time is also important. Aside from seasonal and successional changes in the vegetation,

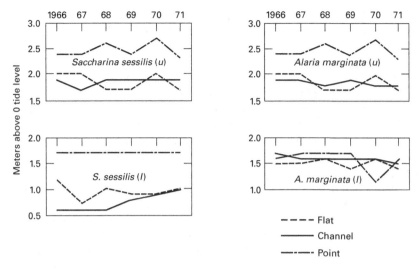

Figure 3.4 Year-to-year changes in upper and lower limits for two intertidal kelp species on three transects at an exposed site on the west coast of Vancouver Island, British Columbia. A gently shelving platform, a rocky point, and a narrow channel are compared. (From Druehl and Green 1982, with permission of Inter-Research.)

the timing of harsh conditions with respect to settling and the timing of the clearing of space with respect to reproduction of potential settlers (and competitors) add to the heterogeneity in the communities; see, for example, Dethier (1984) and Sousa (2001). The patchiness in the pattern is not static. Vertical limits of species vary from year to year (Fig. 3.4), dependent in part on variations in emersion/submersion histories (Dethier and Williams 2009; Pearson *et al.* 2009). Relative abundances and distributions of species that are nearly equal as competitors will change from time to time, and longer-term changes are also known to occur (Lewis 1980; Paine 1994; see sec. 4.3.1). Understanding such long-term changes in communities requires long-term observations, while an understanding of the short-term changes in populations requires extensive data on how various factors affect the critical stages of the life cycles of the algae. Observations of the presence of a seaweed at a given site can be interpreted to mean that conditions there have been suitable for its growth since it settled. Absence, on the other hand, provides little insight: perhaps conditions were unsuitable at some time, or perhaps the reproductive bodies of the species were unable to reach the

site. Unsuitable conditions may have been present always, or only at some brief time in the past, if the seaweed was ever growing there.

The concept of "community" has been long-debated. At one extreme is the view that communities consist merely of coincidental associations of seaweeds, and at the other is the view that seaweed communities are closely integrated units, almost super-organisms (Chapman 1986; see Irving and Connell 2006a). The current concept for terrestrial plants is of interdependent assemblages of species, that interact with one another both positively (e.g. facilitation) or negatively (competition). How plants assemble into predictable communities has been well studied in terrestrial ecosystems (e.g. Wilson 1999), and the application of assembly rules to seaweed communities may be a useful framework for predicting how species associations develop (Irving and Connell 2006a). Also important is how local communities interact with each other. A local seaweed community is not a "closed" system, but it is "open" to the effects of other benthic communities which supply it with algal and invertebrate propagules via water currents, termed "supply-side ecology" (Underwood and Keough 2001). These larger groupings of interconnected local communities are called metacommunities (e.g. Okuda *et al.* 2010). Therefore although seaweed communities are often illustrated as distinct communities growing in homogeneous vertical bands, this illustration belies the fact that communities are dynamic, patchy in space and time, and shaped by processes including the physiology of individuals (Chapters 5, 6, and 7), biotic interactions such as competition and facilitation (Chapter 4), the supply of propagules (Chapters 2 and 4), and biotic and abiotic disturbance (Chapters 4 and 8). In the next sections, we trace the key developments in approaches to studying vertical zonation of seaweeds in the intertidal and subtidal.

3.1.3 Factors controlling vertical zonation

The factors controlling the upper and lower limits of intertidal algae and invertebrates have fascinated researchers for well over a century (reviewed by Russell 1991; Paine 1994; Chapman 1995; Raffaelli and Hawkins 1996; Menge and Branch 2001; Connell and Gillanders 2007; Essay 2). Initially, there was the idea that some factor(s) related to tidal emersion caused vertical zonation, and early research focused largely on physical factors. The concept of critical tide levels (CTLs), that is the positions in the intertidal zone at which there are marked increases in the duration of exposure or submergence, was introduced first by Colman in 1933 and was a popular theory for around 50 years. Colman's (1933) CTLs were based on average percentage exposures, which did not take into account duration and time of exposure. The most widely recognized concept was by Doty (1946) who defined CTLs, using maximum durations of exposure or submergence, which Swinbanks (1982) amplified. Correlations between zone boundaries and CTLs may be expected, because the stress of exposure to the atmosphere increases with the duration of exposure, and seaweeds have differing abilities to recover from the stress (see Chapter 7). However, Underwood (1978) showed that Colman's exposure curve was inaccurate, and when more rigorous experiments were conducted, and statistical analyses applied, there was no relationship

between CTL and zonation (Hartnol and Hawkins 1982; see Raffaelli and Hawkins 1996).

Despite CTLs being rejected as *the* cause of zonal patterns, physical factors associated with emersion clearly contribute to shaping rocky shore communities (Gilman *et al.* 2006). The important criterion is not the theoretical exposure time determined from tide tables, but the actual exposure time (Fig. 3.5), however, this has not been measured very often. Druehl and Green (1982) attempted to correlate vertical distribution with actual submergence/emergence durations. Using a "surf-sensor" devised to record submergence events, they examined the submergence histories of three contiguous areas (within 50 m of each other) with differing slopes in a wave-exposed location and the relations between seaweed vertical distributions and corrected exposure times. Actual submersion/emersion curves differed from predicted tidal curves in two ways: (1) The intertidal range was greater, and the lowest emerged level was higher than predicted; a gently shelving transect was closest to the predicted, and a rocky point the most different (Fig. 3.5a). This compares to the general observation depicted in Figure 3.5b. (2) The harmonic pattern of the tidal curves was disrupted, again the most disrupted on a rocky point (Fig. 3.5b, c). Thus waves tend to increase the emersion time. Druehl and Green (1982) found that the upper limits of seaweeds correlated most closely with accumulated time submersed, whereas the lower limits correlated best with the duration of the longest single exposure.

Gilman *et al.* (2006) extended the work of Druehl and Green (1982) by measuring the effective shore level (ESL), a method of standardizing emersion times across wave-exposure gradients, and defined as a given point on the shore "to be equal to the absolute shore level (i.e. the vertical distance above still-water chart datum) with equivalent emersion characteristics (timing and duration) in the absence of waves." (Harley and Helmuth 2003). Using small temperature sensors that were anchored at various positions on the shore, they logged temperature every 15 mins for 2 years to quantify the time of the "first wetting" of each tidal cycle, and also embedded tiny temperature sensors in the shells of intertidal mussels to examine the temperature

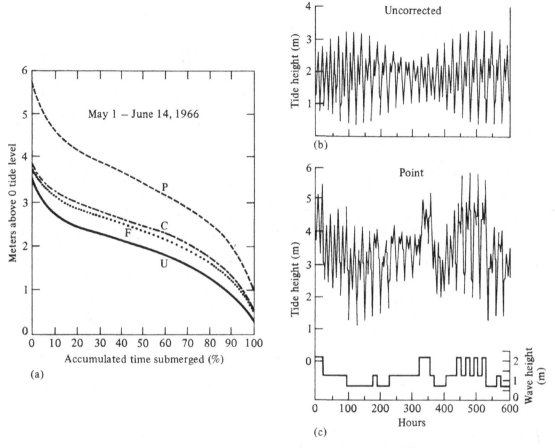

Figure 3.5 Submergence–emergence data from a site on Vancouver Island, British Columbia. (a) Measured accumulated time submerged as a function of elevation above zero tide for a rocky point (P), a channel (C), and a gently shelving rock face (F), all within 50 m of one another, compared with data from 6-min tidal predictions (U). (b) Predicted tide heights over a lunar cycle. (c) Actual tide heights and wave heights at the rocky point. The wave-height data are derived from twice-daily observations. (From Druehl and Green 1982, with permission of Inter-Research.)

experienced at an organismal scale. They found a clear relationship between ESL and the upper limit of mussels at some sites, but not others, once again illustrating that no one factor controls zonation. Summed up by Williams and Dethier (2005), "tidal height is not a perfect proxy for stressful conditions in the intertidal".

In the 1970s to early 1980s, research began to focus on biological explanations for rocky shore zonation, related to the abilities of seaweeds at different shore positions to avoid, tolerate, or recover from environmental stress (e.g. Schonbeck and Norton 1978; Dring

and Brown 1982; see Chapman 1995 and sec. 7.5.1). In their seminal studies on intertidal fucales (Fig. 3.6), Schonbeck and Norton (1978, 1979a, b, c, 1980a, b) conducted reciprocal transplants of *Pelvetia canaliculata* and *Fucus spiralis* from their own zones to higher and lower zones, and found that they grew more slowly when transplanted to a higher zone, but faster when transplanted down shore. They concluded that factors related to emersion (particularly desiccation) controlled the upper limits of each species of fucoid, while the lower limits were set by competition for light and

Figure 3.6 The intertidal seaweed community is not the two-dimensional world it appears to be at low tide. When underwater, the seaweeds form an erect and dynamic canopy. *Ascophyllum nodosum* (center and left) and *Fucus vesiculosus* (right) are buoyed up by pneumatocysts. (Photo taken on Isle of Man, UK by Tim Hill, ©Tim Hill, and reproduced by permission of the photographer.)

space i.e. faster growing species outcompeted slower growing species. Subsequent experiments have shown that this finding is not universal: seaweeds transplanted to higher vertical zones can survive, as long as biological competitors are removed indicating biological control of the upper limits in some cases (see Raffaelli and Hawkins 1996). The current concept is that each species has a "zone of tolerance" for stresses related to emersion, the limits of which can be moved up and down depending largely on competition, although the generality of this in different geographic regions (e.g. tropics vs. temperate) requires testing. The exception is the top of the uppermost zone on the shore, where the marine environment gives way to terrestrial conditions; here the upper limit is ultimately controlled by emersion time (Chapter 7).

In order to test the generality of the concept that upper limits of a zone are more stressful than lower zones, Dethier and Williams (2009) measured the growth rates of *Fucus distichus* growing on the upper, mid and low shores of 14 beaches around San Juan Island, USA. In spring and summer, when low tides occur in daylight, growth rates of mid-shore populations were greater than those of high shore populations, as would be predicted by the "environmental stress model" (Essay 2). In fall and winter, however, when low tides occur in darkness, the high shore environment became less stressful, and growth rates of high shore species remained at a similar level to those of summer, while those of mid-shore populations declined. The lower growth rates of mid-shore populations were also due to higher rates of herbivory in this zone. Their conclusion was that that the "stressfullness" of a particular zone on the intertidal varied seasonally and was dependent on the timing of low tide.

Essay 2 A journey towards rocky intertidal meta-ecosystem dynamics

Bruce A. Menge

Rocky intertidal communities long have served as "model systems" in the sense of contributing to advances in ecology, both through empirical and modeling approaches. The pioneering insights provided by the work of Lewis (1964) and of Stephenson and Stephenson (1972) on regional and global patterns of community structure, including the striking zonation so often seen on rocky shores, are two early examples. The studies of Connell (1961a, b) and of Paine (1966) demonstrated the power of field experimentation in revealing important processes underlying community pattern. For ecology in general, these works were ahead of their time, but the experimental approach eventually overtook and replaced the purely observational approaches used to infer dynamics that were the dominant mode of study in terrestrial ecology in the 1960s through the 1970s (e.g. Robles and Desharnais 2002).

In 1965, I joined R. T. Paine's lab at the University of Washington in Seattle. This was an exciting time, and with my lab mates and mentors, I quickly became steeped in the big ideas of ecology, including the Hairston *et al.* (1960) "green world" hypothesis (hereafter HSS), the modeling approaches of MacArthur and others on issues such as community stability (MacArthur 1955), species diversity (MacArthur 1965; Pianka 1966), limiting similarity and species coexistence (MacArthur and Levins 1967), island biogeography (MacArthur and Wilson 1963), and species interaction strength (MacArthur 1972). These were highly controversial ideas, especially HSS, but in the 1960s and 1970s, the evidence needed to test these ideas was extremely limited or nonexistent. Although I did not expect to singlehandedly fill this gap, I settled on two career goals: carrying out experiment-based field research to investigate the dynamics of rocky intertidal communities, focusing on the interplay between species interactions and environmental context and how this might shape a community, and using the resulting data to test and modify prevailing theory on how communities were organized.

After dissertation and postdoctoral studies focused on population characteristics of the predatory intertidal sea star *Leptasterias hexactis* and its competitive interaction with the iconic keystone sea star *Pisaster ochraceus* (e.g. Menge 1972a, b; Menge and Menge 1974), I moved to Boston, Massachusetts, where I initiated an investigation of rocky intertidal community dynamics along the New England coast. I adopted the approach employed by P. K. Dayton (e.g. Dayton 1971, 1975), which used a method later dubbed the "comparative-experimental" approach (Menge *et al.* 1994). This technique involves carrying out replicated, identically designed and implemented field experiments at multiple sites varying in physical conditions ("environmental stresses") such as wave force and associated factors like desiccation and temperature. Dayton (1971) used this approach to demonstrate that predation by whelks and sea stars, and limpet grazing were powerful influences in structuring communities along a wave exposure/desiccation gradient.

In New England, which has a harsher and more variable coastal environment (with colder winters and hotter summers), and a much less diverse community, similarly designed studies showed that in the mid-intertidal zone, strong effects of predation (by whelks, *Nucella lapillus*) occurred in wave-sheltered areas, but on more wave-exposed areas, consumers were ineffective in controlling sessile organisms (Menge 1976). Investigation of the mechanisms underlying this result indicated that the foraging activity of the whelk was inhibited by wave forces (Menge 1978a, b). In the low intertidal, parallel studies revealed similar dynamics, with the exception that predation was due to a group of predators, not just whelks (Lubchenco and Menge 1978). Sea stars and crabs also exerted strong effects in wave-sheltered areas (Menge 1983). Similar trends were found for herbivorous littorine snails, which were completely excluded from occupying wave-exposed headlands by strong wave action (Lubchenco and Menge 1978; Lubchenco 1983, 1986). Thus, predation/herbivory were the dominant structuring forces in wave-sheltered areas and, in the absence or ineffectiveness of predators and herbivores, competition for space was the dominant structuring force in wave-exposed areas.

Based on these results, John Sutherland and I proposed a modification and expansion of the HSS model (the "Environmental Stress" model) that we suggested was appropriate when trophic complexity varied (Menge and Sutherland 1976, 1987). The HSS model assumed a three-level trophic structure, with predators eating herbivores which ate plants (Hairston *et al.* 1960). We had observed systems in which predators, or predators and herbivores were absent or rendered ineffective by physical conditions, indicating that systems existed in which there were only two or even just one trophic level. We noted that at least in aquatic systems, four trophic levels sometimes existed. We also noted that

Essay 2 (cont.)

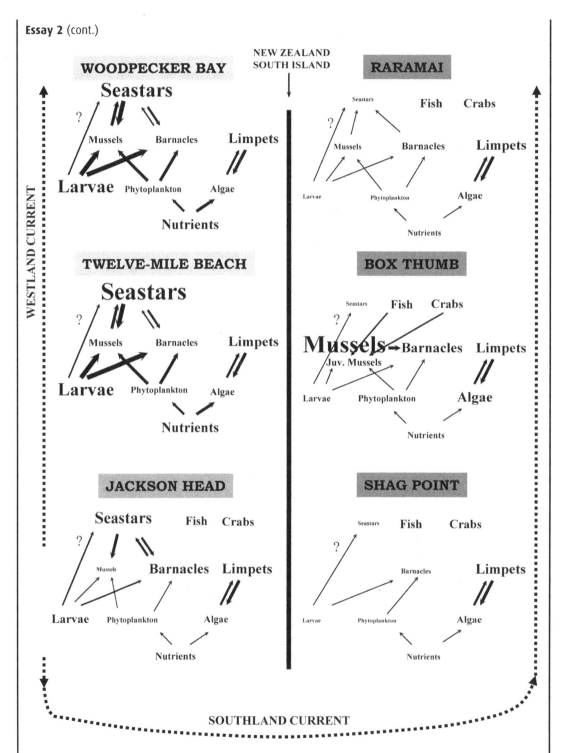

Figure 1 Diagram of interaction webs (i.e., the subset of ecologically important links between components of a food web) for a set of six sites on the east and west coasts of the South Island of New Zealand, an example of a meta-ecosystem (based on

Essay 2 (cont.)

Menge *et al*. 2003). Field experiments (of predation rates and grazing effects) and measurements (of sessile invertebrate recruitment, phytoplankton, nutrients, and abundances) are the basis of the interaction web diagrams. The thick central line represents the South Island of New Zealand. Webs to the left of the line are for sites on the west coast, and webs to the right of the line are east coast sites. From north to south on the west coast, distances among sites were Woodpecker Bay to Twelve-Mile Beach to Jackson Head, 50 and 295 km, respectively. On the east coast, distances were Raramai to Box Thumb to Shag Point, 145 and 260 km, respectively. The dotted lines on the side represent the Westland Current and Southland Current and their general directions. The Westland Current periodically reverses, generating alternations between weak upwelling (northward flow) and relaxations or downwelling (southward flow). The Southland Current generates persistent downwelling conditions on the East Coast. Jackson Head is in the region where the eastward flowing Tasman Current collides with the South Island, and likely also experiences upwelling and downwelling depending on meanders in the Tasman Current. In the webs, upward-pointing arrows indicate ecological subsidies (recruitment, nutrient flow, phytoplankton, and algal biomass flow) and downward-pointing arrows indicate top-down effects. Line thickness qualitatively indicates the magnitude of the effect or flow, and font size reflects relative abundances of each group. Question marks indicate the actual magnitude of the arrow is unknown; arrows shown are estimates based on observations of numbers of sea star juveniles. Note that top-down effects tend to be strongest where ecological subsidies are greatest (i.e. site names with light gray shading, which are sites with intermittent upwelling), and top-down effects tend to be weakest where ecological subsidies are smallest (i.e. site names with dark gray shading, which are sites with persistent downwelling). Factors driving sea star abundance are unknown, but likely involve variable dispersal, which appear high on the west coast and extremely low on the east coast. Limpets form an important exception to the idea of top-down effects dependent on bottom-up inputs; their effects tend to be strong at all sites, regardless of oceanography and algal abundance (Freidenburg 2002; Guerry 2006). (Courtesy of Bruce Menge.)

---

omnivory, thought to be rare, was actually likely to be common in many systems. Omnivory could short-circuit the strict pattern of one trophic level feeding only on the level below. For example, omnivory was common (14 of 51 consumers or 27%) in the Panama rocky intertidal food web of the Pacific coast (Menge 1986).

In the 1980s, my research focus turned to the diverse and relatively unstudied, at least from a community perspective, Oregon coast. Because of the lack of understanding of community dynamics in this system, my initial focus was on a single site, Boiler Bay (BB), with some supplementary studies at a second site, Strawberry Hill (SH), some 65 km south. I soon realized that patterns of community structure at these sites contrasted strongly in several ways. Although both had basically the same biota and same trophic structure, and in this sense were "typical" of US west coast rocky shores, I observed and then documented that BB tended to be more macrophyte dominated, while SH tended to be more sessile-invertebrate dominated (e.g. Menge 1992; Menge *et al*. 1994). Predators and herbivores also were more abundant at SH than at BB. Why the difference?

Throughout much of the period of development of modern rocky intertidal ecology, most considered that the primary influence of the ocean was to deliver waves, which often dislodged sessile organisms, leading to an emphasis on disturbance as a primary determinant of community structure (e.g. Dayton 1971; Menge 1976; Connell 1978; Paine and Levin 1981). A second, later emphasis was on the ocean as a habitat for larvae and on its role in delivering larvae to the adult habitat (e.g. Underwood and Denley 1984; Gaines and Roughgarden 1985). The influence of the ocean as determinant of variation in food (detritus, phytoplankton) and nutrients in rocky intertidal systems, however, was assumed to be of minor importance.

The comparison between BB and SH forced a reconsideration of this assumption. To determine the cause of the differences between these two sites, we embarked on efforts to evaluate the relative influences of environmental factors (e.g. wave forces, temperature, nutrients, phytoplankton), larval supply (recruitment of dominant sessile invertebrates), organism performance (growth of mussels, barnacles, and seaweeds), and species interactions (competition, predation, herbivory, facilitation). In particular, we were intrigued by the role of phytoplankton and detritus as possible determinants of community structure and dynamics through bottom-up effects on growth and survival of sessile invertebrates, which were the prey of sea stars and whelks.

Satellite imagery had shown that the California Current varied in distance from shore, from nearer off BB to farther off SH, a difference that corresponded with variation in the width of the continental shelf. From this, I hypothesized that oceanographic

Essay 2 (cont.)

conditions varied along the coast, generating differences in phytoplankton, nutrients, larval transport, and detritus (Menge 1992). I proposed that all these factors but nutrients were generally higher at SH, thereby providing a higher supply of food (phytoplankton, detritus) for filter-feeding invertebrates, including mussels and barnacles, and that this led to faster growth rates and thus higher secondary productivity. I further proposed that differences in currents favored higher recruitment of mussels and barnacles at SH, thereby explaining the higher abundance of these organisms at SH and through bottom-up effects, to higher densities of predators. At BB, we suggested that stronger upwelling led to higher nutrients, which favored higher abundance of macrophytes.

Starting in the 1980s, ecologists became increasingly aware of the need to expand the spatial and temporal scales of their studies (e.g. Dayton and Tegner 1984a; Wiens 1989; Levin 1992), and consideration of scale is standard in most of modern ecology. This need raised a new challenge, however. Diamond (1986) and others had argued that the limitations of field experiments were precisely that their scales were too small to be relevant to communities and ecosystems as a whole, especially when examined across a landscape. The problem was that as noted above, studies based strictly on observation had also been found to fall short of what was needed to understand system dynamics, so what were we to do? One answer was to use macroecological approaches (Brown 1995), which involved assembling massive data sets on factors thought to be important to the dynamics of global ecosystems, and study them using regression and other approaches for insights into causal relationships. This approach, however, also has limitations. No experimental component is included in such analyses, and spatial variation such as that seen along environmental gradients is largely ignored (e.g. Paine 2010).

Another answer involved the marriage of the comparative-experimental method (e.g. Menge 1991; Menge *et al.* 1994; Paine 2010; see above) and a new conceptual framework for considering how ecosystem dynamics might vary called "meta-ecosystem ecology" (Loreau *et al.* 2003). Meta-ecosystem ecology, defined as a set of ecosystems linked by flows of energy, materials, and propagules (Loreau *et al.* 2003; Leroux and Loreau 2008; Gravel *et al.* 2010a, b), is a recent conceptual advance in efforts to understand the importance of scale in ecosystem dynamics. In this approach, multiple local ecosystems are embedded in an environment in which most effects are difficult to manipulate (e.g. waves, nutrients, currents) but of which modern technology and sampling methods allow quantification. Performance of identically designed local-scale experiments at multiple study sites in this environment can provide important insights into the causes of pattern from local to large scales, such as those of Large Marine Ecosystems.

The California Current Large Marine Ecosystem (CC-LME) ranges from southern Vancouver Island through Baja California, and like other major Eastern Boundary Current ecosystems, is dominated oceanographically by summer upwelling. Starting with the work of Paine (1966), a large body of investigation of the dynamics of populations, communities, and ecosystems in this system has accumulated. Until the early 1990s, however, most of this research was spatially and temporally limited, occurring at one to a few study sites usually located within a feasible distance of university-based researchers, and focused primarily on local-scale processes such as species interactions and disturbance. The research done at BB and SH, though still involving just two study sites relatively close together, incorporated additional elements of recruitment and bottom-up subsidies, two defining characteristics of meta-ecosystems. Research initiated at these sites in the 1980s was expanded in the 1990s and 2000s, and now involves >40 sites along the coastline of the CC-LME. A key stimulus for this expansion was the observation of coastal phytoplankton blooms, and evidence that variation in such blooms was associated with variation in recruitment, growth rates of sessile invertebrates, and abundance and impact of predators.

An example of a meta-ecosystem scale investigation is in Menge *et al.* (2003). The goal of the study was to understand how the relationship between the sea star *Pisaster ochraceus* and its primary prey *Mytilus californianus* varied among sites and between regions with differing upwelling regimes. An earlier hypothesis proposed that with increasing upwelling, the role of local interactions decreased and the influence of larval transport (reflected in recruitment rates) increased (Roughgarden *et al.* 1988; Connolly and Roughgarden 1999). While some results from 14 sites ranging across 1200 km nested within three oceanographic regimes were consistent with model predictions (e.g. patterns of phytoplankton and recruitment), others were not (predation rate was only weakly associated with upwelling). This system showed important variation at all scales, from local, to regional, to latitudinal, and a strong coupling between oceanographic conditions and ecological pattern and process, but the relationship was far more complex than predicted by the coupled benthic–pelagic model. Although the investigation

Essay 2 (cont.)

of meta-ecosystem dynamics at the appropriate spatial scale is still in its infancy, examples are accumulating (Navarrete *et al.* 2005; Witman *et al.* 2010). Figure 1 shows a diagrammatic summary of the structure and dynamics of a meta-ecosystem at wave-exposed rocky intertidal sites in New Zealand (Menge *et al.* 2003). In my view, this approach will be increasingly adopted as ecologists, conservationists, and oceanographers continue to grapple with the problems of understanding ecosystem dynamics at scales appropriate to those of coastal marine environments.

Bruce Menge is a Distinguished Professor of Zoology and the Wayne and Gladys Chair in Marine Biology at Oregon State University, USA. He is a marine ecologist who has spent his career studying how ecological communities are structured. His current research focuses on determining the relative influences of oceanographic conditions, connectivity, and species interactions in structuring coastal ecosystems, and how ocean acidification will alter these systems.

The search for the factors shaping intertidal communities continues on a range of levels. At an organismal level, the physiological mechanisms of stress tolerance are being elucidated, including the application of "omics" (see Essay 5, Chapter 7). Molecular tools allow insights into the genotypic and phenotypic responses of populations to acclimate and adapt to different vertical zones (Pearson *et al.* 2009). Hays (2007) found a genetic component in the ability of seaweeds to resist stress: the progeny of *Silvetia compressa* from the high shore were more resistant to atmospheric exposure than the progeny from low-shore populations, indicating a "home-height advantage". The modeling approach of Johnson *et al.* (1998) generated 13 new hypotheses on factors controlling seaweed zonation that require experimental testing (see their Table 5). For example, the boundary between zones will be less distinct (more diffuse) on shores with lower levels of light (hypothesis c3; Johnson *et al.* 1998). On the community level, Benedetti-Cecchi (2001) demonstrates that the effect of horizontal factors in shaping vertical distributions needs to be taken into account (see Fig. 3.14). In summary, while biological zonation patterns are found on intertidal shores from the tropics to the poles, elucidating the physical and biological processes underpinning these patterns requires the study of processes on scales ranging from cellular to community to biogeographic.

3.2 Subtidal zonation on rocky shores

The vertical zonation apparent on the rocky intertidal extends underwater, with community assemblages changing with depth. Seaweeds occupy the infralittoral zone of the subtidal (sublittoral, Fig. 3.3a) and the ultimate depth of this zone is dictated by light availability. The zone below (circalittoral) is occupied by invertebrates (Witman and Dayton 2001; Fig. 3.7). Seaweeds in the shallow subtidal, i.e. above the surface mixed layer (thermocline), may encounter high, and fluctuating, irradiances and temperatures which can result in physiological stress, similar to that observed in intertidal zone (Graham 1997), but subtidally there is no desiccation stress. Seaweeds growing below the thermocline will experience a more stable environment, with lower temperatures and higher nutrients than in the shallow subtidal (Witman and Dayton 2001). Local features such as wind mixing of the water column that causes turbidity, or the upwelling of cold, deep water, modify these general patterns. Water motion decreases with depth, and while subtidal populations can experience substantial drag forces during storms (evidenced by mounds of beach-cast material – Essay 3, Fig. 1, Chapter 5) they do not experience the extreme forces associated with breaking waves in the low intertidal (Denny 1988; Chapter 8). Consumer pressure is greatest in the shallow subtidal and declines with depth (Witman and Dayton 2001).

The long history of research apparent for intertidal ecology is lacking for subtidal ecology because until quite recently these communities were not accessible. Even though underwater research is now commonplace using scuba diving and underwater cameras, in the 1990s, three times more papers were published on intertidal communities than subtidal (Witman and

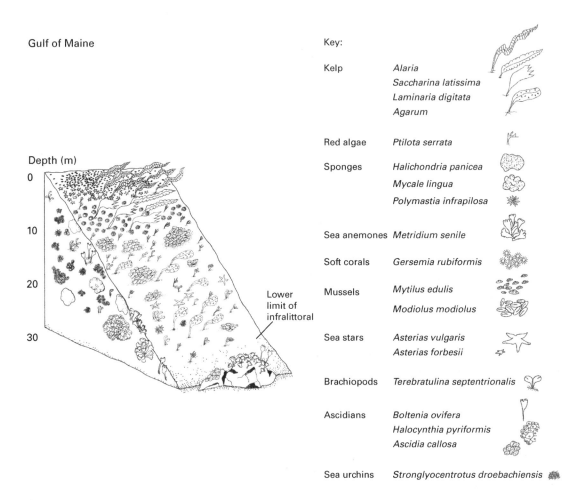

Gulf of Maine

Key:

Kelp	*Alaria*
	Saccharina latissima
	Laminaria digitata
	Agarum
Red algae	*Ptilota serrata*
Sponges	*Halichondria panicea*
	Mycale lingua
	Polymastia infrapilosa
Sea anemones	*Metridium senile*
Soft corals	*Gersemia rubiformis*
Mussels	*Mytilus edulis*
	Modiolus modiolus
Sea stars	*Asterias vulgaris*
	Asterias forbesii
Brachiopods	*Terebratulina septentrionalis*
Ascidians	*Boltenia ovifera*
	Halocynthia pyriformis
	Ascidia callosa
Sea urchins	*Stronglyocentrotus droebachiensis*

Depth (m)
0
10
20
30

Lower limit of infralittoral

Figure 3.7 Rocky subtidal zonation in a cold-temperate zone. A typical wave-exposed site in the Gulf of Maine, such as Murray Rock, Monhegan Island, or Boone Island, showing communities on a sloping rock shelf (front slope) versus a vertical rock wall (left-side diagram) down to a depth of 33 m. The lower limit of the algal dominated infralittoral zone extends to ~ 25 m, as indicated by shotgun kelp *Agarum clathratum* (previously *Agarum cribrosum*) on the slope. A shallow bed of laminarian kelp is truncated at about 10–15 m depth by sea urchin grazing. Dense carpets of blue mussel recruits can cover the substrate down to ~5 m depth. Not shown in the upper 10 m of the slope community is an understorey of red algal turf. Beds of horse mussels, *Modiolus modiolus*, are a conspicuous feature of slopes at 10–20 m depth. Note that macroalgae are absent from vertical rock walls, which are covered by soft corals, sponges, ascidians, and aggregations of the sea anemone *Metridium senile* in the deeper zone (not to scale). (From Witman and Dayton 2001, reproduced with permission.)

Dayton 2001). The majority of early experimental studies were of seaweed–herbivore interactions (Schiel and Foster 1986), a trend that continued into the late 1980s, at which time the focus changed to recruitment, and, to a lesser extent, disturbance. There are still relatively few studies on the relationships between subtidal seaweeds and abiotic factors, or of seaweed–seaweed interactions (Witman and Dayton 2001). Much of the autecological work has been on large laminarians, initially on sporophytes which dominate the vegetation in kelp beds, but more recently on microscopic gameto-phytes (see Graham *et al.* 2007a; Bartsch *et al.* 2008). Studies have often been limited to sites of particular interest (e.g. urchin barrens), so that the generality of the effects seen are often poorly known (Schiel and Foster 1986; Connell and Irving 2009).

A general division of the infralittoral into zones dom-inated by different functional groups was proposed by Vadas and Steneck (1988): an uppermost zone of thick, leathery macrophytes, a middle zone of coarsely branched, sheetlike, or filamentous algae, and a deep zone of crustose algae. The smaller functional groups occur at all depths; the dominance and zonation come about because of the successive elimination of larger forms with depth. Vadas and Steneck compared litera-ture on tropical sites (e.g. Littler *et al.* 1986a) with their study in the Gulf of Maine to show that this pattern is widespread. They also noted that within the genus *Hali-meda*, the depth distribution of seaweed habits parallels the functional-form groups, with a gradation from erect to low-lying species (Hillis-Colinveaux 1985). Those studies included very deep algae, down to the limit of algal growth. The full development of this zonation depends on the availability of substrate, while the depths of the zones depend on water clarity (see sec. 5.2.2).

Vegetation patterns on the rocky infralittoral are dependent on the slope of the substratum. In temperate regions, seaweeds dominate the gently sloping shelves while sessile invertebrates predominate on steep walls; this pattern is related to the preferential settlement of invertebrates on shaded, steep walls that accumulate less sediment, and for seaweeds there is a greater avail-ability of light for photosynthesis on the less-steep shelves (Fig. 3.7). In the tropics, a similar pattern exists but here corals dominate the zones that in temperate and polar regions are occupied by seaweed. Horizontal patterns of zonation are also seen subtidally. If a com-munity is studied at one depth along a line running parallel to the shore, then seaweed communities occur in horizontal "habitat patches" of various scales, for example, from a small kelp patch to a kelp bed that may be many kilometers long. "Rocky subtidal commu-nities may be viewed as a mosaic of habitat patches that changes somewhat predictably with depth" (Witman and Dayton 2001, p. 343).

3.3 Seaweed communities

We now describe the main abiotic attributes and dom-inant seaweed assemblages of the three major biogeo-graphic zones (tropical, temperate, and polar), followed by an overview of seaweed communities classed by habitat type. The large-scale biogeographic patterns of seaweeds are shaped by water currents and dispersal, temperature, and paleoclimatic events (Gaines *et al.* 2009; Huovinen and Gomez 2012; see sec. 7.3.8), and the characteristic features and seaweed assemblages of each region are synthesized by Lüning (1990).

3.3.1 Tropical

Tropical seas are characterized by warm waters (>22°C) that are thermally stratified year round, and seawater above the thermocline has low concentra-tions of inorganic nutrients (i.e. oligotrophic). Incident photon flux density is high, light penetrates deep into the water column because there are few phytoplankton blooms, and seaweeds can be found growing as deep as 268 m (Littler *et al.* 1985, sec. 5.2.2). Coral reefs and seagrass beds are both tropical habitats in which sea-weeds are an important component (reviewed by Mejia *et al.* 2012). These communities are character-ized by high productivity, with nutrients being tightly recycled within the system (Mejia *et al.* 2012). Tropical regions have higher rates of herbivory compared to temperate and polar regions, because fish grazing activity is controlled largely by temperature (Floeter *et al.* 2005).

In coral reefs, the corals (with their dinoflagellate symbionts) are the habitat-forming foundation species (see Essay 6, Chapter 9), whereas seaweeds tend to be understory. This contrasts with temperate and polar regions, where large, canopy-forming brown seaweeds are the structural elements. The seaweeds of coral reefs have been well studied, because of their import-ance in reef primary production, nutrient cycling, sub-stratum stabilization, and also because they compete with corals for the same resources (substratum, light, nutrients). Coral reefs are fragile ecosystems that are susceptible to anthropogenic impacts (Connell 2007b;

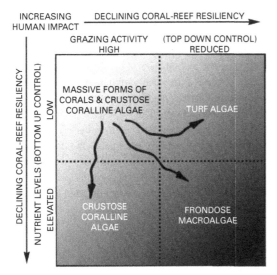

INCREASING
HUMAN IMPACT
DECLINING CORAL-REEF RESILIENCY →

GRAZING ACTIVITY   (TOP DOWN CONTROL)
HIGH   REDUCED

DECLINING CORAL-REEF RESILIENCY
NUTRIENT LEVELS (BOTTOM UP CONTROL)
LOW
ELEVATED

MASSIVE FORMS OF
CORALS & CRUSTOSE
CORALLINE ALGAE

TURF ALGAE

CRUSTOSE
CORALLINE
ALGAE

FRONDOSE
MACROALGAE

Figure 3.8 The relative dominance model. All of the four sessile functional groups depicted occur under the conditions in every compartment of the model; however, the RDM predicts which groups will be predominant under the complex interacting vectors of eutrophication and declining herbivory (most often anthropogenically derived). Crustose coralline algae are posited to be competitively inferior and dominate mainly by default; where frondose algae are removed by herbivores and coral are inhibited by nutrients. The dotted lines represent tipping points where the external forcing functions of increasing nutrients and declining herbivory reach critical levels that reduce resiliency to phase shifts. Light to dark shading indicates declining desirability of each functional group from a management perspective. Hypothetically, one vector can partially offset the other (e.g. high herbivory may delay the impact of elevated nutrients, or low nutrients may offset the impact of reduced herbivory). We further posit that such latent trajectories can be activated or accelerated by large-scale stochastic disturbances such as tropical storms, cold fronts, warming events, diseases, and predator outbreaks; events from which coral reefs have recovered for millions of years in the absence of humans. (From Littler *et al.* 2006, Elsevier, reproduced with permission.)

see Essay 6, Chapter 9) and if the balance between light, nutrients and herbivory changes, then a phase shift from coral to seaweed-dominated ecosystems can occur (Fig. 3.8; but see Bruno *et al.* 2009; sec. 6.8.5 and Essay 6, Chapter 9). Coral reef seaweeds are extensively reviewed by Littler and Littler (2011a, b, c), Fong

and Paul (2011) and Mejia *et al.* (2012). The early works of D.S. Littler and M.M. Littler provided the historical knowledge needed for frameworks that allow an assessment of how coral reefs are responding to human activities (Littler and Littler 1988, 2011a; Littler *et al.* 2006; Fong and Paul 2011). Although not discussed here, cyanobacteria are an essential part of coral reef functioning, including the nitrogen-fixing components of the algal turf (Littler and Littler 2011b, d; Fong and Paul 2011). Coral reef seaweeds can be broadly categorized into three groups: calcifiers, turfs, and upright foliose or leathery.

Calcifying seaweeds include encrusting and upright corallines (Rhodophyta) and *Halimeda* (Chlorophyta) which co-dominate on the wave-exposed slopes of the fore-reef, and corallines also dominate the reef crest (Littler and Littler 2011c; Fig. 3.9). These calcifiers have relatively low rates of productivity, but are resistant to grazing and have significant roles in sediment/sand production; *Halimeda* contributes approximately 8% of global carbonate production (Hillis 1997). Crustose coralline algae are the "glue" that consolidates and stabilizes reefs, and they are important facilitators of reef development because they release chemical attractants that trigger the settlement and metamorphosis of invertebrates, including the animal symbiont of the coral (Heyward and Negri 1999; Harrington *et al.* 2004), and they can suppress the growth of some fleshy algae (Vermeij *et al.* 2011). *Padina*, a brown seaweed, is lightly calcified. *Halimeda* produces rhizoids, and along with some fleshy siphonous green seaweeds, is also able to grow on the cobble and sandy substratum of the reef lagoon (Fong and Paul 2011).

Turfing communities are defined as "sparse to thick mats of diminutive and juvenile algae that are <2 cm tall" (Littler *et al.* 2011b). Mats typically contain 30–50 species, in some cases with >20 species per 1 cm^2 (Diaz-Pulido and McCook 2002; Fricke *et al.* 2011a, b). The seaweed component of turfs in the early recruitment phase are typically filamentous and polysiphonous genera such as *Feldmannia*, *Sphacelaria*, *Acrochaetium*, *Anoctrichium*, *Herposiphonia*, and *Cladophora*, and there are also crust-forming species such as *Pringsheimiella* and *Jania*, plus many species of cyanobacteria and diatoms. In the later successional

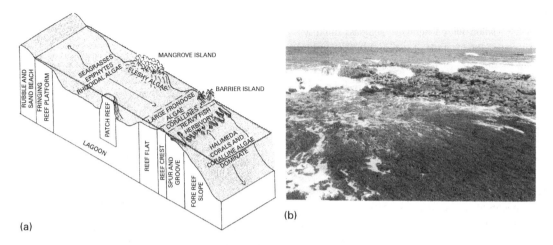

Figure 3.9 Coral reefs. (a) Sectional view through a tropical continental shelf containing characteristic barrier reef and mangrove systems. Dominant macrophyte groups are indicated for the various habitats. (b) Photograph of *Porolithon–Lithophyllum* ridge at Pago Bay, Guam. (Part a from D.S. Littler *et al.*, 1989, with permission of Cambridge University Press; b by María Schefter, ©María Schefter.)

stages, juvenile and adults of fleshy species such as *Lobophora* and *Dictyota* are seen (Diaz-Pulido and McCook 2002). Turfs are highly productive due to their fast growth and rapid turnover rate (4–12 days), but they are also heavily grazed by fish and urchins so that their standing stocks are low; their growth rate is just fast enough to escape deleterious herbivore grazing (Adey and Goertemiller 1987; Klumpp and McKinnon 1989; Williams and Carpenter 1990). They are present in all regions of the reef, prefer low nutrient conditions and tend to overgrow if herbivory levels are reduced (e.g. Kopp *et al.* 2010).

Foliose seaweeds include the brown genera *Lobophora* and *Dictyota*, green seaweeds from the families Caulerpaceae and Udoteaceae, and red seaweed from the genera *Gracilaria*, *Laurencia*, *Asparagopsis*, and *Halymenia* (Fong and Paul 2011). These are common on reefs although they are not always apparent as they can be tucked in among the corals. Visually apparent are the tropical fucoids, which are leathery, and include *Sargassum*, *Turbinaria*, and *Cystoseira*. These are able to tolerate wave action and are thus found on the fore-reef, and also in some regions of the reef flat. With anthropogenically induced changes to coral reefs, some fleshy and leathery seaweeds are becoming

more apparent, and they compete with corals for primary substratum (Birrell *et al.* 2008; Fong and Paul 2011; Fig. 3.8; Essay 6, Chapter 9).

Seagrass beds grow in soft sediment environments in both temperate and tropical regions but the highest species diversity is in the tropics (reviewed by Hemminga and Duarte 2000; Williams and Heck 2001; Mejia *et al.* 2012). Within seagrass beds, some rhizophytic algae e.g. *Caulerpa* and *Halimeda* grow attached into the sediments, whereas seagrasses have roots that allow them to strongly anchor to the sediment even in areas of relatively high water motion; they have an important role in sediment stabilization via roots and attenuating water flow (Williams and Heck 2001). Epiphytic seaweeds, including corallines and filamentous species, are an important component of seagrass beds and are responsible for up to 40% of net primary production, and also 40% of the aboveground biomass (see Mejia *et al.* 2012). They also increase the structural heterogeneity of the community, and increase space available for mobile and sessile invertebrates. Because they have a higher nitrogen content (including the N of the invertebrates that grow on them) epiphytic seaweeds are preferred by grazers compared to the underlying seagrass (Mejia *et al.*

2012). Small epiphytic coralline seaweeds benefit from their association with seagrasses: seagrass photosynthesis causes the local pH of seawater to rise, and this enhances the rate of calcification of the corallines (Semesi *et al.* 2009). Detached algae can get trapped in seagrass beds and may grow vegetatively, although excessive drift algae can have a negative effect on the seagrass (Mejia, *et al.* 2012).

3.3.2   Temperate

The temperate regions of the northern and southern hemispheres are characterized by progressive seasonal cycles of light and temperature that result in the water column becoming seasonally stratified. During summer, surface waters are warm and nutrient deplete, while in winter they are well-mixed and nutrient replete, but cool. Intertidal and subtidal regions are often dominated by large, canopy-forming, brown seaweeds of the Orders Fucales and/or Laminariales which are foundation species, forming the structural elements of the system (see Figs. 3.6 and 3.10 and sec. 4.1). While species differ between biogeographic regions (Lüning 1990), they all have similar functional roles in habitat creation and the supply of energy to higher trophic levels. These canopies modify the local environment substantially, by both reducing light and increasing the heterogeneity of the underwater light environment via the generation of light flecks (Wing and Patterson 1993). They also reduce seawater velocity, creating a relatively quiescent environment in which seaweed and invertebrate propagules can settle and recruit (Reed *et al.* 2006). Space is a primary resource in the infralittoral, and seaweeds compete with invertebrates for rock substratum (sec. 3.2), but they also provide a home for mobile and colonial invertebrates, both on their blades and within their holdfasts (Witman and Dayton 2001; Christie *et al.* 2009). The grazers of temperate systems include sea urchins and mesograzers such as amphipods (sec. 4.3). Many brown seaweeds have chemical defenses against grazing (sec. 4.4.2) and rather than being eaten directly, their tissue enters the detrital food web (see Fig. 5.30). The dominant seaweeds and their ecological roles are reviewed by Witman and Dayton (2001);

Figure 3.10 *Macrocystis pyrifera* fronds form a complex underwater forest that provides a habitat for many other organisms and greatly affects the physico-chemical conditions within. Here, the heterogeneity of underwater light is evident (Photo courtesy of John Pearse, © 1983, John Pearse.)

Steneck *et al.* (2002); Connell (2007a); Bartsch *et al.* (2008); Graham *et al.* (2007a); Huovinen and Gómez (2012); Flores-Moya (2012).

Members of the Order Laminariales are found in temperate regions of all continents. The two tallest subtidal species, *Macrocystis pyrifera* and *Nereocystis leutkeana*, co-habit in the NE Pacific (Steneck *et al.* 2002; Graham *et al.* 2007a; Springer *et al.* 2010). They are typically >15 m although *Macrocystis* has a more complex morphology with many fronds creating a multifarious habitat throughout the water column (Fig. 3.10) whereas *Nereocystis* has a single, thin stipe and blades lie on the water surface (Figs. 2.10 and

8.13). *Macrocystis* has been particularly well studied, and has a widespread distribution in the northern and southern hemispheres. This genus did consist of four, or five, species, but it is most likely just one species that has considerable phenotypic, reproductive and physiological plasticity (Graham *et al.* 2007a). Other well-studied genera are *Laminaria* and *Saccharina* (many species of the genus *Laminaria* are now treated as *Saccharina*; Lane *et al.* 2006). These form smaller canopies in temperate coasts worldwide, except for the west coast of South America, Australia, and New Zealand (Lüning 1990; Huovinen and Gomez 2012). On the NE coast of the Pacific there is a rich diversity of Laminariales, whereas New Zealand and Australia have only a few Laminarians but a high diversity of Fucales, including *Durvillaea* spp. which are also found in Chile (Huovinen and Gomez 2012). Other kelp genera that form smaller canopies include *Ecklonia, Lessonia, Pterogophora,* and *Undaria*.

Within the kelp canopies are other layers of seaweeds, that form progressively smaller canopies. Understory seaweeds include a high diversity of bladed e.g. *Rhodymenia* sp. and *Callophyllis* sp. and filamentous reds e.g. *Polysiphonia, Ceramium,* and each layer further attenuates light and water velocity. The rock surface is often dominated by crustose and erect corallines (Nelson 2009). The conditions inhabited by understory and rock-surface seaweeds is therefore so highly modified by the layers of canopy above, that environmental conditions there (i.e. light, nutrients, and water motion) are very different from those of the upper water column.

When discussing intertidal zonation (sec. 3.1) we focused on the Fucales of the North Atlantic, which have been extensively studied, but in many temperate regions it is red seaweeds that dominate particular vertical zones. *Chondrus crispus* and *Mastocarpus stellatus* occupy the mid- and low-intertidal of the NE Atlantic (Dudgeon *et al.* 1990), and in southeastern New Zealand the mid- and high intertidal zones are dominated by *Bostrychia arbuscula* (previously *Stictosiphonia arbuscula*), *Apophlaea lyallii* and a range of *Pyropia* species, with only one common high shore brown seaweed, *Scytothamnus australis* (Adams 1994). In the NE Pacific, where there is a high diversity

of Laminariales, the mid- and low-intertidal is dominated by *Saccharina sessilis* (previously *Hedophyllum sessile*), *Egresia menziesii*, and *Alaria marginata* (Druehl 2001).

3.3.3 Polar

Polar regions are found from around 60° N and S to the poles, and the most northerly and southerly limits of benthic algae are 80°N and 77°S. The abiotic environment of the north and south poles is reviewed by Zacher *et al.* (2009). Annual solar irradiance is up to 50% lower than that of temperate and tropical regions, and some seaweeds may withstand 8 months of darkness (see sec. 5.2.2). At the lower latitudes of 60°N and S, there are day–night cycles with 5 h of day in winter and 20 h in summer. The light available subtidally depends on the thickness of sea ice, which forms in fall and winter and melts in spring. Under the ice, irradiance is reduced to <2% of incident and PFDs of 6.5 µmoles m^{-2} s^{-1} have been measured below 1 m of ice in the Arctic in June. When the ice melts, the underwater light environment changes, with 600 µm photons m^{-2} s^{-1} being recorded at 4 m. Melting ice also causes a decline in surface salinity which can drop from 34.5 S_A to as low as 27 S_A, although in deeper waters (15 m) there is little variation in either temperature or salinity. Surface seawater temperature is generally low ranging from -1.8 to +2.2°C off the Antarctic Peninsula, and up to 6.5°C in the Arctic. In the Arctic, seawater nutrients follow a seasonal pattern similar to temperate regions, but in the Antarctic, inorganic nitrogen is available in high concentrations (14–33 µM) year round (Zacher *et al.* 2009). Polar seaweed communities are also shaped by physical disturbance by ice scouring and iceberg stranding. The adaptations of seaweeds to the unique environmental conditions of polar regions are discussed in Chapters 5 and 7. A special issue of *Botanica Marina* was dedicated to polar algae (Wiencke and Clayton 2009) and readers are directed to the 13 papers therein, and also texts of Lüning (1990), Wilce (1990) and Thomas *et al.* (2008).

The contrasting cold-water histories as well as prevailing hydrographic conditions have resulted in the formation of distinct seaweed floras in the polar regions of both hemispheres. The geological history of Antarctica started some 90–130 million years ago with the separation of the Gondwana landmass and its southward migration. The positioning of a giant landmass at the southern Pole, in line with pronounced seasonality and low solar inclination, resulted in particularly low temperatures. However, the distinct seaweed flora of Antarctica would not have been formed without the additional peculiarities in hydrography of the surrounding waters: The Antarctic Circumpolar Current (ACC) biogeographically isolated the water masses surrounding Antarctica from oceanic waters and contributed to their very stable temperature regime, with sea surface temperatures hardly exceeding 2°C even under summer conditions. The cooling of water masses around Antarctica started about 15 million years ago. Both the permanent low-temperature exposure and the isolation by the ACC, have resulted in a high degree of endemism, and an almost impassable barrier for the arrival of cold-temperate species from the surrounding sub-Antarctic region (Huovinen and Gomez 2012).

For the Antarctic, 33% of the recorded seaweed species are endemic. The greatest diversity is found on the Antarctic Peninsula, which extends further north than other regions of the Antarctic and the higher diversity is thought to correlate with the higher photon flux densities (Wulff *et al.* 2009). Intertidal seaweed communities are few and restricted to sheltered habitats, including crevices. These are dominated by *Pyropia endiviifolia* (previously *Porphyra endiviifolia*), *Prasiola crispa*, and *Ulorspora pennicilliformis*, while the green seaweeds *Ulothrix, Enteromorpha bulbosa*, and *Acrosiphonia* dominate in tide pools. Subtidally, however, there are extensive macroalgal assemblages dominated by large, canopy-forming, brown seaweeds: *Desmarestia anceps* down to 10 m, *D. menziesii* from 15–20 m, and *Himanothallus grandifolius* from 35–40 m. These brown seaweeds are functionally similar to the Laminarian kelps of the Arctic and temperate coasts: the Antarctic is the only continent with no members of the Laminariales. The non-acidic Desmerestiales are thought to have arisen in the Antarctic, and radiated northwards (Peters *et al.* 1997). The fucoid *Durvillaea antarctica* occurs here as an

annual (it is perennial in New Zealand and Chile). A rich understory community of green, red, and brown seaweeds, are also zoned subtidally. Biomass and diversity decrease closer to the South Pole, and the Ross Sea is the southern-most region where seaweeds grow, with just 17 species (Wulff *et al.* 2009).

In sharp contrast to the Antarctic, Polar regions of the northern hemisphere are geologically young, display less stable environmental conditions, and have a low degree of endemism. The onset of cooling of the northern Polar regions started about 2.7 million years ago. The Arctic regions as we know them nowadays have been formed progressively with the retreat of the ice cover after the last glacial maximum about 12 000 years ago. Due to ice coverage, and, thus, light exclusion, no habitat was available for seaweeds at that time. Subsequently, with the retreating ice, organisms encountered new habitats, and migrated northwards. The marine flora of the Arctic is thus regarded as a depleted descendent of the cold-temperate flora.

Arctic seaweeds originate from both the north Atlantic and Pacific coasts (Wulff *et al.* 2009; Hop *et al.* 2012). On the Norwegian Island, Svalbard, there are 70 species reported and many of these are familiar European genera, e.g. intertidal, *Fucus evanescens, Ulothrix*, and *Urospora* and on wave-exposed sites, *Chordaira flagelliformis*. Subtidally, species diversity increases and is dominated by the Laminariales: *Saccharina latissima* (*L. saccharina*) *Alaria esculenta*, and *Laminara digitata*, with a rich understory of foliose and calcareous red seaweeds whose distribution extends to the deepest subtidal zones. A similar but impoverished flora exists on the Russian Arctic and Greenland coasts. The Canadian high Arctic has many species in common with Svalbard, but also species that are of Pacific origin.

3.3.4 Tide pools

Tide-pool communities are difficult to characterize because of the intrinsic variety in pool conditions but conditions in most differ from both the adjacent exposed intertidal and subtidal areas. Physical conditions within a pool will differ from those of the adjacent seawater depending on the size of the pool (especially surface: volume ratio) and its height on the shore (thus duration of exposure), as well as on the extrinsic factors of atmospheric conditions (e.g. Zhuang 2006; Martins *et al*. 2007).

High shore pools can be very stressful because of rapid changes in temperature, salinity, pH, nutrients, and oxygen concentration (see secs. 6.8.5, 7.2.3). They tend to be populated by single tolerant species such as the greens *Chaetomorpha aerea, Ulva intestinalis*, the browns *Ralfsia verrucosa*, and reds e.g. *Hildenbrandia* (Wolfe and Harlin 1988a, b; Kooistra *et al*. 1989; Metaxas *et al*. 1994; Kain 2008). Communities in mid- and low-shore pools, where physical conditions are progressively less harsh, are more diverse, resembling the communities of the low littoral and sublittoral fringe. As on the exposed rocks, there are seasonal changes in the flora, which are nonetheless character- istic for each pool (Dethier 1982; Wolfe and Harlin 1988a; Metaxas *et al*. 1994). In deep pools there may be zonation within the pool (Kooistra *et al*. 1989); community composition may depend on herbivory, disturbance, nutrient availability, or the presence of allelopathic interactions (van Tamelen 1996; Master- son *et al*. 2008; Atalah and Crowe 2010). The upper zones in deep pools are affected by high temperatures, a phenomenon related to aerial exposure, whereas the lower zones are more strongly regulated by irradiance, as in the sublittoral. Competitive interactions are found in all zones within a pool (sec. 6.8.5).

3.3.5 Estuaries and salt marshes

Estuaries and salt marshes are characterized by hori- zontal gradients of salinity (Raffaelli and Hawkins 1996; Kaiser *et al*. 2011). Estuaries occur where a river flows into the sea, they are typically tidal and charac- terized by a "salt-wedge" in which the low-density freshwater floats on top of the denser seawater. At the head of the estuary where the freshwater flows in, the water column profile is that of freshwater, and at the mouth of the estuary, the profile is that of seawater. In the mid-regions of an estuary, there is a surface freshwater layer, a mixing layer where seawater is entrained by the flowing freshwater, and beneath this is seawater. For a seaweed that is located at the

mid-point of an estuary, therefore, the rise and fall of the tide means that it will experience freshwater, sea- water and the range of salinities inbetween on a daily basis. Temperature differences are also associated with river water versus seawater, and with tidal action that results in periodic exposure to the atmosphere (e.g. Mathieson and Penniman 1991). Estuaries are typically turbid, with strong attenuation of light, because of the high sediment loading and mixing. The substratum is often soft sediment, and dominated by benthic micro- algae, and some filamentous or branching seaweeds that grow detached, or can withstand burial within the sediments, for example unattached varieties of *Fucus* and *Ascophyllum* (Mathieson *et al*. 2001) and *Graci- laria vermiculophylla* (Abreu *et al*. 2011). Estuarine communities typically have the richest flora towards the mouth of the estuary, and they become progres- sively poorer with more stress-tolerant species towards the head. They are ideal habitats in which to study stress associated with salinity and pH fluctuations (Chapter 7).

Salt marshes are distributed worldwide in intertidal regions where water motion is sufficiently slow that sediments can accumulate. These very productive wet- land communities are dominated by terrestrial plants that are able to tolerate waterlogged, anoxic soils and regular variations in salinity. The dominant plants vary with geographic location, and best known are *Spartina* spp., *Juncus* spp., and *Salicornia*. They form a barrier between the terrestrial and marine environments, pro- viding coastal protection and other important eco- logical roles such as filtering sediments and nutrients from the water column, and supporting fisheries (Pen- nings and Bertness 2001; Valiela and Cole 2002). Simi- lar to estuaries, the substrate is composed of sediment, which can be sand, mud, or silt depending on the salt marsh. The seaweed floras of salt marshes are much less diverse than other intertidal communities, but include some very physiologically tolerant species such as *Bostrichia*.

A particularly interesting algal component of salt marshes in the northern Atlantic are the dwarf and unattached *Fucus* species which have generated considerable interest since early work in the UK and Ireland (Cotton 1912; Baker and Bohling 1916).

Salt-marsh *Fucus* are typically smaller than their rocky intertidal counterparts, tend to reproduce vegetatively, may have lost the ability for sexual reproduction, and have extensive proliferation of their thalli. Many of these species/varieties/ecads have no holdfast: *F. cottonii*, "Dwarf *Fucus*", *F. vesiculosus* ecad *volubilis*, *F. veisculosus* f. *mytili*, *F. spiralis* ecad *lutarius*, while holdfasts are present on *F. vesiculosus* var. *spiralis*, *F. vesiculosus* f. *gracillimus*. Unattached *Fucus* are placed in three categories depending on their lifestyle: (1) loose lying on the sediment surface; (2) entangled forms whose thalli grow in a spiraling pattern that facilitates them tangling in salt-marsh plants such as *Spartina* stems, thereby weakly attaching themselves to surrounding vegetation; (3) embedded forms for which part of the thallus is embedded in the sediment (Norton and Mathieson 1983). In Maine, USA, the various salt-marsh *Fucus* ecads arise from both morphological transformations when an attached *Fucus* becomes detached and from hybridization between different *Fucus* species (Mathieson *et al.* 2006).

3.3.6 Deep-water seaweeds

While deep-water seaweeds communities (>30 m) are well known for clear tropical waters (sec. 5.2.2), the full range of their extent and diversity in temperate waters was not realized until the late 1990s. Technological developments, including extended dive times at depth using enriched-air Nitrox SCUBA or closed-circuit rebreathers, and Remotely Operated Vehicles (ROVs) equipped with video cameras, have enabled quantitative studies (Spalding *et al.* 2003; Leichter *et al.* 2008). One of the first such surveys was in central California, where Spalding *et al.* (2003) found communities that were "surprisingly abundant and diverse". They found three distinct communities that are zoned with depth, and they named each zone after the visually dominant components: The "Pleurophycus" zone is from 30–45 m, within which are *Desmarestia* species, coralline algae, and a few foliose red and green seaweeds. The "Maripelta" zone (40–55 m) is dominated by foliose red seaweeds, while the deepest 55–75 m zone "Nongeniculate coralline algae" had far fewer species and were dominated by patches of encrusting corallines.

While 93% of the species recorded are also found at <30 m, the study revealed new records for this area, e.g. the green seaweed *Palmophyllum umbracola* was found in all zones between 35 and 54 m. Most surprising were the lush, subtidal beds of *Pleurophycus gardneri*, because it had never been recorded in drift material around central California: currents are thought to wash the detached kelp drift to deeper water submarine canyons.

Manipulative experiments to assess interspecific competition, and seaweed rate processes (e.g. photosynthesis, sec. 5.7), previously conducted only in the intertidal and shallow subtidal, have now been extended to deep-water sites. On a tropical reef near Eilat (Israel), Brokovich *et al.* (2010) found that grazing by fish was lower at 50–65 m where just 20% of the algal turf was consumed compared to 40–60% consumption at 5–30 m. Growth and photosynthetic rates of seaweed turf were also lower at the deep site, but because the decline in grazing pressure with depth was greater than the decline in algal growth, then the seaweed standing stock was estimated to be higher at the deeper site. A study of the population dynamics of the Fucalean *Cystoseira zosteroides* at 54 m in the NW Mediterranean revealed very slow growth rates of 0.5 cm per year, and an average mortality rate of only <2% (Ballesteros *et al.* 2009). Further such experiments will provide insights into the productivity of deep-water seaweeds and the assessment of factors controlling the lowest limits of seaweed distributions (see sec. 5.2.2).

3.3.7 Floating seaweeds

The Sargasso Sea is famous for its floating populations of *Sargassum*, but other seaweeds that become detached from the shore during storms also raft and some can travel long distances (reviewed by Thiel and Gutow 2004, 2005; Rothäusler *et al.* 2012). *Durvillaea antarctica*, for example, is thought to have rafted from New Zealand to Chile, where it established (Fraser *et al.* 2009). Rafting seaweeds can also connect communities that are otherwise geographically isolated, such as *Fucus vesiculosus*, *Macrocystis pyrifera*, and *Durvillaea antarctica* (Muhlin *et al.* 2008; Hinojosa *et al.* 2010; Collins *et al.* 2010). For *Macrocystis*, dispersal by floating rafts may cause the continuous recolonization of remote sites and explain the relatively low genetic diversity between populations (Macaya and Zuccarello 2010). Most rafting seaweeds are Fucales that have buoyant blades, including *Fucus* and *Ascophyllum* that have pneumatocysts and *Durvillaea antarctica* that has a uniquely inflated medulla (Rothäusler *et al.* 2012). Some red and green seaweeds also raft but they do not have buoyant structures so dispersal is more limited. The movement of rafts is dependent on tides and currents, and they are often found concentrated in convergence zones of coasts or the open ocean (Hinojosa *et al.* 2010; Rothäusler *et al.* 2012). A striking example is of a rafting *Ulva* bloom in the Yellow Sea, China, in 2008. Liu *et al.* (2009) used satellite images to show how the world's largest seaweed bloom happened to occur. The bloom formed over a 3-month period, from a number of *Ulva* rafts that had travelled hundreds of kilometers from the source population. Hydrographic conditions resulted in these large rafts converging offshore to form a massive raft that was 1200 km^2, and this was deposited in Qingdao Harbour; inconveniently, the raft's arrival coincided with the sailing events of the 2008 Olympic Games (see also sec. 10.10).

Rafting seaweeds do not travel alone. Thiel and Gutow (2004) report that up to 1200 species travel by rafts, including cyanobacteria, algae, protists, and invertebrates. The invertebrates that are associated with rafts are adapted physiologically and behaviorally to a rafting lifestyle. In a survey of seaweed rafts in the Southern Ocean, it was estimated that around 70 million *Durvillaea antarctica* rafts were drifting at any one time, of which 25% had holdfasts that contained a wide diversity of invertebrates (Smith 2002). The growth rate of floating seaweeds is affected by biotic and abiotic factors (Rothäusler *et al.* 2009, 2012). *Macrocystis* rafts grew fastest at low temperatures (<15°C) than at medium (15–20°C) and high (20°C) ones, because grazing by associated amphipods was lower at the lowest temperate. In the high temperature treatment, raft degradation was greatest, and such degradation could prevent rafts crossing the tropics (Rothäusler *et al.* 2009).

3.3.8 Other seaweed habitats and communities

Virtually any substratum in salt or brackish water is habitat for seaweeds. Rocks, wood, sand, glass, shell, and other non-living substrata are most typical, along with the surfaces of other algae and other living submerged plants and the shells of living mollusks. Endolithic algae are important in reef processes (Littler and Littler 2011d). Some remarkable seaweed habitats have been reported: the tissues of sea pens, where some *Desmarestia* gametophytes grow (Dube and Ball 1971); the beaks of parrotfish, which provide a "moving reef" for pioneer species such as *Polysiphonia scopulorum* and *Sphacelaria tribuloides* (Tsuda *et al.* 1972); a symbiotic association with a sponge (Price *et al.* 1984); the face and belly of the Hawaiian monk seal (Kenyon and Rice 1959); and the necks of green turtles (Tsuda 1965). Kelp gametophytes grow endophytically within the cell walls of filamentous and bladed red seaweeds (Garbary *et al.* 1999; sec. 4.5.2) and as epibionts (sec. 4.2.2).

3.4 Invasive seaweeds

The number of marine organisms introduced into new environments is growing, due mostly to human activities of aquaculture, the aquarium trade, and shipping vectors (Fig. 3.11). There are concerns over the ecological impacts of invasive seaweeds on the native benthic communities of coral reefs, seagrass beds, and temperate communities, and this topic has received considerable attention since the mid-1990s. See the special issue of *Botanica Marina* (Johnson and Chapman 2007) and reviews by Inderjit *et al.* 2006; Schaffelke *et al.* 2006; Williams 2007; Williams and Smith 2007; Andreakis and Schaffelke 2012. At least 277 seaweed species have been involved in 408 introductions, and are represented by most orders of the red, brown, and green seaweeds (Fig. 3.11). Some introduced species (also termed adventive or non-indigenous marine species (NIMS)) maintain low numbers in their new environment, but others become invasive, with the potential to outcompete the native

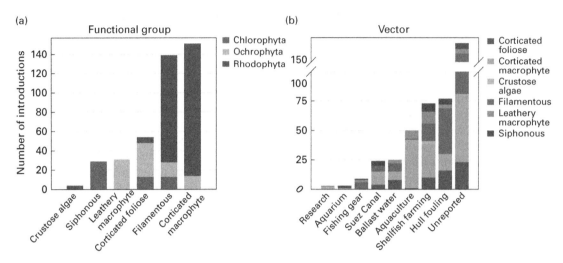

Figure 3.11 Invasive seaweeds. The number of seaweed introductions that (a) fall into different functional groups (Steneck and Dethier 1994) and phyla and (b) are accounted for by different vectors or modes of introduction. (From Williams and Smith 2007. © 2007 by Annual Reviews.)

flora. The ecological consequence of seaweed invasions is in restructuring nearshore food webs. There are numerous examples of invasive seaweeds, but some species stand out as being particularly aggressive "high-profile" invaders. The "top five" invaders based on their abilities to disperse and establish, and on their ecological impact, are *Codium fragile* ssp. *fragile* (=*tomentosoides*) originally from the North Pacific which now has a worldwide distribution, the laminarian kelp *Undaria pinnatifida*, originally from Korea, also with a global distribution, *Caulerpa taxifolia*, *Aspargopsis armata*, and *Grateloupia doryphora* (Nyberg and Wallentinus 2005).

Following its arrival at a new location, an introduced species must establish itself. There are three key features of establishment listed by Schaffelke *et al.* (2006). First, the rate of infection: a critical number of invaders is required to start a population, and a lag phase in population growth is typically observed. Second, the biotic and abiotic characteristics of the environment must both be suitable. For example, the Canadian coasts of the northwest Atlantic has well-established populations of *Codium fragile* but it has not established on the UK coast of the northeast Atlantic, despite their similar abiotic conditions. This difference is explained by differences in grazing

pressure, grazer identity and diversity between the northeast and northwest Atlantic (Chapman 1999). An ability to rapidly colonize disturbed areas was considered a key factor in the establishment of *Undaria pinnatifida* in Tasmania (Valentine and Johnson 2004). Communities that have a higher species diversity are thought to be more resistant to invasions than low-diversity habitats (biotic-resistance hypothesis, e.g. Bulleri and Benedetti-Cecchi 2008). Third, the ecophysiological characteristics of the arriving NIMS must match those available in the new environment. Morphological and physiological plasticity seem key traits of invasive seaweeds. For example, *Caulerpa prolifera* actively "forages" for nitrogen-rich areas and exhibits considerable morphological plasticity depending in its nitrogen and light environment (Collado-Vides 2002b; Malta *et al.* 2005). Many invasive species have high growth rates and therefore they can easily become dominant in eutrophic conditions. They are also tolerant to a range of environmental stresses such as temperature (Gacia *et al.* 1996), desiccation (Bégin and Scheibling 2003; Schaffelke and Deane 2005), or wave action (Russell *et al.* 2008).

Once established in a new area, the invader spreads, expanding its geographic and habitat range (Lyons and

Scheibling 2009; Sorte *et al.* 2010). A key attribute for successful dispersal is a range of reproductive modes which facilitate both short- and long-distance dispersal. *Codium fragile* ssp. *fragile* in the northwest Atlantic off Canada spreads short distances via parthenogenic release of swarmers which quickly (within minutes to hours) settle close to the adult population. They also reproduce vegetatively via fragmentation of the thallus, by producing "buds", or whole adults can be dislodged, and these vectors then travel from meters to kilometers, and are capable of attaching to a new substrate (Watanabe *et al.* 2009). *Undaria pinnatifida* sporophytes release massive numbers of spores which are viable for 5–14 days which enable short-range (<10 m) dispersal whereas long-range dispersal is most likely from detached sporophytes that float with a current. Although *Undaria* sporophytes cannot reattach as can *Codium*, they can rest on the shore and seabed, and release spores (Forrest *et al.* 2000). *Caulerpa taxifolia* is a well-known invader of the Mediterranean Sea and Californian coast that can spread via fragmentation and via its association with benthic filamentous algal mats, which can float (Smith and Walters 1999; Chisholm *et al.* 2000). Once established, *Caulerpa taxifolia* modifies its local environment by trapping silt, which favors the formation of an algal turf instead of the canopy of bladed seaweeds that grew prior to the invasion (Bulleri *et al.* 2010).

Invasive seaweeds are well established on coral reefs too, but there have been far fewer ecological impact studies compared to temperate regions (Williams and Smith 2007). The Hawaiian islands, for example, have had 21 seaweed introductions and several highly aggressive species have overgrown reefs and severely degraded the system. Key invaders are the red seaweeds *Acanthophora spicifera*, *Gracilaria salicornia*, *Eucheuma* spp., and *Kappaphycus* spp. Some of these were intentionally introduced to establish a phycocolloid aquaculture industry, but they now cover up to 50% of corals in some reefs in Kane'ohe Bay (Conklin and Smith 2005).

The ecological impacts of seaweed invasions are not well known, for example, whether the diversity of the invaded ecosystem increases or declines, or if the food web becomes re-structured. Feeding experiments, in which a range of seaweeds are fed to grazers (sec.

4.3.2), can help identify if invaders are being consumed. For example, the invasive *Sargassum muticum* is the least preferred of 13 seaweed species tested in multiple-choice feeding experiments, and a low grazing pressure was thought to add to its successful invasion of the Portuguese coast (Monteiro *et al.* 2009). Invasive species that are structurally or chemically defended may not be grazed directly but instead contribute to the detrital food webs. After a while, some invasive seaweeds may become a dominant component of an ecosystem, and a new community develops. In the Isle of Shoals, Maine, USA, *Codium* established in 1983, and community surveys over the next 20 years have revealed that the *Codium*-based ecosystem is becoming increasingly complex, with animals and other algae adapting to the new environmental conditions (Harris and Jones 2005).

Being able to detect invasions, and track the progression of establishment and spread, are important tools in ecological impact assessment (Meinsesz 2007). Detecting shifts in seaweed community structure is greatly aided if there are good historical records of the seaweed community (e.g. Mathieson *et al.* 2008; Bates *et al.* 2009). In regions where the seaweed flora is poorly characterized, detecting introduced seaweeds can be more difficult particularly for cryptic and uncommon species, although this problem has been alleviated to some extent by the use of molecular tools that can help determine the strain and origin of a species (Booth *et al.* 2007; Mathieson *et al.* 2008; Andreakis and Schaffelke 2012). How might a manager of a coastal ecosystem predict whether an introduced species will become invasive? What are the characteristic traits of successful invaders, and why do some introduced species become invasive while others do not? Nyberg and Wallentinus (2005) categorized species traits of 113 seaweeds introduced to Europe (of which 26 were known to be invasive) into "dispersal", "establishment", and "ecological impact" categories and they ranked each species' probability of becoming invasive. This technique worked well, predicting 15 of the 26 invasive species in the "top 20" known invaders.

Early detection and eradication is key in removing newly established invasive seaweed because once they have spread, eradication becomes much more difficult and there are very few success stories. In California, USA, *Caulerpa taxifolia* was eradicated by hand collection and bleaching the rock to remove gametes and spores (Williams and Schroeder 2004). The "super-sucker", essentially an underwater vacuum cleaner, was developed to suck invasive seaweeds off Hawaiian coral reefs, although in this case it is used to control spread rather than for eradication (Smith *et al.* 2008). However, re-introductions via the same original vector can be problematic and management tools have to include measures to prevent this (Meinesz 2007).

The concerns over the impacts of seaweed invasions on native benthic communities, and of the effects of other anthropogenic stressors such as UV radiation (sec. 7.6), ocean acidification (sec. 7.7), sedimentation (sec. 8.3.2) and pollution (Chapter 9), highlight the importance of knowing what species are present in a native community, their functional role within the community, and their population dynamics. Such an understanding requires the temporal and spatial analysis of seaweed communities, to which we now turn our attention.

3.5   Community analysis

3.5.1   Vegetation analysis

Zonation and community descriptions of the types given in sec. 3.3 provide only a superficial view of the habitat of the most conspicuous and dominant organisms. To properly understand population and community structure and the controlling forces, quantitative data and methods of vegetation analysis are needed. As Schiel and Foster (1986, p. 286) noted, "the structure of a community is essentially a numbers game". The abundance of an organism may be measured in several ways. The percentage of the substrate covered by the species or association is an appropriate measure where the vegetation is dense and uniform, as, for example, with algal turf or encrusting species. The number of individuals may be counted if they are clearly separate, but this may not be appropriate for some clonal-modular seaweeds for which ramets can

spread over several meters (see sec. 1.2.3). The biomass (usually dry weight) or energy content of the population may be measured. Each method can be difficult or impossible with some species. For example, one cannot count the numbers of individuals in a turf (or even in an *Ascophyllum* bed), because each of the overlapping seaweeds sends up numerous erect shoots; on the other hand, it is difficult to determine the biomass of encrusting species or to ascertain the percentage cover of widely separated individuals. Biomass estimates are useful for energy studies, but are of no use for population biology or demography (Schiel and Foster 1986).

The basis for any quantitative study of an area is a sound sampling procedure (see texts by DeWreede 1985; Underwood 1997; Kingsford and Battershill 1998; Murray *et al.* 2006). The samples must be representative of the population as a whole and must, among other things, reflect the heterogeneity of the population. Samples are most commonly chosen by placing quadrat frames in the study area and employing some means for ensuring the randomness required for statistical analysis (e.g. random-number tables). The vegetation in the quadrat can be counted or photographed, which will allow assessment of the same area on subsequent dates, or it can be collected for one-time biomass and species determination. Achieving randomness in placing the quadrat can be difficult because the terrain usually is irregular and the vegetation patchy. "Targeted" sampling, where quadrats are placed in a position selected by an investigator, violates the underlying assumptions of independence of errors needed for ecological studies (Murray *et al.* 2006; Fig. 3.12). As an alternative to mathematically random sampling, sampling units can be placed "systematically" at fixed intervals along lines (transects) stretched horizontally or vertically across the shore, or in a grid (Russell and Fielding 1981; Murray *et al.* 2006; Fig. 3.12). Another practical consideration is the size of the quadrat. The quadrat size appropriate to the population or community to be studied and to the type of substrate can be assessed through preliminary work. Table 5.3 in Murray *et al.* (2006) has useful examples of quadrat sizes that have been used to estimate the abundance of particular species in intertidal studies, and

(a) Random

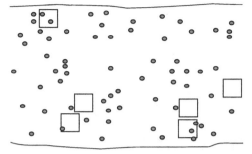

(b) Systematic

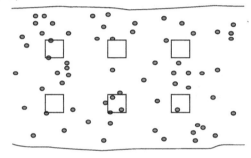

(c) Targeted

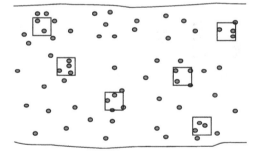

Figure 3.12 Schematic drawing illustrating (a) random, (b) systematic, and (c) targeted approaches to locating sampling units. Random quadrat locations were determined using a random-number generator; targeted quadrat locations were determined "by eye". In this example, the true density is 8.3 individuals m^{-2}; random is 9.3, systematic 11.1, and targeted 33.3 individuals m^{-2}. (From Murray *et al.* 2006, reproduced with permission.)

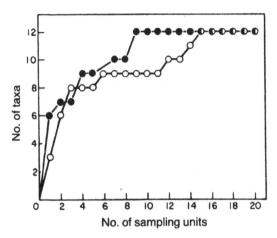

Figure 3.13 Effect of area sampled on number of taxa counted microscopically. Numbers of epiphytic blue-green algal taxa at two localities are related to the numbers of sampling units. The sampling unit was the first 1 cm of an average shoot of *Cladophora prolifera*; area about 2.5 mm^2. (From Wilmotte *et al.* 1988, with permission of *British Phycological Journal*.)

Kingsford and Battershill (1998) suggest a quadrat size of 1 m^2 for general subtidal surveys. In general, the number of taxa found in a quadrat will increase to a plateau as quadrat area increases (Fig. 3.13). Too small quadrats will underestimate species diversity, and too large quadrats will generate unnecessary work and reduce the number of replicates possible (Murray *et al.* 2006).

Seaweeds are not uniformly distributed over the areas they occupy; they are patchy at all scales (e.g. Chapman and Underwood 1998; Sousa 2001; Benedetti-Cecchi 2001; Coleman 2002; Fraschetti *et al.* 2005). Great variation between transects in a given area is also evident and sampling designs must take this into account (Murray *et al.* 2006; Connell and Irving 2009). However, the variance itself may also be an important driver of community structure (Benedetti-Cecchi *et al.* 2005; Burrows *et al.* 2009). Also, spatial and temporal changes in patterns are not independent of scale, that is the patterns measured on a small scale may not be apparent on a larger scale and vice versa (Benedetti-Cecchi 2001; Fraschetti *et al.* 2005; Fig. 3.14; Essay 2). Most studies on distributional patterns of shores have been at the local scale which provides detail specific to

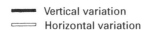

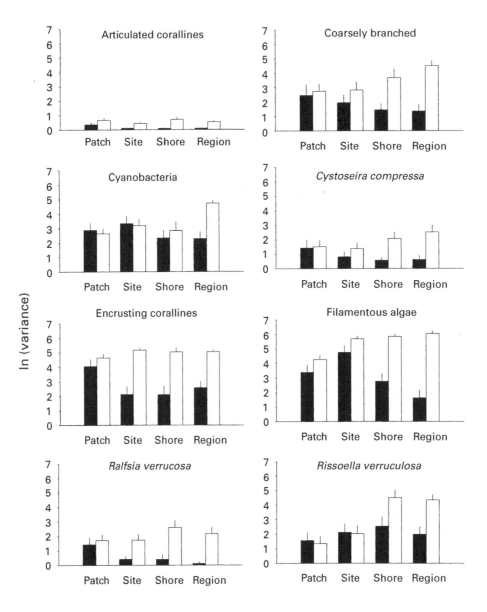

Spatial scale

Figure 3.14 Mean (+SE, = 18) values of vertical and horizontal variability in abundance (in log form) of algae at four spatial scales. (From Benedetti-Cecchi 2001. © Inter-Research 2001.)

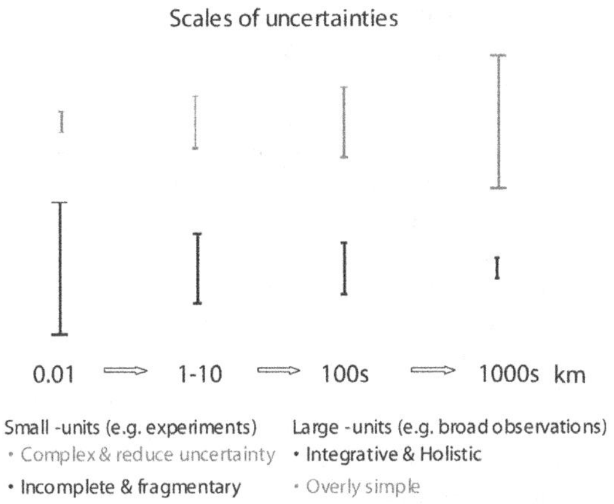

Figure 3.15 Conceptual diagram of the magnitude of uncertainty (size of bar) associated with spatial scale and methodology (small-scale experiments versus broad-scale observations). Broad-scale descriptions may oversimplify local patterns (but provide clear regional contexts for their interpretation) and local experiments may provide incomplete and uncertain outcomes for understanding generality (but provide unambiguous and precise information at local scales). A challenge is to integrate both approaches. (From Connell and Irving 2009, reproduced with permission.)

that site, whereas large-scale patterns over latitudinal ranges can provide a generalized, holistic view (Schoch *et al.* 2006; Blanchette *et al.* 2008). There are advantages and disadvantages to both approaches: large-scale ecological syntheses risk missing the important biotic and abiotic drivers in a local system, while studying small-scale patterns and processes may overlook landscape-scale changes (Paine 2010; Fig. 3.15). Nevertheless, both approaches are needed for a full understanding of coastal ecosystem functioning (Connell and Irving 2009).

Determining the diversity of species within seaweed communities has been a strong research focus because of the potential impacts of anthropogenic changes to coastal habitats, and the concern that communities will become less diverse and more homogeneous (Benedetti-Cecchi 2005; Worm *et al.* 2006). The effects of pollutants on species composition and diversity are discussed in Chapter 9. A major theme is the extent to which diversity and ecosystem function are linked, and whether a diverse community is more resilient to environmental change, or more productive, than a less diverse community (e.g. Arenas *et al.* 2006; Stachowicz *et al.* 2007; Boyer *et al.* 2009; Duffy 2009). "Diversity" is dependent on the richness (i.e. the number of species in a given area) and evenness (i.e. the relative abundance) of a species within a community (see Scrosati and Heaven 2008). Knowledge of both these factors in assessing the impacts of disturbance on species diversity is useful, because a disturbance (including an anthropogenic disturbance) may have no effect on species richness but could cause a major change in the relative abundance of each species. The various indices that are used to quantify diversity are discussed by Murray *et al.* (2006), and the use of these in assessing responses of communities to disturbance is discussed by Svensson *et al.* (2012; see sec. 8.3.2).

The term "biodiversity" is used rather loosely, but Coelho *et al.* (2010) break it down into four components: (1) Genetic diversity within a species, and this will govern the ability of each species within a community to respond (i.e. regulate, acclimate, and adapt) to an environmental change. (2) Species diversity. (3) Ecosystem diversity which is the range of communities within an area and their interactions with one another (e.g. recruitment, carbon subsidies), and includes non-living features such as substratum type. (4) Functional diversity, which is the range of biological processes, such as light harvesting and nutrient uptake, within an ecosystem. Knowledge of seaweed (bio)diversity can aid in the detection of long-term changes to communities (e.g. Steinbeck *et al.* 2005), including anthropogenically induced changes (Helmuth *et al.* 2006), such as ocean warming (Schiel *et al.* 2004; Halpern *et al.* 2008; Hawkins *et al.* 2009) and sedimentation (Balata *et al.* 2007). Diversity indices are also useful indicators of ecosystem health (Juanes *et al.* 2008), and can be used to help make decisions on where to establish marine protected areas (Palumbi 2001; Claudet *et al.* 2008) and on which seaweed species are rare and therefore require protection (Brodie *et al.* 2009). They also provide essential information for the restoration of seaweed communities in degraded habitats, for example, *Macrocystis* communities in southern Australia (Gorman and Connell 2009).

In some situations, broad-scale surveys of seaweed habitats are useful, and remote sensing offers a method that can be less labor intensive than the habitat surveys described above. Satellite and aerial images can be useful in detecting long-term (tens of years) changes of, for example, the extent of surface canopies of *Macrocystis* (Cavanaugh *et al.* 2010). Ocean color maps have been used for many years to assess global patterns of phytoplankton productivity, and a similar technique, airborne hyperspectral remote sensing, has been applied to underwater benthic communities in temperate rocky shores and tropical reefs. This technique relies on the detection of different colors emitted by different organisms, and should be able to distinguish between red, green, and brown seaweeds but in turbid coastal waters the signal may be lost (Kutser *et al.* 2006). Gagnon *et al.* (2008) integrated a hyperspectral approach with digital bathymetry and were able to detect *Codium*-only and *Codium*-kelp communities with 75% accuracy. Seasonal shifts between *Ulva* and *Ceramium* spp. as community dominants were detected in a eutrophic estuary in California, USA (Nezlin *et al.* 2007). The productivity of a habitat can be estimated by combining ocean color mapping with laboratory measurements of primary production (Dierssen *et al.* 2010). Remote methods can also estimate seaweed biomass with a view to harvesting, and tracking seaweed invasions (André-fouët *et al.* 2004).

Communities are not static: succession is a normal process in communities that must be taken into account in an assessment of community structure (Foster and Sousa 1985; Bruno and Bertness 2001; Sousa 2001). Successional stages probably rarely lead to a climax community; disturbances of many kinds clear space on small and large scales (sec. 8.3.2). A three-step research program for studying successional patterns (Table 3.1) allows one to evaluate the mechanisms of temporal change. Connell and Slatyer (1977) proposed three models of succession: (1) the facilitation model, in which earlier colonizers make the habitat more suitable for subsequent settlers and less suitable for themselves (sec. 4.1); (2) the tolerance model, in which later stages simply grow up more slowly through the earlier settlers, without any positive

Table 3.1 An approach to the investigations of successional patterns and mechanisms of species replacement.

Step 1: Observe the natural regime of disturbance to which an algal assemblage is subjected; make quantitative measurements if possible

Step 2: On the basis of these observations, design and conduct a multifactorial experiment to reveal the patterns of succession that occur under a variety of realistic regimes of disturbance that differ in

1. Intensity
2. Areal extent
3. Frequency of occurrence
4. Season of occurrence
5. Various combinations thereof

Step 3: Formulate and test specific hypotheses concerning mechanisms of successional-species replacements; this may involve studies of

1. Interspecific competition
2. The impact of grazing
3. The tolerance of species to physical stress

Source: Foster and Sousa (1985), with permission of Cambridge University Press.

effects of the earlier stage; and (3) the inhibitional model, in which early settlers prevent succession for example. These models are classical and highly cited, and Connell and Slatyer provided a cornerstone for testing and further developing ecological theory on community succession in the marine environment (Sousa 2001). Experimental research has provided support for each of these three models, and the predominant model that operates within a community varies with the location of the community (biogeographic and zonation), and can change as the community develops (see Sousa 2001).

3.5.2 Population dynamics

In seaweed vegetation analyses, there are relatively few studies of seaweed population dynamics, or demography, as distinct from biomass or productivity. Demography is the study of changes in the *numbers of organisms* in populations and of the factors influencing them (Russell and Fielding 1981). Methods for demographic study have been reviewed by Chapman (1985, 1986), Schiel and Foster (2006), and Murray *et al.* (2006). Seaweeds pass through several stages in their lives that span 4–5 orders of magnitude of size (see Chapter 2; and Fig. 1 of Schiel and Foster 2006). For a complete population-dynamics description, we need to know about propagule pool size, recruitment, germling and adult mortality, age at first reproduction, reproductive life span, proportion of individuals reproducing at a given time, fecundity and fecundity-age regression, and reproductive effort versus growth and predator defense (Chapman 1986). Demographic processes of terrestrial plants have been much better studied than algae, and there has been a tendency to apply terrestrial plant models to seaweeds. However, such applications are inappropriate because seaweeds have two life stages (gametophyte and sporophyte) that are exposed to, and affected by, the environment, whereas for terrestrial plants it is the dynamics of the diploid phase alone that is key; also spores are not equivalent to seeds (McConnico and Foster 2005; Schiel and Foster 2006). The development of demographic frameworks that are specific to seaweeds is important, because "understanding seaweed population biology critically underpins an ecological understanding of the communities they dominate" (Schiel and Foster 2006).

Obtaining demographic information for seaweeds is "enormously complicated" because they have complex life cycles and many inhabit wave-exposed and subtidal environments that are difficult to access (Schiel and Foster 2006). Directly aging seaweeds is problematic for most species, with a few exceptions: *Laminaria* species, *Pterogophyora californica*, *Ascophyllum nodosum*, and the red seaweed *Constantinea subulifera* (Klinger and DeWreede 1988; Cheshire and Hallam 1989; Murray *et al.* 2006; Bartsch *et al.* 2008). Tagging or marking individuals and following survivorship is a good method of assessing the age of seaweeds that have robust thalli e.g. the Laminariales and Fucales, but it is impossible for soft, fleshy, and filamentous algae (Chapman 1985; Murray *et al.* 2006). Acquiring accurate information on early life-history stages, and microscopic stages, in the field is

also difficult, although seaweeds with larger propagules e.g. fucoids have been better studied (e.g. Creed *et al.* 1996; McConnico and Foster 2005; Schiel and Foster 2006). Tools that can aid demographic studies include matrix models (Åberg 1992; Ang and deWreede 1993; Chapman 1993; Engelen and Santos 2009).

Nevertheless, there have been an increasing number of demographic studies, and a greater focus on the early life-history stages. Schiel and Foster (2006) synthesize most demographic research on the Fucales and Laminariales and other studies include: *Nereocystis luetkana* and *Costaria costata* (Maxell and Miller 1996), *Sargassum lapazeanum* (Rivera and Scrosati 2006), *Sargassum ilicifolium* and *Turbinaria triquetra* (Ateweberhan *et al.* 2005, 2006, 2009), and *Macrocystis* (Buschmann *et al.* 2006). There are far fewer studies on green seaweed demographics: e.g. *Codium bursa* (Vidondo and Duarte 1998) and *Halimeda incrassata* (van Tussenbroek and Barba Santos 2011). Red seaweeds are particularly difficult with their triphasic life histories and in some cases indistinguishable isomorphic generations. There have therefore been relatively few demographic studies for red seaweeds (Faes and Veijo 2003), with some notable exceptions e.g. *Mazzaella splendens* (Dyck and DeWreede 2006 a, b). In the next paragraphs, we highlight what is known about some of the key processes in seaweed population dynamics.

The "investment" a species makes in reproduction compared to vegetative growth can be assessed from its "reproductive effort" (RE), often estimated from the mass of the reproductive structures relative to the mass of vegetative material (see Åberg 1996). Reproductive effort will depend on whether a seaweed is semelparous (single reproductive event) or iteroparous (multiple reproductive events) (Brenchley *et al.* 1996). For iteroparous seaweeds, RE ranges from 4% for *Macrocystis* (Klinger and DeWreede 1988) to 12.7% for *Fucus distichus* (Ang 1992), and 40–50% for *F. serratus*, while at the other extreme, the RE of the semelparous *Himanthalia* is 98% (Brenchley *et al.* 1996). In terrestrial plants, resource-allocation theory suggests that there is a "cost" associated with reproduction, or a "trade-off" between vegetative growth and reproduction, but there have been few such studies on seaweeds (e.g. Åberg 1996) and in most cases research suggests few costs to reproduction (Santelices 1990). For example, *Fucus disctichus* allocates 12.7% of its dry weight to reproduction, but no "cost" in terms of mortality and longevity was observed (Ang 1992). Similarly, Ateweberhan *et al.* (2005, 2006) found cost of reproduction for *Turbinaria triquetra* or *Sargassum ilicifolium*. The reason may be that seaweed reproductive structures can photosynthesize and are not as physiologically specialized as terrestrial plants (Klinger and DeWreede 1988; reviewed in Dyck and DeWreede 2006b). However, for a *Macrocystis pyrifera* population in California, a trade-off between reproduction and growth was recorded. *Macrocystis* can release spores year round, but following a grazing event in which amphipods removed most of the kelp blades, reproduction stopped for 7 months, and resources were allocated to vegetative growth. This re-allocation of energy from reproduction to vegetative growth may be a mechanism to aid the recovery of the established sporophyte population (Graham 2002).

The dispersal distance of spores or zygotes that are broadcast into the water column is short, especially when compared to invertebrates and fish, because seaweed propagules have shorter life spans (reviewed Santelices 1990; Reed *et al.* 2006; Gaines *et al.* 2009). Dispersal distance will depend largely on local oceanic conditions, but also on the longevity of the propagule and its phototactic ability: the longer it lives, the greater the chance of long-distance dispersal, and if positively phototactic it will stay in the surface for longer (Gaylord *et al.* 2002; Reed *et al.* 2006; see sec. 2.5.1). Reed *et al.* (1988) studied variations in spore settlement by collecting spores on glass slides at different distances from parent seaweed stands and then culturing the collection for several days to allow settled spores to grow into recognizable gametophytes (of kelps) or germlings (of filamentous brown algae). They found that whereas kelp dispersal indeed occurred over a very short range most of the time, storms greatly increased dispersal. In contrast, *Ectocarpus* regularly showed long-range dispersal; its spores, unlike those of the kelps studied *(Macrocystis, Pterygophora)*, are phototactic and tend to remain in the water column

longer, rather than settling. In a further study, Reed *et al.* (2006) showed the effect of a *Macrocystis* bed on spore dispersal. From an individual that was growing in isolation, spore dispersal was around 15 m for 70% of those studied, but 5% of spores travelled 2000 m. However, when kelp were grouped together into a kelp bed, there was less variation in dispersal distances, with most spores travelling between 80–500 m. The effects of kelp beds on dispersal are likely to be complex, as the kelp themselves generate small eddies that might increase suspension, and bending of the kelps in current will change the dynamics of water exchange within and outside of the kelp bed (Gaylord *et al.* 2002; see Chapter 8).

Seaweeds release millions or even billions of spores or microscopic stages into the water column (Santelices 1990). For example, *Macrocystis* releases up to 800 spores mm^{-2} of sorus (Buschmann *et al.* 2004). The majority of spores released die either before they settle (pre-recruitment) or in post-settlement events: densities of adult populations do not reflect the enormous number of propagules released. Chapman (1984) estimated for *Saccharina longicruris* (previously *Laminaria longicruris*) that from a pool of 8.89×10^6 microscopic sporophytes m^{-2} y^{-1} just one macroscopic sporophyte would develop. For *Ascophyllum nodosum*, field experiments in which the fate of newly settled zygotes was followed for over 1 year revealed a 99.9% loss after 400 days, explaining why populations consist almost entirely of mature adults (Dudgeon and Petraitis 2005). Schiel and Foster (2006) compiled and replotted published data on survivorship of microscopic and macroscopic (>1 cm) phases of Laminarales and Fucales (Fig. 3.16). Mortality rates of all species are higher in the microscopic stage compared to the macroscopic stage, and survivorship of kelp spores was several (2–3) orders of magnitude lower than that of fucoids. Other examples of high juvenile mortality are *Halimeda incrassata* (van Tussenbroek and Barba Santos 2011) and *Palmaria palmata* (Faes and Viejo 2003). Reduction in numbers by inter- and intraspecific competition, herbivory, and abiotic agents such as desiccation or wave action is therefore an inevitable part of the pre-and post-recruitment period (Reed 1990a, b).

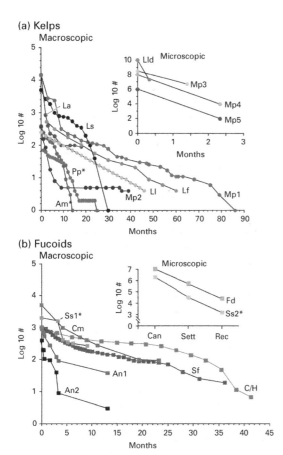

Figure 3.16 Survivorship curves for cohorts of kelps (a) and fucoids (b). Macroscopic cohorts began with juveniles 1–10 cm long for kelps and 1–3 cm for fucoids. Microscopic kelp cohorts began and ended with microscopic stages described below for each curve. Microscopic fucoid cohort abundance plotted against events rather than time: zygotes/embryos released at canopy height (Can), number settled (Sett), and number of recruits (Rec, 1–3 cm long). *denotes annual population. Kelps: Am, *Alaria marginata*; La, *Laminaria angustata* (currently *Saccharina angustata*); Lf, *L. farlowii*; Ll, *L. longicruris* (currently *Saccharina longicruris*); Lld, *L. longicruris* (currently *Saccharina longicruris*) and *L. digitata* (average, spores in water to spores settled); Ls, *L. saccharina* (currently *Saccharina latissima*); Mp1, *Macrocystis pyrifera*; Mp2, *M. pyrifera*; Mp3, *M. pyrifera* (settled spores to microscopic sporophytes); Mp4, *M. pyrifera* (high density of settled spores to visible sporophytes); Pp, *Pelagophycus porra*. Fucoids: An1, *Ascophyllum nodosum* (best survivorship); An2, *A. nodosum* (worst survivorship;); C/H, *Cystoseira*

Mortality is not always highest for the microscopic stages, however. In the annual *Leathesia marina* (previously *Leathesia difformis*) Chapman and Goudey (1983) found higher mortality among adults, due to increased crowding as they grew larger. Juveniles entered an environment essentially free of intraspecific competition. *Codium bursa* had high mortality for new recruits (<2 cm diameter) and a similar mortality rate in large (>12 cm), mature individuals, while intermediate sized seaweeds (3–12 cm) appeared to escape losses from grazing and physical disturbance, resulting in comparatively low mortality rates (Vidondo and Durarte 1998).

For some seaweeds, the microscopic stages that successfully recruit may serve as a "seed bank" (perhaps better termed "propagule bank") that is an available "pool" of microscopic stages that can be recruited into a population under certain environmental conditions and/or developmental cues (see Schiel and Foster 2006). For example, Creed *et al.* (1996) estimated that there were 37 000 individuals m^{-2} of *Fucus serratus* that were <1 mm tall growing beneath the mature canopy. In a wave-protected fjord in southern Chile, *Macrocystis* populations are unusual in that they have an annual life cycle. Here the sporophyte generation disappears in fall for 4–5 months, but it reappears in spring, most likely recruiting from a "gametophyte bank" (Buschmann *et al.* 2006). Another example is in California, USA, where *Macrocystis* gametophytes undergo "delayed development" making them available for recruitment whenever a window of suitable environmental conditions arises (Carney 2011). The identity of species within the propagule bank directly affects species' recruitment and thus competitive interactions in the early successional stages of community development. This in turn

---

osmundacea (currently *Stephanocystis osmundacea*) / *Halidrys dioica* (currently *Stephanocystis dioica*); Cm, *Carpophyllum maschalocarpum*; Fd, *Fucus distichus*; Sf, *Silvetia fastigiata* (= *Silvetia compressa*); Ss1 and Ss2, *Sargassum sinclairii*. (From Schiel and Foster 2006. © 2006 by *Annual Reviews*. See their figure legend for a list of the original studies from which the plotted data were obtained.)

influences the identity of consumers that establish, and hence the subsequent community structure (Worm *et al.* 2001).

The very high mortality rates typically seen for new recruits are caused by the interactions of many biotic and abiotic factors including grazing (sec. 4.3), facilitation (sec. 4.1), competition (sec. 4.2), temperature extremes (sec. 7.5), UV-R (sec. 7.6), and the intensity of wave action, the size of the cleared patch available for colonization, and on sedimentation (sec. 8.3.2). The relative importance of each of these factors in affecting the survivorship of recruits is site specific and will also depend on the time of year at which recruitment takes place. With such inherent complexity, predicting the succession of species and community that develops upon bare rock that has become available at a particular site, seems unfeasible. However, models, including functional groupings, such as the three classical ones described below (and see Connell and Slayter 1977, above), help researchers to frame questions so that manipulative experiments can be designed to test the generality or responses, and predict how communities might develop in different systems.

In an environment populated to its carrying capacity, the level of competition will be high, and the demand for resources will approximate the supply (see Paine 1994). Natural selection there will tend to favor competitive ability (including grazer resistance), with a slow growth rate and delayed reproduction as the price. In unstable areas, where there is a low level of competition because resources exceed demand, selection is predicted to favor rapid growth, early reproduction, and short life spans. In an unpredictable environment, where a population or community may be suddenly removed, species with high growth rates (opportunistic species) will have the advantage and thus become primary settlers. However, populations with very high growth rates tend to overshoot the carrying capacity (there is a delay in the feedback that controls growth), and this typically results in a population "crash". This crash, in turn, allows slower growing species to take over. Such late successional species fare best in stable environments. There is, however, a continuum of characteristics between these extremes, discussed in sec. 8.3.2.

The attributes that hypothetically should increase the fitness of opportunistic and late-successional macroalgae were formulated by Littler and Littler (1980) (Table 3.2) and the costs and benefits of each attribute compared (Table 3.3). By stating the hypotheses in this form, they were able to test some of the predictions; see also Littler *et al.* (1983a). As expected, they found that thin, rapidly growing, short-lived algae were characteristic of unstable (temporally fluctuating) environments, whereas coarse, slower growing, long-lived algae were characteristic of stable environments. Some species, however, through morphologically or ecologically dissimilar alternate phases, have attributes of both extremes (Santelices 1990). For example, *Mastocarpus papillatus*, *Scytosiphon* (Fig. 2.2), and *Petalonia* all have a crustose stage (late-successional) and an erect phase (opportunistic). In terrestrial environments, stress-resistant plants tend to have late-successional characteristics, but among marine algae the more stress-resistant species, such as *Ulva*, are opportunistic species (Littler and Littler 1980).

Grime's (1979) theories on the factors that constrain terrestrial plant establishment and growth are a cornerstone of modern ecology, and have also been applied to seaweed communities. *Stresses* are more or less continuous suboptimal conditions that restrict plant productivity, such as shortages of water or nutrients, or suboptimal temperature or salinity (discussed in sec. 7.1). *Disturbances* are discontinuous events, such as mechanical or chemical destruction of tissue (see sec. 8.3.2). Plants and seaweeds can be categorized as (1) competitors, which occupy habitats of low stress and low disturbance, (2) stress tolerators, or (3) ruderals (opportunists), which occupy highly disturbed areas. There is no successful strategy for areas where stress and disturbance are both high and therefore no seaweeds can grow in these areas.

This concludes our survey of the fundamental details of seaweeds as individuals (Chapters 1 and 2), and the habitats and communities in which they live. In Chapters 4–8, we will consider how seaweeds respond to the biotic and abiotic factors that shape seaweed communities. In Chapter 9, we discuss how individual seaweeds and communities respond to anthropogenic pollution, and finally, in

Table 3.2 Attributes that would seem, a priori, to improve the fitness of opportunistic macroalgae (representative of young or temporally fluctuating communities) versus late-successional macroalgae (characteristic of mature or temporally fluctuating communities).

Opportunistic forms	Late-successional forms
1. Rapid colonizers on newly cleared surfaces	1. Not rapid colonizers (present mostly in late seral stages); invade pioneer communities on a predictable seasonal basis
2. Ephemerals, annuals, or perennials with vegetative shortcuts to life history	2. More complex and longer life histories; reproduction optimally timed seasonally
3. Thallus form relatively simple (undifferentiated), small, with little biomass per thallus; high thallus SA:V	3. Thallus form differentiated structurally and functionally, with much structural tissue (large thalli high in biomass); low thallus SA:V ratio
4. Rapid growth potential and high net primary productivity per entire thallus; nearly all tissue photosynthetic	4. Slow growth and low net productivity per entire thallus unit because of respiration of non-photosynthetic tissue and reduced protoplasm per algal unit
5. High total reproductive capacity, with nearly all cells potentially reproductive; many reproductive bodies, with little energy invested in each propagule; released throughout the year	5. Low total reproductive capacity; specialized reproductive tissue, with relatively high energy contained in individual propagules
6. Calorific value high and uniform throughout the thallus	6. Calorific value low in some structural components and distributed differentially in thallus parts; may store high-energy compounds for predictably harsh seasons
7. Different parts of life history have similar opportunistic strategies; isomorphic alternation; young thalli just smaller versions of old	7. Different parts of life history may have evolved markedly different strategies; heteromorphic alternation; young thalli may possess strategies paralleling those of opportunistic forms
8. Can escape predation because of their temporal and spatial unpredictability or by means of rapid growth (satiating herbivores)	8. Can reduce palatability to predators by complex structural and chemical defenses

Source: Reprinted with permission from Littler and Littler (1980), *American Naturalist*, vol. 116, pp. 25–44, © 1980, The University of Chicago Press.

Chapter 10, we discuss the various ways in which humans utilize seaweeds.

3.6 Synopsis

Seaweeds live in complex communities in tropical, temperate, and polar regions in which they respond to a wide variety of ever-changing biotic and abiotic factors. The rise and fall of the tides results in seaweeds of the rocky intertidal zone being subjected to rapid fluctuations in light, temperature, and nutrients. The ability of intertidal seaweeds to recover from environmental stresses associated with exposure to air is a major determinant of the shore position that they occupy. In the subtidal, seaweeds also grow in vertical bands, and here light availability is a key factor determining their zonal position. In the tropics, seaweeds are a natural component of both coral reef and seagrass ecosystems. However, in some coral reefs anthropogenic change is affecting the balance between seaweeds and corals, favoring seaweeds. In temperate and polar regions, large brown seaweeds form the structural elements of the community, providing a habitat for other seaweeds and animals, and supplying energy to higher trophic levels. Antarctic seaweeds have been isolated from other biogeographic regions for ~100 million years, which has resulted in a high

Table 3.3. Hypothetical costs and benefits of the attributes listed in Table 3.2 for opportunistic and late-successional species of macroalgae.

Opportunistic forms	Late-successional forms
Costs	
1. Reproductive bodies have high mortality	1. Slow growth and low net productivity per entire thallus unit result in long establishment times
2. Small and simple thalli are easily outcompeted for light by tall canopy formers	2. Low and infrequent output of reproductive bodies
3. Delicate thalli are more easily crowded out and damaged by less delicate forms	3. Low SA:V ratios relatively ineffective for uptake of low nutrient concentrations
4. Thallus relatively accessible and susceptible to grazing	4. Overall mortality effects more disastrous because of slow replacement times and overall lower densities
5. Delicate thalli are easily torn away by shearing forces of waves and abraded by sedimentary particles	5. Must commit relatively large amounts of energy and materials to protect long-lived structures (energy that is thereby unavailable for growth and reproduction)
6. High SA:V ratio results in greater desiccation when exposed to air	6. Specialized physiologically, and thus tend to have a narrow range of morphology
7. Limited survival options because of less heterogeneity of life-history phases	7. Respiration costs high because of maintenance of structural tissues (especially during unfavorable growth conditions)
Benefits	
1. High productivity and rapid growth permit rapid invasion of primary substrates	1. High quality of reproductive bodies (more energy per propagule) reduces mortality
2. High and continuous output of reproductive bodies	2. Differentiated structure and large size increase competitive ability for light
3. High SA:V ratio favors rapid uptake of nutrients	3. Structural specialization increases toughness and competitive ability for space
4. Rapid replacement of tissues can minimize predation and overcome mortality effects	4. Photosynthetic and reproductive structures relatively inaccessible and resistant to grazing by epilithic herbivores
5. Can escape from predation because of their temporal and spatial unpredictability	5. Resistant to physical stresses such as shearing and abrasion
6. Not physiologically specialized, and tend to have a broader range of morphology	6. Low SA:V ratio decreases water loss during exposure to air
	7. More available survival options because of complex (heteromorphic) life-history strategies
	8. Mechanisms for storing nutritive compounds, dropping costly parts, or shifting physiological patterns permit survival during unfavorable but predictable seasons

Source: Reprinted with permission from Littler and Littler (1980), *American Naturalist*, vol. 116, pp. 25–44, © 1980, The University of Chicago Press.

degree of endemism and physiological adaptations to polar conditions. In contrast, Arctic seaweeds have a cold-water history of just 2.7 million years and the flora is very similar to the temperate seaweeds of the north Atlantic and Pacific. Lush populations of deep-water seaweeds (>40 m) have recently been discovered. Some seaweeds are buoyant and form floating rafts, and salt marshes harbor a range of unattached species. Many seaweed species have been spread by humans from their native range to new locations, and some of these adventive species have become invasive, altering the ecosystem functioning of their new habitat.

Seaweed communities are not static, and community composition varies in both space and time. Analysis of seaweed communities requires a rigorous sampling design that incorporates random sampling. Succession takes place on any newly exposed surface, and seaweeds and animals interact competitively, often facilitating or inhibiting later stages of succession. Vast numbers of propagules are produced by seaweeds, but only small numbers survive to maturity. Competition takes place within and between species, and grazing and physical factors account for much of the mortality among juveniles. Each environment has a carrying capacity, and populations are maintained generally at or below that capacity. Some species experience rapid growth and have short life spans; these tend to be pioneer or opportunistic species. Other species grow more slowly, reproduce later, and put more energy into maintenance and defense; these tend to dominate later in the succession and to form persistent canopies. Highly competitive seaweeds can occupy habitats in which stresses and disturbances are both low. Opportunistic species are found where disturbances are frequent but stress is low. Some slow-growing seaweed can tolerate stress, but cannot survive if there are also frequent disturbances.

4

Biotic interactions

The environment of an organism includes both biotic and abiotic (physiochemical) factors. Communities of marine organisms encompass not only the seaweed communities but also the animal communities, of which the benthic grazers and their predators are most important to seaweed ecology. Thus, the biotic interactions of seaweeds include not only competition with other seaweeds (both within and between species) and with sessile animals but also predator–prey relations at several trophic levels, and facilitation; the mix of such interactions will change as the individual changes with age and environmental history.

Biotic interactions are complex, and their study often requires large-scale and long-term observations and manipulations in the laboratory, as well as in the field. Interactions can be positive (e.g. facilitation, mutualism, and commensalism), negative (e.g. competitive exclusion, consumption) or neutral, where there is no effect of one species on another. Studies on biotic interactions in the marine environment have traditionally focused on competition but more recently facilitation has been recognized as an important way in which biota interact. The minireviews of Olson and Lubchenco (1990), Carpenter (1990), Paine (1990), and Maggs and Cheney (1990) remain useful frameworks, as are the more recent syntheses found within *Marine Community Ecology* (Bertness *et al.* 2001) and *Marine Ecology* (Connell and Gillanders 2007).

4.1 Foundation species and facilitation

"Habitat modifiers" are organisms whose presence alters the physical and biological environment (also called autogenic ecosystem engineers; see review of Bruno and Bertness 2001). However this loose definition could be applied to most species, and so the terms "foundation species" and "keystone facilitators" are used to help rank the mode and impact of environmental modification (Bruno and Bertness 2001). Foundation species "have a large effect on community structure by modifying environmental conditions, species interactions, and resource availability through their presence (e.g. kelps, seagrasses, mussels and corals) and not their actions" (Bruno and Bertness 2001). *Macrocystis pyrifera* is an excellent example of a foundation species, as it creates a tall (6–68 m), three-dimensional biological habitat for mobile and sessile invertebrates, fish, and epiphytes and it modifies light and water motion (Schiel and Foster 1986; Graham *et al.* 2007a; Fig. 3.10).

Facilitation is a positive interaction between two or more species and keystone facilitators are organisms that have a "large, positive effect on a community through their actions" (Bruno and Bertness 2001). In the intertidal, facilitators generally ameliorate adverse physical conditions; for example, they provide shade, reduce evaporative water loss, moderate temperature, reduce water motion, and increase heterogeneity, thereby enhancing propagule settlement (e.g. Bertness and Callaway 1994; Bulleri 2009; sec. 2.5.1). In Italy, the invasion of *Caulerpa racemosa* has been facilitated by native algal turfs. These turfs are topographically complex and facilitate *Caulerpa* by trapping its vegetative propagules and providing a substratum into which the stolons can penetrate and anchor (Bulleri and Benedetti-Cecchi 2008). The survivorship of

Ascophyllum germlings under desiccation stress is greater when it grows in a mixed-stand with *Fucus serratus* germlings than in a monoculture of *Ascophyllum* germlings (Choi and Norton 2005a). In this case, the faster growing and desiccation-tolerant *Fucus* germlings provide a moister environment for the slow-growing, desiccation-intolerant *Ascophyllum* germlings. A further development is "indirect facilitation", whereby one species facilitates another via an intermediate species (Thomsen *et al.* 2010). For example, *Macrocystis* canopies may facilitate benthic sessile invertebrates through their action of shading understory seaweeds, thereby slowing their growth and reducing their competitive ability for primary substratum (Arkema *et al.* 2009).

In terrestrial communities, facilitation research has been framed within the "stress gradient hypothesis" (cf. environmental stress model; see Callaway 2007), and the applicability of this model to the marine environment is evaluated by Bulleri (2009) (and see Essay 2, Chapter 3). The prediction is that in physically stressful habitats, such as the high intertidal zone, there should be greater facilitation between species thereby ameliorating environmental stress. In the subtidal, which is more benign in terms of physical stress but can have greater biotic stresses such as herbivory, then facilitative mechanisms should serve to reduce consumer pressure. Bulleri's quantitative review of the literature provided general support for this model. Amelioration of environmental stress and release from consumer pressure were facilitative mechanisms in both environments but in the intertidal ~50% of studies reported amelioration of environment as the key facilitative mechanism compared to ~20% in the subtidal. The interactions are likely to vary depending on species composition in different sites, and along horizontal gradients of water motion. While there is an increasing number of examples of facilitation in marine communities, in most cases the underlying mechanisms have yet to be elucidated (Bulleri 2009). Furthermore, most interactions may not be strong. For example, in examining over 3000 interactions of intertidal species in New Zealand and the west coast of North America, Wood *et al.* (2010) found that the vast majority had very low interaction strengths.

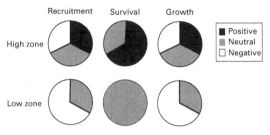

Figure 4.1 Facilitation. The effects of the intertidal, canopy-forming alga *Ascophyllum nodosum* on invertebrate species at high- and low-tidal levels. Data are the percentage of one-way interactions for each demographic process that are positive, negative, and statistically neutral and are pooled responses for nine invertebrate species that are associated with *Ascophyllum* in northern New England, USA. (Redrawn from Bruno and Bertness 2001, with permission from the authors.)

An early test of how facilitation varies along an environmental stress gradient was conducted by comparing the community responses to clearing within high- and mid-shore *Ascophyllum* beds in a wave-protected estuary in Maine, USA (Bertness *et al.* 1999; Fig. 4.1). Importantly in this study, the physical environment was monitored at the scale of the organisms, which is key to assessing the mechanisms of facilitation (Bulleri 2009). The *Ascophyllum* canopy ameliorated the physical conditions at both shore levels by shading the underlying rock, reducing evaporative water loss, and maintaining rock-surface temperatures that were 5–10°C lower than bare rock (see also sec. 7.2.3). The response of understory organisms to the removal of this habitat modifier was "somewhat idiosyncratic" and depended on the shore position and species examined. For example, in the high shore, the presence of the canopy generally enhanced the recruitment of mussels and crabs, negatively affected the recruitment of herbivorous snails, and had no influence on barnacle recruitment. In general, the canopy had positive effects on high shore organisms and species' interactions, compared to negative or neutral canopy effects in the low shore. The results lend support to the environmental stress model but also demonstrate the complexity of biotic interactions, and that habitat

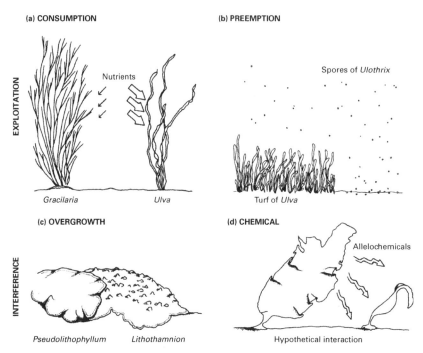

Figure 4.2 Schematic representations of some mechanisms of competition between seaweeds. (a) Consumption of nutrients mediates the interaction between *Gracilaria tikvahiae* and *Ulva* species in the field. (b) A dense turf of *Ulva intestinalis* preempts space, preventing colonization by spores of *Ulothrix flacca* (previously *Ulothrix pseudoflacca*) in culture. (c) *Pseudolithophyllum muricatum* overgrows *Lithothamnion phymatodeum* in the field on a smooth, artificial surface in the absence of grazers. (d) An alga uses allelochemicals to exclude a competitor (hypothetical interaction). (Drawing by Charles Halpern, from Olson and Lubchenco 1990, with permission of *Journal of Phycology*.)

modifiers influence different species at various life stages, even within a 2 m vertical zone on the shore.

4.2   Competition

The term "competition" implies that a common resource is potentially limiting (Denley and Dayton 1985; Carpenter 1990; Paine 1994; Edwards and Connell 2012). Competition will be most important in a community structure where abiotic stresses and disturbances are reduced (Carpenter 1990), but the competitive abilities of organisms will, in general, be affected by abiotic conditions (see sec. 4.3.1). Two kinds of competition have been recognized, although they are not mutually exclusive (Paine 1994; Fig. 4.2):

Interference competition results from interactions between organisms that may not relate directly (if at all) to any limiting resource; examples include preempting resources, whiplash, and allelopathy. *Exploitative competition* involves a scramble for a limiting resource (e.g. space, light, nutrients) without direct antagonism between organisms. If interference competition is taking place, however, exploitative competition must also be potentially possible (Denley and Dayton 1985).

4.2.1   Interference competition

Primary rock substratum is a limiting resource in many marine habitats and an ability of one species to grow over another is a useful way of securing both space and other primary resources, such as light, from the overgrown species. Competition for space is most obvious among crustose species and between turf-forming species and erect algae. Crustose coralline algae of a given species may fuse and lose their individual identities when their thalli meet, or they may form "minimal borders", whereas between species, distinct borders are formed, with one species sometimes overgrowing the other (Paine 1990). Encrusting corallines are often overgrown by other encrusting corallines, or turf-forming erect corallines and fleshy seaweeds. They are thought to be competitively inferior to the overgrower and suffer reduced growth and fecundity (reviewed in Airoldi 2000). However, the wide diversity and persistence of crustose coralline seaweeds in most intertidal and subtidal habitats worldwide indicates mechanisms of tolerating overgrowth (Dethier and Steneck 2001). One mechanism is the transport of photosynthates and nutrients from uncovered parts of the thallus to the overgrown parts, explaining why overgrowth does not always cause competitive exclusion (Bulleri 2006; Underwood 2006). Other mechanisms for tolerating overgrowth may include an ability for rapid wound healing, frequent recruitment to the population, slow growth, and low metabolic demands (Dethier and Steneck 2001).

"Pre-emptive competition" is the process by which an individual procures a limiting resource (e.g. space, nutrients) thereby making it unavailable to others (Menge and Branch 2001; Edwards and Connell 2012). Turf-forming algae often are able to preempt space in the intertidal zone. For example, in central Chile, *Codium dimorphum* forms a thick, spongy crust and is able to overgrow and exclude most other lower-intertidal algae during fall and winter. During spring and summer, irradiance and temperatures increase, and the occurrences of low tides shift to the daytime. The *Codium* crust is bleached in the lower intertidal and killed in the mid-intertidal (i.e. it becomes non-competitive), and new borders are created, along which grazers can attack. During fall, winter, and early spring, *Codium* reinvades the mid-intertidal (Santelices *et al.* 1981).

In southern California, a low-intertidal turf comprising *Chondracanthus canaliculata* (previously *Gigartina canaliculata*), *Laurencia pacifica*, and *Neogastroclonium subarticulatum* (previously *Gastroclonium subarticulatum*) outcompetes the brown alga *Egregia menziesii* (previously *Egregia laevigata*). The kelp recruits only from spores and only at certain times of the year, whereas the red algae expand vegetatively at all seasons via prostrate axes, encroaching on any space that becomes available. During one study, a 100-cm^2 clearing in the middle of a *C. canaliculata* bed was completely filled in 2 years. The turf traps sediment, which fills the spaces between the axes and prevents the settlement of other algal spores on the rock (Sousa 1979; Sousa *et al.* 1981). *E. menziesii* may settle if clearings become available at the right time, but by the time it becomes adult and reproductive (it lives only 8–15 months), the turf will have encroached all around and will prevent the kelp from replacing itself in that area. In the mid-intertidal zone, a turf of *Corallina* is persistent and dominant; Stewart (1989) showed how its life history, physiology, and morphology adapt it to weather, tides, and sand (sec. 8.3.2).

In Nova Scotia, Canada, populations of the invasive seaweed *Codium fragile* ssp. *tomentosoides* were kept in check by competitive dominance of the native canopy-forming seaweeds *Laminara* and *Desmarestia*, until another invader, the bryozoan *Membranipora membranceae*, decimated the *Laminaria* populations. This freed up space which *Codium* preempted, and kelp assemblages have not been able to re-establish (Scheibling and Gagnon 2006). In southern New Zealand, turfing communities of coralline seaweeds can prevent the re-establishment of fucoids over many years (Schiel and Lilley 2011).

Another form of interference competition is whiplash, by which canopy-forming seaweeds clear space around their holdfasts (Irving and Connell 2006b). For example, clear space is maintained around *Lessonia* on wave-swept shores in Chile by the sweeping effect of the blades (Santelices and Ojeda 1984). In dense beds, scouring of the shore by these blades completely inhibits recruitment of juveniles. In pools, from which adults are absent, urchin grazing eliminates recruitment. However, in patchy *Lessonia* beds, juveniles can be recruited into openings that are large enough to ensure reduced whiplash and yet small enough to feature reduced grazing pressure. Similarly, in Australia the abrasion of the substratum by *Ecklonia radiata* creates an understory of tolerant encrusting corallines, whereas in the gaps that are free from whiplash, filamentous turfs and erect corallines dominate (Irving and Connell 2006b).

The examples above are of seaweeds competing with other seaweeds for space, but they also compete for space with sessile invertebrates, such as filter-feeding barnacles and mussels. The outcome of this competition will depend on at least three factors: irradiance, the presence or absence of grazers, and the presence or absence of predators seeking sessile animals (Foster 1975). Under very low irradiance, sessile animals will dominate, whether or not they face predators, but under moderate irradiance competition for space will increase and the algae may predominate over the sessile animals if predators reduce the numbers of animals (Witman and Dayton 2001). The lower limits of sublittoral algal growth may be partly regulated by the inability of the algae to compete against sessile animals, rather than by the inadequacy of the light flux to the algae per se (Foster 1975; see review by Graham *et al.* 2007a)

Competition for space takes place among *Postelsia*, mussels, and a red algal turf (chiefly *Corallina officinalis*) on the northeast Pacific coast (Paine 1988). Mussels will outcompete the turf that becomes established in clearings in the mussel bed. If clearings occur when nearby *Postelsia* are fertile, new sporophytes will grow up on the rock, but they are annuals and will be replaced by the turf. Although *Postelsia* can attach to turf, it cannot persist on that substrate, and apparently it relies on waves to remove mussels and provide new bare rock on which it can re-establish. In other mussel beds, such as those in Chile, various algae coexist with the shellfish. The mussels ingest algal spores (some of which may survive), but they protect germlings from desiccation and probably fertilize them. Algal populations are also regulated by small grazers among the mussels (Santelices and Martinez 1988). In all these examples, the influences of external factors, especially abiotic factors, are apparent.

Herbivorous fish and limpets that "garden" seaweed provide interesting examples of interference competition. In California, the ocean goldfish, *Hypsypops*

rubicunda, establishes nests of filamentous red algae on poorly lit vertical walls dominated by sessile animals. It does this by systematically removing the animals from the area, with the result that the algae can successfully compete for space (Foster 1972) (sec. 4.3.2). A classical example of "gardening" by an invertebrate was reported by Branch (1975), who showed that the limpet *Patella longicosta* in South Africa establishes *Neoralfsia expansa* (previously *Ralfsia expansa*) "lawns" by grazing away the established crustose corallines. It then grazes off competing algae and grazes the *Neoralfsia* in strips, so that productivity is actually enhanced because of edge growth.

4.2.2 Epibionts and allelopathy

In the "space race", one solution is to grow as an epibiont, an organism that lives on the surface of another organism. Epibionts on seaweeds include macroscopic organisms such as filamentous seaweeds, small coralline crusts such as *Melobesia*, a wide range of sessile invertebrates including encrusting bryozoans such as *Membranipora*, colonial hydrozoa such as *Obelia*, calcareous polychaete worms such as *Spirorbis*, and microorganisms including diatoms and bacteria. Epiphytism (in this case an alga growing on another alga) is a common way of life, although relatively few seaweeds are obligate epiphytes. In some locations, most of the algae are epiphytes. Such a place is the intertidal zone in southern California, where two species of *Corallina* occupy over 60% of the total rock substrate, and at any given time and site some 15–30 species of seaweeds are epiphytic on the coralline turf (Stewart 1982).

Epibiosis, whether by macroscopic or microscopic organisms, has the fundamental effect of modifying the basiphyte's body surface and creating a new interface with its environment (Wahl 2008). Epiphytes are often viewed as having negative effects on the basiphyte (host or substratum species) and the language used to describe epiphyte–basiphyte relations reinforces this view of a "battle" for resources: epiphytes "challenge" the host seaweed that may respond by "defending" itself from further "colonization". However, while there are many examples of negative impacts on the basiphyte, positive effects have also been reported, and it is important to remember that epiphytes themselves can be productive members of ecosystems (Ruesink 1998). The strength and sign (+ve or –ve) of interactions are both species and environment specific (Wahl 2008). Furthermore, the level at which an epiphytic relationship is examined can influence the conclusions drawn. In laboratory studies, the diatom *Isthmia nervosa* caused reduced growth rate and increased drag forces on the red seaweed *Odonthalia floccosa*, but these effects were not found at a population level because *Odonthalia* escapes the severe impacts of diatom epiphytism by completing its growth and reproduction prior to colonization by the epiphyte, which occurs in late summer (Ruesink 1998). To unravel the details of epiphyte–basiphyte interactions, investigations at the level of both the individual and community are needed.

The negative effects of macroscopic epibionts are generally to shade the basiphyte (Cancino *et al.* 1987; Kraberg and Norton 2007; Rohde *et al.* 2008), impede gas and nutrient exchange (Hurd *et al.* 1994a) and thereby decrease its growth rate (Honkanen and Jormalainen 2005). Epibionts can also increase drag on the fronds and cause blades to be more susceptible to crack-propagation (Krumhansl *et al.* 2011; see sec. 8.3.1), which both result in blade/frond loss. Moreover, heavy encrustation of kelp blades by bryozoans can lead to loss of the blades to predatory fishes, which cannot get the animal prey without also taking the alga; small *Macrocystis* beds have been destroyed in this way, especially during times of nutrient stress in kelps (Bernstein and Jung 1979). Epiphytes can also influence trophic interactions (Karez *et al.* 2000). In the Baltic Sea, seaweed epiphytes reduce the light levels at the surface of *Fucus vesiculosus*. This causes a reduction not only of the basiphyte's growth rate but also in its ability to allocate resources to chemical defense, thereby making it more palatable to the isopod herbivore *Idotea*. Here, epiphytism had a doubly negative effect on the seaweed but a positive effect on the grazer (Karez *et al.* 2000).

There are examples of positive interactions between epiphyte and basiphyte. Dense epiphytes on the *Corallina* turf described earlier may help alleviate desiccation stress (Stewart 1982), and epiphytism may allow shade-loving seaweeds to grow in sunnier locations, by creating shade (e.g. *Cladophora rupestris*; Wiencke and Davenport 1987). In some cases, the shaded basiphyte can increase pigment content (shade adaptation) thereby compensating for reduction in incident light (Muñoz *et al.* 1991). The bryozoan *Electra pilosa* increases CO_2 availability to *Gelidium corneum* (Mercado *et al.* 1998). The transparent, colonial hydrozoan *Obelia* positively influences *Macrocystis* growth, perhaps via the provision of ammonium (Hepburn and Hurd 2005). Benefits to the animals include carbon exuded from the seaweed (De Burgh and Frankboner 1978; Muñoz *et al.* 1991), a renewable substratum and increased feeding efficiency because kelps extend above the benthic boundary layer (see Hepburn *et al.* 2006).

Some relations between basiphyte and epiphyte are exclusive. The green epiphyte *Pilinia novae-zealandiae* (previously *Sporocladopsis novae-zelandiae*) settles specifically on sori of *Lessonia nigrescens*, and will not settle even on the close relative *L. trabeculara*. This relationship appears chemically mediated, with rhizoid-like anchoring structures of the epiphyte being induced by chemicals released by host sorus tissue (Correa and Martínez 1996). Another example is the larvae of the bryozoan *Lichenopora novae-zelandiae* that settle preferentially on young, meristematic regions of the kelp *Agarum fimbriatum*. As the kelp grows, the bryozoans are slowly moved distally and new meristematic tissue is available for further cohorts of *Lichenopora* recruits. The herbivorous snail, *Tegula pulligo*, preferentially eats the old *Agarum* tissue but the presence of adult bryozoans on older tissue deters *Tegula* grazing. This relationship between epiphyte and basiphyte is symbiotic, with the seaweed providing a substratum for the bryozoans, and the bryozoan some protection against grazing (Durante and Chia 1991). Other examples of mutualisms between epiphytes and basiphyte are explored in sec. 4.5.1.

The surfaces of many seaweeds are conspicuously clean and epiphyte-free, indicating that they have mechanisms to remove the epiphytes, prevent their settlement or to kill them. Many seaweeds slough their

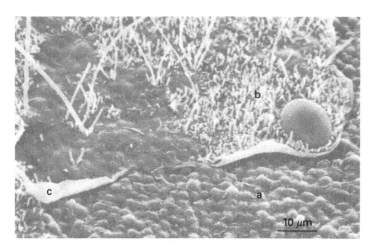

Figure 4.3 Sloughing of outer cell wall layers, with removal of epiphytes. SEM view of the surface of *Chondrus crispus* showing the cuticle (c), with bacteria (b) sloughing away, leaving a clean algal surface (a). Scale = 10 μm. (From Sieburth and Tootle 1981, with permission of *Journal of Phycology*.)

outer layers, which has the effect of ridding the thallus of epiphytes (Fig. 4.3). *Ulva intestinalis* continuously produces new wall layers and sloughs off the outer layers of glycoprotein (McArthur and Moss 1977). In *Halidrys siliquosa* (Moss 1982), *Himanthalia* (Russell and Veltkamp 1984), and encrusting corallines *Sporolithon ptychoides* and *Neogoniolithon fosliei* (Keats *et al.* 1997) the old walls of the epidermal cells are shed as "skins" following production of a new wall underneath. Another coralline, *Hydrolithon onkodes*, sheds its epithelium in a non-synchronous manner, by which epithelial cells degenerate individually (Keats *et al.* 1997). The whole outer epidermal layer may be shed by *Ascophyllum nodosum* (Fillion Myklebust and Norton 1981), and similar mechanism of "cuticle peeling" was reported for the bladed red seaweed *Dilsea carnosa* (Nylund and Pavia 2005). The distal parts of *Laminaria* species are continuously eroded and as a result epiphytes are restricted to a few that can grow and reproduce before they reach the end of the production belt (Russell 1983). "Blade abandonment" is employed by the Caribbean siphonous green seaweed *Avrainvillea longicaulis* to remove heavily epiphytized portions of the thallus. Here, the cytoplasm is removed from the heavily epiphytized sections, which are then abandoned, and at the same time, there is a rapid

proliferation of new blades. Blade abandonment/proliferation is a low-energy mechanism of removing epiphytes without losing photosynthates (Littler and Littler 1999).

Some epiphytes, however, have means of defeating the host's attempts to slough them off. *Ascophyllum nodosum* can become heavily epiphytized by *Vertebrata lanosa* (previously *Polysiphonia lanosa*), which attach in wounds and lateral pits, to avoid removal when the outer cell layer is sloughed (Lobban and Baxter 1983; Pearson and Evans 1990; Longtin and Scrosati 2009). *Vertebrata* can reduce incident light by 40% but *Ascophyllum* may compensate for the infection by re-allocating reproductive output to non-epiphytized lateral pits, so that receptacle biomass in non-infected lateral pits is increased (Kraberg and Norton 2007).

The process of the production and release of chemical compounds that inhibit epibionts is termed allelopathy (Harlin 1987). These compounds are typically secondary metabolites, ones that have no known internal roles in primary metabolic processes such as photosynthesis, respiration, and nitrogen metabolism, although the phlorotannins of brown seaweeds do have primary roles (Amsler and Fairhead 2006; sec. 1.3). Compounds include small, non-polar, volatile molecules that are highly soluble and diffuse rapidly. These may be non-halogenated, such as the terpenoids of *Dictyota* spp. and *Laurencia*, or halogenated with bromine or iodine, for example the brominated furanones of *Delisia* (Steinberg *et al.* 2001, 2002; Viano *et al.* 2009; Potin 2012). In fact, an enormous number of bioactive compounds have been identified from algae (e.g. Blunt *et al.* 2007; see sec. 4.4.1). In their review, Gressler *et al.* (2009) list 295 non-halogenated volatile organic compounds that have been discovered from 31 algal species, and a single species of *Gracilaria* contained nine volatile halogenated organic compounds (VHOCs) (Weinberger *et al.* 2007). The motivation for screening and identifying these compounds has come largely from their potential use as "natural" antifoulants to replace chemicals that are toxic to the environment, such as tributyl tin (TBT) and copper (Nylund *et al.* 2005; sec. 9.5), and also as pharmaceuticals (Gressler *et al.* 2009; Chapter 10). Their putative physiological and ecological roles, however, have received comparatively little attention.

Five criteria should be met to demonstrate that a bioactive chemical has an allelopathic function (Dworjanyn *et al.* 2006a). These are: (1) the seaweed population under observation should be (mostly) epiphyte-free. (2) The chemicals should be bioactive at the concentrations that occur within the boundary layer at the basiphyte's surface, because measuring concentrations in the seaweed tissue will give misleading results. (3) The mechanism of presentation of the chemical to the epiphyte must be known. (4) The mechanism(s) of delivery from internal storage organelles to the seaweed surface must be known. (5) Chemicals released by the epiphyte might interact with those released by the basiphyte and such interactions should be elucidated. The ability to demonstrate these five criteria is methodologically difficult, but a succession of studies on the red seaweed *Delisea pulchra* were the first to demonstrate allelopathy, unequivocally, by a seaweed (de Nys *et al.* 1995; Maximilien *et al.* 1998; Dworjanyn *et al.* 1999, 2006a).

Delisea pulchra synthesizes four furanones, which are brominated compounds that differ from one another in the substation of functional groups (Fig. 4.4a). They are contained in a central vesicle,

which delivers the furanones to the seaweed surface. At concentrations found at *Delisea*'s surface, the furanones inhibit the settlement of *Ulva* and *Ectocarpus siliculosus* gametes, *Ceramium* tetraspores, and *Vertebrata* carpospores (Dworjanyn *et al.* 2006a). However, each of the four furanones has different bioactive properties. For example, at very low concentrations of 10–100 ng cm^{-2}, furanone 3 inhibits propagule settlement whereas the bioactivity of furanone 2 is an order of magnitude lower and unlikely to function alone as an antifoulant. The furanones of *Delisea* are presented in a lipid matrix, which may spread the furanones evenly over the seaweed surface and/or prevent their degradation and dissolution. These furanones also inhibit bacterial settlement on *Delisea*'s surface. However, the surface concentration of furanones is not sufficient to kill bacteria directly. Instead, they affect both the ability of the bacteria to attach to the seaweed surface, and their swarming and swimming behavior (Maximillian *et al.* 1998). The furanones are structurally similar to bacterial signaling molecules, acylated homoserine lactones (AHLs), and interfere with the AHL regulatory system (Givskov *et al.* 1996; Steinberg and de Nys 2002). In two other red seaweeds, *Asparagopsis armata* and *Laurencia obtusa*, the bioactive compounds are delivered from internal storage structures (gland cells and *corps en cerise*, respectively) via a tubular structure to the seaweed surface (Paul *et al.* 2006 a, b; Salgado *et al.* 2008; Fig. 4.4b).

The phlorotannins (polyphenolics) of brown seaweeds (sec. 1.3) have often been cited as antifoulants but evidence for this is equivocal (see also 4.4.1). For *Ecklonia radiata*, the levels of phlorotannins reported at the seaweed surface in the field were much lower than those needed to cause allelopathy (Jennings and Steinberg 1997). There is, however, strong evidence that the non-polar secondary metabolites dictyol E and pachydictyol A of *Dictyota menstrualis* have allelopathic activity (Schmitt *et al.* 1998). These chemicals caused death, abnormal development, or reduced growth to the larvae of hydrozoans at concentrations that were 5% lower than those found at the seaweed surface.

Compared to red and brown seaweeds, relatively few studies have been done on allelochemicals

(a)

R₁ R₂

(1) H   Br
(2) H   H
(3) OAc H
(4) OH   H

Furanone

(b)

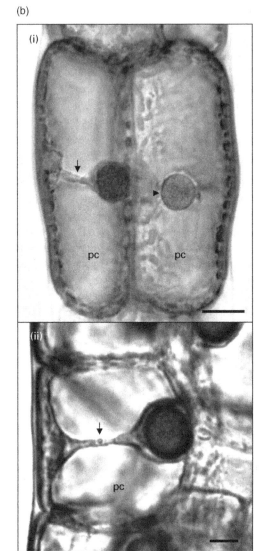

Figure 4.4 (a) *Delisea pulchra*. Structures of four major furanones (1 to 4). (b) *Asparagopsis armata*. Light micrographs of (i) the tetrasporophyte and (ii) gametophyte of *A. armata*. (i) Gland cells (arrow) within pericentral (parent) cells (pc) of the tetrasporophyte and connective structure (arrow head). (ii) Lateral branch of the gametophyte with a gland cell within the parent cell (pc) showing a structure connecting the gland cell to the outer cell wall material may be seen on this structure (ii, arrow). (Part (a) from Dworjanyn *et al.* 2006a; (b) from Paul *et al.* 2006b. © Inter-Research 2006.)

released by green seaweeds. *Ulva reticulata* and its epiphytic bacteria *Vibrio* both released highly soluble macromolecules that had antifouling properties against a polychaete and bryozoan (Harder *et al.* 2004a) and an allelopathic effect of *Ulva* on *Fucus* zygotes and oyster larvae has been reported (Nelson *et al.* 2003). Particularly interesting is an allelopathic effect that *Ulva* has on phytoplankton (Jin and Dong 2003; Jin *et al.* 2005; Tang and Gobler 2011). Phytoplankton blooms cause shading and they take up essential nutrients, thereby reducing seaweed growth rates (Kavanaugh *et al.* 2009). Allopathic chemicals released by *Ulva lactuca* caused lysis or reduced growth of seven species of phytoplankton and this may be a mechanism by which benthic algae reduce phytoplankton numbers and thereby increase light penetration (Tang and Gobler 2011). Candidate allelochemicals include polyunsaturated fatty acids (Alamsjah *et al.* 2008).

Competition between corals and seaweeds can be important in structuring reef communities, but allelopathy has been little studied (see Fong and Paul 2011). In the Caribbean, Foster *et al.* (2008) showed that direct contact of *Dictyota* with the coral *Montastraea annularis* caused a decline in coral fecundity, with smaller and fewer eggs being produced. Subsequently, Rasher and Hay (2010) found that the coral reef seaweeds *Halimeda*, *Dictyota*, and *Lobophora* release lipid-soluble chemicals that can kill corals upon contact. Both studies indicate that if numbers of herbivores on reefs decline (for instance due to harvesting) and seaweed populations increase, allelopathy by seaweeds may become an additional stress on coral reefs.

Another chemical mechanism of deterring potential epiphytes is that of an oxidative burst (see sec. 7.8). Here research has focused on the ability of seaweeds to defend themselves against pathogens (e.g. Potin 2008, 2012; Cosse *et al.* 2008; Goecke *et al.* 2010). For example, bacteria are the important first colonizers of seaweed surfaces and the founders of biofilms which themselves can attract and deter other settling organisms (e.g. Steinberg *et al.* 2001; Matz 2011; Friedrich 2012), but some are pathogenic. Seaweeds need to recognize that a pathogen is present at its surface, so that they can rapidly respond to the "challenge". Pathogenic bacteria cause damage to the seaweed cell wall by enzymatically degrading it. The degradation products are oligosaccharides (e.g. oligoagars and oligocarageenans of red seaweeds, and oligoalginates of brown seaweeds), and these act as elicitors, signaling that the host cell wall has been breached. In *Gracilaria conferta* (= *G.* sp. *dura* in Weinberger *et al.* 2010), the oligoagars elicit an oxidative burst, producing hydrogen peroxide in quantities large enough to kill surface bacteria (Weinberger and Friedlander 2000). Oxidative burst is common in other members of the Gracilariaceae, but not all the species utilize the same biochemical pathways (Weinberger *et al.* 2010). Within the brown seaweeds there is a range of responses to pathogen attack. Küpper *et al.* (2002) found that the sporophytes of all the Laminariales they tested

responded to pathogen attack with a strong oxidative burst, as did *Desmarestia dudresnayi* and *Pylaiella littoralis*. However, *Dictyota dichotoma* had only a weak oxidative burst, and no oxidative burst was detected for gametophytes of *Laminaria digitata*. For the Fucales they tested there was no oxidative burst, but they did release large amounts of H_2O_2, indicative of a constitutive (rather than inductive) response.

Gametophytes of *Chondrus crispus* are much less susceptible than sporophytes to infection by the endophyte *Acrochaete operculata* and this is explained by differences in the sulfation of the carrageenan (Correa and McLachlan 1991; sec. 5.5.2). Following an attack by *Acrochaete*, oligo-λ-carrageenans and oligo-κ-carrageenans are released from the cell walls of the sporophyte and gametophyte, respectively. Oligo-λ-carrageenans released by the sporophyte cause enhanced protein synthesis and the production of polypeptides by *Acrochaete*, making the endophyte more pathogenic. However, the oligo-κ-oligocarrageenans released by the gametophyte increase its ability to recognize *Acrochaete*, and the result is reduced levels of infection. Sporophytes and gametophytes are also differently able to defend themselves against *Acrochaete* attack, because *Acrochaete* triggers a strong oxidative burst in the gametophyte, but not the sporophyte (Bouarab *et al.* 1999, 2001). The signaling molecules involved in eliciting these reactions are oxylipins, oxygenated derivatives of fatty acids that are also involved in immune responses of plants and animals (Bouarab *et al.* 2004).

Finally, some algae including *Ulva* and *Cladophora*, may avoid epiphytism simply because of their very rapid growth and their changes in pH at the thallus surface caused by a rapid metabolic rate (den Hartog 1972). When the growth of such species slows, epiphytes soon cover them. This idea has been used to explain the competitive dominance of *Ulva intestinalis* in Swedish rock pools (Björk *et al.* 2004).

4.2.3 Exploitative competition

Intraspecific exploitative competition is manifest in the effects of density on plant or seaweed size and survival. Populations of newly recruited juveniles may be very dense, whereas by the time they mature there will be far fewer of them (but biomass is greater) (Fig. 3. 16); part of the decline is due to intraspecific competition (i.e. self-thinning). The ability to predict changes in biomass based on population density is a powerful predictive tool for population studies. Early workers in terrestrial systems focused on a "one line fits all" approach, which resulted in the -3/2 self-thinning law and this law was also applied to seaweeds. However, when the -3/2 self-thinning model was tested for benthic seaweeds (Schiel and Choat 1980), it became apparent that a "one line fits all" model was not appropriate (Scrosati 2005). This is because other factors such as grazing, nutrient supply and abrasion affect survivorship, and the effect of these factors is not necessarily dependent on density (Reed 1990b; reviewed by Schiel and Foster 2006). Scrosati (2005) presents a critical assessment of self-thinning studies on seaweeds, including a long list of possible pitfalls in both methodology and data interpretation. Importantly, to test if self-thinning occurs for a particular species, both the biomass and density of a seaweed stand need to be followed through time. This is because self-thinning occurs when seaweed populations are actively growing and in periods of slow or no growth, it will not be detected (Rivera and Scrosati 2008). The -3/2 law has been largely abandoned by both terrestrial and marine ecologists, and the modern concept is that of "dynamic thinning lines" which vary intraspecifically according to abiotic and biotic conditions (cf. Reed 1990b; Scrosati 2005).

As seaweeds within a population grow, any differences in the sizes of germinated zygotes become progressively magnified, with the larger individuals outcompeting the smaller ones for resources (particularly light). Once self-thinning starts, however, this size inequity tends to decrease, as the smallest size-classes die leaving mostly large, similar-sized individuals. The Gini-coefficient (G') is a measure of the skewedness of the size distribution of individuals within a population, and can be used to explore the changes in population size distribution over time (Creed et al. 1998; Arenas and Fernández 2000; Rivera and Scrosati 2008).

The density at which zygotes settle is variable, as is the timing of germination and therefore the abiotic environment which the early recruits encounter (Schiel and Foster 2006). Field experiments, such as those by Reed (1990 a,b), in which microstages are "seeded" at different densities onto bare substratum, and the density followed over time, can reveal natural changes in population demography. However, detecting the particular abiotic factor(s) responsible for changes in populations can be difficult in the field. Factorial laboratory experiments are a useful method of teasing apart the relative effects of various abiotic factors on population dynamics. New recruits can be grown at a range of settlement densities and exposed to different combinations of abiotic conditions (Creed et al. 1997; Steen and Scrosati 2004). For *Fucus serratus*, nutrient enrichment had a marked effect on the size distribution of the populations. After 76 days in culture, high nutrient treatments had, on average, much larger individuals with an uneven size distribution (high G') compared to the unenriched treatment in which individuals were smaller and population size distribution was normally distributed (lower G'). Settlement density and irradiance also influenced the outcome of experiments, but nutrient concentration was clearly the most important controlling factor in these experiments (Creed et al. 1997). In similar experiments, germlings of *F. serratus* and *F. evanescens* showed different responses to temperature (7ºC and 17ºC), nutrient enrichment, and settlement density (Steen and Scrosati 2004; Fig. 4.5). Nutrient enrichment at 17ºC caused reduced survivorship of *F. serratus* at the highest density treatment. For *F. evanescens*, density alone reduced survivorship at the highest temperature and there was no effect of nutrient addition.

Non-coalescing clonal and coalescing clonal seaweeds do not necessarily follow the same self-thinning lines seen for unitary seaweeds such as *Fucus serratus* and *F. evanescens* described above (reviewed by Scrosati 2005). Many clonal seaweeds grow by spreading laterally, producing vegetative ramets (e.g. *Pterocladiella capillacea*; sec. 1.2.3). For these, population density is assessed by counting the number of fronds rather than the number of individuals because identifying an individual is very difficult (Scrosati and DeWreede 1997; Scrosati 2005). Self-thinning has not

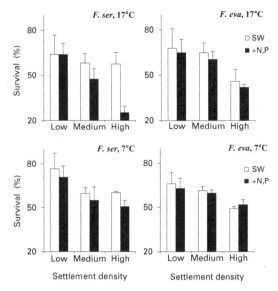

Figure 4.5 *Fucus serratus* (*F. ser.*) and *F. evanescens* (*F. eva.*). Mean survival of germlings as a function of settlement density after 90 days of cultivation at 7°C and 17°C (*SW*) and N-P-enriched seawater (*+N,P*). Error bars denote upper 95% confidence limits (*low*, *medium*, and *high* 10, 50, and 250 germlings cm⁻², respectively). (From Steen and Scrosati 2004, reprinted with permission.)

been detected for most clonal red seaweeds examined including *Gelidium corneum* (previously *Gelidium sesquipedale*) and *Chondrus crispus* (Santos 1995). The frond dynamics are thought to be determined by intra-clonal regulatory processes, because the ramets are genetically identical and physiologically integrated, so may not be in direct competition with each other (Scrosati and DeWreede 1997). Some clonal seaweeds, such as *Asparagopsis armata* and *Sargassum lapazeanum*, are structurally more similar to unitary seaweeds, with a large frond relative to a small holdfast, and for these species self-thinning has been recorded (Flores-Moya *et al.* 1997; Rivera and Scrosati 2008).

Interspecific exploitative competition between seaweeds has been studied in the laboratory only a few times (Russell and Fielding 1974; Enright 1979; Steen 2004; Choi and Norton 2005b). The lack of this type of experiment seems surprising as they are commonly used in terrestrial plant studies on interspecific

competition (Steen 2004). One reason may be the difficulties of maintaining larger seaweed in culture beyond a small size. Two classical experimental designs have been used to examine interspecific competition involving two species (see Steen 2004). (1) The replacement design (de Wit replacement series; de Wit 1960) in which the density of the population is held constant but the relative proportions of each species is changed. The competitive abilities in co-culture are then compared to their growth abilities in monoculture. (2) The additive design in which an initial density of one species is held constant, and the competitors are added. In both designs there are confounding effects (reviewed by Steen 2004). The de Wit replacement series cannot separate the effects of inter- and intraspecific competition (Underwood 1986) and such ratio diagrams give no information about competitive interaction in the field, where initial densities are not experimentally constrained (Inouye and Schaffer 1981). Nevertheless, additive and replacement designs can provide useful information on growth and survivorship at different densities, but it is important to understand the nature of the confounding effects so that the results are not inappropriately attributed to competition (Choi and Norton 2005a, b).

Steen (2004) used a factorial design (that avoided the confounding effects described above), to examine the effects of settlement density, nutrient supply, and temperature on the competitive interactions between *Fucus serratus*, *F. evanescens*, and *Ulva compressa*. The result, that the competitive ability of *Ulva* increased with increasing temperature and nutrient supply, helps explain why *Ulva* tends to dominate in eutrophic conditions, and why *Fucus* outcompetes ephemeral seaweeds at low temperatures. Choi and Norton (2005b) explained the competitive dominance of *Fucus serratus* and *Himanthalia elongata* by differences in their morphology at different times in their life. Their windows of reproduction are similar, but in the field they mostly occur as monospecific stands. During its early life stages, *Himanthalia* forms a hat-like (syncytial) morphology, followed by a button-like morphology; both morphologies create rigid canopies that are virtually impenetrable to light, and thereby exclude *Fucus*. Under a less dense *Himanthalia* canopy, *Fucus* may be able to survive until it is taller than *Himanthalia*, and then it will form a dense canopy thereby out competing *Himanthalia*.

4.3   Grazing

4.3.1   Impact of grazing on community structure and zonation

Every community has herbivores, which play a critical role in transferring energy from primary producers to higher trophic levels. Different grazers dominate in different geographic locations (Poore *et al.* 2012). For example, on the eastern Atlantic coast of Europe, patellid limpets are the dominant intertidal grazers while on the western Atlantic of the USA, these limpets are absent and the periwinkle *Littorina littorea* dominates (it was introduced to this region in 1858), and in the north Atlantic of Iceland, neither grazer is present (Jenkins *et al.* 2008). Herbivorous fish occur chiefly in warmer waters (40°N to 40°S) and are especially significant in the tropics (Floeter *et al.* 2005; sec. 3.3.1). On temperate subtidal reefs, sea urchins and gastropods are the main large grazers, although there are examples of herbivorous fish exerting influence. For instance, in southern New Zealand, *Odax pullus* grazes extensively on juvenile *Durvillaea antartica* thereby controlling its distributional limits (Taylor and Schiel 2010). Another important group are mesograzers, including amphipods, isopods, gastropods, and polychaetes, which can occur in high densities (Brawley 1992b; Duffy and Hay 2001). Such differences in grazer identity, richness and abundance means that the modulation of seaweed communities by grazers varies with geographic region (Poore *et al.* 2012).

Herbivore impact on seaweed populations can be dramatic. In temperate subtidal regions of the North Atlantic and Pacific, and the South Pacific, urchins create and maintain "barrens", in which large kelps are removed and the community is dominated by coralline seaweeds (Andrew 1993; Hagen 1995; Wright *et al.* 2005; see review by Uthicke *et al.* 2009). In Tasmania, for example, the East Tasman Current system has moved south resulting in a rapid regional

seawater warming and a poleward migration of the barren-forming urchin *Centrostephanus rodgersii* (Ling and Johnson 2009; Johnson *et al.* 2011).

The causes of the "catastrophic" changes in ecosystem structure associated with the urchin–kelp relationship in Nova Scotia, Canada and Maine, USA, have been controversial, because the change from a productive kelp-based ecosystem to urchin barrens has consequences for loss of ecosystem services to humans (e.g. Breen and Mann 1976; Elner and Vadas 1990; Lauzon-Guay and Scheibling 2010). However, long-term studies show that the cycles are natural and the decadal switches between "alternate stable states" are driven by the size, density, depth distribution, and feeding behavior of the green urchin *Strongylocentrotus droebachiensis* (Lauzon-Guay and Scheibling 2010). Populations of green urchins undergo "boom and bust" cycles with rapid population increases followed by mass mortalities (Scheibling and Hatcher 2007; Uthicke *et al.* 2009). Recovery of a stable kelp ecosystem after an urchin mass mortality event (caused by amoebic disease) is fast (months–years), whereas the reverse transition to an urchin barren is gradual (years–decades). The mechanism by which barrens are created depends on the substratum that occurs below the kelp zone. Where rock substratum is available, deep-water urchins form feeding fronts of up to ~400 individuals m^{-2} that move up into the kelp beds, clearing them as they go. The situation is different when the substratum is sand. In this case, patches occurring randomly within a kelp bed are expanded by resident urchins, which may form aggregations of up to ~150 m^{-2}, until the patches coalesce into a barren. Other factors affecting urchin grazing behavior are sedimentation, temperature, and wave action (Lauzon-Guay and Scheibling 2007; sec. 4.3.2 and 8.3.2).

A classic example of how sea urchins trigger the change between alternate stable ecosystems is seen in Alaska, USA (Estes *et al.* 1998, 2004). Here, urchin populations are regulated directly by predatory sea otters and indirectly by apex carnivores, notably humans who hunted otters to local extinction in the 1800s (Estes *et al.* 2004). The removal of otters caused a shift from a highly productive kelp-dominated ecosystem to less productive urchin barrens. Studies of islands in the Aleutian archipelago, with and without otters, illustrate the dramatic changes in the energy flow through the ecosystem in the absence of kelp (Duggins *et al.* 1989; Fig. 5.30). Carbon from kelp production supports the entire food web and on islands with little kelp, Duggins *et al.* (1989) showed that growth rates of secondary producers, mussels and barnacles, in the intertidal and subtidal were much reduced. It was reported that the effects flow up through the different trophic levels of the food web, and even affect bald eagles, benthic-feeding sea ducks, and gulls (Estes *et al.* 2004).

The extent to which urchins control *Macrocystis* populations in California has been the subject of much debate (e.g. Foster *et al.* 2006; see reviews by Graham *et al.* 2007a; Foster and Schiel 2010). In Palos Verdes and Point Loma, Southern California, the purple urchin, *Strongylocentrotus purpuratus*, and the large red urchin, *S. franciscanus* created urchin barrens during the 1950s to 1970s (Leighton *et al.* 1966), considered at the time a catastrophic collapse of the ecosystem. The overfishing of sea urchin predators, including lobsters, sheephead wrasse, and sea otters, and/or urchin competitors (abalone) is the most common explanation, but evidence for such a strong top-down control is weak. These urchins preferentially consume drift algae, but will actively graze living kelp when the supply of drift is insufficient; such a switch from foraging to grazing behavior may explain the 1950s to 70s decline in kelp (reviewed by Foster and Schiel 2010). In time, these kelp beds recovered, but to a lower density, which can be explained by the increased sediment loading in this system and hence less bare rock for holdfast attachment (Fig. 4.6). Other dramatic declines in the 1980s were due to El Niño events, but again the populations recovered. Foster and Schiel (2010) conclude that while urchin grazing and El Niño contribute to declines, improving water quality (i.e. reducing sedimentation and sewerage) is the best management tool for conserving kelp beds in southern California. These kelp beds are controlled by complex interactions between biotic and abiotic events, and urchin grazing pressure is only one factor contributing to kelp bed declines.

The effects of urchin grazing in the tropics was dramatically demonstrated by a mass mortality in the Caribbean in 1983, when 93–100% of the urchin *Diadema antillarum* died throughout the region (de Ruyter van Steveninck and Bak 1986; Hughes *et al.* 1987; Carpenter 1988; Lessios 1988, 2005). There was an ecosystem-wide response which included reef degradation, and a phase change to a macroalgal-dominated system in which the productivity was severely reduced (Carpenter 1990; Hughes 1994). This change has lasted 20 years and in Panama where the mass mortality was first observed there has been no recovery of *Diadema* populations (Lessois 2005). In Jamaica and St Croix, however, there is evidence of local recovery of *Diadema* populations which appear to be spreading in an east–west direction, facilitating the recruitment of scleractian corals; this spread may represent the onset of a reverse phase shift to coral reefs in the Caribbean (Carpenter and Edmunds 2006).

The examples from Alaska and Nova Scotia whereby the action of a single apex predator *indirectly* affects the community of primary producers are classic cases of a "trophic cascade". In Tasmania too, the overfishing of rock lobster *Jasus edwardsii* likely contributed to the formation of urchin barrens (Johnson *et al.* 2011). Such an effect was also seen in Leigh marine reserve, northern New Zealand. Here, fishing of predators such as *J. edwardsii* and the fish snapper (*Pagrus auratus*) was stopped in the 1960s. Over time, populations of the urchin *Evechinus chloroticus* decreased and those of the kelp *Ecklonia radiata* recovered (Shears and Babcock 2003; Babcock *et al.* 2010). In the intertidal, an example of a trophic cascade was reported from the Aleutian archipelago, following the introduction of the Norway rat, which triggered an ecosystem switch from seaweed domination to domination by sessile invertebrates. The rats consume foraging seabirds with the result that the numbers of the birds' prey, grazing herbivores, increased. These in turn removed more algal biomass and freed up space for sessile invertebrates such as barnacles (Kurle *et al.* 2008; Fig. 4.7). Another intertidal example of a trophic cascade is that of the starfish *Pisaster ochraceous* in the northeast Pacific, studied by R.T. Paine (references cited in Menge *et al.* 1994). Here, in the low intertidal zone of

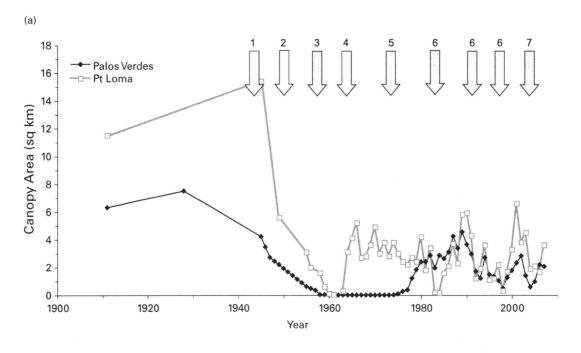

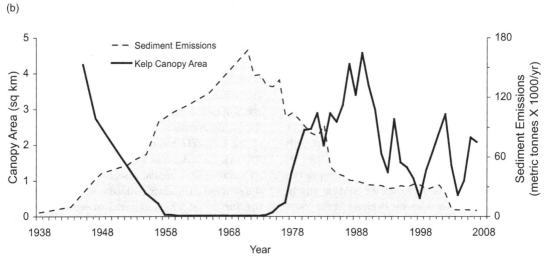

Figure 4.6 (a) Giant kelp canopy area of the Palos Verdes and Point Loma kelp forests. Events indicated by arrows: 1, installation of shallow sewage outfall at Palos Verdes; 2, sediment discharge from Mission Bay and sewage discharge from San Diego Bay at Point Loma; 3, 1957–59 El Niño; 4, sewage discharge moved offshore at Point Loma; 5, beginning of sea urchin fishery in southern California and advanced sewage treatment at Palos Verdes; 6, El Niño events after 1959; 7, record southern California rainfall and runoff. (b) Palos Verdes giant kelp canopy area and mass emission rates of suspended solids from ocean sewage discharges off Whites Point. Refer to the original article for sources of information used in these figures. (From Foster and Schiel 2010, reproduced with permission.)

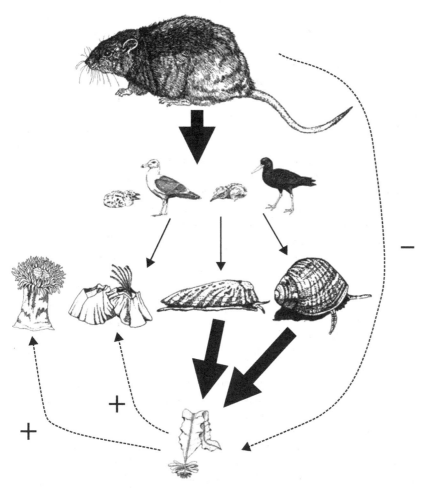

Figure 4.7 Introduced Norway rats indirectly alter the intertidal community in the Aleutian Islands through direct predation on birds that forage in the intertidal. Dotted arrows indicate indirect effects, whereas solid arrows indicate direct effects. Rats keep glaucous-winged gull and black oystercatcher numbers low, which releases intertidal invertebrates such as barnacles and herbivorous snails and limpets from foraging pressure. Greater numbers of grazing invertebrates leads to a significant decrease in algal cover, which allows more settling space for sessile invertebrates. The marine rocky intertidal is altered from an algae-to an invertebrate-dominated system. (From Kurle *et al.* 2008. © 2008 by The National Academy of Sciences of the USA. Illustration by Gena Bentall.)

wave-exposed headlands *Pisaster* preys on mussels, thereby creating space for a seaweed community, whereas the zone above where *Pisaster* does not forage is dominated by mussels. However, further experiments by Menge *et al.* (1994) showed that in neighboring wave-sheltered shores, *Pisaster* did not act as a

keystone predator and there was no trophic cascade. Here, the community control was "diffuse" whereby the overall grazing pressure is still strong but it is by a group of predators each of which independently exert little effect. Keystone versus diffuse control of primary producers may also explain the different effects of urchin removal on kelp beds in Alaska (trophic cascade) versus southern California (no cascade), discussed above.

Urchins, with their sometimes dramatic effects on kelp beds, have received much attention, but another important group is the diminutive mesograzers, whose influence on community structure has been neglected because they are small and highly mobile, making field experimentation difficult (Poore *et al.* 2009; Newcombe and Taylor 2010). In their meta-analyses, Poore *et al.* (2012) found that mesograzers can have an effect on seaweed abundance that is similar to that of macrograzers (see their Fig. 3b). Amphipods can cause large-scale defoliation of seaweeds (Tegner and Dayton 1987; Graham 2002; see sec. 3.5.2). Mesograzers can also positively affect their hosts by grazing epiphytes, by providing nitrogen via excretion, and by facilitating the release and dispersal of spores (Brawley and Fei 1987; Buschmann and Bravo 1990; Duffy 1990). Micro- and mesocosms have been employed to help overcome the difficulties associated with field experimentation. Brawley and Adey (1981) studied the effects of gammarid amphipods in a coral reef microcosm and they found that filamentous pioneer algae such as *Ectocarpus rallsiae* (previously *Hincksia rallsiae*), *Bryopsis hypnoides*, and *Centroceras clavulatum*, among others, were quickly grazed down, allowing a chemically defended species, *Hypnea spinella*, to dominate. Grazing by caprellid and gammarid amphipods prevented epiphytic overgrowth of *Gracilaria*, even in the presence of fish predation on the amphipods (Brawley and Fei 1987).

Severe defoliation of seaweeds by mesograzers is thought to be a rare event in nature because mesograzer populations are usually kept in check by predatory fish (Poore *et al.* 2009). There are several experimental examples of trophic cascades in which the magnitude of fish predation on mesograzers controls benthic community structure (Duffy and Hay 2000; Davenport and Anderson 2007; Newcombe and Taylor 2010). In North Carolina, USA, a mesocosm experiment showed that brown algae dominate because omnivorous fish preferentially feed on red and green seaweeds, and also prey on the brown-seaweed-loving amphipods (Duffy and Hay 2000). In a field experiment in California, USA, Davenport and Anderson (2007) used cages to exclude predatory fish from *Macrocystis* beds. In the absence of the predatory fish, mesograzer populations increased substantially, causing a decline in *Macrocystis* fitness as it produced fewer fronds and apical meristems. Mesocosm experiments in New Zealand showed that epiphyte loading on seaweeds was lower in the presence of a small predatory wrasse (*Notolabrus celidotus*) which consumed the mesograzers (Newcombe and Taylor 2010). However, a rather different result was observed by Poore *et al.* (2009) who used a slow-release insecticide to kill amphipods in the field, thereby avoiding the potential experimental artifacts associated with grazer-exclusion cages and micro/mesocosms. The insecticide killed up to 93% of amphipods but this had no detectable effect on either the growth rate of the host seaweed, *Sargassum linearifolium*, or the abundance of epiphytic seaweeds on *Sargassum*'s surface. It is possible that small predatory fishes naturally suppressed amphipod populations at Poore *et al*.s' study site so that the experimental removal of amphipods had little further effect, and/or other grazers such as gastropods took over the role of the absence of amphipods, termed "redundancy". The study of Poore *et al.* (2009) clearly illustrates that the ecological role of mesograzers is still not fully understood.

In the intertidal zone, grazing pressure can influence the outcome of competition between algae, and algal diversity. In New England, USA, intertidal rock pools tend to be dominated by either *Chondrus crispus* or *Ulva*. *Ulva* is the competitively dominant species, but is also the preferred food of the dominant grazer, *Littorina*. *Chondrus* is successful only when high densities of *Littorina* keep the rapidly growing *Ulva* in check. Removal of the snails from a *Chondrus*-dominated pool resulted in a takeover by *Ulva*. At an intermediate density of *Littorina*, species diversity within the pools was greatest, because grazing allows the competitively inferior species to exist as they are "released" by grazing from competition with *Ulva* (Lubchenco 1978). In the same experiments but on an emersed site, the bimodal effect seen in tide pools was not observed. Instead, algal species diversity decreased linearly with increasing *Littorina* density.

This is because on emersed shores, *Littorina*'s preferred food *Ulva* (and other fast-growing ephemeral species) is competitively inferior to canopy-forming fucoid algae in the mid-intertidal and *Chondrus* in the low intertidal. In this case, increased grazing pressure results in reduced diversity as ephemeral species are grazed, and the unpalatable (i.e. chemically defended) canopy formers remain. Consumers can modulate species diversity and control the outcome of competition, but the strength of this effect will vary in different environments.

Also in the intertidal, Coleman *et al.* (2006) were the first to conduct hypothesis-driven experiments on the effects of limpet exclusion across a latitudinal gradient (54o5'N, Isle of Man, UK to 37o41'N, southern Portugal) using identical methodologies and experimental timing across the study. The exclusion of limpet grazers triggered higher algal biomass at all sites but latitudinal trends emerged. In the northern sites, limpet removal consistently (i.e. deterministic) resulted in the establishment of a fucoid canopy in the mid-intertidal mussel/barnacle zone, and the spatial variability in algal cover decreased (i.e. a less patchy distribution). In southern Portugal, the community response to limpet removal was inconsistent (i.e. stochastic) and varied between sites. This is because compared to northern sites, there are very few or no dominant canopy-forming algae but a higher diversity of turfing species that are able to rapidly colonize free space. At this site, therefore, limpet removal caused increased small-scale spatial variability. The general pattern was that in a physically "harsh" environment (southern Portugal) limpet exclusion was unpredictable compared to the predictable colonization in the relatively benign northern sites.

In sec. 3.1.3, we considered how processes related to the physiological ability of seaweeds to withstand emersion control (in part) their zonal positions on the intertidal zone, and this theme is further explored in Chapters 5, 6 and 7. However, the upper and lower vertical limits of seaweeds can also be modified by grazers (reviewed by Raffaelli and Hawkins 1996; Essay 2, Chapter 3). The upper limit of a red *Laurencia-Gigartina* belt in parts of Britain is controlled by mid-shore *Patella* populations (Lewis 1964). In central Portugal and southern England, grazer exclusion caused seaweeds to increase their upper distributional limit by 0.5 m (Boaventura *et al.* 2002). Grazer exclusion extended the lower distribution of *Mazzaella parksii* (previously *Mazzaella cornucopiae*) on Tatoosh Island, Washington State, USA (Harley 2003). Noël *et al.* (2009) found that grazing pressure in rock pools in southern England was twice as high as that in neighboring emergent sites because at high tide, the limpet *Patella vulgata*, moves into the rock pools to graze. In South Africa, the exclusion of the mid-intertidal limpet *Cymula oculus* over a 5-year period caused the community diversity to change. When limpets were present, the mid-intertidal had one seaweed species (the herbivore-resistant crust *Ralfsia verrucosa*) and nine invertebrates, compared to nine seaweeds and 19 invertebrates in grazer-free plots (Maneveldt *et al.* 2009). Remarkably sparse grazers are capable of preventing the establishment of foliose algae (Underwood 1980). Removal of the grazers from plots in the animal-dominated middle-to-upper-intertidal zone in Sydney, Australia, allowed foliose algae to establish, but in this case the initial upward range extension did not persist as most algae did not survive as adults owing to the harsh physical conditions of this intertidal shore (Underwood 1980).

An interesting difference in seaweed–animal interactions between North American and southeastern Australian rocky intertidal zones has been pointed out by Chapman (1986). In North America, low-shore algal-dominated zones occur only where carnivory by whelks and seastars impacts the competitively superior barnacles and mussels. In Australia, in contrast (Underwood and Jernakoff 1981), seaweeds competitively displace limpets, which feed only on microscopic algae/stages, and there is a dense lower-intertidal belt of algae. The algal zone ends abruptly in the mid-intertidal, where it is superseded by barnacles or grazing molluscs (depending on wave action).

Zonation can also be influenced by a combination of grazing and competition. For instance, in southern Chile, a low-intertidal zone of crustose corallines and *Ulva rigida* normally separates a lower zone of *Ahnfeltiopsis furcellatus* from an upper zone of *Mazzaella laminarioides* (previously *Iridaea boryana*). But this band is maintained by mollusk grazing; when the herbivores were removed, *Gymnogongrus* and *Mazzaella* (*Iridaea*) outcompeted the slow-growing corallines and ephemeral *Ulva*, so their zone limits changed (Moreno and Jaramillo 1983). Crustose coralline algae compete for space by overgrowing each other (Fig. 4.2c), but limpets influence the outcome in tide pools in Washington. Thicker crusts overgrow thinner crusts. In high pools, frequent but low intensity grazing rasps away the surface cells of *Lithophyllum impressum*, but deeply wounds *Pseudolithophyllum whidbeyense*, which lacks a multilayered epithallus. *Lithophyllum* wins. In lower pools, meristems of both crusts are injured, but *Pseudolithophyllum* rapidly overgrows its wounds and the neighboring *Lithophyllum* (Steneck *et al.* 1991).

Interactions among grazing, competition, and wave action have been extensively studied in the Gulf of Maine, USA (Essay 2, Chapter 3). Lubchenco and Menge (1978) showed that on wave-exposed coasts, mussels dominate because wave action reduces the activity of mussel predators, while on wave-protected shores *Ascophyllum* dominates due to higher levels of predation on mussels. These results suggest that the communities are controlled by consumers (i.e. top-down control), and the effect is dependent on wave action. In a third habitat type, "wave-sheltered bays and estuaries", *Ascophyllum/Fucus* beds and mussel beds are found in mosaics, with patches of each community type growing next to one another, although here too *Ascophyllum* dominates areas of slow flows and mussels are found in high-flow sites. Two research groups have used this habitat type to examine how communities develop following simulated disturbance by ice scour. Bertness *et al.* (2002) created patches of different sizes within each habitat, and used grazer-exclusion cages to manipulate consumers. When the consumers were present, neither mussel beds nor fucoid canopies returned over a 3-year period, whereas in the absence of consumers the original community re-established. They proposed a model in which these communities are under consumer control, modulated by water motion and the supply of propagules, and both communities are highly deterministic and self-replicating, so that following a disturbance

event, the same community will develop, and positive feedbacks serve to retain that community (Fig. 4.8).

In a separate series of experiments spanning 9 years, also in sheltered bays along the coast of Maine, Petraitis *et al.* (2009, and see references therein) tested the hypothesis that mussel beds and *Ascophyllum* beds represent alternate stable states, rather than consumer-controlled self-replicating communities. To do this, they cleared patches of differing sizes to simulate ice scour, but only within the wave-sheltered *Ascophyllum* beds. They argued that if mussel beds can develop in a cleared site within an *Ascophyllum* bed and in the presence of active consumers, then this is evidence that disturbance can cause the community to switch to an alternate stable state. For cleared sites that were >2 m diameter, all but one site became dominated by either *Fucus vesiculosus* (60%) or mussels (37%), but *Ascophyllum* did not recover. This effect was scale-dependent, and within the 1 m cleared patches *Ascophyllum* populations re-grew. Petraitis *et al.* (2009) make the case that the large plots simulate severe ice scour, the sort of dramatic event that could tip an ecosystem from one stable state to another. They propose a model that incorporates the site-specific, top-down control proposed by Lubchenco and Menge (1978) and Bertness *et al.* (2002), but they suggest that the community that develops in the sheltered sites will strongly depend on the identity of the colonizing species, and this in turn will depend on the size of the cleared patch (Fig. 4.9). Large patches tend to be disconnected from neighboring communities, and therefore succession patterns will depend on the identity of early settlers. *Fucus vesiculous*, mussels, and barnacles are competitively superior at colonizing large spaces, whereas in a small cleared plot, *Ascophyllum* is the superior colonizer. These long-term studies of the same communities are very valuable, and illustrate clearly the complexities of interactions between seaweeds, grazers, water motion, and the scale of a disturbance. The effects of physical disturbances on seaweed communities is considered further in sec. 8.3.2.

The focus of this chapter is biotic interactions, but it is clear from many of the above examples that abiotic factors can influence the outcome of competition. The

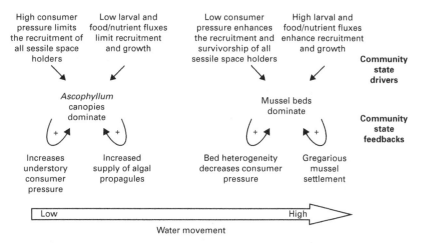

Figure 4.8 Conceptual model of the deterministic generation and maintenance of *Ascophyllum* canopy–mussel bed alternate community states on Gulf of Maine rocky shores. At low-flow sites, intense consumer pressure by snails and crabs leads to habitats dominated by an unpalatable seaweed canopy, whereas at high-flow sites, low consumer pressure and high recruit supply lead to mussel bed habitats. Positive feedbacks in each community type lead to its expansion and persistence. If stochastically determined alternate stable states occur in this system, they are predicted to occur at intermediate-flow sites where consumer control and the predictability of larval recruitment are relaxed. (From Bertness *et al.* 2002. © 2002 by the Ecological Society of America.)

terms "top down" and "bottom up" have been coined to describe if the major controlling factor of a community is consumers or resource supply (typically inorganic nutrients), respectively (Lotze *et al.* 2001, Hillebrand *et al.* 2007; Korpinen *et al.* 2007). Several examples of top-down control were given in the previous sections, including trophic cascades. Lotze *et al.* (2001) propose that the relative effects of bottom-up versus top-down factors on seaweed assemblages will depend on: (1) the nutrient status of the system; (2) seasonal variations in both nutrient supply and grazing pressure; (3) whether seaweeds are annual or perennial; and (4) the age of the seaweed, as new recruits are likely to respond differently from mature seaweed. It is clear from this long list of factors that top-down versus bottom-up control will be site specific (Hillebrand *et al.* 2007).

Strong top-down control on primary producers is predicted in nutrient-poor environments, while consumers will have less control when nutrients are abundant (Lotze *et al.* 2001). Guerry *et al.* (2009) showed that top-down control strongly regulated an upper shore community dominated by limpets and low-lying

algae in southern New Zealand. Overall, there was little effect of nutrient enrichment on this community which was dominated by encrusting and corticated seaweeds, although the abundance of foliose seaweeds increased when nutrient supply was enhanced and grazing pressure reduced. Bottom-up effects are thought to take longer to exert an influence on the system compared to grazer effects (e.g. Nielsen and Navarrete 2004) and Guerry *et al.* (2009) wondered, if their experiment had run for >1 year, whether a different pattern might have begun to emerge. In rock pools in Oregon, USA, nutrient fertilization affected seaweed community structure and primary production, and the effect depended on water motion. However, while nutrient enrichment affected seaweed productivity, there was no obvious flow-on effect to the higher trophic levels (Nielsen 2001). Korpinen *et al.* (2007) tested the effects of herbivory and nutrients along a depth gradient. In shallow sites, nutrient enrichment increased seaweed density but the grazers were able to keep up, and top-down control dominated. In summary, the interactions between resource supply and consumer pressure are not easy to predict,

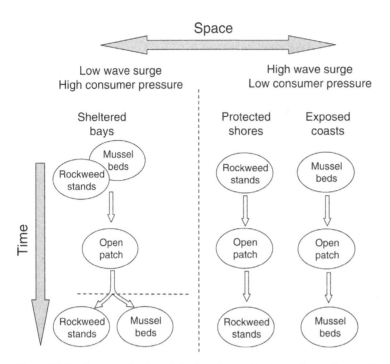

Figure 4.9 Development of rockweed stands and mussel beds over time and space. On open coastal shores there is a gradient in wave surge from protected to exposed shores with predators of mussels showing an increase in activity as wave surge diminishes. Work by Lubchenco and Menge (1978) suggested that rockweeds are confined to protected shores where predators are active enough to eliminate mussels. In sheltered bays, both mussels and rockweeds occur as a mosaic of patches and our work suggests that development depends on which species becomes established first. The *horizontal dotted line* shows where a developing patch is "committed" to becoming a mussel bed or rockweed stand. *The vertical dotted line* shows the hypothetical boundary between the two kinds of development, but transition could also be gradual or not linked to wave exposure alone. (From Petraitis *et al.* 2009, reproduced with permission.)

and the relative importance of top-down versus bottom-up control depends on the spatial and temporal variation in grazers and nutrient enrichment, and these controls will vary along environmental gradients. Furthermore, nutrient addition experiments in the field often produce equivocal results, probably because of the numerous other factors that affect population and community dynamics. The bottom-up effects of nutrients are discussed further in secs. 6.8.5 and 9.6.

4.3.2 Seaweed–herbivore interactions

In the previous section we discussed how herbivore grazing affects community structure, but seaweeds are

also affected on an individual level. The expected herbivore damage has three components: (1) the probability that an individual seaweed will be encountered by a herbivore; (2) the probability of the seaweed being eaten, at least partially, if it is encountered; and (3) the cost, in terms of fitness, of the grazing damage (Lubchenco and Gaines 1981). Minimum damage will occur to seaweeds that are out of contact with grazers, are unpalatable, or are able to sustain damage without significant cost. The effect of grazing on the seaweed will depend on the amount and kind of tissue lost and on the timing, particularly with respect to reproduction. Grazing may sometimes increase productivity by reducing self-shading in algal turfs; Williams and Carpenter (1990) suggested this as the reason that tropical turfs grazed by urchins are 2–10 times more productive than turfs grazed by fish or mesograzers. Loss may also be counter-balanced by the survival and dispersal of algal spores during digestion by herbivores. Santelices *et al.* (1983) showed that the spores of opportunistic species survive much better than those of late-successional species. At one extreme, great damage to *Macrocystis* can be done by urchins, which eat little of the seaweed but cause the loss of entire fronds by chewing through the bases of the stipes (e.g. Leighton 1971; see Schiel and Foster 1986). At the other extreme, some crustose coralline algae are susceptible to grazing by limpets but do not suffer much harm from it. The weakly calcified epithelial layer is removed, but it is replaced by the meristem beneath, which survives. Indeed, in the absence of limpet grazing, the epithelial layer of some corallines becomes too thick and the seaweed is overgrown by epiphytes and dies (Steneck 1982).

The means available to seaweeds for minimizing consumption are collectively known as "defenses" and several schemes have been used to categorize seaweed defenses (Littler and Littler 1984, 1988; Duffy and Hay 1990, 2001; Iken 2012). The various mechanisms available essentially fall into three broad categories: escape, tolerance, or deterrence. Escapes may be in space, time, and size, or there are associational escapes in which one organism grows in close association with an unpalatable or territorial species. Consumption may be tolerated via rapid growth involving the replacement of vegetative and reproductive tissues such as occurs in e.g. turf-forming algae of coral reefs (sec. 3.3.1). Potential consumers may be deterred by structural and morphological features, and/or by using chemicals that are unpalatable to the consumer (Nicotri 1980; Duffy and Hay 2001).

Temporal and spatial escapes. Not all places at all times feature high grazing pressure. Even on tropical reefs, where herbivory is generally high, the grazing pressure varies markedly in time and space (Hay 1981a, b). Vertical migrations of herbivores occur (Vadas 1985), and there are places that herbivores cannot reach. For example, in temperate waters, germlings of *Fucus vesiculosus*, which are more susceptible to periwinkle grazing than are older seaweeds, escape predation when they are in crevices that the snails cannot reach (Lubchenco 1983).

Crustose or boring stages (or crustose parts of heterotrichous seaweed) are more resistant to attack by grazers (but see below), being either out of reach of the grazers or not readily removed by them. Shell-boring algae, such as gomontia and conchocelis stages, are especially well protected against grazing. Tropical reef carbonate is a major habitat for boring, filamentous green and red algae. Numerous species are present (certainly not all are alternate stages of macrophytes), making up over 90% of the algal biomass within the coral (zooxanthellae contribute relatively very little) (Littler and Littler 1984, 2011d). Grazing by parrotfish does, nevertheless, reach these algae. Upright stages, in contrast, may be completely removed through damage to their relatively narrow stipes. The disadvantage to the crustose or boring stage is the increased susceptibility to being overgrown, but this problem may be mitigated by grazers removing the epiphytes.

A refuge may occur by virtue of being too far away from grazer shelters. Visually striking halos occur around the edges of clearings in mussel beds on the west coast of North America (Fig. 4.10) because grazing limpets are limited in how far they can roam from their hiding places during low tide (Paine and Levin 1981). A similar phenomenon occurs around tropical patch reefs, which provide shelter for urchins during the day (Ogden *et al.* 1973). In the first case, desiccation avoidance drives the limpets back; in the case of the urchins, the driving force is predatory fish.

Refuge habitats also include areas with too much wave action for grazers. For example, in laboratory experiments, the feeding rates of large specimens of the sea urchin *Stronglyocentrotus nudus* on *Laminaria* spp. decline when water velocities are greater than 0.25 m s^{-1}, and feeding stops at 0.4 m s^{-1}. These results support field observations that there are more seaweeds and fewer sea urchins in the shallow-subtidal zone in the Iwate Prefecture, Japan, whereas in deeper water there were more urchins and fewer seaweeds (Kawamata 1998, 2010). Strong wave action on tropical reef crests reduces grazing pressure from fish and urchins and allows a "climax" coralline algal community (*Lithophyllum congestum* and *Hydrolithon*

Figure 4.10 Halos around the edges of algal patches within mussel beds on the exposed Washington coast are caused by grazers (limpets and chitons) that shelter among the mussels during low tide and graze only 10–20 cm from shelter. They are unable to forage in the middle of the large patch, but can graze the whole of the small patch (the two patches are the same age). The fleshy algae in the middle of the larger patch are *Porphyra* (center), surrounded by *Halosaccion*. The halo is occupied by barnacles and *Corallina vancouveriensis*. (Photo courtesy of Robert T. Paine, © 1991 by Robert T. Paine.)

pachydermum (previously *Porolithon pachydermum*) in the Caribbean) to build intertidal algal ridges (Adey and Vassar 1975) (see Fig. 3.9b).

Lubchenco and Cubit (1980) and Slocum (1980) have advanced the hypothesis that heteromorphic algal life histories may allow a response to grazing. Lubchenco and Cubit (1980) worked with several winter ephemerals *(Ulothrix, Urospora, Petalonia, Scytosiphon, Bangia,* and *Porphyra)* in areas where the grazing intensity varied seasonally. When grazers were removed experimentally, upright forms of the algae were found at times of the year when normally only the small phase would occur. Slocum (1980) found that the relative amounts of crusts and blades of *Mastocarpus papillatus* depended on grazing intensity, and Dethier (1982) concluded that seasonal changes in abundance may be largely or entirely due to herbivore abundance and feeding rates. If variations are unpredictable, the population survival will be highest with continuous production of both morphs, although only one will survive at a time (Lubchenco and Cubit 1980). "Bet-hedging" (as Slocum calls it) maximizes the total progeny of each such generation. However, Littler

et al. (1983b) suggested that although crustose stages are more grazer-resistant, they are also more resistant to sand scour and burial, wave shock, desiccation, and temperature, thus representing a phase appropriate for various "harsh" conditions. The situation is further complicated because the reproduction of some species or populations of *Porphyra* and *Scytosiphon* is known to be cued by light or temperature, and one would not expect to find erect forms developing out of season. These different ideas on whether or not the heteromorphic life histories, apparent for many seaweeds, are a strategy for avoiding periods of intense grazing pressure still requires experimental testing (sec. 2.2). It must also be recognized that these are not mutually exclusive alternatives.

Associational escapes. Protection may be gained from proximity to unpalatable algae or even toxic animals (Hay 1988). The dictyotalean brown, *Stypopodium zonale*, is one of the most toxic seaweeds, and palatable algae such as *Acanthophora spicifera* suffer much less grazing damage when close by *S. zonale* (Littler *et al.* 1986b, 1987). Even plastic models of *S. zonale* have significant salutary effects on neighboring algae. A parallel example was found in North Carolina, where understory *Gracilaria tikvahiae* was protected by *Sargassum* (Pfister and Hay 1988). The costs of competition with the neighbor were outweighed by the benefit of grazer protection and survival was at the expense of a slower growth rate. In Brazil, the rate of herbivory on *Dictyota* sp. is lower when it grows in association (as a facultative epiphyte) with *Sargassum furcatum* than when grown separately, because *Sargassum* is unpalatable and avoided by herbivores (Pereira *et al.* 2010). In California, USA, seaweed recruitment is greater in the presence of the anemone *Corynactis californica*, because the anemone deters urchins. Both the red and purple urchins used in experiments became stressed when placed onto, or next to, the anemone; they retracted their podia and their movement slowed. This is an example of an associational refuge for seaweeds that is caused by an anemone-triggered change in urchin behavior (Levenbach 2008).

The case of the ocean goldfish, *Hypsypops rubicunda*, nests discussed in sec. 4.2.1 as an example of interference competition is also an example of the benefits of falling into association with a territorial animal. This temperate-water damselfish is a carnivore, but it drives away all other fish, including herbivores (which otherwise would try to eat its nest). More complex behavior is shown by a group of tropical damselfish collectively known as "farmer fish". These fish maintain algal assemblages inside their territories that differ markedly in species abundance and species composition from those of surrounding areas (Fig. 4.11a) (Lassuy 1980; Hinds and Ballentine 1987; Klumpp and Polunin 1989; reviewed by Ceccarelli *et al.* 2001). Contiguous territories can cover large areas of reef flats (Fig. 4.11b) and contribute substantial productivity, most of which is consumed by the fish (Brawley and Adey 1977; Klumpp *et al.* 1987; Polunin 1988). Farmer fish are selective regarding which algae they will allow to grow, weeding out certain species.

While most studies have focused only on the seaweed component of the algae in the farms, the diet of the fish (in Papua New Guinea) was recently shown to consist 30–80% of epiphytic diatoms on the turf (Jones

(a)

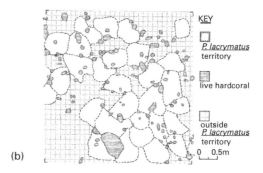

(b)

KEY

☐ *P. lacrymatus* territory

▧ live hardcoral

☐ outside *P. lacrymatus* territory

0   0.5m

Figure 4.11 Farmerfish territories. (a) A common modification in the growth form of massive *Porites* (coral) colonies in territories of *Stegastes nigricans* on the Great Barrier Reef, Australia. The damselfish maintains dense algal turf on the sides of the lobes (1.15 cm across in this photograph). (b) Diagrammatic map showing the distribution and sizes of adult territories of *Plectroglyphidodon lacrymatus* on a section of reef on Matupore Island, Papua, New Guinea. (Part a from Done *et al.* 1991, with permission of Springer-Verlag, Berlin; b from Polunin 1988, with permission of Elsevier Science Publishers.)

et al. 2006). There is often a dense, multi-layer diatom assemblage around the seaweed filaments, which the fish apparently can selectively remove, but there may be regional differences in turf communities even with the same fish species (reviewed by Lobban and Jordan 2010). Hata and Kato (2004, 2006) presented evidence that in Okinawa, Japan, one species of *Vertebrata* has an obligate relationship ("cultivation mutualism") with the dusky farmerfish, *Stegastes nigricans*. These fish maintain small territories with this species as a "virtual monoculture" and this *Vertebrata* species (determined using molecular techniques) occurs only in these farmerfish territories. In more tropical localities (Guam, Papua New Guinea) the same fish has larger, more diverse algal farms, but no one has looked yet for that species of *Vertebrata* there.

Our discussion has focused on mechanisms by which seaweeds escape consumption by association, but seaweeds themselves can form associational refugia for small mesograzers. For some amphipods, their seaweed host provides both food and a refuge. The chemically defended seaweeds with which they are associated offer protection against predators, but there are also trade-offs because the food quality of defended seaweeds can be lower than that of non-defended seaweeds. Also, the amphipods must not consume too much or they will lose their home (Cerda *et al.* 2010). Ampithoid amphipods build nests on their seaweed host using silk threads to curl the seaweed around themselves or gluing pieces of seaweed together (Cerda *et al.* 2010; Fig. 4.12). The nests are built on the meristematic regions and the amphipod continuously extends its nest in the opposite direction to kelp blade growth, so that the nest stays in position on the meristem. It can be difficult to determine whether the seaweed is more important as a refuge or as food source, but Sotka *et al.* (1999) achieved this by studying the effect of a non-herbivorous amphipod *Ericthonius brasiliensis* on its specialist habitat, the tropical seaweed *Halimeda tuna*. While *Erichonius* does not eat its host, it still had a negative impact because small predatory fish feed preferentially on colonized seaweed and damage the meristematic tissue. In New Zealand, the amphipod *Aora typica* selects its seaweed habitat based on the quality of the refuge that it, offers, rather than on the nutritional value of the seaweed to the amphipod (Lasley-Rasher *et al.* 2011).

Structural and chemical deterrents. Many herbivores show preferences for certain food types, and, in some cases, they are able to detect their preferred species. Sea urchins, for example, will move upstream in a current of water that has passed over a kelp, but their feeding behavior is influenced by both preference and availability (Vadas 1977). Food choice by herbivores will have strong flow-on effects to higher trophic levels, and there

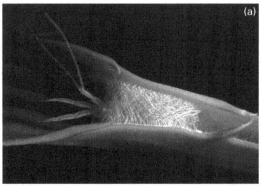

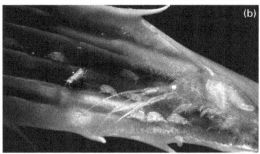

Figure 4.12 Tubiculous nests of *Peramphithoe femorata* on *Macrocystis pyrifera* blades. (a) Weblike amphipod-silk produced by pereopod glands. (b) Amphipod female accompanied by its juvenile offspring. (From Cerda *et al.* 2010. © 2010 Marine Biological Laboratory.)

has been much interest in the reasons that herbivores select one food over another. Structural properties of the seaweed, e.g. morphology or toughness affect grazer choice, as does the chemical composition, or in some cases, both (Nicotri 1980). The roles of structural defenses received early research attention, but it is the chemical ecology of seaweed–herbivore interactions that has become a major research area (sec. 4.4).

The probability of the grazer feeding on a seaweed once an encounter has taken place will depend on the form and palatability of the seaweed. Steneck and Watling (1982) predicted the susceptibilities of seaweeds to molluskan grazers (not including opisthobranchs) by arranging both into functional groups. Algal groups were based on morphology and toughness, similar to those of Littler and Littler (Table 1.1). Some algae, of course, pass through two or more such groups as they grow, and species with heteromorphic life histories straddle two groups. Molluskan grazers were grouped according to the type of radula and grazing action they employed: "brooms" (e.g. keyhole limpets), "rakes" (e.g. *Littorina littorea*), "shovels" (e.g. true limpets), and "multi-purpose tools" (e.g. chitons). Algal groups 1 and 2 are readily scraped off the substrata by grazers with broom like or rake like radulae, whereas the largest or most expansive algae (groups 5 and 7) are consumed primarily by the other two groups of herbivores, which can occupy the alga and gouge out tissue. Intermediate-size algae, of moderate toughness (groups 3, 4, and 6), appear to have a size refuge from these molluskan grazers, being too large to be grazed off the substratum and too small to be occupied.

The use of functional groupings as indicators of grazing susceptibility has been questioned (Padilla and Allen 2000; sec. 1.2.2). The problem lies in the fact that form is only one of a suite of factors that influence seaweed susceptibility to consumers. Even though seaweeds may be chemically defended and their morphology can be highly plastic, grazers can select what they eat based on food quality (e.g. Nicotri 1980). Clements *et al.* (2009) suggest that focusing on the properties of seaweeds that deter herbivorous fish has led to a lack of understanding of how fish select nutritious food. Padilla and Allen (2000) suggest that seaweeds with similar forms may have very different mechanical, material, and structural properties, each of which may affect grazer choice. Nevertheless, Poore *et al.*'s (2012) meta-analysis of 613 grazer-exclusion experiments suggests that seaweed traits, such as functional group and phylogenetic identity, are important in determining their susceptibility to grazing (Fig. 4.13). The Poore *et al.* (2012) analysis supports the commonly held view that coralline seaweeds are among the least preferred by grazers; but as for the idea that coralline seaweeds calcify as a defense mechanism against grazing, it should be remembered that calcifying seaweeds evolved 100 million years prior to their invertebrate herbivores (Steneck 1992).

To identify the factors driving a herbivore's preference for some seaweeds over others, feeding choice assays are used. Such studies often involve presenting a grazer with a single seaweed to eat (no-choice experiment) and/or in a separate experiment, a choice between two or more seaweed species (choice experiment), and measuring the amount of each that is eaten. The utilization of a choice and/or no-choice assay will depend on the question being asked, and the organisms being tested. A choice assay may be ecologically more relevant for a mobile herbivore that in nature has a range of food sources. For a less mobile invertebrate that in nature has little food choice (e.g. a gastropod living on an epiphyte-free seaweed), a no-choice assay may be appropriate. A combination of both assays can be used to determine if low nutritional quality is driving herbivore preference. For example, in a choice experiment, a herbivore is presented with two seaweeds, A and B, and the result is that B is preferred. A no-choice experiment is then run in which the seaweeds A and B are presented separately to the herbivore. If A is eaten more than B in this no-choice experiment, it suggests that A was of a low food quality (e.g. low protein) and when the herbivore has no choice, it ate more of seaweed A to compensate for its low food value (Cruz-Rivera and Hay 2000). If, however, in the no-choice experiment seaweed A was still preferred less than seaweed B, then the reason for this can be further investigated further by testing for structural versus chemical defenses. To do this, seaweeds A and B are separately dried, ground to a fine powder and then mixed with an agar solution to make texturally uniform seaweed "jello" (Hay *et al.* 1998). The seaweed–agar mixtures are then fed to the herbivores in choice experiments. If A was structurally defended (e.g. a tough thallus), then the herbivore should show no preference between agar containing ground seaweed A or B. If seaweed A was chemically defended, then the grazer should eat less of the seaweed A agar compared to seaweed B agar. In this latter case, seaweed A can be further tested for bioactive chemicals using a bioassay-guided experimental approach (e.g. Deal *et al.* 2003). Here, candidate bioactive chemicals are sequentially purified from a crude extract of seaweed A. These extracts are then mixed with a ground-up, palatable seaweed, which is then presented to the herbivore in agar, and the amount eaten is compared to agar containing just the palatable

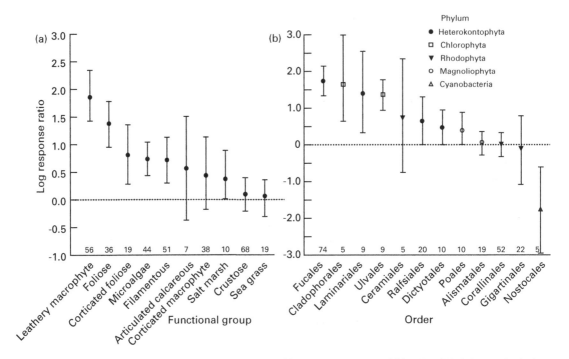

Figure 4.13 Variation in the effects of excluding grazers among (a) functional groups, and (b) orders (labeled by phylum) of primary producers. Means and 95% confidence intervals are derived from linear mixed models of log response ratio against predictor variables. The dotted line at zero is the effect size expected if there is no effect of removing herbivores. Sample sizes for the means are shown above the z-axis. Orders and functional groups with fewer than five replicate observations are excluded. (From Poore *et al.* 2012. © 2012 Blackwell Publishing Ltd/CNRS.)

seaweed. If the herbivore is deterred by a particular purified compound, then this is likely to be the deterrent.

There are several potential pitfalls in the interpretation of feeding choice experiments (Vadas 1985; Cronin and Hay 1996a; Van Alstyne *et al.* 1999, 2001). First, the age of the alga may affect its palatability. *Littorina* snails readily eat *Chondrus* sporelings less than a few weeks old, but they have little taste for older seaweed. Second, there may be intraspecific resource partitioning such that preferences differ among individual herbivores, such as individuals of different ages. Finally, the amount of alga eaten in a feeding experiment may not reflect the importance of that alga in the field.

In choice tests, calcifying seaweeds are often low on the preference list (reviewed by Duffy and Hay 2001).

This is often attributed to a structural defense, with the calcium-carbonate impregnated cell walls (see sec. 6.5.3) making them hard and difficult to bite. However, calcium carbonate by itself can act as a defense to some consumers, and some calcifying seaweeds also contain chemicals that deter grazers. Hay *et al.* (1994) tested the relative importance of calcium carbonate and secondary metabolites as feeding deterrents for three calcifying tropical green seaweeds (*Halimeda goreaui*, *Udotea cyathiformis*, and *Rhipocephalus phoenix*) and three herbivores, an urchin, amphipod, and parrot fish. They made artificial foods using agar, to which they added calcium carbonate and/or secondary metabolites (see below) that had been extracted from the seaweeds. Food choice experiments were conducted in which consumers were given control agar, and experimental treatments of agar plus calcium

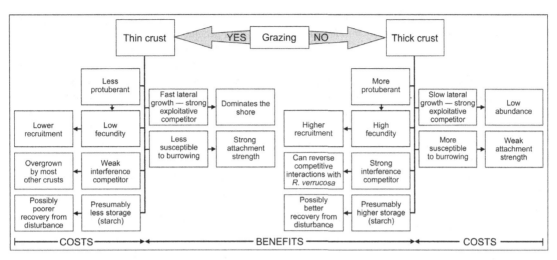

Figure 4.14 A schematic diagram of the benefits and costs to *Spongites yendoi* resulting from the presence and absence of grazing by *Scutellastra cochlear*. (From Maneveldt and Keats 2008. © NISC Pty Ltd.)

carbonate or secondary metabolites, or both. Calcium carbonate alone reduced feeding by the amphipod, was ineffective on the parrotfish, and reduced feeding by the urchin only when agar food quality was reduced by reducing the organic content. The secondary metabolites, in contrast, reduced grazing by all the consumers. When calcium carbonate and secondary metabolites were presented together, however, grazing was even further reduced. A likely explanation is that calcium carbonate changes the pH of the invertebrates' gut, making them more susceptible to the effects of the secondary metabolites.

Although calcifying seaweeds are unpalatable to some herbivores, they are the preferred food of others (Fig. 4.14). In South Africa, the crustose coralline seaweed *Spongites yendoi* provides up to 85% of the diet of the gardening limpet, *Scutellastra cochlear*, while non-calcifying (fleshy) seaweed contribute just 7% to their diet. Surprisingly, the food quality of *Spongites* and the fleshy seaweed was similar. In regions where *Spongites* and *Scutellastra* co-occur, *Spongites* is thin, with a high rate of lateral growth and a strong attachment to the substratum, but it is also more susceptible to overgrowth by other crustose coralline seaweed and has lower fecundity. In regions where the limpet *Scutellastra* is absent, *Spongites* is thick and a better

competitor with other crustose coralline seaweeds, but it has a slow rate of lateral growth and is only weakly attached to the rock. This interesting study illustrates the costs and benefits that a limpet consumer has on the coralline seaweed's fitness. It may be an example of facultative mutualism because the species are not dependent upon one another, but there are advantages to the seaweed in terms of range expansion when it grows with the limpet, and the limpet benefits from a food source (Maneveldt and Keats 2008).

Some grazers are generalists and others are specialists (Sotka 2005; Sotka and Reynolds 2011). Specialization for single food sources is usually found in invertebrates whose individuals are small enough to occupy and gouge the larger algae, which are, from the small grazer's vantage point, abundant and long-lived food supplies. The sacoglossan mollusk *Elysia hedgpethi* is small and slow-moving, and yet selective in choosing only siphonous green algae such as *Codium* and *Bryopsis* (Greene 1970). Its relative in Scotland, UK, *Elysia viridis*, used to feed on *Cladophora ruprestis*, but its preference switched following the introduction of *Codium fragile* spp. *tomentosoides* upon which it is now a specialist feeder (Trowbridge and Todd 2001). Another gastropod, *Aplysia californica*, is able to become less specialized as it grows and its mouth-parts toughen; new recruits grow only on *Plocamium* (Pennings 1990).

Mesograzers are often considered to be generalist feeders, but there are several examples of them showing specialist tendencies (also see above). Amphipods tend to live on their food, and it may be that their grazing induces chemical defenses in the seaweed, and the seaweed then provides a better refuge from omnivorous fish, i.e. the relationship has evolved over time to become mutually beneficial. Sotka *et al.* (2003) showed that where the amphipod *Ampithoe longimana* co-occurs with *Dictyota*, it is tolerant to the chemical defenses, and the seaweed provides them with a refuge. However, amphipods of the same species that do not co-occur with *Dictyota* are intolerant of *Dictyota*. They further tested the effects of *Dictyota* on the feeding preferences of different full-sib families of the amphipod, and found a range of phenotypic preferences, with some families being more strongly deterred than others. This suggests a phenotypically plastic response, and that specific traits can be inherited over time.

4.4   Chemical ecology of seaweed–herbivore interactions

The chemical ecology of seaweed–herbivore interactions has been one of the most active areas of seaweed physiology and ecology since the late 1980s. This is evidenced by the sizeable number of reviews published on the topic (e.g. Amsler and Fairhead 2006; Ianora *et al.* 2006; Paul *et al.* 2006c; 2011; Hay 2009; Amsler 2012; Iken 2012), and the textbooks *Marine Chemical Ecology* (McClintock and Baker 2010) and *Algal Chemical Ecology* (Amsler 2008) in which seaweeds feature strongly. Chemical defenses are apparent in temperate (Jormalainen and Honkanen 2008), tropical (Pereira and da Gama 2008), and polar (Amsler *et al.* 2008, 2009; McClintock *et al.* 2010) seaweeds. Apparent from the literature is that chemical ecology is still a relatively new research area, and a large number of initial studies have given rise to a wide range

of results, that sometimes appear contradictory. A second generation of experiments is now underway that includes the application of new methodologies, such as genomics and metabolomics (Pelletreau and Targett 2008; Prince and Pohnert 2010). Here we outline the chemicals that are considered to be bioactive against grazers, review the two main models that have been used as frameworks to test predictions on how seaweeds might regulate their defenses against herbivores, and select a few key examples that illustrate current ideas and highlight future directions.

4.4.1   Bioactive chemicals

The general biosynthetic pathways by which the various classes of "secondary metabolites" are made are outlined by Maschek and Baker (2008; Fig. 4.15), but within each of these classes there are an enormous number of compounds with specific functionality. The specific compounds are created by tailoring enzymes, that make small modifications to molecules, thereby imparting their unique bioactivity. More than 1500 secondary metabolites have been reported for red algae, including all classes of seaweed secondary metabolites except phorotannins (polyphenols). Red seaweeds are particularly rich in compounds that are halogenated with bromine or chlorine, although most of these do not have a known ecological function (Blunt *et al.* 2007; sec. 4.2.2). The brown algae also possess a wealth of secondary metabolites (>1100 reported), with *Dictyota* and its >250 terpenes accounting for a third of these. Upon grazing, a blend of bioactive chemicals is thus released from *Dictyota*, to which the grazers may respond (Wiesemeier *et al.* 2007). Green seaweeds have the least variety of secondary metabolites, with <300 compounds found from genera including *Caulerpa*, *Halimeda*, and *Udotea* (Maschek and Baker 2008). The effect of these secondary metabolites on the consumer is not an inherent property of the chemical per se, but is the result of a biochemical reaction between the secondary metabolite and the digestive processes of the consumer (see Sotka *et al.* 2009). The consumers themselves have various

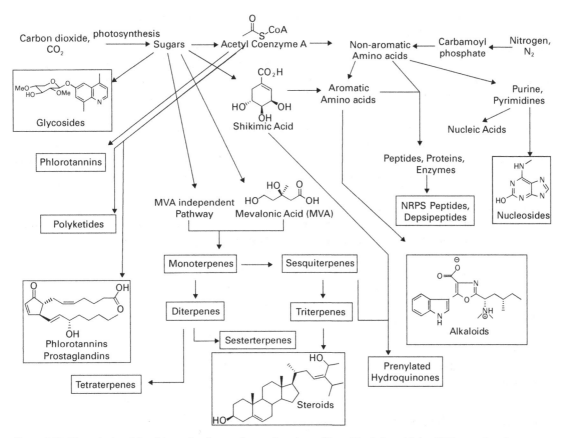

Figure 4.15 Biosynthetic origin of the major classes of natural products. (From Maschek and Baker 2008, reproduced with permission.)

"offensives" against defended seaweeds, including an ability of some to detoxify the secondary metabolites (Sotka and Whalen 2008).

Terpenoids are the largest group of secondary metabolites reported from macroalgae, and are among the most prevalent feeding deterrents in tropical seaweeds (Sotka and Whalen 2008). They are lipids, structurally related to compounds such as carotenoid pigments (Howard and Fenical 1981; Maschek and Baker 2008). The Caulerpales and Dictyotales are particularly rich in terpenoids, as are the red seaweeds such as *Laurencia* (see sec. 4.2.2).

The phlorotannins of brown seaweed can be present in high concentrations (up to 20% of dry weight) and there are many polymeric compounds based on phloroglucinol (Ragan and Glombitza 1986; reviewed by Amsler and Fairhead 2006). Despite their abundance in brown seaweeds, their role in chemical defense against grazing is equivocal (Amsler and Fairhead 2006). Confounding problems are that phlorotannins have many other roles in cells (secs. 1.3, 7.6, 7.8), and their production can be influenced by seawater nitrogen availability (e.g. Peckol *et al.* 1996; Hemmi *et al.* 2004; sec. 4.4.2). Phlorotannins have been the focus of many studies partly because the spectrophotometric assays for bulk concentrations are very simple to conduct (Amsler and Fairhead 2006). However, these bulk extracts contain other bioactive compounds which may act as the deterrent. Using a bioassay-guided approach, Deal *et al.* (2003) showed that a crude extract of *Fucus vesiculosus* deterred grazers, but it was a polar galactolipid that was effective, not the phlorotannins. In order to advance research on phlorotannins, purified phlorotannins are required. This extraction is technically difficult because they are easily oxidized (Amsler and Fairhead 2006), but Audibert *et al.* (2010) undertook the first steps in such a procedure. Amsler and Fairhead (2006) suggest that for the field to move forwards, a change of focus may be required, moving away from feeding experiments or chemical analyses alone and towards combining feeding assays with detailed chemical analyses and fractionation. The mode of action of the phlorotannins also needs to be determined. For example, they may deter consumers indirectly by lowering the food value of seaweed by precipitating proteins (Targett and Arnold 2001). To test for such effects of phlorotannins, and other secondary metabolites, Sotka *et al.* (2009) suggest a pharmacological approach, in which the effect of a secondary metabolite on an animal is studied at a physiological level in terms of its adsorption, distribution, metabolism, and excretion (ADME).

4.4.2 Chemical defenses against grazers

Several theories (or models) have been used to frame ecological questions relating to the evolution of seaweed chemical defense against consumers (reviewed by Pavia and Toth 2008). Two stand out as having been most widely tested. The first is a "supply-side model" involving the argument that the ability of a seaweed to defend itself is controlled by the supply of factors that can limit growth. The carbon–nutrient balance model (CNBM) hypothesizes that carbon-based secondary metabolites are produced only when growth is limited by the supply of resources, typically nitrogen, but also light. When nutrient supply is not limiting, then seaweeds will allocate fixed carbon to growth, whereas in low nutrient conditions, growth slows, and fixed carbon is available in excess and allocated to defense. This theory has been extensively tested on the Fucales and phlorotannin production, but with mixed results (see Table 7.3 in Pavia and Toth 2008). Furthermore,

the light levels used in experiments can influence the outcome, as light will affect growth and thus the C:N balance. In summary, it is unclear whether changes in secondary metabolite production that might occur in relation to resource supply represent a "defensive strategy" or are simply a by-product of a changing metabolism (e.g. Jormalainen *et al.* 2003). For terrestrial plants, the CNBM has been largely rejected for a variety of reasons (Hamilton *et al.* 2001). However, as evident in the ensuing discussions, the CNBM, with its resource-allocation focus, contains elements of the more popular "optimal defense theory", and testing these together can be instructive (Pavia *et al.* 1999; Pavia and Toth 2008). A third model, the growth–differentiation balance model (GDBM) is also a supply-side model, but it has not been tested very often because of experimental design difficulties (see Cronin and Hay 1996b for an example) but it is considered by some to be a more "mature" model than the CNBM (see Pavia and Toth 2008).

Optimal Defense Theory (ODT) is a demand-based model in which the costs of allocating resources to defense are balanced against the benefits of the defense, and was first developed for terrestrial plants (Rhoades 1979). Ragan and Glombitza (1986, pp. 225–6) were among the first to suggest using this model for seaweed. Predictions include:

1. organisms evolve defenses in direct proportion to their risk from predators and in inverse proportion to the cost of defense, other things being equal;
2. within an organism, defenses are allocated in direct proportion to the risk of the particular tissue and the value of that tissue in terms of fitness . . . and in inverse proportion to the cost of defending the particular tissue;
3. commitment to defense is decreased when enemies are absent and increased when organisms are subject to attack;
4. commitment to defense is a positive function of the total energy and nutrient budget . . . and is negatively related to energy and nutrients allocated by the organism to other contingencies (cf. the CNBM).

Nested within ODT is the concept of constitutive versus inducible or activated defenses. Constitutive defenses are those that are synthesized all of the time. An "inducible defense" refers to an increased production of an existing chemical following grazing, and the timescale for up-regulation can be hours to months (Cronin and Hay 1996b). An "activated defense" is where an inactive chemical precursor is converted to the bioactive chemical, and this reaction occurs on a timescale of seconds (Paul and Van Alstyne 1992).

The production of constitutive defenses may be a useful strategy in areas where grazing pressure is consistently high, such as in the tropics. Higher levels of constitutive defenses may also be predicted for tissues that are of particular importance to a seaweed's survival, such as specialized reproductive tissue or a structurally important part of a stipe. For *Ascophyllum nodosum*, basal regions have greater levels of constitutive defense than apical regions, and the basal regions also have a much stronger induced response to grazing than apical shoots (Toth *et al.* 2005). The oldest stipes of *Sargassum filipendula* are constitutively defended from amphipod grazing by being tough, whereas grazing of the tissues supporting the apical meristems triggered an induced chemical defense (Taylor *et al.* 2002). The Antarctic brown seaweed, *Desmarestia anceps*, has a high level of tissue specialization and in accordance with ODT, the most "valuable" structure, the primary stem, was more chemically defended than other parts of the seaweed. On the other hand, *D. menziesii*, which has less tissue differentiation, had similar levels of chemical defense throughout its thallus (Fairhead *et al.* 2005).

To demonstrate the presence of an inducible or activated defense, one must demonstrate that levels of the secondary metabolite increase following grazing, and that the induced seaweed tissue has a stronger effect on deterring grazing compared to the non-induced tissue. Simply showing that the concentration of a putative defensive chemical increases following grazing is not sufficient evidence of an induced defense. One of the earliest examples was that of the activated defense system of *Halimeda* spp. Here, injury caused a secondary metabolite, halimedatetraacetate, to be converted into the more potent chemical halimedatrial, and it had a deterrent effect on the herbivorous fish (Paul and Van Alstyne 1992). Activated

defenses are common in green seaweeds, but there are few examples from red and brown seaweeds (Pelletreau and Target 2008). An early example of an induced defense was that of *Dictyota menstrualis* for which levels of terpenoids increased following amphipod grazing, and the seaweeds became less susceptible to grazers (Cronin and Hay 1996b). This laboratory experiment provided support for patterns observed in the field, in which *Dictyota* growing in heavily grazed areas was better defended than those growing in areas of low grazing pressure.

The effect of induced defenses on grazing is species specific because grazers are differentially susceptible to the same chemicals. For example, the pinfish *Lagodon rhomboides* is deterred at concentrations of 0.1% dry mass of seaweed tissue, which are lower than those naturally found in *Dictyota ciliolata*, whereas amphipods and urchins are susceptible at 0.5% (Cronin and Hay 1996b; Fig. 4.16). Inducible defenses were thought to occur only in brown and green seaweeds (Toth and Pavia 2007), until they were reported for *Delesseria sanguinea*, *Furcellaria lumbricalis*, and *Phyllophora pseudoceranoides* (Rohde and Wahl 2008).

Several studies show the importance of grazer identity on inducible defenses. For *Ascophyllum*, grazing over several weeks by the gastropod *Littorina obtusata* caused phlorotannin levels to increase, whereas the isopod *Idotea granulosa* had no effect, and nor did mechanical clipping that was used to simulate grazing. These laboratory results were supported by field observations that *Ascophyllum* grazed by *Littorina* have higher phlorotannin content than ungrazed (Pavia and Toth 2000). Defenses of the kelp *Ecklonia cava* were induced by *Littorina* grazing in summer, but not fall, and there was no effect of grazing by the abalone *Haliotis discus*, showing that in addition to grazer identity, the season can affect induced responses (Molis *et al.* 2006).

If a defense is to be induced, the seaweed needs to be able to detect an "attack" by a grazer (cf. pathogen attacks in sec. 4.2.2). Pavia and Toth (2000) showed that physical damage alone (i.e. clipping) did not cause an induced response in *Ascophyllum*, and suggested that a chemical signal must be involved.

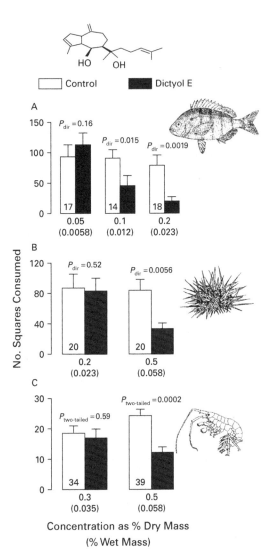

Figure 4.16 The effect of dictyol E in choice assays on the feeding behavior of (a) *Lagodon rhomboides*, (b) *Arbacia*, and (c) *Ampithoe*. Sample size is given at the bottom of each control bar. The tested concentrations of dictyol E increase from left to right and are given below each pair of bars as both % DM and% WM. Bars are the mean (±1 SE) amount of food consumed per replicate (i.e., individual pinfish, urchin, or group of 4–6 amphipods). *P* values are from paired-sample *t* tests. Natural concentrations of dictyol E are 0.020–0.045% WM. (From Cronin and Hay 1996b. © 1996, the Ecological Society of America.)

Coleman *et al.* (2007) confirmed this idea when they demonstrated that α-amylase, an enzyme present in snail saliva, triggers an inducible defense (Fig. 4.17). They grew *Ascophyllum* for 4 weeks under four pre-experimental conditions: control, control + α-amylase, simulated grazing + α-amylase, and herbivore grazing. They then placed the seaweed in experimental tanks with littorinid snails and quantified seaweed consumption, snail movement, and phlorotannin content. The pattern observed for phlorotannin content was: control < control + α-amylase < simulated grazing + α-amylase < herbivore grazing (Fig. 4.17a). Approximately the reverse pattern was seen for amount of tissue eaten; the herbivores ate less of the *Ascophyllum* grown in the presence of snails and more of the control seaweed (Fig. 4.17b). Herbivore behavior was also altered (Fig. 4.17c). Snails given *Ascophyllum* that had been subjected to snail grazing were more mobile, had smaller meals, and ate less overall. When snails are presented with a less palatable food, they are thought to spend more time foraging (for more palatable food). The inducible response seen in *Ascophyllum* was probably triggered by an oxidative burst, as the α-amylase will break down cell wall carbohydrates into oligosaccharides (sec. 4.2.2). It is interesting that herbivore grazing had a greater inductive effect than simulated grazing plus α-amylase, as it suggests that another signaling chemical is also involved.

Chemicals released by seaweeds following grazer damage may trigger their neighbors to up-regulate their own defenses. This idea that seaweeds can detect grazer-related waterborne cues from their neighbors has been tested, with varying results. For *Sargassum filipendula* grazed by *Ampithoe longimana*, there was no evidence of waterborne cues affecting neighboring, ungrazed seaweeds (Sotka *et al.* 2002). However, waterborne cues did induce defenses in *Ascophyllum nodosum* and *Fucus vesiculosus* (Toth and Pavia 2000a; Rohde *et al.* 2004). For *Laminaria digitata*, providing seawater that has been in contact with its conspecific neighbors resulted in an enhanced sensitivity to a defense elicitor, alginate oligosaccharide (see sec. 4.2.2) (Thomas *et al.* 2011).

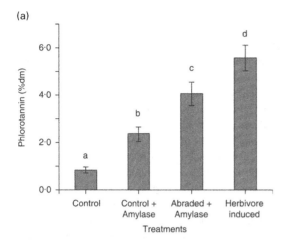

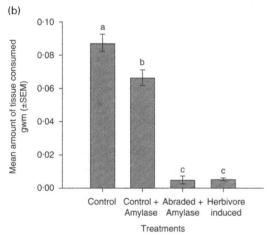

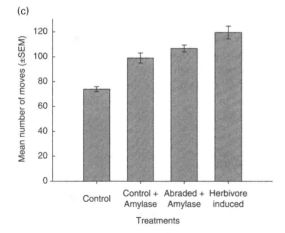

A central premise of the ODT is that the production of chemical defense induces a cost; the allocation of energy to a defense comes as a result of less energy diverted into growth and/or reproduction. However, this has not been tested often because, as highlighted by the study of Dworjanyn *et al.* (2006b), "cost" is difficult to assess. For *Ascophyllum nodosum*, Pavia *et al.* (1999) found a negative relationship between phlorotannin content and growth. In contrast, Rohde *et al.* (2004) found that growth rates of *Fucus vesiculosus* were the same whether or not chemical defense had been induced. In this case, the production of defenses was down-regulated within 2 weeks of the cessation of grazing. Dworjanyn *et al.* (2006b) grew *Delisea* with and without bromine, which is required for making furanones. Seaweeds without bromine grew faster and their thalli were larger than seaweeds grown with bromine, providing direct evidence that furanone production comes with a cost (although the results of their two other tests were equivocal). A benefit of producing chemical defenses was shown by Haavisto *et al.* (2010) who found that when the isopod *Idotea* was fed defended seaweed, it produced fewer eggs.

It has been hypothesized that because the diversity of herbivores is much greater in the tropics where, in addition to invertebrates there are many herbivorous fish compared to temperate systems, then tropical seaweed should be better defended than their temperate counterparts (Gaines and Lubchenco 1982; Paul and Hay 1986). However, evidence to support this

Figure 4.17 (a) Phlorotannin concentrations of *Ascophyllum nodosum* after 2 weeks of exposure to induction treatments (control, application of α-amylase, abrasion followed by application of α-amylase, prior grazing by *L. obtusata*). (b) Consumption of primary shoot of *Ascophyllum nodosum* in grams wet mass (gwm) by *Littorina obtusata* over 100 min after 2 weeks of exposure to induction treatments. (c) Changes in position of *Littorina obtusata* on sections of primary shoot of *Ascophyllum nodosum* over 100 min after 2 weeks of exposure to induction treatments. Similar letters indicate treatments were not significantly different as shown by posthoc SNK tests. (From Coleman *et al.* 2007 © 2006 The Authors. Journal compilation © 2006 British Ecological Society.)

popular theory is mixed and definitive experiments are still required (Pereira and de Gama 2008). In their meta-analysis, Poore *et al.* (2012) found that seaweeds from temperate, tropical, and polar seaweeds are exposed to the same level of grazing, albeit by different grazers in different biogeographic regions. Tropical seaweeds appear to be particularly rich in lipophilic secondary metabolites, which supports the idea of an evolutionary trend in response to the higher diversity of grazers compared to temperate regions (reviewed by Sotka and Whalen 2008) but this idea requires further testing. Cetrulo and Hay (2000) compared the frequency of occurrence of activated defenses in temperate and tropical seaweeds, and found no biogeographic pattern. Pereira and de Gama's (2008) analysis of the literature suggested that temperate seaweeds have a greater range of defenses, perhaps because there are a greater number of seaweed species in temperate seas compared to the tropics (see Kerswell 2006).

Several interesting strategies are used by *Halimeda* in defense against herbivores. Plastid migration (sec. 1.3.2) may protect against surface grazers such as saccoglossan mollusks, which try to capture plastids for their own use (Drew and Abel 1990). Many species of *Halimeda* (but not all) put out new buds at night, when herbivorous fish are inactive. The new growth is unpigmented at first and has high concentrations of halimedatrial. Around dawn, the tissue turns green, perhaps because of plastid migration from the parent segment (Drew and Abel 1990). Calcification begins when photosynthesis begins. After 48 h of photosynthesis, the toxin level decreased, and calcification and morphological toughness should provide adequate defense (Paul and Van Alstyne 1988a). However, not all *Halimeda* species behave this way. New segments of *H. macroloba* remain white throughout the next day and occasionally are produced during the day (Hay *et al.* 1988; Larkum *et al.* 2011).

Some *Desmarestia* spp. have a high sulfuric acid content (vacuolar pH ~1) and this can deter grazers (McClintock *et al.* 1982). For *D. munda* from Washington State, USA, the sulfuric acid content is around 16% of its dry weight (Pelletreau and Muller-Parker 2002). In feeding choice experiments, in which the pH of

seaweed agar was manipulated using sulfuric acid, the green sea urchin (*Strongylocentrous droebachiensis*) was deterred from grazing at pH ≤ 3.5 units. In a community survey in Svalbard, Norway, there were 2–3 times fewer green sea urchins within *D. viridis* beds compared to the nearby urchin barrens (Molis *et al.* 2009). In this study, sulfuric acid was shown to cause a change in the behavior of *S. droebachiensis*. A small volume (500 μL) of pH 7.5 sulfuric acid presented to the green urchins caused them to stop moving, and when the urchins were presented with just 25 μL of sulfuric acid at pH 1, they moved in the opposite direction. Furthermore, when the kelp *Alaria esculenta* was grown in association with *D. viridis*, it was eaten less, indicative of an associational defense (Molis *et al.* 2009).

Some herbivores, far from being deterred by secondary metabolites, actually seek them out and use them for their own defense. An example of this is *Elysia halimedae*, which feeds preferentially on *Halimeda macroloba*. It takes halimedatetraacetate, which is effective against fish feeding, and simply converts one aldehyde group to an alcohol, to produce a deterrent for its own protection, as well as for its eggs (Paul and Van Alstyne 1988b). Similarly, *Aplysia californica* can extract plastids from the red seaweed that it consumes, and use them to synthesize a chemical defense. The initial plastid extraction occurs in the fore-gut of *Aplysia*. Then, the plastids are engulfed in specialized "rhodoplast digestive cells" where the red seaweed pigment r-phycoerythrin is extracted, and then chemically modified to produce a "purple defensive ink", that acts as a defense against predators such as anemones and crabs (Coelho *et al.* 1998). Another example is the decorator crab, *Libinia dubia*, which covers its shell with *Dictyota*, using this seaweed's chemical defenses to deter omnivorous fish (Stachowicz and Hay 1999).

4.5 Symbiosis

At one end of the spectrum of biotic relationships among seaweeds is epiphytism (sec. 4.2.2). Some instances of epiphytism are quite specific; for other seaweeds, epiphytism is simply one solution to the space problem. Symbiosis (literally, "living together") implies a closer relationship than simple epiphytism, but includes a range of mutualistic partnerships and parasitism. Mutualisms form between epibiotic invertebrates and seaweeds, and seaweeds also live in mutual associations with corals. Endophytes are algae that grow within seaweeds, and here commensal and hemi-parasitic relationships have been recorded.

4.5.1 Mutualistic relationships

There are many well-known examples of mutualisms between microalgae and other organisms such as lichens (cyanobacteria + fungi) and hermatypic corals (dinoflagellate + invertebrate) (Yellowlees *et al.* 2008). Until the early 1990s, the main examples of symbiotic relationships between macroalgae and other organisms were with endophytic fungi such as *Blidingia minima* var. *vexata* + *Turgidosculum ulvae, Prasiola borealis* + *Guignardia alaskana, Apophlaea* spp. + *Mycophycias* (formerly *Mycosphaerella*) *apophlaeae, Pelvitia cannaliculata* + *Mycophycias* (e.g. Kohlmeyer and Kohlmeyer 1972; Kohlmeyer and Hawkes 1983; Kingham and Evans 1986; Rugg and Norton 1987; Zuccaro and Mitchell 2005). However, seaweed mutualisms seem much more common than previously thought and recent examples include those with colonial invertebrates (see sec. 4.4.2), bacteria and "whole community mutualisms" (Hay *et al.* 2004; Bracken *et al.* 2007). "The renaissance of interest in positive interactions (mutualisms, commensalisms and facilitation) has highlighted their ubiquity and the critical role they play in ecosystems worldwide" (Stachowicz and Whitlatch 2005; see sec. 4.1).

The association between the ascomycete fungus *Mycophycias ascophylli* and the fucoid *Ascophyllum nodosum* has been studied extensively by D.J. Garbary and co-workers (e.g. Deckart and Garbary 2005; Xu *et al.* 2008). Specimens of *Ascophyllum* >5 mm long are invariably infected with *Mycophycias*, but specimens <3 mm high lack the fungus (Kohlmeyer and Kohlmeyer 1972; Garbary and Gautam 1989). The fungus grows throughout the host thallus, forming a

hyphal network around each seaweed cell (Deckert and Garbary 2005). Fungal fruiting bodies (perithecia) are formed chiefly, but not exclusively, in receptacles of the hosts, where they appear to the naked eye as small black dots. Perithecia form on vegetative apices at about the same time as receptacles are produced, but persist about a month longer (Garbary and Gautam 1989). Infection takes place after *Ascophyllum* has germinated; the eggs apparently are not infected, even though perithecia are present in receptacles. The fungus *Mycophycias* obtains carbon from the host; in axenic culture it can grow on laminarin and mannitol, but not alginic acid (Fries 1979). *Ascophyllum* benefits from the relationship as zygotes infected with the fungus are more tolerant to desiccation than uninfected zygotes (Garbary and London 1995). Infected *Ascophyllum* zygotes are also longer, with more apical hairs and smaller rhizoids than uninfected zygotes, and when they are 8-months old, infected *Ascophyllum* is four-times longer than uninfected *Ascophyllum* (Garbary and McDonald 1995).

On coral reefs, seaweeds may grow in association with sponges (e.g. Price *et al.* 1984; Scott *et al.* 1984) and the associations between the sponge genus *Haliclona* and red seaweeds have provided valuable insights into the nature of these mutualisms. Mutualisms can be facultative, when the partners are able to live apart, or obligate. On the Great Barrier Reef, Australia, the red seaweed *Ceratodictyon spongiosum* grows as an obligate "stable unit" with *Haliclona cymiformis* (Trautman *et al.* 2000). The sponge possesses two signaling molecules that can control the carbon metabolism of the seaweed (Grant *et al.* 2006). The host release factor (HRF) causes the release of photosynthetically fixed carbon from the seaweed partner, making it available for uptake by the animal. A photosynthesis-inhibiting factor (PIF) inhibits carbon fixation. This study represents a pivotal step forward in understanding how sessile invertebrates and seaweeds might communicate with one another. In the Bay of Mazatlan, Mexican Pacific Ocean, the sponge *Haliclona caerulea* grows in a facultative association with the calcareous seaweed *Jania adhaerens* (Enríquez *et al.* 2009). The aposymbiotic form of *Jania* is restricted to the intertidal zone, but when in association with *H. caerulea* its range extents to the subtidal. The sponge provides the alga with structural support that permits a greater degree of morphological plasticity than aposymbiotic *Jania*: the seaweed grows four-times taller in the association and decreases the amount of calcium carbonate by between 29% (transplanted to 1 m depth) and 68% (5 m) (Enríquez *et al.* 2009). An interesting question is how, following reproduction, do seaweeds and sponges form new mutual associations? *H. caerulea* larvae actively select *Jania* as a settlement surface (Ávila and Carballo 2006), but whether this is the same for the obligate symbiosis between *H. cymiformis* and *Ceratodictyon spongiosus* is not known.

Ammonium subsidies from sessile and mobile invertebrates to their seaweed hosts is well established (Taylor and Rees 1998; Wai and Williams 2005; Pfister 2007). In an extension of this idea, Bracken *et al.* (2007) suggest whole-community mutualism between high intertidal rock-pool *Cladophora* and associated invertebrates: *Cladophora* provides food and habitat for the invertebrates which in turn provide nitrogen thereby enhancing the growth rate of their food (see sec. 6.8.5 for further discussion).

4.5.2   Seaweed endophytes

Some seaweeds live endophytically, inside other seaweeds, and there are examples in each of the three seaweed lineages (Potin 2012). Green and brown endophytes are pigmented, but some are hemiparasites and obtain partial nutrition from the host (Burkhardt and Peters 1998). The majority of red seaweed endophytes are obligate parasites, but there are no known green seaweed parasites, and just one brown parasite, *Herpodiscus durvilleae* (Heesch *et al.* 2008). A common feature of endophytes is their simple morphology, which made identification difficult until the advent of molecular taxonomy. Most macroalgal endophtyes are filamentous (Burkhardt and Peters 1998). Some are unicellular, for example the sporophyte generation of the filamentous green seaweed *Acrosiphonia* (Codioales) grows as a unicell inside a range of red and brown hosts including *Mazzaella* sp. and *Sparlingia pertusa*, as well as the crustose species

Ralfsia pacifica and *Hildenbrandia occidentalis* (Sussmann and DeWreede 2002).

Some of the pigmented green and brown endophytes cause symptoms in their hosts, others severe diseases. The Ectocarpalean genus *Laminariocolax* is distributed in temperate coastal waters worldwide and is found growing within members of the Order Laminariales. The degree of infection by the endophyte has been classified as: (1) infected, but no symptoms; (2) moderate (dark spot disease), with symptoms including wart like proliferations, and dark spots on the thallus; and (3) severe when stipes or blades are physically distorted such as stipes growing in a spiral (Peters and Schaffelke 1996; reviewed by Bartsch *et al.* 2008). The filamentous seaweed *Acrochaete operculata* is a pathogenic endophyte of *Chondrus crispus* (sec. 4.2.2). The motile zoospores of *Acrochaete* settle on the surface of *Chondrus* surface and, using extracellular enzymes, digest the cell wall and penetrate the host. *Acrochaete* not only causes direct cellular damage but also facilitates secondary bacterial infection, together causing severe degeneration of the host (Correa and McLachlan 1991, 1994). *Mazzaella laminarioides* also hosts a pathogenic filamentous green alga, *Acrochaete ramosa* (previously *Endophyton ramosum*) (Sánchez *et al.* 1996).

Kelp gametophytes are cryptic and had rarely been observed in the field, until Garbary *et al.* (1999) discovered them growing endophytically within the cell walls of 17 species of (mostly) filamentous red seaweed; the hosts each housed tens to hundreds of gametophytes. The kelp spores settle and germinate on the red seaweed's surface, then burrow into their host. Oogonia are formed at the surface of the host, providing a mechanism for liberation of gametes. By living endophytically, the gametophytes perhaps escape grazing that would occur on rock surfaces, sedimentation, or increased light, and the chemical defenses found in many red seaweeds might provide them with additional protection (Hubbard *et al.* 2004). In the Antarctic, Amsler *et al.* (2009) found that 8 of 13 seaweeds surveyed contained endophytes. The brown endophytes that they isolated grew well in culture, and in no-choice feeding experiments, amphipods much preferred the endophytes over the host seaweed.

Amphipod grazing pressure in the Antarctic subtidal is high and, as a result, few uniserate filamentous seaweeds are seen. Amsler *et al.* (2009) suggest that by living inside another seaweed, these filamentous brown escape amphipod grazing.

Herpodiscus durvillaea is an obligate parasite on *Durvillaea antarctica* in New Zealand. It belongs to the order Sphacelariales, but is unusual because it has an extremely heteromorphic life cycle with a minute, reduced, gametophyte stage. It is considered to be fully parasitic as it contains only small, gray plastids and attempts to grow it in culture have been unsuccessful. The discovery of a functional, brown seaweed rbcl (Large subunit of RuBisCO), however, raises the interesting question of putative photosynthetic ability (Heesch *et al.* 2008).

While there are only a few examples of parasitic brown and green seaweeds, up to 15% of red algae are parasites (Goff 1982; Kurihara *et al.* 2010). Examples of host–parasite pairs include *Gracilariopsis longissima* (previously *Gracilaria verrucosa*) and its non-pigmented parasite *Holmsella pachyderma* (Evans *et al.* 1973), and *Vertebrata lanosa* and its weakly pigmented alloparasite *Choreocolax polysiphoniae* (Callow *et al.* 1979). The traditional view was of two types of red seaweed parasite. Around 80% were considered to be adelphoparasites, those parasites that evolved directly from their hosts and are therefore closely related. The remaining 20% were considered alloparasites that are not closely related to their host. This distinction has been challenged because molecular phylogenies indicate that most red seaweed parasites are in fact closely related to their hosts, although the degree of relatedness does vary (Goff *et al.* 1996, 1997; Zuccarello *et al.* 2004). Zuccarello *et al.* (2004) therefore suggest that all red seaweed parasites are adelphoparasites.

The most recently evolved parasites have strong phylogenetic affinities for their host, whereas ancient parasitic lineages show greater divergence in their small subunit rDNA, indicating greater speciation (Zuccarello *et al.* 2004). *Harveyella mirabilis* is considered an ancient parasite as it has hosts in two families, and is the only red seaweed parasite known to occur in both the Atlantic and Pacific Oceans. It is

thought to have undergone a geographic range extension with its host *Rhodomela confervoides* which is also found in the Atlantic and Pacific. It is also thought to have "switched host" to become parasitic on its second host *Gonimophyllum skottsbergii*. Interestingly, *Harveyella mirablis*' host, *Gonimophyllum skottsbergii*, is also a parasite on *Cryptopleura crispa* – a parasite on a parasite is termed hyperparasitism.

The ability of red algae to form secondary pit connections is considered an important factor in the evolution of parasitic species (Goff and Coleman 1995; Goff *et al.* 1996, 1997). Goff and Zuccarello (1994) revealed the steps following spore germination of two parasites *Gardneriella tubifera* and *Gracilariophila oryzoides*, which grow on *Gracilariopsis lemaneiformis* (Fig. 4.18). Genetic material from the parasites is transferred by "infection rhizoids" that enter the host cells and through which pass the parasite's nuclei. The result is that the parasite "takes control" of the host cells, triggering alterations to the anatomy and physiology of the host. Changes in the host include plastid de-differentiation to proplastids, increased storage of Floridian starch and the production of more mitochondria. This process of parasitic infection is apparently similar to the "normal" developmental steps that give rise to the carposporophytes of *Gracilariopsis*, providing additional support for the evolution of parasites from their hosts.

Finally, little is known about the interactions between marine viruses and seaweeds. One milliliter of seawater contains millions of virus particles. For phytoplankton, virus infections can lead to the termination of a bloom, and they are thought to be an important driver of genetic diversity (Suttle 2007). Given the ubiquity of viruses, it seems likely that they also affect aspects of seaweed physiology and ecology, and this will be an interesting line of future enquiry. Examples of seaweed diseases are given in Chapter 10.

4.5.3 Kleptoplasty

The saccoglossan mollusks (sea slugs; secs. 1.3.2, 4.3.2) *Elysia hedgpethi* and *E. viridis* not only eat

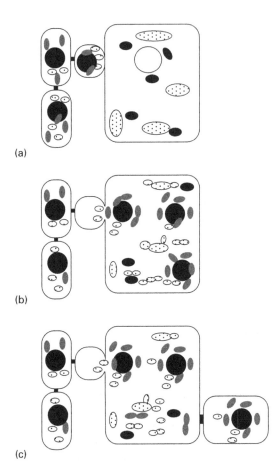

(a)

(b)

(c)

Figure 4.18 Transfer of parasite nuclei, mitochondria and proplastids into a host cell. (a) The parasite cells (left) have formed a conjunctor cell containing a parasite nucleus (black), mitochondria (gray), and proplastids (white with dots). This conjunctor cell fuses with the host cell (host nuclei are white, host plastids are dotted, and host mitochondria are black ovals) delivering the parasite organelles into the host's cytoplasm. (b) The parasite nucleus and mitochondria replicate in the host cell, and the host plastids divide to form numerous proplastids. The host nuclei and mitochondria replicate in the host cell. The host nucleus may disappear or persist. (c) Ultimately, a cell is cut off from the heterokaryotic host plus parasite cell. This cell contains a parasite nucleus, parasite mitochondria, and proplastid derived from the host plastids. (From Goff and Coleman 1995. © American Society of Plant Physiologists.)

the green seaweeds upon which they are specialist feeders, they extract plastids from the seaweed which become integrated with their own metabolism, producing "solar-powered" sea slugs (Rumpho *et al.* 2006, 2008, 2011). Such relationships between an animal and plastid have been termed symbiotic (Greene 1970), but because only the plastids are in the relationship with the animal, the term "kleptoplasty" ("stolen plastids") is typically used (Rumpho *et al.* 2006, 2011). Many species within the sea slug genus *Elysia* are kleptoplasts, and the residence time of functional plastids in their digestive epithelium varies between species. The *Codium* plastids acquired by *E. hedgpethi* remain photosynthetic for about 8 days and the sea slug has to replenish them from its seaweed diet if it is to remain photosynthetic (Greene 1970). The plastids that *Elysia chlorotica* acquires from the Xanthophyceaen alga *Vaucheria litorea* are functional for about 10 months and this species relies on plastid photosynthesis for 80% of its 1-year life (reviewed by Rumpho *et al.* 2006). In primary endosymbiosis, the engulfed cyanobacteria lost all but 5–10% of their DNA to the eukaryotic host nucleus, explaining why plastids are unable to live freely (sec. 1.3.2). Rumpho *et al.* (2008) questioned how, in the absence of seaweed nuclear DNA, plastids could function within *Elysia chlorotica*. Using genome sequencing, they found a functional gene for oxygenic photosynthesis (*pbsO*) within the sea slug that is identical to that within the seaweed nucleus. This gene has been incorporated into the germ line of *C. chlorotica*, possibly acquired by horizontal gene transfer (sec. 1.4.2).

4.6 Synopsis

Biotic interactions include positive interactions such as facilitation and mutualism, and negative interactions such as competition within and between species for space, light, nutrients, or any limiting resource. Predator–prey relations at several levels affect seaweeds directly or indirectly. Interference competition for space takes place among algae and between algae and sessile animals; the outcome can be influenced by both herbivores and carnivores. Exploitation competition for light and nutrients takes place among algae. Epiphytism solves the space problem for the epiphyte, but creates a competition problem for the basiphyte species. Some seaweed produce allopathic chemicals or slough their outer layers to inhibit the growth of epiphytes.

Herbivores that are important in seaweed communities include fish, sea urchins, and mesograzers (diminutive invertebrates with small ranges but high densities). Damage to seaweeds by grazers will depend on the occurrence of an encounter between the two, on how much is eaten or broken off, and on what parts are lost and when, especially as the losses affect reproduction and hence the fitness of the individual. Seaweeds may escape grazing by utilizing temporal or spatial refugia, or growing in association with a chemically defended or territorial organism.

Some seaweeds tolerate grazing, while others are chemically and/or structurally defended. Seaweed chemical defenses may be constitutive, or they may be induced or activated following grazing. Snail saliva and waterborne cues from neighboring seaweed that have been grazed can both induce chemical defense.

Mutualistic relationships have been identified between seaweeds and bacteria, fungi, and invertebrates, including sponges and hydrozoans. Several seaweeds harbor internal fungal symbionts. There are many examples of seaweeds living endophytically, inside another seaweed, some as hemi-parasites. There is one example of a brown seaweed parasite, but many red endophytes are parasitic and these can cause changes to the anatomy and physiology of the host seaweed. Some sea slugs are kleptoplasts and extract seaweed plastids which they use to become a photosynthetic animal.

Light and photosynthesis

In their natural environment, seaweeds grow in exceptionally diverse and dynamic light climates. Water transparency and the continual ebb and flood of tides have profound effects on the quantity and quality of the light that reaches seaweeds at their growth sites, adding greatly to the variation already present in the irradiance at the Earth's surface. The primary importance of light to seaweeds is in providing the energy for photosynthesis, energy that ultimately is passed on to other organisms. In addition, light perceived as a signal also has many photoperiodic and photomorphogenetic effects (see secs. 2.3.1, 2.3.3, 2.6.2). Thus, light is the most important abiotic factor affecting seaweeds, and also one of the most complex.

The principles of photosynthesis are similar in algae and higher plants, and indeed some principles (e.g. the Calvin cycle) were worked out using (mostly unicellular) algae. However, there are several important features of seaweeds and their habitats that stand in sharp contrast to those in higher, and mostly terrestrial plants, and it is on these that we shall focus. Such features include the diversity of pigmentation among marine algae and the diversity of the light climate in the oceans, the nature of carbon supply in the sea, and the diversity of photosynthetic products in different algal classes. This chapter focuses on the processes in eukaryotic algae. Reference is also made to the prokaryotic cyanobacteria, only to highlight evolutionary or functionally important differences or commonalities. It is assumed that the common details of photosynthetic mechanisms and pathways have been covered in introductory courses; they will be reviewed only briefly in the following section. Textbooks on plant physiology and biochemistry offer extensive treatments of all aspects of angiosperm photosynthesis (e.g. Buchanan *et al.* 2000; Raven *et al.* 2005). The accounts of radiation climate, light harvesting, and carbon metabolism presented here with respect to aquatic ecosystems owe much to the detailed books by Falkowski and Raven (2007) and Kirk (2010), which readers should consult for more information and references.

5.1 An overview of photosynthesis

Photosynthesis encompasses two major groups of reactions. The primary reactions, commonly referred to as "light reactions", involve the capture of light energy and its conversion to chemical potential as ATP and NADPH. The primary reactions, in turn, consist of three processes: energy absorption, energy trapping, and generation of chemical potential. The secondary reactions, also termed "dark reactions", include the sequence of reactions by which this chemical potential is used to fix and reduce inorganic carbon. Both reactions do run in parallel. A very illustrative visualization of this general reaction sequence using the example of higher plants and/or green algae in presented in the textbook by Raven, Evert and Eichhorn (see Fig. 7–9 in Raven *et al.* 2005).

Light, as electromagnetic radiation, is defined by its dual character and travels as packets (photons) and as waves. According to the Planck equation $E = h*c/\lambda$, the energy (E) of a photon is inversely related to its wavelength (λ); in other words, blue light (wavelength around 430 nm) contains about double the energy of

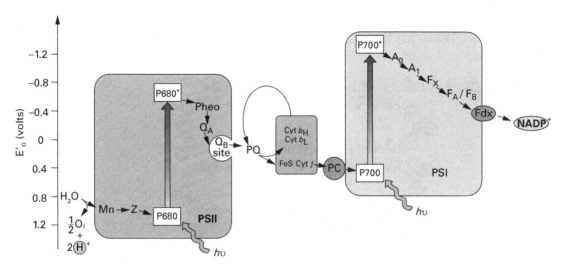

Figure 5.1 The "Z scheme" illustrating the flow of electrons according to their redox potential while passing through the electron transport chain. Electrons once excited by light trapping in PS II show high redox potential (negative E'_0), which will be reduced in each redox reaction within the transport chain. A second excitation of electrons in PS I allows for $NADP^+$ reduction. The gap of electrons in the PS II reaction center chlorophyll will be compensated by electrons generated by water splitting. (From Buchanan *et al.* 2000, with permission of John Wiley and Sons Ltd.)

red light (wavelength around 700 nm). When hit by a photon, pigment molecules absorb this quantum of energy and, thus, are in a chemically excited state for a small fraction of a second before releasing that energy again. In the case of chlorophyll, this energy release from the so-called "first excited state" can be accomplished either by heat dissipation, fluorescence emission, or, most relevant to energy supply for photosynthesis, energy transfer to neighboring pigments. The close packing of pigment molecules into light-harvesting pigment complexes allows most of the energy to be passed from molecule to molecule via the process of resonance energy transfer (also referred to as "exiton transfer", or "Förster transfer", referring to its discovery by the German chemist T. Förster in 1948) until it eventually gets to the reaction center. The significance of fluorescence emission by relaxing chlorophyll molecules as a proxy to estimate photosynthetic performance will be addressed in sec. 5.7.1.

Finally, an electron is boosted out of the chlorophyll in the reaction center (the so-called "special pair"), giving the electron a large redox potential. It is then captured by an electron acceptor and passed along an electron transport chain consisting of several compounds capable of undergoing redox reactions. Some compounds also pick up or pass on protons (H^+) with the electron. The reaction centers and the electron transport chain are arranged in the thylakoid membrane in such a way that a proton gradient develops across the membrane, the membrane being impermeable to H^+. This gradient is relieved through the transmembrane ATPase complex, in which the energy of the gradient is used to phosphorylate adenosine diphosphate (ADP). How the different complexes of the photosynthetic electron transport chain are arranged within the thylakoid membrane is nicely illustrated in Figure 7-12 in Raven *et al.* (2005).

The photosynthetic machinery of plants and algae usually host two photosystems, which are connected by the electron transport compounds. Another helpful depiction of the energetic processes that drive the photosynthetic primary reactions is to arrange the sequence of events according to the redox scale in the "Z scheme" (Fig. 5.1).

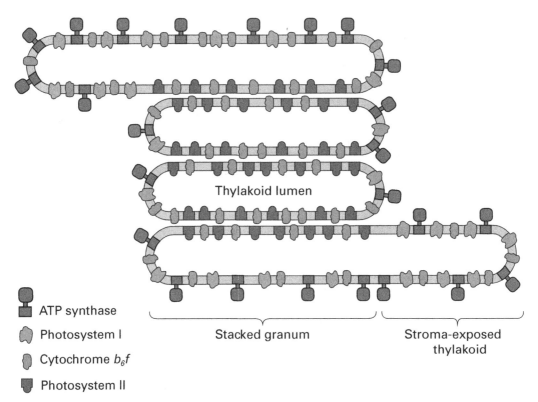

Thylakoid lumen

ATP synthase
Photosystem I
Cytochrome $b_6 f$
Photosystem II

Stacked granum   Stroma-exposed thylakoid

Figure 5.2 Lateral distribution of photosystems and other major complexes of the electron transport chain in the thylakoid membrane of higher plants/green algae. PS II is predominantly located in the stacked membrane regions, while PS I and particularly the ATP synthase complex require interaction with the thylakoid stroma to allow drainage of reduced NADP by the Calvin cycle, as well as phosphorylation (From Buchanan *et al.* 2000, with permission of John Wiley and Sons Ltd.)

Photosystem II (PS II) as the electron donor site also hosts the water-splitting complex, which is also the generation site of oxygen upon the oxidation of water. The electrons generated are passed along the electron transport chain until they are finally used to reduce $NADP^+$ in non-cyclic phosphorylation, which drives CO_2 fixation in the Calvin cycle. Electrons boosted out of the reaction center of chlorophyll *a* in PS II (P680) are given another boost in PS I (P700). One of the proteins in red algae and other "chromophyte" algae that is analogous to green-plant proteins (rather than homologous) is the electron-transfer compound between the cytochrome $b_6 f$ complex and the oxidizing end of PS I. In these algae, cytochrome *c* takes the place of plastocyanin (Raven *et al.* 1990). The two

catalysts are similar in many ways, but cytochrome *c* uses iron (Fe), whereas plastocyanin uses copper (Cu) as the prosthetic group.

Although the two photosystems are held in the membrane, they are not locked together in a 1:1 ratio as the Z scheme implies. Rather, there is a differential distribution of PS I and PS II between stacked and unstacked parts of the thylakoids, and also there are variable numbers of pigment complexes that may pass energy to both reaction centers, rather than to only one. This spatial separation is also referred to as "lateral heterogeneity" (see Fig. 5.2), and describes the thylakoids of streptophyte green algal and higher plant chloroplasts. Lateral heterogeneity is not apparent in brown algal plastids (with three-fold stacked thylakoids) and, due to the absence of thylakoid stacking, lateral heterogeneity is not observed in red algal plastids.

The Z scheme as an electron-flow diagram is far too simple since it shows only how the redox potential drives the process, but it does not show the dynamic aspects of the physical mechanisms. ATP and NADPH are the end products of the light reactions; they contain the energy saved from light and are used to fix inorganic carbon in the Calvin cycle in the stroma of the plastid. The Calvin cycle does not directly require light, although some of the enzymes involved are stimulated by light. The steps in the Calvin cycle are called the "dark" reactions of photosynthesis, but in fact the pathway functions only during the light and, thus, the term "secondary reactions" seems more appropriate. Commonly the reaction sequence of the cycle can be divided into three phases: (1) carboxylation (the fixation of CO_2); (2) reduction (the consumption of ATP and NADPH, and gain of triose phosphate); and (3) regeneration, which constitutes a complex reaction sequence in order to regenerate the CO_2-acceptor molecule ribulose-1,5-bisphosphate for another fixation event (Fig. 5.3).

Basically, the Calvin cycle is the only means of net carbon fixation in most plants; it is called C_3 photosynthesis because the first product of CO_2 incorporation is a three-carbon acid, 3-phosphoglycerate (3-PGA). However, some higher, mostly monocot plants in dry areas and/or areas that receive high levels of irradiance have an additional series of reactions in the cytoplasm that fix CO_2 into 4-carbon acids. However, in these C_4 plants no net CO_2 fixation results from the extra steps, because the C_4 compounds are broken down to give CO_2 again, which is then refixed in the Calvin cycle. This sequence serves to concentrate CO_2 at the plastid, and both pathways function in the daytime. In plants that use the so-called crassulacean acid metabolism (CAM plants), C_4 fixation occurs at night (when stomata can be opened), and release of CO_2 to the Calvin cycle in the same cells occurs during the day, when stomata are closed. C_4 as well as CAM metabolism are efficient means to suppress photorespiration (see sec. 5.4.2). However, in some seaweeds, especially the large browns, there is a different process

called light-independent carbon fixation, which results in net carbon fixation in darkness (see sec. 5.4.3).

Delving into more detail in the Calvin cycle, carbon dioxide (1 carbon) and ribulose-l,5-bisphosphate (RuBP; 5 carbons) react to give two molecules of PGA, the enzyme being, of course, RuBisCO (ribulose-l,5-bisphosphate carboxylase/oxygenase). Subsequent steps, consuming the ATP and NADPH from photosynthetic primary reactions in a pathway that is almost the reverse of glycolysis, convert PGA to glyceraldehyde-P and dihydroxyacetone-P and then put these together to produce fructose-1,6-bis-P. Some fructose-bis-P is siphoned off to produce low molecular weight free sugars or sugar alcohols. Other fructose-bis-P molecules, together with glyceraldehydes-P and dihydroxyacetone-P enter a complex series of carbon transfers that regenerate RuBP. Ultimately, for net production of one hexose molecule, six turns of the cycle and six CO_2 molecules are needed. As we will see below, the different algal groups are characterized by different light-harvesting pigments and photosynthetic products. This diversity together with the different light climate that exists underwater, contrasts markedly with the story of photosynthesis as known from green land plants.

5.2 Irradiance

5.2.1 Measuring irradiance

"Light" refers to the narrow region of the electromagnetic spectrum whose wavelengths are visible to the human eye, not including the ultraviolet and infrared wavelengths. However, there are important differences between the spectral sensitivities of the human eye and plant photosynthetic pigments. Our visual pigment, rhodopsin, has one major absorption peak, in the green region (556 nm), with absorption decreasing on either side. The chlorophylls and other light-harvesting pigments have different absorption peaks, and together they absorb across a broad region of what is called photosynthetically active radiation (PAR) (Fig. 5.4). PAR is usually defined as wavelengths of 400 to 700 nm, but there is some evidence that photosynthetic absorbance extends down to 300 nm in *Ulva*

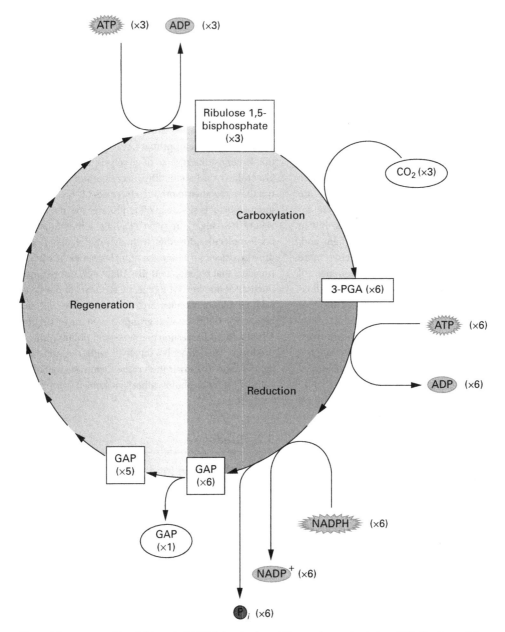

Figure 5.3 Overview of the Calvin cycle, highlighting the three different phases, the timing of CO_2 fixation and the consumption of ATP and NADPH, previously generated by photosynthetic primary reactions. 3-PGA, 3-phosphoglycerate; GAP, glyceraldehyde 3-phosphate. (From Buchanan *et al.* 2000, with permission of John Wiley and Sons Ltd.)

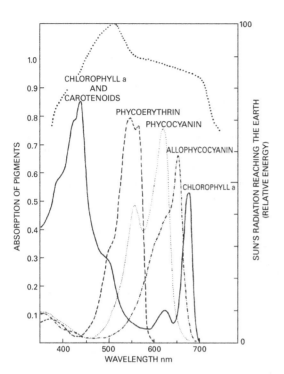

Figure 5.4 Spectrum of solar energy at the Earth's suface (upper dotted curve), and absorption spectra of algal pigments. (From Gantt 1975, © 1975 by the American Institute of Biological Sciences.)

lactuca and the tetrasporophyte of *Bonnemaisonia hamifera* (Halldal 1964).

The light meters used by lighting engineers and photographers are designed with the same spectral sensitivity (bias) as our eyes, and their measure of light is called illuminance. However, they do not give an accurate measure of the light available to plants, even on land. Since the spectral quality of light changes under water, such instruments are particularly inappropriate. Unfortunately, they are still used (as they are inexpensive); they give data in the units of either foot-candles or lux. The meters that are appropriate for such use are uniformly sensitive to all wavelengths, like broad-band radiometers.

Photosynthesis is a quantum process, and so the most appropriate measurement is the number of photons of PAR received by a unit of algal surface. This quantametric measure is called the photon flux density

(PFD). *Irradiance* is, strictly speaking, a measure of the amount of *energy* falling on a flat surface. Underwater, scattering contributes upwelling light, which becomes a significant component when downwelling light is dim (i.e., at depth), or over reflective coral sand (Kirk 2010). Instruments with spherical collectors can measure light from all directions, called "scalar irradiance" if energy is measured, or "photon flux fluence rate" if photons are measured. In any given study in which light is measured, only one of these terms can be accurate. Ultimately, the significant value for algae is the total number of photons they can *trap*, which no measure of incident light can convey, because of reflection and transmission of some photons. On the other hand, simple terms like "light intensity" have been criticized as referring to the light emitted by the source, rather than the light received. We shall use the term "irradiance" to mean the amount of PAR falling on a surface, regardless of whether we are citing data for quanta or energy, and regardless of whether or not upwelling light is included (see Kirk 2010 for further definitions).

The units most commonly used to report PFD are micromole photons per square meter per second (μmol m^{-2} s^{-1}; in the older literature the unit "Einstein" is sometimes used as a synonym for a mol of photons). The energy of light is measured in joules or watts, joules being units of total quantity, and watts units of flux (Smith and Tyler 1974). Thus, 1 W m^{-2} = 1 J m^{-2} s^{-1}. Under "full sun" near noon in temperate areas, the irradiance reaching the surface of the water is approximately 2000 μmol m^{-2} s^{-1} and varies with latitude, albedo, and season.

Nowadays a wide range of radiometers are available in order to record and characterize the light climate with a multitude of application options. Broad-band radiometers provide readings of instantaneous irradiance counting all photons in the photosynthetically active wavelength range; logging of these data over a given time period will provide the dose of radiation. Independent light loggers can be deployed for continuous recordings of the underwater radiation climate within a seaweed community. In addition, submersible spectroradiometers may even provide spectrally resolved readings of the underwater light field and are thus most valuable for photosynthesis studies on seaweeds, i.e. to address the effects of UV radiation (e.g. Hanelt *et al.* 2001; Huovinen and Gómez 2011).

5.2.2 Light in the oceans

Light reaching the Earth's surface already shows variations as a result of scattering and absorption by the atmosphere, but those are relatively small compared with the changes it undergoes in water. The total irradiance is also affected by the sun's angle, being less as the sun approaches the horizon. Four factors that influence light as it interacts with the oceans are: (1) its heterogeneity over time; (2) the effect of the water surface on light penetration; (3) its spectral changes with depth; and (4) its attenuation with depth (Kirk 2010).

Heterogeneity. The underwater light climate varies seasonally, both predictably because of changes in daylength and solar angle and unpredictably because of cloudiness and turbidity from storm waves, runoff, and seasonal plankton blooms. There are few data to illustrate the magnitude of the changes, but Lüning and Dring's (1979) work at Helgoland in the North Sea (54°N) still represents one of the most thorough studies. Instantaneous readings were taken every 20 min and extrapolated to give daily totals (Fig. 5.5a). Approximately 90% of the annual total light in the subtidal zone was received from April through September; during winter, the photic zone was very shallow mainly because of the low angle of the sun relative to the water surface. Large diurnal fluctuations in irradiance also occur, because of changes in clouds, tides, turbidity, and the angle of the sun (Larkum and Barrett 1983, Fig. 5.5b). Finally, there are instantaneous changes in irradiance due to waves and to canopy movement (Fig. 5.5c).

A study on the underwater light climate of an Arctic fjord system demonstrated the even more pronounced seasonality of light availability at high latitudes (Hanelt *et al.* 2001). Apart from the pronounced seasonal changes in the atmospheric light regime at a latitude of approx. 80°N (Polar day and night, lasting for several months each), the underwater radiation climate was substantially affected by sea ice cover as well as fresh water input from calving glaciers and melting snow. This increased the terrigenous particle supply to the fjord and thus strongly reduced water transparency. The input of terrigenous sediment is an important modulating factor of underwater light, which of course is most relevant in proximity to river inflows. However, as anthropogenic pressures on coastal systems are steadily increasing, sedimentation in coastal ecosystems due to large-scale construction activities in the coastal zone has become an issue of environmental concern (Chapter 9). Tropical waters generally are much clearer than temperate waters, but the silt load produced by poor land-management practices is increasingly a problem, particularly during the rainy season.

Water turbidity as a major factor controlling light availability to aquatic autotrophs at a certain depth can be assessed using the vertical attenuation coefficient (K_d, Hanelt *et al.* 2001; Kirk 2010). This parameter is determined by the simultaneous measurements of downward irradiance at two different water depths according to the formula: $K_d = \ln [Ed_{(z2)}/Ed_{(z1)}] \times (z_1 - z_2)^{-1}$ where $Ed_{(z1)}$ and $Ed_{(z2)}$ are the respective irradiances at depths z_1 and z_2. A low K_d of e.g. 0.1 m^{-1} indicates clear water conditions with a 10% light attenuation per meter. Higher turbidity results in higher attenuation and thus higher K_d values, and a value of 1 m^{-1} represents a 63% attenuation per meter (Hanelt *et al.* 2001).

The light climate in the intertidal zone is even more complex than that in the subtidal, but nevertheless it has been modeled by Dring (1987). The complexity stems from three factors: (1) the water type (as discussed later), which may or may not change significantly from time to time at a given site; (2) the tidal range, which has a monthly progression; (3) the timing of high and low tides in relation to the diurnal changes in irradiance, which are functions of both the progression in the times of the tides and the changes in daylength. Dring (1987) found that (for British shores) the lower intertidal receives less light in summer than in spring because the high tide is more likely to occur during the day as daylength increases. He demonstrated that in estuaries with turbid water and large

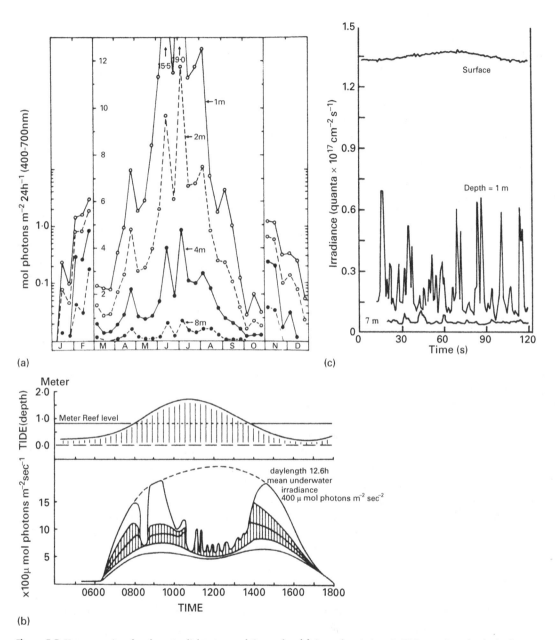

Figure 5.5 Heterogeneity of underwater light at several timescales. (a) Annual variations in PAR at various depths in the North Sea. (b) Diurnal changes in underwater irradiance as a result of solar flux and tide changes at Long Reef, Sydney, Australia. Upper part shows the course of the midday high tide. In the lower part, the upper and lower solid lines show the maximum range of irradiance; the center line is the mean, and the hatched area the 70% range. The broken line shows maximum irradiance on a day with a midday low tide. (c) Momentary changes in irradiance at the surface, at 1 m (in the region of greatest heterogeneity), and at the seabed, 7 m below the surface in a California giant kelp bed. (Part a after Lüning and Dring, 1979; b after Larkum and Barrett 1983; c courtesy of and © 1983 Valerie Gerard.)

tidal ranges, critical low light levels may occur in the intertidal rather than subtidally. For instance, kelp growth was formerly considered to be limited at 1% of surface irradiance, as will be discussed later. In the Avon estuary (UK) this level becomes intertidal, and kelps disappear because they are also limited in upward extent by exposure to air. *Fucus serratus*, which normally occurs into the subtidal zone, has an intertidal lower limit in the upper Avon estuary, again because of inadequate mean irradiance. The time course of the tides' influence on the underwater solar radiation available to seaweeds was also reported for two locations in southern Chile. When the low tide coincided with the maximal daily irradiance at solar noon, PAR levels in the infralittoral in Seno Reloncavi (7 m tidal range) were 40% higher than that during high tide, while at 2 m depth (subtidal), the difference was 30%. On the coast of Valdivia (2 m tidal range), when high tide coincided with the maximal daily irradiance, light was reduced by 15% (Huovinen and Gómez 2011).

Effect of the surface. The sea surface plays a large role in the underwater light climate (Campbell and Aarup 1989; Mobley 1989). Some of the light hitting the sea surface is reflected; the percentage reflected will depend on the angle of the sun to the water, and hence also on the state, or roughness, of the water. Reflection from a smooth sea with the sun near its zenith is only about 4% of the total light (sun plus sky), whereas with a sun altitude of 10°, reflection is about 28%. With an overcast sky, reflection is about 10% regardless of the altitude of the sun. Waves will increase light penetration when the sun is low by increasing the angle between the water surface and the sun, but whitecaps and bubbles in rough seas will increase reflection and can cut the light entering the water by as much as 50%. Under sunny skies, waves can cause considerable heterogeneity (glitter) in the subsurface light field by temporarily focusing the sun's rays to certain spots and away from others, especially in the top few meters of water (Kirk 2010), with significant (though diverse) effects on growth and photosynthesis on seaweeds (*Chondrus crispus*; Greene and Gerard 1990). The effects of wave-induced lightflecks on brown algal photosynthesis has been studied by Wing and Patterson (1993): For intertidal *Postelsia palmaeformis* stands in California the light flecks generated by wave action may significantly enhance the primary productivity of canopy and understory algae (see Chapter 8). Schubert *et al.* (2001) demonstrated that wave-focusing may result in short-term irradiance peaks (<1 s) five times higher than *in situ* irradiances, i.e. in shallow water habitats at the Baltic coastline peaks as high as 9000 µmol m^{-2} s^{-1} have been detected.

Spectral changes with depth. As solar energy penetrates the oceans, it is altered in both quality and quantity. The attenuation results from absorption and scattering, as occurs in the atmosphere. Water itself absorbs maximally in the infrared and far red, above 700 nm; far red (*c.* 750 nm) penetrates only to depths of 5–6 m (Smith and Tyler 1976). Ultraviolet radiation is also readily absorbed by the water column, but UV-B radiation might still adversely affect marine organisms in the upper 5 m (Bischof *et al.* 2006a). Absorption of radiant energy by particles in the water such as phytoplankton, will depend on the pigments they contain. Scattering by particles larger than 2 µm contributes to attenuation by increasing the length of the optical path of quanta once they are in the water, and thereby increasing the opportunities for absorption. Smaller particles and sea salts do not contribute appreciably to attenuation in the visible region (Jerlov 1970, 1976; Kirk 2010).

Attenuation of light by various processes and particles results in seawaters of different optical properties, and these have been classified by Jerlov (1976). The oceans have been divided into two broad categories: green coastal waters and blue oceanic waters, with subdivisions in each of these categories (Fig. 5.6). Jerlov (1976) distinguished five ocean-water types and nine coastal-water types. Figure 5.6 shows that in the clearest water (oceanic type I), the maximum transmittance is at approximately 475 nm. The Jerlov water type that transmits the least solar energy is coastal type 9. In such water the maximum transmittance occurs at about 575 nm (green). The characteristic green color of coastal waters is due to absorption

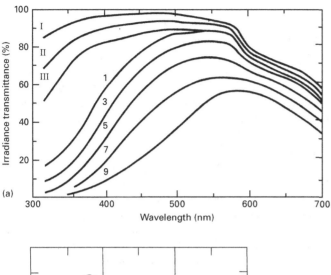

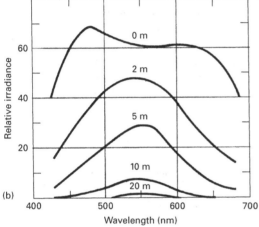

Figure 5.6 (a) Percentage of transmittance downward of irradiance of various wavelengths in Jerlov´s different optical types of water. The types range from clear oceanic (type I) to turbid coastal (type 9). (b) Energy spectra of natural light at various depths in the northern Baltic Sea. (From Jerlov 1976, reprinted with permission from *Marine Optics* © 1976 Elsevier Science Publishers.)

at shorter wavelengths by algal pigments and by yellowish dissolved organic substances (*Gelbstoff*) that absorb strongly in the blue wave bands. *Gelbstoff* comes from terrestrial humic material brought to the seas by rivers, and it is also produced in the sea by algae. However, measurements with sensitive spectroradiometers have shown that Jerlov overestimated light penetration, especially in the blue region and in coastal waters (Pelevin and Rutkovskaya 1977). As a result of the wavelength dependence of transmittance (Fig. 5.6a), both the quality and quantity of light will change with depth in any given water mass (Fig. 5.6b).

Light penetration in very turbid estuaries is even less than that through Jerlov's worst water type, and Dring (1987) extrapolated two additional and theoretical coastal-water types, 11 and 13.

Limits to growth. Irradiance in the sea is reduced with increasing depth. Most of the ocean floor (>90%; Russell-Hunter 1970) is permanently dark and has no algal growth, but what is the depth limit to which seaweeds can grow and what is the minimum light needed for growth? The compensation depth at which irradiance allows enough photosynthesis to just balance respiration (all integrated over time), is an important but elusive number. A value of 1% of surface irradiance (i.e. of "full sun") has commonly been used in oceanography to define the bottom of the euphotic zone (Steemann-Nielsen 1974), though Lüning and Dring (1979) have reported 0.5–1% as the lower limit for Laminariales and 0.05–0.1% as the lower limit for multicellular algae. Lüning (1981a) has suggested that the reason seaweeds can adjust their metabolism to extremely low light is that the flux is more nearly constant and predictable than the light climate for plankton.

Ideally, given the extreme variations in surface irradiance, and especially for studies of long-lived seaweeds, we need to know annual total irradiance. Surface irradiance in the Mediterranean is about 3000 MJ m^{-2} yr^{-1} or about 12.6 $\times$ 10^3 mole photons m^{-2} y^{-1}. Lüning and Dring (1979) measured the annual irradiance as 1.3 MJ m^{-2} yr^{-1} at the lower limit for seaweeds off Helgoland. However, flux measurements usually are made on a per-second basis, with surface irradiance, as mentioned previously, 1500–2000 µmol photons m^{-2} s^{-1} (e.g. Littler et al. 1985). These are midday values, and means for the whole day obviously will be lower. For instance, Osborne and Raven (1986) estimated the mean daily surface irradiance in Britain to range from 1000 µmol photons m^{-2} s^{-1} in June to 75 µmol photons m^{-2} s^{-1} in winter. The deepest known seaweeds, at 268 m on a seamount off the Bahamas, receive a maximum of 0.015–0.025 µmol photons m^{-2} s^{-1}, or a mere 0.0005% of surface irradiance (Littler et al. 1985). These red crustose corallines show a large light absorptance and employ light-harvesting pigments with a high energy cost in their production per unit light absorption rate (Raven and Geider 2003). Raven et al. (2000) outlined that it is difficult to explain algal growth below 0.5 µmol m^{-2} s^{-1} as there are energy consuming reactions (like redox back reactions of reaction center II, the leakage of H$^+$ through thylakoid membranes and the turnover of photosynthetic proteins), which use an increasing fraction of energy when photon flux density decreases. Thus it is yet not entirely clear how these crustose red algae can grow down to such water depth where the average incident photon flux density for 12 h per day does not exceed 0.02 µmol m^{-2} s^{-1} (Raven and Geider 2003).

Evidently, the depth range for seaweeds, where the substratum allows growth, extends much deeper than has commonly been thought and the study by Runcie et al. (2008) addressing photosynthesis of tropical deep-water algae by submersible logging fluorometers was pioneering in this respect. Recently, the existence of deep-water kelp refugia in the tropics has attained much attention. Graham et al. (2007b) established a model that predicts the potential occurrence of extended areas of kelp forests stretching down to 200 m. While occurrence of these extreme habitats still awaits confirmation, the model already resulted in the discovery of an extended deep-water population of *Eisenia galapagensis* off Galapagos, increasing in abundance at depths greater than 60 m. The establishment of such a deep-water population is likely to be supported by regional upwelling of cool, nutrient-rich water (Fig. 5.7). During an El Niño (warm-water) event along Baja California in 1997–98, Ladah and Zertuche-González (2007) found that deep-water *Macrocystis pyrifera* survived while most of the shallow population died, also suggestive of a depth refugia.

The changing radiation climate. In recent years the potential impacts of a changing radiation climate on seaweed ecology has gained much interest. The depletion of the stratospheric ozone layer (Madronich et al. 1998) and the resultant increase in the irradiance of harmful ultraviolet-B radiation (UV-B) reaching the Earth's surface has been found to be a threat to terrestrial and aquatic ecosystems (Häder et al. 1998; Björn et al. 1999). As a result of a 10% loss of stratospheric ozone, the irradiance at 320 nm would increase by 5% at the Earth's surface, but the irradiance at 300 nm (which is in the UV-B range and thus highly energetic) would double. UV-B radiation confers its damaging effects upon its absorption by biomolecules (proteins, lipids, pigments, nucleic acids) and may thus significantly impair physiological processes. In plants, as in seaweeds, the photosynthetic machinery has been shown to be severely harmed by UV-B exposure (Vass 1997; Bischof et al. 2000). As the primary effects on biomolecules and physiological processes may result in alterations in growth, productivity and reproduction, a significant UV-B-induced impairment of

(a)

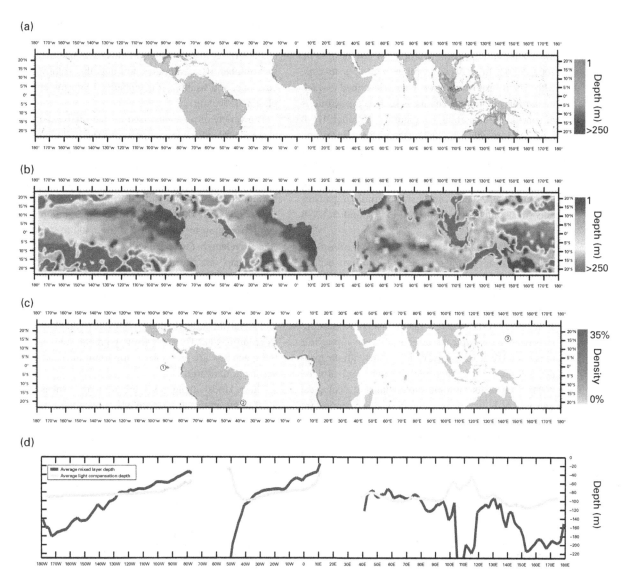

Figure 5.7 Components of a deep-water kelp refuge model for tropical regions worldwide. (a) Distribution of illuminated benthos receiving irradiance above kelp photosynthetic compensation point. (b) Mixed-layer depth. (c) Predicted location of deep-water tropical kelp populations (shading indicates frequency of predicted kelp habitats within 50-km radius). Numbers are locations of known deep-water tropical kelp taxa: 1, *E. galapagensis*; 2, *L. brasiliensis* and *L. abyssalis*; and 3, *L. philippinensis*. (d) Equatorial cross section (averaged between 5°N and 5°S latitudes) showing the mixed-layer depth relative to the kelp photosynthetic compensation point. (From Graham *et al.* 2007b with permission from the National Academy of Sciences.)

ecosystem function has also been revealed (Bischof *et al.* 2006a). With respect to seaweeds, the effects of UV-B exposure have been found to be highly species specific and to also depend on the respective developmental stage (Wiencke *et al.* 2000). Over all, it has been found that UV-B-susceptibility generally affects vertical zonation patterns of seaweeds (Bischof *et al.* 1998) and that the microscopic developmental stages are usually more affected than adult sporophytes (Wiencke *et al.* 2000, 2006). In seaweeds various strategies to cope with UV-B exposure have been described, which include acclimation mechanisms such as the synthesis of specific UV-absorbing compounds. Although UV-B radiation is readily absorbed by the water column, and in particular in coastal waters containing high concentrations of *gelbstoff*, seaweed productivity might still be reduced by UV-B at about 5 m water depth (Bischof *et al.* 2006a). Obviously, intertidal organisms need to be adequately equipped to respond to UV-B exposure.

Ozone depletion is a factor of concern mostly in polar and subpolar regions and these regions experience fluctuating UV-B levels. Tropical areas are not affected from stratospheric ozone depletion but UV-B irradiance is permanently high, thus ecophysiological studies on UV-B-tolerance mechanisms are particularly needed on tropical seaweeds, as even supportive effects of UV-B on recovery of photosynthetic processes have been reported (Hanelt and Roleda 2009). Due to its multiple molecular and ecological effects, UV-B has been identified as an important factor, which contributes to the structuring of seaweed communities (Bischof *et al.* 2006a). We shall elaborate in more detail on seaweed responses to UV-B in Chapter 7.

5.3　Light harvesting

5.3.1　Plastids, pigments, and pigment– protein complexes

Among the different algal phyla there are substantial differences with respect to the fine structure of plastids as well as thylakoid arrangement and respective pigment composition. As Fig. 5.8 shows, stacked granum thylakoids, as found in terrestrial plants as well as in the Chlorophyta, is just one means to increase light-harvesting efficiency. The Phaeophyceae (and the Heterokontophyta in general) usually have appressed thylakoids in groups of three, and a encircling girdle

thylakoid band. The Rhodophyta display single thylakoids bearing structures called phycobilisomes. Other characteristics in which algal plastids may differ include the number of membranes forming the chloroplast envelope, and whether it is embedded into the endoplasmic reticulum.

The variety of pigments used for light harvesting in algae is striking when compared to the low pigment diversity found in terrestrial plants. The need for accessory pigments, which broaden the wavelength range of efficient light harvesting is easily recognized by comparing the transmission and absorption spectra of seawater and chlorophyll *a*. The wide range of minimum light absorption by chlorophyll *a* from 490 to 620 nm (the wavelength range which is also referred to as the "green gap" or "green window") includes the wavelength range of maximal light transmission through seawater (at 465 nm for ocean, 565 nm for coastal waters). Underwater light is particularly rich in blue and green, the range, which is relatively weakly trapped by chlorophyll *a*. Thus, accessory pigments, narrowing the green gap are particularly important for light harvesting underwater.

Three kinds of pigments are directly involved in algal photosynthesis: chlorophylls, phycobiliproteins, and carotenoids (Larkum and Vesk 2003; Mimuro and Akimoto 2003; Toole and Allnutt 2003;). Chlorophyll *a* is universal and essential in the reaction center and is found in all algae. The other pigments, along with the bulk of the chlorophyll *a*, funnel energy to the reaction centers. Some additional chlorophylls occur in seaweeds. Chlorophyll *b* is found in Ulvophyceae (and in other Chlorophyta and higher plants); chlorophylls c_1 and c_2 (Fig. 5.9) occur in Phaeophyceae (and other lines of Chromophyta). All are tetrapyrrole rings with Mg^{2+} chelated in the middle; chlorophylls *a* and *b* each have a long fatty acid tail ($C_{20}H_{39}COO-$) that chlorophylls c_1 and c_2 lack. Chlorophylls *c* absorb blue light more strongly and red light less strongly than do chlorophylls *a* and *b*. In addition to the chemically different chlorophylls, these pigments, especially chlorophyll *a*, bind to proteins in various ways to create even more variety, as seen in their absorption spectra (especially the red peak). Whether

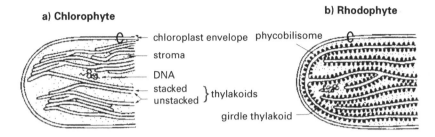

a) **Chlorophyte**

b) **Rhodophyte**

chloroplast envelope   phycobilisome

stroma

DNA

stacked

unstacked } thylakoids

girdle thylakoid

c) **Phaeophyte, Chrysophyceae,
Diatom, Haptophyte, Xanthophyte**

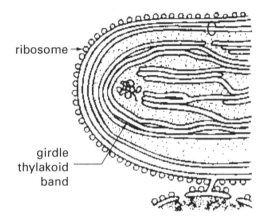

ribosome →

girdle
thylakoid
band

Figure 5.8 Plastid types in (a) Chlorophyta, (b) Rhodophyta and (c) the Phaeophyceae.
(From Larkum and Vesk 2003. With permission of Kluwer Academic Publishers.)

chlorophyll *d* may sporadically be present in some red algae or if this finding was due to contamination of samples with red algal associated cyanobacteria was not resolved for many years (Manning and Strain 1943; Miyashita *et al.* 2003). Indeed, chlorophyll *d* has been found to occur in larger concentrations in a cyanobacterium (*Acaryochloris marina*) from light-depleted environments that are rich in infrared radiation (Miyashita *et al.* 2003). Chlorophyll *d* was not exclusively attributed to light harvesting but also to replacing chlorophyll *a* in the reaction center of particularly PS I (Larkum and Kühl 2005). Thus, two enigmas related to algal photobiology became resolved: (1) the occurrence of chlorophyll *d* in red

algae is most likely due to a contamination of samples with cyanobacteria living in close association; and (2) there are indeed photoautotrophic organisms conducting oxygenic photosynthesis, which host a pigment other than chlorophyll *a* in their reaction centers.

Due to the spectral composition of the underwater light climate, the importance of carotenoids as light-harvesting pigments is much more pronounced in macro- and microalgae than in terrestrial plants. Seaweed plastids contain a wide variety of specific carotenoids, which are not present in higher plants. In addition to light harvesting, carotenoids are important in photoprotection under condition of excessive

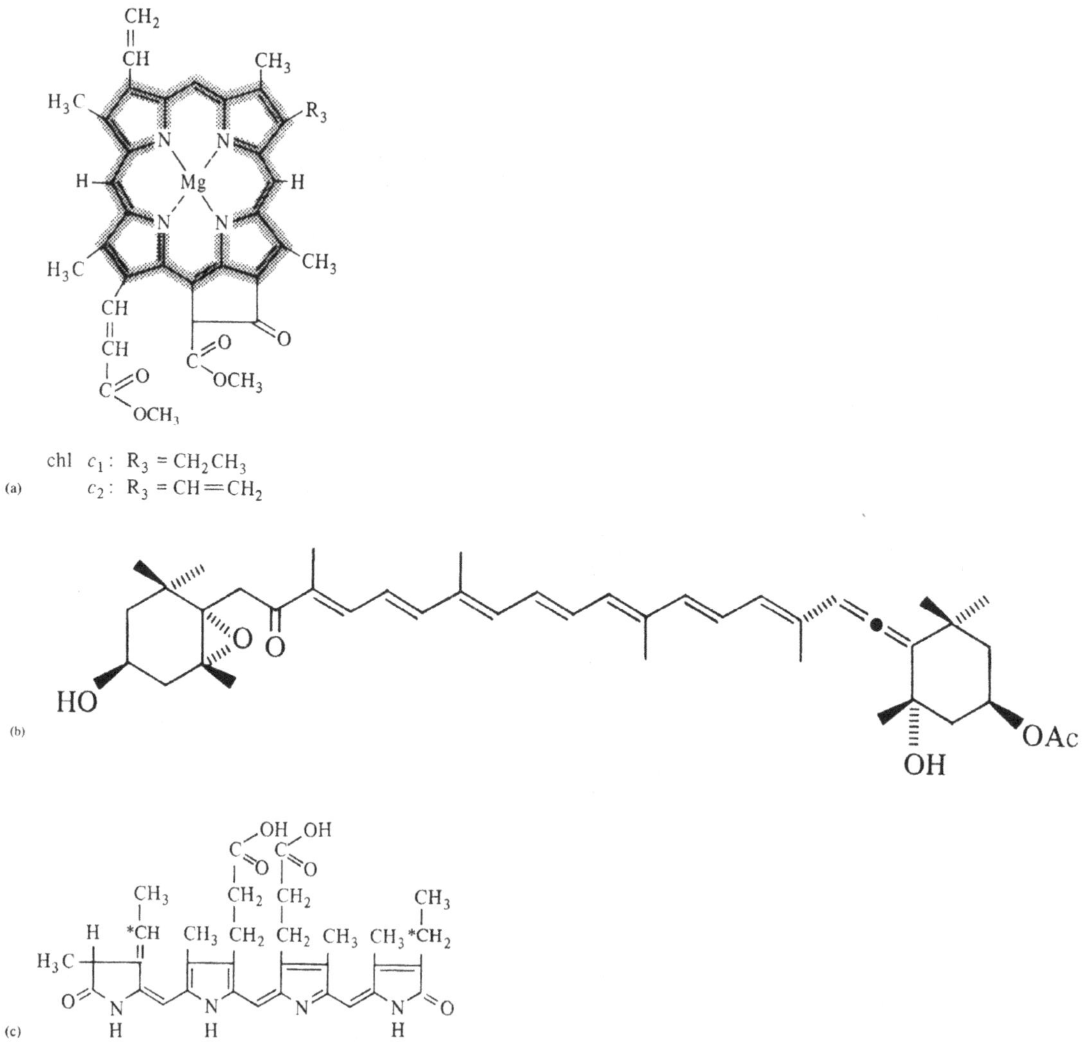

Figure 5.9 Algal pigment types. (a) Chlorophylls (chlorophyll c), showing the conjugated double-bond system (stippled). (b) Carotenoids (fucoxanthin). (c) Phycobilins (chromophore of phycocyanobilin). Sulfide bridges to protein cys residues form at carbons marked*. (Redrawn after various sources.)

irradiance. Carotenoids are C_{40} tetraterpenes, and two groups have to be distinguished: carotenes and xanthophylls. Carotenes are hydrocarbons, and xanthophylls contain one or more oxygen molecules (Fig. 5.9). Larkum and Barrett (1983) distinguished three groups of algae on the basis of the abundance and roles of their carotenoids: (1) Red algae and cyanobacteria have several minor carotenoids. Those associated with thylakoids are confined to the reaction centers, where they have a protective role. The light-harvesting complexes in these algae consist of phyco-biliproteins. (2) In most green algae, carotenoids play a role in light harvesting but are not the predominant antenna pigments. (3) There is a mixed group in which carotenoids play a major role in light harvesting. This group includes the brown algae and other "Chromophyta" with fucoxanthin as dominant light-harvesting pigment (Fig. 5.9b), plus some siphonous Chlorophyta and some deep-water green algae from other orders (e.g. *Ulva japonica, U. olivascens)* with siphonaxanthin. Fucoxanthin, β-carotene, and siphonaxanthin are unusual carotenoids because *in vivo* they absorb into the blue-green region. This, however, also points to their significance to seaweeds as the underwater light climate is particularly rich in green light and poorly exploited by the chlorophylls. Violaxanthin is another major xanthophyll in brown algae (and also present in green algae), but does not significantly contribute to light harvesting. However, as we will see below, violaxanthin plays an important role in photoprotection.

In all groups of seaweeds, some β-carotene is bound in a close and oriented association with PS I, indicating efficient energy transfer to this reaction center, but it probably has only minor significance to energy transfer to PS II. The PS II related function of β-carotene is rather photoprotection, i.e. the quenching of aggressive singlet oxygen (Telfer 2002). Thus, the second major role of carotenoids is in protecting the reaction center chlorophyll from photo-oxidation. For instance, in the xanthophyll cycle (known in green and brown algae), zeaxanthin can capture oxygen to form violaxanthin, which can be reduced again by ascorbic acid (Goodwin and Mercer 1983; Rowan 1989). The central role of xanthophyll cycling in photoacclimation and photoprotection will be discussed in sec. 5.3.3.

Phycobiliproteins are characteristic pigment–protein complexes of red algae and cyanobacteria; they also occur in Cryptophyceae. Unlike chlorophylls and carotenoids, they are water soluble, not lipid soluble, and form the so-called phycobilisomes on the surfaces of thylakoids, rather than being embedded in the membranes. The chromophore of phycobiliproteins consist of pigmented phycobilins, which are linear tetrapyrroles (Fig. 5.9c), covalently bound in various combinations to protein complexes. There are two principal phycobilins: phycoerythrobilin (red) and phycocyanobilin (blue). The protein complexes always have two different polypeptide chains, α and β, usually in a 1:1 ratio, but with as many as six of each chain in the phycobiliprotein. Three main classes of phycobiliproteins occur: Phycoerythrins (PE) absorb in the green region (495–570 nm) and generally have two phycoerythrobilin molecules on each α-chain and four on each β-chain (in two PEs there is a third polypeptide bearing a third phycobilin: phycourobilin). Phycocyanins (PC) and allophycocyanins (APC) generally have one phycocyanobilin on each protein chain. Phycocyanins absorb in the green–yellow region (550–630 nm), and allophycocyanins in the orange–red region (650–670 nm). Representative absorption spectra of these pigments are shown in Figure 5.4. Each of the three types occurs in red algae and cyanobacteria, although the proportions vary, as is evident from the predominant algal colors. Depending on light quality a change in the PE:PC ratio within the phycobilisome has been observed pointing to complementary chromatic adaptation in red algae (Lopez-Figueroa and Niell 1990; Sagert and Schubert 1995).

Pigments are arranged in or on the thylakoid in very particular ways. The reaction centers are special chlorophyll–protein complexes, surrounded by a core antenna of chlorophyll *a*. Most of the chlorophyll *a* and other accessory pigments are organized into light-harvesting complexes in the membrane or, in the red algae, as phycobilisomes on the membrane. These arrangements act as funnels to pass energy to the reaction centers. At each level the non-phycobilin pigments are noncovalently bound to proteins, giving many different combinations with different absorption spectra.

Each of the two photosystems (PS I, PS II) has a unique chlorophyll composition. The reaction center chlorophylls, which are the molecules directly involved in electron transfer (as opposed to energy transfer), are termed RC-I (or P700) and RC-II (or P680). Previously PS I has been most successfully studied, but recently also the understanding of structure and function of water-splitting PS II has significantly advanced (see Buchanan *et al.* 2000). Some of the chlorophyll *a* is closely associated with the reaction center in PS I to form core complex I (CC-I), formerly called P700-chlorophyll-*a*-protein complexes. Core complex I comprises two 84 kDa hydrophobic proteins, hosting 75 to 100 molecules of chlorophyll *a* and 12 to 15 molecules of β-carotene, as the internal antenna system (Kirk 2010). Core complex I can harvest light energy itself, however, the larger share of energy fuelled into RC-I via CC1 is captured by the light-harvesting pigment–protein complex termed light-harvesting complex I, LHC I (see below). Core complex II (CC II) is functional in photosystem II. About 40 chlorophyll *a* molecules and one P680 reaction center form one PS II unit. In CC II, chlorophyll *a* as well as β-carotene is arranged in two pigment–proteins, CPa-1 (approx. 52 kDa in size) and CPa-2 (approx. 48 kDa), also referred to as CP47 and CP43. CPa-1 contains about 20–22 chlorophyll *a* and 2–4 β-carotene molecules. CPa-2 contains 20 chlorophyll *a* and 2 β-carotene (Green and Durnford 1996; Kirk 2010). These complexes form the internal PS II antenna and are closely associated to the reaction center comprising the D1 and D2 reaction center proteins as well as the water-splitting complex (see Buchanan *et al.* 2000 for details).

The light-harvesting complexes of green and brown algae contain most of the chlorophyll *a* and all of the chlorophyll *b*, fucoxanthin, and other light-harvesting pigments (except a small amount of the β-carotene). Presumably LHC I in most of the green algae largely resemble LHC I from higher plants, containing chlorophyll *a* and *b* (in a ratio about 3.5:1) and the xanthophylls, with lutein being an important component. However, in some green algae, e.g. *Codium*, siphonoxanthin is the main carotenoid in LHC I (Chu and Anderson 1985). Most of the light-harvesting pigment–protein complexes characterized so far are related to LHC II. LHC II is the larger of the light-harvesting complexes and can contain as much as half the total pigment in Chlorophyta. Chlorophyll *a* is universally present in LHC II in combination with the respective accessory chlorophyll (*b* for green algae, *c* for brown algae) as well as xanthophylls characteristic for the alga under investigation (Kirk 2010). A chlorophyll-*a*/*b*–siphonoxanthin–protein complex occurs in those greens that have siphonoxanthin

(Larkum and Barrett 1983). Two different sizes of LHC II, not correlated with the presence or absence of siphonoxanthin, have been found in Ulvophyceae (Fawley *et al.* 1990).

Light-harvesting complex II in brown algae has been characterized by Katoh and Ehara (1990) and De Martino *et al.* (2000). In the brown alga *Petalonia fascia*, as well as in the giant kelp *Macrocystis pyrifera*, authors described large supramolecular pigment–protein associations consisting of several to many LHC II complexes characterized as chlorophyll-*a*/*c*2-fucoxanthin–protein and chlorophyll-*a*/*c*1/*c*2–violaxanthin–protein complexes. In *Petalonia fascia* each complex contains 128 molecules of chlorophyll *a*, 27 chlorophyll *c*, 69 fucoxanthin, and 8 violaxanthin (Katoh and Ehara 1990). In *Fucus serratus* the pigment ratio of chlorophyll *a*:chlorophyll *c*:fucoxanthin in LHC II was found to be 100:16:70 (Caron *et al.* 1985, 1988). More recent studies on the kelps *Laminaria saccharina* (now referred to as *Saccharina latissima*) and *Macrocystis pyrifera* suggest that the fucoxanthin-containing LHCs from brown algae are homologous proteins. Thus, discrimination between LHCs specific for either PS I or PS II was not possible, suggesting that both photosystems are either equipped with the same type of antenna or even sharing the very same antenna (De Martino *et al.* 2000). In this way, imbalance in energy distribution between PS I and PS II is not likely to occur.

Most of the data so far still imply that each photosystem has its own antenna. However, it has been suggested by Larkum and Barrett (1983) that rapid changes in underwater light quality and quantity would make it advantageous for algae to control the distribution of energy between photosystems and proposed a mechanism allowing the passage of excess energy from PS II to PS I; a process commonly termed "spillover". The mode of energy partitioning between photosystems is hitherto conceptualized by a phenomenon referred to as "state transitions", referring to the mobility of light-harvesting units between PS II and PS I: by migration of LHC units from one photosystem to the other, the respective optical cross section is altered (Forsberg and Allen 2001). There is now evidence that the underlying mechanism involves a

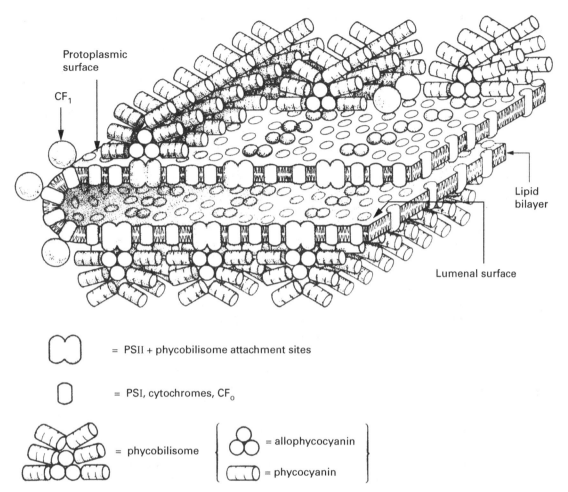

Protoplasmic surface

CF_1

Lipid bilayer

Lumenal surface

= PSII + phycobilisome attachment sites

= PSI, cytochromes, CF_o

= phycobilisome

= allophycocyanin

= phycocyanin

Figure 5.10 A model of the thylakoid membrane from cyanelles of *Cyanophora paradoxa*, showing the phycobilisomes attached to PS II and the CF_1 coupling factor at ATPase attached to a unit comprising PS I, cytochromes, and the CF_1 subunit of the coupling factor. If phycoerythrin were present, it would be on the outer ends of the phycocyanin rods. (From Giddings *et al.* 1983, with permission of the American Association of Plant Physiologists.)

phosphorylation of the light-harvesting PS II protein. The resultant negative charge induces the detachment of LHCs from PS II and the migration to PS I (Allen 2003; Kirk 2010).

Phycobiliproteins form clusters on the surfaces of thylakoids (Fig. 5.10), where they are more amenable to observation (Cohen-Bazire and Bryant 1982; Glazer 1985). Hemidiscoidal phycobilisomes occur in most cyanobacteria and a few red algae, whereas the usual shape in red algae apparently is globular (e.g. Talarico

1990). In field material of *Porphyra umbilicalis*, and perhaps other red seaweeds, both shapes of phycobilisomes occur, which may provide a means of response to fluctuating irradiance (Algarra *et al.* 1990). Phycobilisomes are divided into two domains, the central core (hosting the pigment allophycocyanin) and peripheral rods (hosting phycocyanin in proximity to the core and phycoerythrin at the periphery of the rod). Specific linker polypeptides assist in attaching the phycobilisome to the thylakoid membrane, as well in orientating the complex properly towards PS II.

The arrangement of the pigments parallels the pathway of energy transfer, which can also be traced in the pigment spectra (Fig. 5.4), because energy is inversely proportional to wavelength: PE → PC → APC → chlorophyll. Most phycobilisomes are associated with PS II; PS I receives some light directly from the remaining phycobilisomes, but most via spillover from PS II (Biggins and Bruce 1989).

Hitherto, a number of red algal species have been characterized with respect to their phycobilisome inventory. For example, *Pyropia yezoensis* (previously *Porphyra yezoensis*) hosts hemi-ellipsoidal phycobilisomes with dimensions of 45.1 nm length, 23.1 nm thickness and 34.6 nm height. A density of 770 phycobilisomes per μm^2 thylakoid membrane surface has been determined. As the density of the PS II was found to be considerably higher, a phycobilisome/PS II ratio of 0.6 has been established for most of the red algal species investigated so far (Toole and Allnutt 2003).

5.3.2 Functional form in light trapping

The light-harvesting ability of seaweeds depends not only on the types and amounts of pigments and their disposition within the thylakoid but also on higher levels of organization such as the arrangement of plastids in the thallus and the gross morphology of algae. The arrangements of the plastids in algal cells are quite varied, but in cells with large vacuoles or many small plastids the plastids tend to be distributed around the periphery of the cell. Furthermore, plastid arrangement may change over daily cycles to modulate light capturing in green and brown seaweeds (Hanelt and Nultsch 1990). Pigmented cells tend to be distributed around the periphery of thicker thalli, with medullary cells having few or no plastids. These arrangements obviously facilitate the passage of light to the pigments, but they may also be essential to maintain adequate inorganic carbon fluxes (Larkum and Barrett 1983).

Many seaweeds are constructed of thick, optically complex tissues (Osborne and Raven 1986; Ramus

1990). The packaging of pigments into light-harvesting complexes, plastids, and cells reduces light absorption, as compared with *in vitro* absorption by a pigment solution (the "package effect"; Mercado *et al.* 1996). On the other hand, refraction and reflection increase the light path within a thallus, enhancing the chance of photon capture. Some seaweeds have internal air spaces (e.g. those tubular species of *Ulva*, which were previously referred to as *Enteromorpha)* or internal layers of calcium carbonate (e.g. *Halimeda)* that tend to increase the backscatter of light, in some cases perhaps even acting as "light guides" (Ramus 1978). The theories on light absorption tend to assume a relationship between pigment concentration (especially chlorophyll *a)* and absorption; this relationship holds for absorption of monochromatic light by pigment solutions (Beer's law), but not for the ordered, hierarchical structure of seaweeds in polychromatic light (Grzymski *et al.* 1997). The net effect of pigment and thallus properties can be seen in thallus *absorptance*, the fraction of incident light absorbed. Figure 5.11 shows absorptance spectra for *Ulva* (70 μm thick) and *Codium* (3 mm thick), with different amounts of pigment. Absorptance of *Codium* changes little over a large range in pigment concentration, whereas that of *Ulva* is significantly affected by pigment concentration. Algae such as *Codium*, which absorb virtually all incident light, have been called optically black. Evidently, the functional form of a seaweed (sec. 1.2.2), which we will further explore on in sec. 5.7.2, affects its light-harvesting ability, and Hay (1986) suggested a set of growth forms, reflecting these differences in the capturing of light as follows:

1. Monolayers with flat, opaque thalli, including fleshy umbrella-shaped algae such as *Constantinea*, and also (though Hay did not discuss these) crustose calcareous forms and non-calcareous crusts;
2. Multilayered, translucent thalli, such as *Ulva, Halymenia*, and *Padina*;
3. Multilayered thalli with flat but narrowly dissected blades or terete branches; here we might also include filamentous thalli (which Hay did not consider);
4. Multilayered thalli with midribs supporting thin blades (e.g. *Sargassum)*; here we might also include

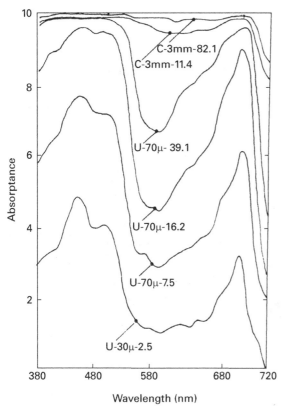

Figure 5.11 Relative contributions of pigmentation and thallus construction to light absorption. Absorptance spectra for *Codium* (C), an opaque seaweed, and *Ulva* (U), a translucent seaweed. Increasing pigment concentration (given in nanomoles per square centimeter) in 70-μm-thick *Ulva* thalli greatly increases absorptance (fraction of incident light absorbed). *Codium* thalli absorb essentially all the incident light, even with pigment concentrations much lower (on a surface-area basis) than those for *Ulva*. (From Ramus 1978, with permission of *Journal of Phycology*.)

Macrocystis, in which the blades are supported by floats along a stipe rather than by a midrib (Lobban 1978a).

As with Littlers' functional-form groups (see secs. 1.2.2 and 5.7.2; Littler and Arnold 1982), a species may fall into more than one category (e. g., by genotypic variation, as shown in *Gracilaria tikvahiae* by Hanisak

et al. 1988). Superimposed on the morphological characters are biochemical/physiological differences in light-harvesting ability (e.g. "sun" vs. "shade" plants, ecotypic variation; Algarra and Neill 1987; Coutinho and Yoneshigue 1990; Gómez and Huovinen 2011).

Hay (1986) hypothesized that in well-lit habitats, multilayered seaweeds (with a high thallus area: substratum area ratio, in higher plants called leaf-area index) would have a growth-rate advantage over monolayered seaweeds with thallus:area ratios near unity, whereas in dimly lighted habitats the reverse should be true. If broad, flat thalli are translucent, they may be multilayered; translucency depends on thin thalli and adequate light, and in very dim light even thin thalli may absorb all the light entering them. Thicker thalli can be multilayered only if their branches are narrow enough to cast little shade on the branches below. Water motion, by constantly rearranging branches and blades, makes even a thick canopy of large blades such as *Macrocystis* effectively multilayered (Lobban 1978a), at least in well-lit habitats. The surface:area ratio of a narrowly branched seaweed with erect thalli, such as *Corallina elongata* was determined to be 1413 m^2 m^{-2} for those from a sunny site and 224 m^2 m^{-2} for shaded individuals (Algarra and Neill 1987), in extreme contrast to the values for kelps and fucoids of 4–7 m^2 m^{-2}.

Peckol and Ramus (1988) criticized Hay's hypothesis when they found that multilayered and not monolayered species were most abundant in a deep, low-light community. Competition for space and nitrogen were also important factors there, and the community presumably reflected the combined results of these opposing selective pressures. The success of seaweeds in the field must always be the result of many factors, and that does not invalidate Hay's hypothesis about the relationship between growth rate and irradiance. Strong support of functional form has been obtained from a study at the virtually atidal west coast of Sweden (Pedersen and Snoeijs 2001). A robust trend of decreasing abundance of filamentous species and increasing abundance of leaf-like species was revealed with increasing water depth. In a survey on 32 seaweed species from the Swedish coast, the dependence of photosynthetic characteristics on thallus morphology

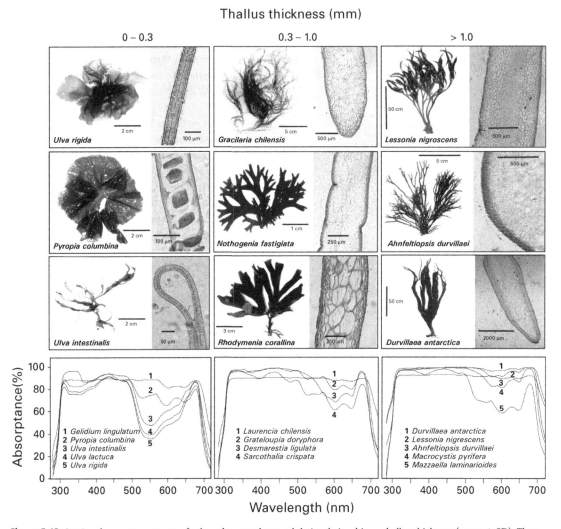

Figure 5.12 *In vivo* absorptance spectra of selected macroalgae and their relationship to thallus thickness (mean ± SD). Three examples of thallus cross sections are shown for each group (From Gómez and Huovinen 2011, with permission of Inter-Research).

was confirmed again. Higher and faster oxygen production rates for thinner and filamentous species, and lower rates for coarser and thicker species were found when normalized to dry weight and the opposite when normalized to the surface area of the respective alga (Johansson and Snoeijs 2002).

Gómez and Huovinen (2011) revisited the functional-form discussion in their study on morpho-functional and zonation patterns of South Chilean seaweeds and showed a relation between thallus absorptance and thickness (Fig. 5.12). However, in the case of intertidal assemblages, the ordination of thallus morphologies in relation to light availability is strongly dependent on other physical contraints (e.g. desiccation, solar UV radiation, wave impact, etc.) and thus, patterns do not neccesarily follow the classical form and function theories (Gómez and Huovinen 2011).

Another complication not addressed in Hay's or the Littlers' theories is the growth habit of turfs. Dense turfs or mats (common on well-lit flats) consist of filamentous algae with a high "leaf-area index", but so crowded that self-shading becomes significant (Williams and Carpenter 1990). Sparse turfs may be closer to the theoretical functional-form group of filamentous algae.

5.3.3 Photosynthesis at a range of irradiances

The rate of photosynthesis depends on the irradiance available, or ultimately the irradiance absorbed. The relationship between photosynthesis and irradiance, the *P-E* curve, is useful for comparing the physiology of light harvesting in different seaweeds. [Note that the symbol "E" is currently used to denote irradiance, and has replaced "I" which was used in most older texts (Falkowski and Raven (2007, p. 240)]. A generalized *P-E* curve is shown in Figure 5.13a. In extremely dim light, respiration is greater than photosynthesis. When photosynthesis balances respiration, the irradiance level is at the compensation irradiance E_c. The rate of photosynthesis increases linearly at first, and the initial slope, α, is a useful indicator of quantum yield. At higher irradiances, photosynthesis becomes saturated (P_{max}), limited by the "dark" reactions. The saturating irradiance, E_k, is defined as the point at which the extrapolated initial slope crosses P_{max}'. Based on these general correlations mathematical models have been formulated to facilitate productivity estimates from *P-E*-curves, to fit functions (with limited data points) as well as to facilitate computation of the critical parameters P_{max}, E_c, E_k, and α. The most common formulas used were originally applied to describe phytoplankton photosynthesis but have widely also been applied to macroalgae. The simple formula proposed by Jassby and Platt (1976) was found to give a particularly good fit:

$$P = P_{max}\tanh(\alpha E_d/P_{max}),$$
with E_d = downward irradiance.

However, different *P-E* models, as Ramus (1990) pointed out (Figure 5.13), assume homogeneous

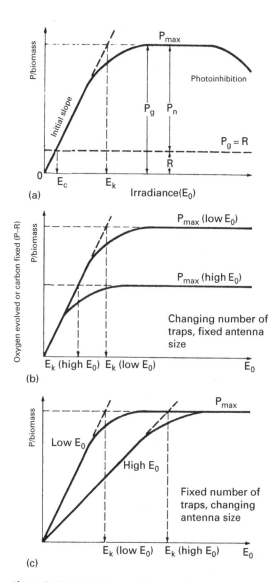

Figure 5.13 Model light-saturation curves for net photosynthesis (*P*) versus incident irradiance (E_0). (a) General model defining P_{max}, maximum photosynthesis; P_g, gross photosynthesis; P_n, net phtosynthesis; R, respiration; E_c, compensation irradiance; E_k, saturating irradiance level. Horizontal dashed line represents zero net photosynthesis $(PR = R)$. (b) Model of adjustment of the photosynthetic unit to extreme low and high 10 by changing the number of PSUs but not their size. (c) Model of adjustment to extremes of irradiance by changing the size of a fixed number of PSUs. (From Ramus 1981, with permission of Blackwell Scientific Publications.)

suspensions of unicells in random motion, in which light scattering is quite different from what takes place within a seaweed thallus. Photon gradients in seaweeds are steep and are greatly affected by the strong backscattering, except in very thin, homogeneous algae like *Ulva* and *Porphyra*. Thus, Ramus argued that for ecological studies of seaweeds we need to relate photosynthesis to *absorbed* irradiance, E_a, not to incident irradiation, E_o, and he defined the "optical-absorption cross section" as a parameter that indicates light-harvesting ability.

Another model, which was originally proposed to describe photosynthesis in tree species has interestingly also been found to be equally applicable to phytoplankton photosynthesis (Webb *et al.* 1974; Peterson *et al.* 1987; Kirk 2010) and should thus also give a good estimation of seaweed photosynthesis:

$$P = P_{max}\left[1 - e^{-\frac{aE_d}{P_{max}}}\right]$$

P-E curves are mathematical models of photosynthetic rates (Jassby and Platt 1976; Nelson and Siegrist 1987; Madsen and Maberly 1990) and the choice of the proper denominator (i.e. units of biomass or chlorophyll) is of crucial importance to the interpretation of results. These models can be used to predict the change in rate if one of the parameters, such as pigment concentration, changes. For instance, an alga may acclimate to low irradiance by increasing the number of reaction centers or the size of the antennae. Ramus (1981) used the concept of a photosynthetic unit (PSU), consisting of a reaction center and its associated light-harvesting pigment complex. We now recognize that there is not a 1:1 ratio between PS l and PS II or between them and their antennae, but the concept of the PSU is still useful. Ramus used the model to predict that if the *number* of PSUs is increased (Fig. 5.13b), P_{max} will increase, and more light will be needed to saturate photosynthesis (higher E_k). On the other hand, if the *size* of PSUs increases, without changing their number (Fig. 5.13c), P_{max} will not change, but less light will be required to saturate photosynthesis; the PSUs in this case will be more

efficient. Some seaweeds have been studied to determine whether they adjust the number or size of the PSUs (or both). Among those that have been studied, *Ulva lactuca* changes its PSU number, whereas *Porphyra umbilicalis* changes both (Ramus 1981).

The levels of irradiance needed to reach compensation are some 2–11 μmol m^{-2} s^{-1} for shallow water seaweeds (Arnold and Murray 1980; Hay 1986), but are much lower in dim habitats. The deepest growing crustose coralline red algae have been found to survive at irradiances as low as 0.01 μmol m^{-2} s^{-1} (Littler *et al.* 1985). While it is evident that low-light-adapted species are to be characterized by high photosynthetic efficiencies and very low compensation and saturation points, it is still difficult to outline algal mechanisms for maintaining growth at irradiances below 0.5 μmol m^{-2} s^{-1} because of the steady operation of energy consuming processes (Raven *et al.* 2000). In polar algae with "shade" adapted characteristics, growth at depths close to 30 m is favored by very low E_k and E_c of 11 and 2 μmol m^{-2} s^{-1}, respectively, which allow these seaweeds to maintain positive carbon balance (net photosynthesis) almost during the complete daylength, and thus to compensate for dark respiration (Gómez *et al.* 2009).

Saturating irradiances show some correlation with habitat, but generally are low compared with full sun, suggesting that seaweeds are all more or less "shade adapted" (Reiskind *et al.* 1989). E_k values in intertidal species are 400–600 μmol m^{-2} s^{-1} (*c.* 20% of full sun), in upper and midsublittoral species 150–250 μmol m^{-2} s^{-1}, in deep sublittoral species <100 μmol m^{-2} s^{-1} (Lüning 1981a), and polar species are even more strongly shade adapted (Weykam *et al.* 1996). A more recent study by Marquardt *et al.* (2010) confirmed the adjustment of E_k in three commercially important red algal species from Chile. Possible mechanisms involve the adjustment of the reaction center ratio, changes in the size of light-harvesting complexes and/or the ratio of light-harvesting versus light-protective pigments. As we will see later, this acclimation to light availability along the depth gradient bears important implications to the carbon balance of seaweeds.

In general, seaweeds respond to low-light conditions, i.e. at increasing water depths, with increasing antenna sizes, as indicated by decreasing chlorophyll a:b ratios in benthic green algae (Yokohama and Misonou 1980). Thus, a larger antenna increases light-harvesting efficiency, however, the alga needs to invest a larger share of energy in pigment synthesis. Thus, when algae are shifted from high to low irradiance, they suffer an "energy crisis" (Falkowski and LaRoche 1991), to which they may respond by diverting macromolecule biosynthesis from lipids and carbohydrates to proteins (for light-harvesting complexes) and then back to lipids (for photosynthetic membranes).

In seaweeds growing at high irradiances in shallow or intertidal habitats, the opposite is observed: smaller antennae reduce the risk of overexcitation, potentially resulting in photoinhibition and photodamage. Under such conditions, the algae have to invest more into the turnover of photosynthetic proteins and enzymes as well as photoprotective mechanisms. At high irradiances, which exceed the demands of photosynthesis, the phenomenon of photoinhibition is observed. Photoinhibition has been defined as the failure of photoprotective mechanisms to mitigate photoinactivation (Franklin *et al.* 2003). In the early literature photoinhibition was mainly attributed to a damaging event on photosynthesis, but more recent studies show that there is also a regulatory and protective component involved. Thus, Osmond (1994) has introduced the term *dynamic* photoinhibition, for a transient reduction in photosynthetic quantum yield in order to protect reaction centers from excessively absorbed energy and suppress reactive oxygen formation by the Mehler reaction, in which electrons (from water) pass to oxygen (see below). In plants and seaweeds applying dynamic photoinhibition, quantum yield rapidly recovers as soon as energy pressure is reduced again (e.g. by an increased water column under high-tide conditions or a reduced solar insulation after noon) and is usually fully restored within a couple of hours. In this sense dynamic photoinhibition is also termed photo*protection* (Franklin *et al.* 2003) and includes all processes that cause a decrease in excitation transfer to the reaction centers. These processes take part in overall photo*acclimation* summarizing all mechanisms involved in the adjustment of structure and function of the photosynthetic apparatus to counteract photoinhibition/photoinactivation. Photoinactivation, previously also referred to as *chronic* photoinhibition, refers only to the slowly reversible inhibition of quantum yield and indicates photo*damage*, e.g. by the photo-oxidation of photosynthetic pigments and proteins mediated by reactive oxygen species, or an increased PS II fragmentation and non-sufficient D1 protein turnover velocity under high light stress (Aro *et al.* 1993). Photoinhibition is commonly observed in seaweeds under desiccation stress in the intertidal zone (Herbert and Waaland 1988; Herbert 1990), but also in submerged seaweeds (Huppertz *et al.* 1990).

The mechanisms of *dynamic* photoinhibition are still not entirely resolved, however, a central role of xanthophyll cycling has been claimed as one important alternative energy sink. By the exergonic epoxidation of violaxanthin to antheraxanthin and further to zeaxanthin, excess energy is drained off the system by heat dissipation (Demmig-Adams and Adams III 1992; Demmig-Adams *et al.* 2008). This process primarily described in higher plants, has also been shown to be operative in green and brown seaweeds (Uhrmacher *et al.* 1995; Bischof *et al.* 2002a). Dynamic photoinhibition is an important feature for shallow water algae in order to cope with high irradiances when low tide coincides with high solar elevation and the respective capability for applying this feature has been shown to reflect the vertical distribution of algae on the shore (Hanelt 1998). However, as Raven (2011) has stated, all mechanisms applied in order to counteract photoinhibition require some kind of metabolic energy, which might result in increased/alternative nutrient and energy demands, ultimately affecting growth.

Inhibition of photosynthesis is also frequently observed when exposing seaweeds to enhanced ultraviolet radiation (UV-R; the impacts of ozone depletion and ultraviolet radiation on seaweeds will be reviewed in Chapter 7). Under field conditions, PAR as well as UV-R may contribute to the inhibition of photosynthetic performance and are, thus, not easily distinguished (Bischof *et al.* 2006a; Fredersdorf and Bischof

2007). However, the mechanisms behind PAR- and UV-R-mediated damage are quite different. Excess PAR favors the generation of reactive oxygen species (sec. 7.1), and thus photo-oxidation of photosynthetic pigments, proteins, and membrane lipids. However, UV-R acts in a more direct way by the absorption of high-energy quanta by UV-chromophores (such as peptide bonds, aromatic residues, etc.), which may directly result in, for example, conformational changes and thus impair molecular functioning, as we will see later. Generally, the process of dynamic photoinhibition does not provide protection against UV-R exposure.

Seaweed photosynthesis is also influenced by changes in spectral composition. These changes typically occur over the depth gradient, within algal canopies, but also over daily and seasonal cycles and can be perceived by algae as environmental signals, which may affect photomorphogenesis, but also pigment composition associated with PS II and I. For example, increases in the ratio of green:red and blue:red will result in increased light absorption by photosystem II and shift phycoerythrin/phycocyanin as well as phycoerythrin/chlorophyll ratios in the rhodophyte *Pyropia leucosticta* (as *Porphyra leucosticta*; Salles *et al.* 1996). Blue light is also known to affect photosynthetic rates and may stimulate photosynthetic capacity, in both the short term and long term. Photosynthesis in red light by *Laminaria digitata* is stimulated by 2 min of blue light, but only for about 1 h (Dring 1989). It is suggested that blue light activates the release of CO_2 from an internal store (Schmid and Dring 1996).

In order to give an autecological perspective of the outlined differences in light utilization, Ramus (1990) developed a model of seaweed production relating growth yield to irradiance. He defined the photon growth yield (PGY) as a quantitative parameter to assess the efficiency of light utilization, where

$$PGY = \mu/E_a$$

with μ as the specific growth rate and E_a the light absorbed. By plotting against both E_o and E_a (Fig. 5.14a, b), he compared *Ulva* and *Codium* and found that the growth of *Codium* saturated quickly and then became photoinhibited (Fig. 5.14a, b), but

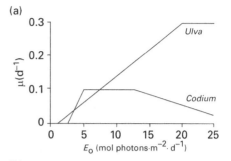

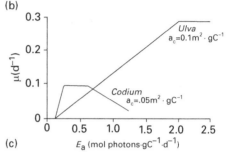

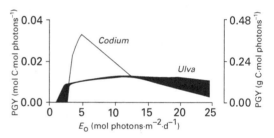

Figure 5.14 Growth-rate–light relations for *Ulva* and *Codium*. (a) Specific growth rate (μ) versus incident light (E_0). (b) Specific growth rate versus light absorbed (E_a), where $E_a = E_0 a_c$ (a_c = carbon-specific absorption cross section). (c) Photon growth yield (PGY) versus E_0, (From Ramus 1990, with permission of Kluwer Academic Publishers.)

that at intermediate irradiances *Codium* used incident light more efficiently than did *Ulva* (Fig. 5.14c).

5.3.4 Action spectra and testing the theory of complementary chromatic adaptation

Each pigment–protein complex has a characteristic absorption spectrum; the sum of all the pigments gives thallus absorption spectra, which, however, are complicated by morphological effects. Narrow wave-band light was used to record absorption spectra in the classic experiments by Engelmann (1883, 1884) and Haxo and Blinks (1950). However, natural light, even in deep water, is broad-band, and this significantly changes the wavelength dependence of photosynthesis. When a small amount of green light (546 nm) supplemented the monochromatic light in Fork's (1963) experiments on red algae, significant increases in photosynthesis occurred below 500 nm and above 650 nm, out of proportion to the amount of extra light, and the action spectrum more closely corresponded to the absorption spectrum. The supplementary wavelength chosen by Fork is close to the peak of the deep-water spectrum.

Engelmann's experiments led to a theory relating seaweed zonation to their pigment complements. Although this theory of complementary chromatic adaptation has been adequately disproved (Ramus 1978; Dring 1981), its "compelling logic" (Ramus 1982) and century-long acceptance have kept it popular as a story in biology textbooks (Saffo 1987). Engelmann's theory proposed that green seaweeds, carrying only chlorophylls, would be restricted to shallow water; the brown algae, with fucoxanthin, could extend farther; and the red algae, with phycobilins filling in a supposed "green window", would have the greatest vertical distribution and would be the only ones in deep (green) water. The theory implied that red algae had evolved phycobilins for survival in deep water. The theory would really apply only where light is limiting, not in the intertidal zone (Dring 1981).

The theory rested on three main assumptions, all false (Saffo 1987): (1) the vertical distribution of red, green, and brown seaweeds would be as just described; (2) light would be the only factor that would affect their zonation; and (3) the pigment complement of red algae would make them more efficient in the submarine light quality. There were many opponents of the theory from the beginning. Oltmanns (1892) particularly insisted that the zonation pattern had been overstated and that total irradiance had a greater role in establishing the vertical distribution of attached seaweeds than did the spectral quality. Lubimenko and Tichovskaya (1928) concluded that the total quantity of pigment, rather than specific pigments, determined an alga's capability to photosynthesize in the low light at a depth of 50 m.

The generalized zonation pattern that Engelmann sought to explain was promulgated in an 1844 treatise on marine biology. Engelmann recognized that red algae could be abundant in shady intertidal habitats, where light would be dim, but not green, but he dismissed that as anomalous (Saffo 1987). Deep-water dredging brings up crustose corallines, but dredging has many limitations, and only by using SCUBA and submersibles can we be sure about the depths at which algae are attached and growing (Runcie *et al.* 2008). In many places, red algae are the deepest growing seaweeds, but we now know that they do not necessarily or exclusively occupy the greatest depths at which seaweeds can be found. Green algae grow at or near the limit of seaweed depths off Malta, Hawaii, and in the Bahamas, for example (Larkum *et al.* 1967; Lang 1974; Littler *et al.* 1985). Some green algae (e.g. *Palmoclathrus* and *Rhiliopsis*) are found *only* in deep water. The numbers of species and biomasses of red relative to green algae do not increase with depth (Schneider 1976; Titlyanov 1976).

Light is by no means the only environmental variable, even in the subtidal zone. Closely related species with similar pigments and functional forms, often have different depth limits. If only the light-saturation parameter for photosynthesis or growth would be considered, most of the seaweeds would be able to colonize deeper habitats than the actual ones. The functional form of a seaweed affects not only its light-harvesting ability but also its nutrient and carbon fluxes (Pekol and Ramus 1988). All these abilities and more contribute to growth and competitive abilities. Sand-Jensen's comment (1987) about carbon uptake applies equally to light harvesting: "Many characteristics have been suggested as adaptations to enhance the external supply of inorganic carbon... The plant, however, is an integrated functional unit that can exploit an array of environmental variables, so rather than looking at those characteristics purely as adaptations to enhance the carbon gain, they can be viewed as general characteristics of the physiology and ecology of the plant."

As far as light-harvesting ability, the particular pigment complement is unimportant if a seaweed is optically black (Ramus 1978, 1981). Thick thalli, or those within which there is much scattering and reflection of light, are the most thorough absorbers, because the optical path is longer and the chances of a photon striking a plastid are greater. Moreover, although green algae absorb relatively poorly in the green region of the spectrum, they do absorb there, and absorptance in the yellow–green region increases disproportionately when the chlorophyll concentration is increased (Larkum and Barrett 1983).

Again, light is not the sole factor in zonation, and the pigmentation of red algae does not better adapt them to subtidal light regimes. Rather, the quantity of light is crucial, and seaweeds harvest dim light by increasing all accessory pigments and making adaptations in thallus morphology and orientation. Nevertheless, changes in pigment complements or ratios under different light qualities, such as phenotypic adjustment (acclimation) to colored light, have been reported, especially in cyanobacteria (Bogorad 1975; also see Ramus 1982; Grossman *et al.* 1989). The effects of light quality must be separated from the effects of low quantity by comparing equal energies of colored lights. Along this line of argument, red algae have been shown to exhibit characteristics of both "light-intensity and light-quality" adapters (Talarico and Maranzana 2000), however the mechanisms behind adaptation to light quality are much more complex than just an adjustment of pigment (phycobilin) ratios. The modulation of spectral components of the light field does rather induce a photomorphogenetic signal cascade, which may i.e. alter phycobilisome structure and attachment to the thylakoid membranes. These signals may ultimately affect algal metabolism and growth in a dynamic light environment (Talarico and Maranzana 2000).

5.4 Carbon fixation: the "dark reactions" of photosynthesis

5.4.1 Inorganic carbon sources and uptake

Seaweeds use inorganic carbon as virtually their sole carbon source. However, some opportunistic seaweeds (*Ulva, Hincksia*) have limited ability to take up and use organic carbon sources, such as glucose, acetate, and leucine, if they are available (Schmitz and Riffarth 1980; Markager and Sand-Jensen 1990) and parasitic seaweeds are at least partially heterotrophic (Court 1980; Kremer 1983).

Seawater has inorganic carbon properties different from those in air or fresh water. Compared with fresh water, its alkalinity is high, its pH is high and stable, and its salinity is generally high. In air inorganic carbon is available only as CO_2, but it diffuses to them rapidly as they take it up. Seaweeds in water can also get CO_2, and the concentration is similar to that in air, but diffusion is 10^4 times slower (see Chapter 8). However, CO_2 in water is part of the carbonate buffer system (Fig. 5.15), and inorganic carbon also is available to many seaweeds as bicarbonate (HCO_3^-). The relative proportions of the forms of inorganic carbon depend on pH and salinity, as shown in Figure 5.16, and also temperature (Kalle 1972; Kerby and Raven 1985). In seawater of

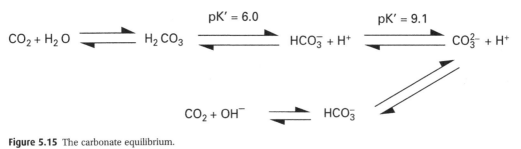

Figure 5.15 The carbonate equilibrium.

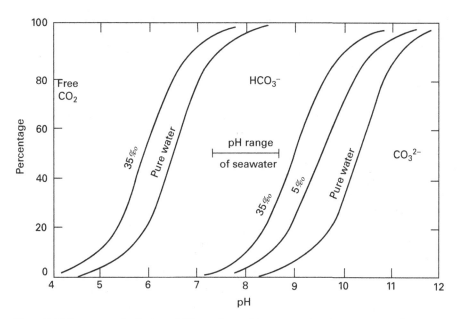

Figure 5.16 Percentage distribution of different forms of inorganic carbon in seawater as a function of pH in three different salinities. (Modified from Kalle 1945, Der Stoffhaushalt des Meeres, with permission of Akademische Verlagsgesellschaft, Geest und Portig KG.)

pH 8 and salinity of 35‰ about 90% of the inorganic carbon occurs as HCO_3^-. Absolute values are approximately 10 μM CO_2 and >2 mM HCO_3^-; this compares with CO_2 at 380 ppm (*c.* 13 μM) in air (Kremer 1981a). Carbon limitation of photosynthesis in the sea had long been considered rare because the concentration of inorganic carbon seems high, and replenishment supplies from carbonates in shells and rocks seem limitless. However, it has been shown that particularly some subtidal red algae, which exclusively rely on CO_2 diffusion into the cells as sole inorganic carbon source, may indeed experience carbon limitation (Raven and Beardall 1981; Maberly 1990).

The capacity to utilize HCO_3^- as the prime inorganic carbon source has been proven for the vast majority of seaweeds, and involves different mechanisms of acquisition (Raven 2010). RuBisCO requires CO_2 as a substrate, and if seaweeds take up only CO_2, they are dependent on molecular diffusion (see sec. 8.1.3) from the relatively low concentrations found in seawater, as compared with the much higher concentrations of bicarbonate. CO_2 diffuses readily across cell and

plastid membranes (or diffusion may be facilitated), but HCO_3^-, like other ions, does not.

The main pathway of HCO_3^- utilization is through the enzyme carbonic anhydrase (CA), which greatly speeds the equilibrium between HCO_3^- and CO_2, and its location in algae can be either extra- or intracellular (Badger 2003). Inorganic carbon acquisition by (external) CA has been confirmed for a large number of seaweeds from contrasting geographic and climatic environments (Gordillo *et al.* 2006; Huovinen *et al.* 2007; Mercado *et al.* 2009; Gómez and Huovinen 2012). In addition, specific porter or proton pump systems (e.g. OH^-/HCO_3^- antiporter), or the activity of an anion exchanging protein have been found to facilitate HCO_3^- incorporation in seaweeds as distinct as *Ulva*, *Laminaria*, and *Cladophora* (Drechsler *et al.* 1994; Klenell *et al.* 2002; Choo *et al.* 2005). Either one or the other mechanism is operative in most seaweeds and both result in an efficient concentration of CO_2 close to RuBisCO, thereby guaranteeing a saturating carbon supply. All strategies related to an increase in inorganic carbon concentration at the fixation site of

RuBisCO are referred to as "carbon concentrating mechanisms" (CCMs). However, any mechanism that concentrates CO_2 inside the cell has to work against the leakiness of the plasmalemma, which allows a backflux/efflux. If HCO_3^- is taken into the cell, it affects the electrochemical potential across the membrane, so it must be either cotransported with a cation (e.g. H^+), or exchanged for another anion (e.g. the OH^- released when HCO_3^- is dissociated; Raven and Lucas 1985). Bicarbonate users typically raise the pH around their thalli, in some cases to >10.5 (Maberly 1990). The expulsion of OH^- relates bicarbonate uptake to $CaCO_3$ precipitation in some calcified algae (Borowitzka 1982; Pentecost 1985; see sec. 6.5.3).

Although most seaweeds tested have can utilize HCO_3^- (Raven 2010), some subtidal red algae cannot (Maberly 1990). Whether or not an effective CCM is operative in a given seaweed species may also be explained by energetic considerations (Kübler and Raven 1994), since a decrease in photosynthetic quantum yield has been observed under conditions likely to induce CCM activity (Raven and Lucas 1985). Kübler and Raven (1994) compared red seaweed species with and without operative CCMs and put forward an outline for an energetic trade-off of utilizing CCMs. *Palmaria palmata* and *Laurencia pinnatifida* both have the potential for employing a CCM. When they were grown under limiting irradiances, they reduced their inorganic carbon uptake activity, whereas under light-saturated growth, their CCM activity increased. This modulation in inorganic carbon uptake efficiency was absent in *Lomentaria articulata*, which is not able to use HCO_3^-. This finding was in line with the general assumption that the presence/absence of CCMs reflects habitat preferences, with seaweeds lacking CCM generally being restricted to shaded/low-light environments (Johnston *et al*. 1992). Again this indicates that higher energetic costs of active CCMs must be compensated by higher photosynthetic rates.

A trend toward a greater use of bicarbonate by intertidal fucoids, as compared with subtidal browns, was tentatively identified by Surif and Raven (1989) and confirmed by Maberly (1990). In general, CCMs in intertidal *Fucus* can be regarded as an efficient strategy to suppress photorespiration (see below; Kawamitsu

and Boyer 1999). Photosynthesis of Fucaceae at light saturation is not carbon limited (Surif and Raven 1989), but the subtidal taxa of browns that they studied were carbon limited at light saturation (i.e., adding more HCO_3^- to the medium increased P_{max}).

Although seaweeds are essentially marine, intertidal species can photosynthesize when emersed, some at rates equal to submerged rates (Madsen and Maberly 1990). Thus, they must use CO_2 because the bicarbonate in the capillary layer of water remaining on the seaweed surface will be quickly exhausted (Kerby and Raven 1985). High concentrations of CO_2 can substitute for an absence of HCO_3^- (Bidwell and McLachlan 1985), and intertidal species can acclimate physiologically to high levels of CO_2 (Johnston and Raven 1990). Surif and Raven (1990) concluded from their study on brown algae that intertidal species are nearly carbon-saturated in air because of their CO_2 uptake ability, in contrast to subtidal species, which show severe carbon limitation in air. In *Fucus spiralis*, photosynthetic parameters such as E_k, Ec, and optimum temperature will change when the seaweed is emersed, and provided that its water content remains high, this high intertidal alga photosynthesizes in air as well as it does in water, if not even better (Madsen and Maberly 1990). However, if we leave the theoretical assumption that water loss is negligible during emersion and apply the more realistic scenario of increasing water loss, a model developed by Maberly and Madsen (1990) shows that the desiccation factor progressively overrides the effect of carbon availability in accordance with the relative position of the specimen on the shore and duration of emersion (Fig. 5.17).

Finally, shell-boring algae, such as the conchocelis stage of *Porphyra/Pyropia*, the gomontia stage of *Monostroma*, and the numerous species endolithic in corals, live in particularly favorable carbon environments. *P. tenera* sporophytes can satisfy most of their carbon and calcium requirements from the shells, presumably by acidifying $CaCO_3$ to give Ca^{2+} and HCO_3^- (Ogata 1971).

As we have seen before, seawater pH readily influences carbon chemistry in the ocean, and recently much concern has arisen on the potential impacts of future ocean acidification, resulting from increasing

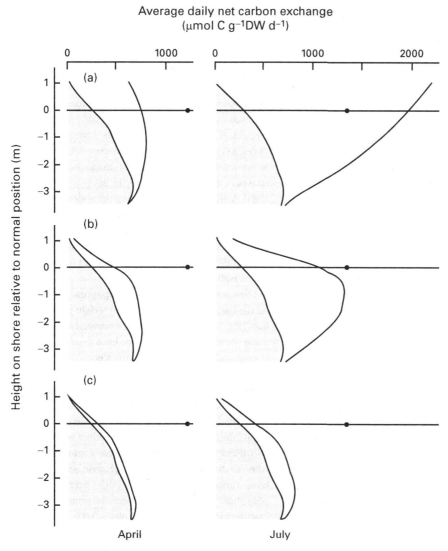

Figure 5.17 Effect of position on the shore on average daily net production in water (stippled area) and air (open area) of *Fucus spiralis* from St. Andrews, UK; total production is the sum of the two areas. Results are depicted for (a) zero water loss, (b) intermediate water loss, (c) extreme water loss in April and July. The horizontal line represents normal height of *Fucus spiralis*. Results indicate that under fully hydrated conditions production in air is equally high as production in water and diminishes with progressing dehydration as a function of shore height and/or time of emersion. (Reprinted from Maberly and Madsen 1990, with permission from Inter Research.)

CO_2 concentrations in the atmosphere. Shifts in seawater carbon chemistry will potentially change the availability of carbon species to seaweeds with or without CCMs, with currently largely unknown consequences for productivity and competitive strength of the respective species. We will elaborate on the ecological meaning of ocean acidification in the essay by Raven in Chapter 7.

5.4.2 Photosynthetic pathways in seaweeds

The primary photosynthetic pathway is variously called the photosynthetic carbon-reduction cycle (PCRC), the reductive pentose phosphate pathway, or most commonly termed the Calvin cycle; all of which basically refer to the process of C_3 photosynthesis, as defined for higher, terrestrial plants. The steps as far as fructose-6-phosphate appear to be common to all algae and higher plants, but from there, characteristic low molecular weight and polymeric carbohydrates are produced in different algae (Craigie 1974; Kremer 1981a). Some of the low molecular weight compounds play important roles in osmoregulation, while other compounds such as some of the sugar alcohols (like mannitol) represent important storage carbohydrates (see below).

Although extracting active enzymes from seaweeds is sometimes difficult because of the presence of mucilages and phenolics, the major enzymes of the Calvin cycle have also been studied in seaweeds (Kerby and Raven 1985; Raven 1997b; Cabello-Pasini and Alberte 2001; Bischof *et al.* 2000, 2002a). RuBisCO, whose three-dimensional structure was established from an angiosperm by Chapman *et al.* (1988), has eight large catalytic subunits of about 53 kDa encoded in the plastid and another eight small subunits of 15 kDa encoded in the nucleus. From the different types of RuBisCO, which have been described in different photosynthetically active organisms, only the type I has been recognized in seaweeds so far (Raven 1997b; Tabita *et al.* 2008). The activity of RuBisCO is modulated by Mg^{2+}, CO_2, light, and a specific activase (Salvucci 1989).

Oxygen is a competitive inhibitor of the carboxylase activity of RuBisCO and acts as a substrate for its oxygenase activity, leading to photorespiration (see below). RuBisCO is a major component of pyrenoids, but many species lack these structures, and in any case, active RuBisCO may be in the plastid stroma.

The principal low molecular weight photosynthetic product in the majority of the green algae examined so far, as in higher plants, is sucrose, a disaccharide of glucose and fructose. There also have been reports of green algae producing glucose and fructose, rather than sucrose, or producing considerable quantities of fructose alone (Kremer 1981a). *Caulerpa simpliciuscula* deposits its hexoses largely in β-linked glucans and sugar monophosphates, rather than in sucrose and starch (Howard *et al.* 1975). The Phaeophyceae are noted for their production of the sugar alcohol mannitol, which can contribute to the macromolecule laminaran (Fig. 5.18). *Himanthalia* also forms altritol (Wright *et al.* 1985), and *Pelvetia canaliculata* produces a C_7 alcohol, volemitol (Kremer 1981a). Mannitol is formed by reduction of fructose-6-phosphate (Ikawa *et al.* 1972), apparently in the plastids (Kremer 1985). Volemitol probably arises in a similar manner from sedoheptulose-7-phosphate, a Calvin cycle intermediate (Kremer 1977). Apart from acting as reserve products, these compounds may fulfill multiple additional tasks in algal metabolism and act as precursors in cell wall formation.

There are three groups of Rhodophyceae, categorized on the basis of their low molecular weight photoassimilates (Fig. 5.18) (Kremer 1981a), however, in recent years many more such compounds have been discovered (Eggert and Karsten 2010). In the Ceramiales, digeneaside is formed, except in *Bostrychia* and the related genus *Stictosiphonia*, in which the sugar alcohols D-sorbitol and (in most species in these general) D-dulcitol are formed instead (Kremer 1981a; Karsten *et al.* 1990). In all other orders, the principal product is floridoside. (Iso-)Floridoside consists of glycerol plus galactose, whereas digeneaside consists of glyceric acid plus mannose. Galactose and mannose come from fructose-6-phosphate by epimerization and are primed by being esterified to a nucleotide phosphate (UTP, GTP) before being coupled to glycerol or glyceric acid. The glycerol/glyceric acid presumably is derived from one of the C_3 compounds early in the Calvin cycle (or glycolysis), such as glyceraldehyde-P. The pathway of dulcitol and sorbitol formation may parallel that for mannitol formation, which was revealed in studies on the mangrove rhodophyte *Caloglossa leprieurii* (Karsten *et al.* 1997). In this species, evidence for the so-called mannitol cycle was found, which in its anabolic phase gives rise to mannitol by the reduction of fructose-6-P to mannitol-1-P catalyzed by mannitol-1-P-dehydrogenase, and the subsequent dephosphorylation by mannitol-1-phosphatase.

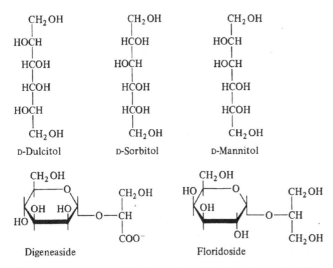

Figure 5.18 Some low molecular weight carbohydrates from red and brown seaweeds.

In the catabolic phase, mannitol is converted to fructose and fructose-6-P by mannitol dehydrogenase and hexokinase.

Photorespiration (or photosynthetic carbon oxidation cycle, PCOC) is a consequence of the oxygenase activity of RuBisCO, in which RuBP is split, with the introduction of O_2 rather than CO_2. One of the early products is the C_2 acid glycolate (Raven *et al.* 2005). In a complex pathway involving peroxisomes and mitochondria, three-fourths of glycolate carbon is salvaged and used in the Calvin cycle (Kerby and Raven 1985). Glycolate oxidase is located in the peroxisomes; as glycolate is oxidized, oxygen is reduced to H_2O_2 (hydrogen peroxide), which in turn is immediately reduced to water by catalase in the peroxisomes. Alternatively, glycolate can be oxidized via glycolate dehydrogenase in mitochondria. Surveys on different algal groups exhibited different glycolate dehydrogenases in diatoms, brown and green algae (Gross 1990; Suzuki *et al.* 1991).

It is suggested that photorespiratory carbon oxidation is commonly found in seaweeds (Raven 1997b), although the number of seaweed species studied with respect to photorespiration and cellular glycolate pathways is very limited so far (Iwamoto and Ikawa 1997). After all, RuBisCO oxygenase activity and photorespiration seem to be suppressed to varying degrees in seaweeds applying carbon concentrating mechanisms and more relevant in e.g. shade-adapted rhodophytes, which do rely more strongly on the diffusive CO_2 supply of RuBisCO (Raven 2010).

5.4.3 Light-independent carbon fixation

Seaweeds, like terrestrial plants, contain enzymes that can interconvert C_3 and C_4 compounds by carboxylation and decarboxylation. Several produce oxaloacetic acid (OAA), the Krebs cycle intermediate that accepts carbon from glycolysis. Because Krebs cycle intermediates are used for biosynthetic pathways, anaplerotic ("filling up") reactions are needed to keep the cycle turning. Several carboxylating enzymes add CO_2 (HCO_3^- may be the substrate of some) to the β-carbon of PEP or pyruvate and hence are called β-carboxylases, and two particularly important ones in seaweeds are PEP carboxykinase (PEPCK; Fig. 5.19a, b) and PEP carboxylase (PEPC):

$$PEP + CO_2 + H_2O \xrightarrow[\text{PEPC}]{Mg^{2+}} OAA + P_i$$

$$PEP + CO_2 + \begin{matrix} GDP \\ IDP \\ ADP \end{matrix} \underset{\text{PEPCK}}{\overset{Mn^{2+}}{\rightleftarrows}} OAA + \begin{matrix} GTP \\ ITP \\ ATP \end{matrix}$$

(a)

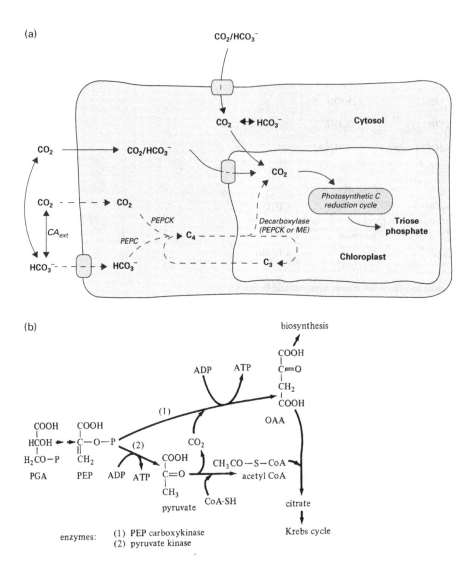

Figure 5.19 (a) Scheme for inorganic carbon transport and CO_2 accumulation in eukaryotic algal cells. (b) Utilization of PEP via PEP carboxykinase (I) in light-independent carbon fixation, or via pyruvate kinase (2) in glycolysis. (Part a from Giordano *et al.* 2005, with permission of *Annual Reviews*; part b modified from Kremer 1981b, with permission of Springer-Verlag, Berlin.)

OAA is rapidly converted to malate or (by transamination) to amino acids such as aspartate. While such reactions are general as anaplerotic components of respiration and biosynthesis, their ability to fix CO_2 has been exploited by some plants. C_4 and CAM vascular plants use β-carboxylases to fix CO_2 onto pyruvate, but as noted earlier, this does not yield net carbon fixation in the long term.

If seaweeds are supplied with labeled carbon in the dark, they fix it into various characteristic products different from those derived from the Calvin cycle. Red and green macroalgae in darkness form only small amounts of amino acids. Brown algae, with considerably higher rates of light-independent fixation, form malate, aspartate, citrate, and alanine, via PEPCK (Kerby and Evans 1983). Although present throughout the brown algae, light-independent carbon fixation (which takes place in the light as well as in darkness) occurs to a significant extent only in young tissues of kelps, fucoids, and Desmarestiales, where it can account for over 13% of the total carbon fixed (Kremer 1981b; Johnston and Raven 1986; Thomas and Wiencke 1991).

The significance of β-carboxylation in kelps and fucoids is seen in two situations. In kelps, photosynthesis is highest in old tissue, and β-carboxylation is most active in meristems. Mannitol is remobilized from mature or old tissue to provide energy and carbon skeletons for growth in the meristems. In particular, many brown and some red algae from highly seasonal environments, in particular the polar regions, display high rates of β-carboxylation, in order to minimize carbon losses during respiration and to sustain growth by salvaged carbon units during the growth season, which in some species is under strongly light limiting conditions (Drew and Hastings 1992; Weykam et al. 1997; Gómez and Wiencke 1998; Wiencke et al. 2009). PEPCK permits salvage of the CO_2 when pyruvate is converted to acetyl coenzyme A (acetyl-CoA) in glycolysis (Fig. 5.19b). There is no difference in energy output, because ATP is generated from PEP whether PEP is used to fix CO_2 or is converted to pyruvate in glycolysis (Kremer 1981b). One mole of mannitol yields 2 moles of PEP and thus can refix 2 moles of CO_2, and the CO_2 fixed into C_4 compounds does not have to be released again to regenerate C_3 substrate, so there can be net carbon fixation in contrast to CAM and C_4 metabolism. This does not mean that dark fixation is greater than CO_2 loss by respiration, and in fact Johnston and Raven (1986) showed that in *Ascophyllum nodosum* it is not, but it can conserve some carbon lost from respiration. In the kelp *Lesssonia nigrescens*, the mechanism of light-independent carbon fixation has also been attributed as a strategy to cope with harmful UV-B radiation (Gómez et al. 2007). While photosynthetic carbon fixation by seaweeds via RuBisCO has been shown to be very sensitive towards UV-B exposure (Bischof et al. 2002a), β-carboxylation is less susceptible, and can thus be considered as a strategy to replenish carbon units under severe abiotic stress (Gómez et al. 2007).

Light-independent carbon fixation obviously shares some similarities with the process of C_4 metabolism in terrestrial, vascular plants, which apply this strategy primarily to increase CO_2 at the reaction site of RuBisCO, in order to suppress photorespiration. However, from what we know so far, most seaweeds apply the strategy of β-carboxylation as an anaplerotic pathway rather than as a CCM (Raven 2010). In contrast, the involvement of β-carboxylation as a tool to suppress photorespiration has been reported from the chlorophyte *Udotea*, for which malate and aspartate were identified as early photosynthetic products from PEPCK (in light) and that these turned over quickly, passing into C_3 compounds (Reiskind et al. 1988, 1989). Those authors inferred that a PEPCK C_4-like pathway is the mechanism reducing photorespiration in this alga. It should be added here, that another potential means for reducing photorespiration is to reduce the O_2 concentration and thus oxygenase activity. Because O_2 is produced by PS II, photosynthesis by PS I could generate ATP without releasing oxygen. This mechanism known from cyanobacterial heterocysts may also be operative in red algal pyrenoids, in which RuBisCO is concentrated. Thylakoids cross the pyrenoid, but phycobilisomes and PS II are absent from these parts of the thylakoids, perhaps yielding a locally low oxygen concentration (McKay and Gibbs 1990; Giordano et al. 2005a).

If a seaweed species utilizes CCMs in inorganic carbon acquisition it can be assessed by their organic carbonate isotope footprint. In particular the $^{13}C/^{12}C$ ratio provides a good indicator for the absence/presence of CCMs, based on the higher carboxylase activity of RuBisCO in fixation of $^{12}CO_2$ than with $^{13}CO_2$ (Raven *et al.* 2002a). A term more frequently found in the literature is the $\delta^{13}C$ value, which ranges widely from -2.7‰ in the green *Codium pomoides* to -35 ‰ in the rhodophyte *Georgiella confluens* (see Raven *et al.* 2002a and citations therein). $\delta^{13}C$ has been found to correlate with the taxonomy and ecology of the seaweeds, with the lowest values being reported from representatives of the red algal class of Florideophyceae. It is further supported that seaweeds with $\delta^{13}C$ values more negative than -30‰ are relying on diffusive CO_2 supply of RuBisCO and thus lack a CCM (Maberly *et al.* 1992; Raven *et al.* 2002b). However, even among species applying CCMs there is lots of variation in $\delta^{13}C$ values, which are influenced by a multitude of physical (diffusiveness, boundary layers) and ecological (growth morphometry, tidal level) factors (Maberly *et al.* 1992; Mercado *et al.* 2009).

5.5   Seaweed polysaccharides

The low molecular weight products of photosynthesis serve many functions, including immediate use in respiration and use as carbon skeletons for functional or structural components of the cells. Some go into storage compounds. Seaweeds are noted for their diversity of storage and structural polysaccharides, which, like pigments, are characteristic of higher taxonomic groups. In recent years, research activities in this area were boosted by the identification of multiple beneficial properties for potential medical applications. Seaweed polysaccharides may have antiangiogenic and antitumoral (as found for *Sargassum*), antioxidant (*Fucus*), and antiviral (e.g. *Cystoseira indica* against herpes simplex, and *Asparagopsis armata* against HIV) capacities (Haslin *et al.* 2001; Ruperez *et al.* 2002; Dias *et al.* 2004; Mandal *et al.* 2007). This section will provide an overview on these polysaccharides.

5.5.1   Storage polymers

Carbon may be stored in monomeric compounds such as mannitol, but much is stored in polymers.

One advantage of polymers for storage is that they have smaller effects on the osmotic potential than would the same amount of carbon in monomeric form. Several characteristic storage polysaccharides are found in red, brown, and green seaweeds and cyanophytes, but most are branched or unbranched chains of glucose units (glucans) (Craigie 1974; McCandless 1981). Most green algae, like higher plants, store starch, a mixture of branched molecules (amylopectin) and unbranched molecules (amylose) (Fig. 5.20). Amylose consists of α-D-glucose units linked $1 \rightarrow 4$; amylopectin also has $\alpha(1 \rightarrow 6)$ branch points. Amylose is insoluble in water, forming micelles in which the molecules are helically coiled, whereas amylopectin is soluble. Dasycladales, such as *Acetabularia*, do not always store starch, but often store inulin, a fructose polymer (Percival 1979). Red algae store primarily floridean starch, a branched glucan similar to amylopectin except for having a few α $(1 \rightarrow 3)$ branch points. Floridean starch forms conspicuous granules located in the cytoplasma. McCracken and Cain (1981) showed that primitive red algae, including five marine species, also have amylose in their starch. Brown algae store laminaran, which, like starch, comprises a branched, soluble molecule and an unbranched, insoluble molecule (these compounds are called simply soluble and insoluble laminaran). The glucose in laminaran is in the β-form, however, and the links are $\beta(1 \rightarrow 3)$ and $\beta(1 \rightarrow 6)$ (Fig. 5.20c). Furthermore, some laminaran molecules, called M-chains, have a mannitol molecule attached to the reducing (C-l) end.

Storage compounds show quantitative changes correlated with season (really with growth), the part of the seaweed, and reproductive condition. These changes have been particularly well documented in commercially valuable kelps, fucoids, and Gigartinales. Studies such as those by Black (1949, 1950) and Jensen and Haug (1956) on *Laminaria* species and fucoids were primarily concerned with fluctuations in the valuable wall matrix polysaccharides and iodine, but they also documented changes in mannitol and laminaran contents. The relationships among growth, storage, and sometimes nitrogen availability have been worked out for *Hypnea*

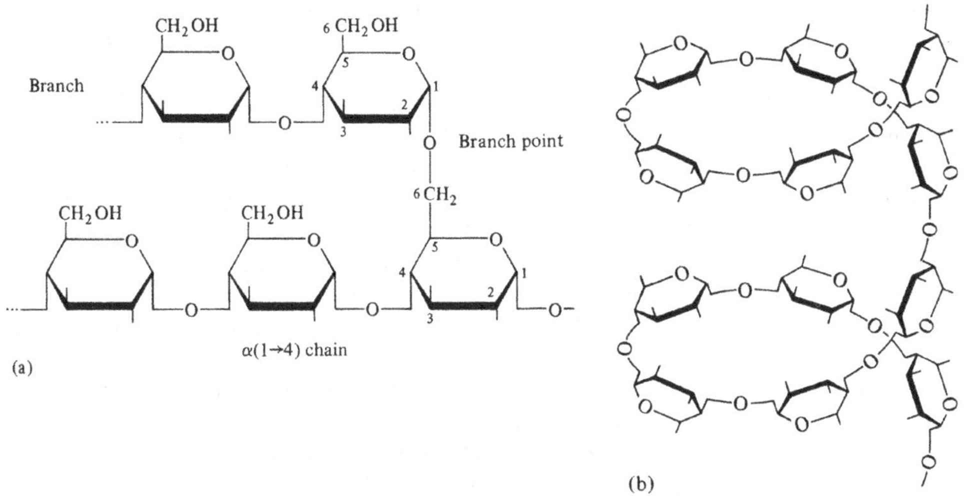

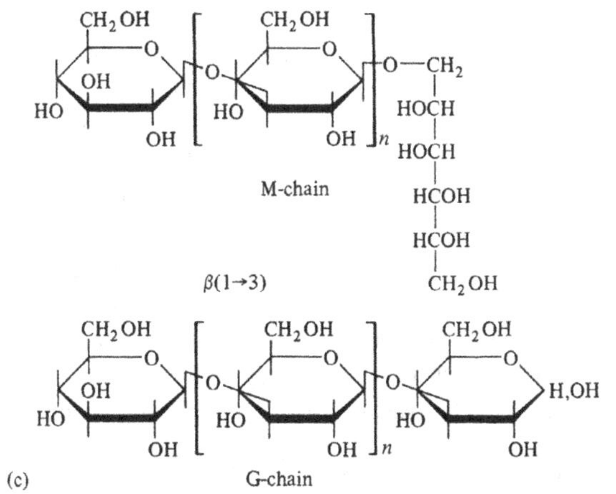

(a)

(b)

(c)

Figure 5.20 Algal storage polysaccharides: (a) amylopectin; (b) amylose; (c) two types of laminaran chains, M with mannitol attached to the reducing end, G with glucose at the reducing end. (Parts a and b from Lehninger 1975, *Biochemistry*. 2nd ed., © 1975, Worth Publishers, New York; c reprinted from Percival and McDowell 1967, *Chemistry and Enzymology of Algal Polysaccharides*. © 1967, Academic Press Inc. (London), Ltd.)

musciformis (Durako and Dawes 1980), *Eucheuma* species (Dawes *et al.* 1974), and *Laminaria longicruris* (Chapman and Craigie 1977; Gagne *et al.* 1982). The buildup of carbohydrate in matrix polysaccharides during periods of low growth or nitrogen starvation has also been shown, as, for instance, in *Chondrus crispus* (Neish *et al.* 1977) and *Eucheuma* (Dawes *et al.* 1977).

5.5.2   Wall matrix polysaccharides

Seaweeds are notable for the large amounts of matrix polysaccharides they produce, and seaweeds (and also seagrasses) are notable for forming sulfated polysaccharides, in contrast to freshwater plants (including algae) and land plants. Each class produces a range of characteristic compounds. Several of these compounds are of commercial value as phycocolloids or have been shown to exhibit antioxidant and antiviral activity. The commercial production and uses of phycocolloids are discussed in secs. 10.3, 10.5, and 10.6.

The matrix polymers present a much more complex picture than either fibrillar polysaccharides or storage compounds. Many are quite variable, so that terms such as "carrageenan" cover a range of similar but not identical molecules. In general, these polymers, like amylose, are thought to form helices that are aggregated in various ways in the gel state (Rees 1975; Kloareg and Quatrano 1988).

As mentioned above, the discovery of multiple commercial applications, i.e. in food sciences and medicine advanced the knowledge of seaweed polysaccharides. Compared to the widely used brown and red algal polysaccharides, chlorophytes are largely understudied in this respect. Chlorophyceae produce highly complex sulfated heteropolysaccharides, and each molecule is made up of several different residues. The major sugars are glucuronic acid, xylose, rhamnose, arabinose, and galactose, made up in several combinations. For *Ulva* sp. the heteropolymer ulvan has been reported as a major cell wall polysaccharide, together with cellulose and small fractions of a xyloglucan and a glucoronan (Lahaye and Robic 2007), making up a complex structural network (see sec. 1.3.1). The term ulvan comprises a huge variety of polysaccharides with varying composition and is now used for all of the complex heteropolymers found in the congener species of *Ulva*. In line with the differences in composition and also species- (and even strain-) specific differences, molecular weights of different ulvan species ranging from 5.3×10^5 to 3.6×10^6 g/mol have been found. Due to its variability in structure and composition, the biosynthesis is not entirely resolved yet. The current state of ulvan research is reviewed by Lahaye and Robic (2007), who have also outlined some of the potential applications. Ulvan has been shown to confer anti tumor and anti coagulant activities, and it seems to suppress plant pathogen growth and increase disease resistance in fish.

Compared to the example of ulvan, the matrix polysaccharides of the brown algae appear relatively simple: alginic acid consisting of mannuronic and guluronic acids, and fucoidan consisting of fucose (Percival and McDowell 1967). However, "fucoidan" is known to cover a wide range of compounds, the simplest being nearly pure fucan (e.g. in *Fucus distichus*). The most complex (e.g. in *Ascophyllum nodosum*) are heteropolymers containing fucose, xylose,

galactose, and glucuronic acid, in which the uronic acid may form a backbone and the neutral sugars form extensive branches (Larsen *et al.* 1970; McCandless and Craigie 1979). Even fucan is complex, having $\alpha(1 \rightarrow 2)$ and $\alpha(1 \rightarrow 4)$ links between fucose units, as well as $\alpha (1 \rightarrow 3)$ branches and varying degrees of sulfation on the remaining hydroxyl groups (Fig. 5.21a). In *Fucus vesiculosus*, more specifically, sulfated fucans were shown to be equally composed of alternating units of 2,3-disulfated, 4-linked and 2-sulfated, 3-linked α-L-fucopryanosyl units (Pomin and Mourao 2008).

Alginic acid consists of two uronic acids (sugars, with a carboxyl group on C-6: $\beta(1 \rightarrow 4)$-D-mannuronic acid (M) and its C-5 epimer α-$(1 \rightarrow 4)$-L-guluronic acid (G) (Fig. 5.21b). In most chains the residues occur in blocks as $(-M-)_n$, $(-G-)_n$, and $(-MG-)_n$ (Kloareg and Quatrano 1988; Gacesa 1988). The strengths of alginates depend on Ca^{2+} binding, with guluronic acid having a much greater affinity for Ca^{2+} than mannuronic acid. This effect depends on the more zigzag conformation of polyguluronic acid, which allows Ca^{2+} to fit into the spaces like eggs in an egg carton. Stiff, guluronate-rich alginates are typical of holdfasts; elastic, mannuronate-rich alginates predominate in the blades of kelps (Cheshire and Hallam 1985; Storz *et al.* 2009). The sequencing of the *Ectocarpus* genome (Cock *et al.* 2010a) has recently opened a new door towards the investigation of brown algal cell wall polysaccharide metabolism and its phylogeny (Michel *et al.* 2010a). Amongst others, authors suggest that brown algal cellulose synthesis "was inherited from the ancestral red algal endosymbiont, whereas the terminal step for alginate biosynthesis was acquired from horizontal gene transfer from an Actinobacterium. This horizontal gene transfer event also contributed genes for hemicellulose biosynthesis. By contrast, the biosynthetic route for sulfated fucans is an ancestral pathway, conserved with animals" (Michel *et al.* 2010a).

Most red algal matrix polsaccharides are galactans in which $\alpha(1 \rightarrow 3)$ and $\beta(1 \rightarrow 4)$ links alternate (Craigie 1990; Pomin 2010). Variation in the polymers comes from sulfation, pyruvation, and methylation of some

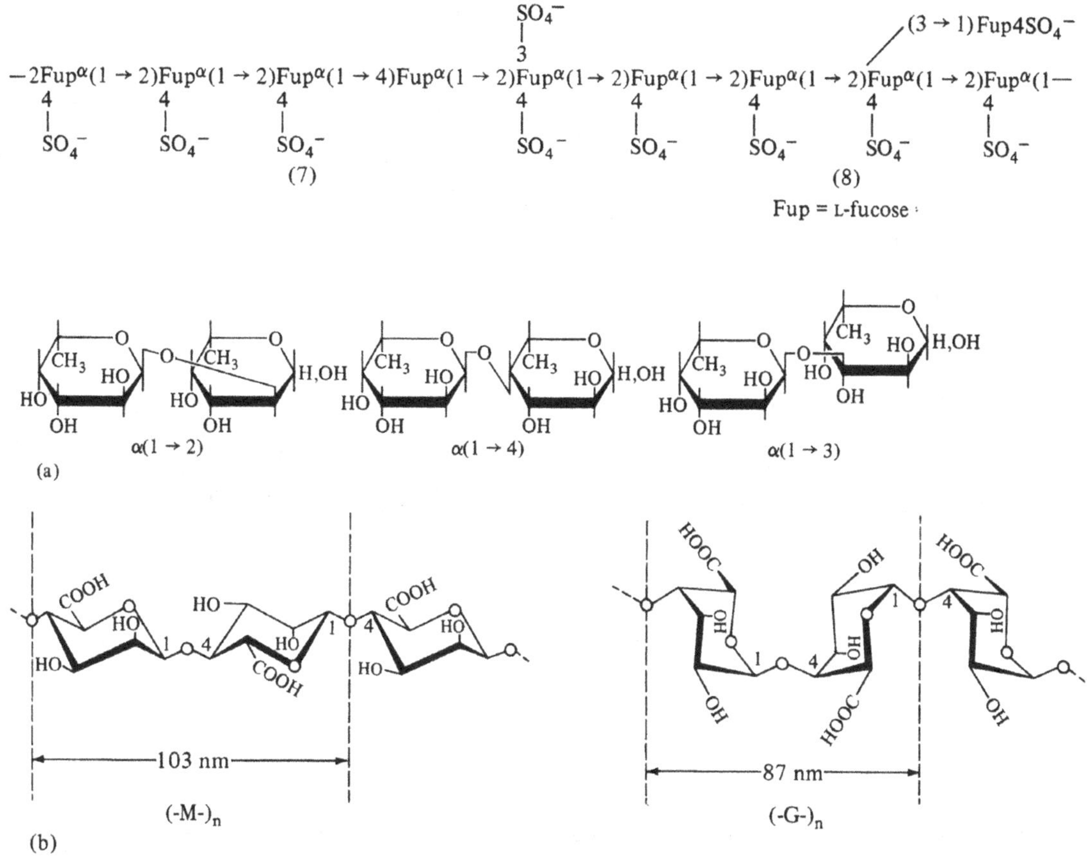

$$\overset{\displaystyle SO_4^-}{\underset{3}{|}}$$

$$-2Fup^{\alpha}(1 \to 2)Fup^{\alpha}(1 \to 2)Fup^{\alpha}(1 \to 4)Fup^{\alpha}(1 \to 2)Fup^{\alpha}(1 \to 2)Fup^{\alpha}(1 \to 2)Fup^{\alpha}(1 \to 2)Fup^{\alpha}(1 \to 2)Fup^{\alpha}(1-$$

with SO_4^- groups at position 4 on residues 1, 2, 3 and 5, 6, 7, 8, 9; and a $(3 \to 1)Fup4SO_4^-$ branch.

(7)   (8)

Fup = L-fucose

(a)

$\alpha(1 \to 2)$ $\alpha(1 \to 4)$ $\alpha(1 \to 3)$

(b)

—103 nm—

(-M-)$_n$

—87 nm—

(-G-)$_n$

Figure 5.21 Brown algal cell wall matrix polysaccharides: (a) fucan, showing overall structure of part of a chain, and details of three kinds of linkage; (b) portions of alginic acid (left, polymannuronic acid; right, polyguluronic acid). (Part a reprinted from Percival and McDowell 1967, *Chemistry and Enzymology of Algal Polysaccharides,* © 1967, Academic Press Inc. (London), Ltd.; b from Mackie and Preston 1974, with permission of Blackwell Scientific Publications.)

of the hydroxyl groups and from the formation of an anhydride bridge between C-3 and C-6 (Fig. 5.22). Still, the most important commercial groups of red algal polysaccharides are the agars and the carrageenans. Agars consist of alternating β-D-galactose and α-L-galactose with relatively little sulfation. The best commercial agar, neutral agarose, is virtually free of sulfate. Some more highly sulfated polymers with the agar structure do not gel and indeed are not referred to as agars (e.g. funoran from *Gloiopeltis* spp.). In carrageenans, β-D-galactose alternates with α-D-

galactose (not α-L-galactose), and there is much more sulfation. The sulfate groups project from the outside of the polymer helix; this polyelectrolyte surface makes the molecule more soluble (Rees 1975). Conversion of a 6-sulfate group to an anhydride bridge, as in λ- and κ-carrageenan, yields a stronger gel. Classically, two carrageenan fractions, λ and κ, were distinguished on the basis of their solubility in KCl. Several varieties are recognized, grouped into κ and λ families based on the presence or absence, respectively, of sulfate on C-4 of the β-galactose residues (Fig. 5.22) (McCandless and

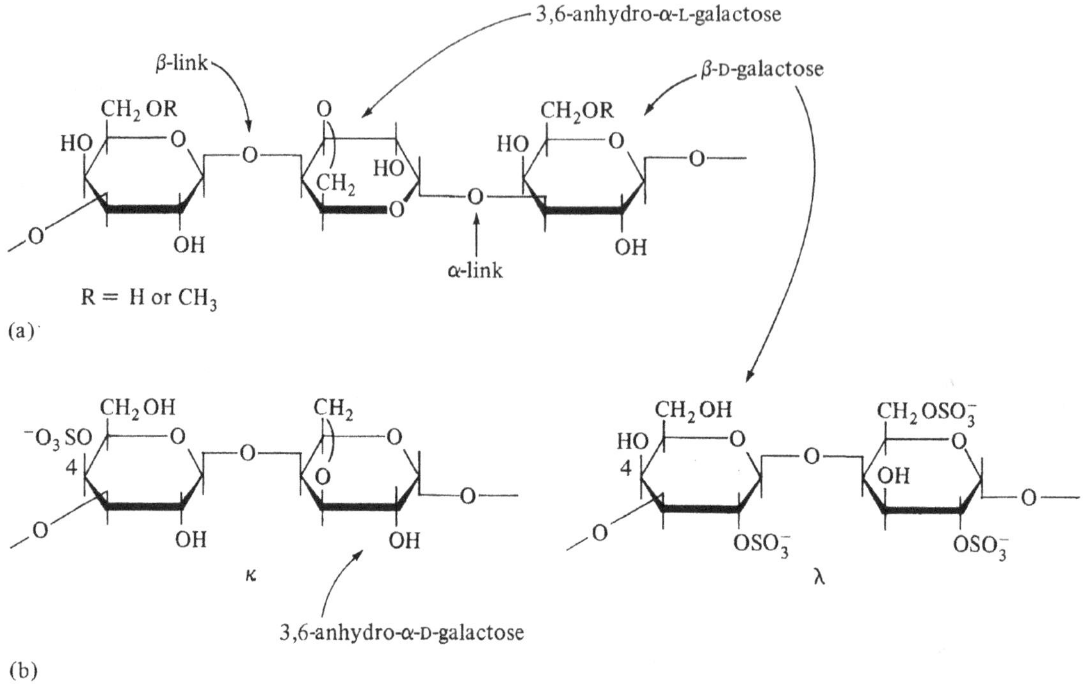

Figure 5.22 Red algal cell wall matrix polysaccharides: (a) agar; (b) κ-and λ-carrageenans showing differences in sulfation, and the C-4 of β-D-galactose used in classifying carrageenans.

Craigie 1979; Kloareg and Quatrano 1988). The more sulfated molecules form stronger gels. The conversion of galactose-6-sulfate to the 3,6-anhydride results in a stiffer gel because it takes a "kink" out of the chain, allowing more extensive double-helix formation and thus a more compact gel (Percival 1979). κ-carrageenan is predominantly found in the gametophytes, and λ-carrageenan in the tetrasporophytes, of *Chondrus crispus* and some other Gigartinaceae and Phyllophoraceae. The significance of this biochemical alternation for the physiology of the seaweeds is not yet clear, especially given the spectrum of carrageenans that are known to exist. Other Gigartinales, such as *Eucheuma*, produce only one kind of carrageenan in both phases (Dawes 1979).

Gel-forming polysaccharides no doubt add rigidity to the cell wall, while perhaps providing a certain amount of elasticity necessary in the intertidal zone. The strength of some gels depends on binding Ca^{2+} or other divalent cations that can cross-link polymer chains. However, the properties of the polymers that tend to increase divalent ion binding are diverse. Gel strength and Ca^{2+} binding of fucoidan depend on sulfation (e.g. in sulfate-free medium, *Fucus* embryos are unable to adhere to the substrate). However, less highly sulfated agars gel better, and sulfate esters interfere with calcium binding in polysaccharides of *Ulva lactuca* (Haug 1976). In *Ulva*, borate becomes complexed with rhamnose residues in the polymers, and Ca^{2+} stabilizes that complex. Sulfate prevents complexing with borate, and hence results in a weaker gel. Although a change in sulfation, or isomerization between uronic acids, could alter wall strength in seaweeds (as might, for example, be useful for spore release), thus far there is little evidence that seaweeds can make such changes once the polymers are outside the cells.

5.5.3 Polysaccharide synthesis

The various cell wall and storage polysaccharides are synthesized at several sites in the cell. Biosynthesis of matrix polysaccharides in general involves the building of a sugar nucleotide precursor in the cytoplasm, polymerization of precursors in the cellular endomembrane systems, export and assembly in the cell wall (Lahaye and Robic 2007). In brown algae, the nucleotide sugar precursors are transported in the Golgi apparatus and there serve as substrate to bound glycosyltransferases. The enzyme UDP-galactosyltransferase, which can transfer galactose to fucoidan, has been located in Golgi bodies of *Fucus serratus* (Coughlan and Evans 1978). Starch granules in the cytoplasm of the red alga *Serraticardia maxima* contain the starch synthesizing enzyme ADP-glucose: α-1,4-glucan α-4-glucosyltransferase (Nagashima *et al.* 1971). In other algae, starch is made and stored in the plastids.

Polysaccharide synthesis seems generally to involve the addition of a nucleotide diphosphate-linked monomer to a primer or an existing chain (Turvey 1978). The route by which mannose is incorporated into alginic acid was postulated by Lin and Hassid (1966) to be D-mannose → D-mannose-6-P → D-mannose-1-P → GDP-D-mannose → GDP-D-mannuronic acid → polymannuronic acid → alginate.

Each different bond type in a polymer is made by a different enzyme. Thus amylopectin synthesis requires one enzyme to make the (1 → 4) links and another to make the (1 → 6) branch points (Haug and Larsen 1974). Complexity seems to be added after polymer synthesis. Alginic acid is made initially as polymannuronic acid; then residues are converted to guluronic acid by epimerization at C-5. Addition of sulfate and methyl groups and the formation of anhydride bridges in various polysaccharides also take place after polymerization (Percival 1979). Wong and Craigie (1978) partially characterized an enzyme from *Chondrus crispus* (also known from other red algae) that forms the 3,6-anhydride bridge on galactose residues sulfated at C-6, with the release of the sulfate. The structure of the resulting κ-carrageenan is shown in Figure 5.22. There is evidence from another red alga, *Catenella*

caespitosa, for extracellular turnover of sulfate ester on λ-carrageenan after deposition. The sulfate appears to be carried by a methylated cytidine monophosphate (de Lestang-Bremond and Quillet 1981; Quillet and de Lestang-Bremond 1981).

5.6 Carbon translocation

In the simplest seaweeds, each cell is virtually independent of the others for its nutrition. However, many seaweeds contain non-pigmented cells in the medulla that evidently are supplied with photoassimilates by the pigmented cortical or epidermal cells. Parasitic algae receive organic carbon via short-distance translocation from their hosts. In those algae with an apico-basal gradient there is clearly movement of growth-regulating substances within the thallus. Such short-distance transport might take place through plasmodesmata, which in some green and perhaps most brown algae traverse the cross-walls and join the cytoplasm into a continuous symplast. In the red algae, pit plugs potentially provide a route between cells, although they usually are bounded on both sides by a membrane (Pueschel 1990; Gonen *et al.* 1996). Translocation of nutrients is covered in sec. 6.6.

Long-distance translocation evolved in higher plants as an adaptation to habitats where light and CO_2 were available in air, whereas water and minerals were available chiefly in the soil. In seaweeds, the whole outer surface is photosynthetic and is involved in nutrient absorption, and so there is no need of translocation for exchange of materials between different regions of the thallus. However, translocation can also serve to redistribute photoassimilates from mature (i.e. non-growing), strongly photosynthetic areas to rapidly growing regions, the so-called "carbon sink" regions. This role is useful only where there is a localized growing region and a relatively large or distant mature region. Kelp sporophytes and fucoids have such translocation structures, and translocation is well established especially in the Laminariales (Schmitz 1981; Buggeln 1983; Moss 1983; Westermeier and Gómez 1996). Transport of mineral ions is well established for Fucales and Laminariales. Long-distance transport of substances has also been documented in Scytosiphonales and Desmarestiales (Moe and Silva 1981; Guimaraes *et al.* 1986; Wiencke and Clayton 1990; see Raven 2003 for review; sec. 6.6).

Translocation in virtually all Laminariales takes place through the sieve elements, also referred to as trumpet hyphae or trumpet cells. These structures lie in a ring between the cortex and medulla throughout the stipe and blades. Longitudinal files of sieve elements branch and interconnect and are connected to cortical cells (Buggeln *et al.* 1985). The structure of sieve elements (in particular, the characteristically perforated sieve plates on their end walls) shows a trend from the smaller, simpler kelps (e.g. *Laminaria*) to the largest, most complex ones (*Macrocystis*). The pores become less numerous but larger in larger kelps, and thus more effective in transport. The sieve elements of *Laminaria* are filled with cytoplasm, organelles, and numerous small vacuoles, whereas those in *Macrocystis* more closely resemble vascular plant sieve elements in having a peripheral layer of enucleate cytoplasm and a very large central vacuole or lumen (Fig. 5.23). The pores in the sieve plates are lined with callose (a β-l,3-linked glucan), which, as in vascular plants can be deposited to block the pores. This mechanism prevents great loss of sieve-tube sap in case of injury and may also regulate routes of translocation. The only kelp known to differ from the description just given above is *Saccorhiza dermatodea* (Emerson *et al.* 1982). Here translocation takes place in the medulla, through highly elongated cells (solenocysts) that are cross-connected by smaller cells (allelocysts). The ultrastructure of solenocysts is similar to that of the sieve elements of other genera. Comparable structures for long-distance transport are also present in the Order Desmarestiales and Ascoseirales.

Kelps translocate the same organic materials that they make in photosynthesis; apparently there is no selectivity in sieve-tube loading. Mannitol and amino acids (chiefly alanine, glutamic acid, and aspartic acid) each account for about half of the exported carbon. In addition, there are numerous other compounds (Manley 1983). These materials move at velocities ranging from less than 0.10 m h^{-1} in *Laminaria* to about 0.70 m h^{-1} in *Macrocystis*, the speed generally correlating with the size of the sieve-plate pores. The *rate* of translocation depends not only on the velocity but also on the amount of material that is moving and hence on the concentrations of solutes and the cross-sectional area of transporting sieve elements. Rates range from 1 g (as dry wt) h^{-1} cm^{-2} in *Alaria* to 5–10 g h^{-1} cm^{-2} in *Macrocystis*. Rates in most vascular plants are 0.2–6.0 g h^{-1} cm^{-2}, although some translocate at much higher rates (Schmitz 1981). Photoassimilates follow a source-to-sink pattern of translocation. Sources include mature tissue; sinks include intercalary meristems and, to a lesser extent, sporophylls and haptera. In species with only one blade, the pattern is simple. Mature distal tissue exports, and meristematic tissue at the blade–stipe junction imports. The pattern becomes complicated in *Macrocystis*, which has numerous blades in various stages of growth and maturity, as well as young fronds developing from old fronds. In this genus, the changing import–export pattern of a blade as it matures is very similar to the pattern found in dicotyledonous angiosperms. Young blades only import, but as they near full size, export begins. Initially export is upward, to the meristem from which the blade developed; later it is also downward, to fronds developing from the base of the parent frond. In *M. integrifolia*, which lives in a seasonally variable habitat, downward export in fall also appears to carry photosynthates to the base of the seaweed for storage (Lobban 1978b, c). When growth stops in kelps, translocation also stops (Lüning *et al.* 1973; Lobban 1978c). A similar pattern of age- and size-dependent changes in carbon translocation has also been confirmed for the southern hemisphere kelp *Lessonia nigrescens* (Westermeier and Gómez 1996; Gómez *et al.* 2007).

The structure of *Macrocystis* sieve elements (Fig. 5.23b) would be consistent with a Münch mass-flow (or pressure-flow) mechanism, whereby sieve-tube loading in the source would cause an osmotic influx of water, which in turn would push the solutes to the sink, where assimilates would be unloaded, as in vascular plants (Buchanan *et al.* 2000). Even the

(a)

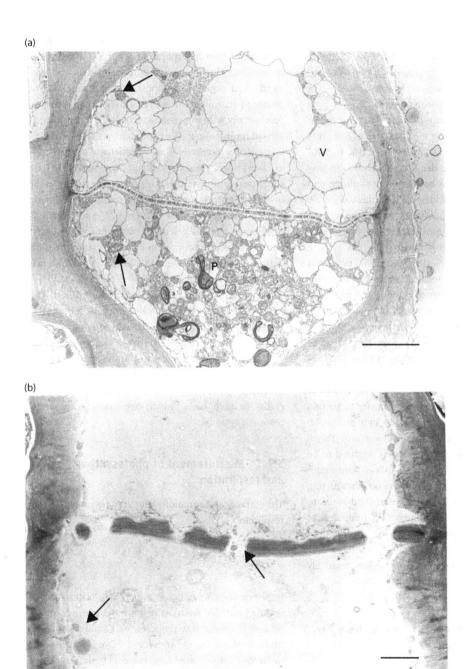

(b)

Figure 5.23 Longitudinal sections through sieve tubes with sieve plates of (a) *Laminaria groenlandica* and (b) *Macrocystis integrifolia*. Mitochondria can be seen in both (arrows); plastids (P) and vacuoles (V) are indicated in *Laminaria*. Note the many narrow pores through the sieve plate of *Laminaria* compared with the few, large pores in *Macrocystis*. Scale = 5 μm. (Part a from Schmitz and Srivastava 1974, with permission of Wissenschaftliche Verlagsgesellschaft mbH., Stuttgart; b from Schmitz 1981, with permission of Blackwell Scientific Publications.)

vesiculate structure of *Laminaria* sieve tubes (Fig. 5.23a) does not preclude mass flow (Schmitz 1981; Buggeln 1983). The phylogenetic gap between kelp and vascular plants suggests that the mechanism in kelps might be worth investigating.

The ecological significance of translocation is that it allows more rapid growth of localized meristems. This is especially important in seaweeds like *Macrocystis*, which may be attached in deep water, where the new frond initials are shaded by both the water column and the surface canopy of blades. In populations of *M. pyrifera* growing in stratified water where the surface layer is poor in nutrients, translocation may also serve to carry nitrogen (as amino acids) to the surface canopy (Wheeler and North 1981). In perennial species of *Laminaria*, translocation serves to move stored carbon from the mature blade to the meristem when new growth begins; this start of new growth is triggered by photoperiod in *L. hyperborea* (Lüning 1986) and by nitrogen availability in several other species (e.g. *L. longicruris* now referred to as *Saccharina longicruris*; Gagne *et al.* 1982). In the Arctic *L. solidungula* seasonal changes in translocation activity relate back to high β-carboxylation activity to sustain growth under prolonged periods of light exclusion under Arctic winter conditions (Dunton and Schell 1986). An analogous strategy has been reported to be active in the Antarctic brown alga *Ascoseira mirabilis*, where medullar structures denominated "conducting channels" show putative translocation, and serve to remobilize photosynthetic products (e.g. mannitol) to power seasonal growth (Gómez and Wiencke 1998). In young individuals, these structures are metabolically active, possess plasmodesmata and contain relatively few physodes (Clayton and Ashburner 1990).

5.7 Photosynthetic rates and primary production

Photosynthesis is a major, easily measured metabolic process and is routinely used as a gauge of environmental effects on seaweeds. It is also the basis of primary productivity, and its accurate measurement is

important for ecological studies. Primary productivity is the rate of net incorporation of carbon into organic compounds. It includes carbon retained in the seaweeds and organic carbon released as exudates or pieces of tissue (or of entire seaweed, if population productivity is being considered). It does not include carbon returned to the environment as CO_2. The development of ecologically sound estimates of primary production is enormously difficult, whether based on short-term photosynthesis or long-term yield. On the one hand, intrinsic variability in photosynthesis and respiration, the complex changes in E_o and total daily irradiance, and the effects of numerous other environmental factors on metabolism are extremely difficult to measure or model. On the other hand, biomass data must account for grazing and other tissue losses (Murthy *et al.* 1986; Ferreira and Ramos 1989). The many generalizations and assumptions required to generate a carbon budget for a seaweed or a community produce estimates with wide confidence intervals. Such estimates are important, however, for resource management at several trophic levels, and much effort has gone into developing models, especially for kelp beds, which are major commercial resource systems.

5.7.1 Measurement of photosynthesis and respiration

The classical equation for photosynthesis and respiration

$$6CO_2 + 6H_2O \underset{R}{\overset{PS}{\rightleftharpoons}} C_6H_{12}O_6 + 6O_2$$

implies some connection between CO_2 fixation and O_2 release (and vice versa), which is much more complex. Several processes take up or release C, and several take up and release O_2; the measured value is a composite of all these, dominated (one hopes) by photosynthetic carbon reduction and water-splitting electron transport. There is a fundamental question about what is being measured in gas-exchange or ^{14}C-fixation experiments.

Whereas photosynthesis consumes CO_2 and produces O_2, respiration uses O_2 and releases CO_2. Because both processes occur when light is available, the gas-exchange rate in the light measures net or apparent photosynthesis. Gross photosynthesis equals net photosynthesis plus respiration in the light, or conversely, net photosynthesis is zero when gross photosynthesis is equal to respiration. However, respiration is usually the dark rate since it is easily measured in the dark. There has been considerable disagreement over the rates of respiratory gas exchange in the light and in darkness, but the weight of data now indicates that respiration is much lower in the light, although the respiratory metabolic pathways operate continuously (Raven and Beardall 1981; Kelly 1989). In higher plants, most of the CO_2 released in the light comes from photorespiration. A small (but unknown) amount of O_2 is consumed in the light by pseudocyclic photophosphorylation (Mehler reaction), generating ATP but resulting in the use of half the O_2 released from water (Raven and Beardall 1981). These processes often are relatively small, but they can become significant, as when photosynthesis is low, or the oxygen concentration is high.

As we will also see below, each method to determine photosynthesis and respiration has its particular advantages and limitations. Still, a widely applied method for measuring photosynthesis and respiration in water is the determination of oxygen release and ^{14}C uptake in light and dark BOD (biological oxygen demand) bottles or other suitable containers. However, due to small volumes this technique imposes constraints on photosynthesis/respiration measurements of seaweeds. At one extreme, turf algae are too small and too entangled with other species (including epiphytes, which become significant at this scale) to be measured individually and in this case one must measure community rates (e.g. Atkinson and Grigg 1984; Hackney and Sze 1988). At the other extreme, very large seaweed must be sampled using relatively small tissue pieces, although large enclosures have also been used (e.g. Hatcher 1977; Atkinson and Grigg 1984). Sampling, especially from a very large kelp-like *Macrocystis*, and the complication of wound respiration are

major obstacles to calculation of accurate rates (Littler 1979; Arnold and Manley 1985). Infrared gas analysis (IRGA) has also been introduced to measure CO_2 exchange of seaweeds in air (Bidwell and McLachlan 1985; Surif and Raven 1990).

The simple equation given earlier also implies a 1:1 ratio between CO_2 fixed and O_2 released. In practice, this ratio is rarely 1.0. Thus, if estimates of productivity (in terms of C) are to be based on O_2 measurement, they must be corrected for the ratio of CO_2 fixed:O_2 released, called the photosynthetic quotient (PQ). Measurements of PQ have frequently been made for phytoplankton and have often shown its value to be approximately 1.2 (Strickland and Parsons 1972). However, there can be considerable variation, depending in part on whether the fixed C is going mainly into carbohydrates (as in the foregoing equation), fats, or proteins and whether or not photosynthetic energy is being directed to uses such as NO_3^- reduction. Few measurements of PQ have been made for seaweeds, but Hatcher *et al.* (1977) reported a range in *Saccharina longicruris* from 0.67 to 1.50. In another study on five abundant seaweed species at the coast of São Paulo, Brazil, PQ values ranging from 0.42–1.01 were determined (Rosenberg *et al.* 1995).

Photosynthetic rates may be expressed on the basis of several denominators, with different results. Ramus (1981) recommended expressing photosynthesis as carbon flux per unit of chlorophyll (also called the assimilation number; Kelly 1989), even while recognizing that the amount of chlorophyll and the rate of photosynthesis do not bear a constant relationship even at saturating irradiances. Respiration probably is best expressed on the basis of total protein, because it is a process taking place in the cytoplasm, rather than on the basis of dry weight, which includes the cell walls. Nevertheless, direct comparison of rates is possible only within carefully controlled groups of experiments, because the complex processes of photosynthesis and respiration are affected by many variables, including irradiance, age of the tissue, nutrient levels, temperature, and pH.

In terms of production estimates probably the method of ^{14}C incorporation is still the most reliable.

However, in recent years substantial new developments have improved photosynthetic measurement techniques, which have increased spatial and temporal resolution, lowered the detection limits and reduced the time for each measurement, yet allowing for improved replication of measurements. This accounts for improvements in the typical Clark-type electrodes, which have been widely used for determination of oxygen evolution. Improved electrode stability, response time, as well as minimized cuvette volumes have shortened the time required for recording *P–E* curves. With respect to oxygen measurements, an enormous improvement has been the development of microsensors for determining oxygen content in water and/or tissues by a needle-like electrode (Revsbech 1989). For example, these electrodes have been used to measure oxygen profiles along depth gradients within *Chaetomorpha* canopies (Bischof *et al.* 2006b) or to monitor microprofiles at the thallus surface of *Halimeda discoidea* (de Beer and Larkum 2001). Recently, a new optical, fluorescence-based technique for measuring oxygen has been introduced in aquatic sciences as the so-called optode system (Gansert and Blossfeld 2008). In brief the measuring principle is that oxygen may act as a selective fluorescence quencher when interacting with a chemical complex (i.e. ruthenium), which is fixed at the sensor tip. Upon excitation with a blue light beam, a red fluorescence signal is emitted, which is quenched depending on the oxygen concentration of the medium. Such measurements are extremely rapid and now they are becoming a state-of-the-art tool in seaweed research, e.g. with respect to estimates of photosynthetic activity in the kelp *Laminaria hyperborea* (Miller and Dunton 2007; Miller *et al.* 2009).

As mentioned above, measurements of photosynthesis by monitoring oxygen evolution/consumption or CO_2 fixation provide absolute readings of rates over time and may thus also allow for production estimates. However, estimates of photosynthetic performance can also be used as some kind of "health indicator" of physiological fitness. This approach has recently fostered more experimentally based seaweed

research on environmental stress physiology (see Chapter 7). Improved methods enable us to compare algal performance under manipulated abiotic conditions over time or across species with a sufficiently high number of replicate measurements. In this respect, a major achievement was the implementation of techniques using the fluorescence signal emitted from chlorophyll to estimate photosynthetic activity. For example, variable chlorophyll fluorescence of PS II has been used to address quantum efficiency of photosystem II, via the widely known PAM (pulse amplitude modulated) fluorescence technique (Krause and Weis 1991; Schreiber *et al.* 1994). The measuring principle is based on the competing pathways of energy relaxation of an excited chlorophyll molecule in the photosynthetic antennae. As initially outlined, the excited electron can be relaxed either by (1) passing the energy to a neighboring chlorophyll to be further processed in photosynthesis, by (2) dissipating energy as heat, or by (3) sending off a fluorescence signal. Under conditions of impaired photosynthetic activity i.e. reduced electron drainage, the first process is restricted and thus a higher share of energy will be drained via the fluorescence and heat-dissipating pathway (Fig. 5.24).

Based on the detection of the fluorescence signal, the measuring devices calculate derived values, such as the optimal quantum yield of photosystem II (also referred to as F_v/F_m, the ratio of variable to maximal fluorescence, with $F_v = F_m - F_0$; F_0 refers to ground fluorescence after a period of dark acclimation and measured under non-actinic irradiances, F_m is the maximal fluorescence yield at supersaturating irradiances), the effective quantum yield ($\Delta F/F_m$) measured under instantaneous light, as well as a variety of quenching parameters e.g. NPQ (non-photochemical quenching), which includes heat dissipation and is often correlated to the de-epoxidation activity of the xanthophyll cycle (Bischof *et al.* 2002b). Measuring the effective quantum yield at a series of increasing irradiances of PAR also allows for the recording of *P–E* curves, which can be used to derive the (maximal) electron transport rate (ETR), as well as the theoretical saturation

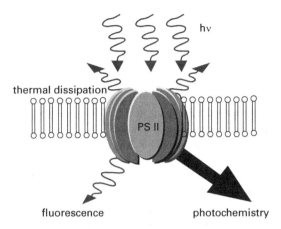

Figure 5.24 Scheme of energy flow in Photosystem II. At normal light conditions the absorbed light energy (hν) is mainly used to drive photochemistry. A small amount of energy is lost by fluorescence, which can be used to calculate quantum yield of photosynthesis. Another share of energy is dissipated as heat. Under excessive light conditions thermal dissipation increases strongly and hence, decreases quantum yield of PS II (From Hanelt and Nultsch 2003, with permission from Springer Publishers.)

point (E_k) of photosynthesis. ETR and effective quantum yield (or "yield") are connected via the formula:

ETR = yield × PAR × 0.5 × AF (with AF being the respective absorption factor of the thallus).

In their studies on higher plants, Genty *et al.* (1989) discovered a good correlation between the electron transport rate determined by fluorescence yield and CO_2 fixation measurements. The chlorophyll fluorescence technique has undergone rapid developments and improvements in recent years and manufacturers offer devices for a multitude of applications (microscopy based, for diving operations, for algal suspensions, fluorescence imaging, etc.). The big advantage of fluorescence-based estimates of photosynthetic performance is their rapidity and ease of application. Thus, quantum yield (in particular F_v/F_m) is nowadays widely used as an overall health assessment parameter in seaweed ecophysiology. Despite the simplicity of conducting the measurements, and apparent ease of

operation, the biophysical bases are complex, and a lack of knowledge of the theory behind the measurement may result in incorrect data acquisition and interpretation (Saroussi and Beer 2007). With respect to data interpretation, the user also needs to be aware that measurements of chlorophyll fluorescence just provide relative estimates of the efficiency of energy transfer from the antenna to PS II and that respiration cannot be addressed. Thus, any kind of production estimate is excluded using solely chlorophyll fluorescence data.

Following the approach of Genty *et al.* (1989), the validity of correlation between fluorescence-derived ETR and CO_2 fixation/oxygen evolution was evaluated for a number of seaweeds, such as the green *Ulva lactuca*, the brown *Fucus serratus*, and *Saccharina latissima* and the red *Palmaria palmata* and *Porphyra umbilicalis* (Hanelt and Nultsch 1995; Longstaff *et al.* 2002; Beer and Axelsson 2004). Overall consensus was that readings from chlorophyll fluorescence may be a good indicator for photosynthetic rates (based on oxygen evolution), but only under low and intermediate irradiances. *In concreto*, effective quantum yields, which decrease at increasing irradiances, have been found to preclude correct determination of ETR below a critical value of 0.1, resulting in the observed mismatch of fluorescence and oxygen data at high irradiances (Beer and Axelsson 2004; Saroussi and Beer 2007). In addition, studies by Nielsen and Nielsen (2008) indicate that thallus morphology (i.e. thallus thickness) of the seaweed species under investigation may also influence the correlation between fluorescence and oxygen data. Taking into account the limitations mentioned above, the introduction of chlorophyll fluorescence still represents a substantial improvement in the rapid assessment of seaweed responses to their environment. While the PAM technique is still the most common tool in seaweed research using chlorophyll fluorescence, other related techniques like the Fast Repetition Rate Fluorescence (FRRF) or Fluorescence Induction and Relaxation (FIRe) are implemented for various applications, sometimes giving different assessments (Röttgers 2007).

5.7.2 Intrinsic variation in photosynthesis

Photosynthetic rates are affected by many abiotic factors besides light and there are also some biotic factors, intrinsic in the individual, that affect photosynthetic rates such as morphology, ontogeny, and circadian rhythms.

Seaweed photosynthetic rates are strongly influenced by thallus morphology (Fig. 5.25), particularly the surface-area:volume ratio and the proportion of photosynthetic to non-photosynthetic tissue (Littler and Littler 1980; also see Littler and Arnold 1982; Littler *et al.* 1983a). In those studies, the sheet group had the highest mean net photosynthesis. This is not surprising, because, in general, all the cells are photosynthetic, have direct access to carbon supplies in the water, and cause little self-shading. On the basis of surface-area:volume ratios, filamentous algae should have had higher productivity, but clumping (which may have advantages in desiccation resistance, etc.) decreases the effective surface area (Littler and Arnold 1982). There was a two-fold decrease in mean rate between each of the first five groups, while the encrusting species had extremely low rates and there was considerable overlap from group to group. For instance, the rate for *Laurencia pacifica* (group 3) was higher than that for *Pyropia perforata* (group 1). In their survey including 32 seaweed species Johansson and Snoeijs (2002) confirmed the strong dependence of photosynthesis on thallus morphology. However, the findings strongly depend on the parameter used for normalization and the use of either dry weight or surface area may result in contrasting findings (Fig. 5.26). Stewart and Carpenter (2003) found that within the same morpho-functional group (and within the same species) morphological responses to differential mechanical exposure (wave action, currents) also affects net photosynthesis (see sec. 8.2.1).

Changes in photosynthetic characteristics take place as tissue or the seaweed ages. In a complex seaweed like *Macrocystis*, ontogenetic gradients occur both along each blade and between blades along each frond. As an example, Figure 5.27 shows net photosynthesis in three categories of blade ages. Steep gradients in the rate per gram of dry weight are seen,

however, the trend changes depending on the denominator. Data expressed on the basis of chlorophyll *a* and those based on area are different because the ratio of photosynthetic meristoderm to cortical/medullary tissues decreases toward the base of the blade. In the thin tips there is relatively more photosynthetic tissue; in the thicker blade bases (also meristematic) there is more respiratory tissue. Obviously a single tissue disc cannot be representative of a blade, nor a single blade representative of the seaweed. Similar ontogenetic changes have been recorded in other kelps (e.g. Küppers and Kremer 1978), *Fucus* (Khailov *et al.* 1978; McLachlan and Bidwell 1978), *Sargassum* (Kilar *et al.* 1989) and *Lessonia nigrescens* (Westermeier and Gómez 1996; Gómez *et al.* 2005). The proportion of non-photosynthetic tissue can also vary between individuals (e.g. wiry versus fleshy specimens of *Gigartina canaliculata*) (Littler and Arnold 1980).

Changes in photosynthetic rate over a day do not necessarily parallel the changes in E_0, partly because of photoinhibition, photoinduction, and diurnal rhythms. The prediction based on instantaneous measurements would be for photosynthesis to parallel the increase in E_0 from dawn until E_0 exceeded E_k, to remain level until E_0 again fell below E_k in the afternoon or evening, and then to decrease toward dusk. However, morning and afternoon P–E curves do not always correspond (Ramus and Rosenberg 1980; Ramus 1981; Gao and Umezaki 1989a, b). There is often a noontime depression of photosynthesis (Hanelt and Nultsch 1995; Bischof *et al.* 2002b), with only partial recovery in the later afternoon, and as much as 70% of daily photosynthesis may take place before noon. The phenomenon of photoinhibition is reviewed in sec. 5.3.3.

Some seaweeds show diurnal changes in photosynthetic rate even under uniform laboratory conditions (Mishkind *et al.* 1979; Ramus 1981). These changes in photosynthetic rates are due to endogenous clocks. Circadian (diel) rhythms affect several aspects of cellular activity, enzymes, or cell division in a number of species (also see sec. 1.3.5). They have periods of 21–27 h (hence circadian) and are entrained to 24 h by light stimuli such as the time of dawn or dusk (Hillman

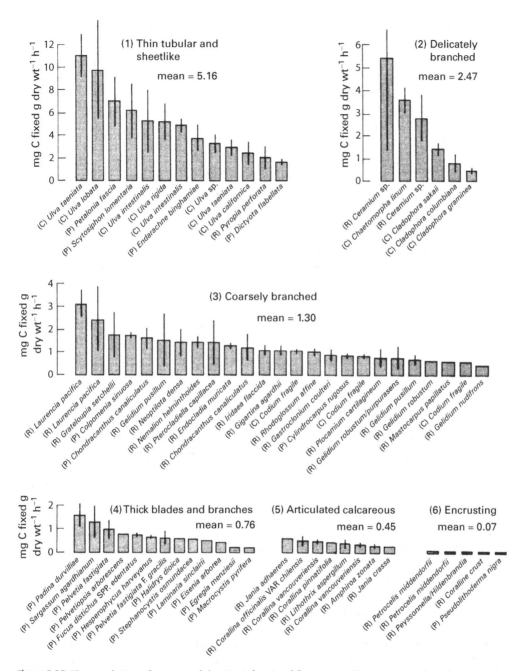

Figure 5.25 Photosynthetic performances of algae in six functional-form groups. Experiments on plants from more than one of six collecting sites, in southern California and Baja California, are shown separately. Mean rate in milligrams of C fixed per gram dry weight per hour is shown for each group. C, Chlorophyta; P Phaeophyta; R, Rhodophyta. (From Littler and Arnold 1982, with permission of *Journal of Phycology*.)

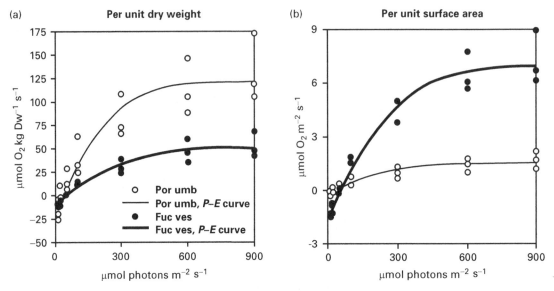

(a) **Per unit dry weight**

(b) **Per unit surface area**

○ Por umb
— Por umb, *P–E* curve
● Fuc ves
— Fuc ves, *P–E* curve

Figure 5.26 Measured O_2 evolution (circles) and *P-E* curves for two species co-occurring in the upper littoral zone of the Skagerrak, *Fucus vesiculosus* (thick lines, filled circles) and *Porphyra umbilicalis* (thin lines, open circles) normalized to (a) dry weight and (b) surface area. (From Johansson and Snoeijs 2002, with permission from Inter-Research.)

1976; Sweeney and Prezelin 1978). The diurnal changes are due to an endogenous rhythm and are not the results of environmental variables (see sec. 2.3.1). This can be shown by transferring seaweeds to continuous (usually dim) light and taking samples for measurement of light-saturated photosynthetic rates at various times parallel to the original light–dark cycle. Characteristically, the rate is highest during a time corresponding to the middle of the light period and lowest in what would have been the middle of the dark period (Fig. 5.28). It has been observed that the circadian rhythm of O_2-based photosynthesis in the tropical alga *Kappaphycus alvarezii* measured during 6 days is affected by different light quantities and qualities, indicating the presence of two photoreceptors in the light transduction pathway (Granbom *et al.* 2001). Over several days of continuous illumination the period of the rhythm gradually changes because there is no entraining stimulus (Mishkind *et al.* 1979; Oohusa 1980). If the light–dark cycle is reversed, the phase of the endogenous rhythm also shifts, as shown in Figure 5.28b.

There are several potential contributors to a cycle in photosynthetic rate. In *Ulva* there is a diel migration of plastids between the sides and faces of the cells (Britz and Briggs 1976), but this does not regulate the diel rhythm of photosynthesis (Nultsch *et al.* 1981; also see Larkum and Barrett 1983). Rather, the probable site of circadian control is the rate-limiting step in electron transport, probably one of the steps between plastoquinone and PS I (Mishkind *et al.* 1979). Further points of control may lie in certain enzymes. Yamada *et al.* (1979), using the brown alga *Spatoglossum pacificum*, showed rhythms in several enzymes of the Calvin cycle, including RuBP carboxylase, fructose-1,6-bisphosphate phosphohydrolase, mannitol-1-phosphate phosphohydrolase, and ribose-5-phosphate isomerase. However, the relationships of these enzyme activities to the circadian clock are likely to prove complex. For example, RuBP carboxylase (in higher plants) is activated by light-driven Mg^{2+} fluxes (Jensen and Bahr 1977). Yet these control points are not the mechanism of the rhythm, but rather its expression.

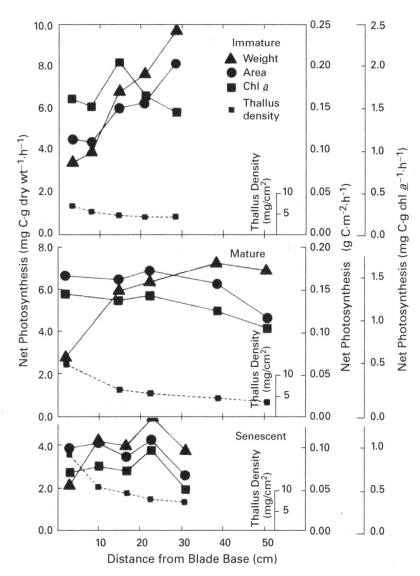

Figure 5.27 Variations in photosynthesis along *Macrocystis* blades of various ages, also showing differences due to normalizing to dry weight (triangles), area (circles), and chlorophyll *a* (large squares). The thallus density (milligrams dry weight per square centimeter) is presented (small squares) for each disc position. (From Arnold and Manley 1985, with permission of *Journal of Phycology.*)

5.7.3 Carbon losses

Exudation has been shown to be the source of considerable carbon loss in seaweeds, which may amount up to 30–40% of net assimilation (Khailov and Burlakova 1969; Sieburth 1969). Kelps have been shown to be particularly active exuders and *in extremo* organic carbon release of about 62% of net primary production has been reported from the phaeophyte *Ecklonia cava* (Wada *et al.* 2007, 2008). However, the rate of

(a)

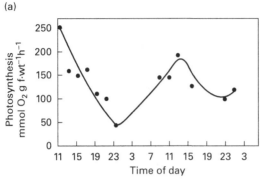

(b)

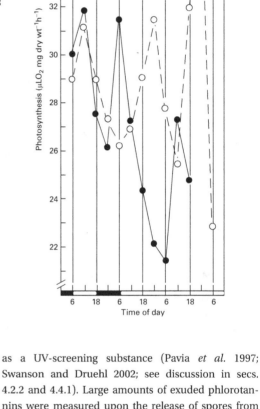

Figure 5.28 Photosynthetic rhythms. (a) Light-saturated photosynthetic rate for *Ulva lactuca* kept in continuous dim light (2 mW cm^{-2}). Along the abscissa, Eastern Standard Time and the hours of darkness in the collection environment. (b) Diurnal changes in photosynthetic capacity of *Pyropia yezoensis*. The solid line shows plants transferred to continuous light after the first day (light–dark cycles shown on lower abscissa). The broken line shows plants transferred to a reversed light–dark cycle (upper abscissa) and shows the reversal of the endogenous rhythm. (Part a from Mishkind *et al.* 1979, with permission of the American Society of Plant Physiologists; b from Oohusa 1980, with permission of Walter de Gruyter and Co.)

exudation is highly species dependent and additionally determined by a variety of abiotic and intrinsic factors, season, and also the level of stress exposure (Abdullah and Fredriksen 2004; Hulatt *et al.* 2009). Recently, also substantial progress has been made in identifying the functions of some of the exuded materials. For example, the compounds released may play a role in fouling control (Harder *et al.* 2004a) and as feeding deterrents (Abdullah and Fredriksen 2004). In recent years, the function of brown-algal-specific phlorotannins (polyphenols) have attracted particular attention, as they may act as a feeding deterrent as well as a UV-screening substance (Pavia *et al.* 1997; Swanson and Druehl 2002; see discussion in secs. 4.2.2 and 4.4.1). Large amounts of exuded phlorotannins were measured upon the release of spores from different species of Laminariales. This exudation was further stimulated by UV-B exposure. It is hypothesized that this excretion of phlorotannins from sporogenic tissue of sporophytes can be considered as a parental investment in order to create UV-B-depleted refugia in the water column and thus to protect UV-susceptible zoospores (Müller *et al.* 2009a; see also Chapter 7).

Tissue loss is more easily documented and understood than exudation. Tissue loss can result from extrinsic and intrinsic causes. Extrinsic causes include direct or indirect grazing damage, physical abrasion, and microbial degradation (e.g. breakdown of old *Laminaria* tissue by the ascomycete *Phycomelaina laminariae*; Schatz 1980). Surprisingly large quantities of tissue are abscised by *Fucus* and *Ascophyllum* following reproduction. In the former, not only the receptacles but also the internodes beneath are shed (Knight and Parke 1950). *Ascophyllum* receptacles can account for half the biomass of the seaweed (Josselyn and Mathieson 1978). On the other hand, kelp spore production, while amounting to 3×10^{10} spores per year, accounts for only 0.17% of *Ecklonia* annual production (Joska and Bolton 1987). *Laminaria* blades continuously lose old tissue from the tips, and the entire blade can be turned over 1–5 times per year (Mann 1972). *Laminaria hyperborea* differs in that a distinct new blade is produced in late winter at the base of the old blade. Carbon, and perhaps nitrogen, will be salvaged from the old blade for the new growth before the old blade is shed. Lüning (1969) demonstrated that the new blade could form in darkness with the old blade present, but if the old blade was cut off, there was virtually no new growth, even in light. Subsequently, Lüning *et al.* (1973) demonstrated translocation of newly fixed carbon from old to new blades, but no one has yet shown with tracers that *old* carbon or amino acids are salvaged. The analogy with senescence of angiosperm leaves suggests that salvage might occur in kelps, because there is translocation within the seaweed.

5.7.4 Autecological models of productivity and carbon budgets

The degree of difficulty in estimating productivity is contingent on the complexity of the seaweed or population. The productivities of *Laminaria/Saccharina* and other linear kelps have frequently been modeled from length increments, because the meristem produces a "moving belt" of tissue (Parke 1948; Mann 1972; Dieckmann 1980; Gagne and Mann 1987). However, simple length–weight regressions (Mann 1972) are not

adequate because the relationship changes over the growth cycle. Kain (1979) found a 10-fold difference when she calculated annual production from the data of Mann (1972) and Hatcher *et al.* (1977) for the same population of *Laminaria (Saccharina) longicruris*. The data of Hatcher *et al.* (1977) came from a carbon budget based on photosynthesis measurements. Various formulas have been used to improve the accuracy of biomass estimates. Gagne and Mann (1987) tested four models for *S. longicruris* and concluded that the best estimates were obtained simply by multiplying linear growth by the weight per unit length of a section from the uniformly wide part of the blade. In this species and some other kelps the tapering meristematic region is a small portion of the overall strap-shaped blade. In more triangular thalli, more complex calculations must be used, and in genera such as *Ecklonia* (Mann *et al.* 1979), *Eisenia*, and *Macrocystis*, this method becomes cumbersome or impossible.

Ferreira and Ramos (1989) wanted to combine short-term estimates from photosynthetic rates with long-term estimates from biomass. For three estuarine species, they measured biomass monthly and ran their calculations for each month based on the starting biomass and adding productivity modeled from irradiance estimates (taking account of hourly tide height and water turbidity) and *P*-*E* relations (including an assumption about the photosynthesis of *Fucus* in air). They simplified their calculations by having a constant day length of 16 h (the study was done in Portugal) and including zero net productivity for the 8 h of darkness (not accounting for respiratory losses). Although the model did not account for losses during a month, it did not carry over any error into the following months, because they started each month's calculation with the actual biomass.

For the agarophyte *Gelidium sesquipedale*, a particularly sophisticated model was set up, allowing for improvements in aquaculturing methods of this seaweed of high commercial interest (Duarte and Ferreira 1993, 1997). This model rests on the previous definition of size classes to simulate individual frond weight and size, biomass, and density of the algal population. This model, which included biomass, integrated

(a)

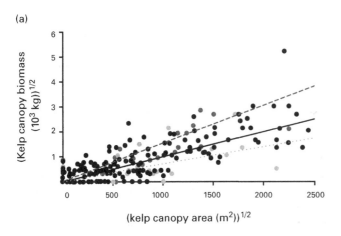

(b)

(c)

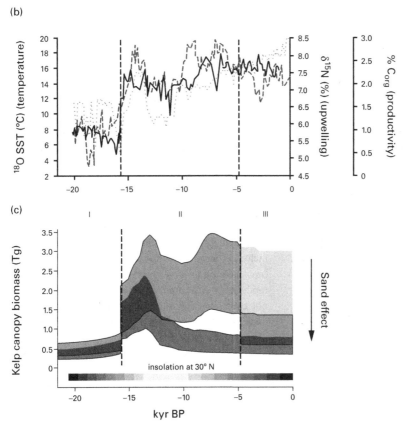

Figure 5.29 Reconstruction of kelp forest productivity (change in biomass) since the last glacial maximum. (a) Annual maxima in giant kelp canopy biomass (wet weight, kg) versus giant kelp canopy surface area (m^2) in southern California under prevailing El Niño (dotted line), La Niña (dashed line) and normal ocean climates (solid line). (b) Palaeo-oceanographic conditions in the Southern California Bight: sea surface temperature (SST; black), upwelling (δ^{15}N; dashed line) and biological productivity

(*cont. over*)

depths, gross productivity, and respiration, exudation, frond breakage, and mortality, finally allowed for seasonal growth predictions and the determination of optimum harvesting time.

Modeling the productivity of a *Macrocystis* bed involves also modeling the large, three-dimensional environment it occupies, because the huge seaweeds significantly affect the physical environment (light, currents) and the chemistry of the water moving through the bed, just as they are affected by these factors. A model has been developed by Jackson and co-workers in a series of steps, including modeling of the physical environment (Jackson 1984), the relations among fronds (Jackson *et al.* 1985), and the relation between irradiance and yield (Jackson 1987). The 1987 model gave results for biomass and production that were "in general agreement" with field measurements, but with differences in the predicted timing of growth maxima and minima. The simple seasonal cycle predicted on the basis of irradiance is complicated by other factors, with nutrients probably most important. Jackson (1987) listed five aspects that could be incorporated into future models, including the adaptations of blades to lower irradiances, the tissue losses due to grazing and decay, the effects of clouds on the light climate, and the effects of nutrients. A computer model of sporophyte densities as controlled by temperature/nutrients (correlated) and irradiance has been developed by Burgman and Gerard (1990).

Graham *et al.* (2010) set up a model on *Macrocystis* stands allowing for estimations of changes in kelp productivity in the Southern California Bay since the last glacial maximum (Fig. 5.29). They integrated survey data on kelp biomass with data on temporal

changes in rocky substrate distribution, sea level, sea surface temperature (SST), presence of nutrient-rich sub-photic water (δ^{15}N) and surface water productivity (%C$_{org}$), the latter three derived from sediment proxies. Pronounced changes in kelp productivity are suggested based on changes in oceanography, i.e. upwelling intensity. In this model, the effects of changing climatic conditions in the past are considered as a base for predictions on the result of future climate change.

5.7.5 Ecological impact of seaweed productivity

Carbon or energy fixed by seaweeds is eventually passed to other trophic levels. Depending on the location, seaweeds can account for more than 50% of the total fixed carbon in coastal habitats (Gattuso *et al.* 2006). The carbon fixed is consequently passed on in various ways and consequently fuels secondary production. The exudates will become available to heterotrophic bacteria as dissolved organic carbon (DOC) in the water column. Particulate matter as seaweed debris can be deposited along the shore (Duggins *et al.* 1989) or even transferred into deep sea habitats (Fischer and Wiencke 1992). Detrital seaweed material may then serve as food for an enormous variety of invertebrates, such as filter feeders, snails, sea urchins, etc. In this way carbon originally fixed by seaweeds will be passed on in two directions: to higher trophic levels by secondary grazers on the detritivores or to lower trophic levels by remineralization of detritus or grazer-derived feces by bacteria and fungi (Fig. 5.30a). Direct grazing on seaweeds by herbivores can be an important (top-down) factor controlling productivity

Caption for Figure 5.29 (*cont.*) (% organic carbon; dotted line; value for most recent 10 000 yr BP is an upper bound). Left vertical reference line at 15 600 yr BP indicates the approximate timing of the Bolling–Allerod warming and marks the beginning of the high productivity period, whereas right vertical reference line at 4750 yr BP marks the approximate onset of mainland sand inundation. (c) Late Quaternary variation in giant kelp canopy biomass (wet weight, Tg) for southern California islands (dark shading) and total (Islands Mainland; light shading). Period I indicates low-productivity conditions; periods II and III indicate high productivity conditions, before and after the beginning of mainland sand inundation. The large arrow and shading during period III indicates total giant kelp canopy biomass after removal of the "sand effect" (70% reduction equivalent to modern difference in kelp occupancy between islands and mainland). At the bottom is an index of Northern Hemisphere solar insolation (Berger and Loutre 1991), with light gray as highest (520 W m^{-2}) and black as lowest (460 W m^{-2}). (From Graham *et al.* 2010 with permission from the Royal Society.)

(sec. 4.3). Sea urchins grazing on kelp beds can be destructive and to a large extent preclude seaweed recruitment (Sjøtun *et al.* 2006). The importance of grazing (by herbivorous fish) is most obvious from tropical reef sites, as herbivores prevent macroalgal growth from taking over slow-growing corals (Mantyka and Belllwood 2007) and coralline algae (Hoegh-Guldberg 2007).

Kelp beds in particular can contribute to temperate nearshore secondary production well beyond the kelp bed ecosystem. Duggins *et al.* (1990) found kelp-derived carbon throughout the nearshore food web in Alaska (Fig. 5.30b). The flow of energy from primary producers into other organisms is important in the dynamics of ecosystems, especially for ecosystem and resource management. Significant progress has been made in the application of trophic markers to track the flow of seaweed biomass. Two particularly valuable approaches are the analyses of fatty acid composition as well as stable isotopes ($\delta^{13}C$, $\delta^{15}N$). While fatty acid analyses allow tracking and distinguishing particular species-specific signals through the food web (Graeve *et al.* 2002), stable isotope analyses facilitate quantitative analyses for ingested seaweed-derived carbon (Rossi *et al.* 2010). In a recent study by Balasse *et al.* (2009) on sheep from Orkney Islands, this powerful tool has been used to estimate and compare the portion of seaweeds consumed as fodder by domestic animals during the neolithicum, iron age, and today. Hanson *et al.* (2010) suggested that always a combination of fatty acid and isotope analyses should be used for future food-web studies. The utility of biomarkers to study the fate of seaweed productivity is elaborated upon in the essay by Glenn Hyndes.

Essay 3 Tracking the fate of seaweed production through food webs using biomarkers

Glenn A. Hyndes

The accumulation of drift algae and seagrass along shorelines (Fig. 1) is often considered unsightly and smelly by the casual observer, but provides a classic example of connectivity among coastal ecosystems. As ecologists, we often focus on the small-scale interactions between primary producers and consumers within habitats, yet understanding those interactions throughout seascapes is crucial, allowing us to "fit the pieces of the jigsaw together" for a broad-scale understanding of ecological interactions at the seascape scale. The late Gary Polis and his co-workers re-invigorated work in this area in the 1990s, when they discussed the role of material moving from one ecosystem "subsidizing" production in another (Polis *et al.* 1997). Their focus was predominantly on seabirds enhancing production on arid islands with little *in situ* production, but the movement of other material across ecotones is also likely to play important "subsidizing" roles. I believe there is growing evidence that seaweed from reefs plays a critical role in enhancing production in a range of coastal ecosystems, a role that is currently underestimated.

Figure 1 Accumulation of red and brown algae and seagrass on a beach. (Courtesy of Glenn Hyndes.)

Essay 3 (cont.)

I first became fascinated by the role of algae and seagrass in driving production while carrying out my doctoral work in southwestern Australia in the 1990s. Using a beach-seine net to sample fish communities, I regularly had to battle the accumulations of drift algae and seagrass in the surf zone! This unwittingly led to my focus on understanding the role of seaweed (and seagrass) in influencing production beyond the borders where it was produced, though it took some years to be in a position to pursue that focus.

Algae, and particularly kelp, are a prominent feature of temperate reefs around the world, and southwestern Australia is no exception. In fact, algae are highly diverse on reefs (Wernberg *et al.* 2003) and as epiphytes in seagrass meadows (Lavery and Vanderklift 2002) in the region. The kelp *Ecklonia radiata* is the dominant canopy-forming macroalga on reefs in the region (Wernberg *et al.* 2003). Furthermore, large quantities of this kelp are dislodged from reefs during storm events and, along with dislodged seagrass leaves, accumulate on the beaches (Lenanton *et al.* 1982) and subtidal habitats (Wernberg *et al.* 2006) in the region (Fig. 2). Clearly, this kelp has to play a significant ecological role in a region whose marine waters are oligotrophic. Given a prevalence of *E. radiata* in New Zealand, Australia, and South Africa (Huisman 2000), it may well play a significant role in other temperate regions. So, what is the fate and importance of kelp in the region, and how do we tackle that question?

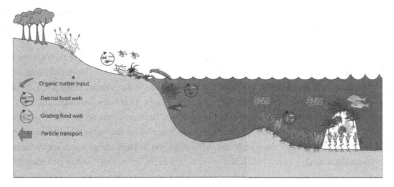

Figure 2 Conceptual model of the flow of kelp from reefs to other coastal ecosystems. (Courtesy of Glenn Hyndes.)

Before discussing those questions, it's worth noting the conundrum regarding brown algae in food webs. Despite the algae having high biomass in many regions, they are considered to contain large quantities of secondary metabolites that are considered to deter grazing (Jormalainen *et al.* 2001; Kubanek *et al.* 2004). However, brown algae have been shown to be consumed by a range of amphipods and isopods (Duffy and Hay 2000; Pennings *et al.* 2000). This may be explained by the secondary metabolites being reduced when the algae are desiccated, decomposed or exposed to ultraviolet light (Cronin and Hay 1996a; Pennings *et al.* 2000), which is likely to occur when brown algae dislodge and accumulate on beaches. With this in mind, *Ecklonia radiata* has particularly high levels of phlorotannins (Steinberg and van Altena 1992), but partially decomposed kelp was shown by Robertson and Lucas (1983) to be consumed by the main amphipod (*Allorchestes compressa*) in surf zones. However, we showed that this amphipod was equally capable of consuming fresh kelp (Crawley and Hyndes 2007). This is perhaps explained by the mesograzers being adapted to high levels of secondary metabolites, an hypothesis proposed by Steinberg and van Altena (1992) for Australian grazers. Regardless, those laboratory experiments showed that kelp can be consumed by mesograzers, but can those results be extrapolated to a significant role in the food webs? To answer this, a food-web biomarker approach is needed.

Traditionally, food-web studies have relied on direct examination of the diets of consumers. This approach is limited due to the difficulty in identifying and quantifying organisms that are macerated or rapidly digested. It is also often difficult to differentiate between ingested and assimilated material. Food-web biomarkers have become a commonly used tool to overcome these issues, as they are elements or compounds that can be traced through the food web (e.g. Canuel *et al.*

Essay 3 (cont.)

1995; Kharlamenko *et al.* 2001; Moncreiff and Sullivan 2001; Cook *et al.* 2004). Thus, if sources of production such as kelp have distinct biomarkers, the fate of this material can be traced through food webs.

Stable isotopes have become a commonly used biomarker in trophic studies, which typically compare the natural abundances of isotopes in food sources and consumers. Carbon ($^{13}C/^{12}C$) and nitrogen ($^{15}N/^{14}N$) are commonly used isotopes, and are generally expressed as $\delta^{13}C$ and $\delta^{15}N$, respectively. For stable isotopes to be used effectively, different food sources need to exhibit distinct signatures, and they need to shift in a predictable way through each trophic level of the food web. The shifts in stable isotope values through each level, known as discrimination, results from physiological processes favoring the incorporation of one isotope over the other into the consumer's tissue. Generally, $\delta^{13}C$ is considered to display minimal change between trophic steps (DeNiro and Epstein 1978; Peterson and Fry 1987; Michener and Schell 1994). This isotope is therefore often used to trace the source of production through food webs. In comparison, $\delta^{15}N$ exhibits discrimination levels of 3–5‰ (Minagawa and Wada 1984; Owens 1987), and is typically used to establish the position of consumers in food webs. These levels of discrimination are, of course, a generality, and can alter among consumers and food sources (Adams and Sterner 2000; Vanderklift and Ponsard 2003). For example, our study showed discrimination levels in amphipods fed on seagrass and algae, respectively, of approximately -10 and 2 to 4‰ for $\delta^{13}C$ and -3 and -1 to +1‰ for $\delta^{15}N$ (Crawley *et al.* 2009). This has obvious implications to the interpretation of stable isotope results, and the examination of the fate of material through food webs, and an issue that is still widely ignored.

Our studies using stable isotopes soon showed another issue with detecting the fate of algae through food webs using bulk stable isotopes. While seagrass displayed distinct $\delta^{13}C$ values compared to algae, different types of algae, i.e. red, green, and brown macroalgae, phytoplankton, and benthic microalgae, all displayed similar values, making it difficult to differentiate algae through the food web (Hyndes and Lavery 2005; Smit *et al.* 2005, 2006). This issue has been raised in a number of other studies (Lepoint *et al.* 2000), and has led to the use of other biomarkers. In some cases, sulfur ($^{34}S/^{32}S$) has helped overcome this issue (Connolly *et al.* 2004), and this isotope has been particularly useful for distinguishing between benthic and pelagic producers since $\delta^{34}S$ in the water column is often high due to ^{34}S-enriched sulfates compared to sediments with ^{34}S-depleted sulfides from anaerobic reduction (Fry *et al.* 1982). However, in my experience, this isotope provides little value in open marine systems.

The adoption of lipid biomarkers, particularly fatty acids, has helped clarify the flow of algae through the food webs. Fatty acids are fundamental components of cellular material. The value of these biomarkers in food-web studies is based on the premise that they cannot be produced by animals *de novo* and move through the food web unaltered (Graeve *et al.*, 1994; Khotimchemko 2003; Sanina *et al.* 2004). Those fatty acids that are diagnostic of particular food sources are, therefore, potentially useful biomarkers. Marine macrophytes have been shown to be rich in n-3 and n-6 polyunsaturated fatty acids, with clear differentiation between red and brown algae and seagrass using fatty acids, and particularly polyunsaturated fatty acids (Graeve *et al.* 2002; de Angelis *et al.* 2005; Alfaro *et al.* 2006; Richoux and Froneman 2008). Fatty acids have been shown to be a valuable tool in resolving complex interactions in marine food webs, particularly when used in combination with stable isotopes (Kharlamenko *et al.* 2001; Alfaro *et al.* 2006; Jaschinski *et al.* 2008).

Our study on wrack-associated communities in surf zones showed that amphipods were characterized by fatty acids that were diagnostic of macroalgae, but particularly brown algae (Crawley *et al.* 2009). The combined use of stable isotopes and fatty acids allowed us to trace brown algae through the food web of surf zones, highlighting the importance of this allochthonous food resource to this ecosystem. This extended nicely on the feeding and habitat experiments in the field and laboratory, clearly showing the flow of brown algae to the primary mesograzers in the system (Crawley and Hyndes 2007). Furthermore, the combined biomarker approach allowed us to trace the flow of material through to fish (Crawley *et al.* 2009), which are major predators of the amphipods in the surf zone (Crawley *et al.* 2006).

The importance of brown algae, particularly kelp, is becoming increasingly more evident as we extend this combined biomarker to trace the fate of food resources in other coastal food webs. Given the high biomass of kelp on reefs in

Essay 3 (cont.)

southwestern Australia, it is perhaps not surprising that we have traced kelp through to various consumers on reefs. For example, stable isotopes and fatty acids showed the consumption of particularly kelp by the urchin *Heliocidarus erythrogramma* and the large turbinid gastropod *Turbo intercostalis*, which are both common grazers on reefs along southern Australia (unpubl. data), and supports the findings of Guest *et al.* (2009) in Tasmania using the same approach. What can't be detected by this biomarker approach is the reliance of *H. erythrogramma* on kelp material that is derived from other reefs, rather than *in situ* production (Vanderklift and Wernberg 2008). This was only detected through clearly thought-out field experiments and highlights the importance of using a multi-dimensional approach to tease apart the complex processes that are driving ecosystem function in marine systems.

What has surprised us is the apparent importance of kelp to consumers in bare sand and seagrass habitats adjacent to reefs. Biomarkers allowed us to detect a high contribution of kelp to sedimentary detritus, and in turn, its importance to the sea cucumber *Stichopis mollis*, a detritivore. Perhaps, more surprisingly, kelp was shown to contribute to the gastropod grazers *Pyrene bidentata* and *Cantharidus lepidus* in seagrass meadows, which also rely on other macroalgae as well as periphyton which are produced *in situ* in seagrass meadows. Since kelp from reefs can accumulate in seagrass meadows (Wernberg *et al.* 2006), we hypothesized that kelp could play a significant role in supporting or subsidizing production in seagrass meadows. Certainly, our feeding experiments showed that both gastropod species had no preference for kelp, red algae, or periphyton (Doropoulos *et al.* 2009), demonstrating a potential pathway for the uptake of nutrients from kelp into seagrass food webs. By adopting a relatively new and novel approach, where we enriched kelp with ^{15}N, we were able to detect the assimilation of kelp by *C. lepidus* in seagrass meadows (Hyndes *et al.* 2012). This approach also allowed us to detect the uptake of nitrogen by seagrass and epiphytes either directly as dissolved organic nitrogen, or dissolved inorganic nitrogen through bacterial activity. Either way, nitrogen from kelp can be traced through primary producers and consumers in seagrass ecosystems.

The combined use of stable isotopes and fatty acids has certainly helped in tracing algae through the food web, and has been particularly useful in establishing the importance of allochthonous resources in marine ecosystems that may not appear obvious. As discussed earlier, stable isotopes have limited capacity to trace material through food webs with multiple potential sources of production. Furthermore, fatty acids tend to be qualitative through detecting diagnostic fatty acids in consumers. I believe the value of biomarkers will be increased markedly by manipulating stable isotope values by enriching or depleting ^{15}N or ^{13}C to experimentally track the flow of material through food webs. Though in its infancy, this approach has been used successfully in coastal food webs (Deegan *et al.* 2002; Mutchler *et al.* 2005). Compound-specific stable isotopes provide an alternative approach, and are increasingly being used in food-web studies (Fantle *et al.* 1999; Cook *et al.* 2004). Rather than examining the stable isotopes of bulk tissue, as in the general approach described above, stable isotopes of specific compounds within an organism's tissue (e.g. fatty acids and amino acids) are examined.

Tracing algae, or any other potential food source, through marine food webs requires a multi-dimensional approach. Food-web interactions can be complex, and sometimes counterintuitive. Food-web biomarkers, such as stable isotopes and fatty acids, have certainly increased our ability to trace sources and fate of production. However, they should not be considered the only tool to be used. Manipulative experiments can provide great insight into the interactive processes within and among ecosystems, and can highlight interactions that may otherwise remain undetected using biomarkers.

*Glenn Hyndes is an Associate Professor in the School of Natural Sciences at Edith Cowan University in Western Australia. He has a broad interest in marine and fisheries biology, but focuses particularly on trophic interactions in coastal ecosystems. His work uses a range of experimental and mensurative approaches, including stable isotopes and other biomarkers, to explore the mechanisms that result in trophic connectivity across ecosystems in coastal seascapes.

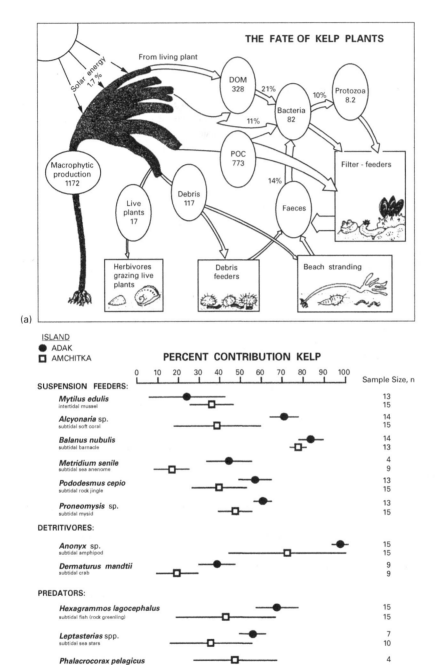

(a)

(b)

Figure 5.30 Fate of seaweed productivity in the nearshore ecosystem. (a) Kelp production and consumption (grams C per square meter per year). Percentages on the arrows indicate conversion efficiencies between different components. (b) Percentages of carbon photosynthesized by kelps found in tissues of consumers at Alaskan islands with extensive subtidal kelp beds. Kelp-derived carbon was identified by its δ^{13}C signature. (Part a from Branch and Griffiths 1988, *Oceanogr. Mar. Biol. Ann. Rev..* with permission of Aberdeen University Press, Farmers Hall, Aberdeen AB9 2XT, UK; b from Duggins *et al.* 1989 © 1989 by the AAAS.)

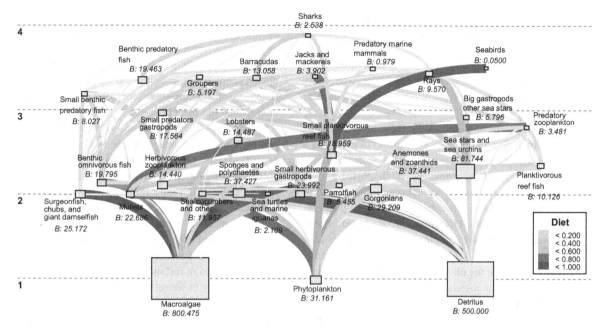

Figure 5.31 Biomass budget schematic for the Bolivar Channel Ecosystem, located between the islands of Isabella and Fernandina, Galapagos archipelago (B, Biomass in t km^{-2}). The graph shows all flows and biomass into the system, the size of each box, which represents a functional group is proportional to the biomass it represents, and the shading and width of the lines connecting functional groups represent the relative proportion of the preys on the diet of a predator. The y-axis shows the relative trophic level of the functional group; the functional group of benthic primary producers comprise seaweeds and benthic diatoms, referred to as "macroalgae and others". (From Ruiz and Wolff 2011, with permission of Elsevier.)

Simulations of the dynamics of an ecosystem (by ecosystem modelers) requires the inputs of production, biomass, and energy flow in and between all trophic levels. Ruiz and Wolff (2011) have recently characterized the trophic flows in the Bolivar Channel, which represents one of the most productive areas of the Galapagos archipelago (Fig. 5.31). The flow diagram shows the prominent role of macroalgae in terms of biomass and flows to the second trophic level. It also highlights that the contribution of phytoplankton only amounts to approx. 4% of the benthic biomass (macroalgae and benthic diatoms). However, detritus is shown to be an important vector for energy transfer from the first to the second trophic level. As seen by the diagram, half of all model compartments are on or near the second trophic level making use of the particularly high primary food production. This information is intended to provide a reference model for future simulations of the effects of different fishing and climate regimes on the system (Ruiz and Wolff 2011).

5.8 Synopsis

Photosynthetically active radiation covers the waveband from 350 or 400 nm to 700 nm. The quantity of light arriving per time, or the flux, is called irradiance and is measured in micromole of photons per square meter per second or microwatts per square meter. The quality and quantity of light in the oceans is highly variable. Irradiance changes at several timescales, and waves strongly affect the penetration of light into the sea. Once in water, light is attenuated by scattering and absorption, with the red end of the spectrum being more strongly reduced and the blue-green wavelength range penetrating deepest. Ultraviolet radiation is absorbed rapidly but may impair growth of seaweeds in shallow waters. Several ocean-water types have been defined on the basis of differences in their light quality parameters.

Photosynthetic pigments in algae always include chlorophyll *a*. In addition, each group of algae has a characteristic array of accessory pigments, including other chlorophylls, carotenes, xanthophylls, and in red algae (as well as cyanobacteria), phycobilins. Chlorophylls and phycobilins occur in pigment–protein complexes. Phycobilins are clustered in phycobilisomes on the thylakoid membrane surface; other accessory pigments (apart from the outer antenna) and the rest of the photosynthetic apparatus are integral parts of the membrane.

The arrangements of thylakoids within the plastids differ among algal phyla, but thylakoid membranes and the processes of photosynthesis are essentially the same in all seaweeds. The photosynthetic apparatus consists of two different reaction centers with slightly different red absorption peaks. Reaction centers are linked to core complexes, and these in turn to light-harvesting complexes; at each level, pigments are bound to proteins. The accessory pigments and most of the chlorophyll *a* are arranged in antennae, with those absorbing at longer wavelengths thought to be closer to the reaction center.

The amount of light harvested by a thallus (its absorptance) depends on both pigment concentration and thallus morphology. Significant morphological aspects are the thickness and the arrangement of branches relative to self-shading. The rate of photosynthesis is strongly dependent on irradiance level. At the compensation irradiance, gross photosynthesis equals respiration and therefore net photosynthesis is zero. At the saturation irradiance, photosynthesis is maximal. At very high irradiances, the photosynthetic rate may decline again because of photoinhibition. Extreme irradiances are harmful and cause photodamage, such as bleaching of pigments and the mechanism behind often involves the generation of reactive oxygen in photosynthesis.

Seaweeds can acclimate to differences in light quality and quantity. The quantity of pigment or the density of photosynthetic units can be increased, and sometimes the ratio of accessory pigments to chlorophyll *a* changes. However, the old idea that seaweeds change color to complement the color of the light in their habitat has been discredited.

Carbon fixation constitutes the "dark reactions" of photosynthesis. Carbon is available in seawater chiefly as HCO_3^-. There are small amounts of CO_2, the proportion depending on pH and salinity. Parasitic algae rely on carbon from their hosts, but other seaweeds have little or no ability to use exogenous organic carbon, particularly at environmental concentrations.

CO_2 is fixed by RuBP carboxylase/oxygenase (RuBisCO) in the Calvin cycle. HCO_3^- is converted to CO_2 with the aid of carbonic anhydrase. In the brown algae, especially in young tissue of kelps and fucoids, light-independent carbon fixation occurs, in light and dark, via PEP carboxykinase, which also uses free CO_2. RuBisCO also acts as an oxygenase, leading to photorespiration, but most seaweeds seem able to suppress this by various means of concentrating CO_2 near RuBisCO.

Various principal products of photosynthetic carbon fixation are found among the seaweeds. Green algae form chiefly sucrose; brown algae form mannitol; red algae form chiefly floridoside and digeneaside. In addition, all groups accumulate some amino acids.

In the kelps, the low molecular weight compounds can be translocated from mature regions to meristematic regions. The driving force probably is a water potential gradient set up by the loading of organic molecules into the sieve elements in the source and their unloading in the sink.

The initial products of photosynthesis may be polymerized into polysaccharides for storage or cell walls. As with low molecular weight products, each seaweed phylum has a characteristic group of compounds. Storage compounds are starch (green algae) or floridean starch (red algae), and laminaran in the brown algae. Seaweed cell walls have a fibrillar layer consisting of cellulose, mannan, or xylan. Various characteristic mucilaginous polysaccharides are also found in seaweed walls. These include some of commercial value, such as agars, carrageenans, and alginates. Chemical changes in these polymers can make them weaker or stiffer gels. The native structures of these compounds are not well known. Different stages in the life history may have different wall polysaccharides.

Photosynthesis is the basis of primary productivity, and its measurement also is often used to indicate seaweed (stress) responses to environmental variables, because it is affected by temperature, pH, circadian rhythms, age of the tissues, and irradiance. Photosynthetic rates can be measured by following CO_2 uptake or O_2 release by illuminated tissue, but account must be taken of respiratory use of O_2 and release of CO_2 in the light. Respiration is usually measured in darkness, but it also occurs to a much smaller extent in the light. As a relative measure of photosynthetic quantum yield

and an indicator of general fitness, the determination of variable chlorophyll fluorescence represents a non-invasive, rapid, and easily applied tool.

The estimation of primary production requires measurements not only of photosynthesis, respiration, and photorespiration (if any) but also of tissue loss and organic carbon exudation rates. Seaweeds can be divided into functional-form groups, such as crustose forms and thin sheets, which have characteristic levels of productivity. Carbon budgets and energy flow models show how seaweed productivity is passed to higher trophic levels. In environments with large seaweeds such as kelps, carbon from primary production may be widely dispersed. In others, such as coral reefs with generally small and microscopic producers, production may be tightly cycled chiefly within the system.

Nutrients

Seaweeds require inorganic carbon, water, light, and various mineral ions for photosynthesis and growth. Natural nutrient sources include vertical wind mixing of the water column, tidal mixing, and release from sediments, while anthropogenic sources include sewage, fertilizers, animal manures, and atmospheric deposition. This chapter will examine the mechanisms of uptake, the nutrient requirements, and the metabolic roles of essential nutrients (excluding C, H, and O). The importance of nutrient uptake and growth kinetics will be discussed in terms of their effects on chemical composition, growth, development, and distribution of macroalgae. Particular emphasis will be placed on nitrogen, because it is the element most frequently limiting to seaweed growth, although P has recently been reported to be limiting in some areas. The nutritional requirements of seaweeds and phytoplankton are generally similar, and therefore some discussion of phytoplankton nutrition is also included when little or no information exists for seaweeds as a possible suggestion for future research.

Seaweeds are important primary producers in shallow coastal and estuarine ecosystems. Seaweed biomass per unit area may be 400 times greater than phytoplankton in these shallow areas and annual production of dry matter per unit area is equivalent to grasslands and even rain forests (Rees 2003; sec. 5.7). Globally, seaweeds are responsible for 5–10% of marine primary productivity, even though they occupy only a fraction of the surface area of phytoplankton. Another interesting comparison between seaweeds and phytoplankton involves their chemical composition. Phytoplankton have a higher protein content than seaweeds (50% vs. 15% dry wt) and a lower C:N ratio (6.7 vs. ~20 for seaweeds) (Atkinson and Smith 1983; Durarte 1992). Therefore, since the growth rate of seaweeds is lower than phytoplankton, their mass specific N demand will also be lower than phytoplankton. However, on an ecosystem basis, their lower N demand per unit biomass compared to phytoplankton is offset by the fact that seaweed biomass can be much higher than phytoplankton in shallow areas.

6.1 Nutrient requirements

6.1.1 Essential elements

The development in the 1960s of defined culture media for growing algae axenically allowed the testing of a variety of elements to determine which are essential and required for growth. A nutrient or element is essential when a deficiency of the element makes it impossible for the alga to grow or complete its vegetative or reproductive cycle and the requirement cannot be replaced by another element. C, H, O, N, Mg, Cu, Mn, Zn, and Mo are considered to be required by all algae (DeBoer 1981); S, K, and Ca are required by all algae, but can be partially replaced by other elements; Na, Co, V, Se, Si, Cl, B, and I are required only by some algae. All the major constituents of seawater, except for Sr and F are required by macroalgae (DeBoer 1981).

Up to 21 elements are required for the main metabolic processes in plants (Table 6.1), but more than double that number are present in seaweeds. The presence of an element in seaweed tissue is not proof that the element is essential, nor is the amount present

Table 6.1 Functions and compounds of the essential elements in seaweeds.

Element	Probable functions	Examples of compounds
Nitrogen	Major metabolic importance in compounds	Amino acids, purines, pyrimidines, amino sugars, amines
Phosphorus	Structural, energy transfer	ATP, GTP, etc., nucleic acids, phospholipids, coenzymes (including coenzyme A), phosphoenolpyruvate
Potassium	Osmotic regulation, pH control, protein conformation and stability	Probably occurs predominantly in the ionic form
Calcium	Structural, enzyme activation, cofactor in ion transport	Calcium alginate, calcium carbonate
Magnesium	Photosynthetic pigments, enzyme activation, cofactor in ion transport, ribosome stability	Chlorophyll
Sulfur	Active groups in enzymes and coenzymes, structural	Methionine, cystine, glutathione, agar, carrageenan, sulfolipids, coenzyme A
Iron	Active groups in porphyrin molecules and enzymes	Ferredoxin, cytochromes, nitrate reductase, nitrite reductase, catalase
Manganese	Electron transport in photosystem II, maintenance of plastid membrane structure	
Copper	Electron transport in photosynthesis, enzymes	Plastocyanin, amine oxidase
Zinc	Enzymes, ribosome structure(?)	Carbonic anhydrase
Molybdenum	Nitrate reduction, ion absorption	Nitrate reductase
Sodium	Enzyme activation, water balance	Nitrate reductase
Chlorine	Photosystem II, secondary metabolites	Violacene
Boron	Regulation of carbon utilization(?), ribosome structure(?)	
Cobalt	Component of vitamin B_{12}	B_{12}
Bromine[a]	{Toxicity of antibiotic compounds(?)	{Wide range of halogenated compounds, especially in
Iodine[a]		Rhodophyceae

[a] Possibly an essential element in some seaweeds.
Source: DeBoer (1981), with permission of Blackwell Scientific Publications.

indicative of the relative importance of the element. Generally, essential (and non-essential) elements are accumulated in algal tissues to concentrations well above their concentrations in seawater, giving rise to concentration factors of up to 10^3 (Phillips 1991) (Table 6.2). Some elements are absorbed in excess of an alga's requirements, whereas others are taken up but not utilized. The elemental composition of the ash in macroalgae is similar to that of phytoplankton.

6.1.2 Essential organics: vitamins

Generally, seaweeds and phytoplankton both require the same vitamins. In contrast most higher plants synthesize their own vitamins and do not depend on environmental sources. The three vitamins that are routinely added to culture media are B_{12} (cyanocobalamin), thiamine, and biotin. Of these, B_{12} is the most widely required by seaweeds and it is present in seawater in lesser amounts (*c.* 1 ng L^{-1}) than are thiamine (*c.* 10 ng L^{-1}) and biotin (*c.* 2 ng L^{-1}). One Chlorophyceae, one Phaeophyceae, and ten Rhodophyceae require vitamin B_{12} (DeBoer 1981). No requirement has yet been found for thiamine and biotin, but very few seaweeds have been studied. The development of HPLC techniques for measuring vitamins may stimulate more vitamin research in the future.

Table 6.2 Concentrations of some essential elements in seawater and in seaweeds.

Element	Mean concentration in seawater (mmol kg^{-1})	(μg g^{-1})	Concentration in dry matter Mean (μg g^{-1})	Range (μg g^{-1})	Ratio of concentration in seawater to concentration in tissue
Macronutrients					
H	105 000	10 500	49 500	22 000–72 000	2.1×10^0
Mg	53.2	1 293	7 300	1 900–66 000	1.8×10^{-1}
S	28.2	904	19 400	4 500–8 200	4.7×10^{-2}
K	10.2	399	41 100	30 000–82 000	1.0×10^{-2}
Ca	10.3	413	14 300	2 000–360 000	2.9×10^{-2}
C	2.3	27.6^{a,b}	274 000	140 000–460 000	1.0×10^{-4}
B	0.42	4.50	184	15–910	2.4×10^{-2}
N	0.03	0.420$^{a\,c}$	23 000	500–65 000	2.1×10^{-5}
P	0.002	0.071	2 800	300–12 000	2.4×10^{-5}
Micronutrients					
Zn	6×10^{-6}	0.0004^a	90	2–680	4.4×10^{-5}
Fe	1×10^{-6}	0.00006^a	300	90–1 500	1.0×10^{-5}
Cu	4×10^{-6}	0.0002^a	15	0.6–80	1.7×10^{-4}
Mn	0.5×10^{-6}	0.00003^a	50	4–240	2.0×10^{-5}

^{a} Considerable variation occurs in seawater (Bruland 1983).
^{b} Dissolved inorganic carbon.
^{c} Combined nitrogen (dissolved organic and inorganic).
Source: DeBoer (1981), including concentrations of elements in seawater from Bruland (1983), with permission of Blackwell Scientific Publications.

6.1.3 Limiting nutrients

More than a century ago Liebig's law of the minimum was formulated which stated that the nutrient available in the smallest quantity with respect to the requirements of the plant will limit its rate of growth, if all other factors are optimal. Nitrogen is the most frequently limiting nutrient followed by phosphorus and occasionally Fe. Their concentrations in seawater vary considerably due to biological activity, and the concentrations of these elements in tissues are 10^4 to 10^5 greater than their concentrations in seawater (Table 6.2). The concentration of a nutrient will give some indication if the nutrient is limiting, but the nutrient's supply rate or turnover time in relation to the alga's growth rate are more important in determining the magnitude or degree of limitation. For example, if the concentration of a nutrient is limiting, but the supply rate is only slightly less than the uptake rate, then the algae will be only slightly nutrient

limited. The possibility of growth limitation by two nutrients simultaneously, or dual nutrient limitation, has been suggested. For example, N limitation may reduce P uptake since N is required for certain enzymes and proteins and that may be involved in P uptake (Harpole *et al.* 2011). However, the ratio of two nutrients (e.g. N:P) required by one algal species may be quite different from the ratio required by another species. Consequently, one species may be N limited, while another species may be P limited, illustrating limitation by a single nutrient for each species which are competing for different nutrient resources.

6.2 Nutrient availability in seawater

The concentrations of various elements in seawater can differ by up to six orders of magnitude (Table 6.2). Those in the nanomolar (nM) range (<0.01 μg g^{-1} in

Table 6.2) are considered micronutrients or trace elements (e.g. Fe, Cu, Mn, Zn) for nutritional purposes. Elements occurring at higher concentrations frequently are referred to as macronutrients (e.g. C, N, P). Generally, marine scientists use micromolar (μM) concentrations, although microgram atoms per liter (μg-at L^{-1}) was used in the early literature. Freshwater nutrient concentrations generally are expressed as micrograms per liter (μg L^{-1}) or parts per billion (ppb). To convert from micrograms to micromoles, divide by the atomic weight of the element (e.g. 1 μM = 14 μg L^{-1} for NO_3 or NH_4, but 28 μg L^{-1} for urea since it contains two N atoms per urea molecule). The methods for measuring nutrients (e.g. nitrate, ammonium, and phosphate) have been reviewed by Parsons et al. (1984) and Wheeler (1985). For most pristine temperate areas, nitrate, nitrite, ammonium, and phosphate concentrations vary from undetectable to about 30, 1, 3, and 2 μM, respectively, for surface coastal water. Organic N sources are often measured as a mixture of dissolved organic nitrogen (DON) (Antia et al. 1991) and only uptake of urea, and some amino acids have been studied individually.

Nutrient cycles have been reviewed by Libes (1992) and Herbert (1999) and only a few basic principles will be reviewed here. The important features of the nitrogen cycle are summarized in Figure 6.1. Processes that bring nitrogen into the euphotic zone (Fig. 6.1) where it can be used for seaweed growth include the following: (1) vertical mixing and upwelling primarily in the form of nitrate from below the nutricline; (2) atmospheric input of ammonium and nitrate either via rain or via sewage inputs and/or animal farming (Paerl et al. 1990); (3) N_2 fixation by bacteria and cyanobacteria; and (4) inputs of nitrogen from land drainage, sewage, and agricultural fertilizers. The input of nitrogen from these four mechanisms above is termed "new nitrogen", while ammonium released from particulate organic matter during decomposition is termed "regenerated nitrogen". Regeneration of nitrogen in the water column occurs as a result of two largely separate processes; one involves bacterial decomposition and the other results from excretion by marine fauna, particularly ammonium by herbivores such as zooplankton and/or benthic grazers.

Ammonium is regenerated from bacterial decomposition of organic matter in sediments, the magnitude of which may be enhanced by the activity of burrowing animals. In some small, shallow eutrophic estuaries, decomposition of extensive mats of Ulva or Chaetomorpha may dominate the nitrogen cycle for short periods in summer (Sfriso et al. 1987; Lavery and McComb 1991a).

Macro- and microalgae may utilize somewhat different nitrogen sources. Nitrogen uptake rates for the kelp Ecklonia maxima and for phytoplankton were compared in an upwelling area off South Africa (Probyn and McQuaid 1985). The kelp took up nitrate and ammonium, but not urea, and nitrate uptake did not saturate at concentrations >20 μM. In contrast, phytoplankton took up all three nitrogen forms, but preferred ammonium and urea. The fact that nitrate was the most abundant and most highly utilized nitrogen resource in this upwelling area indicates that most (80%) of the yearly kelp productivity is based on "new" nitrogen (NO_3^-) rather than on recycled nitrogen (NH_4^+ and urea).

Another suggested source of nutrients, especially nitrate, is submarine discharge of ground water in coastal areas (Capone and Bautista 1985; Lapointe 1997). Because the concentration of nitrate in groundwater may be very high (50–120 μM), a small amount of discharge can significantly enrich nitrogen-impoverished coastal waters in the summer. The availability of groundwater nitrate to Sargassum filipendula and Ulva intestinalis (previously called Enteromorpha intestinalis) has been confirmed by the induction of high nitrate reductase activity (Maier and Pregnall 1990). Regenerated nutrients from the sediments could also be important sources of nitrogen enrichment since ammonium and phosphate concentrations may reach 10 and 1 mM, respectively. If the circulation of seawater is such as to entrain some fraction of the interstitial-sediment water into the surface layers, then significant nutrient enrichment of the water column will result from these regenerated nutrients. Smetacek et al. (1976) observed a seldom recognized mechanism whereby nutrients are "squeezed" out of the sediments by an intrusion of high salinity water displacing lower salinity interstitial water, and resulting in an ~10-fold

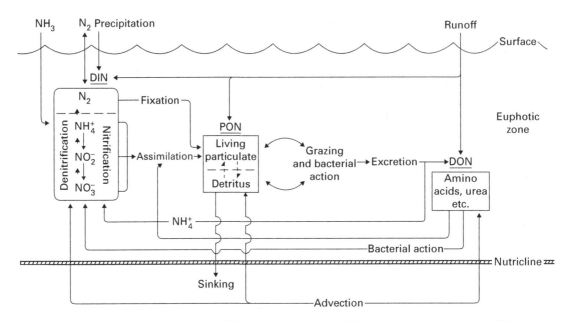

Fig. 6.1 Schematic of the marine nitrogen cycle. PON, particulate nitrogen; DON, dissolved particulate nitrogen; DIN, dissolved inorganic nitrogen. Sources of nitrogen include precipitation, runoff, regeneration and excretion, advection and the sediments. (From Turpin 1980, with permission from the author.)

increase in ammonium, phosphate, and silicate to the water column.

Red, green, and brown seaweeds are unable to make direct use of nitrogen gas (N_2) even though it is 20 times more abundant than nitrate. N_2 fixation associated with *Codium decorticatum* (Rosenberg and Paerl 1981) and *C. fragile* (Gerard *et al.* 1990) has been attributed both to N_2-fixing bacteria (e.g. *Azotobacter*) and to heterocystous cyanobacteria (e.g. *Calothrix*) that occur as epibionts. Gerard *et al.* (1990) found no evidence that the host seaweeds obtain significant amounts of this fixed nitrogen, but Philips and Zeman (1990) indicated that certain species of cyanobacteria (e.g. *Oscillatoria*) are thought to contribute significantly to *Sargassum*.

At the pH of seawater, phosphorus exists primarily as three ionic species. At pH 8.2 and 20°C, HPO_4^{2-} accounts for 97%, PO_4^{3-} for <1%, and H_2PO_4 for 2.5% (Turner *et al.* 1981). Orthophosphate ions (PO_4^{3-}) can form metallophosphate complexes (e.g. with Ca^{2+} and Mg^{2+}), or they can combine with organic compounds, and therefore free orthophosphate represents <30% of

the total inorganic phosphate in seawater. The standard technique for determining inorganic phosphate with the molybdate reaction tends to overestimate the true concentration, because some forms of organic phosphorus are hydrolyzed by reagents in the phosphate analysis (Cembella *et al.* 1984).

6.3   Pathways and barriers to ion entry

6.3.1   Membrane structure and ion movement

Algal and higher plant plasma membranes consist of lipid bilayers interspersed with various proteins (Raven *et al.* 2005). Most substances required by cells are polar and require transport proteins to transfer them across the membrane. Each transport protein is highly selective and they provide a pathway for specific solutes to cross the membrane without the solute coming in contact with the hydrophobic interior of the lipid bilayer. Ions usually have a much lower permeability than uncharged molecules such as CO_2 or urea. The charge on the ion makes it difficult for the ion to penetrate a membrane that is electrically polarized and contains charged groups that either repel or attract (immobilize) the ions. In addition, ions usually are strongly hydrophilic, and their particle sizes frequently are increased by a substantial layer of the water of hydration. Both of these properties tend to decrease the rate of diffusion (Glass 1989). Transport proteins are divided into three groups: pumps, carriers, and channels. For general aspects of the membrane structure and ion movement see Raven *et al.* (2005) and Buchanan *et al.* (2000).

6.3.2 Movement to and through the membrane

Ions enter a cell by moving across the concentration boundary layer (CBL), reaching the cell surface, then through the cell wall and plasmalemma (cell membrane) into the cytoplasm. The thickness of the CBL may affect the uptake rate of an ion, because if turbulence around the thallus is low, the CBL can become thick and uptake may be limited by the rate of diffusion (see sec. 8.1.3). The cell wall, unlike the plasmalemma, does not generally present a barrier to ion entry. When a macroalga is placed in nutrient medium, there may be an initial rapid uptake that does not require energy (i.e. is independent of metabolism) and usually lasts less than a minute. This observation is generally attributed to diffusion into the apparent free space that is exterior to the plasmalemma. Ions can readily be removed from the apparent free space by washing the alga in nutrient-depleted medium. Some ions, especially cations, may not reach the plasmalemma because they become adsorbed to certain components of the cell wall. Polysaccharides and proteins have sulfate, carboxyl, and phosphate groups from which protons can dissociate, leaving a net negative charge on these compounds in the cell wall (Kloareg *et al.* 1987). In effect, these macromolecules act as cation exchangers, and consequently large numbers of cations can be adsorbed but not actually enter into the cell. The concentrations of Ca, Sr, and Mg in brown algae are largely the result of ion exchange between seawater and the acid polysaccharide alginate in the cell walls; the amounts of these ions in the cytoplasm and vacuole compose a relatively small portion of the total in the whole thallus.

6.3.3 Passive transport

The direction of transport of uncharged molecules is determined by the concentration gradient on each side of the membrane and they diffuse down a free energy or chemical potential gradient and hence the term "downhill transport". The rate of diffusion varies with the chemical potential gradient or the difference in activities (approximately equivalent to the concentrations) across the plasmalemma. Many important gases (e.g. CO_2, NH_3, O_2, and N_2) cross lipid bilayers by dissolving in the lipid portion of the membrane, diffusing to the other lipid–water interface, and then dissolving in the aqueous phase on the other side of the membrane (Glass 1989). Uncharged molecules such as water and urea are also highly mobile. However, molecules such as NH_3 may be trapped inside cells when they are converted to ions (Reed 1990c). The pK_a for NH_3/NH_4 is 9.4, and in seawater with a pH of 8.2, only 5–10% of the total ammonium is present as NH_3 (hence, it is preferable to use the term "ammonium" rather than "ammonia" when referring to concentrations in seawater). At a higher pH, which may be found in dense cultures or restricted tidal pools, the percentage of NH_3 can increase to 50% (at pH 9.4) or more, allowing rapid diffusion into the cell. Because the pH of the cytoplasm is only 7–7.5, most of the NH_3 that enters is protonated to NH_4 and cannot diffuse back across the membrane. NH_3 uptake can account for approximately 10% of net uptake at high pH (Glass 1989). Passive diffusion occurs without the expenditure of cellular metabolic energy and therefore temperature has little effect on the rate; however, the electrical gradient that may drive passive cation movement is the result of cellular metabolism. In addition, no carriers or binding sites are involved in diffusion, and therefore it is non-saturable.

6.3.4 Facilitated diffusion

Facilitated diffusion resembles passive diffusion in that transport occurs down an electrochemical gradient, but frequently the rate of transport by this process is faster than diffusion. It occurs via either carrier proteins or channel proteins and exhibits properties similar to active transport in that: (1) it can be saturated, and transport data fit a Michaelis–Menten-like equation; (2) only specific ions are transported; and (3) it is susceptible to competitive and non-competitive inhibition. However, in contrast to active transport mechanisms, any energy expenditure required for transport must be indirect (see Glass 1989; Raven *et al.* 2008 for further details).

6.3.5 Active transport

Active transport is the transfer of ions or molecules across a membrane against an electrochemical potential gradient which requires energy, hence the term "uphill" transport. Because external concentrations of inorganic nutrients (e.g. NO_3^- and PO_4^-) typically are in the micromolar range, and intracellular concentrations are in the millimolar range (i.e. a difference of $1000\times$), passive diffusion along an electrochemical gradient is not important (Reed 1990c). To demonstrate active transport conclusively, as opposed to free or facilitated diffusion, the following criterion must be satisfied: Active uptake is energy-dependent. A change in the uptake rate should occur after the addition of a metabolic inhibitor (e.g. dinitrophenol) or a change in temperature, because both of these factors influence energy production. Other properties of active transport, such as unidirectionality, selectivity of ions transported, and saturation of the carrier system (exhibiting Michaelis–Menten kinetics), are not definitive criteria for active transport because they are also characteristic of facilitated diffusion.

6.4 Nutrient-uptake kinetics

The kinetics of nutrient uptake will depend on which uptake mechanism is being used. If transport occurs solely by passive diffusion, then the transport rate will be directly proportional to the electrochemical potential gradient (external concentration) (Fig. 6.2a). In contrast, facilitated diffusion and active transport will exhibit a saturation of the membrane carriers as the external concentration of the ion increases. The relationship between the facilitated diffusion or active uptake rate of the ion and its external concentration is generally described by a rectangular hyperbola, similar to the Michaelis–Menten equation for enzyme kinetics (Fig. 6.2b). K_s (equivalent to K_m) is called the half-saturation constant, and it is the substrate concentration at which the uptake rate is half its maximum rate. The lower the value of K_s the higher is the affinity of the carrier site for the particular ion. The transport capabilities of a particular macrophyte are generally described by the parameters V_{max} and K_s. The slope of the initial linear part of the hyperbola (i.e. the ratio $V_{max}:K_s$), represents the affinity for the substrate and is a more useful parameter for comparing the competitive abilities of various species for a limiting nutrient than K_s (Duke *et al.* 1989; Harrison *et al.* 1989). This is similar in concept to the use of the slope, α, in the P–E curve (Fig. 5.13). One of the serious problems in using K_s is that its value is not independent of V_{max} (i.e. when V_{max} decreases, the value of K_s will also decrease, even though the initial slope of the hyperbola remains the same) (Fig. 6.2c). The Michaelis–Menten curve can be fitted with a non-linear regression using various computer programs. For a summary of kinetic parameters (V_{max}, K_s, $V_{max}:K_s$) for nitrate and ammonium for various green, red, and brown seaweeds, see Appendix 1 in Rees (2003). In a comparison of K_s and V_{max} values, the average V_{max} and K_s was 185 nmol cm^{-2} h^{-1} and 17.5 µM, respectively, for NH_4 uptake for 56 macroalgae and 93 nmol cm^{-2} h^{-1} and 9 µM, respectively, for NO_3 uptake for 44 macroalgae. For phytoplankton, the average K_s was much lower at 0.7 µM for NH_4 and 4 µM for NO_3 (Rees 2003, 2007). Since V_{max} and K_s values for ammonium are higher than for nitrate, ammonium may be a more important N source than NO_3 for macroalgae. The $V_{max}:K_s$ (affinity of uptake) for NO_3 and NH_4 was 9 and 22, respectively, for phytoplankton and 10 and 12 for macroalgae, surprisingly similar (Taylor *et al.* 1998).

Active uptake may not follow simple saturation kinetics. Uptake systems in higher plants may be biphasic (Crawford and Glass 1998). In the case of biphasic kinetics, a plot of V versus S reveals two rectangular

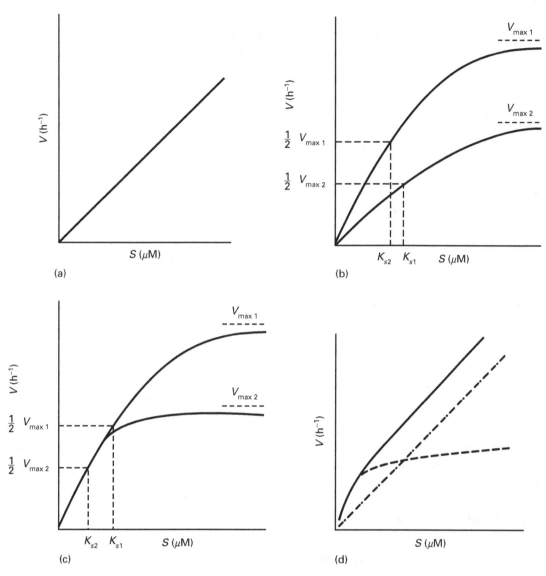

Fig. 6.2 Hypothetical plots of nutrient uptake rates (V) and concentrations of the limiting nutrients (S) for:
(a) passive diffusion, (b) facilitated diffusion or active transport, where $V_{\text{max 2}} = \frac{1}{2}\,V_{\text{max 1}}$ and consequently $K_{s1} < K_{s2}$.
(c) Since $V_{\text{max 2}} = \frac{1}{2}\,V_{\text{max 1}}$ for two species, the decrease in V_{max} by 50% resulted in a decrease in K_{s2} (for species 2), even though the slope (the affinity for the nutrient) remained the same for species 1 and 2. Therefore, the slope (V_{max}: K_s) is a better measure of the competitive ability of a species at low/limiting nutrient concentrations. (d) Passive diffusion plus active transport (solid line) and active transport (dash line) with the passive diffusion component (dot-and-dash line) subtracted.

hyperbolas, which frequently are referred to as the high and low affinity systems. At low substrate concentrations, the high-affinity system operates, exhibiting a high degree of ion specificity and a low value for K_s, whereas at high concentrations the low-affinity system is operative, and it exhibits much less ion selectivity and a very high value for K_s. To date there is no strong evidence for biphasic uptake kinetics in macroalgae, but Collos *et al.* (1992) clearly demonstrated biphasic kinetics of nitrate uptake in two marine diatoms. They concluded that since the intracellular nitrate pool was in the millimolar range, uptake rates at 200–500 μM must have been active in order to transport nitrate against this concentration gradient. Deviation from saturation kinetics may also occur if active uptake and diffusion occur simultaneously. Diffusion is not likely to be important at low substrate concentrations, but it may be significant at concentrations well above environmental concentrations. In this latter case, the total uptake rate is composed of active uptake plus diffusion and does not exhibit saturation kinetics (Fig. 6.2d) and it has been shown to occur for ammonium uptake in *Macrocystis*, *Gracilaria tikvahiae*, and *Agardhiella subulata* at ammonium concentrations >25 μM (Haines and Wheeler 1978; D'Elia and DeBoer 1978; Friedlander and Dawes 1985). Under conditions where the nutrient is limiting, an uptake rate may not follow a Michaelis–Menten hyperbola, because uptake may be controlled by the internal nutrient level rather than by external nutrient concentrations (Fujita *et al.* 1989).

6.4.1 Measurement of nutrient-uptake rates

Nutrient-uptake rates may be expressed in four commonly used units. Uptake may be normalized to surface area ($μmol\ cm^{-2}\ h^{-1}$), wet weight ($μmol\ g_{ww}^{-1}\ h^{-1}$), dry weight ($μmol\ g_{dw}^{-1}\ h^{-1}$), or the nutrient content in the seaweed, which simplifies to a specific uptake rate (h^{-1}). If conversion factors are not given, nutrient uptake data that are expressed in different units cannot be accurately compared. There are two main techniques for measuring nutrient-uptake rates: (1) radioactive or stable isotope uptake (Glibert *et al.* 1982; Naldi and Wheeler 2002); and (2) disappearance of the nutrient from the medium measured

colorimetrically (Harrison and Druehl 1982; Harlin and Wheeler 1985; Harrison *et al.* 1989). The advantages and problems of the ^{15}N technique have been reviewed by Glibert and Capone (1993). Isotope measurements have high sensitivity, allow short incubation times, determination of inorganic N efflux by isotope dilution (O'Brien and Wheeler 1987) and can determine rates of N assimilation into specific organic molecules. Problems of using isotopes include: (1) different parts of the thallus will accumulate the isotope at different rates, and samples from different areas of the thallus should be taken and averaged in order to obtain a whole thallus uptake rate (the most useful measurement for ecological purposes). (2) Often not all of the isotope can be accounted for at the end of the experiment (i.e. the missing isotope problem), usually due to the release of inorganic or organic N into the medium, resulting in isotope dilution (see Glibert *et al.* 1982; Naldi and Wheeler 2002; Mullholland and Lomas 2008 for technical details). When comparing the stable isotope method with the nutrient disappearance method, O'Brien and Wheeler (1987) found good agreement. That contrasts with the findings of Williams and Fisher (1985), who reported that more ammonium disappeared from the medium than could be accounted for by the incorporation of ^{15}N in *Caulerpa*; isotope dilution of the ammonium pool was ruled out. They suggested that either: (1) a secondary ammonium sink, such as wall sorption or bacterial uptake, reduced ammonium concentrations; or (2) ^{15}N was lost as labeled dissolved organic nitrogen or was volatilized during ^{15}N sample preparation. Other studies support the release of both inorganic and organic N (mainly as amino acids) (Naldi and Wheeler 2002; Tyler and McGlathery 2006). Hence, estimates of new N assimilation based solely on growth rate and tissue N content may underestimate the actual N uptake by up to 100% due to released DON (Tyler and McGlathery 2006). In all of these techniques, very short incubation times of <10–15 min likely yield an estimate of gross uptake rate (influx), whereas long incubation times (>6 h) give rates that would approximate net uptake, taking account of efflux of the nutrient (inorganic and organic forms) from the thallus back into the medium.

Nutrient uptake rates for seaweeds are measured in the laboratory by incubating epiphyte-free tissue discs or preferably whole seaweeds in filtered natural seawater to which nutrients (except the one under study), trace metals, and vitamins have been added at saturating levels. In order to eliminate the effect of rapid diffusion into the apparent free space, seaweeds frequently are pre-incubated in saturating nutrients for 1–2 min and then placed in appropriate experimental concentrations of the nutrient; see Reed and Collins (1980), Harrison and Druehl (1982), and Harlin and Wheeler (1985) for further discussion. Adequate stirring is important to minimize the thickness of the boundary layer that could otherwise reduce uptake rates. On the other hand, stirring by vigorous aeration may cause the loss of ammonia since at high biomass density, active growth, and a pH near 9.4, 50% of the total $NH_4 + NH_3$ will be present as the gas (NH_3) which is easily released during aeration as discussed in sec. 6.3.3 (Pereira *et al.* 2008).

There are two basic approaches to determine nutrient uptake parameters by following the disappearance of the nutrient from the culture medium (Harrison *et al.* 1989). The first approach is to spike the culture with the nutrient of interest and follow the nutrient disappearance for several hours until nutrient exhaustion occurs (i.e. the perturbation experiment); this method gives the time course of the nutrient decrease. In Figure 6.3a, the decrease is linear with time and is typical if the thallus is not nutrient limited. However, if the thallus is nutrient limited then a non-linear decrease in the nutrient occurs (Fig. 6.3c) as is frequently the case when ammonium (Probyn and Chapman 1982; Rosenberg *et al.* 1984; Thomas and Harrison 1987) or nitrate (Thomas and Harrison 1985; Thomas *et al.* 1987a) is limiting. In this case, a second method involving a short incubation period must also be used to correctly estimate the slope $V_{max}{:}K_s$, and the maximal uptake rate, V_{max} (Harrison *et al.* 1989). The second method involves the use of many containers with different concentrations of the nutrient, and each with a different piece of thallus (i.e. the multiple flask measurement). The incubation period is usually short (10–30 min) but constant for all concentrations (Fig. 6.3b). Because uptake rates for

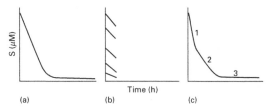

Fig. 6.3 Hypothetical time series showing nutrient disappearance from the medium using the following methods: (a) the perturbation method, where saturated uptake is linear with time; (b) the multiple-container, constant-incubation-time method; (c) the perturbation method, where saturated uptake is non-linear with time. Phase 1 (V_s) is the "surge" uptake, probably due to filling of internal pool(s); phase 2 (V_i) represents an assimilation rate under internal control; phase 3 (V_e) is the depletion of the nutrients when concentrations are limiting (i.e. external control). (From Harrison and Druehl 1982 with permission of Walter de Gruyter and Co.)

N-limited seaweeds frequently vary with time, Harrison *et al.* (1989) suggested putting a time superscript on V_{max} values (e.g. $V_{max}^{0-10\ min}$ means maximal rates measured over the first 10 min of the incubation). In summary, a time course of nutrient uptake rates should always be run before deciding which method to use in measuring nutrient uptake rates in order to determine if the species is nutrient limited.

Nutrient-limited algae exhibit three phases of uptake, especially for ammonium, as follows: (1) the initial surge uptake (V_s) which represents pool filling; (2) internally controlled uptake (V_i) which is a slower uptake rate due to feedback inhibition from filling intracellular pools on the uptake system; and (3) the externally controlled uptake rate (V_e) when the nutrient concentration is low enough to slow the uptake rate (Harrison *et al.* 1989). N-limited *Ulva lactuca* showed the same three phases as phytoplankton (Figs. 6.4 and 6.5) and Pedersen (1994) found that the decrease in ammonium was not linear over time (Fig 6.4a) and ammonium uptake decreased with time (Fig. 6.4b).

There are several reasons why the uptake rate varies. Under nitrogen limitation intracellular nitrogen pools may be low and the initial enhancement in uptake rate over the first 0–60 min represents a pool-filling phase

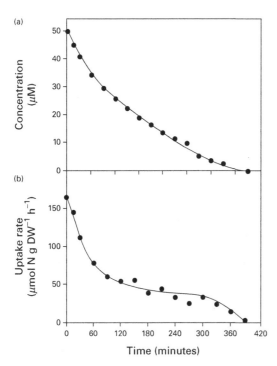

Fig. 6.4 Time course of: (a) ammonium depletion, and (b) ammonium uptake during a perturbation experiment where 50 μM ammonium was added to N-limited *Ulva lactuca* (From Pedersen 1994, with permission from the Phycological Society of America.)

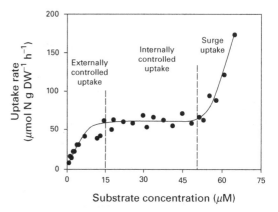

Fig. 6.5 Ammonium uptake when 70 μM ammonium was added to N-limited *Ulva lactuca* during a perturbation experiment. Note the three phases of uptake with "surge" uptake at the beginning of the experiment (From Pedersen 1994, with permission from the Phycological Society of America.)

(Fujita *et al.* 1988; Harrison *et al.* 1989; Pedersen 1994). Lartigue and Sherman (2005) showed that in addition to an increase in tissue nitrate during surge uptake of nitrate, nitrate reductase increased, but tissue ammonium and free amino acids remained constant (Fig. 6.6) The first surge phase of uptake (V_s) can help an alga overcome its N debt since uptake is several fold faster than the growth rate. As the pools fill, the decrease in uptake rate may be due to feedback inhibition on the uptake system (Harrison *et al.* 1989). Therefore, this latter uptake rate does not represent the true transmembrane transport that is free from feedback inhibition. The perturbation method where the nutrient is followed until it is depleted from the medium is not recommended for estimation of V_{max} and K_s because the nutritional status of the thallus is changing with time (i.e. becoming less N deficient).

Nevertheless, the perturbation method is useful in determining the assimilation rate or V_i, the rate at which intracellular nitrate or ammonium is incorporated into amino acids and proteins (Fig. 6.3c see section 2 of the curve and Fig. 6.5).

In order to distinguish between uptake and assimilation, the protonophore, carbonyl cyanide *m*-chlorophenylhydrazone (CCCP), was used to measure NH_4 assimilation (Rees *et al.* 1998). Tissue was incubated with ^{14}C-methylammonium which acted as a tracer for uptake and accumulation of intracellular NH_4. CCCP completely inhibited NH_4 uptake and allowed the release of unassimilated ammonium and methylammonium and thus it inhibits active uptake and assimilation, but not passive uptake. It was concluded that for *Enteromorpha*, the V_i phase is a reliable estimate of the maximum rate of assimilation. This method works well for green algae, but there is interference with the ammonium assay for red algae. Therefore, it is easier to measure assimilation rates during the V_i phase of a perturbation experiment as described by Harrison *et al.* (1989) and Pedersen (1994).

Whereas the limiting nutrient (especially ammonium and phosphate) may enhance uptake rates upon

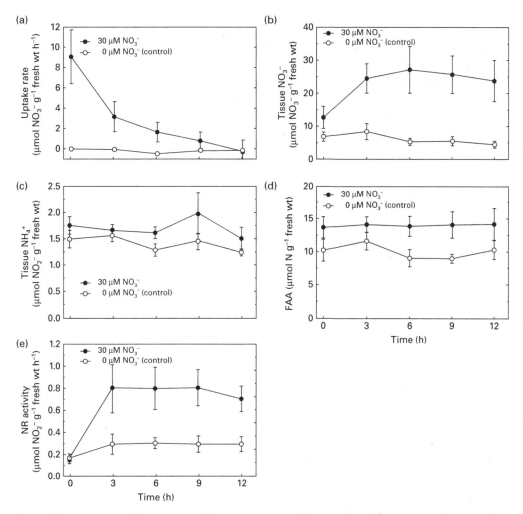

Fig. 6.6 (a) Nitrate uptake; (b) tissue nitrate; (c) tissue ammonium; (d) free amino acids (FAA); and (e) nitrate reductase activity following a 30 μM pulse of nitrate to N-limited *Ulva*. (Redrawn from Lartigue and Sherman 2005, with permission from the Inter-Research Science Center.)

re-addition to NH_4^+ or P-limited algae, C fixation is often shut down (Gordillo 2012). For microalgae, this depression in C fixation may last for minutes (Turpin and Hoppe 1994), while in macroalgae it may take 24 h for the photosynthetic rate to recover (Williams and Herbert 1989). In addition to uptake by the thallus, epiphytes (microscopic algae and bacteria) on the surface of the thallus may also contribute to nutrient uptake. These epiphytes are not easily removed. Antibiotics such as streptomycin and penicillin have been employed to inhibit bacterial uptake, but it is unlikely that there is an antibiotic concentration that can completely inhibit the bacterial uptake of nutrients without also affecting the alga.

6.4.2 Factors affecting nutrient-uptake rates

Physical factors. Light affects nutrient uptake indirectly through photosynthesis, which: (1) provides energy (ATP) for active transport; (2) produces carbon skeletons that are necessary for incorporation of nutrient ions into larger molecules (e.g. amino acids and proteins); (3) provides energy for the production of charged ions that establish chemical potentials; and (4) increases the growth rate and thus increases the need for nutrients. There is strong evidence for the effects of irradiance on uptake rates for nitrate (Fig. 6.7a). The nitrate-uptake data roughly fit a rectangular hyperbola for *Macrocystis*, but the curve intersects the *y*-axis, indicating that nitrate uptake in the dark is substantial. In contrast, ammonium uptake in *Macrocystis* was independent of irradiance (Fig. 6.7a). Photoperiod affects nitrate uptake, possibly because of the diel periodicity in activity and synthesis shown by the nitrate reductase enzyme (Berges 1997). The metabolic cost of N assimilation is considerable and diel cycles may minimize these costs (Turpin and Huppe 1994). The maximum rate of NH_4 assimilation was maximum in the early part of the day since the rates of processes involved in the early processes of assimilation are greater in the early part of the day, possibly to coincide with peak nitrate reductase activity and growth rates (Gevaert *et al.* 2007). Diel periodicity of nitrate uptake is usually more pronounced for nitrate-sufficient than for N-limited algae (i.e. dark uptake is a higher percentage of light uptake when algae are N limited). Raikar and Wafar (2006) observed surge uptake of ammonium over 4 min for three coral atoll macroalgae and surge uptake in the dark was ~80% of the light, indicating that the macroalgae could readily take up pulses of ammonium that was being excreted by reef animals during the night. It is surprising that few studies have been conducted on the effects of UV-A and UV-B on nutrient uptake and assimilation since high intertidal seaweeds are exposed to high doses of UV light for extended periods of time during low tide and possibly a second stress in the form of N limitation during summer.

Temperature effects on active uptake and general cell metabolism approximate a Q_{10} value of 2 (i.e. a 10°C increase in temperature leads to a doubling of the rate). For a purely physical process, such as diffusion, temperature has less effect, and the Q_{10} value is 1.0–1.2. Several studies have indicated that the effect of temperature on ion uptake is ion specific and is

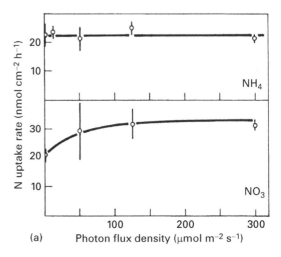

(a)

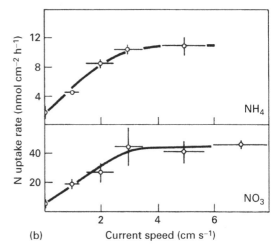

(b)

Fig. 6.7 Nitrate and ammonium uptake rates for *Macrocystis pyrifera* mature blade discs as functions of (a) irradiance and (b) current speed. (From Wheeler 1982, with permission of Walter de Gruyter and Co.)

dependent on the algal species (Raven and Geider 1988). For example, a marked decrease in nitrate-uptake rate was observed for *Saccharina longicruris*, but not for *Fucus spiralis* (Topinka 1978), as the temperature was lowered. This apparent discrepancy may be explained by the fact that these two species have different temperature optima for nutrient uptake. Because temperatures can fluctuate daily, it would be interesting to study how rapidly uptake rates can respond to changes in temperature. An important dual stress that emersed seaweeds must tolerate is the combination of desiccation and high summer temperatures, especially high intertidal seaweeds (Dudgeon *et al.* 1995).

Water motion is another factor that is important in the movement of ions to the surface of the thallus and in helping to decrease the thickness of the boundary layers around the thallus (Hurd 2000) (see sec. 8.1.3). In areas of low turbulence, or in unstirred laboratory cultures, transport across the concentration boundary layer is limited by the rate of diffusion, and not necessarily by the concentration of the nutrient in the medium (Hurd 2000). The reduction in uptake rate is pronounced in nutrient-limited waters, and especially if the thallus is thick rather than filamentous (hence a lower SA:V ratio). The giant kelp *Macrocystis* encountered transport limitation for carbon and nitrogen when the current over the fronds was <3-6 cm s^{-1} (Fig. 6.7b). Water velocity increased ammonium uptake in summer when the rhodophyte *Adamsiella* was N limited, but surprisingly there was no influence on nitrate uptake (Kregting *et al.* 2008a).

Tolerance to desiccation is related to the shore position of species in the intertidal zone. Exposure to air during a low tide frequently results in loss of water from the thallus, depending on the season. Mild desiccation (10–30% water loss) enhances short-term (10–30 min) nutrient-uptake rates when they are submerged in nutrient-saturated seawater (Fig. 6.8) (Thomas and Turpin 1980; Thomas *et al.* 1987b). This enhanced uptake response occurred when growth was nutrient limited and the thallus had been exposed to repeated periodic desiccation for several weeks. The relative degree of enhancement of the nitrogen uptake rate, the percentage desiccation that produced maximal uptake rates, and the tolerance to higher degrees of desiccation were positively related to tidal height for five intertidal macroalgae (Thomas *et al.* 1987b). Low-intertidal species such as *Gracilaria pacifica* showed no enhancement of nitrogen uptake following desiccation. In contrast, two high intertidal species, *Pelvetiopsis limitata* and *Fucus distichus*, showed a two-fold enhancement of nitrate and ammonium uptake

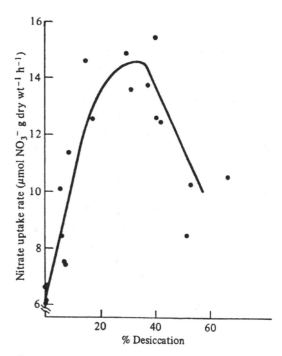

Fig. 6.8 Nitrate-uptake rate as a function of percent desiccation of *Fucus distichus*. Plants were desiccated to different degrees and then placed in medium containing 30 μM NO$_3$ for 30 min. (From Thomas and Turpin 1980, with permission from Walter de Gruyter and Co.)

following desiccation >30% and continued uptake even following severe desiccation (50–60% of original fresh weight); the period of enhanced nitrate uptake was much longer (20–60 min) than that for enhanced ammonium uptake (10–30 min). Using *Gracilaria pacifica*, Thomas *et al.* (1987a) showed that when seaweeds from the low intertidal (1.0 m) were transplanted to high intertidal sites (1.8 m), they developed enhanced nitrogen uptake rates following desiccation, whereas high intertidal seaweeds transplanted to the low intertidal did not maintain their enhanced nitrogen uptake abilities. This study also demonstrated that both intraspecific and interspecific adaptations were dependent on intertidal height. Further studies on the influence of intertidal zonation revealed that species growing at the highest shore position generally had saturable and higher nitrate and urea uptake compared to low-shore species and ammonium uptake did not saturate even at >100 μM ammonium additions (Philips and Hurd 2004). Species growing highest on the shore had higher total tissue N and larger internal pools with NH_4 pools > NO_3 (Philips and Hurd 2003). Hurd and Dring (1991) found that the degree of tolerance to desiccation (i.e. how quickly species recovered their maximal phosphate uptake rates after losing 50% of their water through desiccation) increased with increasing shore height of the fucoid algae.

Chemical factors. Chemical factors such as the concentration of the nutrient being taken up and the ionic or molecular form of the element will affect uptake rates. For example, ammonium is often taken up preferentially and more rapidly than nitrate, urea, or amino acids in many seaweeds (DeBoer 1981). Uptake rates can also be influenced by the concentrations of other ions in the medium. Ammonium may inhibit nitrate uptake by as much as 50% in many seaweeds (DeBoer 1981) and microalgae (Dortch 1990). In contrast, *Gelidium*, *Macrocystis*, and *Saccharina* take up nitrate and ammonium at equal rates when they are supplied simultaneously (Haines and Wheeler 1978; Harrison *et al.* 1986; Ahn *et al.* 1998). Surprisingly, however, the kelp *Nereocystis* preferred nitrate over ammonium (Ahn *et al.* 1998). There is a seasonal influence on the preference of nitrogen forms. All three N resources were utilized simultaneously in winter in the order of $NH_4 > NO_3 >$ urea, while in summer the order was $NH_4 = NO_3 >$ urea (Philips and Hurd 2003). Intracellular ion concentrations in the cytoplasm and vacuoles will also influence uptake rates. Wheeler and Srivastava (1984) found that the nitrate-uptake rate was inversely proportional to the intracellular nitrate concentration in *Macrocystis pyrifera* (previously *M. integrifolia*). Pedersen (1994) followed the decrease in tissue N in *Ulva lactuca* during 25 days of N starvation and found that ammonium uptake peaked at low tissue N, but as tissue N declined further, ammonium uptake decreased, suggesting that there is an optimal N stress condition for the maximum enhanced uptake (Fig. 6.9).

Biological factors. Biological factors that influence uptake rates include the surface-area: volume (SA:V) ratio, hair formation, the type of tissue, the age of the

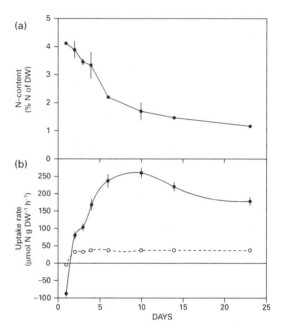

Fig. 6.9 (a) Decline in tissue N content, and (b) short-term ("surge") uptake (filled circles) and longer term uptake ("internally controlled") uptake (open circles) during 24 days of N starvation of *Ulva lactuca*. Surge uptake peaked after 10 days of N starvation. (From Pedersen 1994, with permission from the Phycological Society of America.)

seaweed, its nutritional history, and inter-seaweed variability. Uptake determinations on whole thalli are preferable for ecological measurements, but if a thallus is too large, then a portion must be used. Cutting the thallus to produce tissue segments may alter the uptake rate, either through a wounding response and increased respiration or by elimination of the translocation system or active-sink region. A comparison of excised sections of *Macrocystis* tissue with whole blades showed a marked decrease in uptake rate by the cut sections (Wheeler 1980), suggesting that caution is needed when extrapolating from cut discs to whole seaweed uptake rates.

Many seaweeds are perennials and therefore their natural populations often consist of different age classes. The nitrate and ammonium uptake rates for three age classes of *Saccharina groenlandica* (previously *Laminaria groenlandica*) decreased with increased age; the uptake rate (per gram of dry weight) for first-year seaweeds was three-times greater than that for second- and third-year seaweeds (Harrison *et al.* 1986). Distinct differences in nutrient-uptake abilities also occur between early life-history stages and mature thalli of the same species. Ammonium and nitrate-uptake rates for *Fucus distichus* germlings were 8 and 30 times higher, respectively, than those for the mature thalli (Thomas *et al.* 1985). The germlings showed saturable uptake kinetics, but the mature thalli did not, indicating that the uptake kinetics of germlings are more like the kinetics of phytoplankton than mature seaweeds. The presence of ammonium inhibited nitrate uptake in mature seaweeds, but not in germlings. These characteristics of nutrient uptake indicate that the germlings are better adapted for procurement of the limiting nutrient than are mature seaweeds. This sizeable difference in uptake abilities probably is due to the large proportion of storage and support tissues in the adult seaweeds, tissues that do not actively require nitrogen. It is also interesting that older fronds and stipes retain some ability to take up ammonium, but entirely lose their capacity to take up nitrate. Young tissue, which is metabolically active, appears to need both nitrate and ammonium to meet its greater nitrogen requirements.

Different parts of a seaweed may also take up nutrients at different rates. The stipe has the lowest nitrogen-uptake rate in *Fucus spiralis*, in keeping with its low metabolic activity (Topinka 1978). Davison and Stewart (1983) found that mature regions of *Laminaria digitata* took up and translocated nitrogen to meristematic sink regions. They demonstrated that 70% of the nitrogen demand by the intercalary meristem in this species was supplied by transport of nitrogen assimilated by the mature blade, probably in the form of amino acids. In *Macrocystis*, Gerard (1982a) determined that mature blades deeper in the water column may serve an important role by taking up nitrogen from the relatively nutrient-rich deep water and translocating it (as amino acids) to the growing blades in the nutrient-poor surface water. Members of the Caulerpales inhabit the soft bottom of oligotrophic tropical waters and they have well-developed rhizoidal holdfasts that grow into the sediments. Williams and Fisher

(1985) found that *Caulerpa cupressoides* not only is adapted to use ammonium in interstitial waters, but also meets virtually all of its nitrogen requirements from the sediments. Significant translocation of ^{15}N label occurred from the rhizoid to the blade, indicating that ammonium taken up by the rhizoids in the sediments was available for light-driven organic nitrogen production in the blades (Williams 1984).

The uptake rate of ammonium is a function of the nutritional past history of the seaweed that can be evaluated by the tissue C/N ratio. When *Gracilaria foliifera* and *Agardhiella subulata* were grown under conditions in which they were N limited (i.e. the molar C/N ratio in the thalli was >10), the seaweeds showed higher rates of ammonium uptake at a given ammonium concentration than did seaweeds that were not N limited (C/N ratio <10) (D'Elia and DeBoer 1978).

The SA:V ratio and the shape of the thallus are also important factors influencing nutrient-uptake rates since nutrient uptake occurs through the surface membrane of the thallus. Therefore uptake rates have been correlated to the SA:V ratio. Rosenberg *et al.* (1984) found that for both nitrate and ammonium, the maximal uptake rate (V_{max}) and the initial slope of the uptake substrate concentration curve, V_{max}:K_s which is an index of the affinity for the limiting nutrient, were positively related to the SA:V ratio for four intertidal seaweeds. Unfortunately, their choice of seaweeds, *Ulva* (a short-lived, opportunistic alga with a high SA:V ratio) and *Codium* (a long-lived, late successional alga with a low SA:V ratio), probably overemphasized the importance of this ratio. In a more extensive study of 17 macroalgae, Wallentinus (1984) measured higher uptake rates for nitrate, ammonium, and phosphate in short-lived, opportunistic, filamentous, delicately branched or monostromatic forms *(Cladophora glomerata, Ulva procera,* etc.) that had high SA:V ratios. The lowest uptake rates occurred among late successional, long-lived, coarse species with low SA:V ratios *(Fucus vesiculosus, Phyllophora truncata,* etc.). An increase in nutrient uptake has also been found to parallel an increase in the SA:V ratio associated with an increase in the number of hairs protruding from the thallus of *Ceramium virgatum* (DeBoer and Whoriskey 1983) (see sec. 6.8.1). Hairs increased the thallus surface area by 180% in *Gracilaria pacifica* and 50% in *Gelidium vagum* (Oates and Cole 1994). Taylor *et al.* (1999) found that when expressed per unit biomass, the maximum rate of ammonium assimilation (V_i) together with surge uptake (V_s) and storage capacity was positively correlated with SA:V ratio for nine species. In contrast when expressed per unit surface area, these parameters were independent of SA:V, suggesting that ammonium metabolism is confined mainly to the outermost layer of cells (Taylor *et al.* 1998).

When uptake rates of phytoplankton and macroalgae are compared on a SA:V basis, phytoplankton always outcompete macroalgae. Hein *et al.* (1995) examined size-dependent (SA:V ratio) N uptake for a large number of micro- and macroalgae (Fig. 6.10). The SA:V ratio and the average uptake rate of microalgae was about 10-fold higher than macroalgae. The tissue N is considerably higher in microalgae (1–14% N of dry wt) than macroalgae (0.4–4.4% of dry wt) (Duarte 1992). Hence, microalgae have significantly higher N requirements because of their high growth rate and high N content and often dominate in high N environments. Therefore, their V_{max} and $V_{max}:K_s$ values were higher and the K_s was lower than in macroalgae and hence all sections of the V versus S curve are higher for microalgae than macroalgae (Fig. 6.11). Because of the slower growth rate (i.e. decreased N demand) of macroalgae versus microalgae, their stored nutrients may last longer in N-limited environments, and therefore macroalgae do better in low-nutrient environments. Rees (2007) examined the comparison between micro- and macroalgae and noted that there is a 10^{17}-fold range in dry weight (DW) and a 10^4-fold range in SA:V ratio. One of the difficulties in comparing these two groups of primary producers is obtaining scaling data with common units. Metabolic rates for macroalgae are commonly expressed as DW or fresh weight, while phytoplankton rates are either per cell (i.e. a transport rate) or as a N-specific rate (units of time^{-1}). However, it is possible to express phytoplankton rates per g DW (Hein *et al.* 1995). Surface-related metabolic processes such as uptake may constrain growth, especially in multicellular seaweeds with a low SA:V ratio. Rates of NO_3 and

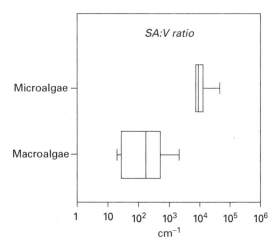

Fig. 6.10 Distribution of relative surface area:volume (SA:V) ratio for micro- and macroalgae, showing the 25 and 75% quartiles, and median. Bars represent maximum and minimum values. (From Hein *et al.* 1995, with permission from Inter-Research Science Center.)

NH_4 uptake rate per unit surface area at low N concentrations of ~1 μM were remarkably similar for both micro- and macroalgae (9–12 nmol NH_4 cm^{-2} h^{-1} and 8–9 nmol NO_3 cm^{-2} h^{-1}) regardless of N source (Rees 2007). The boundary layer may have an adverse affect on uptake rates with a more pronounced reduction in uptake rates for larger organisms such as seaweeds with a low SA:V. Therefore, water motion/stirring is more important for seaweeds than phytoplankton.

Another way to compare nutrient demand by various seaweeds is to compare the ratio of maximum nutrient uptake rate to the uptake rate at the maximum ambient concentration of a nutrient. Rees (2003) termed this ratio the "safety factor" and it provides a simple estimate of the amount of surplus capacity of a nutrient uptake system. (Note that the term "safety factor" is also used in seaweed biomechanics but the meaning is different – see sec. 8.3.1). Hence, a low safety factor means that the uptake system is operating close to the maximum rate. Rees (2003) found that safety factors were low for nitrate and phosphate uptake by seaweeds and ammonium uptake by phytoplankton, while seaweeds had a high safety factor for

(a)

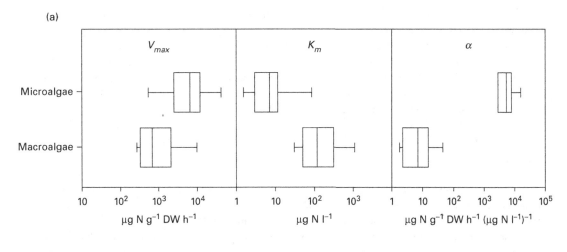

(b)

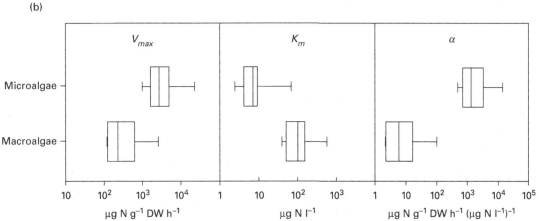

Fig. 6.11 Distribution of: (a) ammonium, and (b) nitrate-uptake kinetic parameters for micro- and macroalgae showing the 25 and 75% quartiles and median. Bars represent maximum and minimum values. (From Hein *et al.* 1995, with permission from Inter-Research Science Center.)

ammonium uptake (see Appendix 1 in Rees 2003). Furthermore, there was a three-fold higher capacity for ammonium uptake compared to nitrate uptake (i.e. the N uptake rate at 5 μM NO_3 = uptake rate at 1.5 μM NH_4). Comparing nutrient-uptake rates from an environmental perspective is also useful. Philips and Hurd (2004) have suggested comparing uptake of the three main N sources (NO_3, NH_4, and urea) at a standard concentration of 2 μM. The advantage of this suggestion is that it clearly demonstrates the importance of NH_4 uptake (Rees 2003).

6.5 Uptake, assimilation, incorporation, and metabolic roles

Consistent definitions are recommended for the processes of nutrient acquisition (Berges 1997). Uptake refers to transport across the plasmalemma. Assimilation refers to a sequence of reactions in which inorganic ions are incorporated into small organic molecules such as amino acids and may be a measureable index of growth rate. Incorporation includes the processes where small organic molecules are combined into macromolecules such as proteins and nucleic acids (Berges 1997; Mulholland and Lomas 2008). Nitrogen uptake and assimilation in macroalgae have been reviewed by Hanisak (1983) and Gordillo (2012) and by Mulholland and Lomas (2008) for phytoplankton. The key metabolic roles of the various essential elements are summarized in Table 6.1 and are discussed in the following sections.

6.5.1 Nitrogen

Uptake. This section will cover uptake rates and mechanisms, and values for uptake kinetic parameters for nitrate, nitrite, ammonium, urea, and amino acids. Because no macroalgae are known to fix N_2 (except those containing symbionts), the inorganic nitrogen sources for uptake are nitrate, nitrite, and ammonium. Most studies focus on NO_3 and NH_4 uptake. Only a few studies have examined urea uptake (Phillips and Hurd 2004) and there is almost no work on nitrite. The uptake rate for ammonium is generally higher than for nitrate at ecologically relevant concentrations. Ammonium does not seem to be toxic to most seaweeds since ammonium additions of 100 to 500 μM yield normal ammonium uptake rates. It is important to note that most of the ammonium uptake measurements have been at a range of concentrations that are up to 10-fold higher than natural concentrations of 0 to 5 μM. In some seaweeds such as *Scytothamnus australis*, ammonium shows saturation kinetics, suggesting active transport in the summer, but linear uptake in winter (Phillips and Hurd 2004). For other species, the ammonium uptake increases linearly with ammonium concentration (Taylor *et al.* 1998; Phillips and Hurd 2004) (Fig. 6.12). This linear increase in uptake rate at high ammonium concentrations may represent a second transport mechanism, perhaps diffusion of ammonia (NH_3) via ion channels. Linear ammonium uptake rates over a range of ammonium concentrations up to 100 μM have also been reported for *Fucus distichus* (Thomas *et al.* 1985), for the kelp, *Saccharina groenlandica* (Harrison *et al.* 1986), for red algae *Gracilaria pacifica* (Thomas *et al.* 1987a) and *Stictosiphonia arbuscula* (Philips and Hurd 2004), and for chlorophytes such as *Ulva* and *Chaetomorpha* (Lavery

and McComb 1991b) and five seaweeds from New Zealand (Taylor *et al.* 1998). While uptake (the V_s phase) may not saturate, assimilation (the V_i phase) generally saturates with increasing NH_4 concentrations yielding K_s concentrations of 20–40 μM NH_4 for the V_i phase (Pedersen and Borum 1997; Taylor *et al.* 1998, 1999).

Uptake of nitrate generally exhibits saturation kinetics (DeBoer 1981; Phillips and Hurd 2004; Rees *et al.* 2007). However, there have been several reports of a linear increase in nitrate uptake with increasing concentrations for *Saccharina groenlandica* (Harrison *et al.* 1986), *Chaetomorpha linum* (Lavery and McComb 1991b), and *Gracilaria pacifica* (Thomas *et al.* 1987a). Concentrations in intracellular (cytoplasmic and/or vacuolar) pools that are 10^3 times greater than that in the surrounding seawater, suggest that uptake is active, but the plasmalemma transport mechanism is not known. Ammonium generally inhibits nitrate uptake in phytoplankton, but for several macroalgae such as *Pterocladiella capillacea* and *Xiphophora chondrophylla*, 20 μM NH_4 did not inhibit NO_3 uptake (Rees *et al.* 2007). However, the habitat and the nutritional status may determine whether NH_4 inhibition occurs. *Ulva intestinalis* from a low N environment showed inhibition of NO_3 uptake by NH_4, but *U. intestinalis* from a high N environment did not exhibit NH_4 inhibition of NO_3 uptake. Therefore, Rees *et al.* (2007) suggested that there are at least two nitrate transport systems, a constitutive transporter that is insensitive to NH_4 and another transporter that is sensitive to NH_4 inhibition and down-regulated by NH_4. In addition to inhibiting NO_3 uptake, ammonium can also inactivate nitrate reductase (NR) activity.

Nitrite uptake has been studied in only a few cases. In *Codium fragile*, the maximal uptake rate for nitrite was found to be similar to that for nitrate, but lower than that for ammonium (Hanisak and Harlin 1978; Topinka 1978). Brinkhuis *et al.* (1989) observed rapid nitrite uptake lasting only a few minutes, followed by a release of some nitrite back to the medium, and then a sustained uptake rate for several hours. Nitrite uptake was linear at concentrations that were much higher than environmental concentrations in *Palmaria palmata* (Martinez and Rico 2004).

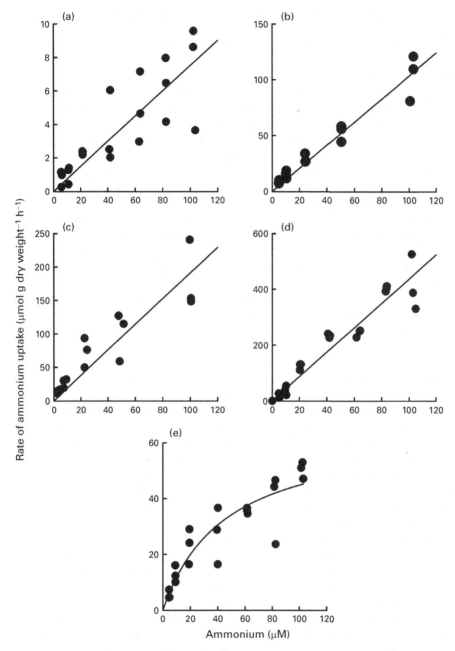

Fig. 6.12 Effect of ammonium concentration on the rate of ammonium uptake by NE New Zealand seaweeds. (a) *Xiphophora chondrophylla*, (b) *Ulva* sp., (c) *Pyropia* sp., (d) *Ulva* sp., and (e) *Pterocladia capillacea*. (From Taylor *et al.* 1998, with permission from the Inter-Research Science Center.)

There have been several reports of growth on dissolved organic nitrogen compounds, such as urea. Urea is an excellent source of nitrogen for many seaweeds (Probyn and Chapman 1982; Thomas et al. 1985), but poor growth has been reported for other species (DeBoer et al. 1978). Phillips and Hurd (2004) found that urea uptake was lower than nitrate and ammonium uptake for four seaweeds, and that urea uptake was higher in summer (e.g. 1-5 μmol g_{dw}^{-1} h^{-1} and K_s = 8-50 μM) than winter, but still 5-10 times slower than NH_4 or NO_3. Assessment of urea as a source of nitrogen for seaweeds in bacterial contaminated culture studies is complicated because the bacteria may break down the urea to ammonium, which can be utilized by the seaweed. Urea uptake is generally considered to be active because it generally exhibits saturable uptake kinetics (Phillips and Hurd 2004), although it should have a high rate of diffusion because it is a small, uncharged molecule.

There has been only one definitive study on the uptake kinetics of amino acids. Schmitz and Riffarth (1980) examined the uptake of 17 different amino acids by the filamentous brown alga *Hincksia mitchelliae* and found the highest uptake rate for L-leucine. The uptake rate for leucine was observed to be light-independent, with a very low affinity $(K_s = 30-120 \mu M)$ and a low maximal uptake rate (0.03 μmol g_{dw}^{-1} h^{-1}). They suggested that exogenous amino acids contribute <5% of the nitrogen demand by this alga.

As discussed above, uptake parameters vary with species and with a wide range of factors. Examples of values obtained for nitrogen-uptake kinetic parameters are summarized in Table 6.3 (see also Appendix 1 in Rees (2003) and Table 1 in Phillips and Hurd (2004) for additional values). The K_s values are similar for nitrate and ammonium and generally high (2-40 μM), up to 10 times higher than for phytoplankton (Fig. 6.11; Hein et al. 1995). The V_{max} values range from 4 to >200 μmol g_{dw}^{-1} h^{-1} for both nitrate and ammonium. Because specific uptake rates (where the rate is normalized to the nitrogen content of the thallus yielding the unit h^{-1}) generally are not reported for seaweeds, comparisons with phytoplankton are seldom possible, although Hein et al. (1995) made the comparison by transforming known phytoplankton uptake rates to the

unit μmol g_{dw}^{-1} h^{-1}. It is important to recall from sec. 6.4.1 that V_{max} frequently varies with the length of the incubation period (Harrison et al. 1989). Therefore, V_{max} values for ammonium measured over 0-10 min may be much higher than values obtained over 0-60 min, In contrast, there may be a lag in nitrate uptake by N-limited algae and nitrate uptake may increase only after 1-2 h, as the added nitrate induces more rapid uptake (Thomas and Harrison 1987).

Assimilation. Nitrate must be reduced intracellularly to nitrite and then to ammonium before incorporation into amino acids. The reduction of nitrate to nitrite is generally assumed to be the rate-limiting step for N incorporation into amino acids (Syrett 1981; Berges and Mulholland 2008). The reduction occurs in two main steps. The first step is the reduction of nitrate to nitrite, catalyzed by nitrate reductase (Berges 1997). Nitrite is transported to the plastids for reduction to ammonium and reduced to NH_4 by nitrite reductase (NiR) (Fig. 6.14). NiR has an iron prosthetic group and uses ferredoxin as a cofactor to supply the electrons. NR and NiR are closely linked which presumably prevents the build-up of intracellular pools of nitrite which may become toxic (Berges and Mulholland 2008). In most cases, NiR activity is greater than NR which suggests that the rate-limiting step in the reduction of nitrate to ammonium is nitrate reduction by NR. The NiR molecule contains five times more iron than NR and therefore iron limitation reduces NiR activity more than NR (Milligan and Harrison 2000). The assay procedure for NiR is more complicated that NR and this may explain why there have been almost no studies on NiR activities in macroalgae, but it certainly deserves attention.

Nitrate reductase has been isolated and purified from many species of microalgae (Berges 1997; Berges and Mulholland 2008), and it is also present in seaweeds (Thomas and Harrison 1988; Brinkhuis et al. 1989; Young et al. 2007, 2009). NR is a relatively large molecule that has three functional domains, such as molybdopterin, cytochrome b_5 and flavin adenine dinucleotide (FAD). The electron donor usually is NADH, but there are some microalgae in which the donor may be NADPH as well (Berges and Mulholland 2008). The enzyme is thought to occur in the

Table 6.3 Examples of nitrogen-uptake kinetic "constants" for seaweeds.

Alga	Temperature (°C)	NO_3^- $K_s \pm SE$ (µM)	V_{max} (µmol $g_{dw}^{-1}h^{-1}$)	NH_4^+ $K_s \pm SE$ (µM)	V_{max} (µmol $g_{dw}^{-1}h^{-1}$)	References
Phaeophyceae:						
Fucus spiralis	5	6.6 ± 0.9		6.4 ± 2.0		Topinka (1978)
	10	6.7 ± 0.8		5.4 ± 2.0		
	15	7.8 ± 1.4		9.6 ± 2.6		
	15[a]	12.8 ± 3.5		5.8 ± 1.8		
Fucus distichus	15	1–5	25	3–5	60	Thomas *et al.* (1985)
Saccharina longicruris	15[b]	4.1	9.6			Harlin and Craigie (1978)
	10[c]	5.9	7.0			
Macrocystis pyrifera	16	13.1 ± 1.6	30.5	5.3[f] ± 1.0	23.8[f]	Haines and Wheeler (1978)
	6–9			50[g]	23.6[g]	Wheeler (1979)
Chordaria flagelliformis	11	5.9 ± 1.2	8.8		20	Probyn (1984)
Saccharina groenlandica	13		20[d]		6	Harrison *et al.* (1986)
			6[e]			
Rhodophyceae:						
Gracilaria foliifera	20	2.5 ± 0.5	9.7	1.6[f]	23.8[f]	D'Elia and DeBoer (1978)
Agardhiella subulata	20	2.4 ± 0.3	11.7	3.9[f]	15.9[f]	D'Elia and DeBoer (1978)
Hypnea musciformis	26	4.9 ± 3.9	28.5	16.6 ± 1.8		Haines and Wheeler (1978)
Chondrus crispus	17			35.5	62	Amat and Braud (1990)
Pyropia perforata	12		15		40–50	Thomas and Harrison (1985)
Chlorophyceae:						
Chaetomorpha linum	25		60		230	Lavery and McComb (1991b)
Ulva prolifera	12–14	2.3–13.3	75–169	9.3–13.4	39–188	O'Brien and Wheeler (1987)
Ulva spp.	15	16.6	129.4			Harlin (1978)
Codium fragile	6	1.9 ± 0.5	2.8	1.5 ± 0.2	13.0	Hanisak and Harlin (1978)
	24	7.6 ± 0.6	9.6	1.4 ± 0.2	28.0	
Ulva lactuca	20			40.7 ± 8.5	138 ± 78	Fujita (1985)
Ulva rigida	25	20–33	60–90		60	Lavery and McComb (1991b)

[a] Dark. [b] Summer tissue. [c] Winter tissue. [d] First-year plants. [e] Second- and third-year plants.
[f] Mechanism-1 uptake (high-affinity system). [g] Uptake of methylamine, an analog of NH_4^+.

cytoplasm, although there is some evidence that it may be associated with plastid membranes or plasmalemma (Berges 1997; Berges and Mulholland 2008). Much of the work on NR has been done on microalgal chlorophytes and there is a need for further work on macrophytes, particularly investigations on NR genes (Falcão *et al.* 2010), nitrate transport proteins, nitrite reductase, and downstream enzymes involved in N incorporation (Berges 1997; Berges and Mulholland 2008).

The *in vitro* assay for NR is widely used now. Techniques for measuring NR activity *in vitro* for microalgae (Berges and Harrison 1995) have been modified for seaweeds (Thomas and Harrison 1988; Brinkhuis *et al.* 1989; Hurd *et al.* 1995; Young *et al.* 2005). There are two issues to consider in enzymatic assays: (1) The enzyme should be completely extracted and in good condition. Measurement of NR in seaweeds is often difficult, because tough rubbery thalli are difficult to grind, but freezing the thallus in liquid nitrogen before homogenization generally resolves this problem (Hurd *et al.* 1995: Young *et al.* 2005). The presence of phenolics (Ilvessalo and Tuomi 1989; Berges 1997) may inactivate the enzyme and interfere with the protein determinations that are used to normalize NR activity. The problems are particularly pronounced in the Phaeophyta, but they have been overcome by adding polyvinylpyrrolidone or bovine serum albumin to bind the phenolics (Thomas and Harrison 1988; Hurd *et al.* 1995), or by purification steps to remove alginate, which also interferes with enzyme activity. Proteases may also attack the enzyme, but this can be stopped by adding the protease inhibitor chyostatin (Berges and Harrison 1995: Berges and Mulholland 2008). Young *et al.* (2005) found that extraction efficiency was significantly improved with 1% Triton X-100, rather than the 0.1% used by Hurd *et al.* (1995). (2) Optimal assay conditions must be used and these are often species specific (Young *et al.* 2005). For example, when the optimum pH for NR activity in *Pyropia* was used for *Ulva*, NR activity was reduced by 50% (Thomas and Harrison 1988).

In vivo NR assays (or the "*in situ*" assay) have some drawbacks. They involve incubating the algae in medium with a very high nitrate concentration (20 mM) and *n*-propanol (to permeabilize the cell membrane) in the dark and measuring the appearance of nitrite in the medium. Some researchers (Davison *et al.* 1984; Corzo and Niell 1994) have recommended the "*in vivo*" assay because it preserves all systems in their natural state, and thus it should give a more realistic estimate of activity. However, because *n*-propanol causes disruption of membranes and lipids, this is not truly an "*in vivo*" assay. Thomas and Harrison (1988) did further studies and detected a major drawback to the *in vivo* assay because activity varied with the length of time the seaweed had been incubated in the *n*-propanol medium, indicating that permeability was limiting the reaction rate, not NR activity. There also are potential problems of diffusion of nitrate into the tissue, especially if too much n-propanol is used (causes bleaching of tissues). Brinkhuis *et al.* (1989) warned that nitrite, which is released into the medium, might be taken back up by the thallus. Thus, the *in vitro* rather than the *in vivo* NR assay is the method of choice (Hurd *et al.* 1995; Berges and Mulholland 2008). Young *et al.* (2005; see their Table 1) listed maximum NR rates for 22 species from the Phaeophyta, Rhodophyta, and Chlorophyta and found that NR rates varied by two orders of magnitude from 2 to >200 nmol NO_3 min^{-1} g^{-1}_{FW}, where FW = frozen weight. Other scaling or normalization parameters for NR activity are protein, tissue N and C, and chlorophyll (Berges and Mulholland 2008).

NR has been used to explore the metabolic activities of various tissues. In the kelp *Laminaria digitata*, the highest NR activities occurred in the mature blades, and activities declined toward the basal meristematic region (Davison and Stewart 1984). These observations are consistent with the suggestion of maintenance of meristematic growth by internal transport of organic nitrogen from the mature blade to the meristem (the sink). The higher NR activity in the outer, highly pigmented meristoderm tissue was probably associated with the high level of photosynthetic activity that occurs there. Activities in the stipe and holdfast were low.

NR activity is regulated by many factors such as light/dark, N status, ambient N sources, temperature, type of seaweed tissue, seasons, etc. Chow and de Oliveira (2008) found that the addition of nitrate rapidly (2 min) induced NR activity, suggesting fast post-translational regulation in the rhodophyte *Gracilaria chilensis*. In contrast, a nitrate addition to starved algae stimulated rapid uptake without concomitant induction of NR activity, suggesting that uptake and assimilation are regulated differently and may respond on different timescales. The addition of ammonium or urea stimulated NR activity after 24 h and then decreased after 72 h. During the dark, NR activity was low, but a 15 min light pulse induced NR activity. They concluded that post-translational regulation by phosphorylation and dephosphorylation is rapid, while regulation of RNA synthesis coupled to *de novo* NR protein synthesis is slower. Young *et al.* (2007) found that NR in *Laminaria digitata* showed a strong diel pattern (low at night and high midday) and the pattern was not controlled by a circadian rhythm, while three *Fucus* species showed no diel pattern. An addition of ammonium suppressed NR activity by >80% in *Laminaria*, but not in *Fucus* (Young *et al.* 2009). Therefore, *Fucus* is relatively unique since NR activity is not regulated by either light or ammonium and it was suggested that *Fucus* may have different isoenzymes compared to *Laminaria*. In nature, *Fucus* species occur higher in the intertidal zone and experience more prolonged desiccation and more control of environmental factors by tidal cycles than *Laminaria* that is submerged most of the time.

An alternative to reducing nitrate to nitrite in the cytoplasm is to store nitrate in the vacuole (Fig. 6.14). Analysis of intracellular nitrate pools indicates that substantial amounts of nitrate may accumulate in the cytoplasm/vacuole in some intertidal macroalgae, especially when NR is relatively inactive (Thomas and Harrison 1985; Naldi and Wheeler 1999). Several studies have followed the time sequence of N uptake, assimilation, and incorporation into storage pools after a macroalga has been given a pulse of N. A 30 μM NO_3 pulse was given to N-depleted *Ulva* sp. and 6 h later, NO_3 uptake decreased due to internal nitrate pool filling (Lartigue and Sherman 2005). Since the peak NR activity was 11-fold lower than the NO_3 uptake rate, and the pools of NO_2, NH_4, and free amino acids remained constant while the NO_3 pool increased

(Fig. 6.6), suggesting that NR activity was the rate-limiting step in assimilating the NO_3 pulse (Lartigue and Sherman 2005). Naldi and Wheeler (1999) added saturating NO_3 or NH_4 concentrations to two N-sufficient macroalgae for 9 days in order to determine where the N was stored. Protein-N ranged from 700 to 2300 μmol N g dw^{-1} (43–66% of TN) and contributed the most to the total nitrogen (TN) increase (41–89%). The free amino acid (FAA) pool accounted for 4–17% of TN (70–600 μmol N g dw^{-1}). In *Ulva fenestrata* NO_3 represents a transient storage pool of up to 200 μmol N g dw^{-1} (7% of TN) and was larger than the FAA pool. Ammonium pools were always <3% of TN. In contrast, *Gracilaria pacifica* has a small NO_3 pool, while the phycoerythrin N pool was ~6% of TN (Naldi and Wheeler 1999). Chlorophyll was not a significant N storage pool. In a similar study, McGlathery *et al.* (1996) followed N pools in the chlorophyte *Chaetomorpha linum* in N sufficiency (TN ~4.6% of dry wt) and N deficiency (TN ~1.2% of dry wt). The nitrate pool was larger than ammonium, while the residual organic N pool (a proxy for FAA) was up to 50% of TN and increased faster than the protein pool. N storage allowed the alga to grow for two generations (14 days). Surprisingly the protein pool did not decrease much, suggesting that the protein pool contributes little to N storage that is used during periods of N starvation.

On the basis of earlier studies on phytoplankton, NR had been thought to provide a good index of nitrate uptake. However, more recent work has shown a poor correlation between uptake and NR enzyme activity that may underestimate uptake by a factor of 10 (Berges and Mulholland 2008). Storage of nitrate in the vacuole, or enzyme instability caused by proteases, may account for this poor correlation. It is important to realize that enzyme activity can serve only as a measure of the maximum capacity of the step, not as a measure of an instantaneous rate.

Assimilation is the incorporation of NO_3^- and NH_4^+ into amino acids and proteins and is a good index of growth rate (Harrison *et al.* 1989; Pedersen 1994; Rees *et al.* 1998). While uptake may exhibit linear kinetics, assimilation may show saturation (Fig. 6.13) but with high K_s values, especially for NH_4 (Harrison *et al.* 1989; Taylor and Rees 1999). The difference between K_s

values for uptake and assimilation can be shown by a notation such as $K_{s\text{-}Vs}$ and $K_{s\text{-}Vi}$. Assimilation can be easily measured during the V_i phase of uptake (usually after 60 min or longer) and integrates several factors

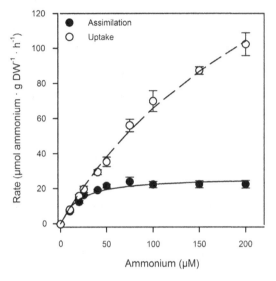

Fig. 6.13 Effect of ammonium concentration on rates of ammonium uptake and ammonium assimilation by *Ulva* sp. (From Taylor and Rees 1999, with permission from the Phycological Society of America.)

such as enzyme activities and the availability of carbon skeletons, ATP and reductant (Turpin and Huppe 1994). Glutamine synthetase (GS) is the primary NH_4 assimilation enzyme and Taylor and Rees (1999) found that rates of GS activity matched maximum rates of NH_4 assimilation. The assimilation rates versus NH_4 concentration shows saturation and K_s concentrations for *in vivo* assimilation range from 20 to 40 μM NH_4 for several macroalgae (Fig. 6.13; Taylor and Rees 1999; Gevaert *et al.* 2007). Therefore, ammonium assimilation will never approach maximum rates in natural populations and will be limited by ambient NH_4 concentrations (Taylor and Rees 1999). At an external NH_4 concentration of 40 μM, the internal NH_4 pool was estimated to be ~11 μmol $(g\ DW)^{-1}$ or ~7.5 mM for *Ulva*. Assuming passive uptake of NH_3, the intracellular pH that would account for this accumulation is ~6, suggesting that most of the ammonium is in an acidic compartment such as the vacuole and not in the cytosol or plastid where GS is located (Taylor and Rees 1999). The high K_m for GS in *Ulva* suggests that internal pooling of ammonium is necessary for GS to operate efficiently.

The first product of ammonium incorporation is glutamine catalyzed by GS, and the second reaction forms glutamate which is catalyzed by glutamine-oxoglutarate

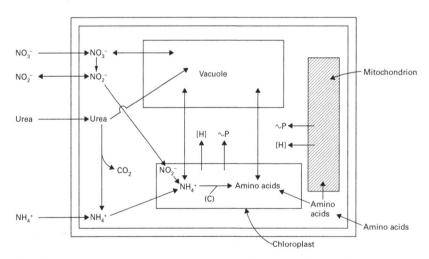

Fig 6.14 Schematic of nitrogen uptake and assimilation in an algal cell. (From Syrett 1981, reproduced by permission of the Ministry of Supply and Services, Canada.)

aminotransferase (GOGAT). Recent labeling and enzyme-activity studies have shown that the GS-GOGAT pathway is the primary route for ammonium incorporation into amino acids in the fronds of *Hincksia mitchelliae* (Schmitz and Riffarth 1980) and *Laminaria digitata* (Davison and Stewart 1984). An investigation of plastids isolated from the siphonous green alga *Caulerpa simpliciuscula* revealed that both pathways of ammonium assimilation (i.e. GS-GOGAT and GDH) were present (McKenzie *et al.* 1979). Studies on phytoplankton have suggested that under conditions of high external ammonium, the GDH pathway is operative; it has a low affinity for intracellular ammonium (K_m = 5–28 mM) and does not require ATP, but it does use NAD(P)H and therefore energy (Syrett 1981; Berges and Mulholland 2008). After an NH_4 or NO_3 perturbation, the amino acid composition changes in *Ulva intestinalis*. Both glutamine and asparagines increased 10-fold while glutamate decreased, as expected in the assimilation of ammonium in the GS pathway (Taylor *et al.* 2006). It was concluded that the glutamine level is a good indicator of recent nutrient history, while tissue N content provides a more integrated longer term index of the ambient nutrient history (Fong *et al.* 1994). Barr and Rees (2003) showed that glutamine was an excellent indicator of recent nutritional history and nutrient loading and better than the glutamine:glutamate ratio and the total level of amino acids (Naldi and Wheeler 1999).

Urea is an organic nitrogen source that must be converted to ammonium, similar to nitrate. It is assimilated by first being broken down into carbon dioxide and ammonium via the nickel-containing enzyme urease, and then the free ammonium is incorporated into amino acids via the GS-GOGAT pathway, as reviewed by Antia *et al.* (1991). A urease enzyme assay exists for phytoplankton, but it has not been applied to macroalgae, except for a brief account of urease in *Ulva lactuca* (Bekheet *et al.* 1984).

Incorporation. Incorporation of nitrogen by N-limited cells is limited by the rate of protein synthesis (Syrett 1981; Mulholland and Lomas 2008). This has been substantiated recently by evidence for the accumulation of internal pools of nitrate, ammonium, and free amino acids after the addition of nitrogen to

N-limited cultures of seaweeds (Haxen and Lewis 1981; Thomas and Harrison 1985) and phytoplankton (Dortch 1982). Such pools would not accumulate if the rates of protein synthesis were equal to or greater than the rates of membrane transport and subsequent metabolism to amino acids. Several species of *Saccharia* accumulate nitrate, and the tissue levels of nitrate account for a significant portion of the total nitrogen (Chapman and Craigie 1977; Young *et al.* 2009). However, in many seaweeds, internal nitrate levels are low and never account for more than 5% of the total tissue nitrogen (Wheeler and North 1980; McGlathery *et al.* 1996; Naldi and Wheeler 1999). Thomas and Harrison (1985) found that intracellular nitrate and ammonium pools in *Pyropia perforata* composed only 10% of the pools of free amino acids. When seaweeds were starved of nitrogen, the nitrate decreased to undetectable levels in 5 days, while the ammonium and amino acid pools decreased by only 50%. Hence, storage of nitrogen as nitrate is not widespread in seaweeds. Amino acids (especially alanine) and proteins appear to form the major nitrogen-storage pools in *Gracilaria tikvahiae* (Bird *et al.* 1982), *G. foliifera* (Rosenberg and Ramus 1982), and *Macrocystis pyrifera* (Wheeler and North 1980). In contrast, citrulline and arginine were abundant in *Gracilaria secundata* (Lignell and Pedersen 1987). The form of nitrogen (NO_3 vs. NH_4) influences the kind of amino acid produced. Jones *et al.* (1996) found that a nitrate pulse to N-limited *Gracilaria* produced an increase in glutamic acid, citrulline, and alanine, while an ammonium pulse produced citrulline, phenylalanine, and serine. High amounts of citrulline and the dipeptide citrullinylarginine have been reported to play major roles as nitrogen-storage compounds in *Chondrus crispus* (Laycock *et al.* 1981), *Gracilaria*, and other red seaweeds (Laycock and Craigie 1977). Protein bodies from 1–10 μm in diameter were observed in the endemic arctic kelp *Laminaria solidungula* (Pueschel and Korb 2001).

The amount of inorganic and organic nitrogen storage can vary considerably in closely related seaweeds or in a given species growing in different localities. *Gracilaria tikvahiae* can store enough nitrogen to allow it to grow at maximal rates for several days without nitrogen (Lapointe and Ryther 1979; Fujita 1985), whereas *Gracilaria secundata* has very limited nitrogen-storage abilities (Lingell and Pedersen 1987). The difference between these two species may correlate with their habitat; *G. secundata* grows in a eutrophic environment, whereas *G. tikvahiae* may be subjected to various periods of nitrogen limitation. A comparison of the storage capacity of five species of macroalgae and phytoplankton by Pedersen and Borum (1996) demonstrated that seaweeds can survive periods of N limitation better than phytoplankton. Because many seaweeds can store nitrogen, this characteristic has been used to advantage in controlling epiphyte growth in aquaculture systems. Weekly pulses of 0.5 mM NH_4 controlled epiphyte growth in *Gracilaria conferta* growing under low irradiances and yielded the highest growth rates for this macroalga (Friedlander *et al.* 1991).

Little is known about catabolism and turnover of cellular protein. Recycling of nitrogen may occur via the photorespiratory pathway (Syrett 1981; Singh *et al.* 1985). Pigments and associated proteins such as phycoerythrin may serve as nitrogen-storage compounds in the Rhodophyceae (Bird *et al.* 1982), but Naldi and Wheeler (1999) found that phycoerythrin was only a minor N storage component in *Gracilaria*.

6.5.2 Phosphorus

Phosphorus is mainly available as the inorganic ions PO_4^{3-} and $H_2PO_4^-$ (hereafter denoted as P_i) and some smaller organic P compounds are bioavailable after enzymatic cleavage of PO_4 from the organic molecule by phosphatase enzymes. Generally, phosphorus is not the main limiting nutrient for macroalgal growth. However, P_i limitation occurs in the Caribbean near Jamaica (Lapointe *et al.* 1992; Lapointe 1997, 1999), some coastal areas with high anthropogenic N inputs such as China (Harrison *et al.* 1990), and Oslofjord (Pedersen *et al.* 2010). In shallow waters, carbonate sediments near coral reefs act as a sink for phosphate and thereby reduce the availability of P for macroalgae (McGlathery *et al.* 1994). Phosphorus plays key roles in nucleic acids, proteins, and phospholipids (the latter are important components of membranes) (Table 6.1).

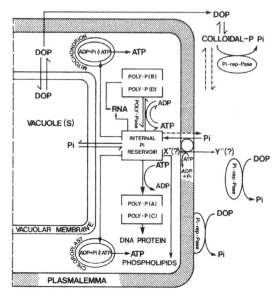

Fig. 6.15 Main features of phosphorus uptake and assimilation in a microalgal cell. DOP, dissolved organic phosphate; P_i, inorganic phosphate. (From Cembella *et al.* 1984, reprinted with permission from *Critical Reviews of Microbiology* 10:317–91, © 1984 CRC Press, Boca Raton, Fla.)

However, its most important role is in energy transfer through ATP and other high-energy compounds in photosynthesis and respiration (Fig. 6.15) and in "priming" molecules for metabolic pathways. For an overview of the ecophysiological role of P in phytoplankton, see Cembella *et al.* (1984) and Wagner and Falkner (2001).

The kinetics of P_i uptake, assimilation, and incorporation are quite similar to N. P_i uptake kinetics can be measured by the uptake of radioactive ^{32}P (Runcie *et al.* 2004) which provides an estimate of flux into the cells (i.e. the influx over a few minutes). In contrast, uptake of stable P_i provides an estimate of the net P_i uptake (i.e. influx minus efflux) and therefore uptake rates from the two methods are not directly comparable. When a macroalga is P_i-starved then "surge" uptake occurs over a few minutes, while surge uptake is largely absent in P_i-sufficient macroalgae. Active P_i uptake occurs by two or more transports systems (e.g. high- and low-affinity systems) in macroalgae (Raghothama 1999). In higher plants and in some algae, there are H^+/P_i and Na^+/P_i co-transport systems (Rausch and Bucher 2002).

A two-compartment model can be used to examine uptake, assimilation, and storage. By using exponential decay models to describe the time course of $^{32}P_i$ release, one can infer the sizes and fluxes from a rapidly exchanging compartment/pool (the cytoplasm) and a slowly exchanging compartment/pool (the vacuole) (Runcie and Larkum 2001). Incorporation and storage can be assessed by determining compounds that are in the rapidly exchanging compartment and soluble (e.g. P_i, sugars, lipids, nucleotides, short chain polyphosphates), or in the slow exchanging compartment and are insoluble in trichloroacetic acid (TCA) (e.g. nucleic acids, long-chain polyphosphates, and phosphoproteins). The emerging research area of functional and structural genomics will allow targeted gene expression, serving as molecular determinants of P_i acquisition and adaptation to P_i stress, to specific tissues under the control of regulated promoters. The complete sequencing of several algal genomes will help to advance the determination of molecular mechanisms of the signal transduction pathway leading to transcriptional regulation of P_i acquisition and acclimation to P_i stress.

Differences in uptake, assimilation, and storage in various algae determines their success in various P environments. *Ulva lactuca* is a cosmopolitan, opportunistic annual with a high nutrient uptake capacity and hence a high growth rate, while the late successional red epiphyte *Catenella nipae* is a slow-growing species (Runcie *et al.* 2004). For both species, the K_s value was 3.7 μM. When both species were P_i-sufficient, most of the $^{32}P_i$ was accumulated as free PO_4. In contrast, when P_i-starved, *C. nipae* incorporated more P_i into the TCA insoluble pool, indicating that it had a greater storage capacity than *U. lactuca*. The rapidly exchanging compartment (i.e. the cytoplasm) had a short half-life of 2–12 min, while the slower exchanging compartment (i.e. the vacuole) has a half-life of 4 and 12 days for *Ulva* and *Catenella*, respectively. *Ulva* took up P_i and transferred it into tissue phosphorus and TCA-soluble P in a few hours. However, *Catenella* took up P more slowly and stored much of it in less mobile TCA insoluble forms. Runcie

et al. (2004) suggested that the long-term storage of refractory forms of P_i makes *Catenella* a useful bioindicator of P_i availability over a ~7-day period, while the P status of *Ulva* may only reflect conditions over a few hours or days. Hence *Catenella* may be a better integrator of highly varying P_i concentrations in estuaries and hence a more useful bioindicator of P_i availability than *Ulva*.

There have been only a few studies on P_i uptake kinetics in seaweeds, despite the fact that P_i concentrations are also frequently near the limit of detection when nitrogen is exhausted from the seawater. Preliminary studies indicate that P_i is actively taken up and saturation kinetics have been examined in the red alga *Agardhiella subulata*, yielding a V_{max} of 0.47 μmol g_{dw}^{-1} h^{-1} and a K_s of 0.4 μM (DeBoer 1981). The kinetic parameters for two chlorophytes, *Ulva* and *Chaetomorpha* that dominate a eutrophic estuary were considerably higher; K_s values for *Ulva* and *Chaetomorpha* were 3.5 and 10 μM, respectively, and V_{max} values were 8.5 and 20.8 μmol gdw^{-1} h^{-1}, respectively (Lavery and McComb 1991b). In *Gracilaria tikvahiae*, the P_i uptake rate showed three phases of uptake, two saturation phases at lower concentrations (0–0.2 and 0–2 μM) and a linear phase (0–11 μM) (Friedlander and Dawes 1985). *Palmaria palmata* exhibited biphasic uptake kinetics with a saturation response at low concentrations and linear uptake at high P_i concentrations that were much higher than environmental concentrations (Martinez and Rico 2004). Two uptake systems occur in *Chara corallina*; a high-affinity P_i transporter expressed in P_i-starved thalli with a K_s of 4 μM and a low-affinity system expressed in P_i-sufficient thalli with a K_s of 220 μM (Mimura *et al.* 1998).

P_i uptake varies with shore position and season. Hurd and Dring (1990) studied P_i uptake in relation to zonation and season in five intertidal fucoid algae. They found an initial (30 min) rapid P_i uptake followed by almost no uptake (30 min; probably due to P_i release back to the medium), and then intermediate rates over several hours in *Pelvetia* and two species of *Fucus*. *Ascophyllum* had a constant slow uptake rate over 6 h. The fucoid algae were divided into two distinct groups: *Pelvetia* and *Ascophyllum* took up small amounts of phosphate over a tidal cycle, and the ratio of their uptake rates was 6:1 *(Pelvetia: Ascophyllum)*. Their uptake abilities were related to their positions on the shore *(Pelvetia* was in the highest zone), and to their low requirements for nutrients (e.g. low growth rate). The three species of *Fucus* formed the second group. They were able to take up larger amounts of P_i during a tidal cycle, the order of uptake being *F. spiralis* > *F. vesiculosus* > *F. serratus*, and their rates were directly related to their positions on the shore. Hurd *et al.* (1993) observed that all species of *Fucus* produced hyaline hairs on the apical region and upper mid-thallus in late winter and shed them in fall. These hairs were shown to enhance nutrient uptake. A reduced uptake rate for *F. spiralis* in July was due to desiccation damage and the grazing of hairs by littorinid snails.

P_i limitation can be assessed by P_i addition bioassays. Growth enrichment studies by Lapointe (1986) utilizing *in situ* cage cultures and a shipboard flowing seawater culture system were conducted in the summer with populations of pelagic *Sargassum natans* and *S. fluitans* in the western Sargasso Sea. P_i enrichment doubled their growth and photosynthetic rates compared with nitrogen enrichment and no enrichment. Further enrichment studies and tissue analysis with *Gracilaria tikvahiae* in the Florida Keys demonstrated that *Gracilaria* was P_i-limited in summer, and N- and P_i-limited in the winter (Lapointe 1985, 1987). O'Brien and Wheeler (1987) also suggested that *Ulva prolifera* off the coast of Oregon may have been P_i-limited in November. Greatly elevated C:P and N:P ratios further confirmed P_i limitation (Lapointe 1987, 1997). Ratios of C:P > 1800 and N:P > 120 for unenriched *G. tikvahiae* were three-fold higher than the normal average 550 and 30 values, respectively, reported by Atkinson and Smith (1983).

Susceptibility to P limitation may depend on the lifeform of members of the green algal Order Bryopsidales (= Caulerpales). They are important calcifying agents in tropical reefs, and they comprise two different lifeform groups: (1) epilithic species with limited attachment structures and (2) psammophytic forms that have extensive subterranean rhizoidal systems. Because the shallow water habitats of the former have relatively low N:P ratios, as compared with the pore waters of the carbonate-rich sedimentary substrata in which the latter are anchored, Littler and Littler (1990) hypothesized that epilithic forms should tend to be nitrogen limited, whereas psammophytic species should be phosphorus limited. This hypothesis was subsequently confirmed by nutrient addition bioassays (Littler *et al.* 1988; Littler and Littler 1990). The epilithic forms, *Halimeda opuntia, H. lacrimosa,* and *H. copiosa*, increased their photosynthetic rates when nitrogen was added, but not when phosphorus was added. In contrast, the psammophytic forms, *Udotea conglutinata, Halimeda monile, H. tuna,* and *H. simulans*, were more stimulated by P addition.

Some seaweeds can use some organic forms of phosphate, such as glycerophosphate, by producing extracellular alkaline phosphatase (APA) enzyme. The ability of cells to enzymatically cleave the ester linkage joining the phosphate group to the organic moiety is due to phosphomonoesterases (commonly called phosphatases) at the cell surface. Two groups of these enzymes have been distinguished on the basis of their pH optima, their phosphate repressibility, and their cellular locations (Cembella *et al.* 1984). Alkaline phosphatases (e.g. phosphomonoesterase) are P_i-repressible, are inducible, have alkaline pH optima, and generally are located on the cell surface (cell wall, external membrane, or periplasmic space) or are released into the surrounding seawater. Acid phosphatases are P_i-irrepressible, are constitutive, have acidic pH optima, and generally are found intracellularly in the cytoplasm. Both types may be found simultaneously in algal cells, with APA aiding in the uptake of organic P compounds, and acid phosphatases playing crucial roles in cleavage of intracellular compounds. APA assays have been conducted using *in vivo* (whole thallus, under field or laboratory conditions), or *in vitro* (following partial purification of the enzyme) approaches. The methodology of measuring APA activity and whether to attempt to simulate field or optimal conditions has been reviewed (Hernández *et al.* 2002). Five isozymes of APA were detected in *Ulva lactuca* (Lee *et al.* 2005). It has been suggested that APA activity in macroalgae could be used for environmental monitoring P status of macroalgae and P_i availability in the field (Hernández *et al.* 2002).

Species should show a clear response to P stress, be widespread, easy to identify, have a high SA:V ratio, and show a strong response to internal P_i concentrations. Examples of some suitable genera are *Gelidium*, *Cladophora*, *Polysiphonia*, and *Stypocaulon*.

Many factors affect APA activity. No correlation was found between the position of *Fucus spiralis* on the shore, but the highest APA occurred in emersed seaweeds at low tide since they cannot take up P_i. APA correlated with whole tissue P and the highest APA occurred in the meristematic zone at the tip of the thallus (Hernández *et al.* 1997). Therefore APA varies due to complex biotic and environmental factors such as stress due to emersion, growth period, P tissue content and N:P ratio, and the region of the thallus. The essential feature of phosphatases that allows them to participate efficiently in cellular metabolism is their ability to be alternately induced and repressed, depending on metabolic requirements. When external P_i concentrations are high, synthesis of APA is repressed and cells exhibit little ability to utilize organic P compounds. Generally, after the external P_i has been exhausted, intracellular P from stored polyphosphates and P_i are used up quickly (Lundberg *et al.* 1989), followed by increased APA activity. Weich and Graneli (1989) found that APA activity in *Ulva lactuca* was stimulated by P limitation and by light. The magnitude of the increase was species specific and depended on the availability of organic phosphates and the degree of P limitation experienced by the cells (Cembella *et al.* 1984; Hernández *et al.* 2002).

When P_i is transported across the plasmalemma, it enters a dynamic intracellular P_i pool. It is incorporated into phosphorylated metabolites (Chopin *et al.* 1990), or stored as luxury phosphorus in vacuoles, or in polyphosphate vesicles (Chopin *et al.* 1997) (Fig. 6.15). P_i-deficient algae possess the ability to incorporate P_i extremely rapidly, and the amount taken up usually exceeds the actual requirements of the cell for growth. The excess is incorporated into polyphosphates by the action of polyphosphate kinase (Cembella *et al.* 1984).

An important difference between P metabolism in vascular plants and that in algae is the formation of these polyphosphates. Seaweeds known to form polyphosphates include species of *Ulva*, *Ceramium*, and *Ulothrix* (Lundberg *et al.* 1989). They used high-resolution NMR with ^{31}P and found relatively short (6–20 P_i units) polyphosphates stored in the vacuole in *Ulva lactuca*. In contrast, they were surprised to find that the brown alga *Pilayella* stored phosphorus mainly as phosphate (not polyphosphate) in its vacuole. Polyphosphate cytoplasmic granules were found in *Chondrus crispus*, mostly in the medullary cells, along the plasmalemma, especially near pit plugs (Chopin *et al.* 1997). The storage compounds are classified into cyclic and linear polyphosphates. These two types cannot be easily separated by simple extraction procedures in cold TCA, but they can be divided into four categories (A–D, Fig. 6.15) on the basis of sequential extraction techniques (Cembella *et al.* 1984). In linear polymers, orthophosphate residues are linked through energy-rich phosphoanhydride bonds similar to those containing terminal phosphate bonds in ATP and hence energy can be released during their hydrolysis.

Green seaweeds of the order Bryopsidales (e.g. *Caulerpa* sp.) are abundant in tropical oligotrophic waters presumably due to their ability to absorb nutrients from sediment pore waters through their extensive rhizoid root-like system (Williams 1984). High P_i inputs in some areas and seaweeds growing in siliciclastic sediments on the west coast of Florida, have been shown to inhibit biomineralization in the rhizophytic green alga *Halimeda incrassata* (Delgado *et al.* 1994; Demes *et al.* 2009b). In contrast, more biomineralization of $CaCO_3$ occurred when seaweeds were growing on P_i-limited carbonate sediments in the Florida Keys. Reduced biomineralization of algae has important implications for higher trophic levels since it serves a role in protecting against herbivory (Hay *et al.* 1994). In addition, ambient P_i concentrations or >2 μM have also been shown to inhibit coral calcification as discussed in the next section on calcium.

6.5.3 Calcium and magnesium

Calcareous seaweeds are a very diverse group that incorporate calcium carbonate (essentially limestone) onto or into their thalli. About 5% of all macrophytes calcify (Smith *et al.* 2012). Calcification has evolved independently in the three major phyla of macroalgae (Rhodophyta, Chlorophyta, and Phaeophyta) and there are more than 100 genera of calcareous algae (Nelson 2009). Calcareous red algae have contributed to the fossil record that extends back to Precambrian times. About 360 million years ago, calcified crusts evolved. They are mostly found in the subtidal and low-intertidal zones because they are susceptible to desiccation (i.e. they contain only 30% water). Bleaching is mainly due to desiccation, while light and temperature are minor contributors to bleaching (Martone *et al.* 2010b). It is amazing that some corallines have been collected at 260 m, the deepest of any known benthic algae, despite being covered with $CaCO_3$ that is nearly opaque to light (Littler *et al.* 1985). They are the slowest growing macroalgae, and growth bands in corallines indicate that they can live to over 700 years, making them possibly the longest-lived marine organism (Steneck and Martone 2007). They tend to thrive where herbivory is the most intense, and there is an interesting positive relationship between limpets which graze epiphytes on the coralline that would shade or smother the calcified thalli. In addition, corallines release compounds that aid in the settlement and morphogenesis of many animals including corals (Nelson 2009). Branched, erect corallines may be heavily grazed by some fish. Some corallines are rhizophytic since they produce rhizoids that can take up nutrients from the sediment and help them colonize sandy substrates in shallow lagoons and possibly facilitate the succession from bare sand to seagrass beds. There has been a renewed interest in coralline algae due to the increasing threat of ocean acidification.

The mode and extent of calcification varies widely. The brown alga *Padina* has a thin white calcified coating, in contrast to the most heavily calcified green alga *Halimeda* and red algae such as *Corallina* (Steneck and Martone 2007). Calcified segments of *Halimeda* may contribute a major portion of the sediment (i.e. white "sand") on an atoll, and thus it plays a role in atoll formation (Barnes and Chalker 1990). The red algal order Corallinales is the best known group of calcified seaweeds and they are either geniculate (jointed or articulated) or non-geniculate (typically crustose). These crusts that look like "pink paint" on hard surfaces have been largely ignored in research, probably due to taxonomic problems since the crust varies with age and environmental factors. They attach to hard substrates including hard-shelled animals such as mollusk shells. Encrusting coralline red algae help to cement reefs together, stabilize them against wave action, and form an algal ridge of up to 10-m thick in the wave-exposed section of a tropical reef. They can produce up to ~10 kg $CaCO_3$ m^{-2} y^{-1} in areas such as the Great Barrier Reef and this calcification rate would produce an upward reef growth of ~7 mm y^{-1} (Chisholm 2000). The articulated corallines form erect fronds that overcome the biomechanical limitation of calcification by producing flexible genicula (like joints) which evolved convergently in the green and red calcifiers (Steneck and Martone 2007). There are free-living non-geniculate corallines called rhodoliths that accumulate in thick beds called maerl that can be mined as a source of $CaCO_3$ that is added to soil to increase the pH (Nelson 2009).

Calcium is deposited as calcium carbonate ($CaCO_3$), sometimes along with small amounts of magnesium and strontium carbonates. Calcium carbonate occurs in two crystalline forms, calcite (hexagonal rhombohedral crystals) and aragonite (orthorhombic), which never occur together under natural conditions. Biomineralogical control of Mg and Ca deposition is partially overridden by the ambient Mg/Ca ratio. The present Mg/Ca ratio in seawater is ~5.2 and therefore corallines contain higher amounts of Mg/Ca in their thalli now (i.e. more aragonite) than during the Cretaceous when the Mg/Ca ratio in the seawater was only ~1 (i.e. more calcite calcification) (Ries 2010; Stanley *et al.* 2010). One explanation why *Halimeda* is presently relatively abundant is that laboratory experiments reveal that its linear growth rate is much slower in calcite seawater (e.g. similar to the Cretaceous period) that had a Mg/Ca ratio of 1 (Ries 2010; Stanley *et al.* 2010). Aragonite is the most soluble form of $CaCO_3$ and the most commonly deposited especially in the crusts of corallines, making them sensitive to pH changes (Fig. 6.16). It is the form that precipitates abiotically because of the high Mg/Ca ratio of 5.2 in seawater and may contain up to ~15% Mg. Aragonite is

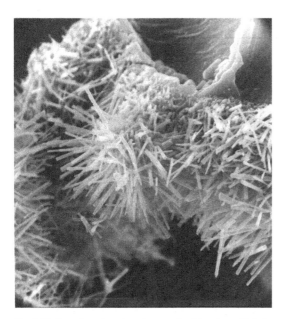

Fig. 6.16 Scanning electron micrograph of the extracellularly mineralizing calcareous alga *Halimeda* sp. showing the thin, unoriented crystals of aragonite (×2880). A portion of the cell wall is visible in the upper right-hand corner. (From Weiner 1986, reprinted with permission from *Critical Reviews in Biochemistry* 20:365–408.)

often confusingly referred to as high Mg calcite. Due to the decrease in both pH and carbonate concentration as a result of ocean acidification, aragonite with relatively high Mg is likely to dissolve before calcite since high Mg increases the solubility of $CaCO_3$ in seawater (Smith *et al.* 2012). Erect articulated corallines generally contain little aragonite. The extensive literature on the morphological and physiological aspects of calcification has been reviewed by Borowitzka (1987, 1989), Simkiss and Wilbur (1989), McConnaughey (1991, 1998) and McConnaughey and Whelan (1997). The deposition sites and forms of algal mineralization vary, but there are two basic types, extracellular and intracellular deposition (Table 6.4).

The best studied example of extracellular deposition is aragonite in the intercellular spaces in the green alga *Halimeda* (Fig. 6.17). Its outer surface is made up of vertical, swollen filaments called utricles (Fig. 6.18). Their outermost walls become fused, forming a barrier that prevents the flow of seawater into the intercellular spaces. Aragonite is precipitated by *Halimeda* and other Chlorophyta as crystals of varying shapes, and without a preferred orientation (Fig. 6.16). The crystals begin as small granules on the fibrous material of the intercellular space (nucleation sites), and then they grow until the intercellular space is almost filled. Ca^{2+} and HCO_3^- enter the intercellular space either by diffusion through the cell walls of the fused filaments or by active uptake into the filament and then an efflux into the intercellular space. During photosynthesis, CO_2 is taken up from the intercellular space, resulting in an increase in intercellular pH (i.e. $HCO_3^- \rightarrow CO_2 + OH^-$) and the CO_3^{2-} concentration with the subsequent deposition of aragonite (Fig. 6.17). This hypothesis for calcification in *Halimeda* is not applicable to all aragonite depositors because other seaweeds with apparently suitable morphology (intercellular spaces) do not calcify (e.g. *Ulva*; Borowitzka 1977).

The Rhodophyceae, in contrast to Chlorophyceae, are primarily calcifiers of cell walls, although some do deposit $CaCO_3$ within intercellular spaces as well (Cabioch and Giraud 1986). Calcification of the wall is so extensive that the cells become encased, except for the primary pit connections at the end. In the Rhodophyceae, calcification occurs in Corallinaceae as calcite and in some members of Squamariaceae, Gigartinaceae, and Bangiales as aragonite. Of these algae, members of the Corallinaceae are both the most abundant and the best known. The extracellular calcification in macroalgae is in contrast to microalgal coccolithophores that have intercellular calcification, but their coccoliths ($CaCO_3$ scales) are deposited externally.

The Corallinales deposit calcite containing high levels of magnesium (6% Mg) within the cell walls of their vegetative cells, except for the meristematic, genicular, and reproductive cells. Calcification commences in the middle lamella and rapidly spreads throughout the cell wall. Along the middle lamella the crystals are arranged along the growth axis, whereas near the plasmalemma the calcite crystals are oriented with the crystal axis at right angles to the plasmalemma (Fig. 6.18). The calcite crystals are in close association

Table 6.4 Sites and forms of algal mineralization.

Sites	Forms	Examples of taxa
Extracellular		
Cell wall surface	Concentric bands of fine aragonite needles on cell surface	*Padina* (Dictyotaceae)
	Surface encrustation of calcite crystals	*Chaetomorpha* (Cladophoraceae)
Intercellular	Fine aragonite needles in intercellular spaces in utricles	*Halimeda, Udotea* (Halimedaceae), *Neomeris* (Dasycladaceae)
	Intercellular crystals of aragonite and/or calcite that may form small bundles	*Liagora* (Liagoraceae), *Galaxaura* (Galaxauraceae)
Sheath	Bundles of aragonite needles in external sheath	*Penicillus, Udotea* (Udoteaceae)
	Irregular bundles of needle-like crystals, usually aragonite	*Plectonema* (Oscillatoriaceae)
Within cell walls	Calcite crystals, often clearly oriented	*Lithophyllum, Lithothamnion* (Corallinaceae)
Intracellular	Calcified plates (coccoliths) of various forms, usually calcite; formed within Golgi vesicles	*Emiliania, Cricosphaera* (Prymnesiophyceae)

Source: Simkiss and Wilbur (1989), with permission of Academic Press.

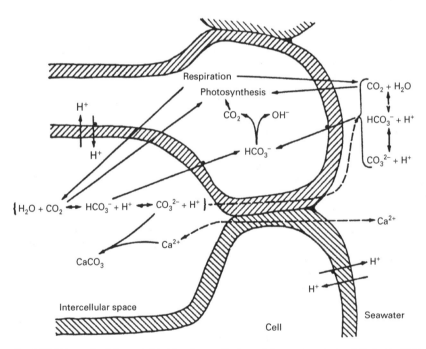

Fig. 6.17 Schematic of the postulated ion fluxes affecting calcium carbonate precipitation in *Halimeda*. A black dot at the plasmalemma indicates that the flux is postulated to be active. (From Borowitzka 1977, with permission of *Oceanography and Marine Biology: An Annual Review*.)

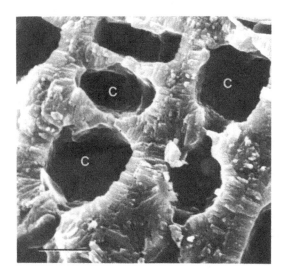

Fig. 6.18 Scanning electron micrograph of a fractured thallus of the crustose coralline *Lithothamnion australe* showing the calcite-impregnated cell walls. This alga has especially large calcite crystals, which can clearly be seen to be oriented at right angles to the cells (c). Scale = 5 μm. (Reprinted from Borowitzka 1982, with permission from Elsevier Scientific Publications.)

with the organic cell wall material, and some cell wall components probably act as a template for deposition and orientation of new crystals. Studies of the compartmentation of exchangeable Ca^{2+} in the thallus of *Amphiroa foliacea* have indicated that there are at least two major compartments for binding and exchanging organic calcium, presumed to be the COO^- and $O-SO_2^-$ O^- groups of the acidic polysaccharide wall compounds (Borowitzka 1979). Because aragonite is the normal crystalline form of calcium carbonate precipitated from seawater, the organic wall material of the corallines is presumed to be responsible for promoting deposition. These results further support the proposed role of some specific cell wall components of coralline algae in influencing the crystallography of the $CaCO_3$ precipitated, in addition to suggesting that they contain compounds (probably polysaccharides) that block crystal growth sites and stop calcification. For example, the genicula (uncalcified joints) are formed when calcified cells decalcify and restructure to create flexible

tissue. In *Calliarthron*, $CaCO_3$ crystals align precisely along an organic matrix, possibly a protein-polysaccharide complex, within coralline cell walls and it is plausible that the xylose-branching of xylogalactans within the intergenicular walls helps define sites for $CaCO_3$ nucleation (Martone *et al.* 2010). In contrast, the highly methylated galactans produced by genicula may not support nucleation, preventing mineral deposition after decalcification and maintaining genicular flexibility for the thallus.

Little new information has emerged since the 1980s and 1990s on the actual mechanism of calcification in the coralline algae except that it is directly proportional to photosynthesis and the CO_3^{2-} concentration, stimulated by light, and highest in young tissue. Two models of calcification in the Corallinales have been proposed by Pentecost (1985) (Fig. 6.19a, b). In one model (Fig. 6.19a), HCO_3^- is taken up and converted to CO_3^{2-} plus a proton (H^+) which is released extracellularly and combines with Ca^{2+} to form $CaCO_3$. The optimum pH for photosynthesis in *Amphiroa* species lies between 6.5 and 7.5. Above pH 8, photosynthesis is greater than would be expected based solely on the utilization of CO_2. This suggests that they take up HCO_3^- at a higher pH (see sec. 5.4.1 for further discussion). The dark calcification rate is approximately 50% of the rate in saturating light at normal pH, and this is considered to be the non-metabolically influenced component of the coralline calcification process (Borowitzka 1989). Nighttime carbonate dissolution can occur due to respiratory acidification of the alga's tissues or microbial activity in mat-like environments (Chisholm 2000). In the second model (Fig. 6.19b), an efflux of H^+ reacts with HCO_3^- in seawater to produce CO_2, which then diffuses into the cells for use in photosynthesis. An efflux of OH^- maintains electrical neutrality in the cell, raises the pH and the CO_3^{2-} concentration in the intercellular spaces, and leads to localized calcification. This second model eliminates a HCO_3^- uptake mechanism since CO_2 is taken up and a high intracellular pH necessary to produce CO_3^{2-}, as required by the first model.

In most calcifying algae, calcification occurs extracellularly in direct contact with the surrounding seawater and is largely determined by physical factors

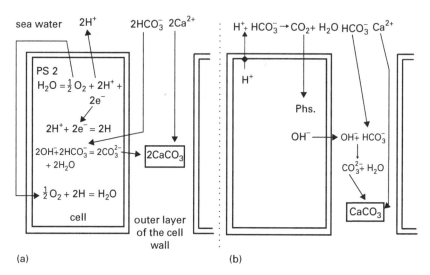

Fig. 6.19 Models of calcification in the Corallinaceae. (a) Scheme proposed by Digby (1977), slightly simplified. (b) Alternative model with localized efflux of H^+ and OH^-. (From Pentecost 1985, with permission of the American Society of Plant Physiologists.)

such as pH, and photosynthesis, respiration, and a light-driven proton pump determines the surface pH. In contrast, in corals, Ca transport is light-driven and the site of calcification occurs intracellularly and hence the coral has direct control over calcification processes (De Beer and Larkum 2001). Biogenically derived calcification is closely coupled to CO_2 uptake during photosynthesis which shifts the carbonate system towards carbonate production and increases the saturation state of $CaCO_3$ and the pH which in turn enhances $CaCO_3$ precipitation and calcification. pH effects can be explained partially by the influence of photosynthesis, respiration, and calcification on the carbonate system. CO_2 uptake for photosynthesis leads to an increase in pH, whereas CO_2 and H^+ release by respiration and calcification leads to a decrease in pH and alkalinity which enhances CO_2 availability. Therefore, both dark respiration and dark calcification lead to a pH decrease which is responsible for the pH decrease at the thallus surface. In the light, CO_2 uptake for photosynthesis must be higher than the combined CO_2 release by calcification and respiration to produce pH values above seawater. Therefore, in *Halimeda discoidea*, calcification is not regulated by the alga, but is a consequence of pH increase during

photosynthesis (De Beer and Larkum 2001). Calcification cannot supply sufficient protons and CO_2 to sustain photosynthesis and hence carbonic anhydrase is required to aid in the utilization of HCO_3^- and thus to alleviate CO_2 limitation.

Data from Ca^{2+}-efflux and stable isotope studies, and the type of $CaCO_3$ crystal isomorph deposited, all suggest that much of the carbon for the production of $CaCO_3$ is of seawater origin. However, long-term pulse-chase studies have indicated that respiratory CO_2 is also incorporated into $CaCO_3$ in *Halimeda*. The precipitation of $CaCO_3$ requires a sufficient Ca^{2+} concentration and a metabolically induced rise in pH. In seawater, this pH shift is easily achieved by photosynthetic CO_2 uptake; however, in fresh water, H^+ efflux or OH^- uptake by the algae may also be necessary. $CaCO_3$ nucleation appears to be also influenced by the organic matter in the seaweed cell wall. Once nucleation is achieved, $CaCO_3$ deposition continues under the influence of localized pH and carbonate changes caused by photosynthetic CO_2 uptake.

Orthophosphate inhibits calcification in corallines, but the mechanism remains unknown. When coralline algae are grown in phosphorus-enriched culture medium (30–150 μM PO_4^{3-}), they are only weakly calcified suggesting that the growth of corallines could be inhibited in phosphate-polluted coastal waters. In addition, the rhizophytic alga *Halimeda incrassata* was significantly more calcified when growing on P-limited carbonate sediments in the Florida Keys than seaweeds from siliciclastic sediments on the west coast of Florida (Delgado and Lapointe 1994; Demes *et al.* 2009b).

Ocean acidification is causing a decrease in pH and an elevation in partial pressure of CO_2 (pCO_2) which is lowering the carbonate saturation state (see also sec. 7.7 and Essay 4). With the predicted reduction of pH to ~7.65, there will be a 300% increase in H_2CO_3, a 9% increase in HCO_3^- and >50% decrease in CO_3^{2-}. These changes are projected to result in a decrease in the abundance of crustose corallines and a shift to more fleshy macroalgae (Kuffner *et al.* 2008). In addition, crustose corallines are generally more sensitive to an increase in pH than corals. Organisms that are able to control the pH at the site of calcification are less likely to be impacted by ocean acidification. Since ocean acidification and global warming co-occur, the effect of an increase in pCO_2 and temperature on calcification should be studied together in factorial designed very long-term experiments (Diaz-Pulido *et al.* 2012). In the crustose coralline *Porolithon onkodes*, warming (26 → 29°C) amplified the negative effects of an increase in pCO_2 such as decreased calcification and even some dissolution and mortality. In contrast, on a local scale, when pH was observed to increase from 7.9 to 8.9 in a tropical seagrass meadow due to rapid photosynthesis, calcification rates of *Hydrolithon* sp. increased several fold (Semesi *et al.* 2009). Hall-Spencer *et al.* (2008) attempted to determine how a decrease in pH would affect corallines growing near a marine CO_2 vent in the Mediterranean on a more evolutionary scale where genotypic adaptation could occur. They found some corallines growing at a pH of 7.6, well below the pH expected in 2100, and showed that some corallines have already adapted to a large pH decrease of 1.5 units. This field observation emphasized the importance of assessing genotypic adaptation rather than phenotypic acclimation.

There are several disadvantages to calcification. Unlike many animals, algae do not require $CaCO_3$ or related salts for skeletal support; in fact, $CaCO_3$ deposits may often be a liability rather than an asset. $CaCO_3$ deposits, especially if they are extracellular, inhibit nutrient uptake by creating diffusion barriers and also limit light penetration into the thallus, thus reducing photosynthesis and inhibiting growth. The high pH near the plasmalemma in calcareous algae could also hinder phosphate uptake, because at high pH, $H_2PO_4^-$ is converted to HPO_4^{2-} and then PO_4^{3-}. Thus algal calcification is best seen as a by-product of photosynthesis, one that many algae have evolved mechanisms to avoid (Simkiss and Wilbur 1989). The main advantage is that the $CaCO_3$ may reduce grazing in corallines. In addition, new *Halimeda* segments are loaded with halimedatriol and other feeding deterrents as they emerge, but once calcification commences, the anti-grazing compounds decrease (sec. 4.4.2).

Calcium has many roles in various cellular functions (Table 6.1). It is important to all organisms for the maintenance of cellular membranes. Calcium has roles in morphogenesis and phototaxis, and evidence for Ca^{2+} as a messenger in light signaling is seen in the role of calcium in the development of cellular polarity in fucoid eggs (sec. 2.5.3 and Fig. 2.23). There is considerable evidence that Ca^{2+}, together with the regulatory protein calmodulin, is an important second messenger in plant cells (White and Broadley 2003). Calcium is required inside the cell by only a few enzymes (α-amylase, phospholipase, and some ATPases). In fact, many enzymes (e.g. PEP carboxylase) are inhibited by Ca^{2+} but it may activate ATPase in some calcareous algae.

Magnesium is an essential component of chlorophyll, forming a metalloporphyrin (Table 6.1). It can play a role in binding charged polysaccharide chains to one another because it is a divalent cation. It is a cofactor or activator in many reactions, such as nitrate reduction, sulfate reduction, and phosphate transfers (except phosphorylases). It is also important in several carboxylation and decarboxylation reactions, including the first step of carbon fixation, where the enzyme ribulose-1,5-bisphosphate (RuBP) carboxylase attaches CO_2 to RuBP. Magnesium also activates enzymes involved in nucleic acid synthesis and binds together the subunits of ribosomes. There are several means by which magnesium may act (Raven *et al.* 2005): (1) It may link enzyme and substrate together, as, for example, in reactions involving phosphate transfer from ATP. (2) It may alter the equilibrium constant of a reaction by binding with the product, as in certain kinase reactions. (3) It may act by complexing with an enzyme inhibitor.

6.5.4 Sulfur

Sulfur is a component of polysaccharides and involved in cellular metabolism and indirectly in climate change processes. Sulfur is important in stabilizing protein tertiary structure, and many algal wall polysaccharides are highly sulfated. Sulfate acts as a major vacuolar osmoticum and a high sulfur content may be a deterrent to grazing (Giordano *et al.* 2005b). Some species of *Desmarestia* have so much sulfuric acid in their vacuoles that the pH is close to 1. Most algae can meet all of their sulfur requirements by reducing sulfate, the most abundant form of sulfur (~29 mM) in aerobic seawater (Giordano *et al.* 2008). Two sulfur-containing amino acids, cysteine and methionine, are important in maintaining the three-dimensional configurations of proteins through sulfur bridges, as well as in linking chromophores to the protein in phycobiliproteins (Fig. 5.21). Some seaweeds produce commercially valuable sulfated polysaccharides that are important in thallus rigidity (e.g. carrageenan in red algae) and adhesion (e.g. fucoidan in brown algae) (secs. 1.3.1, 2.5.2, 5.5.2, 10.3, and 10.5) (Coughlan 1977). Sulfur nutrition and utilization in algae have been reviewed by Schiff (1983) and Giordano *et al.* (2005b) and in higher plants by Leustek *et al.* (2000).

6.5.5 Iron

Iron is an important trace element for macroalgae growth. Iron, the fourth most abundant element in the Earth's crust, is one of the least soluble metals in oxygenated waters. For this reason, iron concentrations of ~0.2 nM in some coastal areas may limit the growth of some seaweeds depending on their requirements. However, it is well established that iron is limiting for phytoplankton growth in the central regions of the Southern Ocean and the equatorial and NE Pacific Ocean with ~0.05 nM where the input of iron from atmospheric dust is low (Boyd and Ellwood 2010). The speciation of iron in seawater is complex, and this is a major impediment to our understanding of iron uptake by algae. At the pH of seawater (c. 8.2), ferric ions combine with hydroxyl ions to form ferric hydroxide, which is relatively insoluble $(K_{sp} \sim 10^{-38}$ M). Therefore, much of the iron is maintained in solution only by the formation of complexes with natural chelators or ligands such as fulvic and humic acids in nearshore areas. In coastal waters, iron supply is relatively high and sources include sediment resuspension, urban inputs, biomass burning particulates, rainfall, river inputs, and runoff. Therefore, iron is seldom limiting in shallow coastal areas. Iron chemistry and its availability are complex and influenced by complexation (by organic and inorganic ligands), algal uptake mechanisms, light, temperature, and especially microbial interactions and recycling (see Boyd and Ellwood 2010; Hunter and Strzepek 2008 for further details).

Algae require iron for photosynthesis, chlorophyll synthesis, respiration, mitochondria electron transport, and nitrate reduction, and many of these processes are down-regulated during iron limitation. For example, Cyt*b6f* (6 Fe atoms), PSI (12 Fe atoms) are disproportionately Fe-rich compared to PSII (2 Fe atoms). Many of the major enzymes of nitrogen metabolism are Fe-containing proteins (NR, NiR, GOGAT, and nitrogenase), and thus nitrogen metabolism is extremely sensitive to iron stress. For example, nitrite reductase has five Fe atoms per molecule compared to two Fe atoms per molecule in nitrate reductase. The chlorotic effect of iron deficiency has been observed for some seaweeds (Liu *et al.* 2000). Radioactive ^{59}Fe uptake of two kelps, *Laminaria* and *Undaria*, and the crustose coralline *Lithophyllum* indicated that iron could be limiting for the two kelps in the northern Japan Sea where ambient iron concentrations are extremely low (<2 nM) (Suzuki *et al.* 1995).

6.5.6 Trace elements

The principal roles of Mn, Cu, Zn, Se, Ni, and Mo are mainly as enzyme cofactors and examples of compounds are given in Table 6.1. Natural copper concentrations range from 1 to 4 μM in coastal waters, whereas open-ocean values are about three orders of magnitude lower (0.5–4 nM) (Libes 1992). The free Cu^{2+} ion and some organically bound copper can be used by algae. Similar to iron, copper availability is controlled by organic complexes. Copper may be toxic at very high concentrations as discussed in sec. 9.3.4. Nickel is a component of the enzyme urease that catalyzes the hydrolysis of urea to ammonium. Additions of selenium have been reported to increase the growth of *Fucus spiralis* and the red alga *Stylonema alsidii* (Fries 1982). An important selenium-containing enzyme glutathione peroxidase occurs in the mitochondria and plastids of algae, where it acts to detoxify injurious lipid peroxides and to maintain membrane integrity (Price and Harrison 1988).

The total iodine concentration in the open ocean is ~0.5 mM and the main species are iodate and iodide (Ar Gall *et al.* 2004). The Laminariales are the strongest iodine accumulators among all living systems since they accumulate iodine in their tissues up to 30 000 times (i.e. typically 1% dry wt, but up to 4.7% in juveniles) compared to seawater (Küpper *et al.* 1998). The importance of iodine for thyroid function and the beneficial effect of iodine-rich Laminariales in treating goiter was first recognized in China and later in Europe (Küpper *et al.* 2008). Micro-imaging of *Saccharina* has revealed that iodine is mainly stored as iodide in the extracellular matrix located in the peripheral tissue (Küpper *et al.* 2008; Verhaeghe *et al.* 2008). Iodide readily scavenges a variety of reactive oxygen species and it has been suggested that its biological role is as an inorganic antioxidant (Küpper *et al.* 2008). During the occurrence of oxidative stress, iodide is effluxed to the atmosphere. In coastal regions, Laminariales are a major contributor to the iodine flux to the atmosphere due to the release and volatilization of molecular iodine and hygroscopic iodine oxides that lead to particles that can act as cloud condensation nuclei.

6.6 Long-distance transport (translocation)

Some large kelps are similar to vascular plants since they also move inorganic and organic compounds by translocation. The movement of organic compounds and the anatomical features of sieve tubes, the translocating tissues that form a three-dimensional interconnected system in the medulla in blades and stipes have been discussed in sec. 5.6. The sieve tubes have sieve plates at their ends with pores up to 6 μm in diameter to facilitate passage through them. Little is known about long-distance transport in red algae. Several studies on translocation were conducted several decades ago, but there are very few recent studies (Raven 2003). Structural and physiological data suggest that while the transport in sieve tubes in the large kelp *Macrocystis* might be by a Münch pressure-flow mechanism, symplastic long-distance transport in other brown algae is less likely (Raven 1984, 2003). *Macrocystis* sporophytes absorb nutrients from seawater and translocate photoassimilates from mature non-growing parts to support rapid growth primarily to their basal (frond-producing) and apical (blade-producing) meristems. Analysis of the sieve-tube sap has revealed the presence of mannitol (65% dry weight), amino acids (15%), and inorganic ions (Fe, Mn, Co, Ca, Zn, Mo, I, Ni) (Manley 1983, 1984). Various tracers (^{32}P, ^{86}Rb, ^{35}S, ^{99}Mo, ^{45}Ca, ^{36}Cl) have been used to show the movements of mineral elements in the thallus of *Laminaria digitata*. Phosphorus, sulfur, and rubidium undergo pronounced long-distance transport, whereas chloride, molybdenum, and calcium do not appear to move (Floc'h 1982). Nitrate was found in the sieve-tube sap of *Macrocystis pyrifera* (Manley 1983), and Hepburn *et al.* (2012) showed that labeled ammonium (^{15}N) can be transported from basal *Macrocystis* blades into the stipe, and then to more apical blades, thereby confirming long-distrance nutrient transport in this species.

There is a high demand for phosphorus in growing or meristematic regions. Movement of ^{32}P in Fucales and Laminariales is consistently from the older tissues toward the younger, growing regions. Therefore, a source-to-sink relationship exists, similar to that observed in vascular plants (Raven 1984, 2003). In *Laminaria hyperborea* the older parts of the blade apparently serve as a source of phosphate for the meristematic regions, in much the same way as has been shown for carbon assimilation (see sec. 5.6)

(Floc'h 1982). Evidence of phosphorus translocation is indirect; there is no significant difference in terms of ^{32}P uptake between young and old tissues, but the older, slowly growing tissues probably translocate unused phosphorus to new, actively growing tissues. The midrib in many Fucales is the main pathway for mineral transport, as shown by the intensity of the labeling in autoradiographs (Floc'h 1982). Most of the translocation occurs through the medulla, with some secondary lateral transport from the medulla to the meristoderm (meristematic epidermis) in the stipe.

6.7 Growth kinetics

6.7.1 Measurement of growth kinetics

Various means are available for measuring the rates of growth for seaweeds. Non-destructive methods include measuring changes in wet weight or surface area or (in kelps) measuring the movement of holes punched away from the stipe at the base of the blade (Brinkhuis 1985). For field work, non-destructive sampling is preferable, because the growth of a given seaweed can be followed over time and related to the ambient limiting-nutrient concentration. Destructive sampling includes the determination of changes in dry weight, or tissue carbon or nitrogen. Using a time series of changes in any of the foregoing parameters, the specific growth rate (μ) can be calculated as the percentage increase in the parameter per day. For example, during exponential growth, μ can be calculated from the daily increases in fresh weight according to the following equation: $\mu = [\ln(N_t/N_0)]/t$, where N_0 and N_t is the biomass at the beginning and at time t, respectively.

The most accurate way to determine the relationship between external nutrient concentration and growth rate in seaweeds is to use continuous-flow cultures. In these cultures, an attempt is made to keep the external nutrient concentration constant by using a high dilution rate (flow rate/container volume) and a low seaweed biomass/container volume ratio. The growth rate is measured at a series of limiting-nutrient concentrations when a steady state is achieved. During

the approach to steady state, a reasonably constant biomass must also be maintained by harvesting (i.e. reducing the biomass) at frequent intervals or by increasing the inflowing nutrient concentration to compensate for the increase in biomass. In phytoplankton continuous cultures, this adjustment of biomass is achieved automatically, because cells are removed along with the out-flowing medium. A steady state can also be approximated by using semi-continuous cultures, in which changes in nutrient concentration in the medium are minimized by making frequent changes in the limiting-nutrient medium since the seaweeds will take up nutrients. The time required to reach a steady state will depend primarily on the growth rate and culture conditions. DeBoer et al. (1978) have arbitrarily chosen the criterion that the biomass must increase 10-fold.

The uptake rate is a good approximation of the growth rate when nutrients do not limit growth, or when steady-state growth occurs under conditions of nutrient limitation. However, nutrient uptake rates may greatly exceed growth rates when nutrients are added to an algal culture growing under nutrient limitation (see sec. 6.4.1). The half-saturation constant of uptake, K_s, that is determined during the transient conditions of short-term uptake experiments, may also be greater than the K_s. value for growth (sometimes denoted as K_μ to eliminate confusion).

6.7.2 Growth kinetic parameters and tissue nutrients

Many steady-state studies were conducted in the 1980s, but recently, the focus has been on the influence of nutrients on uptake rates (rather than growth rates). DeBoer et al. (1978) found that the nitrogen growth kinetics for two red algae, *Agardhiella subulata* and *Gracilaria foliifera*, followed typical growth-saturation curves (Fig. 6.20). The values of K_s for growth ranged from 0.2 to 0.4 μM for various nitrogen sources, and growth rates were saturated at concentrations of NH_4^+ or NO_3^- as low as 1 μM. Similarly, the annual brown alga *Chordaria flagelliformis* had low K_s values (0.2–0.5 μM) for the three nitrogen substrates: nitrate, ammonium, and urea (Probyn and Chapman

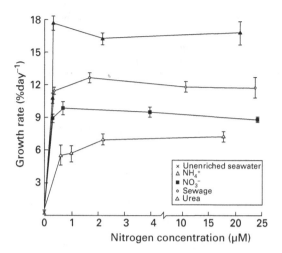

Fig. 6.20 Growth rates of the rhodophyte *Agardhiella subulata* as a function of N concentration from various sources. (From DeBoer *et al*. 1978, with permission of the Phycological Society of America.)

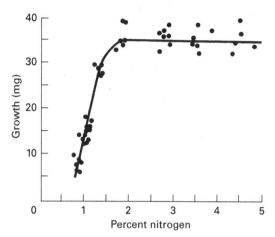

Fig. 6.21 Relationship between growth and internal nitrogen concentration in *Codium fragile*; growth measured as the increase in dry weight after 21 days. (From Hanisak 1979, with permission of Springer-Verlag, Berlin.)

1983). Those very low values were in contrast to other studies, where the growth rate was saturated at 10 μM nitrate for *Saccharina latissima* (Chapman *et al*. 1978; Wheeler and Weidner 1983), between 6 and 15 μM nitrate for juvenile *Macrocystis pyrifera*, and at 30 μM nitrate or ammonium for the estuarine green alga *Cladophora* aff. *albida* (Gordon *et al*. 1981). These latter seaweeds would be poor competitors against phytoplankton, whose growth rate is saturated at 1 μM nitrogen or even less. Growth of juvenile sporophytes of *Macrocystis pyrifera* was found to be saturated at PO_4^{3-} concentrations >1 μM (Manley and North 1984).

Although the growth rate is often related to the concentration of nutrients in the external medium, many early studies with phytoplankton have shown that the growth rate can be estimated more accurately from the concentrations of nutrients within cells (i.e. the cell quota). This basic principle is related to the common agricultural practice of plant tissue analysis, where the critical tissue concentration is where growth begins to saturate. Higher or lower tissue concentrations indicate nutrient storage or deficiency, respectively (Hanisak 1979, 1990; DeBoer 1981).

Representative critical tissue concentrations of nitrogen for a number of macrophytes are given in Table 6.5, along with their minimum tissue concentrations. The growth of *Codium fragile* was found to be more directly related to the tissue nitrogen concentration than to the nitrogen concentration in the surrounding water (Fig. 6.21). Tissue nitrogen ranged from 0.9 to 4.8%, but growth rates remained constant >2%. Whereas tissue nitrogen generally reaches saturating values with increasing concentrations of external nitrogen, phosphorus may not saturate as readily (Gordon *et al*. 1981; Björnsäter and Wheeler 1990; Pedersen *et al*. 2010) (Fig. 6.22). Therefore, the critical tissue phosphorus concentration is more difficult to determine accurately. It took 3 weeks to reach steady-state growth at a range of phosphorus concentrations in juvenile *Macrocystis pyrifera*. In the hyperbolic relationship between external PO_4^{3-} concentration and growth rate, tissue phosphorus ranged from 0.12 to 0.53% of dry weight for PO_4^{3-} concentrations from 0.3 to 6 μM and the critical tissue phosphorus was about 0.2% of dry weight (Manley and North 1984).

In fact, seaweeds do not have just one critical tissue concentration of nitrogen when other factors such as

Table 6.5 Representative maximum, critical, and minimum nitrogen and phosphorus tissue concentrations.

Species	Nutrient	Maximum (% dry wt)	Critical (% dry wt)	Minimum (% dry wt)	References
Phaeophyceae:					
Chordaria flagelliformis	NH_4^+		1.5	0.5	Probyn and Chapman (1983)
	NO_3^-		0.9	0.3	
Saccharina latissima	NO_3^-		1.9	1.3	Chapman *et al.* (1978)
Macrocystis pyrifera	NO_3^-		*a*	0.7	Wheeler and North (1980)
Macrocystis pyrifera	PO_4^{3-}		0.2		Manley and North (1984)
Pelvetiopsis limitata	NO_3^-		1.2–1.5		Fujita *et al.* (1989)
	NH_4^+		0.9		Fujita *et al.* (1989)
	Nitrogen	2.2	1.5	0.86	Wheeler and Björnsäter (1992)
	PO_4^{3-}	0.5		0.27	
Chlorophyceae:					
Chaetomorpha linum	Nitrogen	3.2	0.7	0.3	Lavery and McComb (1991b)
	PO_4^{3-}	0.23	0.04	0.01	Lavery and McComb (1991b)
Cladophora albida	Nitrogen		2.1	1.2	Gordon *et al.* (1981)
Codium fragile	NO_3^-	2.6	1.9	0.8	Hanisak (1979)
	PO_4^{3-}	0.48		0.28	
Ulva intestinalis	Nitrogen	5.1	2.5	2.0	Björnsäter and Wheeler (1990)
	PO_4^{3-}	0.73		0.37	
Ulva fenestrata	PO_4^{3-}	0.6	0.5	0.3	Björnsäter and Wheeler (1990)
	Nitrogen	5.5	3.2	2.4	
Ulva rigida	NO_3^-		2.4		Fujita *et al.* (1989)
	NH_4^+		3.0		Fujita *et al.* (1989)
	Nitrogen	3.2	2.0	1.3	Lavery and McComb (1991b)
	PO_4^{3-}	0.06	0.025	0.02	Lavery and McComb (1991b)

a Linear relationship between μ and Q over the experimental range of Q.

light are also considered. This is because light modifies the nitrogen requirement for maximal photosynthesis and growth by altering biochemical constituents (i.e. pigments, RuBisCO, and nitrogen reserves) that affect the nitrogen level (Lapointe and Duke 1984; Shivji 1985). For example, the nitrogen level for *Gracilaria tikvahiae* at 7% of surface irradiance averaged 25% more than that for seaweeds grown at surface irradiance under similar nitrogen conditions, because of increases in pigments (e.g. phycoerythrin), tissue NO_3^-, and proteins. The ambient concentration of one nutrient may also affect the critical tissue concentration of another nutrient. Phosphorus-limited *Ulva* maintained high levels of tissue nitrogen, but N-limited algae experienced reductions or depletion of

phosphorus (Björnsäter and Wheeler 1990). Therefore, because light and ambient nutrient concentrations vary seasonally, the critical N and P tissue concentrations also vary seasonally (Wheeler and Björnsäter 1992).

Tissue nutrient concentrations have been used to monitor nutrient limitation in the coastal waters of Denmark using *Ceramium virgatum* (previously *C. rubrum*). Monthly monitoring of tissue nitrogen and N and P concentrations, followed by comparisons with laboratory-determined critical tissue concentrations, revealed that nitrogen was limiting for growth, except in May when growth was P-limited (Lyngby 1990).

The maximum to minimum tissue nitrogen ratio has been used to explain species competition and

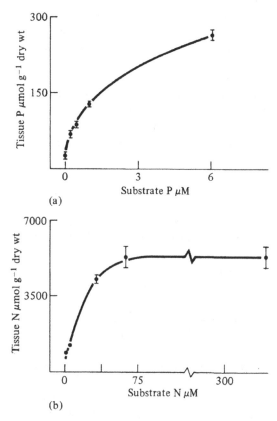

(a)

(b)

Fig. 6.22 Concentrations of total nutrients in *Cladophora* tissues as a function of the nutrient concentration supplied in complete medium: (a) Tissue P when P was supplied at 0–6 μM; nitrogen at 375 μM; (b) Tissue N when N was supplied at 0–375 μM; phosphorus at 12 μM. (From Gordon *et al.* 1981, with permission of Walter de Gruyter and Co.)

subsequent succession. Macroalgal biomass was monitored in the Peel-Harvey estuary in southwestern Australia for two decades (Lavery *et al.* 1991). Macroalgal blooms occurred suddenly in the late 1960s, and *Cladophora montagneana* dominated until 1979. A catastrophic event compounded by a series of unfavorable conditions resulted in the loss of *Cladophora* from the deep areas and its estuary-wide replacement by *Chaetomorpha*, which was more competitive in shallow areas. Periods of high nutrient concentrations favored *Ulva rigida* and *Ulva intestinalis* (previously *Enteromorpha*), whereas

Chaetomorpha resumed dominance during periods of low nutrients. A study of the nutritional ecophysiology of *Chaetomorpha* and *Ulva* highlighted some of the physiological mechanisms that were responsible for the change in their ecological distribution (Lavery and McComb 1991b). Critical tissue nitrogen and phosphorus concentrations revealed that *Ulva* requires about twice as much nitrogen as *Chaetomorpha* (20 vs. 12 mg gdw[-1]) but less phosphorus (0.25 vs. 0.5 mg gdw[-1]). However, *Ulva* takes up ammonium and nitrate at slower rates than *Chaetomorpha*. Thus, *Ulva* frequently is N limited during the spring because of its high nitrogen requirements, its reduced ability to take up nitrogen, and its limited ability to store nutrients over winter. (Its ratio of maximum tissue nitrogen to minimum tissue nitrogen is 2:5; see Table 6.5.) *Chaetomorpha* persists over winter, allowing it to take advantage of elevated nutrient concentrations over the winter and store nutrients (its maximum:minimum ratio for tissue nitrogen and phosphorus is 11 and 23 respectively) that are used to support high growth rates in the spring and summer when nitrogen becomes limiting. In addition, dense accumulations (mats) of *Chaetomorpha* reduce the oxygen levels in the water overlying sediments, and the subsequent anoxia induces phosphorus release from the sediments (Lavery and McComb 1991a). In summary, the maximum to minimum tissue nitrogen ratio is an index of nitrogen-storage capacity (e.g. *Chaetomorpha* stores five times as much nitrogen as *Ulva*) and the critical nitrogen concentration indicates the onset of maximum growth rates.

6.8 Effects of nutrient supply

6.8.1 Surface-area:volume ratio and morphology

Because nitrogen is the primary nutrient that limits seaweed growth, variations in seaweed growth rates should parallel variations in the nitrogen supply. The uptake capacity (V_{max}) is a direct function of the surface-area:volume (SA:V) ratio (Rosenberg and Ramus 1984; Wallentinus 1984; Hein *et al.* 1995), whereas the nitrogen-storage capacity varies approximately inversely with the SA:V ratio (Rosenberg and Ramus 1982; Duke *et al.* 1987). This suggests that the degree of coupling between seaweed growth rates and the nitrogen supply may also be a function of SA:V. High SA:V species, which have high V_{max} values, but low storage capacity, will have growth rates highly correlated with the nitrogen supply. In contrast, low SA:V species with low V_{max} values will store nitrogen and have growth rates relatively independent of the nitrogen availability (Rosenberg and Ramus 1982; Raven and Taylor 2003). For example, *Ulva curvata*, an opportunistic annual with a high SA:V ratio, is capable of utilizing transiently high ammonium concentrations (high V_{max}), and it is capable of high growth rates. In contrast, *Codium fragile*, a persistent perennial, cannot utilize ammonium pulses efficiently, has a slow growth rate, and tends to integrate short-term variability by virtue of its low growth rate (Ramus and Venable 1987). Thus, a seaweed's "functional form" (Littler and Littler 1980) (sec. 1.2.2, Table 1.1) may determine its ability to buffer nutrient variability. In general terms, opportunistic species often are annuals that feature rapid nutrient uptake and rapid growth potentials, high SA:V ratios, low nutrient storage capabilities, and low levels of defense against herbivory (presumably because tissue loss is readily replaced via rapid nutrient uptake and growth), and they typically dominate eutrophic environments (Littler and Littler 1980). On the other hand, perennials tend to be slow growing and can store large quantities of nutrients that can act as a buffer during times of nutrient shortage (Carpenter 1990). In a review of macroalgae that dominate in nutrient-rich estuaries, Raven and Taylor (2003) noted that the "nuisance" algae were ephemeral (i.e. annuals), rapidly growing r-selected species [see Table II, II, and IV in Raven and Taylor (2003) for kinetic parameters for r- and k-selected species]. The nuisance algae are almost exclusively chlorophytes belonging to the genera *Chaetomorpha*, *Cladophora*, and *Ulva*, except for browns such as *Ectocarpus* and *Pilayella*.

A consequence of multicellularity is a decreased SA:V ratio for seaweeds. Some seaweeds appear to respond to nutrient deficiency by the production of hyaline hairs from the thallus surface, akin to the production of root hairs in vascular plants. DeBoer (1981) and DeBoer and Whoriskey (1983) observed that in the presence of low nitrogen concentrations (e.g. $NH_4^+ < 0.5$ μM) and moderate agitation, hair-cell formation was enhanced in *Hypnea musciformis*, *Gracilaria* sp., *Agardhiella subulata*, and *Ceramium virgatum*. Ammonium concentrations >20 μM inhibited hair formation. Interestingly, cytoplasmic streaming occurred in hairs on the apical (meristematic) regions of the thalli, where nutrient uptake rates are high, but not in hairs on the lower part of the thallus, where low uptake rates occur. Ammonium uptake rates for *Ceramium virgatum* with hairs were approximately twice those for those without hairs. DeBoer and Whoriskey (1983) suggested that these hairs may increase the surface area and hence the number of nutrient uptake sites. In *Fucus*, hair formation began in late February, even before phosphate was depleted from the seawater, and formation stopped in October (Hurd *et al.* 1993). In the laboratory, hairs formed more rapidly under phosphorus limitation and these seaweeds were capable of taking up phosphate two- to three-times faster than hairless *Fucus spiralis*. Therefore, the production of hairs may help macroalgae compete with phytoplankton and bacteria for the available supplies of phosphate. Other mechanisms by which hairs might enhance nutrient acquisition are discussed in sec. 2.6.2.

6.8.2 Chemical composition and nutrient limitation

Extensive analysis of the chemical composition of marine plankton has revealed that the ratio relating carbon, nitrogen, and phosphorus is 106:16:1 (by atoms) (i.e. C:N = 7: 1 and N:P = 16:1). This is commonly referred to as the Redfield ratio. Decomposition of this organic matter tends to occur according to the same ratio. However, Atkinson and Smith (1983) reported that benthic marine macroalgae and seagrasses are much more depleted in phosphorus and less depleted in nitrogen, relative to carbon, than are phytoplankton. Therefore, the median ratio C:N:P for seaweeds is about 550:30:1 (i.e., C:N = 18:1 and

N:P = 30:1). Thus, the amounts of nutrients required to support a particular level of net carbon production are much lower for macroalgae than for phytoplankton. In addition, seaweeds should be less prone to phosphorus limitation with their N:P ratio of 30:1 than are phytoplankton, with an N:P ratio of 16:1. The high C:N:P ratios in seaweeds are thought to be due to their large amounts of structural and storage carbon, which vary taxonomically. Niell (1976) found higher C:N ratios in the Phaeophyceae than in either the Chlorophyceae or Rhodophyceae. The average carbohydrate and protein contents of seaweeds have been estimated at about 80 and 15% respectively, of the ash-free dry weight (Atkinson and Smith 1983). In contrast, the average carbohydrate and protein contents of phytoplankton are 35 and 50%, respectively (Parsons *et al.* 1977).

In an extensive survey, Lapointe *et al.* (1992) found that macroalgae from a variety of carbonate-rich tropical waters in the Caribbean were significantly depleted in phosphorus (relative to carbon and nitrogen), as compared with macroalgae from temperate waters (0.07% P vs. 0.15% P on a dry-weight basis). The

mean tissue N:P ratio for tropical macroalgae was 43.4, compared with 14.9 for temperate forms. These data and observations of high alkaline phosphatase activities suggest that these tropical macroalgae may have a higher tendency to be P limited, whereas temperate macroalgae tend to be N limited, except for the study by Wheeler and Björnsäter (1992) discussed below.

The C:N ratio in tissue is often used to determine if a macroalga is N limited using the guide line that C:N ratios >20 indicate possible N limitation. When *G. tikvahiae* was limited by both light and nitrogen, growth rate was a parabolic function of the tissue C:N ratio (Fig. 6.23), because light and nitrogen have opposite effects on the relationship between the C:N ratio and the growth rate (Lapointe and Duke 1984). Under light limitation, NO_3^- uptake is relatively rapid compared with rates of carbon fixation, and that results in a low C:N ratio and accumulation of nitrogen reserves such as phycoerythrin and tissue nitrate. Alternatively, under high light with nitrogen limitation, carbon fixation is relatively rapid compared with the rates of NO_3^- uptake and that results in a high C:N ratio and decreased nitrogen content (i.e. phycoerythrin).

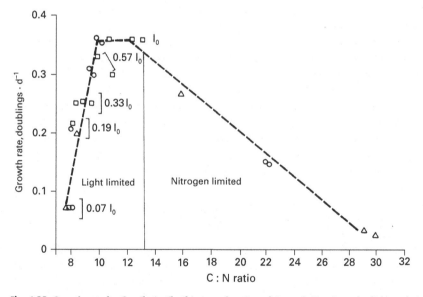

Fig. 6.23 Growth rate for *Gracilaria tikvahiae* as a function of tissue C: N ratio under light and nitrogen limitation. *G. tikvahiae* was grown under identical conditions in outdoor, continuous-flow seawater cultures. (From Lapointe and Duke 1984, with permission from the Phycological Society of America.)

Laboratory studies on *Ulva* showed that nutrient supplies were related to nutrient composition in their tissues (i.e. supply ratios high in nitrogen and low in phosphorus resulted in high tissue N:P ratios as expected) (Björnsäter and Wheeler 1990). Wheeler and Björnsäter (1992) measured seasonal variations in tissue N:P ratios in five seaweeds (range of N:P was 5–22) and compared them with seasonal variations in ambient nutrient concentrations and critical tissue nitrogen and phosphorus concentrations. Surprisingly, they found that phosphorus limitation is more frequent than nitrogen for the five seaweeds off Oregon, USA, that is in contrast to phytoplankton that are always N limited.

Deviations from the Redfield ratio frequently are used to infer which nutrient is limiting the growth of phytoplankton. P-limited phytoplankton have N:P ratios >30:1, whereas N-limited phytoplankton have N:P ratios <10:1. In N-limited seaweeds, C:N ratios are >18:1 because of a decrease in amino acids and proteins and an increase in carbohydrates (Björnsäter and Wheeler 1990). For example, in *Ulva lactuca*, the concentrations of β-alanine and asparagine can decrease 20-fold under nitrogen starvation. The red algae showed greater variability in amino acid composition, whereas the browns had a more uniform composition, with alanine, aspartate, and glutamate being the major amino acids (Rosell and Srivastava 1985). *Chondrus crispus* has a higher carrageenan content in low N seawater than in N-enriched medium and a similar effect is seen in the agar content of *Gracilaria foliifera*. Elevated transient uptake rates of ammonium have also been used to indicate nitrogen limitation in *Gracilaria foliifera* and *Agardhiella subulata* when tissue C:N ratios are >10 (D'Elia and DeBoer 1978).

The chemical composition of many temperate seaweeds varies seasonally, primarily because of the onset of nitrogen limitation in the coastal waters in summer. Wheeler and Srivastava (1984) found that in British Columbia, Canada, tissue nitrate (as ethanol soluble nitrate) and total nitrate paralleled the ambient nitrate levels and showed summer minima and winter maxima (from 0 to 70 μmol gfw^{-1} for nitrate and from 0.9 to 2.9% of dry weight for total N) in *Macrocystis*

pyrifera (previously *M. integrifolia*). In contrast, in California, Wheeler and North (1981) found that neither nitrate nor ammonium accumulated in the tissue of *M. pyrifera* possibly due to the higher light levels in California where slow growth could occur in the winter; free amino acids accounted for a major portion of the soluble nitrogen. Juvenile *M. pyrifera* sporophytes do not appear to store nitrogen.

Higher molecular weight compounds may be involved in nitrogen accumulation and storage. The dipeptide L-citrullinyl-L-arginine can accumulate to high concentrations when *Chondrus crispus* is supplied with nitrate or ammonium at low temperatures (Laycock *et al*. 1981). Much of this reserve can be readily mobilized for growth of the seaweed when it encounters higher temperatures, increased irradiance, and low levels of external nitrogen, which are common during the late spring and summer months. Consequently, rapid growth rates sustained by declining nitrogen reserves persist well after the disappearance of ambient nitrogen. The accumulation of soluble nitrogen reserves appears to be optimized under conditions of low temperature and reduced light (Rosenberg and Ramus 1982). Under these conditions, the rate of accumulation exceeds the requirements for growth. Pigments, or more likely the proteins associated with them, may also play a secondary role in nitrogen storage (Smith *et al*. 1983). There have been several observations of marked decreases in pigment content under nitrogen deficiency (DeBoer 1981). Chlorophyll and phycoerythrin concentrations in *Gracilaria foliifera*, *Agardhiella subulata*, and *Ceramium virgatum* were strongly influenced by the concentrations of inorganic nitrogen in the culture medium.

6.8.3 Nutrient storage and nutrient availability

Nutrient storage interacts with growth rate (the alga's nutrient requirement/demand ratio) and determines low long an alga can continue to grow under nutrient limitation. Pedersen and Borum (1996) tested the hypothesis that fast-growing algae are more impacted by nutrient limitation than slow-growing algae using

Table 6.6 Critical and maximum tissue concentrations of P (mean ±SD), absolute amount of stored P and estimated storage capacity that stored P can support maximum and simulated growth rates, respectively. Parentheses: range of storage capacities obtained by using a range of photosynthetic quotient values to estimate growth rate from photosynthesis (from Pedersen *et al.* 2010, with permission from Inter-Research Science Center).

	Critical tissue P-concentration	Maximum tissue P-concentration (μmol P g^{-1} DW)	Absolute P-store	Storage capacity at max. growth rate (wk)	Storage capacity at simulated seasonal growth rate (wk)
Ulva lactuca	65.5	125 ±20	59.5	2 (2–2)	4 (3–4)
Ceramium virgatum	142.9	186 ±22	43.1	1 (1–2)	5 (4–7)
Fucus vesiculosus	<38.7	173 ±7	>134.3	>10 (10–13)	>12 (11–>52)
Fucus serratus	71.9	161 ±20	89.1	8 (7–11)	11 (11–>52)
Ascophyllum nodosum	48.1	90 ±9	41.9	12 (11–16)	22 (15–>52)
Laminaria digitata	69.4	133	63.6	19 (17–31)	>52 (>52–>52)

phytoplankton, four ephemeral algae (*Ulva*, *Cladophora*, *Chaetomorpha*, and *Ceramium*) and one perennial, *Fucus*. The N requirements per unit of biomass and time were up to 30-fold higher for the fast-growing compared to the slow-growing algae due to the 10-fold faster growth rate and 3-fold higher critical N concentration that was required to maintain μ_{max}. They found no systematic variation with respect to groups or morphology in the size of the storage pools. Therefore, nutrient storage cannot be evaluated from only concentrations because interspecific variations in the critical N concentration and growth rates influence the size of the storage pool and how long the storage pool can last under nutrient limitation. They concluded that low nutrients in summer restricts the growth of phytoplankton and ephemeral macroalgae, while slow-growing perennials such as *Fucus* can maintain near maximal growth rates during most of the summer because of their slow growth rate (nutrient demand) which allows the nutrient pool to last much longer. In a similar study of P limitation in Oslofjord, Pedersen *et al.* (2010) showed that slow- and fast-growing macroalgae had similar pool sizes, but the much slower growth rate of the perennial

macroalgae allowed the P storage pool to be slowly utilized over many weeks (Table 6.6). Therefore, the storage pool should be viewed as a supply and demand scenario, where storage capacity (i.e. how long the pool can sustain growth) represents the supply and demand is represented by the growth rate. They found that when P was limiting during summer, thin, fast-growing species such as *Ulva* and *Ceramium* were P limited because their P storage was quickly used up. In contrast, thicker, slower growing species such as *Fucus* and *Ascophyllum* were less P limited because of greater P storage capacity (Table 6.6; Pedersen *et al.* 2010).

Light and temperature limitations on growth tend to uncouple growth from nitrogen uptake and storage. Low light and temperature lowers a seaweed's demand because growth is slower and hence the storage pool and/or a lower supply of nutrient may be adequate for maximal growth rate for many weeks (i.e. a lower demand is balanced by a lower supply) (Duke *et al.* 1989). This may explain why seaweeds accumulate nitrogen at low levels of light and temperature, because nitrogen uptake may be less limited by light and temperature than is growth.

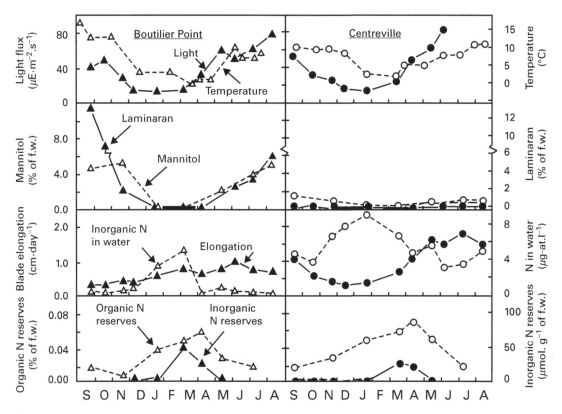

Fig. 6.24 Growth, internal inorganic and organic nitrogen reserves, and laminaran and mannitol content of blades of *Saccharina longicruris* at two sites in Nova Scotia with contrasting light and nitrogen environments. (From Gagne *et al.* 1982, with permission from Springer-Verlag, Berlin.)

6.8.4 Growth rate and distribution

The growth and productivity of seaweeds are controlled in part by environmental factors such as irradiance, temperature, nutrient availability, and water movement. Marked seasonal fluctuations in nutrient availabilities (especially nitrogen) are illustrated in the classical seasonal study of the kelp *Saccharina longicruris* (previously *Laminaria*) off eastern Canada (Hatcher *et al.* 1977; Gagne *et al.* 1982). In these kelp beds, there are two main limiting factors, light and nitrogen availability. Nitrogen is present year round at a coastal upwelling site, while nitrogen is limiting for 8 months in a nearby small bay. The interactions between light and the different nitrogen availability at the two sites gave rise to

marked differences in the seasonal pattern of kelp growth. In the nitrogen-limited bay, the seaweeds grow mainly during the period of nitrogen availability in winter and early spring. The bulky thallus is able to store substantial quantities of both inorganic and organic nitrogen during late fall and winter, to be used for growth in late spring and early summer, when nitrogen is becoming limiting but when light conditions are improving (Fig. 6.24). Thus, they are able to prolong their growth in spring for at least 2 months after nitrogen becomes limiting. During the summer, when irradiance is high, seaweeds in the bay store carbohydrate as laminaran. These carbon reserves are remobilized and used in conjunction with nitrogen (high ambient concentrations) to produce amino acids and proteins for growth in early winter, even though light is limiting for photosynthesis. At the upwelling site, seaweeds do not build up laminaran reserves, and growth rate follows irradiance, being greatest in summer when there is high nitrogen from upwelling (Fig. 6.24). Since these seaweeds do not have carbohydrate reserves, their growth rate declines during early winter as irradiance declines.

A similar example to the upwelling versus small bay comparison above, involves the comparison between Arctic and Antarctic seaweeds. The Arctic shows high seasonal fluctuations and is usually N limited in the summer. The endemic kelp *Laminaria solidungula* exhibited characteristics of a storage specialist with high V_{max} which allows it to take advantage of the high spring nitrate concentrations (Fig. 6.25) (Korb and Gerard 2000). It stops its growth during the N-depleted summer and accumulates carbon skeletons that are used in early spring when light is still low but nitrate is available. It utilizes both nitrate and ammonium equally, allowing it to minimize the impact of seasonally low nitrogen concentrations through luxury uptake and storage. In contrast, the Antarctic is similar to the upwelling site above since nutrients are seldom limiting. The Antarctic endemic kelp-like *Himantothallus grandifolius* (Desmarestiales) has a high V_{max} for ammonium (Fig. 6.25), allowing it to utilize pulses of ammonium and ammonium uptake was reduced very little by prolonged periods of darkness, while nitrate uptake was greatly reduced. Thus, this species shows a strong energy-conserving trait since less energy is required to assimilate ammonium compared to nitrate.

There is always some nutrient supply via regeneration and other processes even when ambient nutrients are at limiting concentrations. Opportunistic species often have high-affinity uptake systems to take up nutrients even at or near limiting concentrations. An annual, *Chordaria flagelliformis*, maintains high growth rates during the summer even when the ambient nitrogen has become exhausted (Probyn and Chapman 1982). *Chordaria* has very low K_s values (0.2–0.5 µM) for nitrate, ammonium, and urea, and therefore this annual can scavenge nitrogen (Probyn and Chapman 1983). The comparatively small

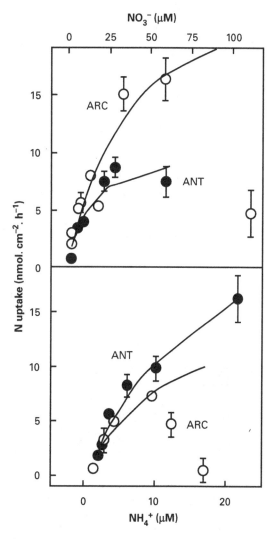

Fig. 6.25 Nitrate and ammonium uptake by Antarctic *Himantothallus* (ANT) and Arctic *Laminaria* (ARC) at different NO_3 and NH_4 concentrations. (From Korb and Gerard 2000, with permission from Inter-Research Science Center.)

intracellular nitrogen pool (0.1–0.4% of dry weight) indicates that *Chordaria* is typical of many opportunistic species since newly absorbed nitrogen is directed into growth rather than storage.

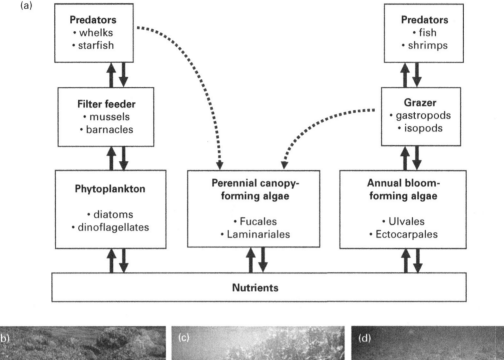

Fig. 6.26 Community food-web interactions. Upward arrows indicate resource ("bottom-up") effects and downward arrows indicate consumer ("top-down") effects. Dotted arrows indicate indirect positive effects. Nutrient enrichment stimulates the growth of phytoplankton and annual bloom-forming algae over perennial algae. Predators and grazers limit the growth of filter feeders and annual algae and may indirectly maintain perennial algae. Excessive eutrophication can override top-down control and favor the development of: (b) mussel beds or, (d) annual algae, both of which displace (c) perennial algal canopies. (From Worm and Lotze 2006, with permission from the Association for the Sciences of Limnology and Oceanography, Inc.)

6.8.5 Effects of nutrient enrichment on community interactions

The complexity of attributing an increase in ephemeral/opportunistic species only to nutrient enrichment is illustrated in Figure 6.26. There are a large number of interacting factors such as three large groups of primary producers, microalgae (phytoplankton and epiphytes), perennial canopy-forming macroalgae, and annual bloom-forming macroalgae, a range of grazers and their predators all interacting with abiotic factors such as light, temperature, wave exposure, type of substratum, etc. Therefore, it is not surprising that there is a diversity of reported community responses to a simple increase in nutrient enrichment and that there are many exceptions to the generalizations (Valiela *et al.* 1997; Worm and Lotze 2006; Russell and Connell 2007; Karufvelin *et al.* 2010). Some examples are provided below for just a few of the many nutrient enrichment effects on community structure studies. In addition, mesocosms that provide essential statistical replication do not completely simulate all field conditions and may yield different results from the natural environment, especially due to differences in the epiphyte load (see also sec. 9.6 on eutrophication).

Much attention has been focused on determining if ecophysiological traits such as nutrient uptake kinetics, or SA:V ratios can explain species dominance patterns in macroalgal blooms. Lotze and Schramm (2000) concluded that the overwhelming dominance of *Pilayella* over *Enteromorpha* in the Baltic Sea could not be explained by interspecific physiological differences. They found that the species distribution was strongly controlled by different overwintering and recruitment strategies as well as selective herbivory. Experimental nutrient enrichment in the field favored each species equally since they had similar nutrient uptake kinetics, however, if grazers were present, *Pilayella* dominated because grazers preferred to consume *Enteromorpha*. Therefore, community interactions must be considered in order to explain species dominance in the field.

Nutrient additions to rocky shore mesocosms yielded remarkably different results when a 2.5-year study was extended to 5 years. Nitrogen and phosphorus were added to concrete mesocosms in the intertidal zone that were dominated by fucoids (Bokn *et al.* 2003). At the end of 2.5 years, fast-growing ephemeral/opportunistic green algae and periphyton increased, while filamentous red algae and large perennial brown algae were largely unaffected. However, on a biomass basis, the fucoids were still dominant. Therefore, established rocky shore communities seemed to be able to resist the effects of increased nutrient loading. Bokn *et al.* (2003) concluded that the combined effects of competition for space and light imposed by the canopy-forming algae, preferential grazing on the opportunistic algae by herbivores, and physical disturbance, prevented the fast-growing algae from becoming dominant after 2.5 years. Additional recruitment studies showed that the opportunistic algae could become dominant when free space became available and if herbivore grazing was also low. They concluded at the end of the 2.5-year study

that the response to nutrient enrichment was a very slow process and may not be enough to stimulate structural changes in some rocky shore communities. Fortunately, they had funding to continue the study for an additional 2.5 years (i.e., total of 5 years) and found surprising results (Kraufvelin *et al.* 2006). During year 4 of the continuing nutrient enrichment, the cover of *Fucus vesiculosus* and *F. serratus* started to decline and by year 5, these canopy species crashed and the opportunistic green algae took over. They continued the study for another 2 years, but used only natural seawater in the mesocosms to determine if the algal community could recover back to the original fucoid-dominated community. In less than 2 years, the algal and animal communities returned to normal. This exceptionally long-term study, indicated that there may be a long-term delay (up to 4 years) in the response to field nutrient enrichments, but the recovery was surprisingly rapid. Therefore, conclusions on community changes based on short-term (e.g. 2 years) studies should be considered with caution and longer term studies are necessary to capture community changes since some processes are slower than others.

In tidal pools, nitrate concentrations are influenced by upwelling, tidal and wind mixing, and microbial activity, while ammonium concentrations are controlled by local-scale excretion by invertebrates (Bracken and Nielsen 2004) and microbial oxidation. Especially at low tide, both forms of nitrogen are available to seaweeds, and Bracken and Stachowicz (2006) found that seaweed species were complementary with respect to their use of nitrate and ammonium; some species prefer nitrate, whereas others prefer ammonium. Thus, when nitrate and ammonium were available simultaneously, uptake by a diverse assemblage was 22% greater than the monoculture average because different species were complementary in their use of different nitrogen forms.

Tidal elevation on the shore can interact with herbivory to influence seaweed community structure and richness, and ultimately nitrogen uptake by seaweed assemblages. Herbivores did not affect algal richness, but low herbivory was correlated with algal species that had higher nitrate-uptake rates (Bracken *et al.* 2011). Species living higher on the shore had higher biomass specific nitrate-uptake rates, particularly at high nitrate concentrations. Grazed seaweeds had lower nitrate-uptake rates when nitrate was low, presumably due to the reduction in the SA:V ratio due to removal of young bladelets and palatable species with high SA:V. Bracken *et al.* (2011) found no relationship between nitrate uptake and seaweed richness when it was considered alone. However, when richness, herbivory and nitrate uptake were evaluated together, all three factors had a strong effect on nitrate uptake at low nitrate concentrations.

Selective grazing can have a top-down effect on macroalgal biomass of certain species by removing young bladelets that in turn has a bottom-up effect by reducing nitrogen uptake. The interactions between grazing and nutrient uptake may dramatically influence the structure and functioning of intertidal communities. When small crustaceans consume large macroalgae, they often target apical regions and small young blades that are more palatable and less chemically defended (Bracken and Stachowicz 2007). These sections of the seaweed usually have high growth rates and a large SA:V ratio which accounts for their disproportionately high nutrient-uptake rates. Bracken and Stachowicz (2007) found that the kelp crab *Pugettia producta* fed selectively on the kelp *Egregia menziesii* and removed thallus parts that had a high SA:V ratio and reduced the kelp's biomass specific uptake of nitrogen by 65% per remaining algal tissue (Fig. 6.27). Grazing is often so intense that morphological changes are dramatic (Fig. 6.28).

In their work evaluating effects of local-scale ammonium excretion on seaweed diversity, Bracken and Nielsen (2004) selected tidal pools of different volume and mussel biomass in order to test the relationship between nitrogen loading and algal diversity. High nitrogen loading due to ammonium excretion by mussels, led to a doubling of the number of species and a higher abundance of faster growing species with high ammonium uptake rates to satisfy their higher nitrogen requirements. Bracken and Nielsen (2004) suggest that an ecosystem has a critical level of nutrient loading, below which small nutrient additions increase diversity when the system is nutrient limited, but above it, diversity declines with excessive enrichment.

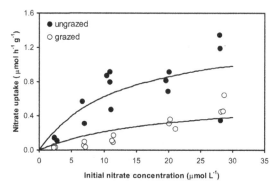

Fig. 6.27 Nitrate uptake by a grazed and ungrazed kelp *Egregia menziesii* as a function of NO_3 concentrations. (From Bracken and Stachowicz 2007, with permission from Inter-Research Science Center.)

(a) Ungrazed thallus　　(b) Grazed thallus

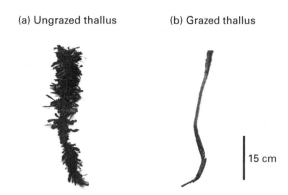

15 cm

Fig. 6.28 (a) Ungrazed thallus of *Egregia menziesii* with many bladelets. (b) Grazed thallus of *E. menziesii* by the kelp crab *Pugettia producta* with only the central stipe remaining. (From Bracken and Stachowicz 2007, with permission from Inter-Research Science Center.)

The reefs of Jamaica and southeast Florida appear to have surpassed the critical level of nutrient enrichment, resulting in pronounced macroalgal blooms (e.g. *Chaetomorpha*, *Sargassum*, and *Codium*). There are different views on the factors responsible for these blooms. Hughes *et al.* (1999) mainly support reduced herbivory, while Lapointe (1999) states that both reduced herbivory and anthropogenic nutrient inputs are important. Lapointe (1997) illustrates how top-down grazing and nutrient availability interact and determine the relative dominance of corals, turf algae,

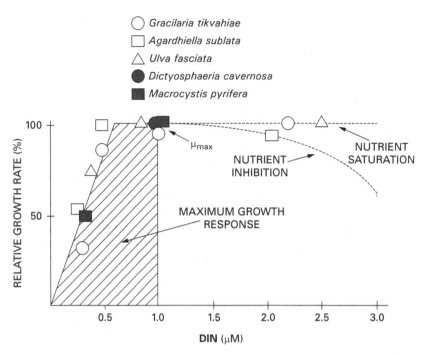

Fig. 6.29 Relationship between relative growth rate (μ:μ_{max}) of several macroalgae and dissolved inorganic nitrogen (DIN). The hatched area indicates N limitation for growth between 0 and ~0.5 to 1 µM. When DIN is >1 µM, growth rates reach μ_{max} or decrease due to epiphytic fouling or toxic effects especially for NH$_4$. (From Lapointe 1999, with permission from the Association of Sciences for Limnology and Oceanography, Inc.)

coralline algae and macroalgae (Fig. 3.8). Lapointe (1999) argued that the increase in the macroalgal turfs was due to a combination of reduced herbivory as a result of overfishing and the widespread die-off of the echinoid *Diadema*, plus the synergistic effects of increased anthropogenic nutrients from sewage and groundwater when natural nutrients inputs are low. Hence, on the north coast of Jamaica with high natural nutrient inputs, macroalgae dominate as shown in the lower right quadrant of Figure 3.8, rather than algal turfs. Water column nutrient concentrations represent the net residual sum of various fluxes/rates such as internal nutrient cycling, algal assimilation, and external inputs, and offer a direct estimate of nutrient sufficiency or limitation for algal growth. The relationship between algal growth rate and DIN concentration for several species indicates that when average concentrations are <0.5 µM on coral reefs, growth rates are limited by DIN

irrespective of flow rate. Under these conditions, an increase in DIN concentrations could lead to an increase in growth rate and the initiation of a macroalgal bloom (Fig. 6.29). For phosphorus, the thresholds are 0.1 to 0.2 µM (Lapointe *et al.* 1992). On the carbonate-rich reefs of Jamaica, evidence that the macroalgae such as *Chaetomorpha linum* were periodically P limited came from the high DIN:P ratios of 85 in the groundwater and the high C:P and N:P tissue ratios of 973 and 45 respectively (compared to the Atkinson ratio of 550C:30N:1P) and high alkaline phosphatase activity. In contrast, macroalgae such as *Codium isthmocladum* on the siliciclastic reefs of southeastern Florida are N limited (Lapointe 1997). Near sites of active groundwater discharge, *Caulerpa brachypus* has invaded during the last few decades (Lapointe and Bedford 2010), somewhat analogous to the invasion of *Caulerpa taxifolia* in the Mediterranean.

6.9 Synopsis

Seaweeds are responsible for 5–10% of the global marine primary productivity and their biomass per unit area may be 400-times greater than phytoplankton in shallow areas. Since the growth rate of seaweeds is slower than phytoplankton, their mass specific N demand is also lower than phytoplankton and hence they can survive more easily during periods of nutrient limitation. The chemical composition of phytoplankton is 106C:16N:1P (by atoms) (i.e. C:N = 7:1 and N:P = 16:1), while the ratio for seaweeds is 550C:30N; 1P (i.e., C:N = 18:1 and N:P = 30:1). Thus, the amount of nutrients required to support a particular level of net carbon production is much lower for macroalgae than for phytoplankton. In addition, seaweeds should be less prone to phosphorus limitation with their N:P ratio of 30:1 than are phytoplankton, with an N:P ratio of 16:1. The high C:N:P ratios in seaweeds are thought to be due to their large amounts of structural and storage carbon.

Seaweeds require various mineral ions for growth, and nitrogen is the most frequently limiting, followed by phosphorus and in a few special cases iron. Elements are taken up as ions (charged particles), which diffuse to the cell surface, through the cell wall and to the cell membrane/plasmalemma. Ions may pass through the plasmalemma passively, either by passive diffusion or by facilitated diffusion; however, they are usually transported by an active process requiring energy. For most ions, the relationship between an ion's uptake rate and the ion's concentration in the seawater can be described by a rectangular hyperbola, with the term V_{max} denoting the maximum uptake rate, K_s the concentration at which $V = \frac{1}{2}V_{max}$, and $V_{max}: K_s$, the affinity for the nutrient. In many cases, saturation uptake kinetics do not occur and the uptake is linear with increasing concentrations. Nutrient-uptake rates can be measured by isotope accumulation in the seaweed tissue or by monitoring the disappearance of the nutrient from the medium. When some N-limited annuals (opportunistic species) are resupplied with a pulse of nitrogen (especially ammonium), the time series of their uptake rate can be divided into phases. "Surge" uptake (V_s) occurs over about an hour during the pool-filling process, followed by a slower rate (V_i) that is a good approximation of the assimilation rate (the incorporation of nutrients into amino acids).

Another way to compare nutrient demand by various seaweeds is to compare the ratio of maximum nutrient-uptake rate to the uptake rate at the maximum ambient concentration of a nutrient. This ratio is called the "safety factor" and it provides a simple estimate of the amount of surplus capacity of a nutrient-uptake system. Hence, a low safety factor means that the uptake system is operating close to the maximum rate. Safety factors tend to be low for nitrate and phosphate uptake by seaweeds and ammonium uptake by phytoplankton, while seaweeds had a high safety factor for ammonium uptake. Therefore, ammonium is potentially a very important nitrogen source for seaweeds since there was a three-fold higher capacity for ammonium uptake compared to nitrate uptake (i.e. uptake rate at 5 μM NO_3 = uptake rate at 1.5 μM NH_4).

Dissolved inorganic nitrogen (nitrate, nitrite, ammonium) and organic nitrogen (urea, amino acids) are the main sources of nitrogen for seaweeds. Ammonium is generally preferred to all other nitrogen forms, except in some kelp. The uptake rate of nitrogen and other ions is influenced by light, temperature, water motions, level of desiccation, and the ionic form of the element. Biological factors that influence uptake include the type of tissue, the age of the seaweed (annual vs. perennial), its nutritional past history, and inter-seaweed variability. Sheet-like or filamentous thalli with a higher surface area:volume ratio tend to have a higher uptake rate than thick bulky seaweeds. After ions are taken up, some (e.g. N, P, S) may be translocated to other tissues within the thallus, especially in large seaweeds such as kelp.

After nitrate and ammonium are taken up, they usually are assimilated into amino acids and proteins. Nitrate is reduced intracellularly to ammonium via nitrate and nitrite reductases. Ammonium is incorporated into amino acids mainly via the glutamine synthetase pathway. Glutamine is the first product, followed by other amino acids via aminotransferase processes. The nutrients may be stored in the cytoplasm and vacuoles, especially if uptake is greater than the growth rate.

There are a few examples of P_i limitation near carbonate reefs in the Caribbean because P_i is bound by $CaCO_3$ in the reefs and in Oslofjord due to various anthropogenic inputs. Seaweeds can take up phosphorus as orthophosphate ions or they may obtain phosphate from organic compounds through extracellular cleavage of P_i using the enzyme alkaline phosphatase. The most important role for phosphorus is in energy transfer through ATP and other high-energy compounds involved in photosynthesis and respiration.

Corallines are an unusual group of macroalgae that deposit $CaCO_3$ onto or into their thallus and crustose corallines make an important contribution to reef formation in the tropics. About 5% of all macrophytes calcify, and they have contributed to the fossil record that extends back to Precambrian times. Corallines are mostly found in the subtidal and low-intertidal zones because they are susceptible to desiccation. It is amazing that some corallines have been collected at 260 m, the deepest of any known benthic algae, despite being covered with $CaCO_3$ which is virtually opaque to light. They are the slowest growing macroalgae, and growth bands in corallines indicate that they can live to over 700 years, making them possibly the longest-lived marine organism. They tend to thrive where herbivory is the most intense since $CaCO_3$ deters grazing. Calcium carbonate occurs in two crystalline forms, calcite and aragonite and the latter form is most commonly deposited, especially in the crusts of corallines. There is considerable diversity in the proposed calcification models for corallines. Generally, calcification is directly proportional to photosynthesis and the CO_3^{2-} concentration, stimulated by light, and is highest in young tissue. During photosynthesis, CO_2 and/or HCO_3^- is taken up from an intercellular space, resulting in an increase in intercellular pH (i.e. $HCO_3^- \rightarrow CO_2 + OH^-$) and CO_3^{2-} concentration with the subsequent deposition of aragonite. Ocean acidification is causing an increase in pH and the partial pressure of CO_2 (pCO_2) which produces a lower carbonate saturation state. Aragonite with relatively high Mg is likely to dissolve before calcite since high Mg increases the solubility of $CaCO_3$ in seawater. This may result in a decrease in the abundance of crustose corallines and a shift to more fleshy macroalgae. In addition, crustose corallines are generally more sensitive to an increase in CO_2 than corals. Since ocean acidification and global warming co-occur, the effect of an increase in pCO_2 and temperature on calcification should be studied simultaneously in factorial designed very long-term experiments to assess genotypic adaptation rather than phenotypic acclimation.

Iron chemistry and its bioavailability is complex and influenced by iron speciation (organic and inorganic), algal uptake mechanisms, light, temperature, and especially microbial interactions and recycling. It is relatively insoluble (~0.2 nM) at the pH of seawater because it exists mainly as insoluble oxides and hydroxides such as $Fe(OH)_3$, but $Fe(III)$ can form organic colloids. Some of the biologically produced organic matter acts as ligands that bind to iron which can increase the solubility by several fold. The organic iron complex can be taken up and released as inorganic iron inside the cell. These marine ligands are similar to siderophores that are produced by soil microorganisms in order to sequester/bind iron. Therefore, up to 99% of the iron in some areas is organically bound, but only some of the organically complexed iron is biologically available. There have been very few studies of iron uptake by seaweeds probably because iron is seldom limiting in nearshore subtidal areas. However, iron uptake in two kelps, *Saccharina* and *Undaria*, and the crustose coralline *Lithophyllum*, indicated that iron could be limiting for the two kelp in the northern Japan Sea where ambient concentrations are extremely low (<2 nM). *Saccharina* and *Undaria* exhibited saturated uptake rates only above 0.3 μM which is 50 times higher that the saturation concentration for *Lithophyllum* which grows even at <0.1 nM iron.

The relationships among external nutrient concentrations and growth rates can be described by a rectangular hyperbola similar to nutrient uptake. An uptake rate will give a good approximation of the growth rate only when nutrients are not limiting growth or when steady-state growth is occurring under conditions of nutrient limitations. The half-saturation constants (K_s) for growth generally are much higher for seaweeds than for phytoplankton, indicating that phytoplankton have a higher affinity for nitrogen and that they probably can outcompete seaweeds when the nitrogen concentration is low. Growth rate is also related to the concentration of nutrients in tissue where the critical tissue concentration is where growth begins to saturate. Seaweeds do not have just one critical tissue concentration of nitrogen because light modifies the nitrogen requirement for maximal photosynthesis and growth by altering N-containing components such as pigments, RuBisCO, and nitrogen reserves. Tissue nutrient concentrations can be used to monitor nutrient limitation by monthly monitoring of tissue nitrogen and N and P concentrations by comparisons with laboratory-determined critical tissue concentrations. The ratio of the maximum to minimum tissue nitrogen is an index of the nitrogen-storage capacity.

A seaweed's "functional form" may determine its ability to respond to nutrient variability. In general terms, opportunistic species often are sheet-like or filamentous annuals with rapid nutrient uptake and growth potential, high SA:V ratios, low nutrient storage capabilities, and low levels of defense against herbivory (presumably because tissue loss is readily replaced via rapid nutrient uptake and growth), and they typically dominate eutrophic environments. On the other hand, perennials tend to be large, thick, slow growing, and can store large quantities of nutrients that can act as a buffer during times of nutrient shortage.

Nutrient storage and the critical N concentration interact with growth rate (the alga's nutrient demand/requirement) and determine how long the storage pool can last and hence, how long an alga can continue to grow under nutrient limitation. Low nutrients in summer restrict the growth of phytoplankton and ephemeral macroalgae, while slow-growing perennials such as *Fucus* can maintain near maximal growth rates during most of the summer because of their slow growth rate (i.e. low nutrient demand) which allows the nutrient pool to last much longer. Therefore, survival during long periods of nutrient limitation is a balance between supply and demand, where storage capacity (i.e. how long can the pool sustain growth) represents the supply, and demand is represented by the growth rate.

In temperate kelp beds, there are two main limiting factors, light and nitrogen availability. The interactions between light and the different nitrogen availability in an upwelling area versus a small bay gave rise to marked differences in the seasonal pattern of kelp growth. In the nitrogen-limited bay, the kelp grew mainly during the period of nitrogen availability in winter and early spring and stored nitrogen was used for growth in late spring and early summer, when nitrogen became limiting. Thus, they were able to prolong their growth in spring for at least 2 months after nitrogen became limiting. During the summer, when irradiance was high, the kelp in the bay stored carbohydrate as laminaran which was used for growth in early winter, even though light was limiting for photosynthesis. At the upwelling site, kelp did not build up laminaran reserves, and growth rate followed irradiance, being greatest in summer when there was high nitrogen from upwelling. Since these seaweeds did not have carbohydrate reserves, their growth rate declined during early winter as irradiance declined.

It is not surprising that there are a diversity of reported community responses to a simple increase in nutrient enrichment and that there are many exceptions to the generalization that an increase in nutrients produces an increase in opportunistic/ephemeral fast-growing seaweeds. There are many interacting factors such as various primary producers, (phytoplankton and epiphytes, perennial canopy-forming macroalgae, and annual bloom-forming macroalgae), a range of grazers and their predators all interacting with other factors such as light, temperature, wave exposure, type of substratum, salinity, etc. In addition, mesocosms are often used because they provide essential statistical replication, but they often do not completely simulate all field conditions and may yield different results from the natural environment, especially due to differences in the epiphyte load that tends to increase in mesocosms. The length of the study may be one factor in the discrepancy between reports. For example, a 2-year study reported little response to nutrient enrichment, but when the study was continued for a total of 5 years, a delayed and dramatic response occurred after 4 years.

The interactions between grazing and nutrient uptake may dramatically influence the structure and functioning of intertidal communities. Selective grazing can have a top-down effect by removing young bladelets that in turn has a bottom-up effect by reducing nitrogen uptake. When small crustaceans consume large macroalgae, they often target apical regions and small young blades that are more palatable. In a tide-pool study, naturally high nitrogen loading due to ammonium excretion by sessile invertebrates such as mussels, led to a doubling of the number of species and a higher abundance of faster growing species with high ammonium uptake rates to satisfy their higher nitrogen requirements. An ecosystem has a critical level of nutrient loading, below which small nutrient additions tend to increase diversity when the system is nutrient limited, but above it, diversity tends to decline with excessive enrichment.

Physico-chemical factors as environmental stressors in seaweed biology

7.1 What is stress?

Any given environmental factor can become a "stress" to a given seaweed species, if it exceeds the upper or lower threshold values of tolerance. Seaweed communities are shaped by the complex interplay of a multitude of external biotic and abiotic factors and the intrinsic responses of the individual seaweed species. These factors are not stable over space or time, requiring frequent metabolic adjustments, termed acclimation (sec. 1.1.3). The genetic frame setting the limitations of acclimation is termed adaptation. Species-specific adaptation and the effectiveness of acclimation to change determine the competitive success of each species in the interaction with other species and thus shape the complex composition of a seaweed community in the field. For example, it is now well established that many biotic interactions of macroalgae (e.g. competition, predation, etc.) are mediated by the environmental stress, and the ways in which they manage it (Menge *et al.* 2003; Essay 2, Chapter 3). Major changes in abiotic factors occur along spatial and temporal gradients: spatially, on a global scale along latitudinal gradients, large changes in temperature, light availability, and seasonality are observed; along the coastline steep gradients in abiotic factors exist stretching from the intertidal to the subtidal zone; but even on very small scales the abiotic environment of seaweeds may change dramatically, e.g. within algal mats (Bischof *et al.* 2006b). Temporally, there are natural fluctuations of abiotic factors due to seasonal events, daily or tidal cycles, and climate variability such as El Niño Southern

Oscillation (ENSO) events. The extent of natural change in the physico-chemical environment to which a seaweed species is exposed can be summarized by the term "habitat stability" and is often tightly linked to vertical zonation patterns along the phytal zone, with intertidal species populating the most demanding, least stable habitat (Davison and Pearson 1996; sec. 3.1). Apparently, the magnitude of the environmental stress along different spatial scales is important to explain the distribution patterns of macroalgae as was reported for rocky intertidal assemblages from Helgoland Island (Valdivia *et al.* 2011). Valdivia *et al.* (2011) indicated that vertical variation in community structure was significantly higher than patch- and site-scale horizontal variation but lower than shore-scale horizontal variation. Most concern and research effort is now directed towards the additional anthropogenic sources of variation in the abiotic environment, from the local to the global scale.

Tolerance levels are naturally species, stage or even specimen specific and Davison and Pearson (1996) stated that "stress must be defined in terms of the response of an individual rather than the value of a particular environmental variable". These authors put forward the very helpful distinction between "limitation stress", which refers to reductions in growth (or other integrative parameters such as reproduction, recruitment) based on an inadequate supply of resources, and "disruptive stress", which include conditions yielding physiological/cellular damage, or requiring metabolic activities to counteract/repair damage. Whether or not an

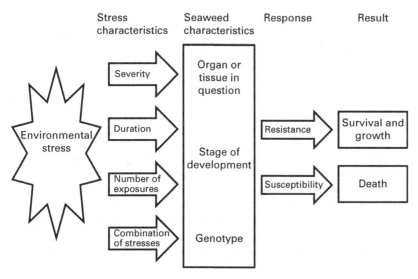

Figure 7.1 The extent of abiotic stress acting on an individual seaweed depends on the complex interplay of external and intrinsic factors. (From Buchanan *et al.* 2000, reproduced with permission.)

environmental variable is generating stress may not only depend on its severity, but also on its duration, frequency and interaction with other environmental factors (see Fig. 7.1).

Physiological stress is often related to "oxidative stress". Under environmental conditions resulting in restricted electron flow through electron transport chains, i.e. in mitochondrial respiration or photosynthesis, electrons might be transferred to O_2, which eventually becomes a reactive oxygen species (ROS) due to this unpaired electron. This applies to the superoxide radical ($O_2^{\bullet-}$), singlet oxygen ($^1O_2$) and the hydroxyl radical ($HO^{\bullet}$). Additionally, hydrogen peroxide (H_2O_2) is commonly also included in the group of ROS. Virtually, any kind of environmental stress, such as high irradiance, temperature and salt stress, malnutrition, pollution etc. may increase the likelihood for ROS formation with subsequent oxidative damage to biomolecules, and physiological processes. This chapter will focus on the outcome of the natural as well as anthropogenic variability of physicochemical factors in shaping seaweed communities in terms of biogeography, vertical zonation, and stress physiology.

7.2 Natural ranges of temperature and salinity

7.2.1 Open coastal waters

The surface temperatures of the oceans vary in two primary ways. First, they decrease toward higher latitudes, from about 28°C in the tropics to 0°C toward the poles, although this trend is markedly affected by ocean currents. Second, the seasonal changes in ocean temperatures are larger at mid-latitudes. In the tropics and at the poles the annual temperature range is often <2°C (Kinne 1970), whereas in mid-latitudes 5–10°C is common.

The new international standard for expressing "salinity" is referred to as "Absolute salinity S_A", according to the TEOS-10 standard, and uses the mass fraction of salt in seawater, based on the dimension of grams of dissolved salts per kilogram of water; thus being very similar to the formerly used "parts per thousand (ppt)" concept (Wright *et al.* 2010). For field applications, salinity can be measured (less accurately, but much more quickly) either with a conductometer, or a refractometer, which measures light refraction, or a hydrometer, which measures density, which is of course temperature dependent.

The salinity of open-ocean surface water is generally 34–37 S_A, though lower off the coasts of areas with high rainfall, and higher in subtropical areas with high rates of evaporation and low rainfall (Groen 1980). Certain seas have markedly higher or lower salinities. The Mediterranean, because there is a high rate of evaporation and little freshwater influx, has salinities of 38.4–39.0 S_A, the Red Sea and in particular the waters of the Gulf of Aqaba may even have salinities as high as 40 S_A; the Baltic, essentially a gigantic estuary, is notably brackish, particularly at the surface, ranging from 10 S_A near its mouth to 3 S_A or less at the northern extreme. In coastal waters, especially those that are partially cut off from the ocean or are subject to heavy runoff, salinity is characteristically 28–30 S_A or lower. Slight latitudinal trends in salinity are overshadowed in coastal areas by freshwater influx. In areas with marked seasonal differences in rainfall, or with winter snow periods followed by spring melt, the salinity, especially of surface water, may change dramatically over the year (Hanelt *et al.* 2001).

Stratification is common in coastal waters, and especially in semi-enclosed water bodies e.g. estuaries, fjords, inlets, etc. Rather sharp temperature boundaries, called thermoclines, may develop between layers of water, especially in sheltered waters or where there is freshwater input. Warm water flowing out of tropical lagoons at low tides does not readily mix with the cooler shelf waters, but forms a buoyancy front as it displaces the water on the shelf. The salinity boundary is called a halocline. Typically but not always, the surface water is warmer and less salty than the deeper water on which it floats. The combined effects of temperature and salinity on seawater density result in a density boundary, or pycnocline, which is a surprisingly strong barrier to mixing. Vertical mixing is thus slow unless wave action is vigorous. In places with moderate or large tidal amplitudes, such pycnoclines (which are often only a meter or two below the surface) will sweep portions of the shallow-subtidal and lower-intertidal zones, causing rapid temperature and salinity changes with each ebb and flood of the tide. Moreover, the importance of thermal stratification to seaweeds is found in two contrasting bays in Newfoundland (Hooper and South 1977). In Bonne Bay, a narrow fjord, a thermocline

forms, warmer water stays near the surface, and Arctic species intolerant of temperatures above 5°C live in the deeper subtidal zone. In Placentia Bay, which is more exposed, there is more turbulence, no thermocline and no Arctic algae, because the water that is too warm for them extends too far down.

7.2.2 Estuaries and bays

The temperature and salinity ranges seen in coastal waters are even more pronounced in bays and estuaries. Shallow bays on Prince Edward Island, in the Gulf of St. Lawrence, freeze over in winter, but warm up to 22°C or more in summer. In Peel Inlet in Western Australia the salinity ranges from 2–50 S_A, the same annual range of salinity that *Cladophora* aff. *albida* can tolerate (Gordon *et al.* 1980). In general with increasing distance up the estuary, the mean temperature will increase, the mean salinity will decrease, and the range of each will become greater.

Salinities in river mouths depend on the proportions of river water and seawater; these proportions depend on the state of the tide and the state of the river, and they change both daily and seasonally. In addition to the gradient along the estuary, the salinity lower on the shore often is more variable than that higher on the shore, because the higher shore tends to be covered only by the surface water (fresher water); whereas the lower shore is alternately under river water and saltwater (Anderson and Green 1980). Because of the slow rate of mixing of layers, even a small freshwater influx can have a pronounced, but very local effect. That is, it can affect individuals as strongly as will a major influx, but it will affect fewer individuals. The course of a freshwater seep across the intertidal zone often can be seen as a bright green path, because *Ulva*, but little else, will be growing in it.

A temperature–salinity (T–S) diagram is a useful way to describe the water climate of an estuary, and it can provide insight into the causes of distribution patterns that mean values conceal. Indeed, Druehl and Footit (1985) have argued that temperature and salinity should not be seen as separate factors affecting seaweed distribution. For instance, the mean annual temperatures and salinities for Nootka and Entrance Island, British Columbia, are nearly identical, but only

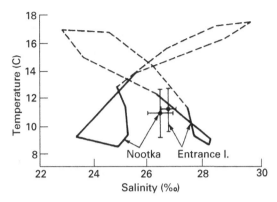

Figure 7.2 Temperature–salinity diagrams for two locations in British Columbia, one of which (Nootka) supports growth of *Macrocystis integrifolia*, the other (Entrance Island, Nanaimo) does not. The large crosses show means and standard deviations for annual temperatures and salinities. The outlines trace monthly temperature–salinity coordinates, with solid lines around winter conditions, and broken lines around summer conditions. The contrast in the two water climates is belied by their annual means. (From Druehl 1981, with permission of Blackwell Scientific Publications.)

Nootka supports growth of *Macrocystis pyrifera* (formerly *M. integrifolia*) (Druehl 1978). Nootka is on the outer coast of Vancouver Island, whereas Entrance Island is in the Strait of Georgia, essentially at the mouth of a large estuary fed by the Fraser River. The *T–S* diagram (Fig. 7.2) shows that for Nootka, when the temperature is high, the salinity is also high; for Entrance Island, when the temperature is high, the salinity is low. In the estuarine zone of the Patagonian Region, the appearance and abundance of the giant kelp *Macrocystis pyrifera* is strongly determined by the gradient in salinity (Dayton 1985). In the presence of high salinity, *Macrocystis* is believed to be able to withstand higher temperatures, although the appropriate physiological tests have not yet been carried out.

7.2.3 Intertidal zone

The principal environmental feature of the intertidal zone is its regular exposure to atmospheric conditions (sec. 3.1). Its temperature regimes are thus much more complex than those of subtidal zones. Myriad microenvironments result from the many factors that affect the local

temperature and the temperatures of resident organisms. Some factors, such as shading, affect the influx of heat to an organism, whereas others, such as evaporation, affect heat efflux. The major source of heat in the intertidal zone during ebb tide is direct solar radiation. Irradiance may be reduced because of shading by clouds, water, other algae, and shore topography (including overhangs, crevices, and the direction of slope). Small-scale topographic features also give shelter from breezes, and hence from evaporative cooling. The important temperature is that of the cytoplasm, which is of course complicated to measure. Two examples of actual algal temperatures recorded during emersion on hot days are given in Figure 7.3. *Endocladia muricata* is a stiff, tufty alga; the temperature at the interior of the clump, which is shaded and yet open to airflow, remains considerably lower than that of the air or the open rock surface (Fig. 7.3a) (Glynn 1965). *Pyropia fucicola*, on the other hand, is flattened against the rock surface like a little solar panel, and on a calm day, such as that illustrated, it becomes much hotter than the air (Fig. 7.3b) (Biebl 1970). These graphs also show the sharp decreases in temperature as the tide covers the seaweed. Notice that the *Pyropia* thallus surface temperature fell from 33°C to 13°C in a matter of minutes after the water reached it.

Other variables that affect the temperatures of intertidal seaweeds are the time of day at which low tide occurs and the extent of heating or cooling due to waves. Seaweeds exposed by a low tide at dawn or dusk will suffer little heating by the sun, whereas if the low tide occurs in the middle of the day, heating (and also desiccation) can be extreme. In summer, the water may be cooler than the exposed rock and algae, as in Figure 7.3b. In winter, seawater frequently is warmer than the air, and so it can thaw algae that have been frozen during exposure to sub-zero air. We will explore the effects of frost in more detail in sec. 7.5.2.

Littoral pools (sec. 3.3.4) are subject to less extreme changes than are open rock surfaces. The longer the exposure and the higher the pool's surface-to-volume ratio, the greater the changes that will take place before the tide refloods the pool. Most of the factors that affect the temperatures of exposed rock pools also affect their salinity. Changes in tide-pool temperatures are more drastic during the day and in the summer (Fig. 7.4a),

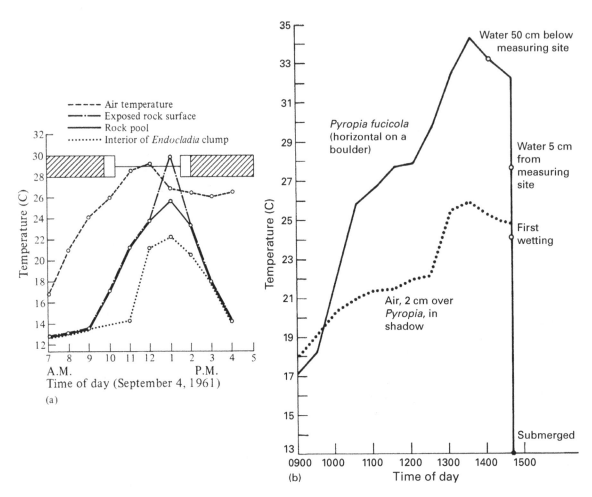

Figure 7.3 Temperature climates of two high shore red algae. (a) Temperature observations of three microhabitats in the high intertidal *Endocladia-Balanus* association at Monterey, California, as related to low-water exposure. Also shown is the air temperature at a nearby weather station during the observation period. The horizontal bar and line at the top of the graph show, for the level observed, the approximate durations of the submerged (cross-hatching), awash (clear), and exposed (line) periods. (b) *Pyropia fucicola* thallus temperature during ebb tide on a calm, sunny day. (Part a from Glynn 1965, with permission of the Zoological Museum, Amsterdam; b from Biebl 1970, with Permission of Springer-Verlag, Berlin.)

because the chief source of heat is solar energy. However, the air temperature may warm or cool such pools (even freeze them). Water added to a pool as runoff, rain, or snow may heat or cool the pool. Tide pools experience little or no mixing and thus may easily become stratified. Temperature stratification that occurs during the day, in the absence of salinity stratification, usually breaks down at night.

Atmospheric variability also subjects seaweeds on open rock surfaces and in tide pools to frequent salinity fluctuations. Evaporation will cause an increase in the salinity of water in the surface film on seaweeds and,

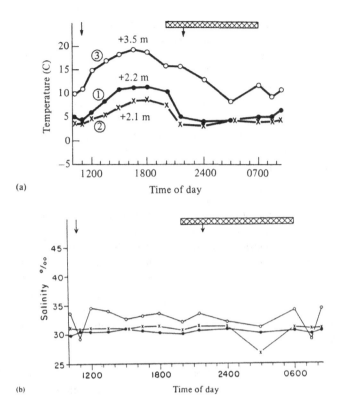

(a)

(b)

Figure 7.4 Temperature (a) and salinity (b) changes in three pools at different heights in the intertidal zone near Halifax, Nova Scotia, 8–9 May 1970. Times of high tide indicated by arrows, and darkness by the cross-hatched bar. Pools 1 and 2, between neap and spring higher high waters, were generally flushed twice daily, except for pool 1 during periods of calm or neap tides. Pool 3, 0.4 m above extreme high water, was washed only during severe storms and contained no perennial macroscopic algae (From Edelstein and McLachlan 1975, with permission of Springer-Verlag, Berlin.)

more slowly, in tidal pools. In contrast, rain, snow, and freshwater streams will cause a reduction in salinity. The community upset that can be caused by a flash flood in the intertidal zone has been documented by Littler and Littler (1987). Because fresh water floats on salt water, and because a long period of evaporation generally is necessary to effect any significant change in the salinity of pool water, salinity will change little in mid-intertidal and low-intertidal pools except during extremely hot days or following torrential downpours. High intertidal pools, which are inundated infrequently or receive seawater only from wave splash, may become very brackish in rainy weather, or strongly hypersaline in hot, dry weather (Fig. 7.4b). Sharp

increases in tide-pool salinities can come about as a result of freezing, because salts are initially excluded from the freezing layer and concentrated in the remaining liquid (Edelstein and McLachlan 1975). Seaweeds are found in pools with salinities from about 0.3 to 2.2 times that of normal (i.e. 35 S_A seawater) -that is, about 10–77 S_A (Gessner and Schramm 1971).

In terms of physiological stress ecology, seaweeds populating the intertidal zone display a remarkable set of adaptations in response to the complex interplay and large amplitude of changing environmental factors. In the following, we will give special attention to intertidal seaweeds when discussing the impact of different abiotic factors.

7.3   Temperature effects

Temperature has fundamental effects on chemical reaction rates. In turn, the metabolic pathways, each the sum of many chemical reactions, are affected by temperatures, but the interactions with other factors become more complex. We shall here examine the effects of temperatures from the chemical level up to the population level and discuss mechanisms for temperature adaptation and acclimation in seaweeds from contrasting environments.

7.3.1   Chemical reaction rates

Temperatures have general effects on chemical reaction rates that are embodied in the concept of Q_{10}, the ratio between the rate at a given temperature t and the rate at t - 10. Typically, $Q_{10} \approx 2.0$; that is, the rate doubles over a $10°C$ increase in temperature, but rates can be much higher. In ice algal assemblages Q_{10} values up to 6 have been described (Kottmeier and Sullivan 1988; Kirst and Wiencke 1995), and for studies on photosynthesis of a number of Antarctic macroalgae a range of Q_{10} from 1.4–2.5 has been determined in the 0–10°C temperature interval (Wiencke *et al.* 1993; Kirst and Wiencke 1995). The effect of temperature is greater on an uncatalyzed reaction than on one that is catalyzed (e.g. by an enzyme) (Raven and Geider 1988). In an enzyme-catalyzed reaction, with Michaelis–Menten kinetics,

$$v = V_{max}[S]/(K_m + [S])$$

with v as the initial reaction velocity and V_{max} as its maximal velocity, $[S]$ as the concentration of the substrate, and K_m as the Michaelis constant, which indicates the substrate concentration at which $v = \frac{1}{2} V_{max}$, both V_{max} and K_m are affected by temperature and $[S]$ may also be influenced via temperature effects on membrane transport affecting substrate availability (Davison 1991). If the rate of an enzyme-catalyzed reaction is measured as it varies with experimental changes in temperature over a broad range, often a peak will be found. This optimum temperature and the sharpness of the peak will also depend on pH and on the purity of the enzyme (*in vitro*). Thermal denaturation of the enzyme will occur above a critical temperature; on cooling, the enzyme may regain its active conformation, or it may be permanently damaged.

Cellular rates, as opposed *in vitro* rates, per unit of enzyme will depend on the amount of enzyme present, as well as on any seasonal changes in its kinetic parameters (Küppers and Weidner 1980; Davison and Davison 1987; Cabello-Pasini and Alberte 1997). Küppers and Weidner (1980) studied six *Laminaria hyperborea* enzymes from diverse metabolic pathways: RuBisCO, phosphoenolpyruvate carboxykinase (PEPCK); malate dehydrogenase (MDH), aspartate transaminase, glycerol phosphate dehydrogenase, and mannitol-1-phosphate dehydrogenase. Enzymes were extracted from the kelp, and their activities were measured under standard test conditions, which included a temperature of 25°C. Seasonal changes were found for all six enzymes, with peaks generally in February–April. In nature, of course, the temperature is not constant (nor as high as 25°C). By determining the effect of temperature on the *in vitro* activity of each enzyme and by recording seawater temperatures the authors were able to calculate what the enzyme activities would have been in the living kelp on the shore. Again they found a seasonal change for each enzyme, but the peaks were in August. An enzyme's activity is the product of turnover rate (temperature dependent) and enzyme concentration. The amount of each enzyme present will depend on nitrogen availability, with RuBisCO (and hence photosynthesis) most sensitive to nitrogen starvation (Wheeler and Weidner 1983). In summer, the kelp has high levels of enzyme activities because of smaller numbers of enzyme molecules with a higher turnover rate; in spring, during the period of nitrogen availability and a high growth rate, a high level of enzyme activity is achieved by an increase in the total amount of enzyme present, although the turnover rate is lower.

Davison and Davison (1987), studying *Saccharina latissima* (as *Laminaria saccharina*), found an inverse relation between temperature and standard (20°C) activity for both RuBisCO and NADP-dependent glyceraldehyde-P dehydrogenase (GAPDH) (another Calvin cycle

Essay 4 Ocean acidification

John A. Raven

Anthropogenic CO_2 emissions to the atmosphere are causing decreases in ocean pH, otherwise known as ocean acidification. The ocean acidification phenomenon involves equilibria between atmospheric and oceanic CO_2 and among inorganic carbon species in seawater; these equilibria are predictable, based on physico-chemical laws (The Royal Society 2005; Doney *et al.* 2008; Hepburn *et al.* 2011; Tyrell 2011). By contrast, the causal relation between increased atmospheric concentrations of CO_2, and other greenhouse gases and global warming, while as physically unassailable as is ocean acidification, is subject to feedbacks that make the temporal trajectory of warming much more difficult to model and predict.

Almost a third of the anthropogenic CO_2 emissions over the last 250 years have dissolved in the ocean, resulting in a decrease in mean ocean surface ocean pH of about 0.1 units, with another 0.4 units of decrease expected by 2100, a total of 0.5 units since 1750 (The Royal Society 2005; Table 1). While "ocean acidification" is a handy term to encompass the phenomena associated with increased CO_2 concentration in the ocean, it is important to recognize that decreased pH is not a driver of the other effects but that it is, with increased concentrations of dissolved CO_2 and of HCO_3^- and a decreased concentration of CO_3^{2-}, itself an outcome of dissolving additional CO_2. Perhaps unexpectedly for the non-specialist, ocean pH is usually not measured directly by characterizing the dissolved inorganic carbon system, but is calculated from measurements of two other components of the system such as total titratable alkalinity and total inorganic carbon.

My involvement with ocean acidification was embarrassingly late. While I had been working on inorganic carbon acquisition in algal photosynthesis since 1967, and recognized the possible effects of increasing atmospheric CO_2 increase on seaweeds before I accepted the invitation in 2004 to chair the Royal Society of London's working group on ocean acidification, it was the production of the report published in June 2005 that made me really focus on the issue. The working group had to produce the report in language that was as accessible as possible to non-experts without over-simplifying the issues, and acknowledging the uncertainties while emphasizing the certainties. These problems of communicating uncertainty without destroying faith in any claims made by scientists come in part from widespread ignorance about the scientific method. The problems experienced by scientists in communicating the generalities of their methodology as well as particular outcomes are not helped if the media focus on a particular conclusion without mention of any uncertainties. To be fair, the media gave a good and balanced account of the 2005 report and always checked with me on what they were going to say.

Table 1 Calculated changes of ocean inorganic carbon chemistry, pH and saturation state of aragonite and calcite at equilibrium with atmospheric CO_2 for pre-industrial CO_2, and for increased, CO_2, relative to 2005 CO_2, for a total alkalinity and a temperature of 18° C. Modified from Table 1 of The Royal Society (2005). In the real ocean there are seasonal and spatial variations in temperature and in the extent of equilibrium of atmospheric CO_2 with dissolved CO_2.

	Pre-industrial (before 1750)	Today (2005)	Twice pre-industrial	Four times pre-industrial	Six times pre-industrial
Atmospheric CO_2	280 ppm	380 ppm	560 ppm	1120 ppm	1680 ppm
$[CO_2 + H_2CO_3]^a$ μmol/kg water	9	13	19	38	56
$[HCO_3^-]$ μmol/kg water	1768	1867	1976	2123	2183
$[CO_3^{2-}]$ μmol/kg water	225	185	141	81	57
[Total Dissolved Inorganic Carbon] μmol/kg water	2003	2065	2136	2242	2296
Surface ocean pH	8.18	8.07	7.92	7.65	7.49
Calcite saturation[b]	5.3	4.4	3.3.	1.9	1.3
Aragonite saturation[b]	3.4	2.8	2.1	1.2.	0.9

[a] The concentration of CO_2 is 400–500 times that of H_2CO_3.

[b] Saturation values are relative to the $[CO_3^{2-}]$ which, with the ocean $[Ca^{2+}]$, is in equilibrium with the specified crystal morph of $CaCO_3$.

Essay 4 (cont.)

Turning to effects on seaweeds, the Royal Society Report (2005) says relatively little about them because there were fewer data for marine macroalgae than for microalgae. Since 2005 more data have become available from experiments on seaweeds (Hepburn *et al.* 2011). Typical experiments involve measurements of growth or metabolism of seaweeds acclimated to seawater equilibrated with CO_2 partial pressures predicted for (say) 2100, compared with seaweeds equilibrated with the present-day CO_2 partial pressure. The simplest case, and the most similar to the well-investigated C_3 embryophytes on land, is that of diffusive CO_2 entry that occurs in a significant fraction of red algae, and especially those in low-light environments, as well as a few green algae. It is likely that these seaweeds will be more competitive in increased CO_2 environments (Hepburn *et al.* 2011). Most seaweeds have inorganic carbon concentrating mechanisms (CCMs) based on active HCO_3^- influx and, probably, active CO_2 influx. These algae may also be more competitive in higher CO_2 if less energy and other resources are used in the CCMs and, at even higher CO_2, the CCMs are replaced by diffusive CO_2 entry (Hepburn *et al.* 2011). By contrast, calcified seaweeds will very likely be less competitive in higher CO_2, since the maintenance of supersaturated conditions at the site of precipitation for crystalline form of $CaCO_3$ produced by the alga will be more difficult in lower-CO_3^{2-} seawater. Furthermore, $CaCO_3$ which has already been precipitated will be subject to dissolution in seawater under-saturated with respect to that crystalline form of $CaCO_3$ (The Royal Society 2012; Doney *et al.* 2009).

Two other considerations may modify these conclusions. One is that other aspects of environmental change, such as warming and the associated shoaling of the thermocline as surface seawater becomes hotter, will interact with the changes in seawater chemistry (Tyrrell 2011). One consequence of the decreased depth of the upper mixed layer is that deeper-growing seaweeds will be exposed to higher nutrient concentrations and those growing higher on the shore, including intertidal algae, will be exposed to a lower nutrient concentration. It is known that nutrient concentration affects CCM expression in algae, although most of the work has been on microalgae (Raven *et al.* 2012). A second consideration is that the experiments have necessarily used extant genotypes of seaweed, so that evolution change (adaptation) has not been examined. Experimental evolution studies have been carried out on microalgae with respect to increased CO_2 (Collins and Bell 2004) and to increased temperature. Such studies would be much more difficult for macroalgae in view of the desirability of hundreds of generations under the new conditions and the longer (at least an order of magnitude) generation times for macroalgae than for microalgae. While any evolutionary adaptations, as is the case for phenotypic acclimations, are constrained by the thermodynamic properties of the inorganic carbon system and other environmental factors, and the evolutionary history of the organisms, this still leaves significant scope for future studies.

*John Raven is Emeritus Professor of Biology at the University of Dundee and an Honorary Fellow at the James Hutton Institute, Scotland. He has 50 years' experience of investigating algal biochemistry, physiology and ecology, covering freshwater and marine habitats and the complete size range of algae. Work on seaweeds has focused on photosynthetic carbon assimilation and how the natural abundance of carbon isotopes can inform our understanding of the mechanism of inorganic carbon entry.

enzyme). They also found an inverse relation between photosynthetic rate and temperature when measured at standard (15°C) temperature, but that relation disappeared when photosynthesis was measured at the growth temperature, indicating that the increase in Calvin cycle compensates for low temperatures and makes *S. latissima* photosynthesis virtually independent of temperature. Indeed a three-fold increase in RuBisCO concentration in *S. latissima* acclimated to low temperature (5°C) compared to specimens kept at 17°C has been shown by Machalek *et al.* (1996).

As a matter of fact, strong temperature dependence does apparently not hold true for the functionality of major proteins involved in maintaining the *primary* reactions of photosynthesis. For the same species Bruhn and Gerard (1996) have found pronounced effects of temperature on the extent and recovery from photoinhibition. It was found that elevated temperatures may enhance protective mechanisms counteracting high light stress, but may impair repair processes, such as *de novo* synthesis and reintegration of D1 protein. In contrast, low temperatures can exacerbate the sensitivity to high light due to an inhibition of the energy transfer from the light-harvesting complexes to the photosystems (Schofield *et al.* 1998), and thus reducing the capacity for recovery (Gómez *et al.* 2001).

In polar seaweeds, seasonal changes in enzyme activity may be triggered by the combined changes in temperature (usually occurring at a rather small amplitude) and light availability. Aguilera *et al.* (2002) have studied the changing activities in three enzymes related to oxidative stress metabolism (superoxide dismutase, catalase, glutathione reductase) in two red (*Devaleraea ramentacea, Palmaria palmata*) and one green seaweed (*Monostroma* sp.) from Spitsbergen. In general, activity changes could be mostly attributed to response to changes in the *in situ* light climate (low light under ice cover in spring, high light exposure after sea ice break-up, high water turbidity in summer due to increased melt water runoff). Also, the respective vertical position of the species was mirrored by overall enzymatic responses. In contrast to *Palmaria palmata* from deeper waters, *Devaleraea ramentacea* maintained high enzyme activities throughout the season. This may reflect its prevalence in a shallow water habitat characterized by low stability, with more pronounced variation in temperature, radiation, and salinity.

As outlined by Raven *et al.* (2002b), temperature may influence carbon acquisition by RuBisCO in many other ways, by enzyme activity, and altered equilibria of CO_2 and C_i, as well as their respective diffusion rates changing with temperature. Furthermore the respective affinity of RuBisCO to CO_2 has been found to differ among the algal classes (Badger *et al.* 1998). Overall it is suggested that low temperatures may favor diffusive CO_2 entry, which, in addition to increased RuBisCO concentration or structural modifications, helps maintain carbon fixation.

7.3.2 Metabolic rates

When rate-versus-temperature measurements are made for complex reactions, such as photosynthesis and respiration, the overall rate is a composite of all the individual reaction rates. If there is a rate-limiting reaction, it will not necessarily be the same one at all temperatures. The effects of a given temperature

change will not be the same on all metabolic processes, because of differing temperature sensitivities of enzymes and the influences of other factors, including light, pH, and nutrients. In photosynthesis, for example, diffusion rates, carbonic anhydrase activity, and active transport of CO_2 and HCO_3^-, all affected by temperature, will determine the supply of substrate to carbon fixation pathways (Raven and Geider 1988; Davison 1991). In general, as Raven and Geider (1988) have pointed out, metabolic processes involve many enzymes and transport processes; so low temperature is likely to limit the overall rate via some particularly sensitive step. Algae can respond by altering the quantity or properties of the limiting component.

Temperature acclimation in enzymes shows up as seasonal changes in the rates of photosynthesis and respiration versus temperature when performances of summer and winter seaweeds are compared under otherwise identical conditions. For instance, Mathieson and Norall (1975) showed that at a given irradiance, net photosynthesis for several algae was maximum at a lower temperature in winter specimens than in summer specimens. More important, the rate of photosynthesis in cold water is higher in winter seaweed than in summer specimens; and summer seaweed can maintain near-peak photosynthesis through warmer temperatures than can winter seaweed. Different thermal traits are also observed in seaweed species growing over extended latitudinal ranges: ecotypic differentiation with respect to prevailing temperature regimes along latitudinal gradients has been identified by conducting "common-environment experiments" in which conspecific isolates from different temperature environments were subjected to identical temperature treatments. Mostly photosynthetic activity and growth were used as response variables pointing to pronounced temperature acclimation, and maybe even adaptation, although studies on the genetic bases of temperature ecotype formation are still scarce (e.g. Johansson *et al.* 2003).

Seaweeds in habitats where the temperature fluctuates seasonally may be better able to acclimate than are seaweeds from stable habitats. For instance, Dawes (1989) compared two species of *Eucheuma*, one from Florida (temperature range 16–28°C) and one from the

Philippines (temperature about 25°C). The Florida sea-weeds were able to acclimate, in stages, to 18°C, but the tropical specimens could not.

Different species, and even different populations of a given species, show diverse responses to temperature changes. Comparisons of macroalgal data from different sources can be facilitated by a mathematical model developed by Knoop and Bate (1990). Their equation relates the photosynthetic rate (P) at a given temperature (T) to the rate of increase at suboptimal temperatures (a) and the rate of decrease at supraoptimal temperatures (b) and to the maximum and minimum temperatures at which photosynthesis occurs (T_{max}, T_{min}):

$$P = a(T - T_{min})\{1 - \exp[b(T - T_{max})]\}$$

The work by Davison and others has allowed us to look more closely into temperature–photosynthesis acclimation and to understand some of the factor interactions. The initial photochemical reactions are temperature independent, but the enzymes of phosphorylation, electron transport, and plastoquinone diffusion are temperature dependent. As a result, light-harvesting efficiency at sub-saturating irradiances (initial slope, α, of a P–E curve) may vary with temperature (Davison 1991). Dark respiration generally increases with temperature. Thus, the amount of light necessary to reach compensation (E_c) increases with temperature. Such short-term effects are not easily related to the long-term responses of seaweeds in the field, however, which are complicated by changes such as acclimation of the enzymes and increases in pigments in seaweeds grown at higher temperatures. *Saccharina latissima* (as *Laminaria saccharina*) grown at 15°C had more chlorophyll *a* (because of more PS II reaction centers and possibly larger photosynthetic units) than did seaweeds grown at 5°C. Because of this, when 5°C-grown seaweeds were assayed at 15°C, E_c and E_k were lower than in 15°C-grown seaweeds (Davison *et al.* 1991). Although E_c increased with incubation temperature in seaweeds grown at both 5°C and 15°C, the effect was canceled out by acclimation effects; E_c and E_k values were similar for seaweeds tested at their growth temperatures. This is important, given that many seaweeds

grow in light-limited environments. *S. latissima* had lower Q_{10} when grown at 15°C (1.05 vs. 1.52); that is, increasing assay temperatures had greater effects on low-temperature seaweed. In general, Q_{10} tends to be higher in winter than in summer (Kremer 1981a). This change appears to be more pronounced in polar species than in temperate species (Drew 1977; Kirst and Wiencke 1995). However, the latter findings imply that patterns in temperature acclimation may be modulated by light availability, and indeed, the most ecological relevant situation is a simultaneous change in both parameters. Thus, Machalek *et al.* (1996) conducted a detailed study on the interaction of thermal acclimation and photoacclimation in *Saccharina latissima*. Specimens were acclimated in a two-factorial design, including two temperatures (5 and 17°C) at two irradiances (15 and 150 µmol m^{-2} s^{-1}). Apart from measurements of photosynthesis, analyses of thallus absorptance, pigment composition, RuBisCO and fucoxanthin–chlorophyll *a/c* binding protein (FCP) concentration, as well as determination of photosynthetic unit size were conducted (Fig. 7.5). At standard temperatures, maximal photosynthetic rates were higher in the cold-water specimens and photosynthetic efficiency (as α) was higher in algae grown at higher temperatures (irrespective of light). At low temperatures, algae kept under low light displayed higher efficiency than under high light. This pattern was also reflected in the differences in thallus absorptance as an outcome of adjustment of pigment composition, PSII reaction center density and FCP abundance.

Interestingly, Antarctic seaweeds, which normally live at constant low temperature but variable light conditions, show photosynthetic acclimation to both factors (Wiencke *et al.* 1993). These studies illustrated the complex metabolic regulation of different physiological processes related to photosynthesis in response to the change of the most prevailing physical drivers, light and temperature.

7.3.3 Growth optima

Numerous studies have investigated the effects of temperature on photosynthesis, respiration, and growth

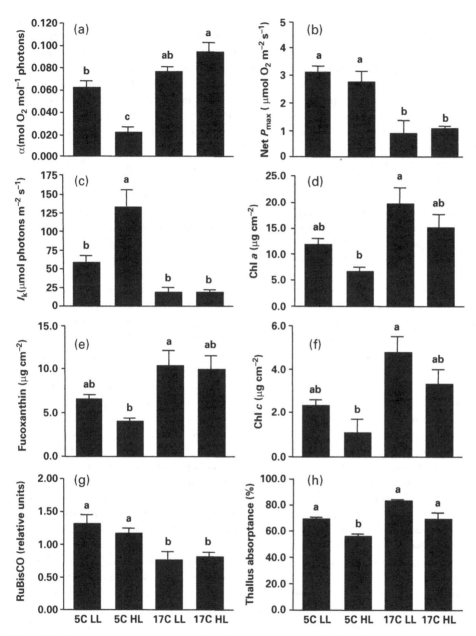

Figure 7.5 Effects of growth temperature and irradiance on parameters related to photosynthesis in *Saccharina latissima* grown at 5 or 17°C and 15 or 150 μmol m^{-2} s^{-1}: (a) photon yield; (b) light-saturated photosynthesis; (c) light-saturation point; (d) chlorophyll *a*; (e) fucoxanthin; (f) chlorophyll *c*; (g) RuBisCO content; (h) thallus absorptance. (Figure compiled from Machalek *et al.* 1996.)

under otherwise uniform conditions. Not surprisingly, maximum rates often have been found to correlate with the temperature regime in an alga's habitat. However, there have been some reports of instances in which the optimum temperature (i.e. that giving the maximum rate) has not been near that for the natural conditions, as found for example for a couple of seaweed species from polar regions (Wiencke and Clayton 2002; Fredersdorf *et al.* 2009). In a classical example, Fries (1966) found the optimum temperatures for growth of three red algae in axenic culture to be 20–25°C, whereas the water temperatures in their habitat, even in summer, rarely rose above 15°C. She speculated that a reason for the discrepancy may have been the absence of bacteria from the cultures. Seaweeds may use growth substances produced by their associated microflora, and marine bacteria grow best at lower temperatures. In other words, there is a physiological optimum for the alga alone, and there is an ecological optimum in nature, where it is interacting with bacteria and fungi and other environmental factors.

Temperature optima vary among species and among strains, as well as between heteromorphic life-history stages (Pereira *et al.* 2011). For example, the sporophyte of the brown seaweed *Desmarestia anceps* that is endemic to Antarctica, showed a sharp temperature optimum for growth at 0°C and a more than 50% decline in growth rate at 5°C. For the male gametophyte a broader tolerance range between 0–5°C was reported, whereas the female gametophyte showed a distinct peak in growth at 5°C (Wiencke and tom Dieck 1989; Wiencke and Clayton 2002; see Fig. 7.9). Changes with the age of the thallus have also been noted. For example, the optimum for cultivated *Pyropia yezoensis* drops from 20°C at the time of conchospore germination to 14–18°C for thalli that are 10–20 mm high, and still lower for larger thalli (Tseng 1981). Moreover, the temperature optimum for an alga under laboratory conditions, where temperature is kept constant, may appear narrower than it is in nature (even if other conditions are equal). This may be particularly relevant when working with organisms raised from stock cultures, which have been kept under laboratory conditions for several

years and, thus, may have even been undergoing genetic alterations. This is an important point for consideration when relating findings from laboratory experiments to field conditions. In this respect, a re-evaluation of the Antarctic seaweeds tested by Wiencke and tom Dieck (1989) and Wiencke *et al.* (1993, 1994) for its light and temperature requirements and kept in stock cultures for more than two decades now would be most insightful.

Age- and stage-specific changes in temperature optima for growth might be further relevant to the invasion ecology of species. Norton (1977) suggested that the very high growth rate of *Sargassum muticum* at high temperatures is what makes it so invasive in relatively warm waters. An example of such an advantage was recorded by Kain (1969), involving *Saccorhiza polyschides* and kelp species. Gametophytes and young sporophytes of *S. polyschides* grow faster at 10°C and 17°C than do those of *Laminaria hyperborea*, *L. digitata*, or *Saccharina latissima* (Fig. 7.6), whereas growth of *Saccorhiza* at 5°C is slower than that of *L. hyperborea* and *S. latissima*. Not only does *Saccorhiza* grow faster in warmer water, but also its cells are markedly larger (at all temperatures), and those in the stipe, especially, elongate greatly, giving the young sporophyte better access to light. At 5°C, *S. latissima* is the most developed after 18 days; *L. hyperborea* does equally well at 10°C and 17°C; and *L. digitata* grows most slowly at all temperatures. *Saccorhiza* has a more southerly distribution than *Laminaria* species in Europe, which correlates with its poor growth in cold water, but correlations of the relative growth rates of the kelp species with their distributions are not so clear.

As stated above, within a given species there may be considerable genotypic variation in temperature tolerance and in optimum temperatures for growth (as for other responses) (Innes 1988; Gerard and Du Bois 1988; Bischoff and Wiencke 1995; Eggert *et al.* 2003). For instance, high- and low-intertidal *Ulva* (formerly *Enteromorpha) linza* have different temperature responses (Innes 1988). Such variation may be great enough for geographically diverse populations to appear as distinct strains or races. Phenotypic variation within each population is also likely.

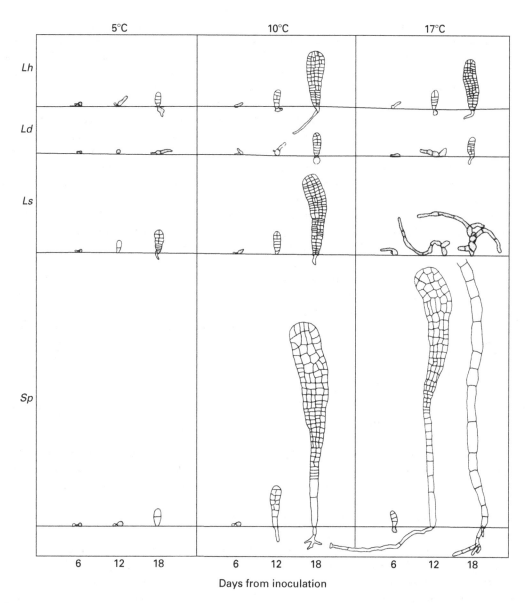

Figure 7.6 Growth and stipe elongation of juvenile kelp sporophytes after 6, 12, and 18 days in culture at 5ºC, 10ºC, or 17ºC. Lh, *Laminaria hyperborea*; Ld, *L. digitata*; Ls, *L. saccharina* (now referred to as *Saccharina latissima*); Sp, *Saccorhiza polyschides*. (From Kain 1969, with permission of Cambridge University Press.)

However, among *Ectocarpus siliculosus*, the populations are not distinct from one another as races (or "ecotypes"), and Bolton (1983) has described their gradual changes as constituting an ecocline. This species is very widely distributed in eastern North America, from Texas to the high Arctic, but its range of genetic variability shifts gradually from population to population along a temperature gradient.

Temperature limits can change, however, or the time of reproduction may change. *Ascophyllum nodosum* reproduces in late winter in Long Island Sound, but in summer in Greenland; gamete release begins at 6°C and ends at 15°C (Bacon and Vadas 1991). West (1972) concluded that strains of *Rhodochorton purpureum* are genetically selected for their temperature dependence of sporulation related to latitude. The very broadly distributed brown alga *Ectocarpus siliculosus* shows phenotypic variation within isolates and also genotypic variation between geographically diverse isolates, with respect not only to temperature optima but also to salinity tolerance (Russell and Bolton 1975). Even a more narrowly distributed species such as *Saccharina latissima* (formerly *Laminaria saccharina*) shows genetic adaptation of temperature optima toward the edge of its range (Gerard and Du Bois 1988). It is also noted that ecotypic differentiation might be manifested differently in the respective life-history stages, as demonstrated for conspecifics of Laminariales from cold-temperate (Helgoland, German Bight) and Arctic (Spitsbergen) environments (Müller *et al.* 2008). More recent molecular genetic studies on thermal ecotype formation may also highlight potential transitions into speciation processes. Ecotype formation in overlapping habitats have been shown for the tropical and subtropical green alga *Cladophoropsis membranacea*, which may have enabled this alga to recolonize habitats after periodic and local extinction, e.g. at the Canary Islands. Thus, the topic of thermal ecotype formation will be highly relevant in the context of globally increasing sea surface temperatures. However, the latter findings also imply that the different populations of *C. membranacea* may rather develop toward the formation of a cryptic multi-species complex (van der Strate *et al.* 2002a).

7.3.4 Temperature tolerance

In general, temperate algae can tolerate cold water at least down to -1.5°C (seawater freezing point). Only 6 species out of 49 did not survive -1.5°C in tests by Lüning and Freshwater (1988) at Friday Harbor,

Washington (where seawater never freezes). Microstages of a few temperate algae, including *Endocladia muricata*, tolerated water as warm as 28°C, but none survived 30°C. Kelps showed the least heat resistance, being limited to 15–18°C (Müller *et al.* 2008). Some Arctic algae in Newfoundland have very low upper limits: *Papenfussiella callitricha* has a limit of 8°C (Hooper and South 1977), and the temperature range for *Phaeosiphoniella cryophila* is about -2°C to +5°C (Hooper *et al.* 1988). In the North Sea, Lüning (1984) found several seaweeds that would tolerate 30°C and one (*Protomonostroma undulatum*) that would tolerate only up to 10°C, with seasonal shifts of up to 5°C in some species (especially *Laminaria* spp. and *Desmarestia aculeata*). However, absolute statements on upper temperature tolerance limits are precluded by the variable responses due to the multitude of interacting factors and from species growing over a wide latitudinal, and thus temperature, gradient which may result in ecotypic speciation. Again, nutrient availability (as NO_3^-) has been identified as an important modulating factor in high temperature tolerance of *Saccharina latissima*: Gerard (1997) has studied two populations that were separated by a distance of approximately 400 km and from habitats with different temperature and nutrient regimes. It was shown that nutrient limitation significantly reduced the thermal stability of the photosynthetic apparatus and the activity of Calvin cycle enzymes, finally resulting even in a negative daily net C fixation. As we will see below, these findings are of great importance to Pacific seaweeds exposed to El Niño conditions.

Most seaweeds will be killed if they become frozen. However, the presence of solutes in water lowers its freezing point, and the high concentration of salts in cytoplasm provides some protection against freezing for intracellular water. Furthermore, the crystallization temperature typically is lower than the freezing point (Spaargaren 1984). Tissue water is not completely frozen until -35°C to -40°C. During progressive cooling, ice crystals form on the outsides of the cells first. This tends to draw water out of the protoplasts, causing dehydration, unless cooling is very rapid, in which case the protoplasts may freeze. Damage is also caused by mechanical disruption of cell components by ice-crystal formation (Bidwell 1979). Damage to the tonoplast will be especially injurious, because toxic materials and inorganic ions stored in the vacuole can thus be released and poison the cell or inhibit metabolic processes in the cytoplasm. Previous studies on the impact of freezing on photosynthesis of marine macroalgae from cold-temperate regions have focused on few model organisms thriving in the intertidal zone, for example *Fucus* spp. (Davison *et al.* 1989; Pearson and Davison 1993, 1994), *Chondrus crispus*, and *Mastocarpus stellatus* (Dudgeon *et al.* 1989, 1990). Damage to *Chondrus crispus* frozen to -20°C was associated with increased plasma membrane permeability and disruption of photosynthetic lamellae, with loss of pigments; but *Mastocarpus stellatus*, which lives a little higher on the shore, was not damaged by that temperature (Dudgeon *et al.* 1989). In most of these early studies, freezing tolerance was mostly characterized by measuring photosynthetic activity before and after freezing exposure. A more detailed study on freezing effects on photophysiology was conducted by Davison and co-workers. The influence of temperature on the biogenic formation and scavenging of reactive oxygen in cold-temperate *Fucus vesiculosus* was studied by Collén and Davison (1999a, b, 2001). They showed for several *Fucus* species from different shore levels, that the degree of oxygen radical production was related to the position of species on the shore (Collén and Davison 1999a, b). As for most other stress factors prevailing in the intertidal, a direct relationship was found between specific freezing tolerance and the position on the shore where the particular species occurs (Davison *et al.* 1989; Lundheim 1997; Fig. 7.7). The latter author sampled nine macroalgal species along their tidal position at Trondhjemfjord in Norway during summer and winter and compared the species-specific mean nucleation temperature (the onset of freezing of tissue samples). Again results confirmed the strong link between freezing tolerance and tidal position. Moreover, seasonal adjustments of freezing tolerance were obvious from difference in nucleation temperature between summer and winter samples.

One might expect a different physiological response for intertidal seaweeds that become frozen while being submerged (thus being trapped in the ice) compared to those that are emerged during low tide. However, it is accepted that in general, frost and drought have similar physiological effects (Pearson and Davison 1994), and in a study by Becker *et al.* (2009) who subjected *Fucus distichus* from Spitsbergen to different freezing treatments, it was shown that the combination of drought and frost had no stronger effects on algal physiology than either factor alone. Moreover, the effect of drought was less important at low temperatures because frozen water is not able to evaporate.

Whereas the effects of freezing are easy to explain, neither the damaging effects of chilling temperatures nor resistances to chilling are well understood for seaweeds. Again, oxidative stress is most likely a prime factor causing photodamage when low temperatures limit electron transport and inhibit photon capture (Davison 1991; Collén and Davison 1999a, b, c, 2001).

Temperature-induced damage to thalli has been seen in both intertidal and subtidal seaweeds. Schonbeck and Norton (1978, 1980a) described the temperature damage to the high intertidal fucoids *Pelvetia canaliculata* and *Fucus spiralis* as consisting of reddish spots of decaying tissue developing some 10 days after thermal stress, along with narrowed apical growth and reduced rates of elongation and weight gain. High temperature damage in both species was less severe at lower humidities (i.e. when the seaweeds were drier). The seaweeds recovered from this damage unless it was extreme. Adverse effects due to unusually high temperatures have been noted in populations of *Macrocystis pyrifera* in California, where warm water promotes black-rot disease (Andrews 1976). We will explore some mechanisms of adaptation and acclimation towards temperature change (including heat stress) in the following section. Nevertheless, almost nothing is known about mechanisms of cryo-protection in seaweeds. In higher plants, the induction of special cryo-protective proteins has been described, such as "cryoprotectin", which stabilizes membranes during freezing (Hincha 2002), or others preventing growth of ice crystals. Designated antifreeze proteins have been identified in the diatom *Fragilariopsis cylindrus* (Armbrust *et al.* 2004), but these molecular mechanisms of

Essay 5 Antioxidant metabolism

Jonas Collén

My PhD thesis project was to work on the physiology of the red algae *Eucheuma denticulatum* and *Kappaphycus alvarezii* in order to understand the metabolism of carrageenan. The aims were to increase yield and control the quality of this economically important polysaccharide. I had the fortune to be able to go to the island of Zanzibar to collect these algae that were grown in recently started commercial cultivations. When I came back to the laboratory the problems started. My previous experiences of growing seaweeds were for different species of *Gracilaria*, which normally grew well in my hands. However, I quickly learned that *Kappaphycus* and *Eucheuma* species are not *Gracilaria*. White parts appeared on the thallus and the tropical seaweed started to fall apart and slowly die. Apparently, it did not like to be transplanted from the warm sunny Indian Ocean to a laboratory at Uppsala University, Sweden. The result of this debacle was that part of my thesis work tried to explain why they died despite all our efforts. For my research career, this also meant that instead of studying carrageenan synthesis and modification I ended up working with stress physiology, and especially oxidative stress since we suspected that reactive oxygen was involved in the demise of my precious algae.

Animals can normally chose to eat or not to eat all the food that is available to them; this choice is more complicated for photosynthetic organisms since plants cannot stop the sun from shining and they have only limited possibilities of reducing the amount of light that reaches their photosynthetic pigments. Just as over indulgence is a problem for humans, absorbing more energy than can be used for photosynthesis is a problem for plants and algae. Excess energy that is not used for carbon fixation can form reactive oxygen species (ROS) like superoxide ions (O_2^-) and singlet oxygen ($^1O_2$) which can later form hydrogen peroxide (H_2O_2) and hydroxyl radicals (OH·). These different ROS have the potential of oxidizing proteins, lipids, and DNA, thereby causing damage to the organism. An imbalance between energy utilization and acquisition can occur when carbon fixation is limited; unfortunately, the dark reactions of photosynthesis are easily disrupted in algae, for example by high or low temperatures, inadequate salinity, or a lack of inorganic carbon. In fact, most stressors will reduce carbon fixation and thus potentially increase ROS production. In addition, most photosynthetic organisms have a maximum carbon fixation rate at a lower light intensity than the maxima experienced. Therefore, photosynthetic organisms try to reduce the energy acquisition by different types of quenching, such as the xanthophyll cycle which uses carotenoids to reduce energy acquisition, or by more or less futile uses of the absorbed energy. These futile uses of energy can sometimes be a two-edged sword; even though they reduce energy, reactive oxygen can be produced as a by-product. As an example, one way for a photosynthetic organism to reduce energy is the water–water cycle (Fig. 1) where oxygen is used as an electron acceptor instead of $NADP^+$ causing the production of O_2^-.

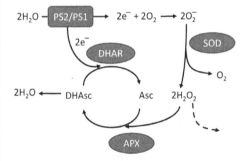

Figure 1 The water–water cycle, simplified. The electrons from the photosystems (PS1/PS2) are shuttled to O_2 forming O_2^- that can be disproportionated by superoxide dismutase (SOD) forming H_2O_2 which, in turn, can be reduced to water by ascorbate peroxidase (APX) during the oxidation of ascorbate (Asc) to dehydroascorbate (DHAsc). In this process both O_2^- and H_2O_2 can potentially damage the alga, or in the case of H_2O_2 escape to the surrounding seawater. (Courtesy of Jonas Collén.)

Essay 5 (cont.)

If the alga does not manage to control the energy, or there is production of ROS by the energy dissipation systems, antioxidant systems are needed to reduce ROS in order to avoid damage. There are two basic types of antioxidants: enzymes that break down ROS, such as catalase, superoxide dismutase (SOD), and ascorbate peroxidase (APX), as well as antioxidants like ascorbate, glutathione, and tocopherols. There is also an interaction between antioxidants and antioxidant enzymes, for example when ascorbate is used as an antioxidant reducing H_2O_2 with the aid of APX (Fig. 1).

I started worrying about ROS metabolism because my seaweeds were dying, but I have continued with the research topic afterwards even if I have moved on to other species that are more respondent to my care, or using algae directly taken from nature. The reason for this is that a large number of studies have shown that the pivotal role of ROS metabolism in terrestrial plants (and humans) is in areas such as growth, signaling, stress, and biotic defense. Much less is known about seaweeds, and considering the phylogenetic diversity and the large evolutionary distance between algae and terrestrial plants, it is likely that seaweeds possess unknown strategies in addition to those already established metabolisms, and therefore represent fascinating research subjects.

Intertidal seaweeds are, to me, especially interesting since they inhabit a high stress environment, with rapid changes in their environment. These alterations are not only caused by weather and seasons, but also through the highs and lows of tides and battling the waves. In addition, in the atmosphere the content of oxygen and carbon dioxide are relatively stable. In tidal and rock pools, the situation is different; during photosynthesis, high concentrations of oxygen can occur at the same time as increased pH and decreased CO_2 concentration. An example is what happens in *Ulva* (Collén *et al.* 1995). In a comparison with experiments in seawater at different pH values, we showed that the excretion of H_2O_2 into the surrounding seawater at pH 9.0 was five times higher than at normal seawater pH 8.2. When I started studying aquatic photosynthesis, the seawater pH was assumed to be 8.2 but with the increasing CO_2-concentration in the atmosphere it is now, *hélas*, 8.1. At the higher pH value, oxygen evolution is lower due to a lack of CO_2; thus, the green seaweed shuttles energy through the water–water cycle and some of the H_2O_2 escapes into the seawater. This phenomenon can cause concentrations of H_2O_2 that can have an impact on other organisms (e.g. an anti-grazing effect) and the organism itself.

There are also some aspects of seawater chemistry that give a twist to oxidative stress in seaweeds, especially the presence of halogens. In many seaweeds, halogen metabolism is very important. Reactive forms of halogens such as bromine and iodine can be formed from their reaction with peroxidases and hydrogen peroxide, producing HOBr and HOI. Another aspect that is different for seaweeds compared to terrestrial plants, is that hydrogen peroxide concentration can be decreased by diffusion into the surrounding seawater, but also be increased if H_2O_2 is present outside the seaweed.

Reactive oxygen metabolism and ecophysiology

In 1996, I started a post-doc with Professor Ian Davison at the University of Maine (Orono, USA). Our hypothesis was that, since abiotic stress causes increased production of ROS in plants, and at the time tentatively in algae, reactive oxygen metabolism plays a pivotal role in general stress tolerance and should therefore be correlated with stress tolerance and environmental parameters. Our principal models were brown algae, particularly different *Fucus* species which inhabit different parts of the intertidal, but we also included some red algae such as *Chondrus crispus*, *Mastocarpus stellatus*, and *Gracilaria tikvahiae*. Our natural laboratory was the intertidal with its different species growing at various tidal heights. We found that different types of antioxidants were present in algae, including antioxidant enzymes such as ascorbate peroxidase, catalase, superoxide dismutase, and glutathione reductase as well as antioxidants such as ascorbate, glutathione, tocopherols, β-carotene, fucoxanthine, and phenolics. We showed that stress tolerance, and thereby potentially their distribution and abundance, was correlated with the content of antioxidants and the enzyme systems that are responsible for defense against reactive oxygen. This was true between species that show different stress tolerance (Collén and Davison 1999a, b, c).

An example is *Fucus spiralis* which is the species found highest on the shore and thus arguably the most stress tolerant; *F. evanescens* is found lower down and *F. distichus* is only found in rock pools and thus in a more stable and arguably less stressful environment. After stress in the form of freezing and desiccation, the production of ROS increased less in the

Essay 5 (cont.)

stress-tolerant species. This was correlated with higher levels of ROS scavenging enzymes in *F. spiralis* and *F. evanescens* compared to *F. distichus*. Similar to our results, Ross and van Alstyne (2007) showed that intertidal *Ulva* had higher H_2O_2 scavenging capability than subtidal *Ulva* in order to be better acclimated to the higher light levels and other more pronounced stressors in the intertidal than in the subtidal. Earlier, we found a similar difference between *M. stellatus* which occurs slightly higher in the intertidal and is more stress tolerant than *C. crispus* and has higher H_2O_2 scavenging capability (Collén and Davison 1999c). In 2001, I moved to the Station Biologique in Roscoff, France, where I have continued with experiments on stress and red algae, focusing on *C. crispus*. This time with a transcriptomic approach, the results reinforced the idea that ROS metabolism plays a key role in stress responses since many of the relevant genes showed increased expression after stress (Collén *et al.* 2007).

Returning to my initial problems with *Eucheuma* culture, they were probably not related to photosynthetically produced H_2O_2. Instead it is likely that the oxidative stress was caused by defense reactions. We know now that bursts of hydrogen peroxide are used as a part of the defense in terrestrial plants, mammals, and algae. This burst of ROS can be triggered by cell wall fragments of either the host seaweed or the attacking microorganisms and is produced by an NADPH oxidase. This defense reaction causes locally increased concentrations of ROS that in itself can have a defensive role; it creates the production of potentially toxic halogen species; it can also induce cell wall strengthening; in addition, it is a signal that the alga is under attack (Potin 2008).

Armed with over a decade of increased knowledge I can look back on my initial experiments with cultivating *Eucheuma* and *Kappaphycus*. Comparing how others are able to successfully grow it today, my guess is that confined in the laboratory instead of in the Indian Ocean the concentrations of bacteria increased, thereby triggering defense mechanisms, including an oxidative burst. The burst would normally help to signal within the alga that an attack was ongoing and serve as an antibacterial substance. However, in the confines of a round flask, the reaction was probably too strong and lethally stressed the algae. Successful laboratory cultivation of eucheumoid species uses antibiotics to reduce the bacterial load and that probably avoids the excessive stress caused by defense reactions.

For my present research, I have taken a step back and am studying the genomes of seaweeds. We have now sequenced the genomes of *Ectocarpus siliculosus* and *Chondrus crispus*. Armed with this new knowledge, I am now returning to the study of stress physiology and ecophysiology using large-scale transcriptomic, metabolomic, and proteomic methods. Another reason for my genomic interest is to be able to gain understanding of carrageenan biosynthesis and physiology, because the initial questions for my PhD thesis are still unanswered, and we cannot have that, can we?

*Jonas Collén is a lecturer at the Station Biologique de Roscoff, France. His research interests are principally on ecophysiology, physiology and genomics of seaweeds. The main projects of the last years have been the sequencing of the genome of the red alga Chondrus crispus *(Irish Moss) and studies on stress responses in intertidal seaweeds.*

freezing tolerance still have to be revealed for seaweeds, in particular in order to understand basic adaptive features in polar environments.

7.3.5 Physiological adaptation to changes in temperature

While species-specific ranges in temperature tolerance, and acclimatory as well as adaptive responses with respect to survival and other integrative parameters such as growth, reproduction and photosynthesis are well described, knowledge of the underlying physiological processes are still understudied in seaweeds, and most mechanistic concepts are derived from studies in terrestrial plants.

With respect to temperature changes, adaptive (long-term) modifications may occur on the enzyme level, either by the adjustment of enzyme concentration, in order to compensate for temperature-dependent variations in activity according to the Q_{10} rule, or by structural modifications altering characteristics such as substrate affinity and temperature

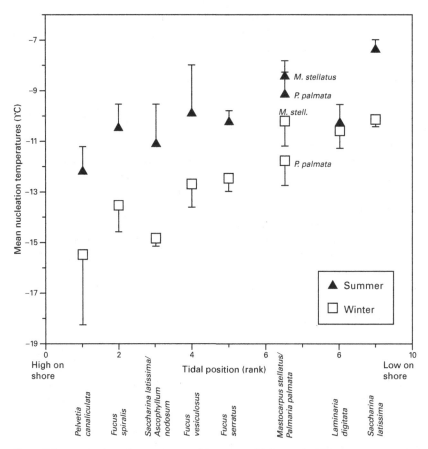

Figure 7.7 Mean nucleation temperatures of fragments of different algal species growing in the littoral zone. Species are ranked by their tidal position, starting with *Pelvetia canaliculata* at the upper eulittoral. Open squares represent nucleation temperatures of fragments from algae collected on 5 March 1996, closed triangles represent nucleation temperatures of fragments from algae collected on 4 July and 6 July 1996. (From Lundheim 1997, reproduced with permission.)

optima. This may apply to the enzymes involved in the Calvin cycle to provide sufficient NADPH drainage under the given environmental conditions (Gerard 1997). This is all-important with respect to the operation of the photosynthetic apparatus under varying light and temperature regimes, as otherwise an imbalance between photosynthetic primary and secondary reactions may occur. While the secondary reactions of photosynthesis are strictly temperature dependent, the primary reactions are not and, thus, low temperatures in combination with high levels of photosynthetically active radiation (PAR) may increase radiation stress to photosynthesis. Due to temperature dependence of

enzymatic reactions, maximal photosynthetic capacity is reduced under low temperatures. The mechanisms behind this include the reduced activity of RuBisCO and also the decreased velocity of the turnover of the D1 reaction center protein in photosystem II, which in addition is also affected by the reduced fluidity of the chloroplast membrane, thus hampering the insertion of *de novo* synthesized proteins (Davison 1991; Aro *et al.* 1993; Becker *et al.* 2010). As enzymatic reactions involved in photosynthetic dark reactions are slowed down at low temperatures, persisting high-energy input due to high irradiances of PAR may further result in increased electron pressure in the electron transport chain. In photosynthesis, absorption of excessive radiation inevitably results in the generation of reactive oxygen species like the superoxide radical and hydrogen peroxide within the Mehler reaction (Polle 1996). In addition, electron transfer from so-called triplet-chlorophyll may result in the generation of extremely reactive and aggressive singlet oxygen (Asada and Takahashi 1987; see Essay 5). The consequences of increased oxidative stress are chronic photoinhibition, bleaching of photosynthetic pigments, peroxidation of membrane lipids, and enhanced degradation of D1 protein (Aro *et al.* 1993; Osmond 1994).

From the mechanisms of temperature-dependent photoinhibition outlined above it can be concluded that an important feature in thermal adaptation might be the respective composition of membrane lipids. Hitherto this aspect has been widely understudied in seaweeds. However, a recent study on the Antarctic rhodophyte *Palmaria decipiens* has revealed flexible adjustments in fatty acid composition (chain length and saturation state) of biomembranes in response to experimental temperature treatments, as well as growth-site-specific differences with respect to tidal levels (Becker *et al.* 2010). Thylakoid membrane fluidity can be maintained via adjustment of fatty acid composition dependent on prevailing temperature. Under low temperature shorter and polyunsaturated fatty acids keep membranes fluid, while at elevated temperatures longer fatty acids with higher saturation states increase rigidity and thus may prevent membrane leakage. The physiological role of fatty acid composition and its molecular control is an important field of research in current seaweed ecophysiology.

However, heat-induced damage to photosynthetic functions may still occur through the failure of thermolabile enzymes. The heat stability of an enzyme seems to be partly related to the temperature at which it was formed. Because there is constant turnover of protein molecules, the enzymatic machinery may thus become gradually acclimated to changing temperatures. Upon temperature stress, the expression of heat shock proteins (HSPs) has been observed in a wide range of organisms, including seaweeds. Heat shock proteins and molecular chaperones (CPNs) are formed in order to protect functional proteins against heat-

mediated denaturation and to maintain operational folding patterns. In addition, CPNs are involved in the folding process of native, newly synthesized polypeptides. Heat shock protein formation under temperature stress has been reported from a wide variety of species of green (*Ulva lactuca, U. intestinalis*), red (*Plocamium cartilagineum, Chondrus crispus, Palmaria palmata*) and brown (*Fucus serratus, F. vesiculosus, F. spiralis, F. radicans, Laminaria japonica, Undaria pinnatifida, Egregia menziesii*) seaweeds, but with pronounced species-specific and site-specific responses depending on spatial gradients, temperature treatment, etc. (Vayda and Yuan 1994; Ireland *et al.* 2004; Collén *et al.* 2007; Fu *et al.* 2009; Lago-Leston *et al.* 2010). For an Antarctic isolate of the rhodophyte *Plocamium cartilagineum*, collected from the field at temperatures between -0.3 and -1.1ºC, high induction of HSP70 encoding mRNA was observed within 1 h of incubation at temperatures between 5 and 10ºC. Longer incubation times and higher temperatures resulted in a reduction of HSP70 transcripts and incubation at 20ºC were lethal to the alga (Vayda and Yuan 1994). The functionality of heat shock proteins was proved by a cloning experiment: various genes encoding for different heat shock proteins were identified in the intertidal rhodophyte *Porphyra seriata*. Under elevated temperature, a significantly higher level of PsHSP70b transcripts was detected. When this gene was introduced and overexpressed in *Chlamydomonas*, heat resistance in terms of survival and growth under elevated temperature increased markedly (Park *et al.* 2011).

It is likely that HSP and CPN induction is not only invoked as a response to temperature stress, but may also reflect a general stress response, as the induction of HSP gene expression was also enhanced under high light conditions in gametophytes of the rhodophyte *Chondrus crispus* (Collén *et al.* 2007). In the chlorophyte *Ulva rotundata*, the cellular content of CPN 60 (also referred to as the RuBisCO binding protein) increased at enhanced UV radiation (under PAR exclusion), however, in this study an additional effect of temperature increase could not be totally excluded (Bischof *et al.* 2002). In all, it is reasonable to argue that induction of HSP and other biomolecules with anti-stress function is complementary or occurs as a response to combinations of factors. In fact, studies examining multiple cDNA microarrays from intertidal algae exposed to different stressors outlined the up-regulation of specific genes for different stresses and additionally, emphasized the importance the gene expression during oxidative stress (Collén *et al.* 2007).

7.3.6 Temperature tolerance in polar seaweeds

As discussed in sec. 3.3.3, Antarctic seaweeds have a cold-water history of around 15 million years compared to just 2.7 million years for Arctic seaweeds. These differences in cold-water history and endemism between the Antarctic and Arctic are clearly reflected by the degree of cold adaptation of the dominant seaweed species in both regions. Cold adaptation in many Antarctic species has been studied by Wiencke and co-workers with respect to photosynthetic and growth performance (Wiencke 1996; Wiencke and Clayton 2002). In species from the stenothermal group of Antarctic Rhodophyta like *Gigartina skottsbergii* or *Georgiella confluens*, growth is already halted at 5°C (Fig. 7.8), so that the respective temperature tolerance range is extremely narrow (Bischoff-Bäsmann and Wiencke 1996). Curiously, in this study similar results were found for *Plocamium cartilagineum*, which is rather a typical component of the cold-temperate flora of the North Atlantic, where it is obviously exposed to a much different temperature regime. Lüning (1984) determined the upper survival temperature of the Helgoland (North Sea, German Bight) isolate of *P. cartilagineum* to be at 23°C. Thus, a remarkable and efficient adaptation and ecotype formation has to be assumed in this species (Vayda and Yuan 1994).

Antarctic seaweeds have adjusted their life cycles to the strong seasonality of their environment. Thus, in accordance with the prevailing temperature regime, different temperature tolerance ranges have been described depending on the respective life-history stage, i.e. for the sporophyte and the male and female gametophytes of *Desmarestia anceps* (Wiencke and tom Dieck 1989; Wiencke and Clayton 2002; Fig. 7.9).

Unfortunately, the amount of data on temperature responses of growth in Antarctic seaweeds is not paralleled by research on the underlying mechanisms of cold adaptation. However, the maintenance of the fluidity of biological membranes by possession of unsaturated fatty acids that prevent rigidification of membrane lipids, molecular adaptations of enzymes in order to maintain sufficient rates of enzyme-catalyzed reactions of key metabolic processes, the evolution of cold shock and antifreeze proteins, and adaptations of the photosynthetic electron transport chain to function at cold temperatures have been regarded as major physiological adaptations of Antarctic algae to temperature (Graeve *et al.* 2002; Becker *et al.* 2010).

Apart from just a few species truly endemic to the Arctic regions, such as the kelp *Laminaria solidungula* (Dunton and Dayton 1995), adaptation to a low temperature environment is largely absent in species populating the Arctic region. In their survey on UV and temperature tolerance in a range of seaweeds from the island of Spitsbergen (at 78°N) Müller *et al.* (2008) and Fredersdorf *et al.* (2009) revealed that most kelps from Spitsbergen display temperature tolerance limits far exceeding the range of the expected temperature increase at their growth site. In *Laminaria digitata*, most temperature-susceptible developmental processes (spore germination, egg release) were not impaired at temperatures up to 12°C (Müller *et al.* 2008). At the study site, current sea surface temperature in Arctic summer may reach up to 5°C (Svendsen *et al.* 2002) and the most pessimistic predictions of Arctic climate change forecast a 5°C increase in atmospheric temperature (IPCC 2007). Most of the specimens studied at Spitsbergen still show similar temperature responses as their conspecifics from more southern locations (e.g. Helgoland, German Bight; Müller *et al.* 2008), though there is some indication of ongoing ecotypic differentiation. For example, in some species like the kelp *Alaria esculenta* spore germination showed an even higher optimum temperature than currently experienced at the Arctic study site. For this species it was also proposed that UV-induced stress might even be compensated by elevated temperatures (Fredersdorf *et al.* 2009). This peculiar

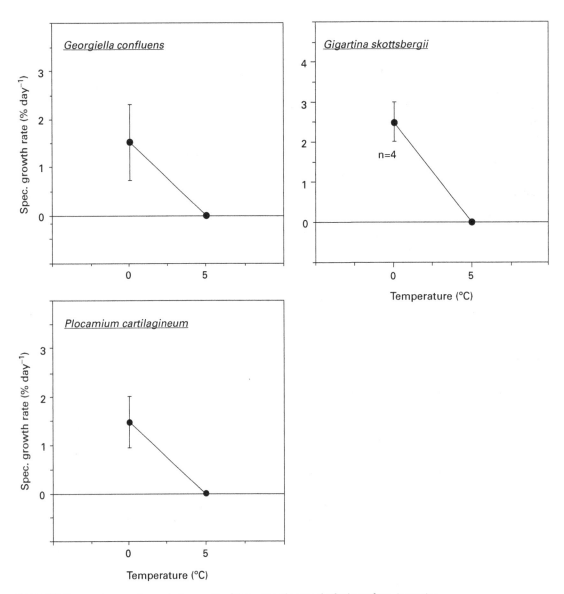

Figure 7.8 Temperature requirements for growth of three stenothermic rhodophytes from Antarctica. (From Bischoff-Bäsmann and Wiencke 1996, reproduced with permission.)

response of *A. esculenta* from Spitsbergen becomes comprehensible when considering the distributional range of the species, which stretches as far south as Brittany in the East and Rhode Island in the West Atlantic.

A number of species like *Plocamium cartilagineum* have been reported to have an amphi-equatorial or even bipolar distribution (Wiencke *et al.* 1994). While an increasing number of allochthonous species are being introduced in previously isolated habitats (e.g.

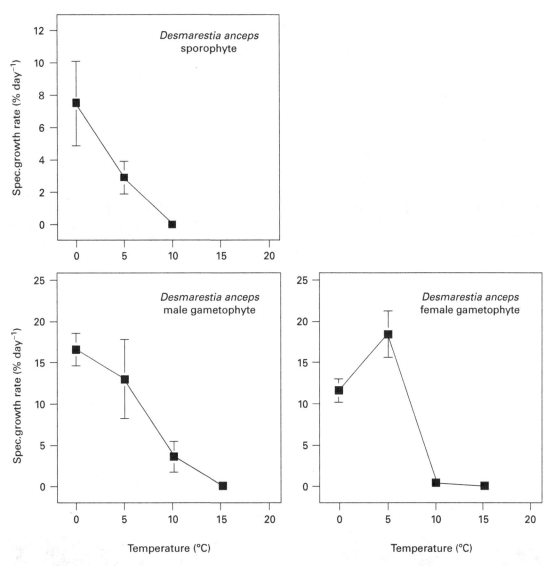

Figure 7.9 Temperature requirements for growth of the sporophyte and the male and female gametophytes of *Desmarestia anceps* from Antarctica. (From Wiencke and Clayton 2002, redrawn from Wiencke and tom Dieck 1989, reproduced with permission.)

by ship traffic via fouled hulls or ballast water), processes related to climatology and hydrography have also contributed to the bipolar spread of species, e.g. during the last glacial maximum (van Oppen *et al.* 1993, 1994). For the green *Acrosiphonia arcta* and the brown *Desmarestia viridis* it is suggested that its upper temperature tolerance limit matches temperatures

prevailing in cold-water gaps at the Atlantic Equator region approx. 18 000 years ago and thus allowed for a floral exchange from one hemisphere to the other (Bischoff-Bäsmann 1997; Fig. 7.10). This hypothesis on the dispersal routes of *A. arcta* is strongly supported by molecular data from van Oppen *et al.* (1993, 1994).

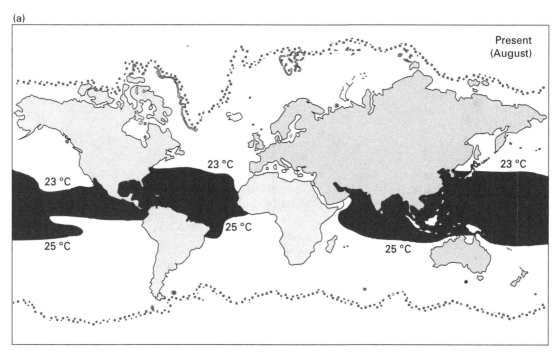

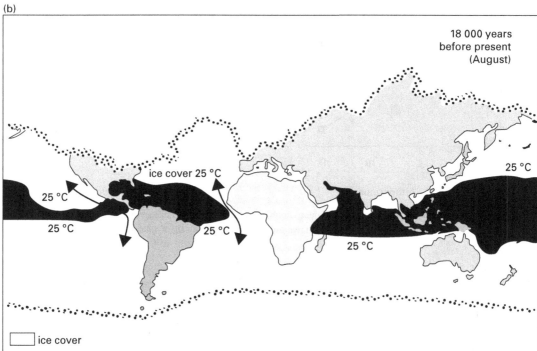

Figure 7.10 Outline for the amphi-equatorial distribution of seaweeds, i.e. the chlorophyte *Acrosiphonia arcta*: (a) The tropics as a present-day warm-water barrier, indicated isotherms correspondent with the boundaries of the tropic waters, as well as with the upper survival temperatures of *A. arcta*; (b) During the last glacial maximum, gaps in the warm-water barrier facilitated floral exchanges between both hemispheres. (Redrawn from Bischoff-Bäsmann 1997, *Berichte zur Polarforschung*, with permission.)

The predictions of the Intergovernmental Panel on Climate Change (IPCC) indicate that the increase in atmospheric and sea surface temperature will be most pronounced in the Arctic. However, the Antarctic Peninsula has also been identified as a region prone to severe environmental change (IPCC 2007) and the potential consequences for the marine biota, including seaweed communities are currently a matter of intense research. However, based on the differences in the degree of cold adaptation found in Arctic versus Antarctic seaweed communities the likely consequences of a temperature increase will differ in both hemispheres. For most seaweed species populating Arctic areas, a 3–5°C increase in the sea surface temperature is not likely to cause lethal effects, but is within the generally broad tolerance range of most Arctic species. It is thus more likely that seaweed communities along Arctic coastlines will change as a result of new invaders coming in from more southern locations, which can, as a consequence of temperature increase, shift their distribution limit even further north. One potential invader might be the kelp *Laminaria hyperborea*, which is already present all along the Norwegian coastline. A potential stepping stone for populating the high Arctic Islands of Spitsbergen might be Bear Island situated in the Barents Sea right in between the northern tip of Norway and Spitsbergen (Müller *et al.* 2009b). In contrast, due to the high degree of cold adaptation present in the most dominant species in Antarctica, elevated temperature may represent a crucial stress factor to seaweed physiology and may thus require new competitive adjustments. In this way strictly cold-adapted endemic species might be replaced by (introduced) subpolar and cosmopolitan species. Interestingly, although Antarctic species commonly live at temperatures <2°C, some retain the capacity to survive at temperatures as high as 19°C (Wiencke and tom Dieck 1989). This raises various questions whether it may reflect a physiological prerequisite to tolerate broad ranges of environmental variability driven by global change. Again, it is an open but a burning question in seaweed stress ecology, if adaptation shown by organisms today can be useful clues to predict environmental change in the future.

7.3.7 El Niño

One of the most important large-scale environmental variations affecting coastal marine ecosystems is the El Niño Southern Oscillation (ENSO). The Pacific region of Central and South America is normally dominated by the cold and nutrient-rich waters of the Humboldt Current Upwelling System (HCS). During an ENSO event, the trade winds, which normally push the warmer surface waters westwards, weaken and a tongue of warm and nutrient-poor water moves eastwards along the equatorial region. Upon reaching the coast, the water moves southward to overlay the cold and highly productive waters of the upwelling system. As a consequence, sea level and surface-water-temperature rise, the thermocline deepens, water transparency greatly increases and nutrient upwelling ceases (Glynn 1988; Chavez *et al.* 1999). The warmer waters first reach the Galapagos Islands, located 1000 km west of the Ecuadorian coast. For this reason, the Galapagos Islands have proved to provide early warning signals for an approaching ENSO (Kessler 2006). The phenomenon of ENSO occurs at intervals of 2 to 7 years with its strongest impacts in the eastern tropical Pacific in the months November to January. An ENSO event is often followed by a period of very strong upwelling, which leads to below normal temperatures. This phenomenon is called La Niña, as the opposite of El Niño. During ENSO, the water temperature is negatively correlated with the nutrient concentration. Sea surface temperatures (SST) in the upwelling area of the HCS are around 13–18°C during a non-ENSO year, but during ENSO may increase to 26°C. Because of the 90 m deepening of the thermocline, the upwelling of nutrient-rich water decreases. The HCS is normally high in nutrients, especially nitrogen (2–6 μM nitrate). But during the strong ENSO in 1982–83, nitrate concentration dropped to <0.2 μM at the coast of Santa Cruz (Galapagos Islands; Kongelschatz *et al.* 1985). Adverse events, such as strong storms, extreme tides, a rise of the sea level, abundant rainfall, etc. are related to ENSO (Suple 1999). In the affected area of the Eastern Pacific and around the Galapagos Islands, ENSO has been associated with the emigration and death of many local species, the collapse of the anjoveta and sardine fisheries off Peru and California, respectively, and an immigration of exotic, tropical invaders (Wolff 1987; Arntz et al. 2006). Even the death of coral reefs (Warwick et al. 1990) also appear directly related to ENSO.

The 1982–83 and 1997–98 ENSOs were the most severe on record, with extensive and protracted increases in sea temperatures (2–5°C) and sea levels (0.2 m or more) in California (on the edge of the affected area). The center of the disturbance was off Peru, where the peak increase in water temperature was 8°C (Glynn 1988). Physical changes were detectable as far north as southern Alaska, though no effects on the biota could be attributed to ENSO that far north (Paine 1986). Mass mortalities of red and green algae in the Galapagos and kelp dieback in northern Chile were among the effects near the center of the El Niño. A reduction of the standing stock and also the death of brown seaweeds like *Macrocystis* sp. were observed in Baja and Central California (Dayton and Tegner 1984b; Carballo et al. 2002; Edwards and Ester 2006) and along the coasts influenced by the Humboldt Current System (Thiel et al. 2007). At the Galapagos Islands the population of *Bifucaria galapagensis* and many species of *Sargassum*, which normally occur in high abundance at the archipelago, disappeared almost completely. The population of *Eisenia galapagensis* was strongly decimated during ENSO (Garske 2002), however, some perennials, such as *Pelvetia fastigiata*, recruited better during the warm-water period. Subtidal seaweeds were also affected by nutrient shortages, because the thermoclines were deeper during El Niño, creating thicker layers of nutrient-depleted surface waters. Growth rates for *Macrocystis pyrifera* (Zimmerman and Robertson 1985) and *Pleurophycus gardneri* (Germann 1988) were reduced. Dean and Jacobsen (1986) were able to show directly that nutrient limitation, not high temperature, had caused the reduced growth of *Macrocystis*. During ENSO, concentrations of <0.5 μM NO_3 were found in the water column. For *Macrocystis*, however, a required concentration of 1 μM NO_3 for maintaining growth and survival was determined (Gerard 1982b). For the kelp *Saccharina latissima*, Gerard (1997) showed that N availability is crucial for thermal acclimation of photosynthesis.

Further interactions between N limitation and other environmental conditions are poorly understood but may be particularly important under ENSO scenarios. During ENSO, water column transparency increases, so high solar irradiance can penetrate deeper, including high irradiances of harmful UV-B radiation, which is characteristic for the tropical and subtropical regions and may contribute in the structuring of seaweed-dominated coastal ecosystems (Bischof et al. 2006a).

The subtidal seaweed ecosystems of the Central and South American coasts are highly productive, hosting abundant populations of macroinvertebrates and fishes (Vasquez et al. 2001). These rich communities may buffer phase shift events and facilitate the recovery from slight ENSO disturbances (Steneck et al. 2002). However, the loss of kelp forests under a severe El Niño event represents a disastrous impact on these coastal ecosystems, since kelps are the main structuring organisms in these systems. Recovery time of perennial kelp forests is very long in contrast to fast-growing opportunistic green algal species and also depends on the biological settings. Edwards and Hernandez-Carmona (2005) suggested that recovery of *Macrocystis pyrifera* beds from El Niño impacts at their southern distributional range depends not only on the stress induced by high temperature and nutrient depletion, but also on substrate availability when competing against populations of species more persistent under El Niño, such as the kelp *Eisenia arborea*. Furthermore, after a complete collapse of the population, recovery must be driven either by connectivity allowing new allochthonous recruits to establish, or by the recruitment from life-history stages, which resist stress exposure either by dormancy or by a stage-specific higher tolerance to stress. Ladah and Zertuche-Gonzalez (2007) proposed that the microscopic diploid stages (the embryonic sporophytes) of *Macrocystis* represent such a resistant stage, which may be regarded as a seed bank analog to allow for recovery of depleted populations at their southern range. The assessment of ecosystem effects of ENSO events based on kelp performance is a highly relevant area of seaweed research because of the extensive impact on the ecosystem.

7.3.8 Temperature and geographic distribution

The knowledge that temperature is a major driver in seaweed biogeography and that temperature regimes change globally imply that new competitive settings and latitudinal shifts in species distributional limits are to be expected in response to climate change (Müller *et al.* 2009b; Wernberg *et al.* 2011a). These studies on the impacts of regional to global temperature shifts build on the classical experiments to test the tolerances of seaweeds and to compare the results with temperature data from distribution limits, with the major researchers being Biebl (e.g. Biebl 1970), van den Hoek and co-workers (e.g. Cambridge *et al.* 1984; Yarish *et al.* 1984) and Lüning (Lüning 1984; Lüning and Freshwater 1988). However, these authors have used widely different exposure times, ranging from hours to months, and survival ability and regeneration capacity was found to decrease with longer exposure times (e.g. Yarish *et al.* 1987). Drawing conclusions from simple stress-response experiments neglects the process of acclimation and adaptation and, furthermore, the recovery time and conditions must be known and standardized. An interpretation of the findings must take into account not only water temperatures but also air temperatures if the algae are intertidal. Moreover, limitations in experimental material (age, stage, reproductive state, etc.) need to be recognized (Lüning and Freshwater 1988).

There is consensus that water-temperature tolerances of different species of seaweeds are at least partly responsible for the patterns of the geographic distribution of the adults, as reviewed by Lüning (1990). Indirect evidence regarding the importance of temperature for seaweed floras can be drawn from the effects of ocean currents, since the floral distributions do not have a strict latitudinal correlation (such a correlation might have been attributable to the amount of light available). The east and west coasts of South Africa are classical examples of the influences of warm and cool currents. The south-coast flora is a mixture of warm- and cold-water species due to the mixing of the warm water of the Agulhas Current and the cold water of the Benguela Current, along with some cold upwelling (Stephenson and Stephenson 1972). Upwelling currents, however, are not necessarily constant throughout the year, and so there can be marked changes in the flora related to the hydrographic "seasons", as off Baja California (Dawson 1951). The discovery of a species of *Laminaria* (a cold-water genus) off the coast of Brazil, only 22–23°S of the equator, is explained by the presence of cold upwelling at the considerable depths where these seaweeds grow (Joly and de Oliveira 1967). Peculiar hydrographic conditions may result in sharp transitions in seaweed biogeography, even on very small scales, e.g. between the Arabian Sea and the Gulf of Oman differing largely with respect to seasonal temperature ranges of coastal waters and species composition (Schils and Wilson 2006). The cold-temperate flora of the southern hemisphere has been modeled by the circulation of the ACC and circumpolar frontal zones which strongly define the diversity, biogeography, and the ecology of seaweeds of the southern South America (Chile and Argentinean Patagonia), the Victoria–Tasmania Region, Southern New Zealand and the sub-Antarctic islands (Huovinen and Gómez 2012).

In a classical approach, Hutchins (1947) defined four critical temperatures as a basis for seaweed biogeographic analysis: (1) the minimum for survival, which might determine the winter poleward boundary for a species; (2) the minimum for reproduction, controlling the summer poleward boundary; (3) the maximum for reproduction, controlling the winter equatorward boundary; and (4) the maximum temperature for survival, determining the summer equatorward boundary. Van den Hoek (1982) added two more potential boundaries, those limiting growth poleward and equatorward. Thus, a species may be restricted by survival temperatures north and south, by reproduction temperatures north and south, or by one survival limit and one reproduction limit. An evaluation of the experimental and phenological evidence led Breeman (1988) to the conclusion that indeed temperature responses do account for most of the geographic boundaries of seaweeds. Restrictions are imposed by high or low survival limits for the hardiest life-history stage (often microthalli in heteromorphic life histories), temperature requirements for reproduction of any stage, and temperature limits on growth and asexual propagation. The lethal limits for macrothalli and their growth/reproduction optima do not necessarily correlate with geographic limits for seaweeds that have heteromorphic life histories (Wiencke and tom Dieck 1989). Species with the same temperature tolerances may still have different geographic boundaries; for instance, rates of minimal growth will be higher in species susceptible to herbivores than in herbivore-resistant species (Breeman 1988). Added stresses will tend to decrease algal resistance to temperature extremes. These additional stressors are likely to be most effective at the edges of the respective distributional ranges. Infrequent periods of unusual cold or heat may wipe out seaweed populations at the edges of their ranges and such brief extremes, not the average warmest or coldest month temperatures, may rather represent the decisive circumstances that control algal distributions (Gessner 1970).

Breeman (1990) has reviewed the potential consequences of predicted temperature increase due to global warming on a number of North Atlantic seaweeds. In a worst-case scenario, a 4°C increase in summer temperatures will result in the northward shift in the southern distributional limit of the canopy-forming *Laminaria hyperborea*, and will deprive extended coastlines (the northern Iberian Peninsula, the Atlantic coast of France, southern British Isles, southern Norway) of this species of utmost ecological significance. Similarly, as outlined by Müller *et al.* (2009b) the northern boundary limit of this species will be extended up to Spitsbergen until the end of the twenty-first century, where *L. hyperborea* might be competing against the Arctic endemic *L. solidungula*. With respect to the future impact of a globally changing temperature regime, there is general consensus that the distributional ranges of seaweeds may change, as determined by the temperature sensitivity of the most susceptible life-history stage. Along latitudinal (and temperature) gradients the species (and their respective life-history stages) forming the seaweed community will be impacted differently by temperature change, implying changes in competition and species composition. Matson and Edwards (2007) have investigated the effect of a temperature increase on various life-history stages of two co-occurring kelp species *Pterygophora californica* and *Eisenia arborea* along the coast of California. This study

again highlights the importance of multiple life-stage studies in environmental stress ecology. While the sporophytes of both species were similarly tolerant to high temperatures, the microstages responded differently, indicating that the southern distributional limit of *P. californica*, currently at northern Baja California, Mexico, is controlled by the susceptibility of its microscopic stages. Ocean warming will, thus, differently change species composition along the California coastline.

Several studies on temperature effects on seaweeds put community responses into focus. Wernberg *et al.* (2011a) have demonstrated that a poleward migration of large numbers of seaweed species along the Australian coastlines has already occurred between the 1940s and today. Extrapolated to global change scenarios, their findings imply the potential retreat to hundreds of seaweed species from the Australian continent (Fig. 7.11). Authors have calculated displacement rates for temperature-induced loss in seaweed species as high as 77 species per 1°C temperature increase (Wernberg *et al.* 2011a). In another study focused on kelp beds, Wernberg *et al.* (2010) have stressed that although some species of kelp at the Australian coastline may be able to respond to increasing temperature simply by metabolic adjustments, resilience after additional impacts (like storm events or pollution) might be impaired. Again, Wernberg *et al.* (2011b) conducted an analysis on four different sites at the Australian west coast and modeled the change in prevailing habitat structures (kelp vs. fucoid vs. mixed canopies, open gaps) along temperature gradients (Fig. 7.12). Although the multiple and complex interactions are hard to interpret with respect to their ecological meaning under global warming, it becomes evident that seaweed communities may undergo profound changes in the future, for example as shown for the declining mixed communities, in which *Sargassum* sp. becomes more competitive at the expense of other fucoid algae. Community changes induced by increasing temperatures and CO_2 may be so drastic that even the term "phase shift" (a shift in the dominating life-form) has been applied. Connell and Russell (2010) have stressed that the expected change in seawater temperature and CO_2 will significantly favor growth of turf algae in a synergistic way. In turn, turf algae have been shown to suppress kelp recruitment, and consequently

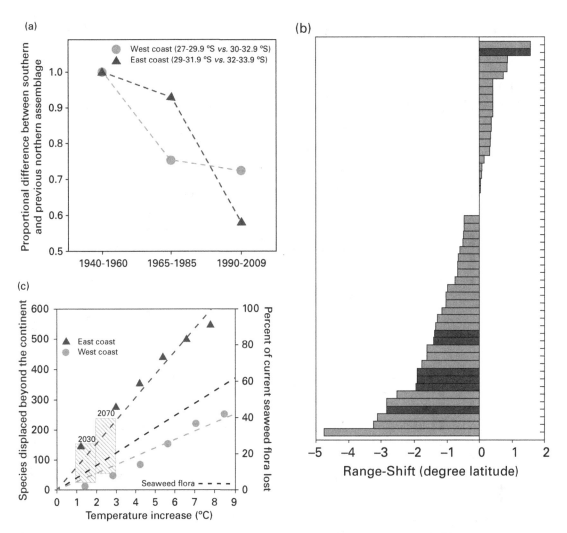

Figure 7.11 Poleward shifts in temperate macroalgae in Australia assessed from herbarium records. (a) Multivariate dissimilarity in assemblage structure (proportional Sorensen dissimilarity) between past northern and subsequent southern floras. (b) Shift in northern range limits of temperate macroalgae in Australia from 1940–60 and 1990–2009. (c) Species displaced beyond the continent (∼44°S and ∼35°S on the east and west coasts – triangle and circle symbols, respectively) given different temperature increases and assuming the median range shifts on each coast are representative for all populations. The gray lines indicate linear regressions through the origin, which yielded species displacement rates of 77 ($R^2 = 0.96$) and 28 ($R^2 = 0.94$) species °C^{-1} temperature increase for the east and west coasts, respectively. The black line indicates the projected relative total species loss (out of 1454 species). The shaded boxes indicate the range of current temperature projections for 2030 and 2070. (Redrawn from Wernberg *et al.* 2011a, reproduced with permission.)

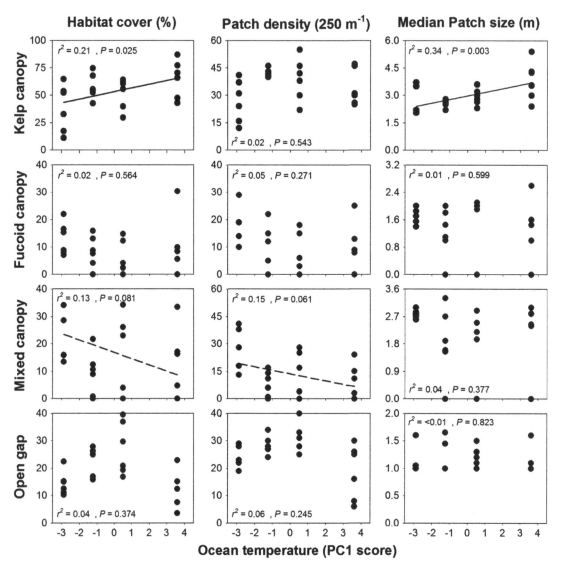

Figure 7.12 Relationship between ocean temperature at four regions (from left to right: Hamelin Bay, Marmion, Jurien Bay and Kalbarri) and characteristics of the biogenic habitat at six independent sites within each region. Kelp canopy refers to monospecific patches of *Ecklonia radiata*. The ocean temperature incorporates temperature variation at multiple temporal scales (From Wernberg *et al.* 2011b, reproduced with permission.)

previously kelp-dominated habitats may shift towards turf-dominated sites. Compared to the sometimes puzzling results from CO_2 perturbation experiments, results from temperature studies are usually quite straightforward. However, as we will see below, temperature and CO_2 increase are directly coupled and it is this kind of bi- to multifactorial studies, which should be conducted with high priority.

7.4 Biochemical and physiological effects of salinity

"Salinity" can be defined simply as grams of salts per kilogram of solution; yet this simple definition belies the physical, chemical, and biological complexities of this "factor". From the physical point of view, the complexity lies in the relationships that seawater density, light refraction, and electrical conductivity have with salinity (and also to temperature) (Kalle 1971). The aspects of salinity that are of biological significance are ion concentrations, the density of seawater, and, especially, the osmotic pressure.

The most important effects of salinity are the osmotic consequences of the movement of water molecules along water potential gradients and the flow of ions along electrochemical gradients. These processes take place simultaneously, and both are regulated in part by the semipermeable membranes that surround cells, and cellular compartments. Water movement is described in the next section, followed by an account of how cells cope with osmotic changes.

7.4.1 Water potential

To understand the physiological effects of salinity, one must understand the basic principles of water potential, which are outlined in plant physiology textbooks (e.g. Buchanan *et al.* 2000). In short, movement of molecules requires free energy. Molecules may acquire free energy from a variety of sources as a result of changes in temperature, concentration, or pressure, as well as that due to gravity and other forces. The free energy of a substance is called its chemical potential, and the chemical potential of water is called water potential, denoted by the Greek letter Ψ ("Psi"). With reference to a cell surrounded by a solution, there are several components to Ψ. Matric potential, Ψ_m, is a measure of the forces that bind water molecules to colloidal material (including proteins and cell walls); it is a minor component in submerged cells, but is important under desiccating conditions on an exposed seashore. Osmotic potential, Ψ_π, is the potential of water to diffuse toward a solution. The osmotic potential of pure water is zero.

Anything dissolved in water will lower its osmotic potential. The more particles there are in solution, the more negative the osmotic potential. Water flows down the potential gradient, that is, toward the more negative Ψ_π. Effectively, water movement results in a dilution of a more concentrated solution. The decrease in Ψ_π is proportional to the number of particles dissolved, regardless of their size. Each dissociated ion of a salt counts as one particle, so that, ideally, a molar solution of sodium chloride has twice the osmotic potential of a molar solution of sucrose. (In practice, a small correction factor, the activity coefficient, must be included in the calculation.) The concentrations of solutions to be used for osmotic measurements are not given as molarity (moles per liter of solvent at 20°C), but as *molality* (moles per kilogram of solvent), because addition of solute molecules dilutes the solvent molecules. By using molality, we refer always to the same number of solvent molecules. Units are (milli)osmols per kilogram: One mole (1 mol) of undissociated solute in 1000 g of water = 1000 mosmol kg^{-1}. Seawater of 35 S_A at 20°C has an osmolality of 1050 mosmol kg^{-1} (Kirst 1988).

As water flows into a plant or algal cell, it pushes against the cell wall and creates a pressure. The tendency of water to move as a result of pressure is called the pressure potential, Ψ_ρ. The pressure potential of water outside the cell is defined as zero at atmospheric pressure, but it will be positive for plants/algae under water, owing to hydrostatic pressure. At equilibrium, when net water flow is zero,

$$\Psi_{\pi e} + \Psi_{\rho e} = \Psi_{\pi i} + \Psi_{\rho i}$$

(Buchanan *et al.* 2000). The pressure potential is a property of the water, but as water presses against the cell wall, the cell wall reacts with an equal and opposite pressure, which is called turgor pressure ($P_i = -\Psi_{\rho e}$). Note that turgor pressure is a property of the cell, not of the water. If the external pressure potential is negligible, turgor pressure at equilibrium will equal the difference between the osmotic potentials inside and outside the cell.

Frequently the overriding component of water potential is its osmotic (or chemical) potential, Ψ_π, and inasmuch as pressures are the important components in osmotic adjustment to salinity changes, turgor pressure and osmotic pressure (π) are useful terms. These hydrostatic pressures are measured in pascals (Pa), the same metric units used for barometric pressure. At equilibrium,

$$\Psi_e = \Psi_i = P_i - \pi_i = -\pi_e$$

or

$$P_i = \pi_i - \pi_e$$

(Reed 1990c). The osmotic pressure of fresh water is minimal, and the turgor pressure in cells is equivalent to the internal osmotic pressure $(P_i = \pi_i)$. In seawater, where $\pi_e \cong 2.5$ MPa, cells must maintain a high internal solute concentration (Ψ_i) to remain turgid.

If the cell is placed in hypertonic solution (a solution of lower solute concentration as compared with the cell), water will flow rapidly out of the cell, because membranes are freely permeable to water. Turgor pressure will be reduced, and at first the cell will become flaccid. Then, as the cytoplasm and vacuole shrink further, the plasmalemma will tear away from the cell wall. The damage to the plasmalemma caused by this process, plasmolysis, is usually irreparable. There are some seaweeds, however, that can survive plasmolysis (Biebl 1962). If the cell is placed in hypotonic solution, water will enter the cell (and ions will leave via channels), causing it to swell, and if the difference in osmotic potentials is great enough, it will burst. Again, such rupture will be fatal. During the first few minutes of submergence in distilled water, seaweed thalli rapidly lose ions from their "free space" (intercellular spaces and cell walls, see also sec. 6.3.1) as the solution in the free space comes to equilibrium with the medium (Gessner and Hammer 1968). Because the osmotic potential of seaweed cells is more negative than that of seawater, and sometimes much more negative, its salinity must be greatly increased before seawater becomes hypertonic. The ability of seaweeds to tolerate high salinity (i.e. their ability to avoid plasmolysis) depends on the difference between the internal and external osmotic potentials and on the elasticity of the cell wall. In the presence of reduced salinity, the turgor pressure will increase. Cells will expand as long as their walls are elastic, but because normal seawater is already hypotonic with respect to the inside of the cell, strain will increase with any reduction in salinity. The strength of the cell walls and the ability of the cells to make their internal osmotic potential less negative will determine their resistance to low salinity.

However, in systems containing charged particles, there is not only a chemical potential of the solutes but also an electrical potential. There is a tendency for the numbers of positive and negative charges to come to equilibrium. Gradients in these two potentials are not always in the same direction. The net passive movement of ions across a cell membrane will depend on the combined electrochemical gradient and will also be complicated by the fact that many molecules cannot freely cross the membrane. Seawater has a very low water potential because of all the salts dissolved in it, but cells maintain even lower potentials (higher concentrations of particles), and the resulting turgor pressure is important for cell growth (Cosgrove 1981).

7.4.2   Cell volume and osmotic control

Control over cell volume clearly is vital to cells that have no walls (including seaweed gametes and spores; Russell 1987), but seaweeds, too, have been reported to alter their internal water potential in response to salinity changes; and these observations have led to the hypothesis that such seaweeds are able to regulate their turgor. By increasing or decreasing $\Psi_{\pi i}$ in response to $\Psi_{\pi e}$ the cells can control turgor pressure. Some studies have reported findings to support this hypothesis, whereas others have indicated that there are exceptions. Moreover, some seaweeds do not regulate turgor (e.g. *Bostrychia scorpioides*; Karsten and Kirst 1989). Algal cells may alter their internal water potential by pumping inorganic ions in or out, or by interconversion of monomeric and polymeric metabolites (Hellebust 1976; Russell 1987; Fig. 7.13). Because membranes leak water freely in both directions, cells cannot change their water potential by pumping water

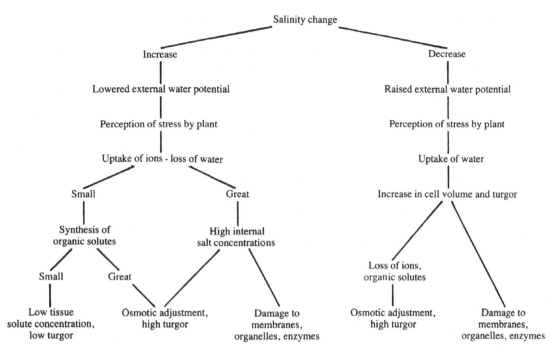

Figure 7.13 Short-term responses of seaweeds to salinity changes. (Redrawn from Russell 1987, with permission of Blackwell Scientific Publications.)

molecules. However, membrane integrated proteins, the so-called aquaporins, for high-performance water transfer between tissues have recently also been described to occur in algae (Anderberg *et al.* 2011; Chan *et al.* 2012b).

Although a change in ion concentration seems an appropriate solution to osmotic acclimation, cytoplasmic enzymes and ribosomes cannot tolerate wide fluctuations in the ionic composition of the cytoplasm; therefore acclimation has to be on the basis of compatible solutes. For instance, inorganic salts such as KCl and KNO_3 have large inhibitory effects on enzymes from *Saccharina latissima* and *Fucus vesiculosus*. Ions are used in the vacuoles of these brown algae for osmotic adjustment and are important for short-term (0–30 min) acclimation in the cytoplasm (Borowitzka 1986). But in the cytoplasm and organelles such as mitochondria and plastids, the water potential is controlled through organic osmolites, in these species the mannitol concentration (Davison and Reed 1985; Reed

et al. 1985). In the cortical cells of *Laminaria* there are large vacuoles, and the tissue contains more K^+, Na^+, and Cl^-, and less mannitol, than meristoderm tissue that has cells with small vacuoles and relatively large proportions of cytoplasm (Davison and Reed 1985).

Compatible solutes must be highly soluble, must have no net charge, and must be retained against their large concentration gradients across the plasmalemma and tonoplast (Blunden and Gordon 1986; Borowitzka 1986). Moreover, they should not be intermediates in any major biochemical pathway, but should be removed by two or three enzymatic steps so that there will be no conflict between osmotic adjustment and metabolism (Borowitzka 1986). Several compounds, such as sucrose, do not meet this criterion; mannitol, although thought of as a major photosynthetic product in brown algae, may in fact serve an osmotic role instead (Reed *et al.* 1985). However, compatible solutes do more than merely serve as particles in the osmotic equation; they interact with enzymes to stabilize them against conformation changes due to water loss (Borowitzka 1986).

Mannitol and sucrose are in the group of low molecular weight carbohydrates, in which the -OH groups are thought to blend well with the structure of cellular water. In particular among the rhodophytes a multitude of such compounds that are active in osmo-regulation have been identified over the past two decades (e.g. floridosides, trehaloses, etc.; see Eggert and Karsten 2010 for review). For the genus *Hypoglos-sum* digalactosylglycerol was identified as a new com-pound with compatible solute function (Karsten *et al.* 2005). This variety of low molecular weight compound forms part of the base for the chemical taxonomy in the red algal class of Florideoyphceae (Karsten *et al.* 2007). Some amino acids, especially proline, may also serve as compatible solute (e.g. in *Ulva intestinalis*; Edwards *et al.* 1987) and also quaternary ammonium compounds and their tertiary sulfonium analogs (Blunden and Gordon 1986; Borowitzka 1986). The tertiary sulfonium compound β-dimethylsulfoniopropionate (DMSP) has been detected in numerous chlorophytes (Kirst 1990; van Alstyne *et al.* 2001), however, the latter study implies that its ecological function might be much more related to anti-herbivory defenses.

Not all seaweeds rely completely on ion and metab-olite concentrations as protection against salinity changes. Some also have morphological defenses and do change volume with salinity. *Porphyra purpurea*, in contrast to many seaweeds, has a nonrigid cell wall composed chiefly of mannan and xylan, rather than cellulose. The cell wall polymers are arranged as gran-ules rather than as ordered microfibrils. *Porphyra umbilicalis*, however, regulates the volume of the protoplasm without the contribution from the vacu-oles, which are solely acting as a deposit for ions with otherwise inhibitory effects on cytoplasmic enzymes (Knoth and Wiencke 1984). Nevertheless, it and other species of *Porphyra* do show changes in ionic compos-ition (especially K^+ and Cl^-) and in metabolite concen-trations in response to salinity changes (Reed *et al.* 1980; Wiencke and Lauchli 1981). Similar phenomena occur in the chlorophyte *Ulva*, and Edwards *et al.* (1987) found that the walls of estuarine *U. intestinalis*

were thinner and hence stretchier than the walls of marine and rock-pool seaweed of the same species, allowing them to swell with the influx of water in low-salinity conditions. In the presence of high salinity, the cell volume decreased, although there were also increases in inorganic ions, sucrose, and proline (Edwards *et al.* 1987). Also, *Ulva* can be repeatedly plasmolyzed and deplasmolyzed without injury to the membranes (Ritchie and Larkum 1987).

7.4.3 Effects of salinity changes on photosynthesis and growth

There tend to be optimum salinities for the processes of photosynthesis, respiration, recruitment, and growth (Fig. 7.14), just as there are optimum temperatures for these processes. Numerous examples of salinity optima are given by Gessner and Schramm (1971), and the effects of hypersalinity were reviewed by Munns *et al.* (1983). More recent studies have exam-ined the interactions among salinity, temperature, and UV-radiation stress (e.g. Fredersdorf *et al.* 2009). Fre-dersdorf *et al.* (2009) found an additional inhibitory effect of reduced salinity on the process of spore ger-mination in the kelp *Alaria esculenta* (Fig. 7.15), which

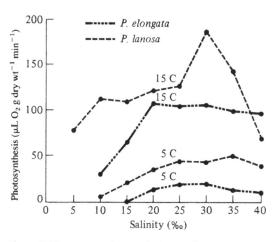

Figure 7.14 Apparent photosynthesis rates for estuarine *Polysiphonia lanosa* and *P. elongata* as functions of salinity at 5°C and 15°C, showing optima near full seawater salinity. (From Fralick and Mathieson 1975, with permission of Springer-Verlag, Berlin.)

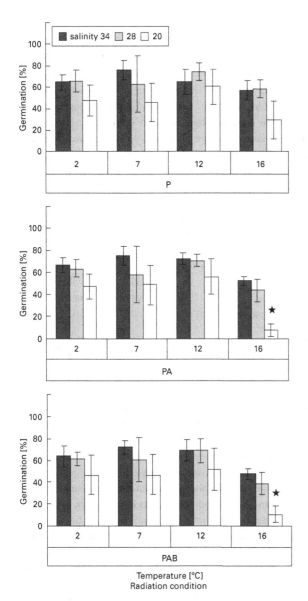

Figure 7.15 Germination rates of *Alaria esculenta* zoospores (expressed in percentage of control), ascertained after 8 h of radiation exposure to photosynthetically active radiation (PAR, P), PAR + ultraviolet (UV)-A (PA), PAR + UV-A + UV-B (PAB) and 7 days of UV recovery below dim light at constant temperature (2, 7, 12, 16°C) and salinity conditions (34 p.s.u. = dark gray bars, 28 p.s.u. = light gray bars, 20 p.s.u. = white bars). SDs are represented by vertical bars (n = 3), and stars indicate interactive effects of temperature and salinity (P < 0.05) (From Fredersdorf *et al.* 2009, reproduced with permission.)

became most pronounced in concert with elevated temperatures. However, as with any kind of stress parameter, the rate of change can be important (Wiencke and Davenport 1987) and, thus, comparisons between different studies are difficult.

Long-term adaptation to different salinity regimes, and consequently, indications for the formation of salinity ecotypes are indicated by the differential growth performance of Baltic and Atlantic *Fucus vesiculosus* (Bäck *et al.* 1992; Fig 7.16). In general, as an intertidal species, *F. vesiculosus* can be characterized as euryhaline, however with marked differences in the respective growth rate at different salinities. As expected, Baltic isolates suffered from up-shock, Atlantic isolates from down-shock salinity treatments. In fact, the Atlantic isolate died at low salinities, under which the Baltic isolate still could maintain active growth. Interestingly, isolates from both locations still exhibited a common optimum salinity, which was S_A 12.

In general, the Baltic Sea gives interesting insights into salinity adaptations over longer times. The Baltic is connected to the North Sea by only a narrow channel. Its post-glacial history is known to have included several major salinity shifts, with the most recent marine period having started about 7500 years ago (Russell 1987, 1988). Its salinity has been low (brackish) for about 3000 years. Various fucoids and other marine algae that occur in the Baltic also occur in the North Sea. Although the populations in the Baltic have consistently lower salinity optima, their thallus organization differs from that of their open-coast counterparts in diverse ways. Some have small thalli but similar-sized cells; some have smaller cells but not smaller thalli; some have only certain cells that are reduced in size (Russell 1987, 1988).

Several researchers have attempted to determine why diluted seawater causes a decline in photosynthesis. There is a sharp drop in the photosynthesis rate for several seaweeds, including *Ulva lactuca*, when they are transferred from seawater to tap water, and a corresponding sharp return to normal when transferred back to seawater. This has been explained as an effect of carbon supply (CO_2 and HCO_3^-) (Gessner and Schramm 1971). Dawes and McIntosh (1981) explained why the rate of

photosynthesis in the red alga *Bostrychia tenera* is temporarily greater in water of certain Florida estuaries than in either full seawater or seawater diluted with distilled water. They found that the estuaries were all fed by spring water, the significant components of which were Ca^{2+} and HCO_3^-. Although *Bostrychia* will die if left too long in water of very low salinity, the increased photosynthesis in spring-water-diluted seawater enables it to survive short periods (a few days) of very low salinities better than in estuaries not fed by spring water. Calcium makes the plasmalemma less permeable to other ions, thus reducing the loss of ions that takes place (in addition to water influx) when cells are placed into dilute seawater (Gessner and Schramm 1971). Yarish *et al.* (1980) found that Ca^{2+} and K^+ were limiting factors for photosynthesis in estuarine red algae. Although Ca^{2+} is unlikely to be absent from brackish water, its concentration may be too low in rainwater, which affects emersed intertidal algae.

7.4.4   Tolerance and acclimation to salinity

When organisms become acclimated to a new range of conditions, they generally lose the ability to perform as well under their previous conditions. This phenomenon has, in the course of evolution, resulted in two nearly separate groups of organisms: freshwater and marine. Very few species or even genera of eukaryotes are able to cross the so-called salinity barrier; three that can are *Cladophora*, *Rhizoclonium*, and *Bangia*. One might suppose that brackish habitats, being mixtures of fresh water and seawater, might be populated by a mixture of freshwater and marine algae, but that is not the case. Brackish waters, even down to salinities of S_A 10 or less, are populated by especially euryhaline marine algae such as *Fucus* and *Ulva*, or typically brackish species such as *Rhizoclonium riparium* and *Vaucheria* species in salt marshes (Nienhuis 1987; Christensen 1988), and *Caloglossa* and *Bostrychia* in mangrove swamps (King 1990).

Intertidal seaweeds generally are able to tolerate seawater salinities of S_A 10–100; subtidal algae are less tolerant, especially to increased salinities, generally withstanding S_A 18–52 (Biebl 1962; Gessner and Schramm 1971; Russell 1987). The former must

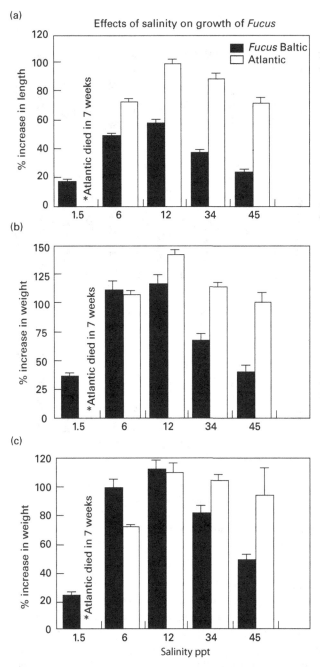

Figure 7.16 The percentage increases (mean + SE) of thallus (a) length (b) fresh weight of Atlantic and Baltic *Fucus vesiculosus* after 11 weeks growth in a salinity range of 1.5–45 ppt (now referred to as S_A). (c) The percentage increase of thallus fresh weight calculated from weight values obtained after one week in culture. *In salinity 1.5 ppt Atlantic material died in 7 weeks. (From Bäck *et al.* 1992, reproduced with permission.)

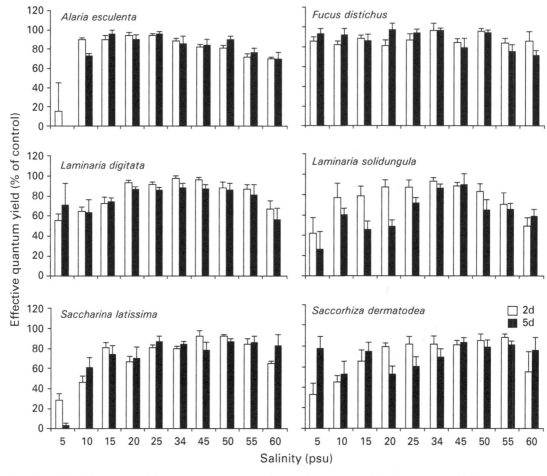

Figure 7.17 The effect of two and five days' treatment with different salinities on the effective quantum yield ($\Delta F/Fm$) in the brown algal species *Alaria esculenta*, *Fucus distichus*, *Laminaria digitata*, *Laminaria solidungula*, *Saccharina latissima* (formerly *Laminaria saccharina*), and *Saccorhiza dermatodea* collected in the Arctic Kongsfjorden. The maximum effective quantum yield of each taxa was set to 100%, and all other values expressed in relation to 100%. Data shown represent mean values $\pm$ SD (n = 5). (From Karsten 2007, reproduced with permission.)

tolerate unpredictable changes in salinity during emersion. Estuarine subtidal seaweeds experience more regular fluctuations. Another interesting habitat in terms of salinity change is the high Arctic coastline. While full saline conditions may prevail during the winter season, the onset of snow melting and glacier calving due to increasing spring and summer temperatures result in pronounced freshwater discharge into

coastal habitats. A survey of six dominant brown algal species from the Arctic Kongsfjord by Karsten (2007) confirmed the link between vertical zonation pattern and salinity tolerance with the shallow water *Fucus distichus* being the most resistant to both hypo- and hyper-osmotic stress, and the Arctic endemic deep-water *Laminaria solidungula* being the least tolerant (Fig. 7.17).

7.5 Further stresses related to water potential: desiccation and freezing

Physically speaking, osmotic adjustments of the individual seaweed under salinity stress represent a response to changes in the water potential of the microhabitat. In that respect, desiccation and frost are also basically related to a change in water potential and thus, the physical drivers are similar for all three stressors and some of the adaptive responses are similar as well. Evidently, adaptation and acclimation to frequent and large-amplitude changes in water potential are required for seaweeds populating the intertidal zone, which may thus be characterized as an extreme environment with extreme changes (Davison and Pearson 1996). At a first glance, the potential danger of desiccation seems to be the most striking factor in the intertidal zone.

7.5.1 Desiccation

Seaweeds are essentially marine organisms, even if they are out of the water more than half the time. Their exposure to the atmosphere is a stress, to which species are more or less tolerant. Except under cool, very humid conditions, loss of water from seaweeds begins as soon as they are emersed; in this way, desiccation is a salinity stress. Removal of seaweeds from seawater also deprives them of their source of nutrients, including most of the inorganic carbon, although postdesiccation enhancement of limiting-nutrient uptake has been reported for *Fucus distichus* (Thomas and Turpin 1980), which would partly compensate for the intermittent availability of nutrients (see also sec. 6.4.2). In a narrow sense, the term "desiccation" is equivalent to "dehydration", but it is used to encompass these nutrient changes as well, because they normally occur together. However, on humid days these other stresses can occur in the absence of dehydration.

Desiccation is related to the surface-area:volume (SA:V) ratio. Small organisms are more susceptible. Sessile organisms can survive harsher conditions as they get larger, but they must be protected from

desiccation when they are small (e.g. by growing in crevices, or by settling when low tides occur at night or in the early morning) (Denny *et al.* 1985). Evaporation rates are affected by temperature; there tends to be more water loss during the day than at night, and more in summer than in winter (e.g. Mizuno 1984). In turn, evaporation relates to salinity through changes in the surface water potential, as we have seen. The seasonal changes in the timing of tides are important, because daytime low tides are much more damaging during hot weather.

A brief exposure to cool, humid air may hardly affect a seaweed, whereas prolonged exposure, particularly during hot summer days, may cause severe stress. The higher up on the shore a species grows, the longer it is exposed to desiccation. The small-scale habitat must be considered, because seaweeds may be partially protected by clumping and by the slope, orientation, and porosity of the shore. Whereas many intertidal seaweeds evidently tolerate desiccation, some largely avoid it by growing under overhangs or under other algae, in crevices, or in other wet areas. For example, *Corallina vancouveriensis* in parts of southern California receives protection where it grows among *Anthopleura elegantissima*, a colonial sea anemone that forms water-retaining carpets (Taylor and Littler 1982). Saccate intertidal algae such as *Colpomenia peregrina* largely avoid desiccation stress by retaining a reservoir of seawater inside the thallus (Vogel and Loudon 1985; Oates 1988). Temperature can have an effect on growth form, which in turn can affect desiccation avoidance, as shown by Tanner (1986) for *Ulva californica*. Above 15°C in culture, and south of Point Conception, California, where water temperatures usually are >15°C, *U. californica* forms densely tufted turfs. These hold water better than does the larger, more foliose growth form found in the cooler waters (and damper air) north of Point Conception, and formed in culture at 10°C. Ongoing studies by Bischof and co-workers further indicate a relation of desiccation rate and irradiance for beach-cast *Laminaria digitata*. This water consumption during photosynthesis (depending on irradiance) may additionally contribute to water loss after beaching.

What takes place during emersion? With respect to photosynthetic responses a peculiar sequence of events has been observed. As soon as a seaweed is removed from water, its photosynthesis rate usually drops sharply, even before any desiccation takes place (e.g. Chapman 1986). This is because the inorganic carbon supply is greatly restricted; a small amount of bicarbonate in the surface film of water on the seaweed is available for photosynthesis, but it is not quickly replenished. CO_2 must diffuse from the air into the water film and dissolve, but the concentration of CO_2 in air is about 10 times lower than the bicarbonate content of seawater. Moderate desiccation, however, has frequently been reported to actually increase photosynthetic rates (Johnson *et al.* 1974; see Davison and Pearson 1996 for review), although continued desiccation leads to further reduction. The reason for the increase seems to be that when the water film has evaporated, CO_2 from the air can penetrate more quickly into the cells. If the relative humidity is experimentally maintained at a high enough level to prevent desiccation, the photosynthesis rate may remain the same over long periods, as found in *Fucus serratus* by Dring and Brown (1982). This is evidence that emersion itself is not detrimental to photosynthesis. A species of *Ulva* in Israel maintained a constant photosynthesis rate over 0–20% water loss and continued positive photosynthesis to about 35% water loss. Beer and Eshel (1983) predicted that these seaweeds would maintain positive photosynthesis for some 90 min after exposure in the morning, but for only 30 min at midday. Only seaweeds in the highest part of the population were unable to maintain positive photosynthesis during low tide. These presumably are saved from starvation by more wetting at other times in the tidal sequence, or through wave splash.

Many experiments have tested the recovery of seaweeds from desiccation stress, often by measuring rates of photosynthesis or respiration upon re-immersion. Characteristically, these experiments consisted in drying out a selection of intertidal algae to measured degrees of water loss, then resubmerging them in water and measuring the rates of gas exchange at various times. Dring and Brown (1982) have correlated stress and recovery with the conditions at the normal positions of different species of *Fucus* on the shore and found similar straight-line relations between photosynthetic rates and water losses (Fig. 7.18) suggesting that the photosynthetic apparatus of these high-shore fucoids is not any more resistant to water loss than low-shore species. Nor is the rate of water loss significantly lower in higher-intertidal species, as one might locally expect. What does differ is the ability of the photosynthetic apparatus (and the cells in general) to *recover* from desiccation stress when resubmerged (Fig. 7.21). Dring and Brown (1982) thus assessed three hypotheses that might explain the effects of desiccation on intertidal seaweeds and on zonation: (1) Species from the upper shore are able to maintain active photosynthesis at lower tissue water contents than are species from lower on the shore; this is refuted by the data in Figure 7.18. (2) The rate of recovery of photosynthesis after a period of emersion is more rapid in species from the upper shore; this hypothesis is also refuted by the available data. (3) The recovery of photosynthesis after a period of emersion is more complete in species from the upper shore. It was furthermore shown by Huppertz *et al.* (1990) that desiccation conserves the state of photoinhibition in *Fucus serratus*, thus recovery from photoinhibition may start from a lesser extent of inhibition than in the more strongly inhibited specimens from lower shore levels. Overall, algae inhabiting the height part of the shore need to increase their time-use efficiency as these species have less time available for productivity or nutrient uptake and hence spend less time in a state of sufficient hydration to photosynthesize (Skene 2004).

Critical steps in photosynthesis upon desiccation are the transfer of electrons from PS II to PS I and the splitting of water (Wiltens *et al.* 1978), and in the desiccation-tolerant rhodophyte, *Porphyra sanjuanensis*, rehydration first led to recovery of the intersystem electron-transfer process, and then recovery of the water-splitting process. Critical cellular components affected by desiccation presumably are biomembranes and the integrity of the plasmalemma can be estimated by the measurement of cellular solute leakage (Hurd and Dring 1991). Upon desiccation and rehydration membrane-bound phospholipids pass from their

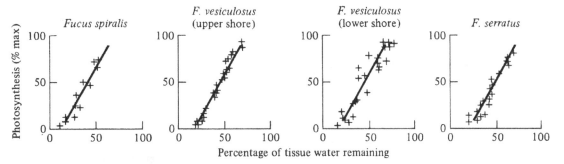

Figure 7.18 Photosynthesis during exposure to air as a function of water content for three fucoids from four heights in the intertidal zone. Linear regression lines for the four populations are not significantly different. (From Dring and Brown 1982, with permission of Inter-Research.)

liquid-crystalline state into a gel-like structure, and back again. Burritt *et al.* (2002) have measured the leakage of amino acids from the rhodophyte *Stictosiphonia arbuscula* after collection from two different shore levels (Fig. 7.19). Again, the extent of leakage was different in the specimens from the different growth sites, indicating high plasticity for acclimation, which was also apparent with respect to the condition of the antioxidant metabolism. The "hardening" of the specimens growing higher on the shore, and, thus being emerged for longer time spans, was evident from a lower degree of membrane leakage, lower production of hydrogen peroxide and a consequently lower degree of membrane lipid peroxidation. All parameters tested related to the antioxidant machinery (glutathione content, activity of reactive oxygen scavenging enzymes) were higher in the upper shore individuals. As we will see below, the generation of reactive oxygen species is a direct consequence of abiotic stress exposure. Hence, studies on the antioxidant metabolism give particular insights into the ecophysiology of species from an environment with a particular stress burden, the intertidal zone.

In order to reveal protective strategies under desiccation conditions the different species and/or different populations of *Fucus* have provided particularly interesting experimental systems (Collén and Davison 1999a,b; Pearson *et al.* 2000; Schagerl and Möstl 2011). For example, *Fucus vesiculosus* is an abundant inhabitant of the intertidal from the North Sea

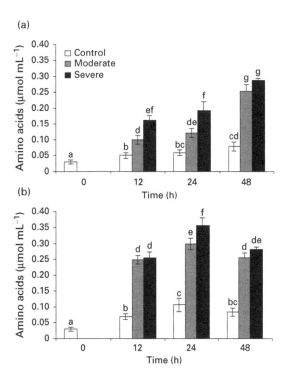

Figure 7.19 Effect of dehydration on amino acid leakage from *Stictosiphonia arbuscula* specimens, during rehydration. (a) High-band community; (b) low-band community. (From Burritt *et al.* 2002, reproduced with permission.)

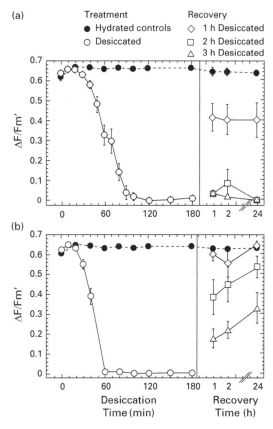

(a)

Treatment
● Hydrated controls
○ Desiccated

Recovery
◇ 1 h Desiccated
□ 2 h Desiccated
△ 3 h Desiccated

(b)

Desiccation
Time (min)

Recovery
Time (h)

Figure 7.20 Effective photochemical yield (ΔF/Fm′) during desiccation and recovery at 25°C of juvenile *Fucus vesiculosus* grown in the central Baltic Sea. (a) Baltic algae; (b) North Sea algae. Controls were kept hydrated in air, and recovering algae were kept in Baltic seawater, both at 25°C. Values are means ± SE (n = 7). (From Pearson *et al.* 2000, reproduced with permission.)

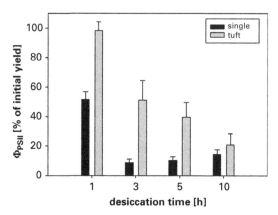

Figure 7.21 Effective quantum yields (%) of tufts (gray) compared to single fronds (black) of *Fucus spiralis* during desiccation lasting 1, 3, 5, and 10 h on a mesh screen exposed to the sun. Values normalized to initial yields (means ± SE; n = 14). (From Schagerl and Möstl 2011, reproduced with permission.)

revealed the protective function related to growth form with tufts being better protected than individual thalli (Fig. 7.21). Evidently, after emersion photosynthetic quantum yield is steadily reduced over time in *Fucus* specimens growing in tufts, while individual thalli suffer a rapid inhibition of quantum yield (Schagerl and Möstl 2011).

7.5.2 Freezing

A further significant stressor in the intertidal zone, at least in polar and cold-temperate regions, is freezing. Freezing of seawater proceeds under the exclusion of dissolved ions, Therefore, the availability of free water becomes reduced, while the concentrations of ions in the unfrozen water are substantially increased, resulting in a further and sometimes drastic, reduction in water potential. In this respect, the physical drivers of freezing, again resemble a salinity stress. We have explored some mechanisms in seaweed cold adaptation in a previous section. With respect to temperature effects on photosynthesis and freezing-induced changes in the osmotic regime, similar mechanisms such as the synthesis of compatible solutes acting as

coastline, whereas its conspecifics from the Baltic only grow permanently submerged. As Pearson *et al.* (2000) have shown, the Baltic population of *F. vesiculosus* has largely lost its ability to withstand desiccation stress (Fig. 7.20)

In *Fucus*, morphological variations may also contribute to desiccation tolerance (Schagerl and Möstl 2011). For *F. spiralis* it was suggested that the thicker and fleshier thalli of low-intertidal individuals are more stress susceptible than the smaller and harder specimens from the high intertidal zone. This study also

(a)

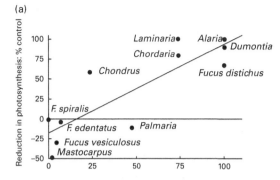

(b)

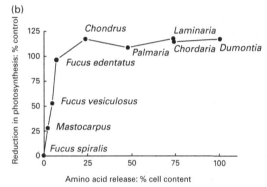

Amino acid release: % cell content

Figure 7.22 Freezing tolerance and vertical zonation in several intertidal brown and red seaweeds: Relationship between percentage cellular amino acid release following 3 h at -20°C and (a) photosynthesis measured 7 d after 12 h at -20°C; or (b) photosynthesis measured immediately after 3 h at -20°C. (From Davison *et al.* 1989, reproduced with permission.)

cryoprotectants also occurs. The synthesis of specialized antifreeze proteins (AFPs) as reported from polar diatoms (Armbrust *et al.* 2004; Bayer-Giraldi *et al.* 2010) has not been confirmed for seaweeds yet. As for osmotic stress, freezing-induced membrane leakage was studied as a proxy for freezing tolerance (Davison *et al.* 1989). As for most stress parameters, freezing tolerance, expressed as photosynthetic impairment and membrane leakage, was generally correlated with the position of species on the shore (Fig. 7.22). Furthermore, it was highlighted by Collén and Davison (1999a,b, c) that the antioxidative metabolism as studied in different species of *Fucus* or in the related red algae *Chondrus crispus* and *Mastocarpus stellatus* from

slightly different shore levels, is a major determinant of freezing tolerance. These authors also stressed, however, that desiccation might be a much more important, as more frequently occurring, driver in shaping vertical zonation patterns of seaweeds.

7.6 Exposure to ultraviolet radiation

Seaweed species populating the intertidal zone and, thus, a habitat of low stability, usually show wide tolerance ranges in relation to the vast variation of the entirety of factors making up their physico-chemical environment. As outlined above, while we expect that intertidal species are well adapted to adequately respond to the large amplitude of variation at their respective growth site, large-scale environmental change, e.g. due to climate forcings, pose new challenges to the organisms. For the last approximately 25 years the increase in solar UV-B radiation reaching the Earth's surface as a consequence of stratospheric ozone depletion, has been discussed as such a stressor that is changing beyond the amplitude of natural variation that the communities have experienced for at least the previous several thousands of years.

The discovery of stratospheric ozone depletion over Antarctica dates back to the 1970s, but recent reports indicate ongoing, though slow recovery of the Antarctic ozone layer (Kerr 2011). Over the Arctic, however, a record loss in stratospheric O_3 has been reported for spring 2011 (Manney *et al.* 2011). Ozone is predominantly generated in the low latitudes, by photolysis of molecular oxygen. In the stratosphere, ozone molecules are subject to UV-R-mediated photolysis and may also be degraded due to the reaction within catalytic cycles with NO, Cl, or Br serving as catalysts (Lary 1997; Bischof *et al.* 2006a). The concentration of these compounds in the atmosphere increases mainly due to anthropogenic emissions, thus leading to ozone depletion. Ozone selectively absorbs UV-B radiation, and a 10% decline in column ozone would result in an approx. 5% increase of surface irradiance at 320 nm (UV-A radiation) while the same decline would be accompanied by a 100% increase at 300 nm (UV-B radiation; Frederick *et al.* 1989).

Due to its high energy content according to the Planck equation

$E = h \times c/\lambda$, (with E = energy, c = speed of light, λ = wavelength, h = Planck's constant),

and to the distinct absorption properties of a number of biomolecules (proteins, nucleic acids), UV-B radiation represents a serious environmental stress factor per se. In general, the effects of UV-B exposure on biological systems are manifold, reaching from the molecular to the ecosystem level. The UV-B-mediated damage to molecular targets may further impair central physiological processes like photosynthesis or nutrient uptake, and ultimately result in reduced growth, production, and reproduction, and altered ecosystem function (see Bischof et al. 2006a; Bischof and Steinhoff 2012 for review). However, as UV-B is an omnipresent component of the physico-chemical environment (at least in terrestrial, intertidal, and shallow water systems), the organisms have developed adequate protection strategies in line with the respective UV-burden in their natural habitat. This becomes apparent when comparing red algal species along a depth gradient with respect to their ability and capacity to synthesize UV-absorbing mycosporine-like amino acids (MAAs) and thus avoid UV-mediated DNA damage (Hoyer et al. 2001; van de Poll et al. 2001). In that regard, Hoyer et al. (2001), put forward the following classification. Type I: These are red algal species, which cannot synthesize MAAs at all, usually these are deep-water species, like *Phycodrys rubens*. Type II are species from the shallow subtidal like *Palmaria palmata* and are able to adjust MAA synthesis to the prevailing ambient radiation conditions. Type III species represent rhodophytes, which permanently maintain high MAA loads, as they usually populate high irradiance environments. As an example, *Porphyra* sp. follows this strategy (Huovinen et al. 2004). A linear correlation between MAA load and seaweed zonation pattern was obtained in the plot of the light-saturation values of different red algal species from Antarctica versus their MAA concentration (Fig. 7.23; Hoyer et al. 2001).

Naturally, in temperate and tropical regions, UV-B irradiances are high due to the high solar elevation. Maximum irradiances are, thus, much higher than to

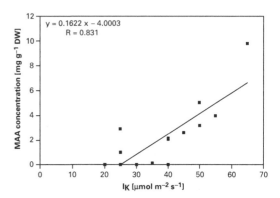

Figure 7.23 Initial light-saturation points (I_k) of Antarctic red algal photosynthesis according to Weykam et al. (1996) plotted against respective mycosporine-like amino acid (MAA) concentration (mg g^{-1} dry wt, DW) in field-collected algae. Regression analysis: y = 0.162x − 4.0, with r = 0.831. (From Hoyer et al. 2001, reproduced with permission.)

be expected in polar regions, even under the most severe ozone reduction scenarios. For example, it has been suggested that UV radiation in the first meters of the water column is a major factor governing the vertical distribution of kelps from Mediterranean regions, which are restricted to depth >10 m (Wiencke et al. 2000). However, in order to gain knowledge on adaptation mechanisms to high UV-B exposure, studies on the photophysiology on tropical shallow water seaweeds would be most insightful, but are, surprisingly, largely lacking. Some of the few studies on (sub-) tropical aquatic macrophytes indeed indicate that UV-B exposure may promote recovery from precedent photoinhibition, indicative of potential adaptation processes in a high UV irradiance environment (Hanelt et al. 2006; Hanelt and Roleda 2009).

The ecological relevance of ozone depletion is strongest in polar regions and a large number of studies on seaweed responses have been conducted in the Arctic, where drastic ozone reductions have recently been reported (Manney et al. 2011). The multitude of physiological effects originating from UV-B exposure are summarized in the reviews by e.g. Franklin and Forster (1997), Vass (1997) and Björn et al. (1999) and conclusions on their ecological implications to seaweeds are presented by Bischof et al. (2006a). There is

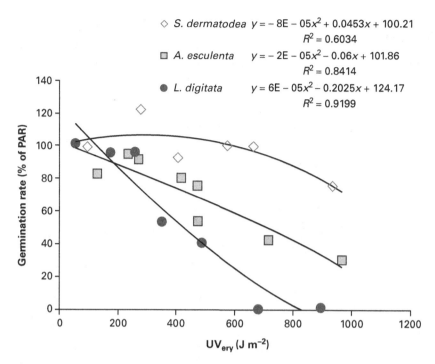

$\diamond$ *S. dermatodea* $y = -8E - 05x^2 + 0.0453x + 100.21$
$R^2 = 0.6034$

$\square$ *A. esculenta* $y = -2E - 05x^2 - 0.06x + 101.86$
$R^2 = 0.8414$

$\bullet$ *L. digitata* $y = 6E - 05x^2 - 0.2025x + 124.17$
$R^2 = 0.9199$

Figure 7.24 Relationship between effective UV-B dose measured as UV irradiance weighted to the minimal erythermal dose function (UV_{ery}) and germination rate of brown algal spores expressed as percentage of PAR. Non-linear regression was used to obtain dose–response relationship. Biological effective doses needed to achieve 50% inhibition of germination BED50 are > 1000 J m^{-2}, 700 J m^{-2}, and 418 J m^{-2} for *Saccorhiza dermatodea*, *Alaria esculenta*, and *Laminaria digitata*, respectively. (From Wiencke *et al.* 2006, reproduced with permission.)

still too few data to assess the ecological impact of an altered UV-B regime on seaweed community structure, but some important insights with respect to polar coastal ecosystems have been recently obtained. The zoospores of Arctic kelp species (and in particular spore germination) represent the life-history stage most susceptible to UV-B exposure, and, thus, developmental cycles might be disrupted by UV-B-induced damage to spores (Roleda *et al.* 2005; Steinhoff *et al.* 2008). However, upon the process of spore release, phlorotannins are exuded from the sporophytes to the surrounding waters, which may decrease the transparency of the water column for UV radiation. This process described for Arctic Laminariales has been suggested as a kind of maternal investment of the sporophyte to protect the most UV-susceptible reproductive stages (Müller *et al.* 2009b). Furthermore, field studies have revealed that

via the pronounced UV-susceptibility of the spores, UV-B radiation contributes to the determination of vertical zonation patterns of Arctic kelp species (Roleda *et al.* 2005; Wiencke *et al.* 2006), by setting the upper distributional limit. The latter authors found a species-specific correlation between spore germination as a function of the UV-B dose received. Again, the species higher on the shore (*Saccorhiza dermatodea*) showed a lower UV-B-susceptibility of germination efficiency than species from lower shore levels (*Laminaria digitata*; Fig. 7.24). Further field experiments, but in an Antarctic intertidal community, revealed that in developing macroalgal communities, species richness is highest when UV-B is excluded from the treatments and that both UV-A and UV-B radiation negatively affected macroalgal succession, highlighting the ecological significance of UV-exposure (Zacher *et al.* 2007).

7.7   Variation in seawater pH and community-based impacts of ocean acidification

Variation in seawater pH in intertidal and shallow waters is frequent and may naturally occur at large amplitudes. Changes in pH and, thus, CO_2 availability are most pronounced in intertidal tide pools where pH can increase as a consequence of photosynthetic activity, or drop when respiration prevails. Thus, this is not

only depending on light availability, but also on the respective community composition (auto- vs. heterotrophs). Currently, concern about climate-driven large-scale acidification of the oceans is increasing (see Essay 4) and we have discussed some potential consequences for photosynthesis (sec. 5.4.1) and calcification (sec. 6.5.3) already. The early studies on ocean acidification, particularly on phytoplankton i.e. coccolithophorids, were mostly characterized by the inconsistent findings, with even strain-specific

(a)

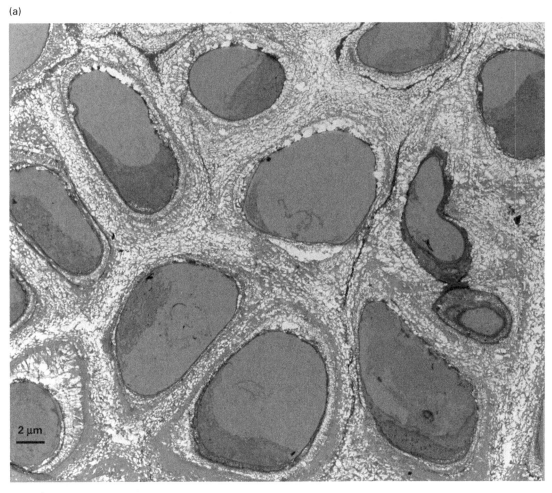

Figure 7.25 Transmission electron microscopy images of longitudinal sections through the youngest tip of each *Corallina officinalis* thalli. Calcite deposition (white material) between the cells is higher in algae grown under ambient (above) versus high (next page) CO_2. (From Hofmann *et al.* 2012a, reproduced with permission.)

(b)

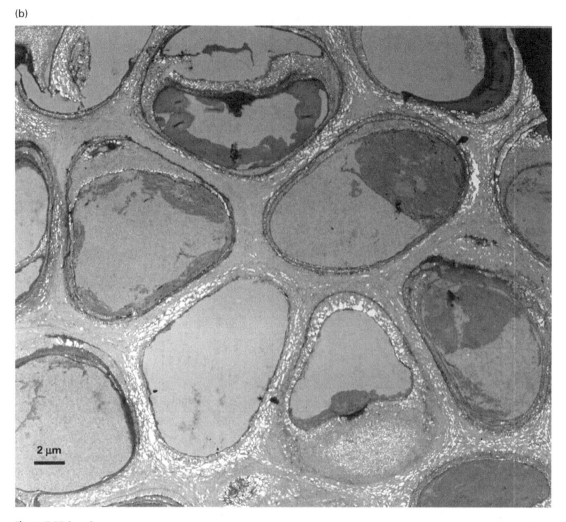

Figure 7.25 (*cont.*)

differences in productivity and calcification (Langer *et al.* 2009). For seaweeds the database for ocean acidification effects is still small, but steadily increasing and effects observed span from negative, neutral to positive in terms of general physiological performance, i.e. photosynthesis and growth. However, results from different studies are still difficult to compare, as effects are species-specific and often specific to the variation in experimental design and conditions, so that a general set of experimental procedures is recommended

(Hurd *et al.* 2009). For calcifying macroalgae, however, recent studies have shown a reduced calcification rate or even decalcification under the CO_2 scenarios forecasted for the next 100 years: In the rhodophyte *Corallina officinalis* the shifting balance of calcification and decalcification results in a net loss of precipitated calcium carbonate (Fig. 7.25; Hofmann *et al.* 2012a). However, the adverse effects of reduced seawater pH are also dependent on water motion, i.e. the thickness of the concentration boundary layer (CBL; sec. 8.1.3),

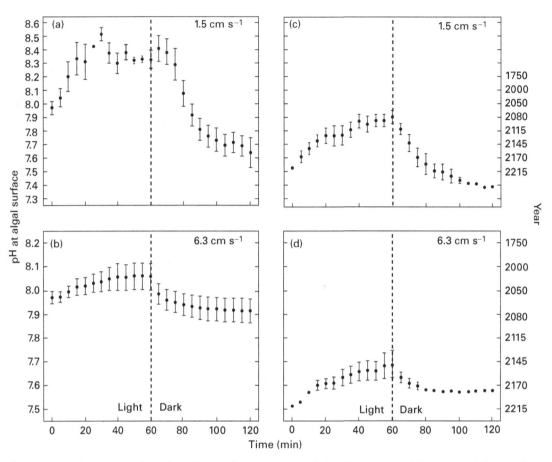

Figure 7.26 pH fluctuations at the surface of the coralline seaweed *Sporolithon durum* measured for 1 h in the light and 1 h in the dark at a mainstream ambient pH of (a, b) 7.97 ± 0.06 and (c, d) 7.50 ± 0.01 at mainstream velocities of (a, c) 1.5 and (b, d) 6.3 cm s^{-1}. Data are also plotted against modeled pH of surface waters projected on centennial timescales, i.e. for the period 1750–2215 (Caldeira and Wickett, 2003). Symbols represent the mean of three replicates (±1 SEM). (Hurd *et al.* 2011, reproduced with permission.)

that should be experimentally controlled. For the coralline alga *Sporolithon durum* high water velocities and thus a reduced CBL helps buffering against both high and low pH at the very thallus surface (Hurd *et al.* 2011; Fig. 7.26). On a community basis, the negative effects on algal calcifiers, resulting in reduced competitive strength in an algal community has been demonstrated in another recent tank experiment on macroalgal communities from the island of Helgoland. The cover of non-calcifying species increased with increasing pCO$_2$ at the expense of the algal calcifiers present (Hofmann *et al.* 2012b; Fig. 7.27). These potential shifts in algal communities have also been reported for a natural kelp forest community in southern New Zealand (Hepburn *et al.* 2011). A first meta-analysis approach on the differential impacts of ocean acidification on different marine organisms confirmed that among the organismic groups tested (seaweeds, seagrass, coccolithophorids, corals, echinoderms, mollusks, etc.) the

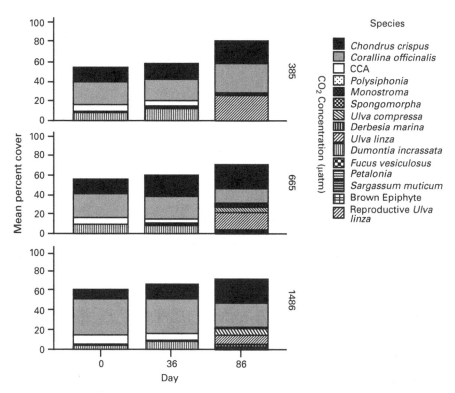

Figure 7.27 Shifts in macroalgal communities from Helgoland (German Bight) subjected to a CO_2 perturbation tank experiment. Mean percent cover of all species of algae present in the experimental communities at days 0, 36, and 86 for three pCO_2 levels are shown (From Hofmann *et al.* 2012b, reproduced with permission.)

calcifying macroalgae are indeed more susceptible (Kroeker *et al.* 2010; Fig. 7.28). Other authors stress that the current focus on the adverse effects of ocean acidification on algal calcifiers neglects the potentially profound effects on the non-calcifying species, with the potential of phase shifts from kelp- to turf-dominated communities supported by rendered carbon availability and altered competitive relations (Connell and Russell 2010).

7.8 Interaction of stressors, oxidative stress, and cross adaptation

All abiotic drivers in seaweed ecophysiology interact, either synergistically or antagonistically. In this respect, studies on the interaction of multiple stressors are of utmost ecological relevance. In the (less frequent) case of antagonism, the variation in one environmental variable may compensate or offset the adverse effects induced by another parameter. This might be observed with respect to the promoting function of enhanced temperatures to increase recovery/repair rates after/during stress exposure. Rautenberger and Bischof (2006) studied two *Ulva* species from Antarctic and sub-Antarctic regions and tested for their susceptibility towards UV-exposure at different temperatures. Overall, the authors found that moderately increased temperatures may compensate for the inhibition of photosynthesis induced by ultraviolet radiation. In the case of kelps from anthropogenically impacted areas in southern Chile, UV responses of

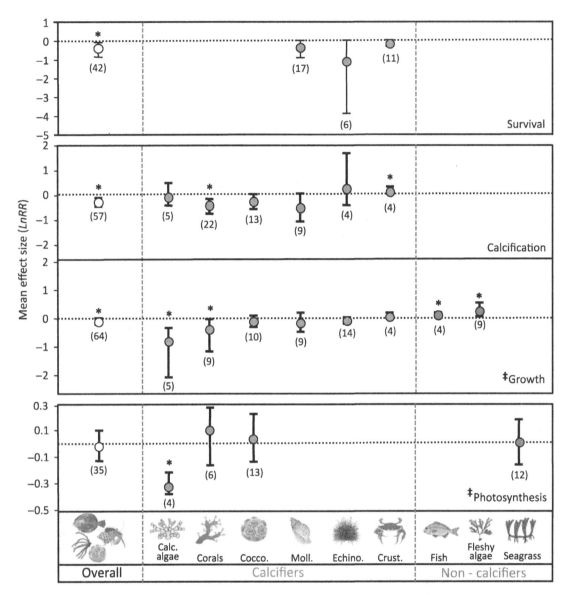

Figure 7.28 Taxonomic variation in effects of ocean acidification. Note the different y-axis scale for survival and photosynthesis. Mean effect size and 95% bias-corrected bootstrapped confidence interval are shown for all organisms combined (overall), calcifiers and non-calcifiers. The calcifiers category includes: calcifying algae, corals, coccolithophores, molluscs, echinoderms and crustaceans. The non-calcifiers category includes: fish, fleshy algae and seagrasses. The number of experiments used to calculate mean effect sizes are shown in parentheses. No mean effect size indicates there were too few studies for a comparison (n < 4). The mean effect size is significant when the 95% confidence interval does not overlap zero (*) Indicates significant differences amongst the taxonomic groups tested. (From Kroeker *et al.* 2010, reproduced with permission.)

seaweeds can be negatively modified by nutrient loads and metals (Huovinen *et al.* 2010).

Because of the complex interaction of stressors, adaptive responses towards one environmental factor can also result in an increased tolerance to another environmental variable, a phenomenon referred to as "cross adaptation". Cross adaptation is most likely to occur in response to different environmental cues, which do, however, result in similar primary physiological effects. An example for such a response is the stress-induced formation of reactive oxygen species (ROS; see Essay 5). The generation of oxidative stress represents a physiological response of an organism to a variety of different stressors, because all stressors finally impair photosynthetic electron drainage. So far, the generation of ROS induced by abiotic stressors prevailing in the natural environment has been documented in seaweeds exposed to high light and UV radiation, changes in temperature and salinity, freezing, desiccation, and is best studied in intertidal species like *Ulva* sp., *Fucus* sp., or *Chondrus crispus* (i.e. Collén and Davison 1999a, b, c; Bischof *et al.* 2002, 2003; Collén *et al.* 2007; Kumar *et al.* 2010a, 2011). Furthermore, ROS formation may also be promoted in algae from polluted areas, i.e. under elevated heavy metal concentrations (Kumar *et al.* 2010b; Ritter *et al.* 2008; see also Chapter 9). Thus, oxidative stress may also be regarded as a kind of general, non-specific stress response. Consequently, as reactive oxygen may also be part of a signaling cascade to perceive environmental stress by the organism (Mackerness *et al.* 1999), non-specific adaptive responses may also be observed in response to various and different stressors. For example, the green alga *Dasycladus vermicularis*, often found in tide pools along the Mediterranean shoreline, synthesizes and excretes the phenolic compound 3,6,7-trihydroxycoumarin in response to different stressors, UV radiation, high PAR, temperature, high salinity (Pérez-Rodríguez *et al.* 2001). Furthermore, it has been shown for the same species that reactive oxygen forms an important part of stress signaling cascades (Ross *et al.* 2005b). It should be noted here, that similar cues are also used in biotic interactions. Seaweeds under biotic stress (i.e. challenged by grazers or pathogens) apply so-called oxidative bursts (the release of hydrogen peroxide) in order to defend against biological assaults (Dring 2005; Potin 2008). These oxidative bursts are generated via a completely different pathway than ROS formation under temperature and light stress, but have been associated with general mechanical stress of seaweeds (Ross *et al.* 2005b). The significance of ROS formation by biological or physico-chemical processes to marine environments has been reviewed by Lesser (2006).

Another type of phenolic compounds, the brown algal phlorotannins, have been suggested to be UV-screening substances and grazing deterrents (Pavia *et al.* 1997); these compounds have been shown to increase in *Ascophyllum nodosum* under elevated UV-B and grazing, thus in response to abiotic and biotic stress. As another example for cross adaptation, heat shock proteins (HSPs) are, as indicated by the name, synthesized under heat stress. However, as their main function is to prevent protein denaturation by unfolding, HSPs might be produced under any kind of stress, which may impair protein function such as high temperature, UV-B exposure, heavy metal and ammonium toxicity. Again, it is likely that in all the cases mentioned reactive oxygen serves as a signaling agent (Dring 2005). In recent years, the involvement of mitogen-protein activated (MAP) kinases in the mediation of stress signals is also suggested for micro- and macroalgae (Jimenez *et al.* 2004). The role of oxidative stress and antioxidative responses with respect to the gradient of environmental stress seaweeds are exposed to along the shore gradient has been extensively studied (see the Essay 5). Again results point to a correlation between the capacity to scavenge reactive oxygen by different mechanisms and the vertical position of a seaweed species in the phytal zone (see Dring 2005 for review; also Collén and Davison 1999a,b, c, for comparisons of reactive oxygen metabolism in *Fucus* species and *Chondrus crispus* vs. *Mastocarpus stellatus* in the intertidal zone). Recently transcriptome analyses on gene expression patterns in two species of *Fucus* and *Chondrus crispus* under abiotic stress underline the stimulation of the reactive oxygen metabolism and heat shock protein synthesis (Collén *et al.* 2007; Pearson *et al.* 2010).

7.9 Physiological stress indicators

As outlined above, stress exposure will affect seaweed biology and processes on a multitude of levels. However, scientists involved in coastal management and environmental protection may aim to detect (any kind of) stress as soon as possible. Therefore, an important direction of recent research in seaweed biology focuses on the establishment of physiological or ecological parameters, which may serve as early warning indicators.

As we have seen so far, most of the abiotic stressors will act on the physiological level, and as such finally result in an impairment of photosynthetic electron transport, with the likelihood of increased reactive oxygen production in photosynthesis (see Essay 5). Thus, for any kind of abiotic stress, reactive oxygen production may be regarded as a general stress response, and thus, the quantification of the level of oxidative stress may serve as a good physiological indicator of algal fitness. Amongst others, oxidative stress in seaweeds has been approached either by measuring the production of malondialdehyde (MDA assay, e.g. Bischof *et al.* 2002) as a by-product of membrane lipid peroxidation, as changes in total antioxidative potential (DPPH essay, e.g. Cruces *et al.* 2012), glutathion turnover (Pawlik-Skowrońska *et al.* 2007) or changes in superoxide dismutase activity (Bischof *et al.* 2003; Rautenberger and Bischof 2006). The use of these assays as easily applicable indicators of stress is limited however, because a well-equipped biochemistry laboratory is required. The photometrical MDA assay has been shown to be a rather inaccurate method, which may be improved by HPLC analysis. The SOD assay is a coupled enzymatic test (McCord and Fridovich 1969; Aguilera *et al.* 2002) and requires advanced laboratory skills. Furthermore, this assay is best suited to detect adaptive traits (e.g. along a depth or latitudinal gradient), and not short-term responses to acute stress, which are less reflected by changes in SOD activity (see also the review by Dring 2005). Overall, such biochemical assays are not applicable as field methods and, thus, their use as easy-to-apply stress indicators is limited.

As abiotic stress ultimately impairs photosynthesis, the development of an easily measurable parameter to assess photosynthetic performance represented a major advance in plant ecophysiology, which could ideally be applied in the field as well (Schreiber *et al.* 1994). As we have seen in Chapter 5, measurements of variable chlorophyll fluorescence are fast and highly feasible under different experimental conditions, although they are less conclusive if physiological knowledge of the underlying processes is lacking (Saroussi and Beer 2007). Still, the determination of maximum quantum yield of PS II, assessed as the ratio of variable to maximal fluorescence (F_v/F_m) can offer a rapid tool to get a first impression on algal photosynthetic fitness under the given experimental conditions (Büchel and Wilhelm 1993; Hanelt *et al.* 2003). With respect to environmental stress ecology of seaweeds, measurements of F_v/F_m have been widely applied, amongst others to assess the impacts of high PAR and UV radiation (Bischof *et al.* 1998; Hanelt 1998), temperature stress (Eggert *et al.* 2006), salinity changes (Karsten 2007), organic pollutants, and heavy metal exposure (Nielsen *et al.* 2003a; Huovinen *et al.* 2010). However, not every change in the physico-chemical environment can be easily detected by F_v/F_m analyses. In attempting to assess physiological stress related to changing pCO$_2$, Hofmann *et al.* (2012a) found that in the calcifying rhodophyte *Corallina officinalis*, changes in calcification, and even decalcification of the cell wall is not accompanied by a change in F_v/F_m.

For most applications in seaweed stress ecology, the measurement of F_v/F_m is a very useful tool for the detection of emerging stress, or at least of changing photosynthetic performance. However, short-term changes in F_v/F_m may instead reflect regulatory mechanisms (e.g. in photoprotection; sec. 5.3.3), which might be an important indicator of short-term algal acclimation strategies, but they do not reflect long-term stress effects. Thus, seaweed stress ecologists may also need to use parameters, which integrate stress responses at a higher long-term level and thus allow for more ecologically relevant conclusions. Such integrative parameters are usually regarded as growth and reproduction. Despite all the molecular and physiological tools, only the integration of multiple stress effects will provide conclusive insights for ecosystem consequences of environmental stress. The significance of any kind of stress indicator has to be further considered with respect to the respective life-history stage under investigation. This has been clearly shown for the responses of different life-history stages of Arctic and temperate kelp to elevated temperature and UV-exposure (Müller *et al.* 2009b). Not only each life-history stage (e.g. gametophyte vs. sporophyte), but also individual developmental steps (e.g. antheridia formation, spore germination) have specific tolerance ranges. Therefore, for dominant seaweed species in a given habitat, stress ecophysiology should attempt to reveal the tolerance ranges of the most susceptible developmental stage or process (Roleda *et al.* 2005; Wiencke *et al.* 2006). The quest for ecological and physiological stress indicators is also valid to predict range expansion potentials of newly invading seaweed species.

On the community level, differences in species-specific stress tolerance will most likely propagate through an alteration of the respective competitive strength and finally be manifested in an altered community structure. A new conceptual approach of monitoring changes in composition of certain morpho-functional groups as indicators for stress exposure of (intertidal) communities was recently proposed for a western Mediterranean macroalgal community in response to varying levels of anthropogenic stressors i.e. pollution and sedimentation (Balata *et al.* 2011). Its general feasibility for application in coastal zone management still needs to be evaluated more thoroughly.

7.10   Synopsis

Seaweeds are exposed to a wide range of abiotic parameters potentially changing at large amplitudes over temporal and spatial gradients of different sizes. This set of the physico-chemical environment requires specific adaptation and acclimation mechanisms, which are furthermore challenged by additional anthropogenic environmental change. Temperature is a major driver in seaweed biogeography and surface temperatures of seawater vary with latitude and ocean currents. The annual range in the open ocean often is only about 5ºC, but in the shallow waters of estuaries and bays it can be much greater. Moreover, intertidal seaweeds are exposed to atmospheric heating and cooling during low tide. Natural salinities in marine and brackish waters range from S_A 10 to S_A 70, but values of S_A 25–35 are most common. Intertidal seaweeds may experience extreme salinity changes because of evaporation or rain/runoff.

Temperatures can have profound effects on seaweeds, owing ultimately to their effects on molecular structure and activity. Biochemical reaction rates approximately double for every 10ºC rise in temperature, but enzyme reactions show peak activities at certain optimum temperatures, above which any changes in tertiary or quaternary structures will inactivate and ultimately denature the enzymes. Photosynthesis, respiration, and growth, being sequences of enzyme reactions, also have optimum temperatures, but the effects of temperature are not uniform across all these processes. These optima vary between and within species. At these more complex levels, other environmental variables have larger effects and may overshadow the effects of temperature. Metabolic rates can become acclimated to gradually changing temperatures. Freezing kills many algae, especially if ice crystals form in the cells. However, many intertidal algae can withstand temperatures well below 0ºC, especially if their cells are partially desiccated. On the other hand, tropical algae will be killed by low temperatures above 0ºC.

In regions of extreme seasonal temperature changes, some seaweeds have life-history events cued by temperature (and also by photoperiod). Through their effects on the life histories of seaweeds and on the temperature ranges that seaweeds can tolerate, temperatures affect the geographic distributions of seaweeds, probably constituting the principal large-scale regulatory factor. Salinity, wave action, and substratum configuration play important but local roles in phytogeography.

The components of salinity that are important to seaweed physiology are the total concentration of dissolved salts and the corresponding water potential, plus the availability of some specific ions, notably calcium and bicarbonate. The internal pressure in the cells of many seaweeds with rigid cell walls is regulated through active movement of ions across membranes or by the interconversion of monomeric and polymeric compounds.

Desiccation stress on intertidal algae is partly a salinity stress, because as water evaporates from the seaweeds, the salt concentration in the remaining water increases. During emersion, seaweeds are also deprived of bicarbonate for photosynthesis. Seaweeds higher on the shore generally are more drought-resistant than those nearer low water.

In general, the intertidal zone represents a highly demanding environment for its permanent change in abiotic conditions, with high irradiances of PAR and UV radiation, large amplitudes of salinity, temperature as well as pH changes. All these stressors finally relate to seaweed physiology, i.e. photosynthesis, and may increase the oxidative stress burden. The ability to suppress oxidative stress is related to general stress tolerance and furthermore to the vertical position of a seaweed species along the phytal zone, reflecting habitat stability. General assumptions on stress tolerance of seaweeds are hindered by differential susceptibilities of different life-history stages. Therefore, the most susceptible state is the decisive stage for the prevalence under the given environmental conditions and priority should be focused on studying the most sensitive life-history stage.

Water motion

The waters of the oceans are in constant motion. The causes of that motion are many, beginning with the great ocean currents, tidal currents, waves, and other forces, and ranging down to the small-scale circulation patterns caused by local density changes (Vogel 1994; Thurman and Trujillo 2004). Hydrodynamic force is a direct environmental factor, but water motion also affects other factors, including nutrient availability, light penetration, and temperature and salinity changes. The forces embodied in waves are difficult to comprehend, unless one has been dangerously close to them; because of the density of water, a wave or current exerts much more force than do the winds. "Imagine a human foraging for food and searching for a mate in a hurricane and you will have only an inkling of the physical constraints imposed on wave-swept life" (Patterson 1989b, p. 1374). The energy amassed from a great expanse of air–ocean interactions is expended on the shoreline as waves break (Leigh *et al.* 1987). Equally difficult to visualize are the microscopic layers of water next to seaweed surfaces where the seaweeds' cells interact with water. Too much water motion imposes drag forces that can rip seaweeds from the rocks, but this also clears patches of "new" space for recruitment. Too little water motion and nutrient concentration gradients form at the seaweed surface which can restrict nutrient uptake, but the same gradients are used by seaweeds to sense how fast the surrounding seawater is moving and thereby cue gamete or spore release.

Studies of seaweed form and function in wave-exposed and wave-protected sites have provided insights into the trade-offs apparent in some species that allow them to maximize resource acquisition in slow flows and minimize drag forces in fast flows. The following texts and reviews provide the necessary background on fluid mechanics: Denny (1988, 1993, 2006); Vogel (1994); Denny and Wethey (2001). "Marine ecomechanics" is an emerging field that uses a "physical framework" to understand the responses of marine organisms on scales from cells to ecosystems (Denny and Helmuth 2009; Denny and Gaylord 2010). We begin this chapter by describing the hydrodynamic environments in which seaweeds grow, and then discuss the mechanisms by which seaweeds can enhance resource acquisition in slow flows and withstand hydrodynamic forces in wave-exposed sites. We finish with a discussion on the effects of wave action and sediments on seaweed communities. The effects of nutrients and pollutants on seaweed communities are covered in Chapters 6 and 9, respectively.

8.1 Water flow

8.1.1 Currents

Currents range in magnitude from the great ocean currents, of interest chiefly in regard to their temperature effects on phytogeography (sec. 7.3.8), to small-scale flows around and over surfaces. Currents attain extreme velocities only in narrow channels, where maximum speeds can surpass 2.5 m s^{-1} (5 knots) on a spring tide (Fig. 8.1). Steady currents of 0.5 m s^{-1} (1 knot) are generally considered strong, but that speed is low compared with the velocities briefly

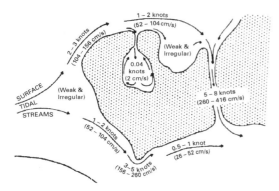

Figure 8.1 Some effects of coastal topography on the velocities of surface currents. (From Hiscock 1983, with permission of Oxford University Press.)

attained in breaking waves. The flow of a current along a shore is complicated by topography, as shown in Figure 8.1. Islands and reefs similarly affect flows and can concentrate planktonic organisms, eggs, and so forth, in slicks at fronts between eddies (Wolanski and Hamner 1988). Moreover, the velocity of a current decreases rapidly close to the seabed, owing to friction and eddy viscosity (the latter a property of turbulence). Thus, at the seabed, water motion will be due as much to turbulence as to the general flow of water (Fig. 8.2).

8.1.2 Physical nature of waves

Ocean waves are generated by wind blowing over the ocean surface. Certain characteristics of waves, such as height, period, and wavelength, will depend on the velocity and duration of the wind and the distance of open water over which the wind has blown. The last of these is referred to as the "fetch". The speed at which waves travel (or propagate) is called their "celerity" to distinguish it from any velocity the water itself may have (Denny 1988; Denny and Wethey 2001). Waves travel far beyond the storms that create them. As they travel, they lose energy, and because longer waves travel faster, they become sorted into groups (wave trains) that are similarly sized and travelling at a similar speed. Longer waves retain more of their energy because they spend less time in transit. Of course, the ocean surface rarely consists of a smooth progression of uniform swells (Fig. 8.3a) – usually it is chaotic (Fig. 8.3b).

As waves approach the shore and interact with the seabed, their celerity will decrease. When the water depth becomes small compared with the wavelength, which is the case for swells near shore, the celerity is proportional to the square root of the water depth; the principle of conservation of energy leads to an increase in wave height for slower waves. Waves tend to approach a beach nearly perpendicularly, whatever their direction was at sea. If a wave approaches obliquely, one end will be in shallower water, and its speed will decrease to less than that of the end that is in deeper water; as a result, the wave front will be curved and will tend to swing around parallel to the shore. The energies of waves tend to be concentrated on promontories and diminished in bays; hence headlands are more exposed to wave action. Some of the momentum of an oblique wave will drive water along the shore, creating a longshore current. Where opposing currents meet, water will flow offshore as a rip current. Both of these currents are important in the transport of spores.

The movement of water in the open ocean as a wave passes is circular (around an axis perpendicular to the wave direction) (Fig. 8.2). The radius of the circle decreases with depth, until there is no effect of the wave at all. In shallow water (less than about 30 m), the circular motion of water is progressively flattened, until at the seabed it is simply horizontal, back-and-forth motion, often referred to as "surge" (Fig. 8.2). On the surface, when the speed of the water at the top of its orbit exceeds the speed (celerity) of the wave, the water begins to spill over the front of the wave, and the wave breaks. As the wave hits the shore, its energy is dissipated by friction. Water flow during the breaking of a wave on the shore consists of a jet of water coming down on the rock from the wave crest, plus a very strong surge parallel to the rock (Fig. 8.4 illustrates the hydrodynamic conditions in the surge component of a breaking wave). For "collapsing" waves, characteristic of gently sloping shores, the impact of the falling water will be relatively great compared with the surge; for "surging" waves, characteristic of steep shores, the surge component is greater.

In metric units, force [mass (kg) × acceleration (m s^{-2})] is measured in newtons (N = kg m s^{-2}); pressure

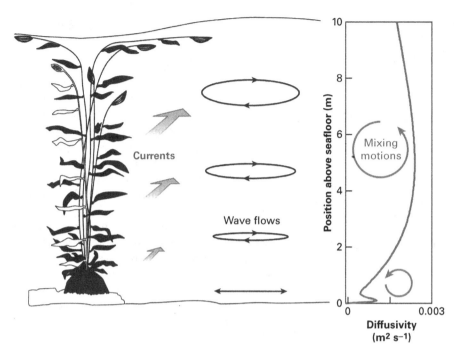

Figure 8.2 Physical factors dictating transport of spores and resultant patterns of dispersal. Schematic showing how current speeds and orbital wave velocities vary with water depth and ensuing patterns of vertical mixing. Length scales of mixing are greater in the midwater column than near the seafloor, but intensities of mixing are higher near the seafloor. These factors produce a non-monotonic mixing profile (eddy diffusivity profile). Eddy diffusivity values were calculated as in Denny and Gaylord (2010) for conditions of 2 cm s^{-1} currents and 2-m waves with 10-s period, both interacting with a sandy seafloor (0.008 m roughness elements) as commonly occurs between forests. (From Denny and Gaylord 2010, © 2010 by *Annual Reviews*.)

[N m^{-2} = Pa (pascal)] is force per unit area. Mechanical energy is the energy available to do work, and it is measured in joules (J = N m). The available energy in an incompressible fluid such as seawater has three components (Denny 1988): (1) *Gravitational potential energy*, as in the falling water of a collapsing wave, is given by the mass of the water multiplied by its acceleration due to gravity multiplied by the distance it falls. (2) *Kinetic energy*, the amount of energy needed to accelerate a given mass of water to a given velocity (which will be lost when the water is stopped) is mass multiplied by (velocity)^2/2. (3) *Flow energy*, as in surge, is pressure × volume. The pressures exerted by plunging breakers may reach 1.5 × 10^6 N m^{-2} (Denny *et al.* 1985). The pressures due to summer wave action usually are around 4 × 10^3 N m^{-2} (Palumbi 1984; Denny

et al. 1985). Wave velocities of 10 m s^{-1} are typical of wave-exposed shores, and velocities of up to 20 m s^{-1} have been measured (Bell and Denny 1994; Denny 2006). On wave-swept shores, intertidal seaweeds survive a pounding by approximately 8600 breaking waves each day and the wave forces they experience are equivalent to wind speeds of 1050 km h^{-1} (Mach *et al.* 2007); on land, a category 6 hurricane-force wind (>280 km h^{-1}) would destroy everything in its path.

When winds are strong enough to cause whitecaps, the spilling water, as in a breaker, has forward movement and can drag the surface canopies of seaweed shoreward, although the actual force is uncertain because the air trapped in the water makes it less dense (Seymour *et al.* 1989). Such wind-generated waves create a problem for *Macrocystis pyrifera*:

(a)

(b)

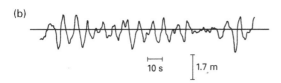

Figure 8.3 Sinusoidal and actual surface waves. (a)
A monochromatic, sinusoidal wave train: wave height, length,
and period are well defined. (b) A record of sea surface
elevations: wave height, length, and period vary. (From *Denny*
1988, with permission of the author.)

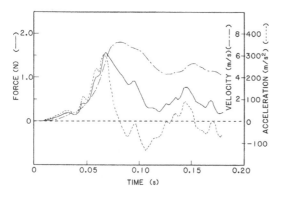

Figure 8.4 Water velocity and acceleration during the initial
portion of a typical wave in the exposed intertidal zone on the
coast of Washington, showing the hydrodynamic force they
impose on an acorn barnacle, as measured in Denny's device.
(From Denny *et al.* 1985, with permission of the Ecological
Society of America.)

because the waves have wavelengths shorter than the
lengths of the fronds in the canopy, the fronds may be
pulled by more than one wave at a time. The destruc-
tiveness of large waves results largely from their direct
hydrodynamic forces, but is increased by the fact that
they can move rock particles, ranging from sand to
cobbles. A severe storm in southern California, January
1988, moved cobbles 200 mm in diameter that were
situated as deep as 22 m – "such missiles are capable
of extensive damage" (Seymour *et al.* 1989, p. 289).

8.1.3   Laminar and turbulent flow over surfaces

To understand the exchange of gases and uptake of
nutrients by seaweeds, we need to know something
about water flow very close to the seaweed surface, for
it is from the water layer immediately adjacent to it
that a seaweed takes up and releases dissolved substances such as CO_2, O_2, H^+, OH^-, nitrate, ammonium,
and phosphate. The following overview is based on
Hiscock (1983); Koehl (1986); Wheeler (1988); Vogel
(1994); Hurd (2000); Denny (1988, 1993); and Denny
and Wethey (2001).

The water in contact with a solid surface is stationary, termed the no-slip condition (Denny and Wethey
2001). Thus, there is a velocity gradient from the unim-
peded "mainstream" or "bulk" seawater to the surface
of the object (Figs. 8.5, 8.6), forming a velocity boundary layer (VBL). The thickness of the VBL is defined as
velocity less than 99% of the free-stream velocity
(Vogel 1994). The gradient is not linear, however:
Within the lower 5% of the layer, the local velocity
already reaches more than 50% of the free-stream
velocity. Because the surface is effectively slowing
down the water, there is a stress set up against the
surface, called fluid (or surface) shear stress, τ_0.
"Shear" refers to a force parallel to a surface, as
opposed to compressive forces, which are applied perpendicularly. One can easily experience shear stress by
holding a hand, parallel to the ground, out the window
of a moving car.

When a current flows over an obstacle, eddies are
set up at the edges, as shown on the right in Figure 8.5.
If the obstacle is a smooth, flat plate, the water flow
over the middle of the plate will be laminar at first.
"Laminar flow" means smooth flow, which we can
imagine as layers (laminae), with velocities decreasing
toward the surface (Figs. 8.5 and 8.6). Shear slows
down the water flow next to the surface, and these
flow layers drag against the current, causing the thickness of the laminar layer to increase (Fig. 8.6); this
happens regardless of whether the surface is smooth
or rough. At some distance along the flat plate, the
laminar layer will become unstable and pass through a
transitional stage to turbulent flow (Fig. 8.6). In

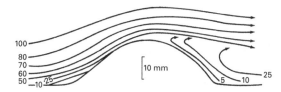

Figure 8.5 Water flow (velocities in cm s^{-1}) over a ridge on the seabed, illustrating laminar flow and an eddy on the downstream side. (From Hiscock 1983, with permission of Oxford University Press.)

turbulent flow there is vertical mixing as well as the horizontal flow. However, even when the VBL is turbulent, there remains at the seaweed surface a region in which turbulence is dampened by viscosity and flow is laminar, termed the viscous sub-layer (Fig. 8.6). In laminar flow, the transport of mass (e.g. nutrients), momentum, and heat are by molecular diffusion, and this region is termed the "diffusion boundary layer". Momentum diffuses along a gradient of velocity, whereas mass diffuses across a concentration gradient (Denny and Wethey 2001).

Whether or not flow at the surface is laminar or turbulent is critical to understanding the flux of dissolved substances to and from the seaweed. The type of flow can be estimated from the Reynolds number (*Re*) which is a dimensionless ratio of inertial to viscous forces, and is related to the size of an object, the friction velocity (u_*) at the surface, and kinematic viscosity (1.06 m^{-2} s^{-1} for seawater at 20°C; Denny 1993). There are many useful applications of *Re* (see Denny 1993). The roughness *Re* (*Re*$_*$), for example, can be used to determine if the structural elements on the surface of an object will cause flow to be laminar or turbulent (Roberson and Coyer 2004; Hurd and Pilditch 2011). *Re*$_*$ < 5 indicates laminar flow, *Re*$_*$ > 60 is fully turbulent with flow being transitional at values in between (Roberson and Coyer 2004). The size of the seaweed, its overall shape and the spacing of branches will affect flow. In unidirectional flows, large bladed seaweeds such as kelps generate turbulence at velocities of 1–3 cm s^{-1} (Hurd and Stevens 1997). For thalli composed of cylindrical branches, as is common in many of the smaller seaweeds such as *Gelidium nudifrons*, the flow is over a series of narrow, closely

spaced, rodlike surfaces. The thallus as a whole will dampen any large-scale turbulence in the water and at low velocities, the water leaving the thallus will exhibit smooth flow. Above a critical velocity, 60–120 mm s^{-1} (depending on the diameter and spacing of the branches), the branches will create microturbulence (Anderson and Charters 1982).

In order for nutrients to move from the mainstream seawater to the seaweed surface where they are taken up they must cross the laminar region of the VBL, as must the metabolic products released at the seaweed surface. Nutrients move by molecular diffusion from a region of higher to lower concentration, creating a gradient at the seaweed surface termed the "concentration boundary layer" (CBL, δ_C; Fig. 8.6). The term diffusion boundary layer is commonly used in this sense (e.g. Hurd 2000), but CBL is useful as it distinguishes the diffusive flux of mass from that of momentum (δ_M) or heat (δ_T; Denny and Wethey 2001; Nishihara and Ackerman 2007). The flux of a dissolved substance across a CBL depends on its molecular diffusion coefficient (*D*), and the difference in concentration between the mainstream seawater and the seaweed surface which, in turn, is dependent on the mainstream seawater velocity and whether flow is laminar, transitional, or turbulent. The flux across the CBL is also dependent on the metabolic status of the seaweed i.e. the rate at which a substance is taken up and/or released by the seaweed surface. Finally, some dissolved substances undergo chemical reactions with other molecules dissolved in seawater and this too can affect the flux across the CBL; this especially applies to CO_2 which reacts with water (Nishihara and Ackerman 2006). For limiting nutrients, if the demand by the seaweed is greater than the supply across the CBL, then "mass transport limitation" of seaweed growth and productivity may occur (Hurd 2000).

The thickness of the CBL at a seaweed surface can be estimated indirectly from Fick's first law of diffusion or the Schmidt number (Denny and Wethey 2001), and directly using microprobes (summarized in Table 2 of Raven and Hurd 2012). Blade morphology and the growth form of the seaweed will affect CBL thickness, because of their effect on generating or dampening turbulence. Small-scale topographic features such as

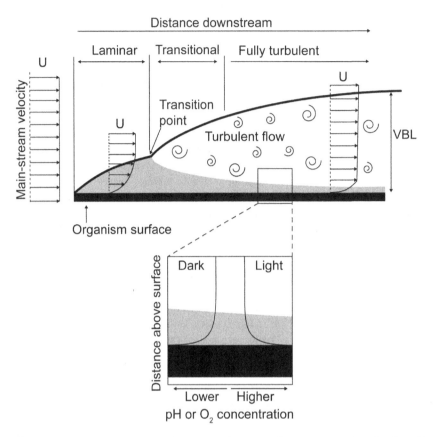

Figure 8.6 Conceptual diagram that illustrates the formation of a concentration boundary layer (CBL) within the velocity boundary layer (VBL) at the surface of a submerged seaweed. Initially, uniform (laminar) mainstream flow (arrows indicate velocity, U) encounters the leading edge of the seaweed blade (idealized here as a flat surface). A laminar VBL forms at the organism's surface, and within this laminar region, a CBL can form if the seaweed undergoes metabolic exchange at its surface. The maximum thickness of the CBL is set by the outer boundary of the laminar region of the VBL (i.e. the shaded region above the organism's surface). The thickness of the CBL decreases with increasing turbulence. Inset: Predicted gradients of pH and O_2 at the surface of a seaweed in the light and dark. In the light, photosynthesis results in both a higher pH and O_2 concentration at the seaweed surface, whereas in the dark, surface values of pH and O_2 concentration are lower due to respiration, and for calcifying seaweeds calcification. (From Hurd *et al.* 2011, *Global Change Biol.*, Blackwell Pub.)

surface corrugations and hairs increase the thickness of the CBL compared to a smooth blade, because they reduce Re_* and trap fluid at the seaweed surface. At mainstream flows of 1.5 cm s^{-1}, hyaline hairs on the surface of *Fucus vesiculosus* cause the CBL to be 0.3 mm thicker than that of hairless blades (Spilling *et al.* 2010; sec. 2.6.2). Some densely branched seaweeds will also trap fluid within their thallus, with the result that the CBL can extend from the base of the seaweed to the top of the canopy, or even beyond. For example, Cornwall *et al.* (2013) detected oxygen and H$^+$ concentration gradients within canopies of an articulate coralline seaweed that were up to 60 mm thick. Furthermore, each "layer" of seaweed within a community (i.e. turf, fleshy understory, kelp canopy) has a CBL associated with it. It is evident that the physical and chemical environment at the surface of a seaweed is very different to that of the mainstream seawater (e.g. Fig. 7.26).

8.1.4 Methods for measuring seawater flow and wave forces

In the subtidal, flow velocity, direction and turbulence can be measured using S4 current meters, acoustic Doppler velocimeters (ADVs) or acoustic Doppler current profilers (ADCPs). Such instruments can be placed within seaweed beds so that the extent to which velocity is dampened can be measured directly (e.g. Stevens *et al.* 2003; Gaylord *et al.* 2007). Figurski *et al.* (2011) developed an "underwater relative swell kinetics instrument" (URSKI), an inexpensive device that incorporates an accelerometer to measure underwater velocities due to passing waves. In laboratory flumes, micro-ADVs can measure velocity and turbulence very close to seaweed surfaces (Hansen *et al.* 2011), and hot film sensors have been used for this purpose in the field (Koch 1994). The rate of dissolution of plaster of Paris from "clod cards" (Doty 1971) and plaster balls (Gerard and Mann 1979) provides a good estimate of mass-transfer rates (Porter *et al.* 2000), and small dissolution blocks can be clipped onto seaweed blades (Koehl and Alberte 1988; Hepburn *et al.* 2007). However, because dissolution rates vary with the mode of water motion (e.g. unidirectional vs. orbital), care must be taken in extrapolating dissolution to velocity (Porter *et al.* 2000), and they are not useful in the intertidal zone as atmospheric conditions will affect dissolution.

The "waviness" of the ocean can be estimated mathematically as the significant wave height (H_S), which is the "average height of the highest third of the waves present at a given time" (Denny and Wethey 2001). Some countries, for example the USA and Japan, have established national networks of wave-rider buoys that continuously record the height of waves approaching the shore, and archived data are available on websites (e.g. Denny and Wethey 2001; Nishihara and Terada 2010). Inexpensive pressure transducers are available to measure wave height locally (Denny and Wethey 2001; Stevens *et al.* 2002). Knowing the size of waves approaching a shore gives an indication of wave exposure, but small-scale local topographies strongly modify water flow such that wave height is not necessarily a good predictor of the hydrodynamic forces experienced by an organism, or the likelihood of a physical

disturbance to a site (Helmuth and Denny 2003). Therefore, it can be useful to measure wave forces on the scale of the organism, or indeed, on the organism itself. In the intertidal zone, the "wiffle-ball" dynamometer developed by Bell and Denny (1994) is an inexpensive method of recording the maximum wave force (Denny and Wethey 2001). A variation of this was used by Boller and Carrington (2006a) to measure forces on intertidal *Chondrus crispus* directly, whereas in the laboratory they measured drag using a more sensitive strain gauge force transducer (Boller and Carrington 2006b). Stevens *et al.* (2002) took advantage of the uniquely large diameter of *Durvillaea* sp. stipes, and their thick blades, to which they attached displacement and forces transducers, and accelerometers. The holdfast was embedded in concrete and the entire "bionic kelp" was anchored to the intertidal where it recorded forces of 300 N on the stipe and blade accelerations of 30 m s^{-2}. Sensors were also used by Gaylord *et al.* (2008) to measure wave forces on *Egregia menziesii* and *Macrocystis pyrifera* in the intertidal and subtidal, respectively. Indirect methods of measuring forces on seaweeds are also available, for example Seymour *et al.* (1989) calculated drag forces on *Macrocystis* canopies from measurements of wave height and frequency.

Wave-exposure indices are derived from the factors that determine wave size and used to categorize the level of wave exposure that a beach experiences (see Burrows *et al.* 2008). The simplest scale is the angle of open water (water unobstructed to some arbitrary distance from the site), as determined from a map (e.g. Baardseth 1970). This method relies on the relation of wave action to fetch, but does not take account of the other factors involved in generating or modifying wave action. A more sophisticated method (Thomas 1986) involves calculating an index based on the velocity, direction, and duration of the wind and the effective fetch; it can be modified for shore slope. Thus, most of the factors affecting wave action are included; the index can be derived from readily available hydrographic and weather data, and it is universally applicable. The advent of geographic information systems (GIS) computer software is further improving such indirect measures (Burrows *et al.* 2008), which can be applied to predicting seaweed percentage cover and biodiversity

(Hill *et al.* 2010). Another indirect approach is the biological-exposure scale, which has the advantage of integrating all the complex components of wave exposure, including the impact pressures of waves, the wetting effects of wind-driven spray, the presence of sedimentation in the intertidal zone, the mobility of loose stones, and the nutrient availability (Lewis 1964; Dalby 1980). Nevertheless, all biological exposure indices are ultimately limited because of the circular argument that underlies them: The calibration or rationalization of the method is dependent on assumptions about the distribution of organisms, which is what the method is designed to explain (Burrows *et al.* 2008).

A variety of laboratory flumes (see Vogel 1994) have been developed to measure the metabolic processes of seaweeds in unidirectional (e.g. Wing and Patterson 1993; Hurd *et al.* 1994b) and oscillating flow (e.g. Carpenter *et al.* 1991; Lowe *et al.* 2005). Flumes can be made sufficiently large to accommodate seaweed communities (e.g. Larned and Atkinson 1997; Rosman *et al.* 2010; Kregting *et al.* 2011). The type of flow (laminar, transitional, turbulent) experienced at a seaweed surface can be visualized using dye tracing or particle tracking techniques (Wheeler 1980; Anderson and Charters 1982; Hurd *et al.* 1997; Fig. 8.7), and

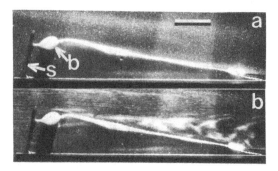

Figure 8.7 Seawater flow around a single *Macrocystis integrifolia* (= *M. pyrifera*) blade at mainstream velocities of (a) 0.5 and (b) 1.5 cm s^{-1}. The stipe (s) and bulb (b) are indicated by arrows in (a). The VBL is laminar at 0.5 cm s^{-1} (a). At 1.5 cm s^{-1}, whole-blade separation occurs, and the VBL is turbulent. Shear and acceleration around the bulb are also apparent in (b). Scale bar = 5 cm. (From Hurd and Stevens 1997; Wiley Pub.)

quantified using an ADV (Hurd and Pilditch 2011). Drag forces on seaweeds have been measured in unidirectional flumes at mainstream velocities of up to 4 m s^{-1} (Bell 1999; Martone *et al.* 2012). However, seaweeds may respond differently in such slow, unidirectional flows compared to the faster, multidirectional flows seen in the field, so that care must be taken if extrapolating laboratory results to the field situation (Bell 1999; Martone *et al.* 2012). To this end, Martone and Denny (2008a) designed a gravity-accelerated water flume that generates velocities of up to 10 m s^{-1}, simulating forces of breaking waves.

8.2 Water motion and biological processes

8.2.1 Function and form in relation to resource acquisition

Metabolic processes such as photosynthesis, respiration, and nutrient uptake in seaweeds are dependent on a flux of dissolved substances to and from the blade surface. As mainstream seawater velocity increases, the thickness of the CBL declines, as does the distance over which substances travel (see Wheeler 1988; Hurd 2000; Fig. 8.6). Physiological rates, including growth, tend to increase hyperbolically with increasing velocity, until a maximum rate is reached, at which point the passage of substances through the CBL is no longer the rate-limiting process (Figs 6.7b, 8.8). The mainstream velocity at which the maximum physiological rate occurs varies with species: growth rates of *Ulva australis* (previously *U. pertusa*) were maximal at 0.5 cm s^{-1} (Barr *et al.* 2008; Fig 8.8); photosynthesis of *Dictyopteris undulata* and *Zonaria farlowii* was saturated at <2 cm s^{-1} (Stewart and Carpenter 2003); for photosynthesis and nutrient uptake by *M. pyrifera* the saturating velocity was 2–6 cm s^{-1} (Wheeler 1980; Hurd *et al.* 1996); ammonium and phosphate uptake were not saturated by 13 cm s^{-1} for coral reef *Dictyosphaeria cavernosa* (Larned and Atkinson 1997); and for a turf community velocities of 20 cm s^{-1} were required for maximal rates (Carpenter *et al.* 1991). The mode of water motion also affects physiological rates. The productivity of coral reef algal turfs was 21% higher

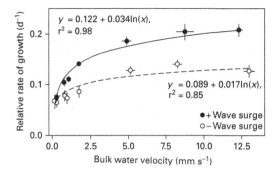

Figure 8.8 *Ulva australis* (previously *Ulva pertusa*). Effect of wave surge on relative growth rates d^{-1} (in g wet weight per day) of specimens maintained at a range of bulk flow rates in outdoor tanks. Symbols represent mean $\pm$ 1 SE (n = 3). Regression lines were fitted by ordinary least squares. (From Barr *et al.* 2008. © Inter-Research 2008.)

in an oscillating-flow chamber, which simulates wave surge, than in a stirred chamber with the same water velocity (Carpenter *et al.* 1991). Similarly for *Ulva australis*, growth and ammonium uptake rates increased by ~1.5 times when wave surge was added (Barr *et al.* 2008; Fig. 8.8). Given that water flow is so important to metabolic processes, laboratory measurements of photosynthesis, respiration, and nutrient uptake, must be made in stirred chambers (Dromgoole 1978; Littler 1979; Harrison and Druehl 1982).

The morphology of many seaweeds varies between wave-sheltered and -exposed sites (synthesized in Table 1 of Koehl *et al.* 2008; Fig. 8.9). Wheeler (1988) proposed that seaweeds had "evolved morphologies that would trip laminar boundary layers to turbulent boundary layers with thinner sublayers" thereby decreasing CBL thickness and increasing the supply of nutrients. In his seminal work, Wheeler (1980) showed that seaweed blades do generate more turbulence than a smooth, flat plate. Hurd *et al.* (1996) argued that if this is an adaptation to enhance nutrient supply, then wave-sheltered seaweeds should generate more turbulence that those from a wave-exposed site, because the latter are not mass-transfer limited. This hypothesis has been tested on *Macrocystis pyrifera*. Typical of many kelps, wave-sheltered blades are wide, thin and flat, with undulations (3–5 cm tall) running

along their edges whereas wave-exposed blades are thick and narrow, with surface corrugations (1 mm tall) running longitudinally (Hurd and Pilditch 2011). Contrary to predictions, rates of nitrate and ammonium uptake were similar for wave-sheltered and wave-exposed blade morphologies (Hurd *et al.* 1996), the transition from a laminar to turbulent boundary layer occurred at the same mainstream velocities (1–3 cm s^{-1}; Hurd *et al.* 1997), and there is little evidence that wave-sheltered blades caused increased turbulence at slow flows compared to wave-exposed blades (Hurd and Pilditch 2011). Nevertheless, blade morphology does affect CBL thickness and nutrient supply, but not necessarily by enhancing turbulence as originally hypothesized. The edge undulations (ruffles) cause blades to flap in slow flows, which results in periodic stripping of the CBL thereby supplying fresh seawater to the seaweed surface (Koehl and Alberte 1988; Stevens and Hurd 1997; Koehl *et al.* 2008; Huang *et al.* 2011). The corrugations of wave-exposed blades are too small to generate turbulence ($Re_* = 1.5$), but they trap fluid at the blade surface (Hurd and Pilditch 2011), as can surface hairs (Spilling *et al.* 2010). This "fluid trapping" may enhance nutrient acquisition by creating a quiescent region for the deployment of extracellular enzymes, such as carbonic anhydrase and alkaline phosphatase, that might be otherwise washed away (Raven 1991; Schaffelke 1999b; Hurd 2000; Enríquez and Rodríguez-Román 2006). The corrugations of wave-exposed blades may reduce frictional drag forces by inducing span-wise vortices (Hurd *et al.* 1997).

Studies on CBLs have focused on their effect on nutrient supply from seawater, but ions and molecules also accumulate at the surface of seaweeds (Fig. 8.6) and these too might affect physiological rate processes. O_2 released during photosynthesis can result in a hyperoxic layer at the seaweed surface which could increase photorespiration and so decrease rates of photosynthesis. This may be the case for *Hydropuntia cornea* (previously *Gracilaria cornea*) in slow flows (Mass *et al.* 2010), but high O_2 concentrations did not reduce the photosynthetic rates of either *Lomentaria articulata* or *G. conferta* (Gonen *et al.* 1995; Kübler *et al.* 1999). For *G. conferta*, the accumulation of OH$^-$

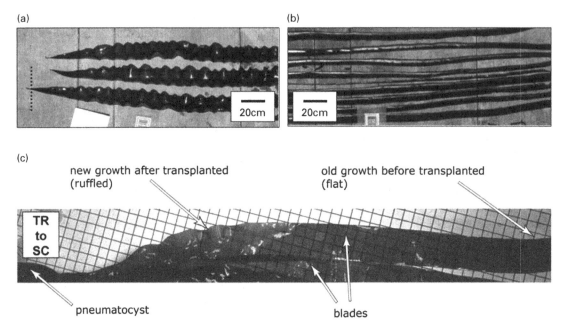

(a)

20cm

(b)

20cm

(c)

new growth after transplanted
(ruffled)

old growth before transplanted
(flat)

TR
to
SC

pneumatocyst

blades

Figure 8.9 A. Ruffled, wide blades from a *Nereocystis luetkeana* collected at a slow-flow site. The dotted line indicates the blade position defined as the "origin" in growth experiments (the position along a blade where the blade first widens from a cylindrical string into a flat blade). (b) Flat, narrow blades from a *N. luetkeana* collected from a current-swept site. (c) Photograph (taken 6 days after the transplant) of the proximal end of some blades of a *N. luetkeana* transplanted from the current-swept site to the slow-flow site (grid marks 1 cm apart). The old, slowly growing distal portion of the blade retained the flat blade morphology that characterized this individual before it was transplanted, while the new, rapidly growing proximal portion of the blade developed ruffles. All individuals (n = 5 kelp) collected from the current-swept site and transplanted back into that exposed site remained flat. (From Koehl *et al.* 2008. Published by Oxford University Press on behalf of the Society for Integrative and Comparative Biology. All rights reserved.)

at the surface was considered responsible for reduced photosynthesis under slow flows (Gonen *et al.* 1995). The hyperoxic layer may be detrimental to mobile and sessile invertebrates living on the seaweed surface (Irwin and Davenport 2002). H^+ concentrations also vary substantially within the CBL and cause large fluctuations in surface pH. Respiration and calcification both cause pH to increase whereas photosynthesis causes pH to decrease and this has implications for how seaweeds respond to ocean acidification (sec. 7.7 ; de Beer and Larkum 2001; Hurd *et al.* 2011; Cornwall *et al.* 2013).

Blade morphology also affects the light-harvesting abilities of seaweeds growing at wave-sheltered

compared to wave–exposed sites (Stewart and Carpenter 2003). The blade flapping described above may generate light flecks, which are known to enhance rates of photosynthesis (e.g. Wing and Patterson 1993). The thin blades of wave-sheltered kelps are also more transparent, which may allow light to be transmitted to blades below (Raven and Hurd 2012). Wave-sheltered blades of *Ecklonia radiata* have greater pigment content per unit blade mass and a greater light harvesting at sub-saturating PFDs (i.e. higher alpha) than wave-exposed blades (Miller *et al.* 2006; Wing *et al.* 2007). Stewart and Carpenter (2003) examined the interactions between photon flux density and flow velocity (2-32 cm s^{-1}) on the photosynthetic rates of wave-sheltered and -exposed morphologies of *Dictyopteris undulata* and *Zonaria farlowii*. At sub-saturating PFD, there was very little effect of flow velocity on net photosynthesis for either species because the supply of carbon and nutrients across the CBL met the demand for photosynthesis. Under a saturating PFD, however, there was a strong effect of water motion on photosynthesis. Rates were similar at seawater velocities between 2 and 20 cm s⁻¹, but then decreased substantially as the thalli reconfigured in high flows. Reconfiguration reduces drag on the thallus (sec. 8.3.1) but also reduces the effective surface area so that self-shading is increased, and thallus compaction means that the supply of DIC to the middle of the seaweed is also reduced.

The morphology of some seaweeds changes when they are transplanted from a wave-sheltered to wave-exposed site, or vice versa. For example, following transplantation to a wave-sheltered site, wave-exposed blades of *Nereocystis luetkeana* develop a morphology that is characteristic of wave-sheltered blades (Koehl *et al.* 2008, Fig. 8.9b). However, this response is not universal. When *Eisenia arborea* was transplanted from both fast- and slow-flow sites into the "kelp forest exhibit" tank at the Monterey Bay Aquarium, USA, they retained their original morphological characteristics for 4 years, suggesting that water motion had caused incipient speciation (Roberson and Coyer 2004).

For those kelps that do show morphological plasticity with water motion, blade development is controlled by the drag force acting on the seaweed. This was first demonstrated by Gerard (1987) who grew juvenile *Saccharina latissima* (previously *Laminaria saccharina*) with and without a weight attached to the distal end. A wave-exposed blade grew in the with-weight treatment, and a wave-sheltered morphology grew in the absence of a weight. Kraemer and Chapman (1991a, b) found the same morphological effect for juvenile *Egregia menziesii*, and also showed that the added force cause a 56% increase in DIC uptake, and that this carbon was allocated to making thicker cell walls. The characteristic edge undulations of wave-sheltered kelps are produced via differential rates of cell growth in the outer blade regions

compared to the midline, termed "elastic buckling" (Koehl *et al.* 2008). These undulations are not rigid, but change their shape and dimensions as fluid flows past them. Mechanical perturbations ·trigger similar morphological changes in terrestrial plants. Niklas (2009) proposed that the mechanism of detecting mechanical stimuli is common to all photosynthetic organisms, and involves receptors in the cell membrane–cell wall interface, which in turn brings about a change in cellulose synthesis and properties of the cytoskeleton that allow the cells to change their size and shape.

8.2.2 Synchronization of gamete and spore release

Concentration boundary layers play a critical role in controlling the timing of gamete and spore release. Observations that a Baltic Sea population of *Fucus vesiculosus* synchronously released gametes in the afternoon only on calm, but not windy, days led to a series of experiments that revealed the physiological mechanisms by which some seaweeds perceive slow flows (Serrão *et al.* 1996; reviewed by Pearson and Serrão 2006). *Fucus* spp. can sense water motion by detecting concentrations of dissolved inorganic carbon (DIC, probably bicarbonate) at their surface, i.e. within the CBL (Pearson *et al.* 1998). In laboratory experiments, gamete release was inhibited by shaking culture flasks or by increasing the concentrations of DIC under no-flow conditions (Fig. 8.10). In calm conditions, the CBL is thicker and DIC at the blade surface is depleted as it is actively taken up by the seaweed. This DIC-detection system is linked to photosynthesis, because a light dose is required. Receptacles become "potentiated" (i.e. ready to release gametes) when given the appropriate triggers of calm conditions and light (Pearson *et al.* 2004). *Silvetia compressa* can be de-potentiated by simply placing them into high-flow conditions (Pearson *et al.* 1998). This mechanism ensures that if local flow conditions in the field became too turbulent, conceptacles will not move onto the second phase of gamete expulsion (see Speransky *et al.* 2001).

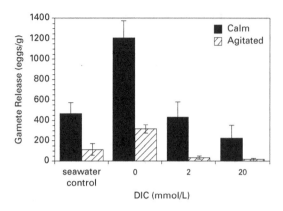

Figure 8.10 The effect of DIC and water motion on gamete release by receptacles of *F. vesiculosus*. Values are means ± 1 SE (n = 4). The experiment was done in unbuffered, DIC-free seawater with or without addition of $NaHCO_3$ (controls in natural seawater). (From Pearson *et al.* 1998, ©1998 by the Ecological Society of America.)

Water motion differentially affects propagule release by different species of seaweed, and different life-history stages of the same species. For the kelp, *Alaria esculenta*, spore release is 2-times greater under shaken compared to calm conditions, while the opposite effect is seen for sperm release by male gametophytes which is 3-fold greater under calm than shaken conditions (Gordon and Brawley 2004). Kelp sporophytes are responsible for the long-distance dispersal of spores, and therefore releasing spores in turbulent conditions is beneficial in terms of broadcasting them further. However, sperm release by the microscopic gametophytes is triggered by a pheromone released by the female gametophytes (sec. 2.4); fast flows would wash the pheromone away from the egg and reduce fertilization success. Hence, for *Alaria*, slow flows are a pre-requisite for successful fertilization. *Ulva lactuca* has an isomorphic alternation of generations and for this species, both spore and gamete release were enhanced in fast flows, but gamete release was enhanced 5-fold compared to just a 20% enhancement of zoospore release (Gordon and Brawley 2004). The coenocytic green seaweed *Bryopsis* also has enhanced gamete release in calm conditions, but for this species

turbulent conditions do not prevent gamete release as seen for the fucoids (Speransky *et al.* 2000).

8.3 Wave-swept shores

8.3.1 Biomechanical properties of seaweeds

Seaweeds growing on wave-swept shores tend to be smaller than the same species growing in wave-sheltered sites, and intertidal seaweeds tend to be smaller than those in the subtidal (Wolcott 2007). "Unlike whales and giant kelp that live in deeper water, intertidal fauna and flora rarely exceed 0.5 × 0.5 m in any dimension" (Martone and Denny 2008a). This is because the forces associated with moving water set the maximum size to which seaweeds can grow, and forces in the wave-exposed intertidal zone are much greater than those in wave-sheltered sites (Denny *et al.* 1985, 1997; Wolcott 2007; Martone and Denny 2008a). The effect of drag (F_D) acting on a seaweed has received the most attention (Mach *et al.* 2007). F_D is directly proportional to seaweed size, which is typically reported as the frontal area (A_F) and quantifies the amount of thallus that is interacting with the flowing seawater (Denny and Wethey 2001). F_D is also proportional to the dimensionless drag coefficient (C_D) which can be viewed as a measure of algal shape (Denny 2006; Boller and Carrington 2006b, 2007; Martone *et al.* 2012). Lastly, F_D is proportional to velocity squared (u^2). Therefore, as a seaweed grows, it will reach a critical large size at which it becomes more prone to dislodgement by wave-induced drag forces. Wave action varies not only from place to place but also seasonally and unpredictably due to storms. Algae may grow large during periods of calm, only to be pruned back, or destroyed, in stormy periods, as is the case for *Fucus distichus* (previously *Fucus gardneri*) and *Pelvetiopsis limitata* in California, USA (Blanchette 1997; Wolcott 2007). Drag forces also induce bending, which can cause thalli to break (Martone and Denny 2008b).

Seaweeds are much more flexible than other sedentary benthic organisms such as corals (Boller and Carrington 2006b). This flexibility allows their thalli to

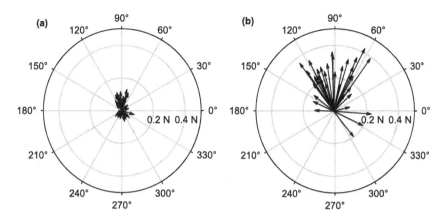

Figure 8.11 Magnitudes and relative directions of peak forces experienced by a representative arborescent alga in the canopy experiment on 7 November 2004. The length of each arrow represents the peak force (N) recorded in a 10 s interval: (a) canopy; (b) solitary. Note that the directions are relative to the arbitrary orientation of the sensor and do not necessarily correspond to any compass bearing. (From Boller and Carrington 2006a, reproduced with permission.)

reconfigure so that their overall shape changes with changing water velocities and direction (Denny *et al.* 1985). Reconfiguration is considered a pre-requisite for survival in highly dynamic fluid environments (Harder *et al.* 2004b). Boller and Carrington (2006b) identified two mechanisms by which *Chondrus crispus* reconfigures to increasing flow. At low velocities (<0.2 m s^{-1}), the stipe bends over, thereby realigning the thallus so that instead of being upright, it lies parallel to the substratum. Such bending reduces hydrodynamic forces on the thallus, and hence there is less force on the holdfast (Koehl 1986). At higher flows, *Chondrus* reconfigured via "compaction" of the thallus, as the branches collapsed together to become more densely packed. Compaction results in changes to both the shape of the seaweed (C_D) and its size (A_F) (Boller and Carrington 2006b, 2007; Martone *et al.* 2012). Seaweeds of very different sizes and morphologies grow together on wave-exposed shores, suggesting that the drag forces that they experience are similar. This idea was explored by Martone *et al.* (2012) who compared the abilities of branched and bladed seaweeds to reconfigure via changes in shape and/or size at high velocities (10 m s^{-1}). For all species tested, reconfiguration occurs via concurrent changes in both shape and area, but the bladed seaweeds changed shape to a greater extent than branched seaweeds, whereas branched seaweeds changed their surface area to a greater extent. The result was that similar drag forces were experienced by seaweeds with different morphologies.

Another way by which seaweeds reduce drag on themselves is to grow in a community. Aggregated organisms (e.g. kelp or seagrass beds) can experience much less severe hydrodynamic forces than solitary individuals (Fig. 8.11). By growing in groups, seaweeds also receive mechanical support from their neighbors (Johnson 2001). In laboratory studies, *Chondrus crispus* growing in the center of an aggregation experienced 83% less drag force than those on the edge (Johnson 2001) and direct measurements of forces in the field showed the same trend, although the percentage reduction (15–65%) was lower (Boller and Carrington 2006a). In southern Australia, *Ecklonia radiata* at wave-exposed sites grows in more densely packed aggregations compared to wave-sheltered sites, and the holdfasts of wave-exposed seaweeds have a greater attachment area and greater forces are required to dislodge them (Wernberg 2005). For the geniculate coralline seaweed, *Calliarthron*, the presence of neighbors reduced frond bending and lessens frond breakage (Martone and Denny 2008a). The sea palm,

Postelsia palmaeformis, may be an exception because the wave velocities experienced by this small kelp are so great that growing as an aggregation is unlikely to protect individuals (Holbrook *et al.* 1991).

Seaweed morphology often varies with water motion (sec. 8.2.1) and this too can affect the hydrodynamic forces experienced by individual seaweeds. For example, the straplike wave-exposed blades of *Nereocystis leutkeana* reconfigure into a streamlined shape and experience lower drag forces than the wave-sheltered blades whose characteristic edge undulations (ruffles) prevent reconfiguration (Koehl and Alberte 1988; Koehl *et al.* 2008). Boller and Carrington (2006a) were the first to measure drag forces on different morphologies of an intertidal seaweed in the field. The flat, little-branched "planar" morphology of *Chondrus crispus* experienced lower drag forces than the highly branched "arborescent" morphology, suggesting that the planar morphology would survive better in wave-swept sites. For the tropical reef seaweed, *Turbinaria ornata*, the production of pneumatocysts is a plastic response to wave action. Those growing on the wave-exposed fore-reef do not produce bladders, but bladders were produced upon transplantation to the wave-sheltered back-reef with the result that they became buoyant (Stewart 2006). By replacing the air within the bladders with water, Stewart (2004) changed the buoyant properties of *Turbinaria* without affecting their morphology. The result, that non-buoyant *Turbinaria* experienced greater forces than buoyant ones, showed that buoyancy also directly affects hydrodynamic forces experienced by seaweeds. Morphological variation does not always result in differences in drag, however. In *Mastocarpus papillatus*, which shows considerable range in morphology (e.g. thallus thickness and degree of branching), drag force on the seaweed was a function of the area of the thallus and not strongly related to morphology (Carrington 1990).

Although drag forces have been the focus of most studies on seaweeds' ability to withstand wave action, other forces such as acceleration, lift, and impingement also act on seaweeds (e.g. Gaylord 2000; Denny 2006; Gaylord *et al.* 2008). Acceleration of breaking waves is not considered important to intertidal seaweeds

because it acts on a spatial scale (1 cm) so small that the seaweeds will not experience it (Gaylord 2000; Denny and Wethey 2001; Denny 2006). Lift and impingement, however, are directly related to sea-water velocity which acts on larger spatial scales, and both can affect the probability of seaweed dislodgement and perhaps their upper size limit (Gaylord 2000; Denny 2006). Lift is analogous to drag but acts perpendicular, rather than parallel, to the flow (Denny 2006). This is why forces are reduced on seaweeds that are able to bend over and lie parallel to the substratum (Koehl 1986). Impingement is the force experienced by an intertidal organism the instant that they are struck by a breaking wave (Gaylord 2000; Gaylord *et al.* 2008). This force acts for only a very brief time, but can increase the forces experienced by some seaweeds by three times compared to drag alone (Gaylord *et al.* 2008). The brief nature of impingement forces also mean that seaweed has very little time to reconfigure to the flow (Martone *et al.* 2012).

The force required to break or dislodge a seaweed, termed tenacity, can be easily measured by clamping the distal end of a seaweed to a spring scale, pulling, and recording both the force and location of the breakage (e.g. Bell and Denny 1994; Shaughnessy *et al.* 1996; Wolcott 2007). Tenacity generally scales to size, with larger seaweeds requiring more force to dislodge them (Thomsen *et al.* 2004), although for *Hormosira banksii* there was no relationship with size, but the direction in which the seaweed was pulled affected its tenacity (McKenzie and Bellgrove 2009). For some species, tenacity increases with wave exposure, for example, *Ecklonia radiata* (Thomsen *et al.* 2004), *Saccharina japonica* (previously *Laminaria japonica*) (Kawamata 2001), and *Chondrus crispus* (Carrington 1990). This was not the case for *Saccharina sessilis* (previously *Hedophyllum sessile*), but tenacity did increase following winter storms showing a seasonal effect (Milligan and DeWreede 2000). The likelihood that a seaweed will be detached in the field can be assessed using a "safety factor" (see Denny 2006) or "environmental stress factor" (ESF; Johnson and Koehl 1994) both of which relate algal strength to the wave force required to dislodge it. (Note that the term "environmental stress" used here = force per unit area, and is not related to physiological stress discussed in Chapter 7). For example, when ESF is <1, a seaweed will be detached from the substratum, but will remain attached when ESF >1, and the higher the value of ESF recorded, the lower the risk of dislodgment (Johnson and Koehl 1994; Stewart 2006; Koehl *et al.* 2008). However, there are some problems with using these simple ratios, because they are based on average values for seaweed strength and environmental stress. To more fully assess the "risk" of dislodgement, the inherent variations in seaweed strength and in wave forces are required (see Denny 2006).

The material properties of seaweeds affect how they will respond to the hydrodynamic forces acting on them. Compared to other biological materials, seaweeds are elastic, extensible, and compliant (i.e. stretchy), but they are weak (i.e. they have a low breaking strength) (Holbrook *et al.* 1991; summarized in Table 1 of Martone 2006). When Martone (2007) plotted thallus strength against the diameter of the thallus at the point where the thallus broke in tensile tests (typically the stipe), he found that seaweeds fell into two groups (Fig. 8.12). The tissues of brown seaweeds are relatively weak, but they had a greater diameter at the breakage point, whereas red seaweeds are stronger but narrower. The results suggest a trade-off between material strength and tissue thickness. At one extreme is the coralline seaweed *Calliarthron cheilosporioides* whose narrow geniculae are made of material that is up to an order of magnitude stronger than other seaweeds, and they are also more extensible, which endows them with surprising flexibility and ability to reconfigure their shape (Martone 2006; Martone and Denny 2008a, b; Martone *et al.* 2010a, 2012). At the other end of the spectrum was *Durvillaea antarctica*, which has the thickest stipe of any seaweed but also the weakest material, a combination of properties that allows it to grow to an enormous size in wave-exposed intertidal sites (Koehl 1986; Stevens *et al.* 2002).

The "pull to break test" is a standard biomechanical technique in which a slice of seaweed tissue (stipe and/or blade) is pulled (i.e. loaded in tension) until it breaks. Based on these tests, some seaweeds appear "over-engineered" for their habitat, as the force

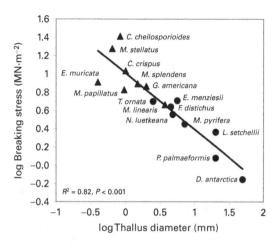

Figure 8.12 Mean breaking stresses of various red macroalgae (triangles) and brown macroalgae (circles) as a function of mean thallus diameters, measured where thalli commonly broke (e.g. stipes). References from which data were extracted are listed in the original figure legend. Species names: *Calliarthron cheilosporioides, Mastocarpus stellatus, Chondrus crispus, Endocladia muricata, Mazzaella splendens, Grateloupia americana* (previously *Prionitis lanceolata*), *Mastocarpus papillatus, Egregia menziesii, Turbinaria ornata, Mazzaella linearis, Nereocystis luetkeana, Macrocystis pyrifera, Fucus distichus* (previously *Fucus gardneri*), *Laminaria setchellii, Postelsia palmaeformis,* and *Durvillaea antarctica.* If diameters were not explicitly reported, mean thallus diameter was estimated by dividing mean breaking force (N) by mean breaking stress (N · m⁻²) and assuming a circular cross section. Diameter measurements of *Fucus distichus* were taken by the author at Hopkins Marine Station (Pacific Grove, CA, USA). Error bars were excluded to increase the readability of the graph. (From Martone 2007, © Phycological Society of America.)

required to break them is much greater than that exerted by a breaking wave (Denny 2006; Mach *et al.* 2007; but see Martone and Denny 2008a). However, this view is at odds with large amounts of drift seaweed seen on beaches after storms (see Essay 3 and Fig. 1 in Chapter 5). By repeatedly applying sub-lethal stresses to seaweeds, the discrepancy between apparent overengineering versus large-scale removal from the substratum has been reconciled. Such repeated loading causes tissue "fatigue" which in turn causes small cracks to form in the seaweed tissue and they propagate until the tissue ruptures. "Fatigue crack growth" may be more significant in determining breakage in seaweed populations than stresses that are applied one time (Mach 2009; Martone *et al.* 2010). Because seaweed tissue is weak, even a very small crack in their surface will propagate quickly, causing the structure to fail. Crack formation is also caused by herbivore grazing and can make seaweeds more prone to dislodgement (Mach *et al.* 2007; see below).

What are the cellular properties of seaweeds that account for the observed differences in material properties between species and with wave exposure? Red and brown seaweeds contain large proportions of cell wall polysaccharides (sec. 5.5) and different forms of these, each with their distinctive gelling properties, are found in different regions of seaweeds (e.g. stipe vs. blade), and for red seaweeds, in different life-history stages (Carrington *et al.* 2001). While it is possible that such biochemical differences might account for the different material properties of seaweeds growing in sites of different wave exposure, support for the idea is weak. Craigie *et al.* (1984) and Venegas *et al.* (1993) found that the blades of wave-exposed *Saccharina longicruris* and *Lessonia trabeculata*, respectively, have greater proportions of polyguluronate (strong, rigid gels) than those of wave-sheltered specimens, but there were no differences between stipes or holdfasts, which are the structures that are crucial for attachment. Blades of *Chondrus crispus* gametophytes were stronger and more extensible than those of tetrasporophytes, which correlated to the gelling properties of the carrageenans, but once again, there were no differences in the polysaccharide properties of the holdfast–stipe junction (Carrington *et al.* 2001). *Egregia menziesii* blades growing in high-energy environments had weaker alginate (less polyguluronate; sec. 5.5.2), which should have led to more flexible blades, and yet in biomechanical tests, the blades were stiffer and stronger than those grown in calmer water (Kraemer and Chapman 1991a). Both Kraemer and Chapman (1991a) and Carrington *et al.* (2001) suggest that properties of seaweed cells other than polysaccharides are responsible for their different material properties.

Their view is supported by Martone (2007) who showed that cell wall thickening in the geniculae of *Calliarthron* increases tissue strength and Demes *et al.* (2011) who show that blade construction affects tissue strength.

In the subtidal, wave forces are much reduced and seaweeds can grow larger than in the intertidal (e.g. Gaylord and Denny 1997). Instead of the rapid reconfiguration of thalli seen for intertidal seaweeds in response to waves, subtidal kelps take a longer time to become fully extended in the flow (Gaylord and Denny 1997). The flexibility of seaweeds also allows them to "go with the flow" and so the velocity experienced by the blade can be much lower than that of surrounding seawater (Stevens *et al.* 2001). Such means of reducing drag are remarkably effective, especially in the stipes of mature *Nereocystis luetkeana*, which become very narrow and elastic at the base (Koehl and Wainwright 1977, 1985; Denny *et al.* 1997). The stipe bases seem incongruously small compared with the holdfast, and yet they can withstand the strain imposed by most waves until they are damaged by grazing or abrasion (Koehl and Wainwright 1977). Two features of *Nereocystis* help the seaweed survive wave action. We can think of the seaweed as a sphere on the end of an elastic string. The maximum acceleration force occurs when a wave first strikes, but the wave force pulls on the stipe only when the seaweed has become fully extended in the direction of the surge. Assuming that the seaweed had become fully extended along the backflow of the preceding wave, it must travel some distance before it is again fully extended (Fig. 8.13). Second, it can stretch a long way before the force will have any chance of breaking the stipe. Extension and stretching take time. Denny (1988) estimates 12–13 s for a mature *Nereocystis* and because the water travels shoreward for only half of a wave period, most waves have too short a period to harm the kelp: "If the stipe is long enough, the sphere never reaches the end of its tether before the water velocity changes direction" (Denny 1988, p. 245). Most seaweeds are much shorter than mature *Nereocystis* and thus will experience the pull of the wave force. Indeed, juveniles of *Nereocystis* are prone to breakage (Denny *et al.* 1997).

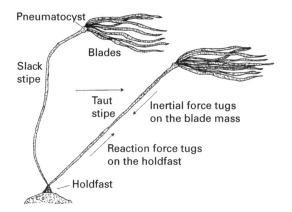

Figure 8.13 The blade mass of a bull kelp (*Nereocystis luetkeana*) is attached to the substrate by a ropelike, flexible stipe. As the blade mass moves with the flow, it may abruptly come to the end of its "tether". The resulting deceleration can apply a substantial inertial force to the stipe. (Redrawn from Denny and Wethey 2001.)

The stipes of some subtidal kelps are comparatively stiff (Gaylord and Denny 1997). Those of *Eisenia arborea* and *Lessonia nigrescens* branch, and each branch has a bundle of blades arrayed at the tip. When the water is calm, this morphology holds the blades spread out for maximum light harvest. When there is a surge, the blades clump together, and the stipe bends, so that the whole thallus becomes streamlined (Charters *et al.* 1969; Neushul 1972; Koehl 1982, 1986; Gaylord and Denny 1997). *Pterygophora* stipes are six times stronger than *Nereocystis*, but they are still very flexible (Biedka *et al.* 1987). Their resistance to breaking comes partly from different tensile and compressive properties. Stipe tissue is easier to compress than to stretch, so that in bending, the compression of the cells on the inside of the bend relieves part of the tension on the cells on the outside of the bend, and the stipe can bend farther before breaking (Biedka *et al.* 1987). In fact, *Pterygophora* stipes must be bent double and squeezed before they will snap, a situation that obviously never occurs in nature. *Pterygophora* stipes are very tough because the cortex is stiff, strong, and extensible. Yet the stipes are brittle, and surface weaknesses such as cracks or grazing marks will greatly reduce the mechanical strength of the stipes (Biedka

et al. 1987; Mach *et al.* 2007). The effect of cracks is partly mitigated by the extensibility of the tissue, which rounds off the apex of the crack and makes it less liable to fracture further (Denny *et al.* 1989). Cracks may actually help the stipeless kelp *Saccharina sessilis* in wave-exposed areas by allowing portions of the blade to be torn away by unusually strong waves, while saving the basal meristem (Armstrong 1987). A similar response has been observed for *Fucus distichus* which sheds branches when transplanted from wave-sheltered to -exposed sites (Blanchette 1997).

Productivity is very high on wave-beaten shores, on both temperate rocky coasts and on tropical reefs. Can the energy imparted by breaking waves enhance productivity? (Leigh *et al.* 1987). Waves might directly enhance photosynthesis by constantly moving algal fronds and reducing self-shading, and by generating light flecks (Wing and Patterson 1993; Kübler and Raven 1996), or through the action of drag forces stimulating carbon uptake (Kraemer and Chapman 1991b). Seaweed competitors may be removed by whiplash (sec. 4.2.1) and wave action can reduce herbivore grazing (sec. 4.3.2). In the laboratory, physiological and growth rates are typically enhanced with increasing water motion up to a certain level, but does this translate into higher rates of production in the field? Rates of growth and production along wave-exposure gradients have been studied only a few times and no clear pattern has emerged. The bladed red seaweed, *Adamsiella chauvinii*, had lower growth rates in slow flows compared to fast flows, explained by sediment accumulation on blades at the slow-flow site (Kregting *et al.* 2008b). For *Saccharina latissima*, growth rates were higher at the wave-sheltered site because there was a greater supply of regenerated nitrogen from invertebrates compared to the wave-exposed site (Gerard and Mann 1979). Growth rates of *Macrocystis pyrifera* were higher at wave-exposed sites during fall, but not in any other season (Hepburn *et al.* 2007). For *L. hyperborea*, growth rates of 2- and 3-year-olds was controlled by density-dependent factors unrelated to water motion, but for 4-year-olds, growth rates were higher at wave-exposed sites (Sjøtun *et al.* 1998). These studies focused on single species in the subtidal, but the effects of breaking waves on community productivity of intertidal sites are needed to assess if, and how, wave action enhances productivity (Leigh *et al.* 1987).

8.3.2 Wave action and other physical disturbances to populations

A "disturbance" can be defined as a factor that causes the removal of biomass from a population through mortality or damage of one or more individual. This is distinct from a "stress" which reduces the growth potential of an organism (Sousa 2001, cf. Grime 1979; see secs. 3.5.2 and 7.1). Physical disturbances include water motion, sedimentation, ice scour, earthquakes, and landslides (see Table 4.1 in Sousa 2001) and pollution (Chapter 9). The extent to which a disturbance affects a population or community will depend on its intensity or magnitude, that is the proportion of the population removed or the extent of the damage incurred on any remaining seaweeds, and on the frequency and duration of the event (Connell *et al.* 1997; Sousa 2001 p. 91). Disturbance regimes have been categorized as a "disaster" or a "catastrophe" (Harper 1977; Paine 1979), but marine ecologists now tend to use the terms "acute" and "chronic" (Connell *et al.* 1997; Sousa 2001). Acute disturbances are "discrete", short-term events such as a particularly large wave that clears a patch of space within a stand of seaweeds, whereas a chronic disturbance is a major event from which the community will take a much longer time to "recover". Acute and chronic disturbances are two extremes of a gradient of disturbance, although if an acute disturbance occurs often enough, its cumulative action can be similar to that of a chronic disturbance (Connell *et al.* 1997). In experimental manipulations, the term "pulse" is used to describe experiments in which there is a one-time removal of biomass (acute) and "press" to describe repeated biomass removal simulating a chronic disturbance (Bender *et al.* 1984; Lilley and Schiel 2006). The effects of disturbance on a population can be direct, as in the creation of new space, or indirect by modifying the physical and biological environment (Sousa 2001).

Water motion is a major source of disturbance to seaweed populations, causing damage or mortality at

Figure 8.14 *Pterygophora* washed ashore with their substrate on the west coast of Vancouver Island, Canada, testify to the force of moving water.

all stages of growth. The windrows of seaweeds cast onto beaches testify to the power of waves to pull up seaweeds and animals, in some cases still attached to rocks (Fig. 8.14). Storms overturn boulders and move sand onto and off beaches. In many areas, storms constitute a regular seasonal phenomenon, although their intensities may vary markedly (e.g. Seymour *et al.* 1989). Physical disturbances that are less frequent include earthquakes, which can cause uplift of intertidal communities. For example, in Chile, mega-earthquakes occur on a ~10-year cycle and these, together with the tsunamis they generate, have devastating effects on intertidal communities (Castilla 1988; Castilla *et al.* 2010). A 2009 earthquake in the Caribbean destroyed 10 of 21 reefs that were part of a long-term monitoring program (Aronson *et al.* 2012). In some areas, cyclones (hurricanes or typhoons) create unpredictable disturbances. In 1980, in Jamaica, Hurricane Allen created waves up to 12 m high and devastated the shallow fore-reef communities, both by the direct force of the water and by the impact of entrained projectiles (e.g. broken coral; Woodley *et al.* 1981).

All of these events are disturbances that destroy some organisms and create space for others. The extent to which populations "recover" to their previous state, following a disturbance, has been the focus of many studies. The population that develops will depend on the intensity and spatial scale of the disturbance. If a disturbance causes relatively little damage to foundation species, then there may be regrowth from the existing population, and little change to the community diversity. In cases where a substantial proportion of the biomass is removed from an area, "patches" of bare space are created within the original system. In this case, the development of a new community will depend more strongly on the supply of planktonic propagules, so that biofilms are formed on the rock surface, and succession ensues (see sec. 2.5). The size of patches created varies greatly, and the patch size itself affects how the population might recover (see the discussion of *Ascophyllum* in sec. 4.3.1). For small patches, there is often recruitment from neighboring seaweeds whereas in large patches, the middle of the patch is too far from the surrounding seaweeds so it tends to be colonized by planktonic propagules. The greater the change in the physical environment, the slower the recovery. Aronson *et al.* (2012) estimated that the widespread damage to the Caribbean reef system described above, will take 2000–4000 years to recover to its pre-earthquake state.

In the following paragraphs, we give three classical examples of how disturbance affects seaweed-dominated communities before discussing two models that are being employed to test the generality of responses. On exposed northeastern Pacific rocky shores, the mussel *Mytilus californianus* is a competitive dominant in the mid-intertidal zone (Paine and Levin 1981). When patches of mussels are ripped out, there is a succession that starts with a biofilm and tends to give way to a restoration of the mussel bed, although this process takes 7–8 years (see Sousa 2001). Within this community, one alga is particularly

dependent on disturbance, the kelp *Postelsia palmae-formis* (Dayton 1973; Paine 1979, 1988; Blanchette 1996). *Postelsia* lives in the mid-intertidal zone on only those headlands most severely exposed to wave action. Small sporophytes develop equally on bare rock, on animals (barnacles attached to mussels), and on other algae, but only those on bare rock persist to maturity (Paine 1988). Commonly, clumps of *Postelsia* and the barnacles/mussels to which they are attached are dislodged by waves and in conjunction with direct wave action on the mussels, creates a bare space. Dayton (1973) suggested that *Postelsia* spores (from nearby individuals) settle onto the new bare patch and grow to maturity. Blanchette (1996) questioned this idea because clear patches are typically created during winter storms, whereas *Postelsia* releases spores in summer when the patches would already be colonized by other sessile organisms. She found that sporophyte recruitment to cleared patches was greatest in winter, most likely from cryptic gametophytes that were growing on the rocks, underneath the mussels. Increased light levels caused by the removal of the mussel canopy would trigger sexual reproduction (cf. *Macro-cystis* and *Laminaria*, sec. 2.3). The mussels may also have a facilitatory effect on the *Postelsia* gametophytes, reducing desiccation stress and protecting them from limpet grazers.

Disturbances are also important to the maintenance of *Ecklonia radiata* kelp beds in western Australia, but here the kelp is the dominant organism in a subtidal habitat (Kirkman 1981; Kennelly 1987a, b; Toohey and Kendrick 2007). *Ecklonia* can maintain its dominance as long as disturbances are small. Shade-tolerant juveniles are present in the understory in a state of arrested development; when small clearings are created, these quickly grow up and replace the mature individuals that have been lost. The rate of survivorship of the juveniles depends on the season in which patches are cleared, with higher recovery in late summer compared to early summer (Toohey and Kendrick 2007). A different pattern is seen for larger-scale disturbances. Toohey *et al.* (2007) created 18 bare patches, each 314 m^2, within *Ecklonia* canopies and compared the diversity of the successional species to that of 18 undisturbed control sites over 34 months.

The diversity in the cleared patches initially increased (after 7 months) compared to control sites as a filamentous turf and foliose seaweed community developed. Next (11-22 months), a *Sargassum* canopy developed in many cleared plots, and by 34 months *Ecklonia* beds were beginning to recover. In some of the cleared plots, however, the filamentous turf remained dominant. The diversity was higher in the cleared plots than the controls throughout the study illustrating that *Ecklonia* is a competitive dominant that locally shades understory algae.

Intertidal boulder fields are subjected to periodic disturbances in winter when storm waves overturn boulders. In southern California, small boulders in areas with short intervals between disturbances, support only early successional *Ulva*-barnacle communities. Large, infrequently disturbed boulders have a red turf dominated by *Chondracanthus canaliculata* (previously *Gigartina canaliculata*). In the mid-intertidal zone, early successional algae competitively inhibit a mid-successional association of red algae, which in turn competitively inhibit the late-successional *Chondracanthus* association (Sousa 1979). In its turn, the *Chondracanthus* turf outcompetes the kelp *Egregia laevigata*, which recruits only from spores and only at certain times of the year. The red algae expand vegetatively throughout all seasons via prostrate axes, encroaching on any space that becomes available. A 100-cm^2 clearing in the middle of a *C. canaliculata* bed was completely filled in 2 years. The turf traps sediment, which fills the spaces between the axes and prevents the settlement of other algal spores on the rock (Sousa 1979; Sousa *et al.* 1981). *Egregia menziesii* (previously *E. laevigata*) may settle if clearings become available at the right time, but by the time the kelp has matured (it lives only 8–15 months), the turf will have encroached all around, thus preventing the kelp from replacing itself.

The intermediate disturbance hypothesis (IDH) proposes that following a disturbance event, essential resources (e.g. space and light) are freed up, thereby allowing competitively inferior species e.g. turfs and foliose seaweeds to establish and subsequently increase diversity (Sousa 2001; Svensson *et al.* 2012). The term "intermediate" is used because if there is too

little disturbance, a climax community of a competitive dominant (e.g. kelp bed) will develop, whereas if a disturbance occurs too frequently, then the kelp beds will not have time to recover (Sousa 2001). However, when testing the IDH, it must be remembered that "diversity" encompasses both species "richness" and "evenness" (sec. 3.5.1) and each of these components of diversity responds differently to disturbance (Svensson *et al.* 2012). The use of different indices of diversity by different researchers has therefore confounded the interpretation of studies that set out to test the IDH, and when testing the IDH, separate hypotheses are required for richness and evenness (Svensson *et al.* 2012).

The IDH predicts that if species richness is plotted against wave exposure a bell-shaped curve should result. Nishihara and Terada (2010) tested this by examining the richness of red, green, and brown seaweeds across Japan (26°N to 33°N). For a total of 437 species, they found that, overall, richness declined with wave exposure. This trend was also seen for red and green seaweeds, but for brown seaweeds, richness increased with wave exposure. They suggested that large brown seaweeds are well adapted to wave exposure and in wave-exposed sites they will outcompete red and green seaweeds. The lack of a bell-shaped curve in the Japan-wide study illustrates that disturbance is only one factor shaping the diversity of species along a water motion gradient (Nishihara and Terada 2010).

Another popular model for testing how wave-induced disturbance affects the diversity and productivity of coastal communities is the dynamic equilibrium model (DEM) which hypothesizes that the productivity of a system will influence how a disturbance regime affects species diversity (Svensson *et al.* 2007, 2010, 2012). The idea is that the more productive a system is, the greater the disturbance that is required to cause the competitive exclusion, by foundation species, of less-competitive species. Svensson *et al.* (2010) tested this using the Ecotone Mesocosm Facility, Sweden. They examined the interactive effects of physical disturbance (five levels of wave action – simulating a pulse disturbance), biological disturbance (with and without grazing by *Littorina littorea* – simulating a press disturbance) and nutrient enrichment (productivity – ambient and nutrient enriched) on seaweed and invertebrate richness. Large boulders were placed in the mesocosms, and the effect of wave action was simulated by rolling the boulders around. Physical disturbance had little effect on algal species richness, whereas invertebrate richness in each of the disturbed treatments was half that of the undisturbed control. For the four dominant seaweeds, the percentage cover of the perennials *Chondrus crispus* and *Fucus vesiculosus* declined with increasing levels of disturbance, whereas percentage cover increased for the fast-growing ephemerals, *Chaetomorpha melagonium* and *Ulva intestinalis*. Also, there was no interaction between physical disturbance and nutrient enrichment. Biological disturbance, however, had a very different effect as it interacted with productivity;

the nutrient-enriched seaweeds grew faster which offset the negative effects of grazing on species richness. This experiment illustrates how the mode of disturbance (physical vs. biological) interacts differently with the productivity of the system, and helps explain why the relationship between wave action and species richness is not necessarily that of a bell-shaped curve.

Sand and sediment are also major agents of disturbance that are associated with water movement affecting seaweed communities directly and indirectly (Fig. 8.15). In wave-sheltered sites, fine sediments accumulate on rock and seaweed surfaces, whereas on wave-exposed sites the rock surface tends to be sediment free but larger sand-particles that are suspended in the turbulent water can scour the rock

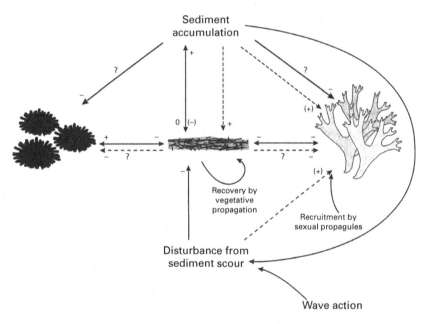

Figure 8.15 Example of complex direct (solid arrows) and indirect (dashed arrows) effects (+, positive; –, negative; 0, no effects) of sedimentation in rocky coast assemblages. Parentheses indicate effects that occur only under certain circumstances, while question marks indicate processes that need further experimental verification. (See original reference for list of sources.) Filamentous turf accumulates sediment. In turn sediment can sometimes reduce turf thickness which does not affect turf cover. Accumulation of sediment is thought to deter grazers and inhibit recruitment of erect algae that compete for space with turf. Severe scour and accumulation can locally remove turf, allowing the temporary development of erect algae. Such positive indirect effects, however, only occur at certain times depending on propagule availability. Turf always recovers quickly from eventual damage by vegetative propagation. (From Airoldi 2003, © R. N. Gibson and R. J. A. Atkinson, Editors, Taylor and Francis.)

surface (Raffaelli and Hawkins 1996; Schiel *et al.* 2006). Seaweed spores that settle on sediment particles are likely to be washed away before long, especially as they grow up into the faster-moving water layers. Also, if sediment settles on top of spores, the spores will be shaded and may be smothered. Devinny and Volse (1978) found that for *Macrocystis pyrifera*, even small amounts of sediment introduced before or along with the spores markedly reduced the percentage of spores able to settle and grow on glass slides. Other studies have similarly found negative effects of sediment on early life-history stages (reviewed by Airoldi 2003). Sediment addition to *Cystoseira barbata*, for example, strongly impeded recruitment and, for settled germlings, caused 83% mortality (Irving *et al.* 2009). Sediment loading into coastal systems is increasing due to deforestation, sewerage, and dumping, so these studies are important to help identify potential effects on the ecosystem services that these systems provide.

The smothering effects of sediments on seaweeds are species-specific, and differentially affect different life-history stages. On the rocky shores of southern New Zealand, *Durvillaea antarctica* dominates wave-exposed intertidal sites while *Hormosira banksii* dominates in wave-sheltered sites. In laboratory tests, ~100% of germlings of both species attached to sediment-free plates and neither species could attach to plates that were covered with 2 mm of sediment. However, under an intermediate covering of sediment, *Hormosira* was better able to recruit than *Durvillaea*, partly explaining its dominance on sediment-covered wave-sheltered sites (Schiel *et al.* 2006). On the Great Barrier Reef, Australia, Umar *et al.* (1998) found that sprinkling sediments onto natural populations of *Sargassum microphyllum* caused declines in recruitment, percentage cover, and growth rates of established seaweed. However, when established *S. microphyllum* fronds were "dusted" to remove natural levels of sediment from the surface, there was no effect on growth rate indicating that the adults could survive some level of sedimentation.

A protective effect of sediment against high light levels was found for mature sporophytes of *Saccharina latissima* (Roleda and Dethleff 2011). The photosynthetic efficiency (F_v/F_m) of blades grown at 220 μmol photons m^{-2} s^{-1} for 7 days under gravel or sand was unaffected, but for a silt + clay treatment, F_v/F_m declined from 0.7 to 0.5. However, this decline was minor compared to the no-sediment controls, for which F_v/F_m was <0.2. A different type of protective effect was seen for *Sargassum* sp. in southern Japan (Kawamata *et al.* 2012). *Sargassum* sp. can recruit onto sediment-covered cobbles, although the mechanisms by which it is able to attach are not known. Since sea urchins graze preferentially on sediment-free sites, the sediment-covered cobbles provide *Sargassum* with a refuge habitat.

Sand scouring and sand burial prevents the growth of many seaweeds but, as for habitats smothered in fine sediments, there are resistant species. Isolated rocks on a sandy beach and sandy areas in largely rocky shores have relatively few species of algae. These species tend to be robust, stress-tolerant perennials such as *Protohalopteris radicans* (previously *Sphacelaria radicans*) and *Ahnfeltia plicata*, or opportunistic ephemerals such as *Chaetomorpha linum*, *Ulva* spp., *Ectocarpus* spp., and colonial diatoms (Daly and Mathieson 1977; Littler *et al.* 1983b). The opportunists are able to settle when scouring is at a minimum and the rocks are bare; they reproduce and disappear before scouring begins again. Many coralline communities in southern California are sand-stressed, but corallines resist sand scour and stabilize the sand (Stewart 1983, 1989). Nevertheless, some coralline communities can be negatively impacted by sediments. In the mainland shore communities of southern California where there is substantial sand scour, the saccate browns (*Colpomenia* and *Leathesia*) grow as epiphytes on turf-forming corallines, apparently unable to anchor directly onto the rock. However, on offshore islands where there is little sand scour, they grow directly on rock (Oates 1989).

Sand movement on beaches typically is seasonal. Sand builds up in spring/summer and is washed into the subtidal in fall/winter (Fig. 8.16). Tolerant seaweeds must survive scouring and even months of burial. Such species include *Ahnfeltiopsis linearis* (previously *Gymnogongrus linearis*), *Laminaria sinclairii*, *Phaeostrophion irregulare*, and *Ahnfeltia* spp. from the

(a)

(b)

Figure 8.16 Habitat of *Laminaria sinclairii* on a sandy beach in Oregon, (a) in April, with little sand present, and (b) in July, with rocks almost buried in sand. Arrows point to the same rock. (From Markham 1973, with permission of *Journal of Phycology.*)

west coast of North America, and a *Polyides–Ahnfeltia* association and *Protohalopteris radicans* (previously *Sphacelaria radicans*) on the east coast of North America and *Mazzaella capensis*, *Ahnfeltiopsis complicata* (previously *G. complicates*), and *Ahnfeltiopsis glomerata* (previously *G. glomeratus*) from South Africa (Markham and Newroth 1972; Markham 1973; Sears and Wilce 1975; Daly and Mathieson 1977; Anderson *et al.* 2008). The characteristics of these algae are listed by Airoldi (2003, Table 1) and include: tough, usually cylindrical, thalli, with thick cell walls; great ability to regenerate, or an asexual reproductive cycle functionally equivalent to regeneration; reproduction timed to occur when seaweeds are uncovered; a sand-tolerant crustose base (e.g. *Mazzaella capensis*) or a strongly heteromorphic life cycle in which the microscopic phase survives sand inundation and the macroscopic phase appears once the sand disappears, e.g. *Porphyra capensis* (Anderson *et al.* 2008). Buried seaweeds require physiological adaptations to withstand darkness, nutrient deprivation, anaerobic conditions, and H_2S. Although the traits of species and communities that are sand tolerant have been identified, the nature of their physiological adaptations has received very little attention (Roleda and Dethleff 2011). These algae have been called "psammophilic" (sand-loving), but as Littler *et al.* (1983b) pointed out, that name implies that the algae do better (higher growth rates or reproductive output) in the presence of sand, whereas in reality they probably do worse, but are not as severely affected as their competitors. This was the case for crustose coralline recruitment to lava boulders in the Galapagos, as sand scour removed the algal turfs that otherwise overgrew the corallines (Kendrick 1991). Perennial, sand-tolerant species may dominate the primary rock substratum, as *Corallina* does in the low-intertidal zone in southern California (Stewart 1982, 1983, 1989), but if they are not completely buried, they may in turn provide a substratum for epiphytic ephemerals (see sec. 4.2.2). In other cases, turfs of filamentous algae dominate in environments subject to high levels of sand scour. In the Mediterranean, for example, the sand trapped by *Wormersleyella setacea* (previously *Polysiphina setacea*) is an integral (96% mass) part of

the community structure (Airoldi 1998; Airoldi and Virgilio 1998).

Some algae, especially the inhabitants of tropical lagoons, such as *Penicillus* and some *Halimeda* species, develop holdfasts in sand and silt. For these algae, sand is a necessary substratum, and the disturbances that they face include burial and uprooting by surge and burrowing animals. For instance, *Caulerpa* species on a soft seabed in the Caribbean frequently were disturbed by conchs, ghost shrimp, rays, and so forth, which "uprooted seaweeds, excavated holes which broke or undermined seaweeds, trampled seaweeds, or caused large-scale sediment redistributions" (Williams *et al.* 1985). Stolons and upright shoots responded to burial by turning upward or forming new erect shoots, but their growth rates were reduced as compared with those of undisturbed seaweeds. Stolon elongation involves very little increase in biomass, and so the cost of recovery from burial may not be great; plastids move into the unburied parts of the siphonous seaweed.

8.4 Synopsis

Water motion fundamentally modifies the physical and chemical environment in which seaweeds grow. Surface waves create a dynamic underwater light environment that can enhance rates of photosynthesis. Breaking waves impart hydrodynamic forces on seaweed which can rip them off the rock, but this also clears new space for seaweeds to colonize. In slow flows, nutrient concentration gradients form at the surface of seaweeds which can restrict nutrient supply.

Seaweed morphology is a trade-off between maximizing surface area available for light harvesting and nutrient uptake, and minimizing drag forces associated with waves. Physiological rates such as photosynthesis and nutrient uptake typically increase with increasing seawater velocity, up to a maximal rate, as the supply of nutrients across the concentration boundary layer (CBL) increases and metabolic waste products are removed. In slow flows, undulations along the edge of kelp blades cause the blades to flap, thereby "shaking off" the and supplying fresh seawater to the surface. Some seaweeds can sense periods of slow flow through the changes in the dissolved inorganic carbon concentration within the CBL, and this acts as a trigger for gamete or spore release.

Breaking waves impart substantial forces on seaweeds, the most important being drag, lift, and impingement. Seaweeds are very flexible, and this allows their thalli to reconfigure to the flow, modifying both their size and shape, and thereby reducing drag. Reconfiguration at high velocities is considered a prerequisite for surviving drag forces, but this also causes self-shading and lower photosynthetic rates at higher velocities. Drag forces on individuals are also reduced by growing in an aggregation. The blades of kelps of wave-exposed seaweeds are more streamlined compared to wave-sheltered sites.

Wave action causes disturbance to seaweed populations, creating patches of space. The frequency and intensity of disturbance will affect community succession, as does the supply of propagules from adjacent areas. Water motion may involve sediment movement. Whereas sediment is generally deleterious to algae, some species are able to tolerate long periods of sand burial, and others gain a competitive advantage in sand-stressed areas.

Pollution

9.1 Introduction

There are several categories of marine pollution. In this chapter, six categories of pollution encountered by macroalgae are discussed: metals, such as mercury, lead, cadmium, zinc, and copper; oil; synthetic organic chemicals, such as pesticides, industrial chemicals, and antifouling compounds; eutrophication (excessive nutrients, such as nitrogen or phosphorus); radioactivity, and thermal pollution. In the 1970s and '80s, marine pollution was a hot topic and therefore some of these important early references have been retained. Over the last two decades, research on metals and eutrophication has been particularly active, followed by oil, antifouling paints, and organic wastes. There has been very little research on thermal pollution, even though it could be a good surrogate at local sites for assessing the potential long-term effects of global warming/climate change. Emerging anthropogenic issues are ocean acidification (see sec. 7.7 and Essay 4) and nanoparticles such as titanium dioxide (TiO_2) (Miller *et al.* 2010, 2012).

9.2 General aspects of pollution

Several general considerations apply to studies on pollutants. Among these are the choice of test organisms, whether to study chronic or acute effects, the level of the effect such as lethal or sub-lethal, the complexities at various levels of organization from physiology to communities, and the issue of what is the biologically available quantity and form or species of the pollutant.

Overall effects of a compound are assessed by acute or chronic exposure. Acute effects are the result of short-term exposure (e.g. 48–96 h) and are determined from the percent survival of an organism over a range of toxin concentrations. Chronic effects are the result of exposure for a relatively long time (e.g. 10% of the organism's life span or longer (Walker *et al.* 2006)).

Assessment of pollution effects is complicated because there are many sources of pollutants that enter the coastal zone and various biogeochemical processes modify their concentration and bioavailability at several organizational levels (Figs. 9.1 and 9.2). At biochemical and physiological levels, effects of a contaminant can result in reduced phenotypic fitness. Pollution studies at the population level have focused on recruitment, mortality, size and age structure, and biomass and population production. Detecting pollution effects at the community level is the most complex because of the great variability in the time and space scales involved (Fig. 9.1) (Underwood 1992; Laws 2000).

To date, two approaches have been attempted at the community level: (1) Descriptive statistics. This involves describing the number of species detected per sample, the abundances of individual species, and the biomass, and then summarizing this information in the form of measures of diversity and species richness. Under conditions of infrequent disturbances, competition between species will result in competitive displacement, whereby a few competitively superior species will dominate the community, and species diversity will be relatively low. If the community is subjected to disturbance by pollution, the competitive equilibrium

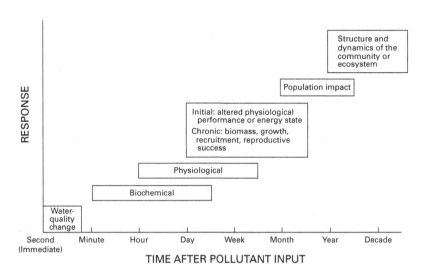

Fig. 9.1 Hypothetical time-related sequence for the potential effects of pollutant input, observed at various levels of biological organization (From Hood *et al.* 1989, with permission of E. W. Krieger Publishing Co.).

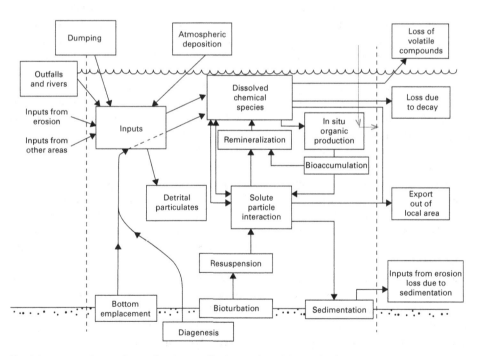

Fig. 9.2 Important biogeochemical processes affecting the fate of chemical pollutants in marine ecosystems. Inputs are from the sediments, horizontal advection, erosion, dumping, and the atmosphere (wet and dry deposition). Losses on the right side are volatilization, decay, export and sedimentation. (Redrawn from Libes 1992.)

will be disrupted, and species diversity will increase. At higher levels of disturbances, species will be eliminated from the community, and diversity will decline. (2) Patterns of species abundance and biomass. Various analyses to rank abundance and diversity have been proposed to measure community effects. One of the most common techniques is multivariate analysis. This procedure attempts to group co-occurring species into multivariate clusters based on a variety of organismal characteristics (Underwood 1992). The difficulty in detecting low-level chronic effects at the community level is well recognized. In most cases, for community effects to be detected, the disturbance is usually a specific event (e.g. oil spill), or polluted effluent from a point source. The need for well-designed impact assessments with repeated sampling and comparison with multiple control or reference sites is essential. This allows natural temporal and spatial variations to be statistically separated from putative impacts (Underwood 1992, 1994).

The adverse effects of pollutants on aquatic organisms generally are identified in terms of their lethal and sub-lethal impacts. Mortality can be readily recognized and quantified, but sub-lethal low-level chronic effects are more difficult to detect. The possible sublethal responses of an organism can be categorized according to the effects on the organism's (1) biochemistry/physiology, (2) morphology, (3) behavior, and (4) genetics/reproduction. There is debate about how to measure responses to sub-lethal concentrations of a pollutant, as well as whether laboratory bioassays give meaningful results (Bayne 1989) and whether the responses observed in the laboratory can be extrapolated to the more natural and varied conditions in the oceans (White 1984). Because of the simplicity of laboratory experimental conditions in comparison with the complexity of the marine environment, there is a need to also make observations directly in the field and take into account seasonal and interannual temporal variability. This could be done, for example, by attaching macrophytes to artificial substrates and putting them in polluted areas. Long-term monitoring of community structural changes in the field is required for a minimum of several years to take into account interannual variability and to assess effects at low pollutant concentrations. Because the bulk of most long-lived pollutants will often end up in the sediments, the mechanisms and dynamics of uptake and release of pollutants from sediments and their transfer to biota needs further investigation (Ahlf et al. 2002). Remote sensing using infrared photography or spectral radiometry is now a powerful new tool to rapidly survey large areas of macrophytes and document major changes over several years. In reality, both laboratory and field measurements are necessary. In order to bridge the gap between the two areas, some laboratory facilities are scaled up and taken into the field to conduct mesocosm experiments (Boyle 1985). Mathematical models may also be used to evaluate data obtained from tests with different endpoints, which take uncertainties of biotest systems into account where the variability and ecological relevance of biotests are incorporated (Keiter et al. 2009).

The *total concentration* of a contaminant may give little indication of its toxicity. Presently we cannot accurately estimate how the biological consequences of the physico-chemical aspects of contaminants will affect the toxicity, because only living systems can integrate the effects of those variables that are biologically important (Philips 1990; Eklund et al. 2010). These aspects include solubility, adsorption, and chemical complexation and speciation (Higgins and Mackey 1987); because of such interactions, only a small portion of the pollutant may be biologically available.

In an evaluation of toxicity, usually a battery of tests is used with relevant representatives for both the primary producers, primary consumers, and secondary consumers. The most commonly used test for the primary producers is growth inhibition of marine microalgae (ISO 10253:2006). Macroalgae represent an important part of the ecosystem and today a number of macroalgae originating from different parts of the world are used for testing single substances, effluent waters, and other complex mixtures. In the USA, the chronic reproduction test with the red macroalga *Champia parvula* has been used since the 1980s (Thursby 1984) and it is a standard within the Americam Society for Testing and Materials (ASTM) (ASTM 2004). The green alga *Ulva pertusa* test is based on the inhibition of sporulation by a pollutant (Han and Choi 2005; Han *et al.* 2008, 2009) and is used in Korea and Australia. The effects on germination of the brown alga *Hormosira banksii* have been proposed for regulatory testing (Seery *et al.* 2006; Myers *et al.* 2007). The growth inhibition test with the red alga *Ceramium tenuicorne*, common in temperate waters in both the northern and southern hemispheres has been developed (Bruno and Eklund 2003) and since 2010, it is a standard test within both ISO (ISO 2010) and Europe.

For algae, growth rate (often estimated by photosynthetic rate, change in dry weight, fluorescence, etc.), germination, or germling growth are used instead of survival (mortality) that is used for animals, since death is difficult to accurately determine in macroalgae. Other endpoints such as sporulation success and reproduction may also be used. In toxicity testing, the response is reported as the concentration where a 50% reduction in growth rate occurs compared to the control and is called the 50% effective concentration (EC_{50}) (Fig. 9.3).

The choice of a test organism is often based on its ease of handling and culturing and the relevance to the investigation (i.e. is it an ecologically important species?). Depending on the objective of the investigation, different strategies may be used. When the aim is to rank the toxicity of a particular chemical, the most sensitive alga should be used. However, if the aim is to make a proper evaluation of possible effects in a certain ecosystem and provide information on the survival of populations, the organism's full life cycle must be considered. The organism's response must ultimately be related to the healthy progression through its full life cycle, including the most sensitive stages such as germination, good growth of both germlings and adults, and successful reproduction. Optimally several ecologically important species should be tested. The recovery process should also be examined, but unfortunately, it is rarely assessed. The possibility of pollution-induced sexual reproduction leading to the development of resistance through genotypic adaptation warrants further investigation in macroalgae.

It has been proposed that because some of the early stages of macroalgae are very sensitive, they can be used for toxicity testing. In the standard battery of

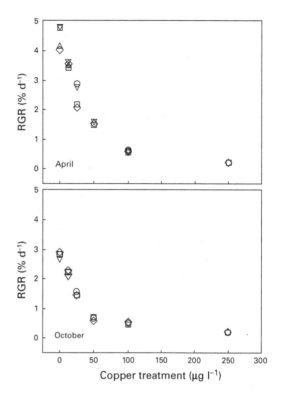

Fig. 9.3 Effects of exposure to Cu on the relative growth response (RGR) based on the length measurements of vegetative apical tips after 7 days exposure and expressed as % day^{-1} for *Gracilariopsis longissima* sampled from five populations in SW England. The effective concentration for a 50% reduction in growth (EC_{50}) was about 30 μg L^{-1}, while 12 μg L^{-1} caused an observable reduction in growth. (From Brown *et al.* 2012, reproduced with permission.)

bioassay tests that use microalgae and various invertebrates, there is no internationally recognized standard seaweed bioassay for toxicants. Han *et al.* (2007, 2008, 2009) have shown that the early life stages of *Ulva* have considerable potential for toxicity testing. They developed a quantitative assay for toxicants in sewage and waste sludge by assessing spore release, germination, and gametophyte growth by *Ulva*. The EC_{50} values for four metals were generally lower (i.e. more sensitive) than the standard Microtox test and five other standard test organisms (see Table 3 and Fig. 2 in Han *et al.* 2007). Since artificial induction of spore release in

Ulva is now routine, testing can be year round, fast (~7 h), cheap, and administered for a wide variety of toxicants. During reproduction, the shedding of motile cells is associated with a clearly visible change in thallus color from yellow-green in the normal vegetative state to dark olive to finally white, reflecting the release of reproductive cells. The inhibition of reproductive cell release can be assessed by estimating the percent area of the thallus shedding reproductive cells (see Fig. 1 in Han *et al.* 2007).

When conducting biotests it is important to be aware of their limitations. For example, the presence or absence of suspended particulate matter is known to profoundly influence the effects of many pollutants through surface adsorption (Ytreberg *et al.* 2011a, b). Other environmental factors such as salinity, temperature, light, nutrient limitation, and the presence of other pollutants may act synergistically or antagonistically with the pollutant (Eklund 2005; Ytreberg *et al.* 2011b).

9.3   Metals

Metal pollution is one of the most active areas of pollution research. Topics that will be discussed in this section include; sources and forms, adsorption,

uptake, biomonitoring, metabolic effects, metal tolerance, interacting factors, and ecological aspects.

9.3.1   Sources and forms

Generally, metals such as Hg and Pb are non-essential for macroalgal growth (Fig. 9.4). Hg may be toxic at only 10–50 µg L^{-1} (Gaur and Rai 2001; Costa *et al.* 2011). However, some metals, such as manganese, iron, copper, and zinc, are essential elements/micronutrients and frequently are referred to as trace metals required for growth (Bruland and Lohan 2004; Sunda 2009) (see also secs. 6.5.5, 6.5.6). They may limit algal growth if their concentrations are too low, but they can be toxic at higher concentrations; frequently the optimum concentration range for growth is narrow (Fig. 9.4). The concentration and speciation of various trace metals in the open ocean are given in Table 9.1 and represent a relatively pristine concentration for comparison with the wide variation found in coastal and more polluted waters.

Metals in minerals and rocks can enter the water naturally through weathering of rocks, leaching of soils and vegetation, and volcanic activity. Therefore, in assessing marine pollution, a distinction must be made between natural sources and those due to human

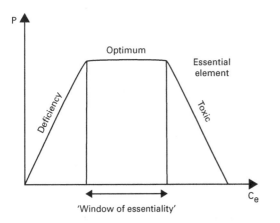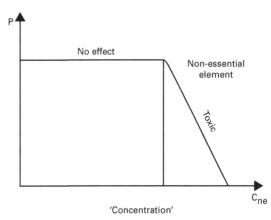

Fig. 9.4 Relationships between performance (P) (growth, fecundity, survival) and concentrations of an essential (C$_e$) or a nonessential (C$_{ne}$) element for macroalgae. At low concentrations, essential elements may reduce growth (i.e. they are deficient). The onset of toxicity usually occurs at much lower concentrations for non-essential compared to essential elements. (Redrawn from Walker *et al.* 2006, reproduced with permission.)

Table 9.1 Range of concentrations and main speciation of various trace metals in various areas of the open ocean. (Modified from Clark 2001.)

Metal		Area	Concentration (μg l^{-1})	Main species in aerated water	
				35% Salinity	10% Salinity
Silver	Ag	NE Pacific	0.00004–0.0025	$AgCl_2^-$	
Aluminum	Al	NE Atlantic	0.162–0.864	$Al(OH)_4^-$, $Al(OH)_3$	
		N. Atlantic	0.218–0.674		
Arsenic	As	Atlantic	1.27–2.10	$HAsO_4^{2-}$	
Cadmium	Cd	N. Pacific	0.015–0.118	$CdCl_2$, $CdCl_3$, $CdCl^+$	Cd^{2-}, $CdCl^+$
		Sargasso Sea	0.0002–0.033		
		Arctic	0.015–0.025		
Cobalt	Co	NE Pacific	0.0014–0.007	$CoCO_3$, Co^{2+}	Co^{2+}, $CoCO_3$
Chromium	Cr	E. Pacific	0.057–0.234	CrO_4^{2+}, $NaCrO_4^-$	CrO_4^{2+}
Copper	Cu	Arctic	0.121–0.146	$CuCO_3$, Cu-organic	Cu-humic, $Cu(OH)_2$
		Sargasso Sea	0.076–0.108		
Iron	Fe	Arctic	0.067–0.553	$Fe(OH)_3$, $Fe(OH)_2^+$	
Mercury	Hg	N. Atlantic	0.001–0.004	$HgCl_4^{2-}$, $HgCl_3^-$	Hg-humic, $HgCl_2$
Manganese	Mn	Atlantic	0.027–0.165	Mn^{2+}, $MnCl^+$	Mn^{2+}
		Sargasso Sea	0.033–0.126		
Nickel	Ni	Arctic	0.205–0.241	$NiCO_3$, Ni^{2+}	Ni^{2+}, $NiCO_3$
		Sargasso Sea	0.135–0.334		
Lead	Pb	Central Pacific	0.001–0.014	$PbCO_3$, $PbOH^+$	
		Sargasso Sea	0.005–0.035		
Antimony	Sb	N. Pacific	0.092–0.141	$Sb(OH)_6^-$	$Sb(OH)_6^-$
Selenium	Se	Pacific and Indian	0.044–0.170	SeO_4^{2-}, SeO_3^{2-}	
Tin	Sn	NE Pacific	0.0003–0.0008	$SnO(OH)_3^-$	
Vanadium	V	NE Atlantic	0.83–1.57	HVO_4^{2-}, $H_2VO_4^-$	
Zinc	Zn	N. Pacific	0.007–0.64	Zn^{2+}, $ZnCl^+$	Zn^{2+}
		Sargasso Sca	0.004–0.098		
		Arctic	0.056–0.225		

activities. Humans contribute metals to the environment during a variety of pursuits such as mining and smelting ores, burning fossil fuels, disposing of industrial waste, and processing raw materials for manufacturing. Most of the dissolved or particulate metal input is transported by water and reaches the oceans via rivers or land runoff. Atmospheric contributions via dry and wet deposition can carry Cd, Cu, Zn, and Pb to the oceans, especially near heavily populated coastal areas. These metals in the atmosphere come from the burning of fossil fuels and human transport systems. Metals in sediments may be reduced or oxidized, primarily by bacteria, and released into the overlying water (Bruland and Lohan 2004).

Metals in an aquatic environment may exist in dissolved or particulate forms. They may be dissolved as free hydrated ions or as complex ions (complexed with inorganic ligands such as OH-, Cl⁻, or CO_3^{2-}) or they may be chelated with organic ligands such as amines, humic and fulvic acids, and proteins. Particulate forms may be found in a variety of situations: as colloids or aggregates (e.g. hydrated oxides); adsorbed onto particles; precipitated as metal coatings onto particles; and incorporated into organic particles such as algae (Gaur and Rai 2001; Sunda 2009). The physical and chemical forms of metals in seawater are controlled by environmental variables such as pH, redox potential, ionic strength, salinity, alkalinity, the presence of

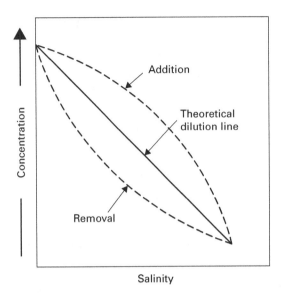

Fig. 9.5 Relationship between the concentration of a dissolved pollutant and salinity (a conservative index of mixing) when the river water concentration of the pollutant is greater than the seawater (solid straight line) and when there is addition or removal (biological uptake or chemical processes) of the pollutant. (Redrawn from Libes 1992, reproduced with permission.)

dissolved organic and particulate matter and colloids, and biological activity, as well as by the intrinsic properties of the metal. Changes in these variables can result in the transformation of the metal's chemical forms and can contribute to their availability, accumulation, and toxicity (Mance 1987; Gaur and Rai 2001; Yterberg *et al.* 2011a, b).

In coastal waters, the concentrations of heavy metals show a linear decrease with distance from river mouths as illustrated by typical mixing diagrams (Fig. 9.5). Deviations below the theoretical linear dilution line are due to removal by biological and/or chemical processes, while deviations above the dilution line are due mainly to an addition of the metal by chemical processes. An example of removal by chemical processes is the salting-out process of high molecular weight fractions and flocculation of inorganic matter as salinity increases; metals may adsorb to these newly formed particles and sink to

the sediments. On the other hand, an addition of the metal could occur when some metals previously attached to particles in the river water are displaced by chloride ions and become potentially available for uptake. For further information on the biogeochemistry of metals see Libes (1992), Laws (2000), and Gaur and Rai (2001).

9.3.2 Adsorption, uptake, accumulation, and biomonitors

Even if metal concentrations are not high enough to affect macroalgal growth, they can concentrate metals several orders of magnitude greater than the ambient environment and this process is called bioconcentration or bioaccumulation since the source of the contaminant comes from water. When the contaminant in the seaweed is a food source for an herbivore, it increases in concentration in the herbivore and higher trophic levels in a process called biomagnification (Gray 2002; Walker *et al.* 2006).

Metals are taken up both passively and actively by algae. Some metals such as Cu, Pb, and Sr, may be passively adsorbed by charged polysaccharides in the cell wall and intercellular matrix (Eide *et al.* 1980; Toth and Pavia 2000b). Other metals (e.g. Zn, Cu, Cd) are taken up actively against large intracellular concentration gradients (Gledhill *et al.* 1997; Burridge and Bidwell 2002). Macrophytes concentrate metal ions from seawater by several orders of magnitude. Good accumulators, i.e. those that do not metabolically regulate metals and therefore demonstrate an almost linear relationship between the uptake and exposure concentration, make good bioindicators (Shimshock *et al.* 1992; Rainbow and Phillips 1993).

There are several reasons to use seaweeds as indicators of metal contamination (Philips 1990; Rainbow and Phillips 1993; Rainbow 1995). The first is that dissolved metal concentrations in seawater are often near the limits of analytical detection and may be variable with time. Metal accumulation by seaweeds integrates short-term temporal fluctuations in seawater. Second, since seaweeds do not ingest particulate-bound metals (as animals do), they will accumulate only those dissolved metals that are

Table 9.2 Concentration of four metals in seawater and brown seaweeds and their tissue concentration factors.

Heavy metal	Concentration in seawater (μg L^{-1})	Fucus vesiculosus		Ascophyllum nodosum	
		Concentration (ppm)	Concentration factor^a ($\times 10^3$)	Concentration (ppm)	Concentration factor ($\times 10^3$)
Zn	11.3	116	10	149	13
Cu	1.4	9	6.4	12	8.6
Mn	5.3	103	19	21	3.9
Ni	1.2	8	6.8	5.5	4.6

^{a} Concentration factor = ppm dried seaweed per microgram of dissolved metal per milliliter of seawater.
Source: Modified from Foster (1976), with permission of Applied Science Publishers.

biologically available (assuming the degree of adsorption is slight. The metals in the algal tissues are easily analyzed by atomic absorption spectrometry and more recently by ICPMS (inductively coupled plasma mass spectrometry) or OES (optical emission spectrometry). Many seaweeds have been used as indicators of trace metal pollution (Phillips 1991; Hou and Yan 1998): browns (e.g. *Fucus*, Bond *et al.* 1999; *Ascophyllum*, Stengel and Dring 2000), greens (*Ulva*, Villares *et al.* 2001, 2005), and reds (*Ceramium*, Eklund 2005; *Pyropia* (previously called *Porphyra*), Leal *et al.* 1997). There is no general agreement as to which species are the best biomonitors. Ho (1990) found that the cosmopolitan green alga *Ulva lactuca* was a good bioindicator for Cu, Zn, and Pb pollution because of its high accumulation capacity. For similar reasons, Forsberg *et al.* (1988) found *Fucus vesiculosus* to be a good detector of metal pollution.

The variation in the background concentration (i.e. no obvious exposure to trace metal pollution) of eight trace elements for brown, red, and green seaweeds is given in Figure 9.6. The concentration factor is the concentration of a metal in a seaweed (as μg dry wt) divided by the metal concentration in seawater and if this factor is constant over months and seasons, then an estimate of the long-term change in the concentration in seawater is obtained. Examples of concentration factors for four metals in two brown seaweeds are given in Table 9.2. The concentration factors for various elements generally range from 10^3 to 10^4. However, there may be considerable variation in the

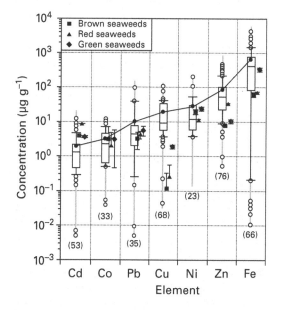

Fig. 9.6 Average concentrations (± 1 SD) of seven trace elements in brown, green, and red seaweeds from areas lacking significant local anthropogenic inputs illustrated using box and whisker plots. Open circles represent values beyond the 90th and the 10th percentiles; gray lines cover the ranges from minimum to maximum values and filled circles represent average concentrations. The numbers in parentheses indicate the number of values included in the calculations of the average metal concentrations. Note the log scale on the *y*-axis. (From de León-Chavira *et al.* 2003, reproduced with permission.)

concentration factor between different species at the same site and even within the same species (Villares *et al.* 2001; Stengel *et al.* 2004). Zn accumulation rates were the most variable in reds compared to browns and some greens, possibly due to their large variation in morphology and surface-area:volume ratio (Stengel *et al.* 2004). The high variation in tissue metals is related to the size, age, part of the seaweed analyzed, reproductive and nutritional state, removal of epiphytes and adhering surface particles (Phillips 1990; Shimshock *et al.* 1992; Villares *et al.* 2001). The high variation in metal concentrations in water may be due to temperature, salinity, suspended load, other pollutants, source of metals, seasonal variation, pH, chelators, dissolved organic matter, etc.

The problems inherent in assessing the magnitude of metal pollution in the field are discussed next. Since seaweeds may *not* always accurately reflect metal concentrations in the surrounding water, direct measurements of trace metal concentrations in the water should also be conducted periodically as a check on the validity of the seaweed biomonitoring data. For further discussion on the necessary precautions in interpreting biomonitoring data and tissue concentration, see the reviews by Phillips (1990), Burridge and Bidwell (2002) and Eklund and Kautsky (2003).

Any factor that tends to alter growth rate such as light, temperature, salinity, nutrient limitation, season, tidal level, turbidity, other pollutants, etc. can alter the alga's effectiveness as a bioindicator or bioaccumulator (Phillips 1990; Eklund and Kautsky 2003; Connan and Stengel 2011). For example, copper uptake was higher in winter than summer, while there was no seasonality in uptake of Pb and Cd in *Pyropia* and *Ulva*, possibly due to different types and amounts of algal exudates present in different seasons (Vasconcelos and Leal 2001).

Wide variation in the results from biomonitoring studies are often associated with the choice of seaweed and its characteristics (i.e. the species, size, age, part of the seaweed analyzed, reproductive and nutritional state, removal of particles and epiphytes adhering to its surface). Generally, old tissue retains metals longer than young tissue when ambient metal concentrations decrease (i.e. the depuration rate is slower). Some

investigators have suggested that the analysis of different aged tissues may be advantageous in determining the contamination history of a site (Stengel and Dring 2000). Higgins and Mackey (1987) found that pretreatment of the kelp *Ecklonia radiata* with an EDTA wash which chelates metals, led to a release of 90% of its total Zn and Cd, 25% of its Cu, and 7% of its Fe.

In turbid waters, metal-contaminated suspended particulates may stick to the mucilaginous surface of the thallus that is difficult to eliminate by washing or brushing prior to analysis (Gledhill *et al.* 1998; Villares *et al.* 2001). This problem is particularly severe for metals that are closely associated with particulates, such as Fe, Pb, and Cr (Barnett and Ashcroft 1985). Epiphytes may also be difficult to remove from a mucilaginous surface and thus this could overestimate the actual metal content attributed to the actual macroalgal tissue since epiphytes also accumulate metals (Stengel and Dring 2000). For example, zinc was found to be mainly associated with extracellular polymers produced by epiphytes and bacteria on *Gracilaria chilensis* and little was incorporated into the macrophyte (Holmes *et al.* 1991). The seaweed may clean its own surface by frequent epidermal shedding in young tissue of *Ascophyllum* and this process periodically lowers the tissue metal concentration and contributes to the high temporal variability that is observed (Stengel and Dring 2000; see sec. 4.2.2). Therefore, there is a need for standard methods for removing particles and attached epibionts (microalgae and bacteria) from different types of thallus tissue without removing some thallus surface tissue. Gledhill *et al.* (1998) compared eight techniques and found that applying a 1:9 ethanol:seawater mixture followed by gentle scraping gave the lowest metal values.

As an alternative to the many problems and uncertainties arising from the passive biomonitoring approach, Brown *et al.* (2012) suggested an "active" approach. They showed that *Gracilariopsis longissima* has constitutive copper resistance and an ability to take up and release copper, making it a good seaweed to use for biomonitoring. The growth rate of seaweeds that were collected in two different seasons from five sites with varying copper contamination was assessed in the laboratory at a range of copper concentrations.

There was no difference in their growth rates despite having been exposed to widely different copper concentrations at the five contaminated sites and their EC_{50} was ~30 μg L^{-1} (Fig. 9.3). Therefore, this species did not evolve copper-tolerant ecotypes expressing different degrees of resistance, suggesting that it has a constitutive copper tolerance. Another ideal characteristic of this species was that it lost 80% of its copper in 8 days when it was transferred to "clean" seawater (i.e. a good depuration response). In addition, it responded well to *in situ* transplant experiments to natural sites with varying copper contamination. In transplant experiments that compared the response of Cu and Cd, Andrade *et al.* (2010) found that large changes in tissue Cu took place in hours, while Cd changes occurred over weeks, suggesting that the transplant biomonitoring approach is more useful for Cu than for Cd. Therefore, the young tissue of an ideal species used in transplant experiments should be able to take up and release metals quickly, but this will vary with the metal. This active biomonitoring approach deserves further comparison to the passive approach that is often not particularly quantitative, given the large number of factors discussed above that affect the final metal concentration factor for a particular species.

9.3.3 Mechanisms involving tolerance to toxicity

There are comparatively fewer studies on the tolerance to metal toxicity conducted on seaweeds, and therefore some general findings from microorganisms and higher plants (Stauber and Florence 1987; Hall 2002) will be discussed to illustrate basic principles (Fig. 9.7). The following section covers extracellular, cell surface, and intracellular mechanisms used by seaweeds/cells to detoxify various metals and reduce toxicity effects. Since Cu is a very common toxic metal and is widely used in antifouling paints, some examples below will focus on Cu.

Detoxification of metal ions outside of the cell or at the cell surface is referred to as an exclusion mechanism, because the metal ions do not cross the cell membrane (Fig. 9.7). Macroalgae may produce

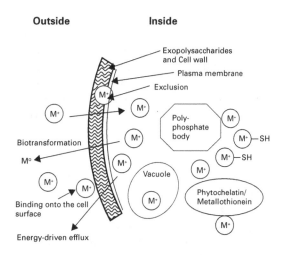

Fig. 9.7 A schematic showing some common mechanisms of metal (M) tolerance in macroalgae. Tolerance may be a result of binding onto the cell surface, biotransformation, efflux, exclusion, binding to polyphosphate bodies, phytochelatin or sulfhydryl groups (SH) (Redrawn from Gaur and Rai 2001.)

extracellular and/or wall-bound chelating compounds (Gekeler *et al.* 1988; Gledhill *et al.* 1999) that can bind to certain metal ions, rendering them non-toxic. Carrageenan from *Kappaphycus striatum* and fucoidan from *Ascophyllum nodosum* have been shown to be capable of effectively binding metals such as Cd, Pb, and Sr through an ion exchange mechanism. The metal binding capacity of carrageenan was found to be correlated with its degree of sulfation. Another exclusion mechanism, not yet observed for macrophytes, is the adsorption or detoxification of a metal ion by surface-living microorganisms that form a biofilm. Possibly epiphytes such as diatoms or bacteria on the surface of the seaweed may act like a biofilter and sequester the metal ion before it reaches the seaweed's membrane (Riquelme *et al.* 1997). Stauber and Florence (1987) found that when trivalent metal ions (Al, Fe, and Cr), or divalent metals (Mn and Co) that could be oxidized by algae to trivalent species, were added to the culture medium, they protected two marine diatoms against Cu toxicity by forming a layer of hydrated metal (III) oxide around the cell. This layer was thought to adsorb Cu ions before they could penetrate the cell. This important study by Stauber and Florence (1987) establishes a mechanism for the previously reported antagonistic effects (Stauber and Florence 1985) of some metals, such as Mn, Fe, and Co, on Cu toxicity (i.e. an increase in Mn decreases Cu toxicity). Nickel, whose chemistry is similar to that of Co, cannot be oxidized in seawater to Ni(III) hydroxide, and therefore it is totally ineffective for protecting against Cu toxicity; similarly, Zn(II) is ineffective. The protective effects of Mn and Co are enhanced by their ability to scavenge damaging superoxide radicals and H_2O_2 (Pinto *et al.* 2003; Lesser 2006; Bischof and Rautenberger 2012).

If a weak extracellular exclusion mechanism is operating, then a metal ion can be transported across the membrane and enter the cytoplasm. If the metal remains as a free ion, then it can exert its toxic effect (Fig. 9.7). Detoxification mechanisms inside the cell include; binding to various compounds with sulfhydryl and carboxyl groups, chelation with intracellular ligands/phytochelatins. Storage in the vacuole is also a commonly used detoxification mechanism. Conversion to different inorganic or organic species is possible, especially for Hg. The tolerance of *Lessonia* spp. to high Cu concentrations was due to the capacity of different seaweed tissues to sequester Cu in the form of precipitates in the cell wall and periplasmalemmal space with minor amounts in the vacuoles (Leonardi and Vasquez 1999). In contrast, in *Enteromorpha flexuosa* (now referred to as *Ulva flexuosa*), most of the copper was found in copper-rich precipitates in the vacuoles (Andrade *et al.* 2004). In upwelling areas where P and Cd concentrations are naturally high, *Macrocystis pyrifera* stored P in polyphosphate bodies (Fig. 9.7) with accompanying high cadmium concentrations (Walsh and Hunter 1992). Phytochelatins (PCs) are a type of metallothionein protein and a metallothionein gene has been identified in *Fucus vesiculosus* that was induced by copper exposure and whose encoded protein product binds cadmium and copper (Morris *et al.* 1999). These metal complexing thiol peptides are produced by some seaweeds that live in metal-contaminated areas (Fig. 9.7). PCs are involved in metal homeostasis and detoxification, and tolerance varies between species growing in the same environment and the variation depends on thallus morphology, polysaccharide composition, intracellular metal accumulation, and the production of the precursor for PC production, glutathione (Pawlik-Skowrońska *et al.* 2007). Glutathione may also be used to reduce oxidative stress and complex with certain metals (Pinto *et al.* 2003). Phlorotannins (brown algal polyphenolics) are strong chelators of metals found in special compartments (physodes) within cells and are potentially capable of detoxifying metals. However, Toth and Pavia (2000b) found that phlorotannin production in *Ascophyllum nodosum* was not induced by high Cu concentrations and concluded that binding of Cu to phlorotannins is probably not an important internal detoxification mechanism in this alga. They suggested that other substances that bind metals such as polysaccharides and/or phytochelatins may be more important for Cu resistance in *A. nodosum*.

In addition to their physiological tolerance to heavy metals, algal resistance to metals is likely to be genetically controlled. Copper tolerance in the ship-fouling alga *Ulva compressa* appears to be genetically determined, because the progeny was also found to be very copper tolerant (Reed and Moffat 1983). Similarly, Nielsen *et al.* (2003a) showed that copper tolerance in the progeny of *Fucus serratus* was at least partially inherited, but this contrasts to the findings of Correa *et al.* (1996) who concluded that the progeny of *Ulva compressa* did not exhibit genetically inherited copper resistance. Since algae can develop resistance to pollutants, the inhibitory effect of a pollutant may be an order of magnitude higher in seaweeds from polluted areas compared to ones from a reference site. For example, Marsden and DeWreede (2000) found that *Ulva* that was living in an acid mine drainage area developed resistance to high concentrations of Cu, but the mechanism was not resolved. Genetic adaptation to heavy metals in a wide range of aquatic organisms has been reviewed by Klerks and Weis (1987) and Hall (2002) for higher plants.

9.3.4 Effects of metals on algal metabolism

The order of metal toxicity to algae varies with the algal species and the experimental conditions, but generally the order is Hg > Cu > Cd > Ag > Pb > Zn (Gaur and

Rai 2001). Even though Hg is the most toxic metal, Cu toxicity has been the most active research area followed by Cd, Zn, and Pb. There have been few investigations of mixtures of metals to evaluate synergism or antagonism and factorially designed experiments investigating the interactions between metal toxicity and environmental factors such as salinity, temperature, light, nutrient limitation, etc., but more of these environmentally realistic combinations need to be examined.

Mercury is the most toxic metal and interacts with enzyme systems and inhibits their functions, especially enzymes with reactive sulfhydryl (-SH) groups (Van Assche and Clijsters 1990). The toxic effects of mercury on algae generally include: (1) cessation of growth in extreme cases, (2) inhibition of photosynthesis, (3) reduction in chlorophyll content, and (4) increased cell permeability and loss of potassium ions from the cell (Gaur and Rai 2001). In an early study, Hopkin and Kain (1978) studied how the different life-history stages of *Saccharina hyperborea* (previously called *Laminaria hyperborea*) were affected and found that the growth of gametophytes was most sensitive. A study of the effects of Hg on increases in length for five intertidal Fucales showed that exposure to a Hg concentration of 100–200 µg L^{-1} for 10 days gave a 50% reduction in growth rate (Strömgren 1980b). There was no effect on growth rate of *Ulva lactuca* at 5 µg L^{-1}, a significant reduction at 50 µg L^{-1} and death at 500 µg L^{-1} (Costa *et al*. 2011). It was suggested that the very rapid uptake rates of Hg by *Ulva*, make it an ideal macrophyte for phytoremediation (see sec. 10.9).

Copper is the second most toxic metal even though it is an essential micronutrient at lower concentrations (Fig. 9.4). Copper toxicity is dependent on the ionic activity (concentration of the free ionic species Cu^{2+}), and not the total copper concentration (Gledhill *et al*. 1997, 1999). The chemical form (speciation) of Cu is an extremely important consideration when conducting toxicity experiments and this issue is well reviewed by Gledhill *et al*. (1997). Therefore, laboratory toxicity testing should be carried out in AQUIL, a widely used chemically defined seawater medium for trace metal studies (Morel *et al*. 1979). However, some organic copper complexes (especially the lipid-soluble ones)

are much more toxic than ionic Cu (Stauber and Florence 1987), because these lipid-soluble complexes can diffuse directly through the membrane. Many algae release Cu-complexing exudates with stability constants varying between 10^6 and 10^{13}, comparable to stability constants for the weak Cu-complexing ligands observed in normal seawater (Gledhill *et al*. 1997). These exudates/ligands likely control the speciation of Cu in seawater and therefore further research on the production of macroalgal exudates is required under a range of environmental conditions.

The cellular effects of Cu are widespread. First, copper affects the permeability of the plasmalemma, causing loss of K$^+$ from the cell and changes in cell volume. The initial Cu binding to the cell may be to carboxylic and amino residues in the membrane protein, rather than to thiol groups, because the Cu-alga stability constant is orders of magnitude lower than the thiol Cu-binding constant. At the cell membrane, Cu may interfere with cell permeability or the binding of essential metals. Following Cu transport into the cytosol, Cu may react with -SH enzyme groups and free thiols (e.g. glutathione), disrupting enzyme active sites. Typical symptoms of Cu-induced oxidative stress in *Lessonia* and *Scystosiphon* are the production of H$_2$O$_2$, superoxide anions, lipoperoxides as well as the activities of the antioxidant enzymes, catalase, glutathione peroxidase, ascorbate peroxidase, dehydroascorbate reductase, and glutathione reductase (Pinto *et al*. 2003; Contreras *et al*. 2009; Bischof and Rautenberger 2012; see also sec. 7.8). The identification of Cu-induced genes in *Ulva* has recently been reported (Contreras-Porcia *et al*. 2011). Cu may also exert its toxicity in subcellular organelles, interfering with mitochondrial electron transport, respiration, and ATP production.

Copper readily inhibits the key cellular process of photosynthesis by uncoupling electron transport to NADP$^+$. As the ionic concentration increases, copper is bound to plastid membranes and other cell proteins, causing degradation of chlorophyll and other pigments. At still higher concentrations, copper produces irreversible damage to plastid lamellae, preventing photosynthesis, and eventually causing death (Küpper *et al*. 2002). DNA replication checkpoints that control spindle alignment and ensure correct cell division in *Fucus* may be principal targets for Cu during algal growth. Nielsen *et al.* (2005) concluded that the inhibitory effects of Cu on photosynthesis are similar to those caused by photoinhibition since very high irradiances affect the photosystem II reaction center and reduce the quantum yield in *Fucus serratus*. Nielsen and Nielsen (2010) found that the susceptibility to Cu toxicity was not dependent on light adaptation and suggested that non-photochemical quenching mediated by the xanthophyll cycle may be involved. Experiments with *Fucus* zygotes showed very strong inhibitory effects of Cu on secretion of cell wall components that are required for cell expansion (Nielsen *et al.* 2003b). The fixation of the polar axis in the developing zygote was inhibited by Cu since it disrupted the calcium signaling process and hence development was curtailed. Since pulse amplitude fluorescence (PAM) (see secs. 5.7.1 and 7.9) reflects the interactions of PSII, membrane degradation, and photosynthetic electron-transfer efficiency, it can be used to assess the damage to the photosynthetic apparatus (Baumann *et al.* 2009). They found that the relationship between internal metal concentration and fluorescence was algal species and metal specific.

Early studies on *Saccharina hyperborea* showed that the effects of copper followed a pattern similar to that for mercury since gametophyte growth was more sensitive than sporophyte growth, but Cu was less toxic than mercury (Hopkin and Kain 1978). The growth of sporophytes of *Saccharina saccharina* was the most sensitive to Cu (>10 μg L^{-1}), followed by the release of meiospores, development of gametophytes ($\sim$50 μ L^{-1}), and settlement and germination of meiospores (500 μg L^{-1}) (Chung and Brinkhuis 1986). At 50 μg Cu L^{-1}, they found that sporophytes of *S. saccharina* showed abnormal growth patterns, haptera-like protuberances, giant cells, and abnormal branching patterns. Bioassays using *Macrocystis pyrifera* revealed that reproduction was three-times more sensitive to Zn than was zoospore germination, but half as sensitive as germ-tube elongation (Anderson and Hunt 1988). Metals may inhibit reproduction by interfering with the ability of the sperm to find the egg, perhaps via a pheromone that is thought to be

involved in this process (Maier and Muller 1986). *Ulva* rhizoid regeneration gave variable results, and the growth of *Ulva* discs was not as sensitive as *Fucus* eggs (Scalan and Wilkinson 1987). The effects of copper were studied on egg volume, fertilization, germination, and development of apical hairs of *Fucus vesiculosus* from the Baltic Sea. Germination was the most sensitive stage and was affected by 2.5 μg Cu L^{-1} (Andersson and Kautsky 1996). They found that a 30 min exposure of the eggs to 10 μg Cu L^{-1} before the spermatozoids were added, resulted in significantly lower fertilization. Nielsen *et al.* (2005) found that inhibition of spore and embryo rhizoid elongation were sensitive endpoints in bioassays which may be affected by prior exposure of the parent seaweed to Cu. In toxicological tests where Cu was applied to spores or zygotes immediately after settlement and prior to germination about 1 h after release, rhizoid elongation was regarded as being a more sensitive endpoint than germination (Anderson *et al.* 1990; Bidwell *et al.* 1998; Bond *et al.* 1999). Recently, a much needed study was conducted by Brown and Newman (2003), where they compared various physiological indicators or endpoints of Cu toxicity in *Gracilariopsis* and found that relative growth rate was the most sensitive, down to 12.5 μg Cu L^{-1} (Fig. 9.8). At 250 μg L^{-1} (i.e. 20$\times$ the inhibiting concentration of growth rate), photosynthetic rate, measured as chlorophyll fluorescence (i.e. photosynthetic efficiency, F_v/F_m) and O$_2$ evolution was impaired at 250 μg L^{-1}. At even higher Cu concentrations (500 μg L^{-1}) ion leakage occurred, phycobilin concentrations decreased, and shrinkage of apical tips occurred (Fig. 9.8).

Copper concentrations in normal oceanic areas are only somewhat lower than potential toxic concentrations. The ratio of the lowest Cu concentration that was toxic for sporophytes of *S. hyperborea* to typical seawater concentrations, was only 3.3, much lower than 200 for Hg and 2000 for Cd. If coastal copper concentrations increase by only a small amount, they will become toxic to this species. Cu at 60–80 μg L^{-1} was found to be somewhat more toxic to the growth of four intertidal fucoids than Hg and far more toxic than Zn, Pb, or Cd (Strömgren 1980a, b). Cu toxicity increases as salinity decreases. The red alga *Ceramium*

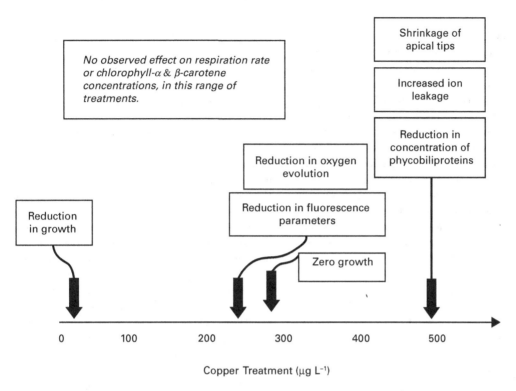

Fig. 9.8 Summary of the comparative effects of Cu concentrations between 0 and 500 µg L^{-1} on the growth and various physiological parameters/endpoints for *Gracilariopsis longissima* following exposure for 7 days. (From Brown and Newman 2003, reproduced with permission.)

tenuicorne showed a 50% growth inhibition at a salinity of 7 at ~3 µg Cu L^{-1} and at a salinity of 20, around 11 µg Cu L^{-1} (Eklund 2005), while *Ulva pertusa* was inhibited at 25–50 µg Cu $^{-1}$ (Han *et al.* 2008). Fielding and Russell (1976) showed that species in mixed culture gave different responses to copper than when grown in unialgal culture and warned that results from unialgal cultures may be misleading because they neglect possible species interactions.

In contrast to Cu, Cd pollution occurs where the concentrations are several orders of magnitude higher than seawater concentrations of ~0.1 to ~5 µg L^{-1} (Table 9.1). High Cd concentrations are found in upwelling areas that accounts for the naturally higher Cd concentrations in macroalgae, phytoplankton, and bivalves. The only known requirement of Cd is as a cofactor for the enzyme carbonic anhydrase and hence

it is sometimes regarded as an essential element. Cd is toxic at very high concentrations and is weakly complexed with organic ligands (Andrade *et al.* 2006). Although the mode of action is largely unknown, its high affinity for sulfhydryl- and oxygen-containing groups results in blocking the essential functional groups of various biomolecules. It can inhibit the uptake and transport of many macro/micronutrients and lead to nutrient deficiencies (Kumar *et al.* 2010b, 2012). Markham *et al.* (1980) reported on Cd uptake and its effects on growth, pigment content, and carbon assimilation in *Ulva lactuca* and *Laminaria saccharina*. The growth rate for sporophytes decreased by 50% at 2000 µg Cd L^{-1} (i.e. several orders of magnitude higher than most polluted areas). Generally, the activities of RuBP carboxylase, PEP carboxykinase, and mannitol-phosphate dehydrogenase were not affected by *in vitro* additions of Cd (Kremer and Markham 1982). They concluded that Cd inhibits one or more steps in protein synthesis and thus leads to enzyme deficiencies and a series of secondary effects. Andrade *et al.* (2010) found that tissue Cu was 200-fold higher in seaweeds from a Cu-contaminated site, but Cd was lower than the control site. In transplant experiments to a Cu-contaminated site, tissue Cu of *Lessonia* increased in a few hours, while a decrease in tissue Cd occurred over weeks. They suggested that antagonistic relationships between metals must be taken into account in the use of tissue metal concentrations in biomonitoring applications. In *Ulva lactuca*, a 0.4 mM Cd addition decreased growth rate and pigments and induced oxidative stress resulting in a twofold increase in lipoperoxidases and H_2O_2, an enhancement in antioxidant enzymes such as superoxide dismutase, ascorbate peroxidase, glutathione reductase, and glutathione peroxidase, and a decrease in catalase (Kumar *et al.* 2010b).

Very little research has been conducted on the less toxic heavy metals, such as Pb and Zn in seaweeds. Zn is similar to Cu since toxicity increased when salinity decreased (Eklund 2005). Significant reductions in growth occurred only at unrealistically high lead concentrations (10 mg L^{-1} as $PbCl_2$). Even though Zn is actively taken up by seaweeds and is an essential trace metal, it has a relatively low toxic effect. In an early study, Strömgren (1979) found that 5–10 g Zn L^{-1} reduced growth by 50% for five intertidal Fucales. In contrast, Cu and Hg toxicities occurred at 1% and Cd and Pb toxicities at 20% of that concentration.

9.3.5 Factors affecting metal toxicity

Adsorption of metals onto particles or complexation with dissolved organics generally will reduce toxicity of a metal by reducing the level of free ions. Exposure to Cu resulted in the rapid release of organic ligands by four species of seaweeds which influenced the bioavailability, bioaccumulation, toxicity, and cellular transport of copper (Andrade *et al.* 2010). It is now possible to divide the ligands into two classes based on their binding strength of Cu (Gledhill *et al.* 1999). When the kelp *Lessonia* was spiked with Cu, it produced large amounts of ligands in a few hours which reduced toxicity and hence, it has been suggested that the re-introduction of this kelp to a Cu-contaminated area might allow the re-establishment of other Cu sensitive seaweeds and invertebrates (i.e. a bioremediation approach) (Andrade *et al.* 2010). Abnormally high concentrations of dissolved organics from dispersed sewage occur near sewage outfalls and probably mitigate metal toxicity in those areas. Because the form in which the metal exists is difficult to characterize, most earlier studies have measured the total concentration of the metal, which does not correlate well with toxicity (Florence *et al.* 1984; Gledhill *et al.* 1997, 1999). This may explain why two studies examining the same *total* concentration of the metal in a particular alga may obtain quite different results. Both the pH and the redox potential can have considerable effects on the availabilities and thus the toxicities of metals (Guar and Rai 2001). At a low pH, metals generally exist as free cations, but at the alkaline pH of seawater, they tend to precipitate as insoluble hydroxides, oxides, carbonates, or phosphates.

The interactions of salinity and temperature with toxicity are not always clear (Munda and Hudnik 1988). Usually the metal concentration of seawater is lower than that of fresh water. Munda (1984) found that the Zn, Mn, and Co accumulations in *Ulva intestinalis* and *Scytosiphon simplicissimus* could be enhanced by decreasing the salinity and this should be considered in biomonitoring in an estuary. This could be associated with surface charge, because phytoplankton and probably seaweeds are negatively charged at low salinities. Salinity had only a minor effect on Cu toxicity in *C. tenuicorne* compared to organic matter concentration that had a significant effect in reducing the bioavailability and hence copper toxicity (Ytreberg *et al.* 2011a, b). Copper bioaccumulation by *C. tenuicorne* showed that this macroalga could access a sizeable fraction of organically complexed copper in addition to Cu^{2+}, when Cu^{2+} concentration to the cell membrane was diffusion-limited. An increase in temperature increased toxicity in some cases, but reduced it in other instances (Guar and Rai 2001). Increased toxicity at higher temperature may be explained by increases in the energy demand, which would result in enhanced respiration.

There can be synergistic or antagonistic interactions between metals and effects of other pollutants on the toxicities of heavy metals. Metal–metal antagonism was observed in a few studies. For example, selenium may relieve Hg toxicity (Guar and Rai 2001), and Mn or Fe may reduce Cu toxicity in various microorganisms (Stauber and Florence 1985; Munda and Hudnik 1986). Significant antagonistic effects were observed for the growth of *Ascophyllum nodosum* following exposure to Cu + Zn and also Hg + Zn (Strömgren 1980c). However, when two highly toxic metals such as Cu and Hg were added simultaneously, generally the toxic effects were additive. There are only a few examples of synergism between metals, such as the effects of Mn and Co on the growth of *Fucus vesiculosus* (Munda and Hudnik 1986).

9.3.6 Ecological aspects

Given the high biomass that is common in temperate intertidal areas, large portions of the non-sediment-bound metals can be associated with the macroalgae, which act as substantial buffers of these elements. Because most of the macroalgal production enters the detrital pool, the decomposition of macroalgal detritus can play a significant role in the cycling of trace metals in coastal waters. Detrital decomposition of the seaweeds may leach substantial amounts of metals, polyphenolics and dissolved organic carbon

such as organic ligands that are capable of forming strong complexes with Cu, Fe, and Zn. Therefore, the high biomass of kelp beds may play a major role in regulating both the concentrations and speciation of heavy metals in nearshore environments. Because of the chelating properties of cell wall constituents such as alginate and fucoidan of brown seaweeds, they could assist with remediation and the recovery of metals (Davis *et al.* 2003; Brinza *et al.* 2009).

Cu can have indirect top-down effects on macroalgae through their grazers. Contamination reduced the colonization of a variety of epifauna on *Sargassum*. The amphipod *Peramphithoe parmerong* showed less preference and lower grazing rates on Cu-contaminated *Sargassum* compared to controls (Roberts *et al.* 2006). Surprisingly, there was no reduction in the amphipod's growth rate. Amphipods store excess Cu that is accumulated from their diet in granules, rendering them non-toxic and then the Cu is gradually excreted. Examples of biomagnification can be found in heavily polluted areas. Studies on coastal waters with very high concentrations of Cd, Zn, and Pb showed that Cd and Zn were accumulated up the food chain. Table 9.3 shows that cadmium is found at relatively low levels in *Fucus* (the primary producer), at higher concentrations in the limpet *Patella* (herbivore), and in greatest concentrations in the carnivorous dog whelk *Thais* (i.e. the pollutant increases as the trophic levels increase).

There is also a human-health concern because of the ability of seaweeds to actively accumulate metals,

Table 9.3 Cadmium concentrations in seawater, seaweeds, and shore animals along a transect from the Severn estuary near the mouth of the Avon River to the Bristol Channel.

Location	Distance from Avon mouth (km)	Seawater (μg L^{-1})	*Fucus* (mg kg^{-1})	*Patella* (mg kg^{-1})	*Thais* (mg kg^{-1})
Portishead	4	5.8	220	550	—[a]
Brean	25	2.0	50	200	425
Minehead	60	1.0	20	50	270
Lynmouth	80	0.5	30	50	65

[a] Not reported.

Source: Butterworth *et al.* (1972); reprinted by permission from *Marine Pollution Bulletin*, vol. 3, pp. 72–4, © 1972 Macmillan Journals Limited.

especially with the increased consumption of commercial seaweeds for various health reasons. In 2004, the Food Standards Agency of the UK issued an advisory against eating the seaweed *hijiki*. Subsequently, Nakajima *et al.* (2006) found that the ingestion of one serving of *Hijikia fusiformis* containing about 100 μg of arsenic in various forms produced concentrations of As in urine that were equivalent to As poisoning. The symptoms of As poisoning are headache, drowsiness, muscle cramps, diarrhea, vomiting, and in severe cases coma and death. Arsenic is an essential component of the pyruvate dehydrogenase complex which oxidizes pyruvate to acetyl-CoA. Besada *et al.* (2009) tested heavy metals in many commercially available edible seaweeds and found that most *hijiki* samples had Cd and As concentrations greater that the recommended limit, and some *Pyropia* sp. had high Zn and Cu concentrations. Seaweeds are also used for fertilizing crops and therefore seaweeds from highly contaminated sites increase the risk of high metal concentrations in the crop plants.

9.4 Oil

The *Torrey Canyon* oil tanker spill off the UK in 1967 and a series of other oil spills in the 1970s set off an awareness of pollution in the marine environment. Today, this is an ongoing problem combined with inputs from tanker operations and a dramatic increase in offshore oil drilling. Over the last five decades there have been over 40 relatively large oils spills with about 25 of them occurring in the 15-year period of 1970–85 (Clark 2001). The *Amoco Cadiz* oil spill (223 000 metric tonnes) off the north coast of France in 1978 was one of the largest and the best studied tanker spills. Evaporation (30%) and stranding on shore (30%) accounted for 60% of the oil spilled. After 3 years, most of the obvious effects had disappeared, but high hydrocarbon concentrations remained in estuaries and marshes that initially had received large amounts of oil (Gundlach *et al.* 1983). Other well studied oil spills are the *Torrey Canyon* off the UK (Hawkins and Southward 1992), the *Exxon Valdez* off Alaska (Preston 2001) and the *Prestige* off northern Spain (Penela-Arenaz 2009). The concern

over oils spills was renewed by the *Deep Water Horizon* drill site leak in the Gulf of Mexico (Camilli *et al.* 2010). In the following sections, the composition, fate, toxicity, and ecological effects will be discussed.

The demand for oil has risen steadily and petroleum accounts for about 40% of the world energy production and >50% of the oil is used in the transportation sector. Tanker accidents and oil well blowouts, which make newspaper headlines, contribute only a small percentage of the total input (~3%), but they can be devastating in local areas. Half of the petroleum hydrocarbon input comes from transportation-related activities, from the discharge of waste oil from industrial and municipal sources and from the routine operations of oil tankers, especially via shipping routes out of the Middle East (Preston 1988; Clark 2001). Since the 1983 International Convention for the Prevention of Pollution from Ships (referred to as MARPOL), all new crude oil tankers have segregated ballast tanks and crude oil washing systems. These requirements eliminated the amount of oil discharged to the ocean by tankers that ballasted their cargo bunkers with water after they had off-loaded the oil. The greatest inputs of oil occur in coastal areas, which often are the most biologically productive. Oil spill detection is now relatively routine with the use of satellite remote sensing (Brekke and Solberg 2005). Oil pollution research was very active in the 1970s and '80s and hence this section includes many of the important earlier references during this period.

Petroleum, or crude oil, is an extremely complex mixture of hydrocarbons with some additional compounds containing O, S, N, and metals such as Ni, V, Fe, and Cu (Preston 1988). The composition of oil varies from one oil field to another and may vary during the lifetime of one oil field. The place of origin (e.g. Nigerian or Kuwaiti crude oil) of the crude oil can be determined by its unique characteristics using gas chromatography (called "finger printing"). Light oil has low sulfur, tars, and waxes, while heavy oil is high in waxes and tars.

Molecular arrangements include straight chains, branched chains, or cyclic chains, including aromatic compounds (with benzene rings) and they are classified into three broad categories:

1. The alkanes (Fig. 9.9a–c; saturated; single bonds; straight-chained or branched), whose general composition is C_nH_{2n+2}, make up 60–90% of the hydrocarbon content of oil. Methane is the simplest hydrocarbon. The presence of only saturated bonds makes them very resistant to degradation. The more branched the molecule (e.g. isobutane), the more difficult it is to biodegrade. Low molecular weight ($>C_6$) alkanes are generally gases (e.g. methane, propane, butane), and high molecular weight alkanes ($>C_{18}$) are solids (e.g. waxes, paraffins). The lower the number of carbon atoms, the more volatile and more water soluble the compound. Alkanes are relatively non-toxic.

2. The cycloalkanes or naphthenes (Fig. 9.9d–f), which are similar to alkanes except that some or all of the carbon atoms are arranged in rings. They are intermediate in their toxicity effects. These compounds have the general formula C_nH_{2n} and account for about 50% of crude oil, the most prevalent being cyclopentane and cyclohexane. Frequently, alkyl groups (e.g. -CH$_3$) are substituted on the cycloalkane ring, forming compounds such as methylcyclohexane. Polycyclic naphthenes are very resistant to microbial degradation.

3. The aromatics (Fig. 9.9g–i) contain one or more benzene rings, and the name comes from the pleasant aroma of these compounds. They are commonly found in crude oil or can be produced during refining; they include benzene, toluene, naphthalene, and phenol. Aromatics usually constitute <20% of the crude oil, but they are very toxic and some polycyclic aromatics (PCBs) are potent carcinogens. They are volatile and readily degraded. Other hydrocarbons, such as alkenes, occur in crude oil in much smaller amounts. They are unsaturated chain compounds possessing double or triple bonds, but without the regular arrangement found in the benzene ring. Examples include ethylene and acetylene which are produced during refining.

Oil pollution involves refined petroleum products as well as the natural crude oil. The refinery technique takes advantage of the fact that the boiling point of hydrocarbons generally increases with increasing molecular size. Therefore, crude oil can be separated

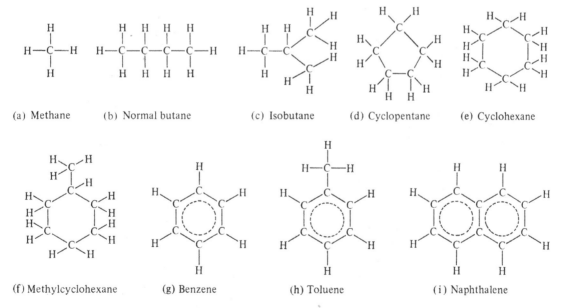

(a) Methane   (b) Normal butane   (c) Isobutane   (d) Cyclopentane   (e) Cyclohexane

(f) Methylcyclohexane   (g) Benzene   (h) Toluene   (i) Naphthalene

9.9 Three groups of hydrocarbons from crude oil; (a–c) alkanes, (d–f) cycloalkanes, or naphthenes, and (g–i) aromatics.

into different refined products by this process of fractional distillation. Refined products (gasoline, benzene, etc.) contain a higher percentage of low molecular weight products and their toxicity tends to be positively correlated with molecular size.

9.4.1   Fate of oil in the ocean

When oil enters the environment it undergoes changes due to weathering, including evaporation, dissolution, emulsification, dispersion, photo-oxidation, and biodegradation. The fate of the oil will depend on the type spilled and where it is spilled. Many refined petroleum products are also spilled, including gasoline, kerosene, fuel oils (no. 2, 3, 4, etc.) and lubricating oils. The main physical, chemical, and biological processes governing the fate of crude oil and weathering process for oil in the ocean are described below and summarized in Figure 9.10 and Table 9.4.

Most oil spills immediately form a thin surface slick (as thin as 0.1 μm) and light oils spread faster than heavy oils. Wind moves oil at about 3% of the wind velocity,

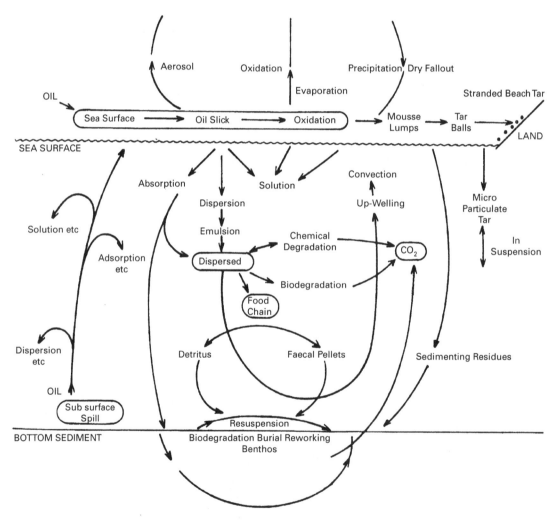

Fig. 9.10 The weathering of a crude oil slick at sea. (From Preston 1988, with permission of Academic Press.)

about the same as a surface drifting object (Preston 1988). Knowledge of the film thickness, coupled with an estimate of the area of the slick, will give an indication of the total amount of oil involved. It is interesting to note that the characteristic rainbow-colored sheen of oil on water is indicative of a film thickness of only 0.3 μm. At this thickness, 350 L of oil could produce a film covering an area of 1 km[2] (Preston 1988).

Many of the hydrocarbons are volatile and begin to evaporate immediately. After 24 h, half of the C_{14} compounds have vaporized, but it takes 3 weeks to evaporate half of the hydrocarbons shorter than C_{17} (Fig. 9.11). Vaporization continues slowly over the next few weeks and it is the most important natural factor in removing

Table 9.4 Contribution (%) of various pathways for the environmental fate of crude oil over various time scales.

Pathway	Time scale (days)	Percentage of initial
Evaporation	1–10	25
Solution	1–10	5
Photochemical degradation	10–100	5
Biodegradation	50–500	30
Disintegration and sinking	100–1000	15
Residue	>100	20
Total		100

Source: Reproduced with permission from Butler *et al.* (1976); © 1976 by the American Institute of Biological Sciences.

oil from the water surface. This time sequence from the initial spill to the formation of tar-like lumps and other resistant products is called weathering. Refined products such as gasoline and kerosene may disappear almost completely, whereas viscous crudes may lose <25% by evaporation (Fig. 9.11).

When the sea surface is roughened by wind, the oil may absorb water up to 50% of its weight and form brown masses called "chocolate mousse" (Laws 2000). Besides the water-in-oil emulsions, oil-in-water emulsions (dispersions) form, especially under the influence of added chemicals (dispersants). Although emulsification and dispersion can give the impression that the oil has disappeared from the surface, it actually continues to exist as tiny droplets, and its potentially poisonous effects persist. However, the toxicity during such dispersion is reduced, because the lighter fractions such as the aromatics and other short-chained molecules evaporate more quickly. On a longer timescale, photochemical oxidation may contribute to the weathering of oil. Through the actions of atmospheric oxygen and solar radiation, the proportions of oxygenated compounds in the slick will increase. For example, aromatics and alkyl-substituted cycloalkanes tend to be oxidized more rapidly, forming soluble compounds and insoluble tars.

Microbial degradation begins to take place only after the oil at the surface has aged and lost some of its highly volatile, toxic components by vaporization. At least 90 strains of marine bacteria and fungi and a few algae are capable of biodegrading some components of petroleum. Oil-decomposing bacteria increase in

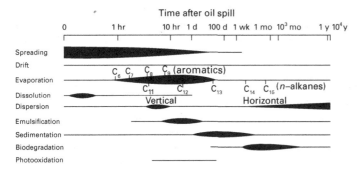

Fig. 9.11 Time course of various factors affecting oil spilled on the sea. Note the non-linear time scale. (Redrawn from Libes 1992.)

number slowly after an oil spill and their growth may be restricted by nutrients. Therefore, N and/or P should be added to speed up degradation (Young-Sook *et al.* 2001). Moreover, many major oil spills occur in temperate waters in winter, when the low temperature restricts bacterial growth rates. Non-toxic dispersants enhance biodegradation by greatly increasing the surface area of the oil (King 1984). Normal alkanes are the most easily degraded, whereas aromatics, cycloalkanes, and branched alkanes are more difficult (Preston 1988). For further aspects of microbial degradation of oil see Gundlach *et al.* (1983) and Preston (1988).

If an oil spill occurs near shore, and if the wind is in the right direction, beaching of the oil may occur. This oil may adhere to rocks, plants, and animals, but large waves during the winter may be strong enough to remove some of the oil from rocks. Oil may be worked into the sediments if a dispersant is used. Penetration into the interstitial spaces between sand grains will result in very slow degradation rates, often because of lack of oxygen in the interstitial water. As a result, oil may persist in sediments for years.

Tar-lump formation and sinking are the final stages of weathering. Oil may adsorb to particles and sink, or it may be consumed by filter-feeding plankton and become incorporated into fecal pellets which sink. Weathered oil may form lumps (usually about the size of peas), which can coalesce and become large enough to form a substratum for sedentary animals such as gooseneck barnacles, again causing the lump to sink (Fig. 9.10).

9.4.2 Effects of oil on algal metabolism, life cycles, and communities

Bioassay investigations have revealed variable effects, depending on the physical and chemical properties of the oils and components tested (e.g. whole oils vs. refined products), the parameters being measured, and the test species employed. Other problems associated with interpreting many bioassay experiments are the lack of details about the manner of preparation of the oil extract, how old the extract was before it was applied (hence what

fraction of the volatiles had been lost), and the total or differential losses of hydrocarbons, especially during long-term experiments. For these reasons, many investigators have chosen to work with individual components of oil whose composition is at least understood and measurable, and then they attempt to extrapolate back to the original oil. The effects of oil on seaweeds, ranging from the initial coating of the thallus, to penetration, impacts on metabolic rates, effects on enzymes, interference of sexual reproduction, and changes in the community, are discussed below.

Coating of the thallus due to the higher molecular weight, water-insoluble hydrocarbons associated with crude oil reduces CO_2 diffusion and light penetration into the seaweed. Reductions in photosynthetic rates often correlate with the thickness of the oil layer. During exposure to the air, the oil reduces desiccation of the blades, allowing photosynthesis to occur for longer than normal, but at a reduced rate. Severely oiled kelp fronds may break because of the weight of oil adhering to the fronds. The loss of thalli by this mechanism is associated primarily with the higher molecular weight, water-insoluble hydrocarbons (Nelson-Smith 1972).

The penetration of oil will depend on the covering on the thallus. The brown algae are thought to be largely protected from oil damage by their mucilaginous coating. This is assumed to be what "saved" the *Macrocystis* beds off Santa Barbarba after an oil well blowout. However, some dispersants may damage this protective mucus layer. The compounds that penetrate the thallus most easily, and hence are most toxic, are the lower molecular weight lipophilic compounds such as aromatics and they may be incorporated into the fatty layer that makes up the interior of cell membranes (Nelson-Smith 1972). The least toxic components and the least water soluble are the long-chain alkanes. The cycloalkanes followed by the long-chain alkanes are intermediate in toxicity. The aromatics and other toxic hydrocarbons appear to exert their toxic effects by entering the lipophilic layer of the cell membrane, disrupting its spacing. As a result, the membrane ceases to properly control the transport of ions in and out of the cell.

Disruption of cellular metabolism usually has been measured through changes in the rates of photosynthesis or respiration and growth. Reductions in photosynthetic rates vary with the type of crude oil, its concentration, the length of exposure, the method of preparation of the oil–seawater mixture, the irradiance, and the algal species. Photosynthetic inhibition occurred at ~7 ppm of crude oil, especially at high irradiance, while other species were unaffected. Crude oil extracts inhibit photosynthesis because of the toxicity of volatile aromatics such as benzene and naphthalene. The actual mechanism of inhibition is seldom investigated because of the difficulty in separating the toxicity effect from the purely mechanical effects of the coating (smothering) of the thallus and the reduction in light reaching the seaweed. Bleaching is commonly observed among red algae and is caused by the breakdown of phycoerythrin by kerosene-related compounds. Lipid-soluble pigments such as chlorophylls may be leached from cells by oil (O'Brien and Dixon 1976). There are only a few studies on the effects of oil contamination on algal respiration. The respiration rate for *Saccharina hyperborea* was inhibited by phenol at 100 ppm immediately (Hopkin and Kain 1978). Oil could interfere with respiration in a number of processes, such as gas diffusion, glycolysis, and oxidative phosphorylation. Mechanical blockage of gas diffusion is thought to be less pronounced for oxygen than for carbon dioxide.

Enzymes and structural proteins may be affected especially by aromatics. Glutathione S-transferase (GST) belongs to a family of enzymes involved in detoxification processes by catalyzing the conjugation of pollutants with glutathione and they are also involved in the detoxification of reactive oxygen species in cells. GST activity in several *Fucus* sp. was investigated as a potential environmental biomarker for petroleum contamination (Cairrão *et al.* 2004). Inhibition of algal DNA and RNA activities has been reported following exposure to high concentrations of crude oil (Stepanyan and Voskoboinikov 2006). Exposures to emulsified oil–seawater mixtures (100–10 000 ppm) for 24 h resulted in decreased DNA in the red algae *Polysiphonia opaca*. Given the fundamental importance of nucleic acids for reproduction

and protein synthesis, this early work provides a start in understanding the mechanisms whereby resistance to damage by oil may be conferred on tolerant species.

Growth rates integrate many cellular processes. In a study in Norway, rocky shore algal communities were kept in 50 m^3 basins, and continuously exposed to diesel oil for 2 years (Bokn 1987). With diesel oil at 130 µg L^{-1} lengthwise growth for *Ascophyllum nodosum* and *Saccharina digitata* was significantly reduced, by about 50% over the 2-year period. At a lower diesel oil concentration (30 µg L^1), there was periodic inhibition of growth, but no overall reduction in length. After 2 years of continuous exposure to oil, the seaweeds completely recovered during the following oil-free growth season and therefore more long-term recovery experiments are needed.

Oil may interfere with sexual reproduction. Several diecious brown algae, including *Ectocarpus* and *Fucus*, secrete olefinic hydrocarbons into seawater as gamete attractants. The possibility that petroleum hydrocarbons could confound recognition of the attractant fucoserratene by *Fucus* spermatozoids, was investigated by Derenbach and Gereck (1980). They found that a combination (rather than a single compound) of petroleum hydrocarbons attracted spermatozoids, but at concentrations about 100 times higher than levels at which fucoserratene is active. The reproductive stages of *Fucus edentatus* and *Saccharina saccharina* are particularly sensitive to oil, especially during gamete or spore release (Steele and Hanisak 1979). There was a marked decrease in secretion of polysaccharide mucilage and the adhesion of settling spores was inhibited at high oil concentrations (Coelho *et al.* 2000). Concentrations of crude or several fuel oils as low as 2 µg L^{-1} blocked fertilization in *Fucus*, apparently because of toxic effects on the sperm. *Saccharina* spores did not germinate above 20 µg L^{-1} of crude oil. Male gametophytes were more sensitive to oils than were female gametophytes, because in both *Fucus* and *Saccharina* they are smaller than females and hence have a higher surface-to-volume ratio, possess fewer stored reserves, and respire at a higher rate, creating a greater energy demand.

9.4.3 Ecological aspects

In the laboratory, the cellular effects of oil that were outlined above are much easier to quantify and document compared to the complex and widely varying community effects in the field. This variability in the field is due to factors such as the type of oil, the amount spilled, the water temperature, the weather conditions, the prior exposure of the area to oil, the presence of other pollutants, and the type of remedial action (e.g. use of dispersants). The impact is full-scale if the spill occurs close to the beach and is quickly washed ashore. The least impact occurs if the oil does not reach the shore for several days, giving time for many of the toxic volatile compounds to evaporate. In this section, examples and challenges of assessing the magnitude of an oil spill and the cleanup process with dispersants are discussed.

Since the early 1960s, various attempts have been made to quantify the effects of large oil spills in different parts of the world on various flora and fauna. The conclusions drawn from these studies have varied considerably, ranging from minimal effects to severe damage. The assessments have varied depending on the ecosystem studied and the community or population observed. Gundlach and Hayes (1978) constructed an "oil spill index" in which different ecosystems were ranked according to their vulnerability. Rocky exposed cliffs are the least vulnerable, whereas salt marshes, mangroves, and coral reefs are extremely vulnerable (Loya and Rinkevich 1980; Sanders *et al.* 1980) (Table 9.5). Since oil is light, it initially affects surface oriented organisms, but later as it weathers, some oil sinks. Communities also have been ranked, with birds and benthic subtidal communities being most vulnerable, and plankton and benthic rocky intertidal communities only slightly vulnerable. Following the death of corals, rapid colonization of algae on the skeletons of the dead corals may be enhanced by oil pollution.

The effects of oil vary greatly depending on the length of time after the spill and the closeness to shore. In cases in which weathering processes have time to eliminate the more volatile toxic components before the oil reaches the shore, the effects of even a heavy oil

Table 9.5 Expected impacts of oil spills on various marine habitats and cleanup recommendations.

Exposed rocky cliffs	In the presence of high-energy waves, oil spill cleanup usually is unnecessary.
Exposed rocky platforms	Wave action causes rapid dissipation of oil, generally within weeks. In most cases, cleanup is not necessary.
Flat, fine-sand beaches	Because of close packing of the sediment, oil penetration is restricted. Oil usually forms a thin surface layer that can be efficiently scraped off.
	Cleanup should concentrate on the high-tide mark; lower beach levels are rapidly cleaned of oil by wave action.
Beaches with medium- or coarse-grained sand	Oil forms thick oil–sediment layers and mixes down to 1 m deep with the sediment. Cleanup damages the beach and should concentrate on the high-water level.
Exposed tidal flats	Oil does not penetrate the compacted-sediment surface, but biological damage results. Cleanup is necessary only if oil contamination is heavy.
Mixed sand-and-gravel beaches	Oil penetration and burial occur rapidly; oil persists and has a long-term impact.
Gravel beaches	Oil penetrates deeply and is buried. Removal of oiled gravel is likely to cause future erosion of the beach.
Sheltered rocky coast	The lack of wave action enables oil to stick to rock surfaces and tidal pools. Severe biological damage results. Cleanup may cause more damage than if the oil is left untreated.
Sheltered tidal flats	Long-term biological damage occurs. Removal of the oil is nearly impossible without causing further damage. Cleanup is necessary only if the tidal flat is very heavily oiled.
Salt marshes and mangroves	Long-term deleterious effects occur. Oil may continue to exist for 10 years or more.

Source: From Gerlach (1982); reprinted with permission of Springer-Verlag, Berlin.

deposition on intertidal flora appear to be largely physical, with injury due to smothering and adsorption of oil. Species that grow between neap and spring high-tide marks, especially those algae near spring high tide, where oil may be stranded for a long time, are most seriously affected by oil coating. Many high intertidal species of Rhodophyceae and Phaeophyceae become oleophilic as their surfaces dry out (O'Brien and Dixon 1976). This strong adsorptive capacity for oil can cause algae to become severely overweighted by adsorbed oil and subject to breakage by waves. For algae with annual basal regrowth, loss of distal blades may be no more debilitating than losses during a winter storm (Nelson-Smith 1972). However, the loss of too many photosynthetic blades during the growing season, when metabolic products are stored, could impair a seaweed's regenerative ability (O'Brien and Dixon 1976).

No clear patterns emerge from the relationships between systematics and the susceptibilities of intertidal algae to oil. Several studies have indicated that Cyanophyceae are particularly resistant to oil (O'Brien and Dixon 1976). Species of Chlorophyceae, in particular, have a remarkable ability to invade areas where other species have been eliminated. The spread of green algae often is due to the die-off of herbivores, which are more susceptible to oil damage than are algae. Early observations suggested that filamentous red algae and corallines were most susceptible to oil and oil–emulsifier blends, possibly because of the destruction of phycoerythrin (Nelson-Smith 1972), but this suggestion requires further confirmation.

The ecological impacts of oil vary with the seaweed's habitat. On rocky shores there may be a slight, short-term impact, but no significant long-term effects on the macrophyte community have been observed (Nelson 1982; Gundlach et al. 1983). After the Exxon Valdez oil spill in 1989, mature Fucus were covered with oil, but did not appear to die due to the oil (Driskell et al. 2001). However, the largest mortality came from the use of high pressure hot water to lift the oil off the rocks in an attempt to improve the negative visual aspect of the oil spill and protect the birds and mammals (Preston 2001). This was a publically visible cleanup process, but it was so harsh that

it killed ~90% of the adult Fucus gardneri and its germlings as well as grazing limpets, periwinkles, mussels, and barnacles (Stekoll and Deysher 2000; Preston 2001). Hence, there was no control of Fucus recruitment by limpet or snail grazing. The recovery of Fucus was slow since propagules only disperse <1 m from the adults. They found a rapid increase in the single cohort of Fucus recruiting soon after the spill to above normal abundance and then a return to near normal abundance about 5 years after the spill. The persistent patterns in size structure and dynamics in Fucus, suggest that full recovery had not occurred by 1996 (7 years after the spill), since there are still considerable oscillations in the population (Driskell et al. 2001). Fucus recovery in the upper shore was slow because recruits are very sensitive to desiccation, especially in the absence of a canopy to protect them (see sec. 4.1). Other effects of the spill included colonization of the upper shore by ephemeral mostly green algae and an opportunistic barnacle. Fucus spread into the lower shore where it prevented the return of red algae. The size distribution of the dominant kelps, Nereocystis, Agarum, and Saccharina was abnormally skewed towards recruits in 1990, implying an adult mass mortality in 1989 (Preston 2001).

Recovery from an oil spill is variable. Rocky intertidal areas that have been cleansed with detergents after oil spills have shown recolonization rates comparable to the rates on control plots. The first macroalgae to recolonize the Cornwall shore after the Torrey Canyon spill in 1967 were Ulva sp. (Hawkins and Southward 1992). They quickly covered the entire area, because the herbivores that usually grazed on them (e.g. limpets and periwinkles) had been killed by the oil. The upper limit of distribution for Saccharina digitata and Himanthalia elongata was higher by as much as 2 m during the first few years of succession. Limpets progressively recolonized, and within 7 years the distribution of seaweeds had returned to normal (Gerlach 1982). Similar observations were made on the Somerset coast of England, where the oil was reported not to adhere to Fucus spiralis, and the percentage of cover of this alga increased from 50 to 100% after the spill (Crothers 1983). No significant effects of the Amoco Cadiz spill were observed for Saccharina, Fucus, or

Ascophyllum (Gundlach *et al.* 1983). Some of the damage to corallines, such as loss of pigments, appears to have been partially or wholly due to the dispersant and its toxic aromatic solvent. Several non-toxic dispersants such as Corexit are available, and therefore toxic effects attributable to dispersants should no longer be a problem.

In the cleanup process dispersants are often used and they are of two main types (Preston 1988). Hydrocarbon or conventional dispersants are based on hydrocarbon solvents; they contain about 20% surfactant and must be pre-diluted and mixed with seawater. Because of the large volumes required to treat even a moderate size slick, these chemicals are more suitable for application from small ships. The second group, the concentrates or self-mix dispersants, are alcohol- or glycol-based and usually contain higher concentrations of surfactant components. Typical dose rates are between a ratio of 1:5 and 1:30 (dispersant: oil), and this makes them more suitable for aerial spraying. Usually the natural motion of the sea is enough to mix these dispersants, and therefore they are much more practical for large oil spills. With both types of dispersants, it is essential to apply the chemical as rapidly as possible (i.e. before mousse formation) for maximum effectiveness. Unfortunately, this traps the more volatile and toxic components that normally would evaporate. Light fuel oils such as gasoline should be left to evaporate, and heavy fuel oils and mousses are not amenable to dispersion. For more information on oil cleanup and bioremediation see Laws (2000).

Dispersants may be toxic as well as the oil. The effects of four dispersants belonging to group 2 above and dispersed diesel fuel and crude oil combinations were assessed by germination inhibition of the brown seaweed *Phyllospora comosa* (Burridge and Shir 1995). Inhibition of germination of the water-soluble fraction of diesel fuel increased when each of the four dispersants were added (i.e. the EC_{50} declined from 6800 to 400 $\mu l \ L^{-1}$). Corexit 9500 produced the largest inhibition of germination, while Corexit 8667 was the least toxic, possibly because it is immiscible in water and its toxicity might be enhanced by factors that increase toxicity. In contrast, for crude oil, the addition of dispersants enhanced the germination rate (i.e. the EC_{50}

increased from 130 to ~3000 $\mu l \ L^{-1}$). However, it may be better not to use dispersants. In cases in which dispersants were not used to aid in the oil cleanup, algal growth generally was less affected. In the case of the San Francisco Bay oil spill of 1971, caused by the collision of two tankers carrying Bunker C fuel oil, pre-spill algal densities were restored within 2 years.

In summary, determining the impacts of oil spills in the field is challenging (Crowe *et al.* 2000). Often there is a lack of data at a particular site before the spill. This is compounded by problems in selecting adjacent control sites that are very similar in terms of the biological community and the physical/chemical setting. It is often difficult to separate the effects of the cleanup (e.g. use of dispersants that may be toxic or high pressure hot water) from the actual effects of the oil. There is often pressure to pursue a cleanup procedure that will be highly visible to the public. It is difficult to persuade regulatory agencies to leave some areas untreated to act as controls for comparison with cleaned areas. Resources tend to be employed in excess during the start of the post spill phase, but as public interest wanes, so does funding for longer term recovery studies. Unfortunately, most oil spills miss the opportunity to scientifically design cleanup experiments to determine which cleanup procedures produce the best results and then refine them for future application (Paine *et al.* 1996; Crowe *et al.* 2000). Large mesocosm studies may be useful to simplify the challenges encountered in field work. For example, Bokn *et al.* (1993) used large outdoor tanks to determine how a low dose of diesel oil affected communities over a 2-year period. While laboratory studies are useful to determine the effects of various concentrations of specific compounds of oil on specific macrophytes, the real but expensive need is to fund long-term (>10 years) field-monitoring programs to document the recovery of the spill.

9.5 Synthetic organic chemicals

Pesticides are classified according to the intended pest target (e.g. herbicides, insecticides, fungicides, etc.). They are used widely in agriculture, forestry, and human-health activities. They enter the marine coastal environment through runoff from the catchment basin, groundwater, rivers, and atmospheric deposition. Coastal lagoons with long residence times may have the highest concentrations, especially if there are intense agricultural activities in the surrounding area. Relatively few laboratory and field studies have examined the effect of pesticides and especially synergistic effects with other stressors on macroalgae, despite the easily detectable concentrations of pesticides in some coastal areas. Since herbicides and insecticides are more of a problem in freshwater than in marine ecosystems, they are not covered here, although there is concern over herbicides and the Great Barrier Reef (Lewis *et al.* 2009). Similarly, polychlorinated biphenyls are being phased out and are not covered here.

Antifouling compounds clearly have an impact on marine organisms. Fouling or the settlement and growth of marine organisms on submerged structures is estimated to have a worldwide cost of >US$3 million annually (Myers *et al.* 2006). Tributyl tin (TBT) was introduced in the mid-1960s and until recently it was the most widely used antifouling agent. The International Maritime Organization's (IMO) banned the use of organotin biocides in antifouling paints in 2008 (Antizar-Ladislao 2008). This ban stimulated the development of new paints containing "booster biocides" like diuron, zineb, seanine, and zinc pyrithione (Myers *et al.* 2006). However, the environmental effects of these booster biocides are still poorly known.

The antifouling property of marine paints is based on the slow leakage of toxic substances such as Cu, irgarol and tributyl tin which prevent the growth and settlement of organisms on the paint surfaces. Since 1970, organotins, particularly the trialkyltin compounds such as triphenyltin (TPT) and tributyltin (TBT), have been widely used as biocides in antifouling compositions for boat hulls and fish-farming gear (Almeida *et al.* 2007). TBT is very persistent and degrades slowly. Even if TBT has been banned for use on pleasure boats for more than 20 years, TBT is still found in high concentrations especially in harbor sediments and still appears to leak out from many boat yards (Eklund *et al.* 2008). Effects on non-target organisms have been recognized at lower levels than were previously anticipated (Langston 1990; Antizar-Ladislao 2008). Shell abnormalities and reduced growth and recruitment in oysters sampled near marinas were the first indication of the TBT problem. Subsequently, effects have been demonstrated in a number of marine and estuarine species (Langston 1990; Antizar-Ladisco 2008). Although TBT and TPT are highly effective against *Ulva*, they are less effective in controlling *Ectocarpus* and the micro-fouling biofilm of bacteria, benthic diatoms, and some green algae that generally precedes settlement by macroalgae (Millner and Evans 1981).

TBT affects all life stages of macrophytes. The EC_{50} for *Ceramium tenuicorne* was 0.49 µg TBT L^{-1} (Karlsson *et al.* 2006). The photosynthetic apparatus of zoospores and the vegetative tissues of *Ulothrix* were found to be relatively insensitive to triphenyltin, compared with those of *Ulva intestinalis* (Millner and Evans 1980). There are questions remaining to be answered: For example, why is *Ulothrix* more resistant to organotins than *Ulva*, even though *Ulothrix* takes up organotins more rapidly? Because TBT has been reported to range from 0.1 to 2 µg L^{-1} in estuaries and especially marinas, these higher concentrations may be inhibiting primary productivity (Hall and Pinkley 1984).

Newer antifouling paints contain booster biocides such as cuprous oxide or cuprous thiocyanate. Zinc oxide is often used in combination with Cu(I) as a booster which increases the toxicity by 200-fold. Because some algae are resistant to the Cu + Zn mixture, other biocides are added such as Cu- or Zn-pyrithione Irgarol 1051 (a triazine herbicide), chlorothalonil, diuron, etc. (as reviewed by Turner 2010). Diuron is one of the most popular biocides since it is less toxic and it has replaced organotin (Konstantinou and Albanis 2004). However, it is still highly toxic to algae and the EC_{50} for growth inhibition of *Ceramium tenuicorne* was 3.4 µg L^{-1} (Karlsson *et al.* 2006). In general, bacteria, crustaceans, and fish are less sensitive than algae with EC_{50} values of >74 mg L^{-1} (Myers *et al.* 2006). Germination of *Hormosira banksii* spores were more sensitive than growth since germination is more related to water concentrations, while growth may be related to diuron concentrations at the substrate interface. Diuron is relatively persistent with a degradation half-life of 14–35 days and water column concentrations range from 0.003–17 µg L^{-1} (Konstantinou and Albanis 2004). Another popular biocide is Irgarol 1051 (an *s*-triazine) and in marinas, concentrations may reach 4 µg L^{-1} (Hall *et al.* 1999). The EC_{50} for growth inhibition of Irgarol on *C. tenuicorne* was 0.96 µg L^{-1} (Karlsson *et al.* 2006). Phytoplankton appear to be more sensitive than macroalgae and bacteria with EC_{50} values ranging from 0.3 to 7.8 µg L^{-1} but only a few macroalgae have been tested (Zhang *et al.* 2008a). Irgarol undergoes photodegradation to another *s*-triazine termed M1 which is less toxic than Irgarol. Since synergistic effects occur, it is important to test not only the active substance in the antifouling paint, but the whole product should be tested to assess the impact on the ecosystem (Karlsson and Eklund 2004), The toxicity of leachates from various antifouling paints were tested on the red macroalga *C. tenuicorne*. Some paints leaked more copper causing the observed toxicity. For some other paints, zinc was toxic and for still others both copper and zinc leaked out in concentrations that could explain the toxicity. For one paint, neither copper nor zinc could explain the observed toxicity. In this case it was hypothesized that zinc pyrethion added as a preservative for the paint could be the reason (Karlsson *et al.* 2010). In a single substance test on zinc pyrethion, an EC_{50} of 3.3 µg L^{-1} was obtained (Karlsson and Eklund 2004).

The effects of antifouling paint particles that are produced by cleaning and scrapping boat hulls have recently been examined (Turner *et al.* 2009). Leachate of Cu and Zn from the paint particles was toxic to *Ulva* at paint particle concentrations as low as 4 mg L^{-1}, an order of magnitude less than the concentrations observed in many poorly flushed harbors. Cu accumulation was largely through adsorption to the cell surface, but significant accumulation of Zn was not observed.

9.6   Eutrophication

In the following sections, the focus will be mainly on the effects of sewage-derived inorganic nutrients on macrophytes, although other anthropogenic nutrient sources such as atmospheric deposition and groundwater discharge are briefly discussed. Near the sewage outfall site, NH_4 and metal concentrations may be high enough to be toxic and reduce growth. The response in temperate areas where macrophytes dominate the seashore, offers a contrast to impacts on coral reefs where macrophytes are normally small, but may become dominant when nutrients upset the delicate balance between corals and their zooxanthellae.

9.6.1   Sewage effluent and impacts of nutrient enrichment on algal communities

Sewage is classified as a complex waste because it contains inorganic nutrients (N and P and possibly toxic NH_4 concentrations), organics, chlorine (from chlorination), and some metals (Camargo and Alonso 2006). There are three main treatment phases, primary, secondary, and tertiary. In primary treatment, the screened sewage is passed to settling chambers, where mainly organic particles settle out and the sludge may be taken to the landfill or incinerated. In secondary treatment, the remaining liquid is aerated to encourage bacterial growth and decomposition of the dissolved organics. In tertiary treatment, nutrients such as N and P are removed by chemical treatment (e.g. precipitation of phosphate by alum) or biological treatment (e.g. conversion of NO_3 to N_2 gas by bacterial denitrification). The discharge of nutritive inorganic and organic wastes into coastal areas with low rates of water exchange may stimulate the growth of algae and produce excessive blooms of phytoplankton and/or macrophytes and create biological, aesthetic, or recreational problems. When eutrophication is extensive, the large volume of algal biomass (both phytoplankton and macrophytes) that was produced from the anthropogenic nutrient load soon begins to decay and seriously depletes the oxygen concentration in the bottom water and this hypoxic (~2 mg O_2 L^{-1}) or anoxic (~0 mg O_2 L^{-1}) condition may stress or kill animals. The sequence of changes in various parameters in a shallow ecosystem during four phases of increasing eutrophication is shown in Figure 9.12. In phase I, perennial macrophytes

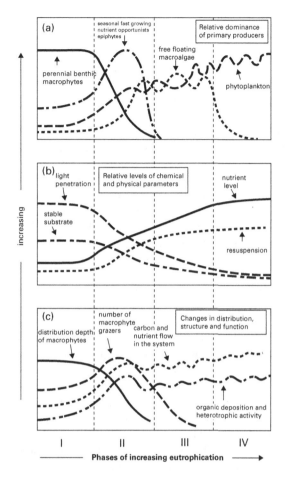

Fig. 9.12 Schematic of typical changes in (a) dominance of primary producers; (b) relative levels of physical, chemical, and biological parameters; and (c) changes in distribution, structure, and function of various parameters during four phases of increasing eutrophication (From Schramm 1999, reproduced with permission.)

dominate, but by phase II, epiphytes increase as well as macrophyte grazers. By phase III, most macrophytes are free floating since light penetration has decreased due to an increase in phytoplankton. In phase IV, there are very few macrophytes with very high phytoplankton biomass and high organic deposition and possible hypoxia in the bottom water, depending on the depth of the water column. The stages of increasing eutrophication and the role of

seaweeds/plants as a nutrient filter in estuaries have been reviewed (Schramm 1999; Grall and Chauvaud 2002; McGlathery *et al.* 2007).

Most studies that have examined the responses of macrophytes to eutrophication have been concerned with a sewage outfall since it is easier to study this small defined area, with a gradient in nutrient concentration away from the outfall. Generally, such studies have been from an ecological point of view, examining changes in community structure and diversity. In an early classical study, Littler and Murray (1975) found 17 fewer species of macrophytes and less cover near the San Clemente Island (California) outfall than in a nearby control area. The outfall flora was less diverse and showed a reduction in community stratification (spatial heterogeneity). They suggested that sewage favors rapid colonizers such as *Ulva* and more sewage-tolerant organisms. Macrophytes near the outfall exhibited relatively higher net productivities, smaller growth forms, and simpler and shorter life histories and most were components of early successional stages. As a follow-up, Murray and Littler (1978) determined experimentally, whether algal communities that are characteristic of sewage-stressed habitats showed high resilience and they found that cyanobacteria, filamentous Ectocarpaceae, and colonial diatoms were the dominant forms during the early successional stages in the cleared areas in both the sewage and control plots. The outfall plots showed rapid recovery by algae such as *Ulva californica*, *Gelidium pusillum*, and *Pseudolithoderma nigrum*, which have a capacity for rapid recruitment. The algal communities in the unpolluted (control) denuded areas did not fully recover, even after 30 months.

The early studies by Murray and Littler (1978) and Kindig and Littler (1980) on the impact of sewage on macrophytes provide excellent examples of the combination of field studies and laboratory studies. They progressed from a community field study to an experimental manipulation (denuded plots) in the field and then to studies of the environmental physiology of important species in the laboratory. Kindig and Littler (1980) studied the responses of 10 macrophytes to various sewage effluents (untreated, primary, secondary, and secondary-chlorinated) during long-term culture studies in the laboratory. *Bossiella orbigniana* and *Corallina officinalis* exhibited increased photosynthesis rates when exposed to primary treated sewage, and in long-term cultures their growth was enhanced. Chlorination of effluent produced only a short-term reduction in growth for the first week of culturing. Three populations of *C. officinalis* with differing pollution histories (pre-exposure to pollution) showed tolerances to sewage corresponding to the extent of their prior exposure. This finding indicates that this species may be able to adapt to sewage stress and suggests that considerable caution must be exercised in the selection of benthic algae as biological indicators of pollution. Downstream and inshore from a domestic sewage outfall 90% of the algal biomass was composed of the corallines *Bossiella* and *Corallina*. Thus, coralline algae appear to be extremely tolerant of high concentrations of sewage.

Excessive growth of green seaweeds in response to sewage effluent is becoming an increasingly common phenomenon in sheltered bays (Reise 1983; Soulsby *et al.* 1985; Tewari and Joshi 1988; Teichberg *et al.* 2008, 2010). An overabundance of *Ulva* on the tidal flats of the Wadden Sea during the summer was attributed to eutrophication by adjacent sewage effluents (Reise 1983). The mats were first composed primarily of *Ulva*, but later others such as *Cladophora*, *Chaetomorpha*, and *Pyropia* appeared as secondary components.

Manipulative field experiments such as transplantation and cleared sites can be helpful to examine species succession near a sewage outfall. The distribution of littoral algae in the inner part of Oslofjord in Norway has been studied over many decades, and *Ascophyllum nodosum* was the dominant alga in the area before 1940 (Bergström *et al.* 2003). When there was a large increase in the sewage load, many species such as *Rhodochorton purpureum*, *Phyllophora truncata*, and *Ascophyllum nodosum* disappeared or became rare (Rueness 1973). He cleared plots in the inner part of Oslofjord near the sewage outfall and in a control area to observe recolonization. In addition, rocks from the control area to which *A. nodosum* had become attached were transplanted to the sewage-stressed area. Regrowth was much faster in the inner

fjord than in the control area. In the inner fjord, the dominant recolonizing species was *Ulva compressa*, followed by *Fucus spiralis*. No *Ascophyllum* germlings were observed. In the cleared control plots, regrowth proceeded more slowly, and green algae were less predominant. The number of species that recolonized was also greater, and after 6 months, the regrowth was primarily dominated by a dense stand of *Pyropia purpurea* in the cleared control area. The *Ascophyllum* transplants into the sewage-stressed area were heavily infested with epiphytes and frequently were overgrown by *Ulva*, *Ceramium capillaceum*, and small mussels (*Mytilus*). Rueness (1973) concluded that the increased competition for substrate and the shading effect of the *Ulva* carpet reduced the chances of the *Ascophyllum* germlings becoming re-established near the sewage outfall.

In oligotrophic areas of the oceans and coral reef communities, significant effects from a sewage outfall have been observed (Pastorok and Bilyard 1985; Laws 2000). Kaneohe Bay is a subtropical embayment in the Hawaiian Islands and the sewage that was discharged into the bay affected the coral reef community in two ways. First, a reduction in water clarity was caused by increased phytoplankton growth. This reduced the amount of light available for the symbiotic zooxanthellae living in the hermatypic corals and thus resulted in reduced coral growth and possibly increase attachment substrate if the corals die. Coral growth may also be affected by a small increase in nutrients that may cause the corals to expel their zooxanthellae since the corals take up the external ambient nutrients rather than deriving the nutrients from the symbiotic relationship (Falkowski *et al.* 1993). Second, the sewage discharge stimulated the growth of the green alga *Dictyosphaeria cavernosa*, commonly known as the bubble alga, which usually establishes itself within a coral head at the base of the frond and then grows outward, eventually enveloping the coral head and killing the coral. This alga was not abundant beyond the sewage-stressed area and therefore it appears to have spread in response to the elevated nutrient concentrations. Eventually the sewage was diverted from the bay, and the recovery of the community has been documented (Laws 2000). Although the inorganic nutrient concentrations have reverted to their pre-sewage levels, the system took many years to fully stabilize, because of the slow release of nutrients from plankton that had sunk and accumulated in the sediment during the sewage discharge period, and because of ongoing non-point-source inputs resulting from urbanization of the shoreline.

Caribbean coral reefs have experienced a loss of coral cover from ~50 to <10% of their substratum in the last 30 years. In many cases, this space had been occupied previously by frondose algae, specifically the brown frondose algae such as *Dictyota*, *Lobophora*, *Padina*, *Sargassum*, and *Turbinaria* (McClanahan *et al.* 2003, 2004). They found that a fertilizer mixed with N and P enhanced the cover and colonization of filamentous green turf algae such as *Ulva prolifera*, but not the erect brown frondose algae and red algae. Therefore, the increase in brown frondose macroalgae in many of the Caribbean reefs does not appear to be solely due to nutrient enrichment. Similar results were observed for a nutrient gradient on the inshore Great Barrier Reef (Fabricius *et al.* 2005). Green and red seaweeds increased with increasing nutrients, while brown seaweeds remained constant.

There is increasing concern over the changes in water quality on coral reefs (see Essay 6). This deterioration in water quality represents a second major stress in addition to ocean acidification (see also sec. 7.7). Cooper *et al.* (2009) have reviewed the assessment of the best bioindicators to detect water quality changes on reefs and made recommendations for improved monitoring programs. Nutrient-rich ground-water impacts macrophyte coral reef communities. The productivity of *Cladophora prolifera* is normally limited by both nitrogen and phosphorus in Bermuda's shallow oligotrophic inshore surface waters (Lapointe and O'Connell 1989). However, seepage of nitrogen-rich groundwater, combined with a high alkaline phosphatase capacity to use organic P, accounts for its recent increase in some areas. The proliferation of *Cladophora* in Bermuda's inshore waters over the past 20 years exemplifies the dramatic ecological changes that occur when oligotrophic marine ecosystems are impacted by nutrient-enriched groundwaters.

The interplay between macroalgal growth and the triggering of phytoplankton blooms was clearly demonstrated in the hypertrophic Venice Lagoon (Sfriso *et al.* 1987). Under aerobic conditions, nutrients were taken up primarily during the spring and summer periods. When there was an imbalance between production and consumption of oxygen, anoxic conditions occurred and large amounts of nutrients were released to the sediments by the decomposition of macroalgae that subsequently triggered a large phytoplankton bloom.

9.6.2 Sewage effluent and toxicity

As outlined in the section above, nutrients may enhance the growth of macrophytes, but in some cases, forms of nitrogen such as NH_4 may be toxic and reduce growth. The ammonium concentrations in discharged sewage can be very high (up to 2200 μM) (Camargo and Alsonso 2006). The disappearance of the sensitive brown seaweeds *Hormosira banksii* and *Durvillaea potatorum* and increase in *Corallina* and opportunistic green macroalgae near the sewage discharge site of Melbourne, Australia, was investigated to determine if ammonia was the main toxicant in the secondary-treated sewage (Adams *et al.* 2008). In aqueous solution, total ammonia exists in two forms (NH_3 (unionized ammonia gas, the most toxic form) and NH_4^+ (ammonium ion) which are controlled mainly by pH (Camargo and Alonso 2006). At a pH of <9.5, there is more NH_4^+ (e.g. seawater of pH 8.2 has ~95% NH_4^+) and at pH >9.5 there is more NH_3. Hence if the pH of seawater is increased to >10, then the ammonia (gas) can be reduced/removed by vigorously bubbling the seawater. Chronic toxicity testing of ammonia was conducted using microalgal growth (*Nitzschia closterium*), fertilization, germination, and cell division of the macroalga *Hormosira banksii* and scallop larval development. The macroalgal tests were the most sensitive with EC_{50} concentrations of 100 μM, followed by scallop larval development (200 μM) and microalgal growth was the least sensitive (800 μM). Sewage may also contain other toxicants and therefore toxicity may not be due to only ammonia. Therefore ammonia must be removed first in toxicity tests by three different

---

Essay 6 Macroalgae play a major role in the future community composition of coral reefs

Britta Schaffelke

Coral reefs are places of outstanding beauty and biodiversity. Owing to their high productivity and their often remote location, they also provide livelihoods for millions of people (Burke *et al.* 2011). When thinking about or visiting coral reefs, seaweeds are not the first thing that comes to mind, but macroalgae provide much of the foundation for the dazzling array of benthic animals and fish that constitutes a coral reef.

I became interested in coral reef macroalgae after studying temperate seaweeds for some years. During my PhD studies I was part of the exciting research group of Prof Klaus Lüning who pioneered the discovery of elaborate endogenous systems that control seaweed growth so that their phases of growth are exactly synchronized to the seasons (Lüning 1994; Schaffelke and Lüning 1994). Then my research focus broadened to the general environmental and ecological drivers of seaweed growth, biomass and abundance in order to understand why the habitat-forming kelps and rockweeds of the order Laminariales and Fucales have been decreasing in the Baltic Sea. A multitude of environmental factors for this decline in seaweeds has since been recognized, ranging from decreasing light availability due to eutrophication, and associated biological consequences such as endophyte infections to blooms of ephemeral macroalgal competitors (Schaffelke *et al.* 1996; Schramm 1999; Lotze *et al.* 2000). The reverse problem of increasing macroalgae lured me to the Great Barrier Reef (GBR) where I started research on the overabundance of habitat-forming seaweeds on inshore reefs due to increased nutrient input.

A large number of environmental factors shape coral reefs and their biological communities, including the seaweeds found on reefs. Macroalgal communities on reefs differ between reef types, change along reef zones (slope, flat, crest, or lagoon), along depth gradients and with seasons, and depend on the successional state of a reef community. For example, coral reefs that have been recently disturbed resulting in coral mortality, generally have higher cover and biomass of seaweeds because these quickly colonize any newly available substratum. On most reefs, a return to coral dominance occurs over several years, however, there are cases where chronic disturbance leads to persistent community changes.

Coral reef macroalgae have important roles in facilitating coral settlement and as primary producers, but an over abundance of seaweeds on some coral reefs, especially of non-calcareous, frondose or foliose growth forms (e.g. the genera *Laurencia*, *Acanthophora*, *Eucheuma*, *Caulerpa*, *Dictyosphaeria*, *Sargassum*, *Turbinaria*, *Lobophora*, *Dictyota*) is often a symptom of environmental change or impact. Coral reefs have always been shaped by disturbances, such as severe storms, coral predation and disease. Over the past decades coral reefs have been exposed to additional acute and/or chronic disturbances such as global climate change, marine pollution and increasing resource exploitation (Pandolfi *et al.* 2003). As a consequence, only 46% of the world's coral reefs are currently considered to be healthy and not under any immediate local or regional pressure (Wilkinson 2008). After a disturbance that leads to coral mortality and/or alters environmental conditions (e.g. substratum, nutrient and light availability, intensity of herbivory), coral reefs either gradually recover or can become unstable and change into undesirable states that are largely dominated by upright macroalgae, suspension feeders or even devoid of larger benthos, and hence of the three-dimensional structure that signifies a healthy coral reef (Figure 1; Mumby and Steneck 2011).

While seaweeds are also affected by acute physical disturbance, the longer term consequences are often beneficial to macroalgae, potentially tipping the competitive balance from corals to seaweeds (Diaz-Pulido *et al.* 2007b). Following coral mortality, the availability of vacant substratum is critical, since macroalgae rarely settle on healthy coral tissue. Seaweeds are generally able to rapidly recruit into available space (Diaz-Pulido and McCook 2002, 2004) and high abundances are commonly observed after hurricanes or cyclones (Done 1992). Many common coral reef seaweed genera are well adapted to regenerate from fragmentation after disturbance, e.g. *Caulerpa*, *Dictyota*, *Gracilaria*, *Halimeda* and *Hypnea*, or from remaining attachment points, such as basal holdfasts or apical "hooks", e.g. *Halimeda*, *Sargassum*, *Turbinaria*, and *Hypnea*. Banks of propagules (see Santelices *et al.* 1995), microscopic recruits or tissue remnants could also facilitate the rapid exploitation of vacant substratum, but the importance of these are unknown for coral reefs.

After colonization, high seaweed abundance is often sustained by a combination of reduced top-down (herbivory) and increased bottom-up (nutrient availability) control, which often co-occur (Littler and Littler 1984; Smith *et al.* 2010; but see Rasher *et al.* 2012). When nutrients are naturally at limiting levels, nutrient enrichment promotes growth and increases

Essay 6 (cont.)

(a)   (b)

Figure 1 (a) Inshore fringing coral reef dominated by *Sargassum* sp.; (b) Fringing reef with high coral cover and no visible macroalgae. (Courtesy of Britta Schaffelke.)

biomass of frondose macroalgae (e.g. Lapointe 1997; Schaffelke and Klumpp 1998a; Stimson *et al.* 2001). This bottom-up control sparked my curiosity because coral reefs are usually believed to occur in oligotrophic waters and by the mid-1990s there had only been a few reports of anthropogenic nutrient enrichment resulting in increased seaweed biomass on coral reefs (Smith *et al.* 1981). What if the habitat-forming brown algae I had seen on the GBR inshore reefs had certain mechanisms to efficiently use the nutrients that are available in this system? I was able to show that *Sargassum* takes up dissolved nutrients during periods of high availability, e.g. during monsoonal floods that carry excess nutrients from the adjacent agricultural catchments (Brodie *et al.* 2012) and store these nutrients in its tissue to sustain increased growth over several weeks (Schaffelke and Klumpp 1998b; Schaffelke 1999a). *Sargassum* spp. also utilize nutrients derived from the high loads of particulate organic nutrients that settle on their thalli to sustain high productivity (Schaffelke, 1999b). Several inshore brown seaweeds (*Chnoospora implexa*, *Hormophysa cuneiformis*, *Hydroclathrus clathratus*, *Padina tenuis*, and *Sargassum* spp.) also have high alkaline phosphatase activity (APA), an enzyme which allows seaweeds to use the organic phosphorus pool in addition to dissolved phosphate (Schaffelke 2001) and may compensate for induced phosphorus limitation in locations subject to nitrogen inputs (Lapointe *et al.* 1992). In contrast, distribution, growth, calcification and photosynthesis of crustose coralline algae (CCA) are negatively affected by increased nutrient, organic matter availability and sedimentation, similar to corals (Björk *et al.* 1995; Harrington *et al.* 2005; McClanahan *et al.* 2005).

Herbivory modifies the nutrient response by removing biomass down to a certain threshold where resident herbivores are not able to consume more macroalgal biomass (Williams *et al.* 2001). The availability of sufficient algal biomass can result in selective grazing of more palatable or more nutrient-rich species and, consequently, to a dominance of species that are not readily controlled by grazing (Stimson *et al.* 2001; Boyer *et al.* 2004; Ledlie *et al.* 2007). On Hawaiian reefs, transitions towards high macroalgal biomass were likely to be initiated by introductions of non-indigenous macroalgae and then sustained by sufficient available nutrients and avoidance/selective grazing by resident herbivores (Smith *et al.* 2004; Conklin and Smith 2005).

The consequences of higher seaweed abundance of coral reefs depend on the seaweed taxa present, their growth forms and life histories. Larger thalli with fast growth rates (mostly brown algal taxa, but also *Halimeda*, *Acanthophora*, *Gracilaria*, *Hypnea*, *Kappaphycus*, and *Eucheuma* species) compete with corals leading to spatial exclusion of coral recruits, reduced coral growth or in extreme cases, partial or total coral mortality (McCook *et al.* 2001; Nugues and Bak 2006; Littler and Littler 2007). In addition, ephemeral taxa, such as *Hydroclathrus*, *Padina*, and *Dictyota* species, can form dense algal mats that physically smother corals underneath and negatively affect the microenvironment for corals by reducing light and water exchange and increasing the concentrations of dissolved nutrients and organic carbon (Hauri *et al.* 2010). The increasing incidences of coral reefs that are dominated by seaweeds are considered to be both a consequence and cause of coral reef

Essay 6 (cont.)

degradation (McCook *et al.* 2001) as seaweeds not only tolerate but thrive in environments that are suboptimal for corals and, once they are established, persist through ecological feedbacks that limit the ability for corals to recover (Mumby and Steneck 2011).

   The global environment is changing and the future does not bode well for coral reefs as they are affected by both global and local or regional environmental pressures. Climate change will further warm the oceans and cause more coral bleaching and coral mortality and lead to higher intensity storms that will damage reefs and sea-level rise. In addition, burning of fossil fuels will increase ocean acidification caused by more dissolved carbon dioxide which will change coral reef communities in ways that we are only beginning to understand (Anthony *et al.* 2011; Fabricius *et al.* 2011) Increasing resource exploitation (fishing, mining) and increased agricultural, urban and industrial development of coasts and catchments can lead to local pressures on coral reefs that facilitate excess seaweed biomass, such as overfishing of herbivores and destructive fishing practices, eutrophication of marine waters (Fabricius 2011) and higher numbers of introduced marine pests due to increasing global shipping trade and aquaculture of non-native seaweeds (Schaffelke and Hewitt 2007; Andreakis and Schaffelke 2012).

   In an assessment of the vulnerability to climate change, Diaz-Pulido *et al.* (2007b) paint a picture of what coral reef seaweed communities may look like in the future: CCA and other calcifying algae will be less abundant and reefs with low herbivory will have high algal cover and biomass, dominated by non-calcified taxa that are often ephemeral and fast-growing, quickly recover/colonize substratum after disturbance and tolerate high temperature and UV light, and fluctuating light (PAR) and nutrient availability and may benefit from an increase in carbon dioxide. Our knowledge of the community dynamics, physiology, and ecology of seaweeds is still limited compared to other major groups of benthic organisms on coral reefs, however, the consequences of this scenario are obvious for the function of the reef ecosystem as a whole. While many of the anthropogenic pressures can be addressed, especially at the local scale, management and conservation efforts to protect coral reefs need to be more effective and comprehensive, including a global effort to reduce greenhouse gas emissions (Burke *et al.* 2011).

* *Britta Schaffelke is a principal research scientist at the Australian Institute of Marine Science, Townsville, Australia. From her seaweed research she has branched out into more general research and management of environmental impacts, especially those related to deteriorating marine water quality. Her current work focuses on the measurement of ecosystem health responses to pollution and coastal development in the Great Barrier Reef.*

methods in order to assess the other toxicants. Ammonia can be removed (taken up) by the ammonia tolerant *Ulva lactuca*, by the use of zeolite (a hydrated aluminosilicate mineral with a negative charge that can absorb NH_4^+), or by aeration (increase pH to 10 and bubble) (Burgess *et al.* 2003).

   The maximum value for ammonium found in the surface waters over the White's Point sewage outfall, off Los Angeles, was 35 µM, and more often the concentrations were 5–10 µM. This represents a dilution of about 100-fold compared with the discharged concentration. Ammonium concentrations of 10–30 µM are not toxic to algae, but at ~200 µM dinoflagellates are more sensitive than diatoms. Chlorophyceae seem to be more tolerant of sewage toxicity than are many phytoplankton species. *Ulva linza* grew well in full-strength sewage effluent, even though the ammonium concentration was 500 µM (Chan *et al.* 1982). *Ulva*

compressa appears to be more sensitive, showing inhibition of photosynthesis when exposed to about 75 µM NH_4. The germination rates for the zygotes of three species of *Sargassum* were 50 and 0% for secondary effluent with and without ammonium (3.5 mM), respectively, but toxicity was reduced by 50% when the pH was <7 (Ogawa 1984); residual chlorine concentrations >3 mg L^{-1} and an anionic surfactant were also toxic. The inhibition of growth that has been observed near sewage outfalls probably is not due to an excessively high ammonium per se, but rather to a combination of high ammonium concentrations and other inhibiting factors such as heavy metals, chlorinated compounds such as chloramine in chlorinated sewage and possibly surfactants.

   In addition to the direct inhibitory effects of sewage on macroalgae, secondary or indirect effects may account for macroalgal decline in eutrophic waters.

There is evidence that some decrease in macroalgal abundance is due to increased growth and shading by epiphytes and filamentous algae associated with beds of macrophytes, as well as increased turbidity in the surface layers because of phytoplankton growth (Schramm 1999; McGlathery *et al.* 2007).

Other components of concern in sewage are detergents (which can cause oxygen depletion because of the organic load), surfactants and sewage sludge with metals and possibly PCBs concentrated in the sludge. Anionic detergents/surfactants account for the bulk of the detergents in household sewage, and three of these, sodium lauryl ether sulfate, sodium dodecylbenzenesulfonate (SDBS), and DOBS 055, were tested on *Laminaria hyperborea* (Hopkin and Kain 1978). SDBS and DOBS 055 reduced the growth of sporophytes at concentrations of 1–10 mg L^{-1}, and gametophyte germination was also inhibited by SDBS. The anionic and neutral surfactants, sodium dodecyl sulfate and Triton X-100 had no effect on tissue incorporation of the metals. The toxicity of detergents and surfactants is attributed to disruption of cellular and intracellular membranes. Indeed, detergents such as Triton X-l00 are used in research to help extract cell components.

9.6.3 Other anthropogenic nutrient sources

There are a large number of other anthropogenic nutrient sources such as fertilizers, animal farming, groundwater, and atmospheric deposition. The use of the two nitrogen stable isotopes ^{14}N and ^{15}N have been used to determine which N source makes the largest contribution to macroalgal growth since various N pollutants have distinguishable $^{15}N/^{14}N$ ratios. By measuring the ratio of ^{15}N to ^{14}N in the dried seaweed tissue and comparing it to a worldwide standard, the relative amount of ^{15}N or $\delta^{15}N$ can be determined (Heaton, 1986; see Essay 3 in Chapter 5). The $\delta^{15}N$ values of various environmental components must also be known and they are used to calibrate and interpret the $\delta^{15}N$ values that are obtained in the algal tissue. For example, the main method of nitrate and ammonium fertilizer production is by industrial fixation of atmospheric nitrogen, resulting in a $\delta^{15}N$ value of ~0‰. In animal or sewage waste,

nitrogen is excreted mainly as urea that is converted to ammonia, some of which is lost to the atmosphere. The remaining ammonia is enriched in ^{15}N and the elevated $\delta^{15}N$ value of about 6–10‰ (depending on the amount of denitrification and ammonification during the treatment process) is distinguishable from fertilizer N with a value of 0‰. The $\delta^{15}N$ of the particulate matter in sewage ranges from 0 to 3‰. Oceanic dissolved inorganic nitrogen (DIN) and particulate organic nitrogen have a $\delta^{15}N$ value of about 3–7‰, while groundwater varies between -2 and 8‰.

By determining these end member $\delta^{15}N$ values (see above) for each site, it is possible to qualitatively determine the contribution of the various N sources for macrophytes' growth. Gartner *et al.* (2002) found that the $\delta^{15}N$ values in three seaweeds corresponded to the distribution of the sewage plume and the faster growing *Ulva australis* had a stronger $\delta^{15}N$ signal than the slow-growing leathery kelp *Ecklonia radiata*. A seasonal trend (higher $\delta^{15}N$) was suggested to be related to faster growth in winter due to the availability of nutrients. They estimated that *U. australis* assimilated ~25–90% of its N requirements from sewage. Similarly, in a global survey of $\delta^{15}N$ values in *Ulva* sp., Teichberg *et al.* (2010) found high values in widely varying areas and concluded that *Ulva* sp. is a good sentinel species for detecting sewage inputs. Cohen and Fong (2005) found that *Ulva* does not fractionate or select for ^{14}N over ^{15}N and suggested that in this alga, the isotope ratio directly reflected what is available in the water. The $\delta^{15}N$ values of the red seaweed *Acanthophora spicifera* (2–6‰) that were incubated for 3 days in the effluent from a shrimp farm indicated that the shrimp farm was the main N source and not agricultural runoff or sewage (Lin and Fong 2008). Stable C and N isotopes were used to assess the recovery from sewage pollution after the closure of a sewage treatment plant in New Zealand (Rogers 2003). A comparison of the $\delta^{15}N$ values for *Ulva lactuca* before the closure and 3 months after the closure of the treatment plant revealed that the $\delta^{15}N$ values became heavier during the 3-month decontamination period and finally reached the same value as the control sites. Very similar results were obtained during the tracking of sewage nitrogen in a gradient away from the sewage discharge site (Tucker *et al.* 1999). They found that *Ulva* became heavier (6‰ and increased to 14‰) about 18 km from the outfall site. The use of $\delta^{15}N$ to assess sewage stress on coral reefs has been demonstrated by Risk *et al.* (2009) where there is about +3‰ increase for each trophic level. Recently, Raimonet *et al.* (2013) questioned whether macroalgal $\delta^{15}N$ values are an indicator of anthropogenic nitrogen inputs or variations in macroalgal metabolism under different nutrient, light, and temperature conditions, and suggested further studies on N fractionation under different environmental conditions.

Anthropogenic atmospheric nitrogen deposition over the past century has mirrored the increases in anthropogenic fossil fuel emissions of NOx ($NO_3 + NO_2 + NH_4$). Since the 1950s, atmospheric deposition of N in the northern hemisphere has increased five-fold and over the last several decades the emission of "new" (in contrast to "recycled" N) nitrogen to the biosphere has doubled (Galloway *et al.* 1994). On a remote island near the Bahamas, episodic deposition of NO_3 (1–137 μM) and NH_4 (2–122 μM) represented a mean deposition rate of 0.2 mg DIN m^{-2} y^{-1} (Barile and Lapointe (2005). They estimated that this atmospheric N provided up to 20% of the new N necessary to meet the growth demands of macroalgae on coral reefs near the Bahamas. Since rain contains more inorganic N than P (molar ratios are 50 to 300 N:P), then P limitation may be induced after a N-rich rainfall event.

9.7   Radioactivity

Wide-scale radioactive contamination in the marine environment has predominately originated from atmospheric fallout associated with nuclear weapons testing programs and is deposited at low activities across large areas of the ocean. In contrast, point-source discharges from nuclear facilities occur at high concentrations and near specific coastal sites. Hence, radionuclides from these two sources show markedly different behavior which influences the uptake by organisms and the transfer through the food chain.

Radioactivity is measured by the radioactive disintegrations in a substance and one becquerel (Bq) is one disintegration per second. It replaces the old unit of activity, the curie (Ci), which is the amount of radioactivity displayed by 1 g of radium (^{226}Ra) and equals 3.7×10^{10} Bq. Some of the metals in seawater are radioactive, and normal seawater has a radioactivity level of ~0.01 kBq L^{-1}. *Fucus* and *Pyropia* normally have radioactivities of 0.2–0.6 kBq kg^{-1} (wet wt), and the levels in mollusks and fish are 0.4–1 kBq kg^{-1} (Gerlach 1982). The levels of radioactivity may be even higher in certain areas, especially near wastes from nuclear power plants.

Seaweeds near nuclear reprocessing plants readily accumulate radioactivity. There are two European nuclear waste reprocessing plants in La Hague (France) and Sellafield (Irish Sea, UK) that discharge various radionuclides such as ^{129}I, ^{137}Cs, etc. Since the amount of ^{129}I released from these two reprocessing plants is 10-fold greater than natural levels, ratios of ^{129}I to other isotopes have been used as a tracer for oceanic currents and transport times. By analyzing archived samples of *Fucus vesiculosus* and *F. distichus*, transport times of isotopes from the Irish Sea and France, into the Baltic Sea and Arctic have been estimated (Kershaw *et al.* 1999; Hou *et al.* 2000; Yiou *et al.* 2002). The impact of anthropogenic ^{129}I on Irish coastal waters revealed that ^{129}I is 100-times higher on the NE coast near the Sellafield nuclear reprocessing plant compared to the west coast, as observed in samples of *Fucus vesiculosus* and in seawater (Keogh *et al.* 2007). Human health is a concern if seaweeds near nuclear plants are consumed. In the vicinity of Sellafield contamination of *Pyropia*, especially with ^{106}Ru, was ~35 times normal. For residents of south Wales, "laver bread" made from *Pyropia* is a food specialty, and by regularly consuming it, they can expose themselves to considerable radiation (Gerlach 1982).

The Fukushima nuclear power plant in northeastern Japan is one of the world's largest plants. A large earthquake on Mar 11, 2011, followed by a tsunami damaged the plant and resulted in the release of radioactivity into the air and ocean. Two of the radionuclides of interest were ^{131}I with an 8-day half-life and ^{137}Cs with a 30-year half-life (Buesseler *et al.* 2011). There was a peak in ocean discharge 1 month after the accident and a rapid decrease by 1000 times in the following month. After four months, ^{137}Cs was 10 000 times normal levels. Presently, dose calculations suggest minimal impact on marine biota and humans due to direct exposure to surrounding waters. In terms of potential biological impacts, radiation doses are generally dominated by naturally occurring radionuclides such as ^{210}Pb. To be comparable to naturally occurring doses of ^{210}Pb, ^{137}Cs levels in fish would need to be 1–3 orders of magnitude higher than what was observed off of Japan (Buesseler *et al.* 2012). However, biological uptake and concentration by food consumption is still under investigation (Buesseler *et al.* 2012).

9.8   Thermal pollution

In the 1970s and early '80s, the effect of thermal pollution was a more active research area but since then, fewer studies have been conducted (Langford 1990; Laws 2000). In this section, some early field examples of thermal impacts on macroalgae are given along with a suggestion that the thermal gradients found near power plants could be viewed as natural long-term experiments to examine the potential effects of global warming/climate change.

Conventional electric power plants account for about 75% of the thermal pollution in the USA, while nuclear power plants and industries such as refineries are minor. The electric power plant uses fossil fuel or nuclear energy to produce steam that is used to drive a turbine. The steam is recaptured in a condenser that is cooled by taking in cool ambient water and discharging heated water. In addition to death from the heated discharge, organisms may be killed or stressed by passage through screens. Some organisms are sensitive to chlorine gas or other biocides that are used periodically to prevent fouling in the piping system. Because there are often multiple stressors, it is difficult to attribute any community effects solely to thermal impacts. Some operations such as cooling the heated effluent to near ambient temperatures by using canals and recycling the cooling water in a closed system operation are used to reduce thermal impacts.

Many northern European countries now recommend that when a new power plant is to be built and is to be cooled by seawater, the temperature increase that will result after mixing cannot exceed 2°C, and in the summer the temperature of the mixed water cannot exceed 26°C. In spite of such restrictions, the species composition of the affected area may change, even in the absence of thermal damage to individuals, because the increase in temperature may affect species competition.

The effects of the thermal discharge on macroalgae can be either deleterious or beneficial, depending on the geographic location, the season, and the species involved. A change in algal species composition may affect the species composition of herbivores and animals higher up the food chain. Other deleterious effects can occur if there is an abrupt plant shutdown, which can deliver a cold shock to the macroalgae. Unlike higher plants, which may be exposed to a wide temperature range (~50°C), seaweeds generally exist in a much narrower range (~10–25°C). Whether macroalgae can survive the increased water temperature will depend on how close they are to their upper limit of temperature tolerance (sec. 7.3.4). This temperature tolerance is not constant for a species and may depend on other environmental factors, such as light, salinity, nutrients, and pollutants (Laws 2000). For example, during summer in temperate shallow bays or tropical areas, the ambient water temperature may already be near the upper range of tolerable temperatures. On the other hand, in spring and early summer in most temperate coastal areas with high rates of water exchange, localized increases in temperature usually will result in increases in growth rates and primary productivity. There have been no studies on the effect of the thermal discharge on reproduction and growth of germlings and young sporophytes for macroalgae.

Three thermal related disorders that have been observed are black rot, tumor-like swellings, and stipe rot. Black rot is a darkening of the blades that usually appears first at the tips, and then spreads toward the base. On the Atlantic coast of North America, a prolonged but intermittent thermal stress was reported to affect the growing (apical) tips of *Ascophyllum nodosum* (Vadas *et al.* 1978) (Fig. 9.13). Significant declines

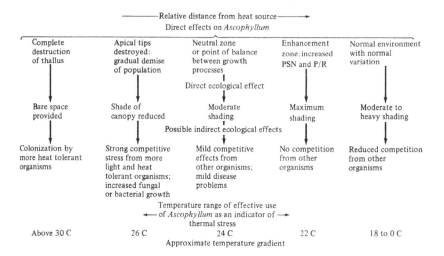

---Relative distance from heat source---→

Direct effects on *Ascophyllum*

Complete destruction of thallus	Apical tips destroyed: gradual demise of population	Neutral zone or point of balance between growth processes	Enhancement zone: increased PSN and P/R	Normal environment with normal variation

Direct ecological effect

Bare space provided	Shade of canopy reduced	Moderate shading	Maximum shading	Moderate to heavy shading

Possible indirect ecological effects

Colonization by more heat tolerant organisms	Strong competitive stress from more light and heat tolerant organisms; increased fungal or bacterial growth	Mild competitive effects from other organisms; mild disease problems	No competition from other organisms	Reduced competition from other organisms

Temperature range of effective use
←--- of *Ascophyllum* as an indicator of ---→
thermal stress

Above 30 C	26 C	24 C	22 C	18 to 0 C

Approximate temperature gradient

Fig. 9.13 Schematic showing potential impacts of elevated temperatures on populations of *Ascophyllum nodosum*, including photosynthesis (PSN) and primary productivity (P/R). (From Vadas *et al.* 1978.)

in percentage cover, biomass, growth, and survival were reported for *Ascophyllum*, but the basal sections of these algae survived, although weakly attached to the substrate. Apical meristems were not initiated in the second spring after the onset of the thermal discharge, but in later years, the population recovered fully. However, the cover of *Fucus vesiculosus* decreased when the thermal discharge began, and it never re-established itself.

Thermal discharge can lead to sub-lethal interactive effects. An increase in the water temperature will increase the respiration rate of organisms, but reduce the solubility of O_2 in water; a 10°C increase in temperature decreases O_2 solubility by 20%. In addition, the heated discharge tends to form a layer on the surface, thereby increasing the thermal stratification and reducing the transport of oxygen to deeper waters and the mixing of nutrients to the surface.

Recently, thermal discharge from power stations has been used as a natural experiment to assess the potential future impacts of climate change and elevated water temperatures (Keser *et al.* 2005). In the thermal plume, enhanced *Ascophyllum nodosum* growth correlated to thermal input from the power plant until the temperature reached 25°C. Above 25°C, growth rates decreased rapidly and mortality occurred when

temperatures were >27-28°C. Since water temperatures reach 22-23°C in eastern Long Island, then a further increase of 4-5°C due to climate change could lead to the demise of *Ascophyllum* (Keser *et al.* 2005). A similar study in California was conducted to gain insights into the effects on a benthic community of a 3°C temperature increase in the thermal plume, possibly similar to a potential increase due to climate change (Schiel *et al.* 2004). They found that the dominant algae, *Nereocystis*, *Pterygophora*, and *Laminaria* declined dramatically in areas of the thermal plume that was about 3°C above ambient. These observations suggest that a 3°C increase due to climate change could have dramatic effects on species composition in algal communities.

9.9 Synopsis

Pollution includes human additions of deleterious materials and energy into the environment. The effects of pollutants on macroalgae can be lethal (acute) or sub-lethal (chronic), and the effects are assessed with bioassay experiments which should be conducted both in the laboratory and in the field. The physico-chemical aspects, such as solubility, adsorption, chemical complexation, and speciation, are extremely important in quantifying the effects of a pollutant. If the contaminant has various chemical forms/species, the total concentration of a contaminant may give little indication of its toxicity since some forms may not be bioavailable. There are obvious limitations to laboratory bioassays, especially because they do not contain natural suspended particulates, which are known to radically reduce the toxicity of pollutants such as heavy metals through adsorption in estuarine areas. Synergistic and antagonistic interactions of the pollutant with salinity, temperature, light, and other pollutants should be investigated. The choice of bioassay organism, its life-history stage, and its potential for long-term recovery are also important in pollution assessments.

The degree of heavy metal toxicity is influenced by the type of metal ion, the amount of particulates in the water, and the algal species. Generally, the order of metal toxicities for seaweeds is Hg > Cu > Cd > Ag > Pb > Zn. Because metal toxicity usually occurs only when the metal exists as a free ion, adsorption of the ions onto particles may be a very significant detoxification process in some environments. Macroalgae have several mechanisms to detoxify the metals or to increase their tolerance. Extracellularly, metals may be detoxified by binding to algal extracellular products. Exclusion of the metal ion may occur at the cell wall via binding to cell wall polysaccharides, or at the cell membrane via changes in the transport properties of the membrane. Intracellularly, metals may undergo changes in valence, or may be converted into nontoxic organometallic compounds. Intracellular precipitation within vacuoles and nuclei has been observed for copper. If significant detoxification does not occur, the metal ion may inhibit the functioning of enzyme systems, eliciting the following responses: cessation of growth due partially to the inhibition of photosynthesis, reduction of chlorophyll content, an increase in cell permeability, and loss of K^+ from the cell. Metals are concentrated several orders of magnitude above ambient seawater and concentration factors may range up to 10 000 times. This process is termed bioaccumulation and further bioaccumulation up the food chain is termed biomagnification.

Generally, metals never constitute a threat to the marine environment other than in estuarine or poorly flushed areas. The difference between the natural concentration and that at which acute effects are observable is normally several orders of magnitude, except for Cu where this difference is smaller. It is the more insidious sub-lethal effects that are most likely to be encountered, and they can occur at concentrations more than an order of magnitude lower than the concentrations that will produce acute effects. The sub-lethal effects are determined using a dose–response curve where the response is measured by relative growth rate. The trend toward increasingly more sensitive biochemical/molecular indices may eventually enable us to detect effects at even lower levels. Sub-lethal effects have been demonstrated under laboratory conditions, but they have rarely been identified under natural and complex field conditions. Sub-lethal and acute toxicities are critically dependent upon the stage of development of an organism. Reproduction and early developmental stages generally are the most vulnerable. Unfortunately, the life-cycle studies that are required to examine the sub-lethal effects as a result of prolonged exposure to a contaminant are complex and expensive, but are needed to determine real effects of trace metal contamination in macroalgal populations.

Petroleum is an extremely complex mixture of hydrocarbons, including alkanes, cycloalkanes, and aromatics. Oil can reduce photosynthesis and growth in macrophytes by preventing gas exchange, disrupting plastid membranes, destroying chlorophyll, and altering cell permeability. In some cases, penetration of oil is reduced by the mucilaginous coating, especially on some brown seaweeds. The components that penetrate the thallus most easily and hence are the most toxic are the lower molecular weight, volatile, lipophilic compounds, including the aromatics. Alkanes are least toxic. In the laboratory, the concentrations at which oil will be toxic will depend on the type of oil, how the extract was prepared, and how soon it was used since the most toxic components are volatile, as well as on the water temperature and the presence of other pollutants or dispersants. Additional factors in the field that can influence oil toxicity include the proximity of the spill to the shore and weather conditions, especially wind which can mix the oil or push it onto the shore. Rocky intertidal areas suffer slight, short-term effects from oil spills, whereas the impacts on salt marshes and coral reefs are severe and longer term. Weathering of oil occurs by a number of processes, the most important of which is evaporation of the most toxic compounds. Herbivores often are more susceptible to oil than macroalgae, and often an increase in ephemeral macroalgal biomass is a response to the reduced grazing pressure.

Antifouling paints are one of the main groups of synthetic organic chemicals that affect seaweeds. Biofouling is the development of biofilms and the attachment of macroalgae to submerged surfaces and it is a very costly problem. New antifouling compounds such as "booster biocides" are being developed because tributyl tin has been banned since 2008. They require further testing to determine their toxicity to all of the life-cycle stages of macroalgae, especially the early stages that are more sensitive. However, tributyl tin continues to leak from boat hulls and marinas, and hence it is still a concern.

Increased supplies of nutrients, especially nitrogen enrichment include, sewage discharge, atmospheric deposition, and groundwater discharge. Nutrient enrichment produces changes in community structure and diversity, especially an increase in fast-growing greens and an increase in epiphytism. The stable isotope of nitrogen ^{15}N is routinely used to determine the relative contribution of N from various N sources that is being used for macrophyte growth. Macrophytes near a sewage outfall tend to show relatively higher net primary productivity, smaller growth forms, and simpler and shorter life histories; most are components of early successional stages. Browns tend to be more sensitive than reds and greens. NH_4 may be toxic close to the sewage discharge site. In tropical waters, corals are very susceptible to nutrient enrichment and reefs degrade to beds of macroalgae if there is chronic eutrophication.

Thermal pollution, originating primarily from the cooling water discharged from power plants, can be stimulatory if the water temperature does not rise above the optimal temperature for growth of a species. Thermal stress on seaweeds has occurred off southern California, and the symptoms include frond hardening, bleaching or darkening, and cellular plasmolysis in *Macrocystis*. Most of the ecosystems that have been studied to date have shown remarkable abilities to recover when the source of the pollutant has been removed. Most of the effects are local and confined to coastal areas, where point sources of pollutants predominate. In many cases, the animals were found to be more sensitive than the macrophytes, resulting in a decrease in grazing and an increase in some species of seaweeds. Recently, several studies have used the gradient in thermal pollution at a power plant to examine the long-term species impacts and genotypic adaptation and these sites could act as a surrogate for global warming/climate change.

Although there is a temptation to generalize about pollutant effects, extreme caution is warranted in view of the large number of environmental and physiological factors that influence toxicity, notably the wide range of tolerances displayed by different organisms. Genetic adaptation to pollutants is important, but there are only a very few well-documented examples to date. Most studies have examined phenotypic acclimation (i.e. short-term responses over a few days or weeks). In addition, indirect effects caused by the elimination of sensitive species could have far greater significance for marine communities than is indicated in toxicity studies with single species.

Seaweed mariculture

10.1 Introduction

As seaweed consumption has increased in the last several decades, seaweed mariculture has filled the gap between wild stock harvest and the present demand. Ancient records show that people collected seaweeds for food starting in about 2500 BP in China (Tseng 1981), and 1500 in Europe (Critchley and Ohno 1998). Presently, the wild harvest of seaweeds is about 1.8 m tonnes y^{-1}, mainly brown seaweeds used for alginates (FAO 2009). In Japan, China, and other Asian countries, where seaweeds have long composed an important part of the human diet, seaweed farming is a major business and over 90% of the seaweed production is from farming for human consumption. Since 1970, the culture of seaweeds has increased at ~8% per year (FAO 2009). Seaweed production from farming nearly doubled from 8.8 to 15.9 million tonnes from 1999 to 2008, with a value of US $7.4 billion (FAO 2010). Most of the world seaweed supply comes from aquaculture and seaweeds were the first to pass the 50% farmed/wild harvest threshold in 1971, compared to fish aquaculture that will exceed the 50% threshold by 2012 (Chopin 2012). About 99% of the farmed production is in Asia and over 70% of the production (10.9 million tonnes) is in China, followed by Indonesia, the Philippines, South Korea, and Japan. Chile is the most important producer outside of Asia with a production of 90 000 tonnes y^{-1} of wild harvested seaweeds. Table 10.1 illustrates the production, value, price and the three main producing countries for the six most important seaweed genera that are grown in aquaculture

systems. Brown seaweeds compose about 64% of the production (67% of the value), reds about 36% (33% of the value), and greens, with ~99% being produced by Asian countries, 0.2% of the production and value (Chopin and Sawhney 2009). There has been a rapid increase in production in the last decade, especially of reds and browns (Fig. 10.1). The largest production (4.6 million tonnes; Table 10.1) is from *Saccharina japonica* (previously *Laminaria japonica*; or kombu in Japan or haidai in China), mainly in China. Korea grows mainly *Undaria pinnatifida* (wakame) with 1.8 million tonnes annually and *Pyropia* (previously *Porphyra*, or nori), while Japan focuses mainly on *Pyropia*. However, due to the high price of nori (*Pyropia yezoensis* and *P. tenera*), Japan is the second most important country in terms of the value of the seaweeds produced (US$1.2 billion) (FAO 2010). Nori (US$16 000/dry tonne) is about twice the price of wakame and about five times the price of kombu (McHugh 2004). Commercial harvesting has increased significantly from a few countries in the 1980s to about 35 countries presently. According to FAO (2010), seaweed aquaculture was 23% of the total world aquaculture production (value of ~$7.4 billion), with 19% represented by mollusks and 50% attributed to fish aquaculture. When considering the world mariculture production (i.e. aquaculture production in seawater), the fact that seaweeds represent 46% of the biomass is a surprise to many. An authoritative reference system on world seaweed resources is available on a DVD (Critchley *et al.* 2006) and it is an update of the book *Seaweed Resources of the*

Table 10.1 Aquaculture production (million tonnes), value (billion US$), price per ton (US$ t⁻¹), and the three main producing countries of the six most important seaweed genera. (Modified from FAO 2009; Pereira and Yarish 2010.)

Genus	Production (million tonnes)	Value (billion US$)	Price per tonne (US$ t⁻¹)	3 main producing countries
Saccharina	4.6	2.89	627	China (98%), Japan, S Korea
Undaria	1.8	0.79	448	China (87%), S Korea, Japan
Pyropia	1.5	1.54	1020	China (58%), Japan, S Korea
Gracilaria	1.1	0.52	490	China (94%), Vietnam, Chile
Eucheuma and *Kappaphycus*	3.2	0.55	174	the Philippines (94%), Indonesia, China

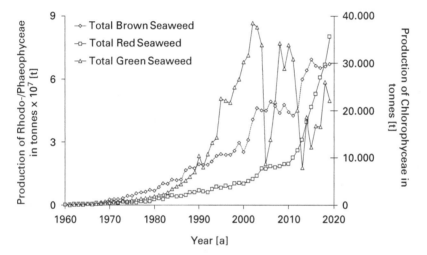

Fig. 10.1 Commercial production of red, brown, and green macroalgae over the last 50 years. (With permission from Buchholz *et al.* 2012.)

World (Critchley and Ohno 1998). Guiry's website (www.seaweed.ie) is also a broad source of information on seaweed production, aquaculture, and their uses.

Seaweeds have a very wide range of uses: from human food, animal feeds, biopolymers (phycocolloids), chemicals, agrichemicals, cosmetics, pharmaceuticals, nutraceuticals to bioenergy compounds. Recently they have been promoted in polyculture systems referred to as integrated multi-trophic aquaculture (IMTA) where they take up nutrients produced by invertebrate and vertebrate animals (Chopin *et al.* 2001; Troell *et al.* 2009; Buchholz *et al.* 2012). The use

of seaweeds for human consumption is very significant in terms of tonnage (8.6 million wet tonnes or 76% of total seaweed usage) and value (US$5.3 billion or 88%), compared to the use by the phycocolloid industry (1.3 million wet tonnes or 11%) and value (US$650 million or 11%) (Chopin and Sawhney 2009). Three genera, *Saccharina* (as kombu), *Pyropia* (as nori), and *Undaria* (as wakame) dominate the edible seaweed market. The value of nutrient bioextraction or phytomitigation (the removal of excessive nutrients in the coastal zone) is difficult to value at present, since there are few guidelines on the discharge and required removal of excessive nutrients into coastal waters. If

Table 10.2 Seaweed phycocolloid production (1000 tons), average price (US$ kg⁻¹) and value (million US$) during the decade 1999–2009. (Modified from Bixler and Porse 2011.)

Phycocolloid	Production (1000 t)		Price (US$ kg^{-1})		Sales (millions US$)	
	1999	2009	1999	2009	1999	2009
Agar	7.5	9.6	17	18	128	173
Alginates	23.0	26.5	9	12	225	318
Carrageenans	42.0	50.0	8	10.5	291	527
Total	72.5	86.1			644	1018

we estimate an average composition for seaweeds of around 0.35% N, 0.04% P ,and 3% C, the nutrient-trading credits (NTC) should be around US$10–30 kg⁻¹, US$4 kg⁻¹ and US$30 t⁻¹ for N, P, and C, respectively. Thus the ecosystem services of cultivated seaweeds (15.8 million tons) are worth at least US$590 million to US$1.7 billion. This is ~23% of their present commercial value (US$7.4 billion; Chopin *et al.* 2012). We all use seaweed products in our daily life. As part of the human diet, seaweeds provide protein, vitamins, and minerals (especially iodine from kelps). In addition, three commercially important phycocolloids, agars, carrageenans, and alginates are extracted from red and brown algae. Their sales volume, value and price in 2009, generally show a significant increase in the last decade (Table 10.2). Agars obtained from *Gelidium* are used extensively in microbiology and tissue culture for solidifying growth media, and more recently in electrophoretic gels. The agar from *Gracilaria* is used mainly in foods. Carrageenans, chiefly from *Kappaphycus*, *Eucheuma*, and *Chondrus*, are widely used as thickeners in dairy products. Alginates, from *Saccharina*, *Laminaria*, *Ascophyllum*, and other kelps or fucoids, are also used as thickeners in a multitude of products ranging from salad dressings to oil-drilling fluids to the coatings in paper manufacture (McHugh 2004; Bixler and Porse 2011).

There are approximately 10 500 known species of seaweeds. Around 500 have been used for centuries for human food and medicine. However, only ~220 species of seaweeds are cultivated worldwide (Chopin and Sawhney 2009). Six genera provide 95% of the seaweed aquaculture production and 98% of its value:

Saccharina (~40% production and 48% value), *Undaria* (~22% production and 18% value), *Pyropia* (~12% production and 23% value), *Kappaphycus/Eucheuma* (~12% production and 2% value), and *Gracilaria* (~8% production and 7% value). Success in seaweed farming is attributed to understanding complex life histories, regenerative capacity of thalli, prolific spore production, and understanding interactions amongst light, temperature, and nutrients. *Kappaphycus/Eucheuma* and *Gracilaria* can be farmed by vegetative propagation, while *Pyropia*, *Saccharina*, and *Undaria* must be propagated from spores. Much of the farming in Asia is in large-scale open-water systems, but some species are cultivated in tanks. Successful mariculture depends on an extensive basic knowledge of the biology and physiology of the seaweeds, strain selection, and how environmental factors that are important to seaweed growth can be manipulated to improve yields. In the following sections, the mariculture practices used in growing six of the most common commercially grown seaweeds, starting with food producers, followed by phycocolloid-producing genera will be summarized. The application of the ecological and physiological principles described in the preceding chapters will be highlighted, including some discussion of the products from these species.

10.2 *Pyropia/Porphyra* mariculture

Pyropia yezoensis (previously known as *Porphyra yezoensis*, Sutherland *et al.* 2011) is one of the most commercially valuable seaweeds. In 2008, about

1.5 million tonnes were produced annually with an annual value of about US$1.5 billion (Table 10.1; FAO 2009; Pereira and Yarish 2010). China produced 800 000 wet tonnes, followed by Japan with 337 000 tonnes and Korea with 224 000 tonnes (FAO 2010). About 90% of *P. yezoensis* is grown in Chinese northern provinces and exported to international markets, while *P. haitanensis* is grown in southern provinces and used domestically in China (FAO 2005). Based on molecular phylogenetics, species that were within the genus *Porphyra* have been split into other genera, with most being treated as *Pyropia* (Sutherland *et al.* 2011; see also sec. 1.4.1). There are more than 50 *Pyropia* species worldwide (www.algaebase.org). Five *Pyropia* species are used in commercial cultivation (*P. yezoensis, P. tenera, P. haitanensis, P. pseudolinearis,* and *P. dentata*), with the first three species accounting for most of the total production (Yarish and Pereira, 2008; Pereira and Yarish 2010). In Japan, *P. yezoensis* f. *narawaensis* has become the most commonly used strain in nori farms because of its vigorous growth rate and more elongate blades (Niwa *et al.* 2009). In the 1980s, a very small *Pyropia* cultivation industry in the state of Washington, USA, emerged after 10 years of development (Mumford 1990). Even though cultivation was biologically feasible, economically viable and the products were of high quality, further development has stopped due to the difficulty in obtaining permits for use of water areas that are also used for recreation and concerns of visual pollution by water-front land owners. Neefus *et al.* (2008) are working on the growth and cultivation of three Asiatic *Pyropia* species that were introduced to the northwest Atlantic.

Pyropia is used primarily in the food industry, especially for the wrapping around sushi and it is known as *nori* in Japan, *gim* in Korea, *zicai* in China, and "purple laver" in Great Britain. Consumption has been expanding worldwide, particularly in North America, over the past decade. This is due in part to Japanese marketing efforts and the increasing consumption of Japanese cuisine. Korean production and exports are increasing and about 25% of their production is shipped to the USA. *Pyropia* has highly digestible protein (20–25% wet weight) and free amino acids (especially glutamic acid, glycine, and alanine), which are responsible for its specific taste. The vitamin C content of *Pyropia* is similar to that in lemons, and it is also rich in B vitamins. It is an excellent source of iodine and other trace elements (Keiji and Kanji 1989; Burtin 2003; MacArtain *et al.* 2007).

10.2.1 Biology

The life cycle of *Pyropia* (and *Porphyra*) involves a heteromorphic (biphasic) alternation of generations (see Fig. 2.7). It is the foliose gametophyte that is eaten. The blade can be yellow, olive, pink, or purple, 1 or 2 cells thick and up to 1 m in length. The blade can reproduce asexually in some species by means of archeospores or aplanospores. In the commercially cultivated species of *Pyropia*, sexual reproduction occurs under the stimuli of increasing day length and rising temperatures. Male gametes are released from the spermatangium and fuse with the female cell (carpogonium). Following fertilization, division of the carpogonium is mitotic, forming packets of diploid carpospores. The released zygotospores develop into the conchocelis phase (the diploid sporophyte consisting of microscopic filaments), which in the wild will bore into shells or other calcareous substrates, where it grows vegetatively. The conchocelis serves as a perennating stage in nature and it can also be maintained in laboratory cultures for long periods of time since it can reproduce asexually/vegetatively. In the presence of decreasing day length and falling temperatures, terminal cells of the conchocelis phase produce diploid conchospores inside conchosporangia. Meiosis occurs during the germination of the conchospore, producing the macroscopic gametophyte.

The great success of the nori industry is due to the discovery of various unknown life-history stages of *Porphyra umbilicalis*. Until 1949, no one knew where the spores came from. Drew's (1949) discovery that the previously described genus "*Conchocelis rosea*" is actually a stage in the *Pyropia/Porphyra* life cycle, transformed the nori industry. It allowed indoor mass cultivation of the filaments in sterilized oyster or clam shells and the "seeding" of conchospores directly onto nets for out-planting in the sea.

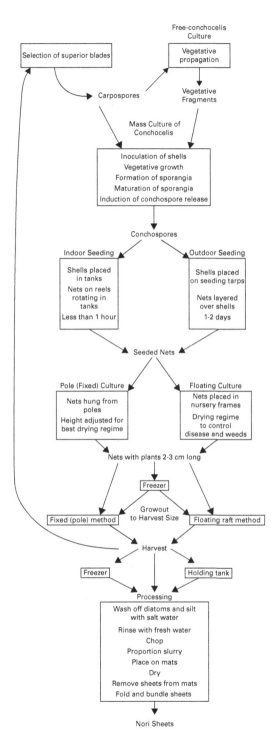

10.2.2 Cultivation

In the beginning, cultivation techniques were very simple. Enhancement of wild stocks was originally achieved by pushing tree branches or bamboo shoots into the mud on the bottom of the bay, or by clearing rock surfaces, so that when conchospores were liberated in early fall, they would have space for attachment and growth. Later, nets were used since they were more easily transported from the collecting grounds to the cultivation areas where they were strung between poles (FAO 2005).

The cultivation technique involves four main steps: (1) culture of the conchocelis and the release of conchospores; (2) seeding of nets with conchospores; (3) nursery rearing of the sporelings; and (4) harvesting. These steps are summarized in Figure 10.2. Mass culture of shells inoculated with conchocelis takes place in tanks in greenhouses (Tseng 1981; Sahoo and Yarish 2005). In February or March, *Pyropia* thalli are induced to release carpospores by being dried overnight and then reimmersed in seawater. Sterile oyster or scallop shells, or artificial substrata treated with calcite granules, are placed in seawater tanks with the fertile *Pyropia* blades, or sprinkled with a suspension of carpospores. It is also possible to take conchocelis filaments grown *in vitro* in a bioreactor, fragment them, and sprinkle them onto shells (Sahoo and Yarish 2005; He and Yarish, 2006). The carpospores germinate and bore into the shells or an artificial substrate under very low light at 10–15°C. Good growth of the conchocelis requires good light, nutrients, and stirring in the culture tanks at 15 to 25°C.

Light and temperature play a critical role in conchospore production. From early July to late August–September, the ambient water temperature increases from about 22 to 28–30°C and then gradually decreases. This is a critical time for formation of conchosporangia and maximal conchospore production, which are dependent on the interaction of

Fig. 10.2 Flow diagram for the production of *hoshinovi* (sheets of *Pyropia*) as practiced in Japan. (From Mumford and Miura 1988, with permission of Cambridge University Press.)

temperature and photoperiod. In the indoor tanks, the irradiance is reduced to ~15 μmol photons m^{-2} s^{-1} in early July and the ambient water temperature continues to rise. In September, when the ambient temperature has fallen to about 23°C, sporulation is encouraged by artificially reducing the photoperiod to 8-10 h per day and dropping the tank temperature to 17-18°C. Conchospores are collected on nets either by running nets through the indoor tanks containing the shells or by placing shells under nets in the field. Conchospore adherence and germination require brighter light, >50 μmol photons m^{-2} s^{-1}, and usually germination is carried out in the sea.

There are certain requirements for the successful seeding of conchospores on nets. The optimal settlement density of 2-5 conchospores mm^2 is achieved in only 8-10 min in a seeding tank, but it can take 1-5 days in the field. The nets are then attached to poles or rafts in the field for nursery cultivation. When the seaweeds reach 2-3 cm in length, they can be left to grow further, cut and repopulated by archeospores, thereby permitting up to 6-8 harvests per growing season. Nets may also be rolled up and frozen for up to 6 months or more, and then brought out to the field later. This allows for multiple harvests during the growing period, especially if the existing crop is destroyed by sporadic fungal or fungal-like parasites.

Several methods are used for suspending nets, depending on the depth of the water and the tidal amplitude (Fig. 10.3). In shallow areas, the nets are suspended from fixed poles (the pillar method), so that the seaweeds are regularly exposed to the atmosphere (Fig. 10.3a). If the tidal range is >2 m, the nets are attached to poles so that they rest just above the bottom at low tide, but float as the tide rises (the floating method) (Fig. 10.3c). This avoids too much shading by the water column. In deep water, *Pyropia* is grown on nets attached to floating rafts near the surface. Intertidal pole cultivation often is preferred because it ensures periodic exposure of the proper duration, which helps to reduce the incidence of disease and the growth of competitive (weed) species, especially epiphytic diatoms (Tseng 1981; Pereira and Yarish 2010). Presently, artificial seeding of

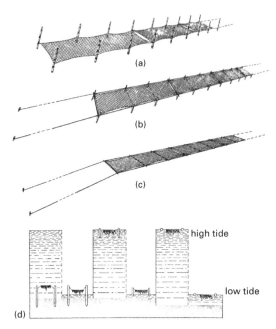

Fig. 10.3 Sketches showing three methods for *Pyropia* cultivation: (a) fixed type of the pillar method: (b) semifloating method: (c) floating method. (d) For each of the methods (a–c), the left side represents the position of the net during high tide, and the right side represents low tide. (From Tseng 1981, with permission of Blackwell Scientific Publications.)

conchospores from hatchery-grown conchocelis is used and the process begins in fall at about 18°C.

The maximal harvest is obtained by the right nutrient and temperature conditions. The optimum temperature for growth decreases as the thallus ages. Thus, the timing of out-planting is important. Delay will slow the growth of the germlings and result in a later initial harvest. The NH$_4^+$ and NO$_3^-$ concentrations must be >3 μM and the best quality nori is obtained when the nitrogen concentration is >15 μM. Fertilizer, such as ammonium sulfate used to be applied as a spray over the beds, or allowed to diffuse from bottles hung on the support poles. However, recent environmental regulations no longer allow this practice. The best quality seaweeds are those harvested from October to December under normal growth conditions. Harvesting is done every 10-20 days, and each net is harvested six or more times using automated harvesters. After the first net has been harvested several times, it may be replaced by another net brought from the freezer when available. This process is repeated at least three or four times until the growing season ends, usually in January or February, because of decreasing quality (Pereira and Yarish 2010).

The discovery that *Pyropia* germlings could survive deep freezing added a new dimension to nori farming. If thalli are allowed to dry to between 20 and 30% of their initial moisture content, and then frozen and stored at -20°C, they can resume normal growth as much as a year later. This practice can be used to extend the useful harvest period into March or April, and it also serves as insurance against failure of the early crop. Freezing the nets allows the farmer to produce more nursery sets than he has grow-out areas and permits flexibility in the control of disease and fouling.

There are several steps in the preparation of nori. The harvested thalli are thoroughly washed in seawater, and all epiphytes and dead tissues are removed. The thalli are chopped and made into a freshwater slurry, then spread over screens and dried. The finished product (in Japanese, *hoshi-nori*) can be eaten directly in sauces, soups, salads, and sushi or can undergo secondary processing. This involves toasting and seasoning the nori sheets for sushi.

10.2.3 Problems in *Pyropia* culture

Just as there are problems with weeds in land agriculture, there are similar problems in *Pyropia* cultivation. In general, there are two kinds of competitive weedy algae: green algae (usually *Monostroma*, *Ulva*, and *Urospora*) and diatoms (e.g. *Licmophora*). Attachment of the spores of these algae is prevented in three ways: (1) Archeospore production by young thalli increases seedling density, so that little space is left for weed spores to attach. The production nets are handled carefully so that the germlings will not be scraped off and no free space is left. (2) Only the distal ends of thalli are harvested, so the nets remain densely covered with nori. (3) If weed algae do establish on the nets, they can be killed off by exposing the net to the air for hours, because weed algae are more susceptible to desiccation than *Pyropia* (Mumford and Miura 1988).

Recently, blooms of the large diatom *Coscinodiscus wailesii* in late fall in the Seto Inland Sea have depleted nitrogen and caused bleaching of *Pyropia* since nitrogen is required for pigment (phycoerythrin) production (Nishikawa and Yamaguchi 2008). In Japan, very dense nets in protected bays can be damaged by a sudden rise in temperature. This often occurs in November when the seawater is very calm and surface waters warm due to the increased stratification. Such a rise in temperature will kill some of the young nori and the nets will become partially denuded, sometimes in large patches, resulting in low production and possible invasion of epiphytes. Another problem is grazing of *Pyropia* by herbivorous fish. If the problem is severe, special nets must be used to protect the crop (Tseng 1981).

Disease remains one of the biggest threats to nori producers, especially when high density farming is employed. Most of the diseases are fungal (Andrews 1976; Neill *et al.* 2008; Gachon *et al.* 2010). *Pyropia* is susceptible to "red rot" or "red wasting" disease caused by an oomycete fungus, *Pythium*, and to "chytrid blight", caused by a fungus, *Olpidiopsis* (Ding and Ma 2005). These two fungi often co-occur and cause mold spoilage and an unacceptable product. Screening for resistance levels to the fungus *Pythium*, which causes "Akagusare" disease in *Pyropia*, has recently been undertaken to determine the host–pathogen interactions involving a susceptible and resistant host (Uppalapati and Fujita 2001). In Japan, farmers have resisted the use of chemicals on this food crop, and many diseases are controlled by drying and/or freezing infected nets, and maintaining healthy seaweeds by good cultivation practices and other innovative methods (Ding and Ma 2005). For example, a green color mutant has been found to have high resistance to red-rot disease that can attack and quickly devastate *Pyropia*. This resistance may be due to the fact that the green mutant has relatively thick cell walls, and there seems to be an inverse relationship between susceptibility and cell wall thickness. Unfortunately, for decades the strain selection in *Pyropia* has been for thin cell walls (better taste and texture). It is doubtful that the green form will be acceptable to consumers because of its color and its thick walls. "Green-spot disease" is caused by pathogenic bacteria such as *Vibrio* and *Pseudomonas* (Neill *et al.* 2008). Each commercially cultivated seaweed is susceptible to one or more diseases, some due to pathogens. However, other perceived "diseases" are actually due to adverse physical conditions (Andrews 1976; Tseng 1981; Gachon *et al.* 2010). "Crown-gall disease" results in tumorous growths, possibly from carcinogenic substances in sewage (Tseng 1981). In Japan, fog has been known to cause serious damage to the nori crops, which are exposed to the atmosphere for part of each day. Sulfites in polluted air account for part of the cause, because they form sulfurous acid when they dissolve in water.

The use of only a few highly selected strains can lead to genetic uniformity and hence to an increased potential for crop failure through either disease or unfavorable conditions affecting all seaweeds similarly. There is a need to produce new conchocelis strains with genetic diversity from wild-collected gametophyte blades that possess better flavor, higher tolerance to disease, and deeper color. Niwa and Sakamoto (2010) demonstrated allodiploidy in natural and cultivated populations and this indicates the potential for development of polyploidy breeding in *Pyropia* in the future. As a safeguard, nets are now seeded with several strains. It has also been suggested that "gene banks" of conchocelis cultures be established to maintain the genetic diversity that is being threatened by widespread use of only a few cultivars (Mumford and Miura 1988). An artificial red mutant (IBY-R1) was isolated by heavy-ion beam mutagenesis and found to have more phycoerythrin (dark red/black – desirable color), higher free amino acids (alanine, and taurine – better taste), but lower yield in comparison to the wild type (Niwa *et al.* 2011). Niwa *et al.* (2008) determined that the superior taste of dried nori sheets was obtained from the first harvest since the blades were thinner and had a higher content of glutamic acid. During subsequent harvests from the same net, the blades became thicker and glutamic acid content decreased and hence the quality was decreased.

10.2.4   Future trends

Traditionally, the typical Japanese nori farmer was largely self-sufficient. The trend is now toward specialization. Now a farmer can buy conchocelis shells from a large firm that grows only conchocelis, or he can buy nets that are already seeded. Lastly, he can sell the raw *Pyropia* to a processor, rather than processing it himself or participating in a cooperative.

The lack of vegetative propagation capacity of the gametophytes makes the farming of *Pyropia* labor intensive and expensive since spores from the conchocelis are needed for every crop. In fact, the few species of *Pyropia* that are farmed are the ones that naturally produce archeospores (formerly known as monospores) from the gametophytes. The gametophytes of some species are able to form blade archeospores, a form of asexual spore that germinates to form a new blade. This allows for a large increase in production and up to 6 to 8 harvests per net (Pereira and Yarish 2010). Strains have been selected for a long, narrow shape, late maturation, and archeospore production. Narrow seaweeds give a greater yield on nets. When seaweeds become reproductive, their rapid growth is partially offset by erosion of the margin. Secondary settlement of archeospores on nets can help overcome an otherwise expensive insufficient initial seeding. Hafting (1999) propagated blades of *P. yezoensis* in suspension culture from archeospores by cutting blades into very small pieces. After ~3 weeks the blade pieces disintegrated, producing archeospores that grew into a dense suspension culture of blades. This finding should speed up production.

Free-living suspension cultures of conchocelis of *Pyropia leucosticta* have been grown in the laboratory at 15°C and moderate light. Conchosporangia formation was induced after four weeks by increasing the temperature to 20°C and decreasing the photoperiod to 8:16 L:D and maintaining it for 6 months (He and Yarish 2006). Release of conchospores was obtained by decreasing the temperature, increasing the light and photoperiod. The conchospores were able to attach, germinate and develop into juvenile blades on twine. Another method was developed by Saito *et al.* (2008) for the dioecious commercially valuable *P. pseudolinearis* which does not produce asexual spores such as archeo-spores. They induced the formation of archeosporan-gium by culturing with allontoin that has cytokinin-like properties to obtain free cells from the gametophyte of *P. pseudolinearis* and then cell wall destruction by mild agitation.

Vegetative propagation of the gametophyte via protoplasts could solve two problems. By elimination of the conchocelis phase, production costs could be lowered, and genetic diversity eliminated. By propagation of vegetative cells from the blade in tissue culture, much greater control can be maintained over desirable genotypes. Protoplast production and fusion techniques (Polne-Fuller and Gibor 1984; Chen 1986; Fujita and Migata 1986; Reddy *et al.* 2008a) have been successful for a number of *Pyropia* species (Polne-Fuller and Gibor 1990; Saito *et al.* 2008). The ability to isolate and regenerate viable vegetative cells allows induction and selection of desired mutations, as well as vegetative cloning of specific isolates. Another benefit of single cell and protoplast technology is the ability to bypass sexuality and thus maintain a pure gene pool.

Recent scientific advances in genetics, physiology, and biochemistry are quickly being applied in new production techniques. The "*Porphyra* Genome" project is underway (Gantt *et al.* 2010; sec. 1.4.1), while part of the genome of *Porphyra umbilicalis* is available (Chan *et al.* 2012a) and the transcriptome of *Porphyra purpurea* have recently been released by the Joint Genome Institute (US Dept Energy, www.jgi.doe.gov/genome-projects/ Program CSP2008: see also www.porphyra.org). The small genome size of about 2.7 Mbp consisting of three chromosomes and the short generation time of 1–3 months, make it suitable for genetic analysis with promising biotechnological and possible genetic engineering applications (Gantt *et al.* 2010).

10.3 *Saccharina/Laminaria* for food and alginates

Saccharina japonica (previously called *Laminaria japonica*) is the most commonly grown kelp and it constitutes ~40% of the world's seaweed production.

The most feasible way to increase production of kelps is through cultivation, because wild harvests have not increased over the past 20 years (Druehl 1988). Wild stocks of *Saccharina* were collected in northern Japan as early as the eighth century, and some was exported to China (Tseng 1981, 1987). Kelp cultivation was initiated in China and Japan in the early 1950s (Kawashima 1984; Brinkhuis *et al.* 1989), and more recently in Korea. China dominates the harvest of *Saccharina*, with over 4 million metric tonnes (wet weight) out of a worldwide harvest of nearly 5 million tonnes (Table 10.1; FAO 2010) since the Koreans prefer *Undaria* over *Saccharina* (McHugh 2004). On an aerial basis, production ranges from 50 to 130 metric tonnes per hectare (Druehl 1988). *Saccharina* is the best concentrator of iodine (10^5 times seawater concentrations) and therefore it is regularly consumed in China to prevent goiter and also used for alginate and mannitol production (Tseng 1987). *Saccharina* (known as *kombu* in Japan), grows best below 20°C and lives naturally in the low intertidal and upper subtidal in cool temperate environments. The harvestable sporophyte alternates with microscopic gametophytes (see Figs. 2.5 and 2.8). Even though the Japanese and Koreans rely primarily on *Undaria* as an edible kelp, the cultivation of *Saccharina* has become increasingly important in meeting the rapid increase in market demands during the last decade (see Fig. 1 in Buchholz *et al.* 2012). In Canada, *Saccharina latissima* (formerly *Laminaria saccharina*) and *Alaria esculenta* are cultivated in association with the development of Integrated Multi-Trophic Aquaculture (IMTA; Chopin *et al.* 2008). In Chile, brown seaweeds such as *Lessonia* spp., *Macrocystis pyrifera*, and *Durvillaea antarctica*, with a combined harvest of 100 000 tonnes y⁻¹, are used for alginate production and abalone feed (Buschmann *et al.* 2008).

Several other kelps are cultivated or harvested from nature for food and alginates. Small-scale cultivation of *Saccharina latissima* is occurring in eastern Canada (Chapman 1987; Chopin and Sawhney 2009), and the eastern United States (Yarish *et al.* 1990). *Alaria esculenta* is also cultivated in eastern Canada (Chopin and Sawhney 2009). Other brown seaweeds include *Ascophyllum nodosum*, harvested from wild populations for its alginate content in eastern Canada (Sharp 1987) and northern Europe (Kain and Dawes 1987), and *Sargassum fusiforme* that is grown for food in Japan (Nisizawa *et al.* 1987). Chile is an important producer of brown seaweeds representing about 10% (250 000 wet tonnes y^{-1}) of the world supply. *Lessonia* spp. with a large natural population and *Macrocystis* with a much smaller wild population are used to feed a rapidly expanding abalone industry and for alginate extraction (Vásquez 2008).

10.3.1 Cultivation

The life cycle consists of an alternation of generations between the microscopic gametophyte stage and the macroscopic edible sporophyte phase (Fig. 2.8). The sporophyte releases zoospores mainly during the summer and fall that settle and germinate into male and female gametophytes. Gametophytes need a low temperature and blue light to become fertile. The male gamete fertilizes the egg and within ~20 days young sporophytes develop. For aquaculture purposes, *Saccharina* can be considered an annual and can be harvested in ~9–11 months.

The cultivation consists of four main steps: (1) collection and settlement of zoospores on seed strings; (2) production of seedlings; (3) transplantation and grow-out of seedlings; and (4) harvesting (Sahoo and Yarish 2005; FAO 2006;Yarish and Pereira 2008). Seed stock is produced from meiospores released from the sori of wild or cultivated sporophytes. The sori are cleaned by vigorous wiping or by brief immersion in bleach and are left in a cool, dark place for up to 24 h. Drying helps to induce the liberation of spores. Spore release usually occurs within 1 h after re-immersion, but they cannot survive above 20°C. The zoospores attach to a substratum such as string within 24 h and develop into microscopic gametophytes. Rearing of seed stock usually is carried out on horizontal or vertical strings. The seed stock is placed in sheltered waters for 7–10 days. The string, with the seed stock on it, is then cut into small pieces and either inserted into the warp of the culture rope or attached by string or tape. A single or double line raft system is used to suspend the young sporophytes in the sea in late fall. The single rope raft

is better, especially in clear water, since it provides better access to nutrients at depth (McHugh 2004). The ropes are checked every few months to thin the seaweed densities and to remove trapped debris and fouling organisms. The seaweeds are kept about 5 m below the surface in the winter to avoid winter storm waves, and 2 m below in the spring and summer to get more light for growth. In May and June, harvestable blades are cut from the ropes, washed to remove epiphytic diatoms, etc., and then dried in the sun.

Ways to shorten the production time have been devised. In the original 2-year cultivation method, seed stock is produced in late fall, when the sporophytes produce their sori. Seed stock is available for outplanting from December to February, and the crop is ready to harvest in 20 months (Kawashima 1984) (Fig. 10.4). In the forced-cultivation method, seed stock of summer sporophytes that spent 3 months in the fall in the field prior to their second growth season behave as second-year seaweeds. This method saves 1 year (Fig. 10.4: Kawashima 1984). The natural cycle has been manipulated even further. By depriving the gametophyte stage of blue light, gametogenesis can be delayed, and sporelings can be produced throughout the year (Lüning and Dring 1972; Druehl *et al.* 1988).

Production can be increased in several ways. Lengthwise growth occurs by production of new tissue in the meristematic zone between the stipe and the base of the blade. Natural shedding of the older distal part of the blade is common in *Saccharina* in summer and this loss amounts to as much as 25% of the total harvest. In China, tip cutting is employed between April and May of the second year to eliminate this loss of harvestable material (Tseng 1987). In northern China, the kelp farms were fertilized in the summer with nitrogen using porous fertilizer cylinders for slow release to overcome nitrogen limitation, but this practice is now reduced due to new environmental regulations.

Saccharina is also subject to disease. Among the more important diseases encountered in *Saccharina* cultivation, three are caused by adverse conditions (Tseng 1987; Neill *et al.* 2008): green-rot disease (irradiance is too low), white-rot disease (irradiance is too high, and nitrogen too low), and blister disease

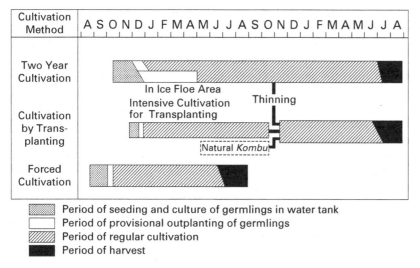

Cultivation Method	A S O N D J F M A M J J A S O N D J F M A M J J A
Two Year Cultivation	
Cultivation by Transplanting	
Forced Cultivation	

In Ice Floe Area
Intensive Cultivation
for Transplanting

Thinning

Natural *Kombu*

☷ Period of seeding and culture of germlings in water tank
☐ Period of provisional outplanting of germlings
▨ Period of regular cultivation
■ Period of harvest

Fig. 10.4 Major seasonal events in the 2-year cultivation method, the forced-cultivation method, and the transplanted-seedstock method of rearing edible kelp. (From Kawashima 1984, with permission of the Japanese Society of Phycology.)

(effects of freshwater from rainfall). There are also two pathogenic diseases. Malformation disease of summer sporelings is characterized by abnormal cell division due to hydrogen sulfide produced by sulfate-reducing bacteria. Swollen stipe, twisted-frond disease has appeared sporadically in China, and the organism causing it is unknown (Tseng 1987). Grazers such as gastropod snails and amphipods may be a problem in addition to diseases.

10.3.2 Utilization and future prospects

Saccharina is widely used in various foods and some use in the alginate industry. Most of the *Saccharina* is dried and eaten directly in soups, salads, and tea, or used to make secondary products with various seasonings (e.g. sugar, salt, soy sauce). *Saccharina* contains about 10–20% protein, and it has been used in China as a dietary iodine supplement (since iodine is low in nori and other red seaweeds) to prevent goiter (Brinkhuis *et al.* 1987). Kelps have been used for medicinal and food purposes for over 1500 years in China, Japan, and Korea (Tseng 1981). *Laminaria hyperborea* and *L. digitata* have been the main source for the alginate industry for some time (Kain and Dawes 1987;

McHugh 2004; Bartsch *et al.* 2008). *Saccharina* spp. (mainly from China) and *Lessonia* spp. (from Chile) are the two main seaweeds that are used for alginate extraction (Bixler and Porse 2011).

Alginates are present in the cell walls of brown seaweeds as the Ca, Mg, and Na salts of alginic acid and are partially responsible for their flexibility. The goal of the extraction of alginates from seaweeds is to convert all of the alginate salts to sodium salts (since the Ca and Mg salts do not dissolve in water), dissolve this in water, remove the seaweed residue by filtration, and recover the alginates from the solution. The most economical way is to precipitate the Na alginates by converting them to Ca alginates and adding acid to convert them into alginic acid (for further details see McHugh 2004). Alginate molecules are long chains that contain two different acidic components, abbreviated here for simplicity as M and G units. The way in which these M and G units are arranged in the chain and the overall ratio of M/G is species specific and hence the properties of alginates from various brown seaweeds vary considerably. Even though *S. japonica* is abundant in China, the low G/M ratio of the alginates extracted from it, yields weakly gelling alginates that are only useful in textile printing and paper coating (Bixler and Porse 2011). Another derivative of alginic acid that is used in the food industry is propylene glycol alginates. For a brief review of alginate chemistry, see Jensen (1995).

The uses of alginates are based on three main properties. First, they increase the viscosity of solutions (i.e. thicken) when dissolved in water. For example, water-in-oil emulsions such as mayonnaise and salad dressings are less likely to separate if they are thickened with alginates. The texture, body, and sheen of yoghurt can be improved with the use of alginates. Second, a gel is formed when a Ca salt is added in order to displace the Na in the alginates. No heat is required and these gels are desirable for many products because they do not melt when heated. For example, they act as stabilizers in ice cream and reduce the formation of ice crystals during freezing, giving a smooth product. They also reduce the rate at which ice cream melts. Alginates increase the amount and the duration of foam for a glass of beer and speed up the clarification of wine. Third, alginates have the ability to form films of Na or Ca alginates that help to preserve frozen food. If the food is frozen in a Ca alginate jelly, the food is protected from the air and therefore rancidity from oxidation is greatly reduced. Many other examples of the use of alginates such as textile printing, the paper industry, and pharmaceutical and medicinal uses are given by McHugh (2004). The main alginophytes are *Saccharina* (Asia), *Laminaria* and *Ascophyllum* (Europe), *Macrocystis* and *Lessosnia* (South America), with much smaller amounts from *Durvillaea* (Asia and South America), and *Ecklonia* (Africa and Australia) (McHugh 2004). The demand and price of alginates has increased in the last decade (Table 10.2).

In China, cultivation has been improved by producing the sporelings by mass cultivation of separate gametophyte clones. The male and female gametophytes are raised in batch cultures and then fertilization takes place under appropriate conditions (Zhang *et al.* 2007, 2008b; Xu *et al.* 2009). The advantage is that large quantities of sporelings with the desirable traits can be produced and this has reduced the need for long-term cultivation in temperature-controlled greenhouses. Thus, the seedling production method using vegetative gametophytes instead of collecting zoo-spores from adult sporophytes is much more cost effective. China is considering offshore cultivation and plans to overcome the problem of nitrogen limitation by using commercial slow-release fertilizer cylinders, if allowed by environmental regulations.

The Chinese have conducted genetic studies of *Saccharina japonica* and have employed methods of continuous inbreeding and selection in developing new strains. They have developed strains that are more tolerant of high temperatures and have high growth rates and high iodine content (Tseng 1987). Currently, microscopic gametophytes, produced from single spores and kept asexual by withholding blue light (Lüning and Dring 1972), can be cloned by fragmentation to provide populations of genetically identical gametophytes. These gametophytes, when crossed, produce genetically identical sporophytes. Using this system, superior strains can be maintained for long periods. The ability to clone the sporophyte directly would provide an optimal system to maintain a superior strain, but this has not been achieved yet.

10.4 *Undaria* for food

All kelps are more or less edible, but the two genera that are most important as an economical food source are *Undaria* and *Saccharina*. *Undaria* has been a foodstuff of great importance and high value in Japan since about AD 700 (Nisizawa *et al.* 1987). At the beginning of this century, the demand for *Undaria* exceeded its production from the wild. Cultivation began in Japan and Korea and later in China. There has been a pronounced increase in production in the last decade (Buchholz *et al.* 2012). In 2007, the annual global production of *Undaria* was about 1.8 million tonnes worth US$0.8 billion, with ~90% coming from China (Table 10.1; FAO 2009). Korea used to be the lead producer, but due to cheaper labor, China is now the leader (FAO 2009). However, about 50% of the total Korean seaweed production comes from *Undaria*. It is becoming more popular in North America and Europe, but that market is still insignificant compared to Asia.

There are three species of *Undaria*, *U. pinnatifida*, *U. undarioides*, and *Underiella peterseniana* (previously *Undaria peterseniana*), but *U. pinnatifida* is the main species cultivated. It is an annual that grows subtidally. It has a life cycle similar to *Saccharina* with an alternation of generations and the large sporophyte is harvested. Young fronds appear in October/November and grow rapidly until spring. In Japan, *U. pinnatifida* (also called *wakame*) is more important than *Saccharina japonica* in terms of the value of the harvest, the amount produced, and its epicurean quality. Several North American substitutes for *wakame* have been developed using *Alaria* and *Nereocystis*.

10.4.1   Cultivation

Cultivation steps and methods are similar to those for *Saccharina*. Typically, twine or rope is immersed in seawater tanks with fertile sporophytes during April and May, when the water temperature is 17-20°C (Sahoo and Yarish 2005; Yarish and Pereira, 2008). Zoospores produced from sporophytes, which previously had been dried for several hours to induce spore release, are allowed to settle on the string until a certain density is reached. The zoospores produce gametophytes and the fertilized eggs develop into young sporophytes. The seedling lines are then lashed to frames submerged in seawater tanks until September or November (McHugh 2004). When the ambient temperature is <15°C, the seedling lines are wound around ropes and positioned on horizontally floating poles at intervals of 1-6 m. Because it is an annual and grows quickly, young kelp sporophytes can be harvested by late winter (only 3-5 months after being transferred to the sea). On shore, the midrib is cut out, and the thalli are put up on lines for drying. The semi-dried product is kneaded and then dried completely and processed.

Undaria is often intermixed with *Saccharina* on floating rafts in Japan. Because *Undaria* has an earlier, shorter growing season than *Saccharina*, it is harvested in March and does not interfere with the growth of *Laminaria*, which is harvested in June (Tseng 1981). This provides the farmers with two crops per year. Because the Chinese prefer *Saccharina* to *Undaria*

many farmers in China do not mix the two species. *Undaria* was accidentally introduced into France in 1971 (Perez *et al*. 1988) and the high demand for *Undaria* in Japan has encouraged the French to explore its cultivation potential. The French use another seeding method, which allows them to obtain great quantity of "seeds" at the gametophyte stage at any time of the year. The gametophytes are kept in the laboratory as vegetative stock, from which a suspension is produced and sprayed onto thread collectors. The collectors are hung under a bank of lights in tanks until the young sporophytes are 3-5 mm, large enough to survive in the open sea (Perez *et al*. 1988). Then the thread is unrolled from the frame, wound around a rope, and placed at a depth of 1 m in the sea.

In the last two decades, pinhole "disease" has increased significantly. The holes are caused by the harpacticoid copepod *Amenophia orientalis* and occur during the growing season from December to April. When the thallus has many pin holes, it cannot command a good price and therefore it is often fed to abalone (Park *et al*. 2008). Other diseases that are common to *Saccharina* also occur in *Undaria* (Neill *et al*. 2008).

10.4.2   Food products and future trends

Undaria is processed into a variety of food products. It is sold boiled or dried and is an ingredient in soybean paste soup and seaweed salad. Raw *wakame* fronds with the midrib removed, are diced and dried, however the fronds fade and soften during storage because some enzymes are still active. To overcome this problem, fresh *wakame* is mixed with the ashes of straw, wood, or briquets, spread on the ground to dry and then packed in a plastic bag in the dark. The alkalinity of the ash inactivates the enzymes and prevents the degradation of alginates, and the frond from softening. It is washed with seawater and then fresh water to remove the adhering ash and salt and dried. This treatment preserves the deep greenish brown color and makes the product more elastic and flavorful for chewing. *Wakame* is used in instant foods such as soups and noodles or dried and marketed as chips.

There have been some recent improvements in the cultivation of *Undaria*. A seeding process called vegetative gametophytes that are grown in flasks is similar to the *Saccharina* technology of growing clonal gametophytes (Yarish and Pereira 2008).

10.5   *Kappaphycus* and *Eucheuma* for carrageenans

Carrageenan production from red algae is mainly centered in Asia and includes eight genera which yielded ~100 000 tonnes of dry seaweeds in 2006 (FAO 2008). Two genera *Eucheuma* and *Kappaphycus* account for over 80% of the global production and ~80% of the raw material for carrageenans comes from the latter (Yarish and Pereira 2008). About 86 000 tons of hydrocolloids were sold in 2009, with carrageenans making up 58%, alginates 31%, and agar 11% (Bixler and Porse 2011). Hereafter, the combination of *Kappaphycus* and *Eucheuma* is termed "*Eucheuma*". Over half of the world's carrageenophytes are produced in the Philippines and Indonesia and minor amounts in China, Malaysia, Western Pacific Islands, and East Africa, where two species *E. denticulatum* (formerly *E. spinosum*) and *K. alvarezii* (formerly *E. cottoni*) are the dominant producers (Ask *et al.* 2002; Bixler and Porse 2011). There has been a significant increase in production in the last decade (Buchholz *et al.* 2012). Some supply of carrageenans comes from natural harvests of *Chondrus crispus* in Europe and eastern Canada. Carrageenan yields in *Chondrus crispus* are higher (40–50%) than those in *Eucheuma* (20–30%); however, costs of labor and faster growth make the Asian carrageenophytes more competitive. Carrageenans are also extracted from natural beds of *Gigartina* and *Mastocarpus* harvested in Chile, Morocco, Mexico, and Peru (McHugh 2004). Some species that have very high yields of carrageenans such as *Kappaphycus striatum* (Madagascar) with 76%, and *Betaphycus gelatinum* and *Kappaphycus alvarezii* (both from the Philippines) with ~70% of their dry weight as carrageenans (Pereira *et al.* 2009, 2012). For more information on production of various countries, sales, and prices in 2009, see the review by Bixler and Porse (2011).

The two main carrageenophytes, *Eucheuma* ("spinosum") and *Kappaphycus* ("cottonii"), are subtropical and tropical red algae, and they occur mainly between 20°N and 20°S latitudes. They grow best in open-ocean waters with high levels of water motion and growth rates range from 0.5 to 5% d^{-1}.

10.5.1   Biology

The life history of "*Eucheuma*" follows the triphasic scheme common for red algae, consisting of the gametophyte (n), carposporophyte ($2n$) and tetrasporophyte ($2n$) (see sec. 2.2). The diploid tetrasporophyte phase ($2n$) produces non-motile meiospores called tetraspores (n). The tetraspores usually produce separate haploid male and female gametophytes (n). A diploid carposporophyte develops *in situ* on the female gametophyte after fertilization. The carposporophyte releases diploid carpospores, which initiate the tetrasporophyte stage again. In contrast to *Pyropia*, the phycocolloid-producing red algae such as "*Eucheuma*" have an alternation of isomorphic macroscopic gametophytic and sporophytic generations and both are harvested.

10.5.2   Cultivation

"*Eucheuma*" farming on a commercial scale began in the early 1970s in Indonesia and the Philippines, which are now the world's largest producers. Farm sites must have the following characteristics: salinities >30; good water movement; clear water at 25–30°C; and a coarse-sand bottom to retain mangrove poles and keep thalli from being covered with silt. At good sites, thalli can double in size in 30 days (e.g. daily growth rates are 3–5%). The bottom is cleared of seagrasses, seaweeds, large stones, corals, and grazers, such as sea urchins. Growth will decrease if thalli become desiccated at low tide. Bottom and floating monoline methods are used. Thalli are cut into small pieces called "seedlings" and tied to the monofilament line at intervals. Invading seaweeds and animals (especially sea urchins) are removed by hand. Seedlings are harvested when they reach ~1 kg in about 2–3 months, and therefore a farmer can obtain up to five harvests per year. A thallus can be harvested whole or pruned back to seedling size. The best seaweeds sometimes are kept as seedlings for the next crop. The seaweeds are washed, air-dried, washed again to remove dead epiphytes and salts, and re-dried (often in the sun).

Fouling of "*Eucheuma*" by other seaweeds and grazing by starfish and urchins can be controlled by regular maintenance (Ask *et al.* 2002). A commonly occurring disease known as *ice-ice* causes a whitening and softening of the thallus that leads to poor-quality carrageenans (Solis *et al.* 2010). It can be caused by high temperature, low salinity and light, and/or marine bacteria of the *Vibrio* and *Cytophaga–Flavobacterium* groups. Solis *et al.* (2010) also found that several marine fungi (*Aspergillus* spp. and *Phoma* sp.) could induce ice-ice symptoms that possessed carrageenolytic and cellulosic activities.

10.5.3 Production, uses, and future prospects

Carrageenans are a family of linear sulfated polysaccharides, obtained by extraction from certain rhodophytes. Chemically, they are comprised of repeating galactose and 3-6-anhydrogalactose units, either sulfated and non-sulfated, joined by $\alpha(1-3)$- and $\beta(1-4)$-glycosidic linkages (see sec. 5.5). Commercial carrageenans are available as stable Na, K, and Ca salts, or often as a mixture of these salts. The carrageenans of commercial interest are called iota (ι), kappa (κ), lambda (λ), and beta (β). Iota forms clear elastic gels with calcium salts, while kappa forms brittle gels with calcium salts and strong, rigid gels with potassium salts. In contrast, lambda forms high viscosity solutions, but no gels. *Eucheuma* produces ι-carrageenans, while *Kappaphycus* produces κ-carrageenans in both haploid and diploid stages. This contrasts with other carrageenan producers, such as *Chondrus*, *Iridea*, and *Gigartina*, in which the gametophyte usually produces κ/ι-carrageenans, and the sporophyte usually produces λ-carrageenans (Dawes 1987; Chopin *et al.* 1999). *Betaphycus gelatinum* from the Philippines and China produces β-carrageenans, with properties intermediate between carrageenans and agars and for this reason sometimes called carragars (Chopin *et al.* 1999).

There are two main kinds of carrageenan extracts and they differ according to the extraction methodology employed: semi-refined and refined carrageenans. In semi-refined carrageenan extraction, the carrageenans are not extracted from the seaweeds. Clean seaweeds are cooked in an alkali solution to improve the gel strength, washed several times leaving the carrageenans and insoluble material behind and then laid out to dry in the sun. Later, the alkali-treated material is ground and sold as seaweed flour that has 9–15% fiber content which causes its gels and solutions to be distinctly cloudy and grainy. This extraction process is much shorter and cheaper than the original refined carrageenan process. However, semi-refined carrageenans have different applications than the refined carrageenans; they are usually used for canning meat and for pet food, and products in which the presence of color and turbidity does not matter. Refined or filtered carrageenans are obtained after a pre-treatment with alkali solution, followed by cooking in water. The seaweed residue is removed by filtration and the carrageenans are recovered from the solution by precipitation with isopropyl alcohol, or by repeatedly freezing and thawing the gel until water is totally eliminated from the carrageenans. Carrageenans are then dried and ground in the same way as semi-refined carrageenans. The resultant product is practically "pure" carrageenans, without any seaweed residue and can be used in the food industry for refined products like desserts, puddings, etc., because it is colorless. For further details see McHugh (2004).

Both ι- and κ-carrageenans form gels with K and Ca salts when the solution is heated to 60°C to dissolve the carrageenans and then the gel forms upon cooling (McHugh 2004). Potassium salts of κ-carrageenans form the strongest gels of all the carrageenans, but they also bleed (loose liquid). Therefore, locust bean gum (from the seeds of the carob tree) is blended to reduce the amount of κ-carrageenans that is used. ι-carrageenans form gels most strongly with Ca salts that are soft and virtually free of bleeding. They can be frozen and thawed without destroying the gel. Most of the market is for ι- and κ-carrageenans that form strong gels. The water extractable ι- and κ-carrageenans are used in instant foods that come in powdered form, in chocolate milk, and toothpaste; λ-carrageenans form a highly viscous, non-gelling, polyanionic hydrocolloid, and it is well suited for instant-mix food products (McHugh 2004). Carrageenans are widely used in the food industry, especially in dairy products where carrageenan additions may be only 0.01 to 0.05%. For example, κ-carrageenans are added to cottage cheese to prevent separation of whey. The cocoa in chocolate milk can be kept in suspension and it produces a weak thixotrophic gel that is stable as long as it is not shaken strongly. Carrageenans may be used to produce nearly fat-free hamburgers (e.g. a McLean burger).

Quality control has emerged as a recent problem for farmers. The quality of the carrageenans has been decreasing recently because of hybridization with native seaweeds in the farming areas in the Philippines. Farmers have reported that their stocks have been growing more slowly than previously, possibly because farmers have been using vegetative cuttings from their present stocks as a source for the succeeding crop (Rovilla *et al*. 2010). To overcome this problem arising from continuous transplantation, periodic infusion of new stock is recommended to prevent this possible genetic deterioration in the quality of the carrageenans and growth rates. Recently, carposporophytes from the wild were made to shed spores in the laboratory and grown in a multi-step culture process until maturity, placed in outdoor tanks and later moved to the sea and long-line for grow out (Rovilla *et al*. 2010). This process offers the possibility of revitalizing stocks with wild strains and avoiding "strain aging".

10.6 *Gelidium* and *Gracilaria* for agar

10.6.1 *Gelidium* production and products

Prior to 1942, *Gelidium* was the primary source of agar, and it was prized for its high-quality agar (high gelling strength and low sulfation). Because of its slow growth, the natural beds have often been over-harvested. In the wild, it grows best in intertidal and subtidal areas with good water movement (McHugh 2004). It is harvested in many countries, especially Spain, Morocco, Japan, and Korea, with smaller amounts from Chile, China, France, Indonesia, Mexico, and South Africa (Critchley and Ohno 1998). In Spain and Portugal, the wild harvesting is mainly from storm-cast seaweed. Harvesting of live material on rocks is done by divers who cut above the holdfast so that rapid regrowth can occur (McHugh 2004). Attempts to increase the production from the natural beds by increasing the areas of the rock substrates, farming *Gelidium* with ropes and rafts, and even cultivating free-floating seaweeds in onshore tanks have met with some success (Santelices 1991; McHugh 2004). Roughly 35% of the world's agar production comes from *Gelidium*, and bacteriological-grade agar is obtained exclusively from this genus, although a very small amount comes from *Pterocladia* (Santelices 1991; McHugh 2004; Buschmann *et al*. 2008). Due to the difficulties in growing *Gelidium*, ~80% of the agar raw material was from *Gracilaria* in 2009, up from 63% in 1999 (Table 10.2; Bixler and Porse 2011; Buchholz *et al*. 2012). Over 90% of the *Gracilaria* is produced by China with a few percent from Vietnam and Chile (Table 10.1).

Agar is extracted by heating the seaweeds in water for several hours to dissolve the agar. The residual seaweeds are removed and the hot solution is cooled and forms a gel that contains about 1% agar. Water is removed from the gel by freeze–thawing or squeezing; the gel is then dried and milled to a uniform particle size and then used in various food processes (McHugh 2004). High pressure membrane presses have greatly improved the dewatering of agar and hence reduced the energy costs. Bacteriological agar (or "Bacto" agar) can only be made from *Gelidium* since the agar has a low gelling temperature (34–36°C) that allows the addition of other compounds to the agar with a minimum of heat damage. In contrast, *Gracilaria* and *Gelediella* produce agars that gel at 41°C or higher (McHugh 2004). Agar, particularly from *Gelidium*, can be fractionated into two components: agarose (the gel-forming component) and agaropectin (a mixture of variously sulfated molecules with low gelling ability). Agarose is produced by separating it from agaropectin and it commands a very high price in the growing biotechnology market and is used in electrophoresis for example (McHugh 2004).

About 90% of the agar that is produced is used in the food industry. Agar forms a gel at 32–43°C, but does not melt until 85°C, unlike gelatin gels (from animals) that melt at ~37°C. They have a different mouth feel from gelatin since they do not melt or dissolve in the mouth as gelatin does. The ability of agar to withstand high temperatures means that it can be sterilized and this is why it is so widely used in bacteriology and pharmacology. It also can be used as a stabilizer and thickener in pie fillings, icings, meringues, and gelled fish and meat products. It is used in vegetarian products such as meat substitutes. In combination with other gums, agar is used to stabilize sherbet and it improves the texture of dairy products such as cream cheese and yoghurt. Unlike starch, agar is not readily digested and adds little calorific value to food, a highly desirable property for the production of low calorie food products.

10.6.2 *Gracilaria* production and products

Since *Gelidium* is slow growing, interest has turned to fast-growing species such as *Gracilaria chilensis* and some harvesting of *Gracilariopsis* occurs in Chile (Santelices and Doty 1989; McHugh 2004; Buschmann *et al.* 2001, 2008). *Gracilaria* grows in temperate to tropical waters (e.g. Chile, Argentina, Brazil, Canada, China, etc.) in the intertidal zone on sandy or muddy sediments that are protected from waves (McHugh 2004; Buschmann *et al.* 2008). Since *Gracilaria* is quite brittle, it often breaks off, leaving a piece to re-grow and the broken piece washes ashore and is picked up. Raking up thalli on the bottom from a boat is also practiced, but care is needed to leave some seaweed material to grow again. The Chileans pioneered the commercial cultivation of *Gracilaria* using *Gracilaria chilensis* that contains a high-quality agar (Buschmann *et al.* 2008). *Gracilaria* is mainly grown vegetatively on the bottom of bays and estuaries on lines, ropes or nets, and in ponds, and occasionally in tanks. Pieces of seaweeds are forced into the sediment; they grow laterally and then vertically. Shoots that are produced can be harvested later (Buschmann *et al.* 1995, 2008;

Yarish and Pereira 2008). Line or rope farming practices are similar to those used for brown seaweeds. Pieces of *Gracilaria* are inserted or tied to a piece of rope strung between two stakes. Pond cultivation is less labor intensive than rope farming and is practiced in many Asia countries. However, the agar from pond-grown *Gracilaria* has low gel strength because of higher sulfate content and is only suitable for food-grade agar (McHugh 2004). Harvesting is possible in 35–45 days and is usually done by scooping it off the bottom and placing it on a floating raft or basket. The remaining thalli produce seed material for the next crop. Epiphytes can be a problem in ponds and are difficult to control since the thalli are on the bottom. Sometimes tilapia or milkfish are used to control the epiphytes.

Gracilaria is harvested from the wild in more than 20 countries, but attempts are being made to farm it and grow it in land-based tanks. *Gracilaria* grows quickly in tanks and adequate supplies of nutrients (NO_3, PO_4, and CO_2) and light are essential. Pond culture is used in Taiwan where wild thalli are collected and broken into 10-cm long pieces. Thalli grow to 10 times their original mass in 3 months. Annual yields are estimated to be 24 dry tonnes per hectare. Raft culture has been employed in China (Ren *et al.* 1984). In St. Lucia, *Gracilaria debilis* and *G. domingensis* are planted together on ropes, similar to the cultivation of "*Eucheuma*" (Smith *et al.* 1984). The faster growing *G. domingensis* sweeps *G. debilis* clean of most epiphytes. Many species of *Gracilaria* are delicate and brittle, but these two species are more robust and are amenable to rope culture. The search for new strains continues, especially for a strain that will produce a strong gelling agar. Normally, harvested agarophytes must be treated with alkali to remove 6-sulfate from L-galactose so that the agar will have adequate gel strength.

Epiphytism is one of the main problems in *Gracilaria* farming in Chile due to the high density of seaweeds that are maintained under monoculture conditions. There is a continuum between epiphytes and endophytes, although the latter is much less common (secs. 4.2.2 and 4.5.2). Leonardi *et al.* (2006) reported five types of host–epiphyte relationships.

At one extreme are the epiphytes that are weakly attached to the surface of the host and therefore do not damage the host tissues (e.g. *Hincksia* sp. and *Ectocarpus* sp.). The fifth type at the other extreme, include genera such as *Polysiphonia* spp., *Hypnea* spp., and *Ceramium* spp. which penetrate deep into the cortex and outer medulla of the host and destroy the host's cells around the infected site. An endophytic amoeba causes whitening, thallus decay and fragmentation in *G. chilensis* by penetrating the cortical and medullary cells and digesting their protoplasm (Correa and Flores 1995). Previous reports attributed very similar symptoms to endophytic bacteria or fungi, but amoeba may also be important.

Strain selection processes often have assumed that sterile clones could be maintained for long periods in diverse environments without major changes. In seaweeds, fragmentation is frequently used to maintain free-floating populations and is the most common method used to farm *Gracilaria chilensis*. Up to now, fragmentation has been considered to be a way to spread a genet into multiple units without modifying its genetic makeup. However, Meneses and Santelices (1999) found that clonal *G. chilensis* exhibited intraclonal variations in performance and such changes may be due to rapid changes in DNA composition associated with growth, via mitotic recombinations; these genetic changes may be enhanced by fragmentation and growth in the field with a variable environment. Therefore, they suggest that the present farming methods of fragmentation which are supposed to select thalli and reduce variability, may in fact be doing the exact opposite by dispersing genetically different units and introducing variability in the farmed population. More recently, Guillemin *et al.* (2008) suggested that the production problems and reduced genetic diversity for farmed populations of *G. chilensis* in Chile is due to the fact that they have been maintained almost exclusively by vegetative (clonal) propagation for the last 25 years. They are harvested by hand and the largest thalli (fastest growing) are retained and replanted. This has resulted in a reduction in genetic diversity, based on the results from six microsatellite markers and may explain why the occurrence of epiphytes and endophytes has become an increasing

problem. The predominance of diploids is an example of heterosis (i.e. a heterozygote excess compared to the typical haplo-diploid wild populations). Thus, the diploid generation has advantageous phenotypic traits (e.g. high growth rates) that have been unconsciously selected by farmers.

10.7   Tank cultivation

Land-based tank cultivation focuses on the production of commercially valuable seaweeds such as *Chondrus crispus* (Irish moss), *Gelidium* sp., *Pterocladia* sp., and *Pyropia* that produce a high value product such as specialty agars or can be used as sea vegetables for niche markets (Craigie and Shacklock 1995; Israel *et al.* 2006; Friedlander 2008; Pereira *et al.* 2012). Tank cultivation is more challenging than the normal open-water systems, but it offers more control of various environmental factors (Titlyanov and Titlyanov 2010). Since the goal is to produce a high-quality, high-value commercial product, small-scale factorial design experiments using controllable environmental factors, such as light, nutrients, and temperature, will determine the optimal combination of these factors for the highest production of the product before the expensive, large scale-up is undertaken (Chopin and Wagey 1999).

To achieve commercially successful tank cultivation of a particular species, many factors must be considered. The success or failure of the commercial venture may be determined by the choice of the site. Since large quantities of seawater are pumped through the tank system, the supply must be of high quality year round, preferably nutrient rich and close to the tanks since large pumps are costly to operate. Tank design (to ensure easy mixing and CO_2 injection), size, and the tank material are critical for successful economic production. Since it is desirable to have production nearly year round, the life cycle of the species is also important. It is more cost effective if the seaweed can be grown asexually in suspension culture since the sexual reproduction phase may require additional nursery tanks and procedures. Strain selection is usually an ongoing task to maximize yield, reduce disease, etc.

Environmental factors such as temperature and nutrients may be expensive to control and in the case of temperature, it may limit year-round operation due to excessive summer or low winter temperatures. Light may be relatively easily manipulated by varying the stocking density and possibly covering with screens during high summer irradiances. The application of nutrient physiology concepts to seaweed aquaculture have been outlined in sec. 6.8.3 and reviewed by Harrison and Hurd (2001). The first elements to become depleted in tanks are C (signaled by a pH increase to 9 or higher) and N. Carbon is usually added as gaseous CO_2 through a fine sparger system since it does not readily dissolve in seawater. The advantages of continual vs. pulsed addition of nutrients has been discussed in sec. 6.8. Pests, weeds (epiphytes), and diseases may destroy the crop and spell failure. Learning the best operating conditions takes time and these will change with the seasons and episodic events such as heavy rain, periods of cloudy weather, etc. For further details on issues to consider for successful tank cultivation of *Chondrus crispus*, see Craigie and Shacklock (1995) and a section in the previous edition (Lobban and Harrison 1994; pp. 294-7). Similarly, Friedlander (2008) has reviewed the recent advances in the tank cultivation of some Gelidiales.

10.8   Offshore/open-ocean cultivation

There has been extensive interest in obtaining energy from seaweeds. *Saccharina* in the northeastern USA (Brinkhuis *et al.* 1987) and *Macrocystis* off California (North 1987) were investigated in pilot offshore culturing for their biofuel potential. Generally, the engineering design could not withstand the storms and these projects were not continued. Floating systems have been suggested since they can withstand storms more easily, but the currents dictate where the floating structure will end up. In addition, since they move with the currents, water and hence nutrient renewal is more limited compared to anchored systems and hence nutrient limitation may occur during some periods.

In contrast to tank cultivation where environmental factors can be fairly well controlled, offshore cultivation depends on the local and rather variable environmental conditions. One of the biggest engineering challenges is designing structures that can withstand storms. One suggestion is to locate them at the same sites as wind farms. The pylons of the wind generators are fixed to the sea floor and they could serve to fasten culture modules (Buck and Buchholz 2005). Another advantage of locating next to wind farms (i.e. a multi-functional area) is that strong legal shipping regulations are enforced and this would avoid destruction of the culture facility by commercial ships.

Offshore mass culture of seaweeds under rough conditions found in the North Sea requires rigid support systems that can withstand storms and also retain the seaweeds. A new offshore ring system was developed by Buck and Buchholz (2004) that can withstand currents of 2 m s^{-1} and 6 m waves, but also permits ease of handling compared to other systems. In addition to the physical engineering challenges, seaweed growth may be slow depending on the surface nutrient concentrations. Some nutrient upwelling pipes have been tested to overcome this problem (Buck and Buchholz 2005).

Saccharina latissima was chosen for testing in the offshore system because it has a flexible stipe that allows it to quickly reorient the thallus and become aligned with the direction of the current (Buck and Buchholz 2005). They found that *S. latissima* grown in sheltered areas had more ruffled margins and wider blades that increased the drag compared to the flat narrow blades of seaweeds farmed under the stress of currents (see sec. 8.2.1). In laboratory tank trials, they found that seaweeds that were grown in the field could withstand high, simulated breaking forces. Their findings indicate that only seaweeds farmed in flowing water develop the best blade morphology for outplanting to offshore systems.

10.9   Integrated Multi-Trophic Aquaculture (IMTA) and biomitigation

Aquaculture production supplies ~50% of the seafood consumed worldwide and production has increased by nearly 10% per year over the last few decades.

Therefore, it is the fastest growing global food production sector (Chopin 2012). Fish and shellfish aquaculture represent 9 and 43% of all mariculture production, while seaweed aquaculture contributes 46% (~15.8 million tonnes) and about 99% of the seaweed production comes from Asia (China, Indonesia, the Philippines, Korea, and Japan) (FAO 2012). In fish aquaculture, only about 25–35% of the food N and P become fish meat and the remaining 65–75% is released into the water (Krom *et al.* 1995). The dominant processes affecting N release within the fish-farming system are excretion by fish as ammonium (~30%), dissolved organic nitrogen (DON) (~30%), and feces-N (~10%), and rapid transformation of DON to ammonium (20–40%). Nitrification (i.e. the conversion of NH_4 to NO_3 by bacteria) is usually small (~10%). Therefore, if these nutrients are not removed, they could contribute to eutrophication and algal/red tide blooms (Fei 2004). They can be removed by bacterial biofiltering processes which are dissimilative with the production of gases such as CO_2 and N_2 and therefore some potential nutrient resources are lost to the atmosphere. In contrast, biofiltration by seaweeds is assimilative with the nutrients being assimilated into particulate components, possibly with commercial value, which can then be harvested. In the 1970s, Ryther *et al.* (1979) developed a trophic level integrated land-based mariculture system and now this concept has been applied to fish farms and tank cultivation.

When seaweeds are grown near fish farms, they take up nutrients from the farm and produce valuable seaweed resources, especially if commercially valuable species such as *Pyropia* and *Saccharina* are used. Therefore, seaweeds could provide biomitigative services, much like terrestrial plants mitigate in the removal of excess CO_2 in the atmosphere. Chopin and Taylor coined the term "Integrated Multi-Trophic Aquaculture" (IMTA) (Chopin 2011) to describe the three additional trophic levels that should be combined with "fed" aquaculture (finfish): (1) suspension organic extractive aquaculture, with invertebrates such as shellfish (e.g. mussels) to recapture the small particulate organic matter from the fish food; (2) suspension inorganic extractive aquaculture, with seaweeds

or other aquatic plants to recapture the dissolved inorganic nutrients; and (3) deposit organic extractive aquaculture, with benthic invertebrates or grazing fish to utilize the large particulate organic matter that sinks to the bottom (Fig. 10.5). Therefore, the wastes from the aquaculture of one species are utilized for growth of another species with the benefit of reducing the potentially polluting effluent. The goal of the multi-trophic system is to move aquaculture to a new level of "ecosystem responsible aquaculture" (Chopin *et al.* 2001). Therefore, this integrated aquaculture arrangement of cultivating fed species along with extractive species could capitalize on the recapturing of the by-products of the fed aquaculture. This mix of different trophic levels (all with commercial value) mimics natural ecosystems, is more profitable than the present standard monoculturing/farming, and should decrease the environmental impacts of fish farms (Ridler *et al.* 2007). Present aquaculture business models do not consider or recognize the economic value of the nutrient biomitigative/ecosystem services provided by biofilters/extractors, because there is no cost associated with aquaculture discharges or effluents in most jurisdictions. However, a nutrient-trading credit system should be considered for some coastal areas that are similar to carbon-trading credit programs (Chopin 2011, 2012). Use of seaweeds is the most cost-effective method of removing/extracting excess nutrients from the environment since they produce harvestable and commercially valuable products (unlike phytoplankton blooms that cannot be economically filtered from the water) (Troell *et al.* 2003; Neori *et al.* 2004, 2007). IMTA programs, on land and in open water, and in different states of development and configuration, are taking place in at least 40 countries (e.g. Israel, Canada, China, and South Africa). They have already provided a proof of concept and are commercially scaling up (Neori *et al.* 2004; He *et al.* 2008; Chopin *et al.* 2008). Of course, they have been used in Asia on a more basic level for centuries. IMTA activities in temperate areas have been reviewed by Barrington *et al.* (2009) and for tropical areas see Troell (2009).

In coastal water, when *Gracilaria* was cultured on ropes about 10 m from the fish cages, it had a ~40%

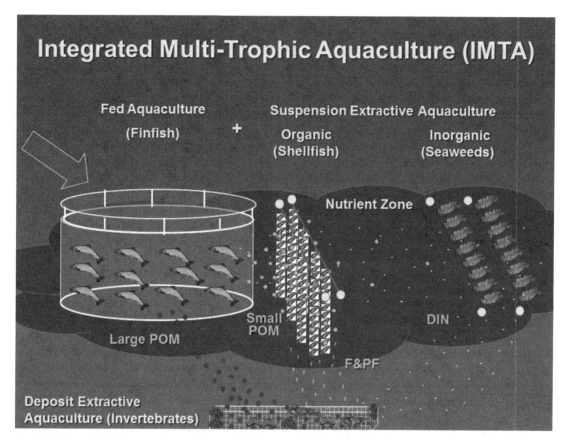

Fig. 10.5 Conceptual diagram of an Integrated Multi-Trophic Aquaculture (IMTA) operation. Finfish (fed aquaculture) and their food can produce three utilizable/extractable components: (1) dissolved inorganic nutrients extracted by seaweeds, (2) suspended organics extracted by shellfish, and (3) particulate organic matter (POM) and feces and pseudo-feces (F and PF) from suspension-feeding organisms extracted by deposit feeders (e.g. echinoids, holothuroids, and polychaetes). The bioturbation on the bottom also regenerates some inorganics and organics, which become available to the seaweeds. (With permission from Chopin 2012.)

higher growth rate (specific growth rate of 7% d^{-1} compared to controls about 1 km away (Troell *et al.* 1999). The N and P tissue content was higher and the accumulated agar production was higher (i.e. even though the yield of agar per unit biomass was lower since more agar is produced when N is limiting; the large increase in biomass more than compensated for the lower yield; Neori *et al.* 2004). Significant nutrient removal has been demonstrated in eutrophic coastal areas in China where a large 300 ha *Pyropia yezoensis* farm reduced NH_4 and NO_2^- up to 90%, while NO_3 was only reduced by~35% and PO_4 by ~60% (He *et al.* 2008). The ambient N:P ratio of the seawater was ~40:1 indicating that the growth of *Pyropia* was P limited and therefore, the average tissue N level was ~6% DW and tissue P was ~1% DW. Thus, *Pyropia* can be used for bioremediation of eutrophic coastal waters in addition to producing a large amount of valuable nori for human consumption (although heavy metal contamination was not assessed). However, in open-water systems, nutrient-uptake efficiency by seaweeds can be relatively low because of the rapid dilution of nutrients from the fish cages by currents (Petrell *et al.* 1993; Troell *et al.* 2003, 2009; Neori *et al.* 2004).

Some pilot-scale IMTA systems are land-based where a 1 ha sea bream – shellfish – seaweed farm has been estimated to produce 25 tons of fish, 50 tons of bivalves, and 30 tons of seaweeds annually (Neori *et al.* 2004). In Israel, a fishpond–seaweed biofilter system was set up where the effluent from intensive sea bream fish culture tanks was sedimented and then drained into *Ulva lactuca* tanks where nutrients were taken up by *Ulva* (Krom *et al.* 1995). Ammonia is toxic to most fish above 100 μM (1.5 mg NH_3-N L^{-1}) and so it must be removed. Nitrification (i.e. bacterial conversion of NH_4 to NO_3) and NH_4 uptake by *Ulva* reduced the high concentration of NH_4 that was produced by the fish to ~20% and PO_4 to ~30% of the incoming effluent. Since *Ulva* does not take up NO_3 very efficiently (Wallentinus 1984), it is ideal for reducing the mainly NH_4 excretion from fish, plus some nitrification due to bacterial conversion of NH_4 to NO_3. To make the seaweed more effective in removing phosphate, the N:P molar ratio in the fish feed must be >16N:1P, otherwise P limits uptake and growth of the seaweeds. In an expansion of the biofilter system described above into a three-stage greenhouse system, Schuenhoff *et al.* (2003) also included abalone (*Haliotis*) and sea urchins (*Paracentrotus*) along with the sea bream. Peak NH_4 excretion by the morning-fed fish coincided with maximal seaweed light-dependent NH_4 uptake. Daytime photosynthesis by *Ulva* took up the CO_2 produced by the fish and helped to balance fishpond pH levels and oxygen demand by the fish was partially met by the photosynthetically generated oxygen. During the day, the pH rises due to photosynthesis and subsequent CO_2 consumption and this potentially increases the toxicity of NH_4 to the fish and decreases their appetite. The protein content of *Ulva* averaged >34% of dry weight and it was fed to the abalone and sea urchins (Schuenhoff *et al.* 2003). Seasonal studies with *Gracilaria* co-cultured with salmon in a tank system removed 50% of dissolved NH_4 in winter and 90–95% in spring (Troell *et al.* 1999).

10.10   Other uses of seaweeds

Seaweeds have a wide variety of uses (see Guiry's (2011) website). With the increased awareness of the high nutritional value of seaweeds, their consumption has been increasing, especially in North America (Pereira *et al.* 2012). The nutritional value including fiber, mineral, fatty acid, vitamin, and protein contents have been reviewed for some of the commonly eaten seaweeds (Keiji and Kanji 1989; Burtin 2003; MacArtain *et al.* 2007; Craigie 2011). The Japanese eat *aonori*, which is a mixture of sea lettuce (*Ulva*) and *hitoegusa* (*Monostroma latissimum*). *Monostroma* is grown commercially and is an important component of *tsukudani*, a green paste made by boiling down this seaweed mixture in soy sauce (Nisizawa *et al.* 1987). In addition, natural products from seaweeds have many medical and pharmaceutical uses and they have been reviewed by Smit (2004) and Løvstad Holdt and Kraan (2011). There is significant commercial interest in the antibacterial and antiviral activities that may partly be responsible for the positive results with tumors and HIV, but getting seaweed products authorized for medical use is challenging (Løvstad Holdt and Kraan 2011).

Brown seaweeds, such as *Ecklonia*, *Ascophyllum*, and *Fucus*, are often available in large quantities on the beach and are used as soil conditioners and fertilizer. They have N and K, but much lower P than traditional manures and typical N:P:K ratios in chemical fertilizers (McHugh 2004). They contain trace elements, but their contribution is small compared to normal seaweed requirements. The large amounts of insoluble carbohydrates in brown seaweeds act as soil conditioners by improving soil structure, aeration, and moisture retention.

Dried *Ascophyllum nodosum* and *Laminaria digitata* are milled to produce a fine powder and added to animal feeds. They contain useful amount of minerals (K, P, Mg, Ca, Na, Cl, and S), trace elements (Fe, Zn, Co, Cr, I, Mo, V, Ni, etc.) and vitamins. Most of the carbohydrates and proteins are not digestible since protein is likely bound to phenolic compounds and therefore only a small addition of the seaweed meal

(~10%) is made to the regular feed (McHugh 2004). *Alaria esculenta* is more effective as an animal meal than *Ascophyllum* since it has better protein digestibility. Feeding trials with *Ascophyllum* indicate that it is beneficial to sheep and cattle in terms of weight gain, but it had no measurable benefit for pigs and chickens (McHugh 2004). Use of *Macrocystis* in the laying hen ration improves the quality of n-3 fatty acid-enriched eggs (Carrillo *et al.* 2008). Fish feeds consist of meat and fish wastes that need to be bound together so that they do not disintegrate when they are in the water. Seaweed meal is used as a binder since it is cheaper than using alginates (McHugh 2004). There is an active market for fresh seaweed such as *Macrocystis, Gracilaria, Palmaria*, and *Ulva* as a feed for abalone (Buschmann *et al.* 2008; Kim *et al.* 2013).

When the price of oil increased dramatically in the 1970s, research into setting up ocean farms for kelp production and the conversion of seaweed to methane by anaerobic fermentation was initiated as an alternative energy source. However, the anaerobic fermentation part of the project was much more achievable than the open-ocean production of *Macrocystis* on large-scale open-ocean farms (McHugh 2004).

Cosmetic body creams and lotions often contain "extract of seaweeds" (alginates and/or carrageenans) that help to keep the skin moist. Thalassotherapy is fashionable in Europe and mineral-rich seawater is used with a variety of mud and algal treatments (McHugh 2004). There are a growing number of products that use seaweeds such as creams, face masks, shampoos, body gels, bath salts, and body wraps, but the efficacy of these products must be determined by the user (Pereira *et al.* 2012).

Seaweeds have the potential to remove N and P from sewage and agricultural wastes. Many seaweeds have the ability to take up and grow at high ammonium concentrations that is often the main N source in sewage. Some of the green seaweeds such as *Ulva* and *Monostroma* are recommended since they are particularly tolerant of extreme environmental conditions (McHugh 2004). It would be ideal to use commercially important species since nutrient wastes could be converted into useful products, but these species are not as tolerant of varying growth conditions. Seaweeds have been promoted as "bioabsorbers" where they are used to remove toxic metals (Cu, Ni, Pb, Zn, and Cd) from industrial wastewater. While there have been many small-scale trials, there are very few reports of actual implementation on a large scale. One natural very large-scale unplanned mitigation (nutrient removal) event occurred in the coastal waters of the Yellow Sea in 2008 just weeks before the Olympic Games sailing events. More than 1 million tons of *Ulva prolifera* (previously called *Enteromorpha prolifera*) was harvested when winds pushed the large pelagic biomass on shore and an estimated 2 million tons of seaweed sank to the bottom (Liu *et al.* 2009). *Ulva prolifera* has presently no commercial value, but the removal of 1 million tons is equivalent to the removal of about 3000 to 5000 tonnes of N, 400 tonnes of P and 30 000 tonnes of C. Thus, there was some biomitigative services rendered during this unplanned natural mega-scale macroalgal bloom worth between US$32 and US$152 million (Chopin 2012). For a critical appraisal of using macroalgae for carbon sequestration, see Chung *et al.* (2006).

10.11 Seaweed biotechnology: current status and future prospects

The development of techniques for the culturing of single cells isolated from seaweeds has led to new opportunities in seaweed biotechnology. Progress in tissue culture and genetic engineering of seaweeds is now making the following technologies possible: (1) large-scale, rapid propagation of genetically uniform seaweeds from superior stock, possibly in photobioreactors; (2) selection of improved varieties using somaclonal variation techniques; (3) development of new hybrids between different cultivars and species by protoplast fusion and cell-culture techniques; and (4) use of recombinant DNA technology to introduce new genetic material into seaweed cells (Evans and Butler 1988; van der Meer 1988; Aguirre-Lipperheide *et al.* 1995; Baweja *et al.* 2009). The advances in macroalgal biotechnology lag far behind the research with terrestrial plants because of reduced interest in seaweeds and their commercial products and because there are substantial gaps in our knowledge of the biochemistry and physiology of seaweeds. In addition, macroalgal genomics have been completed for *Chondrus crispus* (Collén *et al.* 2013) and are ongoing for *Porphyra umbilicalis* (Gantt *et al.* 2010; Chan *et al.* 2012a) and there have been recommendations for genetic analysis of *Saccharina japonica* and *Undaria* (Waaland *et al.* 2004; sec. 1.4.1).

The starting point for applying new developments in technology to seaweeds is successful isolation of viable protoplasts (single cells without cell walls) from diploid and haploid tissues under axenic conditions (Reddy *et al.* 2008a). Reliable procedures for successful isolation and regeneration of protoplasts-to-plant systems are available for a number of commercially important species, including anatomically complex genera such as *Saccharina*, *Undaria*, *Pyropia*, and *Gracilaria* (Reddy *et al.* 2008b). The first steps in successful seaweed protoplast isolation were the pretreatment of the thallus with proteases or plasmolytic solutions and finding the specific cell wall lytic enzyme for the total removal of the rigid and complex polysaccharide cell wall (Polne-Fuller and Gibor 1987; Evans and Butler 1988; Aguirre-Lipperheide *et al.* 1995; Reddy *et al.* 2008b). For some species, protoplast-based methods have been developed for producing stocks of seedlings (micro-thalli) that are capable of surviving for many years in an incubator and still retain their ability to develop into leafy thalli. In addition to providing a seed stock source, protoplasts are being used for physiological and biochemical studies since the single cells of these seaweeds can be grown as easily as microalgae or phytoplankton in the laboratory under various environmental conditions.

Hybridization and even intergeneric hybridization has been documented to occur in the field in natural kelp populations (Druehl *et al.* 2005a) and now it can be implemented in the laboratory. The success in protoplast isolation and regeneration provided the tool and impetus to study somatic cell hybridization (genetic recombination) to develop new and improved strains. Protoplast fusion offers a unique opportunity to produce new hybrids between related but sexually incompatible or sterile species (see Table 3 in Reddy *et al.* 2008b). Interspecific and intergeneric protoplast fusion has employed the polyethylene glycol (PEG) and electrofusion methods. In most cases, the frequency of heteroplasmic fusion products and subsequent recovery of somatic hybrids following protoplast fusion was very low (Reddy *et al.* 2008b). These somatic hybrids should have a combined genome and a novel cytoplasmic combination. The natural colors of seaweeds help to identify heterokaryons, but methods used for terrestrial plants such as differential fluorescent labeling and genetic and physiological complementation for isolating heterokaryons (bi-parental fusion products) should be used.

The formation of callus (undifferentiated cell masses or tissue explants that form at the sites of wounds) can be used as a source of vegetative cells. A callus can be induced to differentiate into whole plant, or it can be stored as germ stock for several years and has been employed for several seaweeds (Evans and Butler 1988; Reddy *et al.* 2008a). However, seaweed calli are generally small and slow growing, and the occurrence of calli in some seaweeds is low and sporadic. Therefore, new thalli are regenerated more easily from protoplasts than from calli. In addition, the role of growth regulators (phytohormones, see sec. 2.6.3) and carbon sources on callus formation in seaweed explants is variable (Reddy *et al.* 2008a). There are limited reports of callus induction by typical terrestrial plant growth hormones such as cytokinins, and auxins. Some different types of cytokinins have been discovered in seaweeds and the next step is to determine if they can increase the induction of calli (Reddy *et al.* 2008a). Another group of plant growth regulators, polyamines (e.g. putrescine and spermidine), have been discovered in seaweeds (sec. 2.6.3). Their concentration is increased by stresses such as high and low salinity and they likely play an important role in cell growth and development in nature.

Gene insertion and genetic transformation in seaweed protoplasts are in the very early stages (Reddy *et al.* 2008b). Efficient expression of foreign genes in transformants depends on a suitable promoter and presently there is a lack of native macroalgal promoters such as seaweed-associated infective viruses and bacteria. Advances in seaweed molecular biology and genetics have been discussed in sec. 1.4.

Presently, it is not possible to produce phycocolloids such as agar and carrageenans cost-effectively by these new biotechnologies, because it is not yet known to what extent improved quality or yield can be achieved. On the other hand, somatic hybridization for producing strains resistant to fungal diseases, for example, is possible (Evans and Butler 1988). Increased research efforts must therefore be devoted to perfecting the underlying biotechnological techniques (protoplast isolation, somatic fusion, callus formation, and thallus regeneration) to realize the potential offered by seaweeds.

10.12   Synopsis

Seaweed mariculture involves large-scale cultivation of seaweeds in the sea or in land-based tanks and the life cycle dictates the farming practice. It is distinct from simple harvesting of wild stock. Wild harvest of seaweeds is about 1.8 m tonnes y^{-1}, mainly brown seaweeds used for alginates. World seaweed aquaculture production is 15.9 million tonnes (valued at ~US$ 7.4 billion) with ~99% being produced by Asian countries. The major seaweed-farming countries are China, Indonesia, the Philippines, South Korea, and Japan (FAO, 2010). Seaweeds frequently are grown for human food, and they provide protein, vitamins, and minerals (especially iodine). Brown seaweeds comprise about 64% of the production (67% of the value), reds about 36% (33% of the value) and greens ~0.2% of the production and value. The commercially important phycocolloids are agars and carrageenans from red algae and alginates from brown algae. The phycocolloid industry utilizes about 1.3 million wet tonnes or 11% of the total seaweed production, while direct food consumption is much higher at 8.6 million wet tonnes or 76% of total production. The phycosupplement industry is growing rapidly and involves soil additives/conditioners, animal feed supplements, pharmaceuticals, nutraceuticals, cosmetics, and natural pigments. The six major genera that are cultivated in Asia are the red algae *Kappaphycus*, *Eucheuma*, *Pyropia*, and *Gracilaria*, and the brown algae *Saccharina* and *Undaria*. Cultivation has two major components:

seed-stock production and field cultivation of the seed stock to a harvestable product.

In the laboratory, during February or March, *Pyropia* thalli are induced to release carpospores, which adhere to shells and germinate into a small microscopic filament called the conchocelis. The conchocelis produces conchospores, which are collected on nets that are then taken to the field, where the spores germinate and grow into the harvestable thallus. Harvested thalli are chopped and made into a slurry, spread over screens, dried, and then pressed into nori sheets. The problems surrounding *Pyropia* culture include fouling of nets by fast-growing weed species, predation by herbivorous fish, and threats from several diseases. The lack of vegetative propagation capacity of the gametophytes makes the farming of *Pyropia* labor intensive and expensive since spores from the conchocelis are needed for every crop. The few species of *Pyropia* that are farmed are the ones that naturally produce archeospores from the gametophytes. The gametophytes of some species are able to form blade archeospores, a form of asexual spore that germinates to form a new blade (formerly referred to as monospores). This allows for a large increase in production and up to six to eight harvests per net.

The two most economically important edible kelps are *Saccharina* (mainly from China) and *Undaria* (mainly from Korea and China). The cultivation consists of four main steps: (1) collection and settlement of zoospores on seed strings; (2) production of seedlings; (3) transplanting and grow-out of seedlings; and (4) harvesting. The kelps have heteromorphic life cycles, and the microscopic haploid stages are reared and manipulated in the laboratory. The sporophytic seed stock is attached to ropes on floating-raft systems. In Japan, *Undaria* is harvested by hand, dried, and made into different types of *wakame*. Most of the *Saccharina* (*kombu*) is dried and eaten directly in soups, salads, and tea, or is used to make secondary products with various seasonings. It is also used for alginate production.

Half of the world's carrageenans are produced in the Philippines and Indonesia, where several species of *Eucheuma* and *Kappaphycus* are cultivated on a large scale. Farming of these seaweeds began in the 1970s, and today it is one of the most notable success stories of the seaweed-farming industry. Floating monoline methods are used. The quality of the carrageenans has been decreasing recently because of hybridization with native plants in the farming areas in the Philippines. Farmers in the Philippines have reported that their stocks have been growing more slowly than previously, possibly because farmers use cuttings from their present stocks as plantlets for the succeeding crop (i.e. vegetative propagation). To overcome this problem arising from continuous transplantation, periodic infusion of new plantlet stock is recommended to prevent possible genetic deterioration in the quality of the carrageenans and reduced growth rates.

Gelidium is the primary source of high-quality agar (high gelling strength and low sulfates content) and bacteriological-grade agar is obtained exclusively from this genus. Because of its slow growth, the natural beds have often been over-harvested. Due to the difficulties in growing *Gelidium*, ~80% of the agar raw material was from *Gracilaria* in 2009, up from 63% in 1999. However, the agar from *Gracilaria* has a lower gel strength and it gels at a higher temperature than *Gelidium* and therefore it is less versatile for various applications. Over 90% of the *Gracilaria* is produced by China with a few percent from Vietnam and Chile. *Gracilaria* is mainly grown vegetatively on the bottom of bays and estuaries on lines, ropes or nets, and in ponds, and occasionally in tanks. Pieces of seaweed are forced into the sediment and it grows laterally and then vertical shoots can be harvested. Line or rope farming practices are similar to those used for brown seaweeds. Pieces of *Gracilaria* are inserted or tied to a piece of rope strung between two stakes.

Integrated multi-trophic aquaculture (IMTA) involves the cultivation of fed species (finfish) with extractive species which utilize/extract inorganic (e.g. seaweeds) and organic (e.g. suspension- and deposit-feeding invertebrates) excess nutrients for their growth. Thus, adding the various extractive trophic levels produces valuable biomass, while simultaneously rendering ecosystem services. IMTA allows some of the food, nutrients and by-products that would normally be lost from the fed component to become a resource that is recaptured and converted

into commercially valuable harvestable seafood. In addition, biomitigative/ecosystem services are performed such as partial removal of nutrients and CO_2 and supply of oxygen for fish respiration. The important criteria for selecting the best seaweeds for an IMTA system are: high NH_4 uptake and growth rate, tolerance of high NH_4 concentrations and capacity for storage of high tissue N (also P to a lesser extent in the marine environment), ease of cultivation and control of the life cycle, resistance to epiphytes and disease, existing or potential market value for the raw material or its derived products, and commercialization not leading to insurmountable regulatory hurdles (Neori *et al.* 2004; Chopin 2011). A local species may be the best choice in order to ensure a good match between its ecophysiological characteristics and the growth environment and one that combines economic value and efficient biomitigation. *Ulva* fulfills all of the criteria above, except that it has presently little commercial value and therefore it is most appropriate if the goal is mainly biofiltration (however, its disposal remains to be addressed). *Saccharina*, *Undaria*, *Alaria*, and *Pyropia* have commercial value, but their life cycles make management of their cultivation more complex and expensive.

The domestication of a seaweed involves the application of ecology, physiology, and genetics to provide a scientific basis for seaweed aquaculture. The domestication of a κ- and ι-carrageenan producer, *Chondrus crispus*, has been well documented over the past three decades. The protocol for domestication involves species selection (i.e. product selection and market survey), site selection, and strain selection (i.e. choosing an individual that will grow the fastest and/or produce the most biomass in the environmental conditions provided). Tank design and water circulation are also critical. Physiological experiments have provided fundamental knowledge used in growing the species efficiently. This land-based system represents the first attempt at large-scale cold-water seaweed aquaculture. The resulting sophisticated and expensive operation system also shifted the original purpose of the cultivation from carrageenans to the production of a high-valued niche market for edible sea vegetables, which is the only possibility to keep the operation economically viable.

Vegetative propagation in which new thalli are produced from fragmented pieces (e.g. *Gracilaria*), or from archeospores produced from the blade (e.g. *Pyropia*), is the most cost effective since the sexual reproduction part of the life cycle can be omitted. Recently, Blouin *et al.* (2007) have seeded nori nets with asexually produced neutral spores produced by *Porphya umbilicalis.* The progeny of asexual reproduction is genetically identical to the parent organism, ensuring that desirable traits of selected clones are retained. However, in Chile, 25 years of vegetative propagation of *Gracilaria chilensis* has led to a dominance of diploids and a marked reduction in genetic diversity.

Seaweed tissue and cell culture are now being used. Protoplasts are cells with their cell wall removed often by enzyme digestion and they can divide into specific cell tissues or germinate into plantlets. Protoplasts can also fuse with other protoplasts (even from other species or genera), resulting in genetic recombination that can produce new desirable traits. Calli are aggregates of undifferentiated cells that are formed at sites of wounds and also in tissue culture. They can be induced to differentiate into plantlets, often with the addition of certain chemicals (growth regulators or hormones). In comparison with calli, new thalli are regenerated more easily from protoplasts and protoplasts are more suitable for suspension culture. While somatic hybridization is well established for terrestrial plants, it is still in the early stage of development for seaweeds. Similarly, work on creating transgenic seaweeds has been initiated, but the potential release of genetically modified seaweeds into coastal waters is a major concern.

References

Section numbers for text citation of references are provided at the end of each reference.

Abdullah, M.I., and S. Fredriksen (2004). Production, respiration and exudation of dissolved organic matter by the kelp *Laminaria hyperborea* along the west coast of Norway. *J. Mar. Biol. Assoc. UK* 84:887–94. [5.7.3]

Åberg, P. (1992). A demographic study of two populations of the seaweed *Ascophyllum nodosum. Ecology* 73:1473–87. [3.5.2]

Åberg, P. (1996). Patterns of reproductive effort in the brown alga *Ascophyllum nodosum. Mar. Ecol. Prog. Ser.* 138:199–207. [3.5.2]

Abreu, M. H., R. Pereira, I. Sousa-Pinto, and C. Yarish (2011). Ecophysiological studies of the non-indigenous species *Gracilaria vermiculophylla* (Rhodophyta) and its abundance patterns in Ria de Aveiro lagoon, Portugal. *Eur. J. Phycol.* 46:453–64. [3.3.5]

Ackerman, J. D. (1998). Is the limited diversity of higher plants in marine systems the result of biophysical limitations for reproduction or evolutionary and physiological constraints? *Funct. Ecol.* 12:979–82. [1.1.1]

Ackland, J. C., J. A. West, and J. Pickett-Heaps (2007). Actin and myosin regulate pseudopodia of *Porphyra pulchella* (Rhodophyta) archeospores. *J. Phycol.* 43:129–38. [1.3.3]

Adams, M. S., J. L. Stauber, M. T. Binet, R. Molloy, and D. Gregory (2008). Toxicity of a secondary-treated sewage effluent to marine biota in Bass Strait, Australia: Development of action trigger values for a toxicity monitoring program. *Mar. Poll. Bull.* 57:587–98. [9.6.2]

Adams, N. M. (1994). *Seaweeds of New Zealand: An Illustrated Guide.* Christchurch, NZ: Canterbury University Press, 360 pp. [3.3.2]

Adams, T. S., and R. W. Sterner (2000). The effect of dietary nitrogen content on trophic level 15N enrichment. *Limnology and Oceanography* 45(3):601–7 [Essay 3]

Adey, W. H., and T. Goertemiller (1987). Coral reef algal turfs: master producers in nutrient poor seas. *Phycologia* 26:374–86. [3.3.1]

Adey, W. H., and J. M. Vassar (1975). Colonization, succession and growth rates of tropical crustose coralline algae (Rhodophyta, Cryptonemiales). *Phycologia* 14:55–69. [4.3.2]

Aguilera, J., A. Dummermuth, U. Kartsen, R. Schrieck, and C. Wiencke (2002). Enzymatic defences against photooxidative stress induced by ultraviolet radiation in Arctic marine macroalgae. *Polar Biol.* 25:432–41. [7.3.1]

Aguirre-Lipperheide, M., F. J. Estrada-Rodriguez, and L.V. Evans (1995). Facts, problems and need in seaweed tissue culture: an appraisal. *J. Phycol.* 31:677–88. [10.11]

Ahlf, W., H. Hollert, H. Neumann-Hensel, and M. Ricking (2002). A guidance for the assessment and evaluation of sediment quality a German approach based on ecotoxicological and chemical measurements. *J. Soils Sediments* 2:37–42. [9.2]

Ahn, O., R. J. Petrell, and P. J. Harrison (1998). Ammonium and nitrate uptake by *Laminaria saccharina* and *Nereocystis leutkeana* originating from a salmon sea cage farm. *J. Appl. Phycol.* 10:333–40. [6.4.2]

Airoldi, L. (1998). Roles of disturbance, sediment stress, and substratum retention on spatial dominance in algal turf. *Ecology* 79:2759–70. [8.3.2]

Airoldi, L. (2000). Effects of disturbance, life histories, and overgrowth on coexistence of algal crusts and turfs. *Ecology* 81:798–814. [4.2.1]

Airoldi, L. (2003). The effects of sedimentation on rocky coast assemblages. *Oceanogr. Mar. Biol.* 41:161–236. [8.3.2]

Airoldi, L., and M. Virgilio (1998). Responses of turf-forming algae to spatial variations in the deposition of sediments. *Mar. Ecol. Prog. Ser.* 165:271–82. [8.3.2]

Alamsjah, M. A., S. Hirao, F. Ishibashi, *et al.* (2008). Algicidal activity of polyunsaturated fatty acids derived from *Ulva fasciata* and *U. pertusa* (Ulvaceae, Chlorophyta) on phytoplankton. *J. Appl. Phycol.* 20:713–20. [4.2.2]

Alfaro, A. C., F. Thomas, L. Sergent, and M. Duxbury (2006). Identification of trophic interactions within an estuarine food web (northern New Zealand) using fatty acid biomarkers and stable isotopes. *Estuar. Coast. Shelf Sci.* 70 (1–2):271–86. [Essay 3]

Algarra, P., and F. X. Neill (1987). Structural adaptations to light reception in two morphotypes of *Corallina elongata* Ellis and Soland. PSZNI *Mar. Ecol.* 8:253–61. [5.3.2]

Algarra, P., J. C. Thomas, and A. Mousseau (1990). Phycobilisome heterogeneity in the red alga *Porphyra umbilicalis*. *Plant Physiol.* 92:570–6. [5.3.1]

Allen, J. F. (2003). State-transitions: a question of balance. *Science* 299:1530–2.

Allen, T. F. H. (1977). Scale in microscopic algal ecology: a neglected dimension. *Phycologia* 16:253–7. [1.1.3]

Almeida, E., T. C. Diamantino, and O. de Sousa (2007). Marine paints: the particular case of antifouling paints. *Prog. Org. Coatings* 59:2–20. [9.5]

Amat, M. A., and J.-P. Braud (1990). Ammonium uptake by *Chondrus crispus* Stackhouse (Gigartinales; Rhodophyta) in culture. *Hydrobiologia* 204/205:467–71. [Table 6.3]

Amsler, C. D. (ed.). (2008). *Algal Chemical Ecology*. London: Springer, 313 pp. [4.4]

Amsler, C. D. (2012). Chemical ecology of seaweeds. In C. Wiencke, and K. Bischof (eds), *Seaweed Biology: Novel Insights Into Ecophysiology, Ecology and Utilization* (pp. 177–88). Series: Ecological Studies, Volume 219. Berlin and Heidelberg: Springer. [4.4]

Amsler, C. D., and V. A. Fairhead (2006). Defensive and sensory chemical ecology of brown algae. *Adv. Bot. Res.* 43:1–91. [4.2.2, 4.4, 4.4.1]

Amsler, C. D., and M. Neushul (1989a). Diel periodicity of spore release from the kelp *Nereocystis luetkeana* (Mertens) Postels *et* Ruprecht. *J. Exp. Mar. Biol. Ecol.* 134:117–27. [2.5.1]

Amsler, C. D., and M. Neushul (1989b). Chemotactic effects of nutrients on spores of the kelps *Macrocystis pyrifera* and *Pterygophora californica*. *Mar. Biol.* 102:557–64. [2.5.1]

Amsler, C. D., and M. Neushul (1990). Nutrient stimulation of spore settlement in the kelps *Pterygophora californica* and *Macrocystis pyrifera*. *Mar. Biol.* 107:297–304. [2.5.1]

Amsler, C. D., and R. B. Searles (1980). Vertical distribution of seaweed spores in a water column offshore of North Carolina. *J. Phycol.* 16:617–19. [1.1.1, 2.5.1]

Amsler, C. D., D. C. Reed, and M. Neushul (1992). The microclimate inhabited by macroalgal propagules. *Brit. Phycol. J.* 27:253–70. [2.5.1]

Amsler, C. D., J. B. McClintock, and B. J. Baker (2008). Macroalgal chemical defenses in polar marine communities. In C. D. Amsler (ed.), *Algal Chemical Ecology* (pp. 91–104). London: Springer. [4.4]

Amsler, C. D., M. O. Amsler, J. B. McClintock, and B. J. Baker (2009). Filamentous algal endophytes in macrophytic Antarctic algae: prevalence in hosts and palatability to mesoherbivores. *Phycologia* 48:324–34. [4.4, 4.5.2]

Anderberg, H. I., J. Danielson, and U. Johanson (2011). Algal MIPs, high diversity and conserved motifs. *BMC Evol. Biol.* 11:110. [7.4.2, 7.10]

Andersen, R. A. (2004). Biology and systematics of heterokont and haptophyte algae. *Am. J. Bot.* 91:1508–22. [1.1.1]

Anderson, B. S., and J. W. Hunt (1988). Bioassay methods of evaluating the toxicity of heavy metals, biocides and sewage effluent using microscopic stages of giant kelp *Macrocystis pyrifera* (Agardh): a preliminary report. *Mar. Environ. Res.* 26:113–34. [9.3.4]

Anderson, B. S., J. W. Hunt, S. L. Turpen, A. R. Conlon, and M. Martin (1990). Copper toxicity to microscopic stages of giant kelp *Macrocystis pyrifera*: interpopulation comparisons and temporal variability. *Mar. Ecol. Prog. Ser.* 68:147–56. [9.3.4]

Anderson, F. E., and J. Green (1980). Estuaries. *The New Encyclopaedia Britannica*. Macropaedia vol. 6, (pp. 968–76). London: Encyclopaedia Britannica. [7.2.2, 7.10]

Anderson, R. J., and J. J. Bolton (1989). Growth and fertility, in relation to temperature and photoperiod, in South African *Desmarestia firma* (Phaeophyceae). *Bot. Mar.* 32:149–58. [2.3.2]

Anderson, R. J., D. R. Anderson, and J. S. Anderson (2008). Survival of sand-burial by seaweeds with crustose bases or life-history stages structures the biotic community on an intertidal rocky shore. *Bot. Mar.* 51:10–20. [8.3.2]

Anderson, S. M., and A. C. Charters (1982). A fluid dynamics study of seawater flow through *Gelidium nudifrons*. *Limnol. Oceanogr.* 27:399–412. [8.1.3, 8.1.4]

Andersson, S., and L. Kautsky (1996). Copper effects on reproductive stages of Baltic Sea *Fucus vesiculosus*. *Mar. Biol.* 125:171–6. [9.3.4]

Andrade, L.R., M. Farina, and G. M. Amado Filho (2004). Effects of copper on *Enteromorpha flexuosa* (Chlorophyta) *in vitro*. *Ecotox. Envir. Safety* 58:117–25. [9.3.3]

Andrade, S., M. H. Medina, J. W. Moffet, and J. A. Correa (2006). Cadmium-copper antagonism in seaweeds inhabiting coastal areas affected by copper mine waste disposal. *Environ. Sci. Technol.* 40:4382–7. [9.3.4]

Andrade, S., M. J. Pulido, and J. A. Correa (2010). The effect of organic ligands exuded by intertidal seaweeds on copper complexation. *Chemosphere* 78:397–401. [9.3.2, 9.3.4, 9.3.5]

Andreakis, N., and B. Schaffelke (2012). Invasive marine seaweeds: pest or prize? In C. Wiencke, and K. Bischof (eds), *Seaweed Biology: Novel Insights into Ecophysiology, Ecology and Utilization, Ecological Studies* 219 (pp. 235–62). Berlin and Heidelberg: Springer-Verlag. [3.4, Essay 6]

Andréfouët, S., C. Payri, E. J. Hochberg, *et al.* (2004). Use of *in situ* and airborne reflectance for scaling-up spectral discrimination of coral reef macroalgae from species to communities. *Mar. Ecol. Prog. Ser.* 283:161–77. [3.5.1]

Andrew, N. L. (1993). Spatial heterogeneity, sea urchin grazing, and habitat structure on reefs in temperate Australia. *Ecology* 74:292–302. [4.3.1]

Andrews, J. H. (1976). The pathology of marine algae. *Biol. Rev.* 51:211–53. [7.3.4, 10.2.3]

Ang, P. O. Jr. (1992). Cost of reproduction in *Fucus distichus*. *Mar. Ecol. Prog. Ser.* 89:25–35. [3.5.2]

Ang, P. O., and R. E. DeWreede (1993). Simulation and analysis of the dynamics of a *Fucus distichus* (Phaeophyceae, Fucales) population. *Mar. Ecol. Prog. Ser.* 93:253–65. [3.5.2]

Anthony, K. R. N., J. A. Maynard, G. Diaz-Pulido, *et al.* (2011). Ocean acidification and warming will lower coral reef resilience. *Global Change Biology* 17:1798–808. [Essay 6]

Antia, N. J., P. J. Harrison, and L. Oliveira (1991). The role of dissolved organic nitrogen in phytoplankton nutrition, cell biology and ecology. *Phycologia* 30:1–89. [6.2, 6.5.1]

Antizar-Ladislao, B. (2008). Environmental levels, toxicity and human exposure to tributyltin (TBT) contaminated marine environment. A review. *Environ. Internat.* 34:292–308.

Ar Gall, E., F. C. Küpper, and B. Kloareg (2004). A survey of iodine content in *Laminaria digitata*. *Bot. Mar.* 47:30–7. [6.5.6]

Archibald, J. M. (2009). The origin and spread of eukaryotic photosynthesis: evolving views in light of genomics. *Bot. Mar.* 52:95–103. [1.1.1, 1.3.2]

Arenas, F., and C. Fernández (2000). Size structure and dynamics in a population of *Sargassum muticum* (Phaeophyceae). *J. Phycol.* 36:1012–20. [4.2.3]

Arenas, F., I. Sánchez, S. J. Hawkins, and S. R. Jenkins (2006). The invisibility of marine algal assemblages: role of functional diversity and identity. *Ecology* 87:2851–61. [3.5.1]

Arkema, K. K., D. C. Reed, and S. C. Schroeter (2009). Direct and indirect effects of giant kelp determine benthic community structure and dynamics. *Ecology* 90:3126–37. [4.1]

Armbrust, E., J. Berges, C. Bowler, *et al.* (2004). The genome of the diatom *Thalassiosira Pseudonana*: Ecology, evolution, and metabolism. *Science* 306: 76–86. [7.3.4, 7.5.2]

Armstrong, S. L. (1987). Mechanical properties of the tissues of the brown alga *Hedophyllum sessile* (C. Ag.) Setchell: variability with habitat. *J. Exp. Mar. Biol. Ecol.* 114:143–51. [8.3.1]

Arnold, K. E., and S. L. Manley (1985). Carbon allocation in *Macrocystis pyrifera* (Phaeophyta): intrinsic variability in photosynthesis and respiration. *J. Phycol.* 21:154–67. [5.7]

Arnold, K. E., and S. N Murray (1980). Relationships between irradiance and photosynthesis for marine benthic green algae (Chlorophyta) of differing morphologies. *J. Exp. Mar. Biol. Ecol.* 43: 183–92. [5.3.3]

Arnold, T. M., N. M. Targett, C. E. Tanner, W. I. Hatch, and K. E. Ferrari (2001). Evidence for methyl jasmonate-induced phlorotannin production in *Fucus vesiculosus* (Phaeophyceae). *J. Phycol.* 37:1026–9. [2.6.3]

Arntz, W. E., V. A. Gallardo, D. Gutiérrez, *et al.* (2006). El Niño and similar perturbation effects on the benthos of the Humboldt, California, and Benguela Current upwelling ecosystems. *Adv. Geosci.* 6:243-65. [7.3.7]

Aro, E. M., I. Virgin, and B. Andersson (1993). Photoinhibition of photosystem II. Inactivation, protein damage and turnover. *Biochim. Biophys. Acta* 1143:113-34. [5.3.3, 7.3.5]

Aronson, R. B., W. F. Precht, I. G. Macintyre, and L. T. Toth (2012). Catastrophe and the life span of coral reefs. *Ecology* 93:303-13. [8.3.2]

Asada, K., and M. Takahashi (1987). Production and scavenging of active oxygen in photosynthesis. In D. J. Kyle, C. B. Osmond, and C. J. Arntzen (eds) *Photoinhibition. Topics in Photosynthesis* 9 (pp. 89-109). Amsterdam: Elsevier Science Publishers. [7.3.5]

Ashen, J. B., J. D. Cohen, and L. J. Goff (1999). GC-SIM-MS detection and quantification of free indole-3-acetic acid in bacterial galls on the marine alga *Prionitis lanceolata* (Rhodophyta). *J. Phycol.* 35:493-500. [2.6.3]

Asimgil, H., and I. H. Kavakli (2012). Purification and characterization of five members of photolyase/cryptochrome family from *Cyanidioschyzon merolae*. *Plant Sci.* 185:190-8. [2.3.3]

Ask, E. I., and R. V. Azanza (2002). Advances in cultivation technology of commercial eucheumatoid species: a review with suggestions for future research. *Aquaculture* 206:257-77. [10.5.2]

ASTM (2004). *Standard Guide for Conducting Sexual Reproduction Tests with Seaweeds.* ASTM E1498-92. [9.2]

Atalah, J., and T. P. Crowe (2010). Combined effects of nutrient enrichment, sedimentation and grazer loss on rock pool assemblages. *J. Exp. Mar. Biol. Ecol.* 388:51-7. [3.3.4]

Ateweberhan, M., J. H. Bruggemann, and A. M. Breeman (2005). Seasonal dynamics of *Sargassum ilicifolium* (Phaeophyta) on a shallow reef flat in the southern Red Sea (Eritrea). *Mar. Ecol. Prog. Ser.* 292:159-71. [3.5.2]

Ateweberhan, M., J. H. Bruggemann, and A. M. Breeman (2006). Seasonal module dynamics of *Turbinaria Triquetra* (Fucales, Phaeophyceae) in the southern Red Sea. *J. Phycol.* 42:990-1001. [3.5.2]

Ateweberhan, M., J. H. Bruggemann, and A. M. Breeman (2009). Seasonal changes in size structure of *Sargassum* and *Turbinaria* populations (Phaeophyceae) on tropical reef flats in the southern Red Sea. *J. Phycol.* 45:69-80. [3.5.2]

Atkinson, M. J., and R. W. Grigg (1984). Model of a coral reef ecosystem. 11. Gross and net benthic primary production at French Frigate Shoals, Hawaii. *Coral Reefs* 3:13-22. [5.7.1]

Atkinson, M. J., and S. V. Smith (1983). C: N: P ratios of benthic marine plants. *Limnol. Ocenaogr.* 28:568-74. [6.5.2, 6.8.2]

Audibert, L., M. Fauchon, N. Blanc, *et al.* (2010). Phenolic compounds in the brown seaweed *Ascophyllum nodosum*: distribution and radical-scavenging activities. *Phytochem. Anal.* 21:399-405. [4.4.1]

Austin, A. P., and J. D. Pringle (1969). Periodicity of mitosis in red algae. *Proc. Intl. Seaweed Symp.* 6:41-52. [1.3.5]

Ávila, E., and J. L. Carballo (2006). Habitat selection by larvae of the symbiotic sponge *Haliclona caerulea* (Hechtel, 1965) (Demospongiae, Haplosclerida). *Symbiosis* 41:21-9. [4.5.1]

Azanza-Corrales, R., and C. J. Dawes (1989). Wound healing in cultured *Eucheuma alvarezii* var. *tambalang* Doty. *Bot. Mar.* 32:229-34. [2.6.4]

Azimzadeh, J., and W. F. Marshall (2010). Building the centriole. *Curr. Biol.* 20:R816-25. [1.3.3]

Baardseth, E. (1970). A *Square-Scanning, Two-Stage Sampling Method of Estimating Seaweed Quantities.* Norwegian Institute of Seaweed Research, Report 33, 41 pp. [8.1.4]

Babcock, R. C., N. T. Shears, A. C. Alcala, *et al.* (2010). Decadal trends in marine reserves reveal differential rates of change in direct and indirect effects. *PNAS* 107:18,256-61. [4.3.1]

Bäck, S., J. C. Collins, and G. Russell (1992). Effects of salinity on growth of Baltic and Atlantic *Fucus vesiculosus*. *Br. Phycol. J.* 27:39-47. [7.4.3]

Bacon, L. C., and R. L. Vadas (1991). A model for gamete release in *Ascophyllum nodosum* (Phaeophyta). *J. Phycol.* 27: 166-73. [7.7]

Badger, M. (2003). The roles of anhydrases in photosynthetic CO_2 concentrating mechanisms. *Photosynth. Res.* 77:83-94. [5.3.4]

Badger, M. R., T. J. Andrews, S. M. Whitney, *et al.* (1998). The diversity and co-evolution of Rubisco, plastids, pyrenoids and chloroplast-based CO_2 concentrating mechanisms in the algae. *Can. J. Bot.* 76:1052-71. [7.3.1]

Baker, S. M., and M. H. Bohling (1916). On the brown seaweeds of the salt marsh. Part II. Their systematic relationships, morphology, and ecology. *J. Linn. Soc. Bot.* 43:325-80. [3.3.5]

Balasse, M., I. Mainland, and M. Richards (2009). Stable isotope evidence for seasonal consumption of marine seaweed by modern and archeological sheep in the Orkney archipelago (Scotland). *Environ. Archeol.* 14:1-14. [5.7.5]

Balata, D., L. Piazzi, and L. Benedetti-Cecchi (2007). Sediment disturbance and loss of beta diversity on subtidal rocky reefs. *Ecology* 88:2455-61. [3.5.1]

Balata, D., L. Piazzi, and F. Rindi (2011). Testing a new classification of morphological functional groups of marine macroalgae for the detection of responses to stress. *Mar. Biol.* 158:2459-69. [1.2.2, 7.9]

Ballesteros, E., J. Garrabou, B. Hereu, *et al.* (2009). Deep-water stands of *Cystoseira zosteroides* C. Agardh (Fucales, Ochrophyta) in the Northwestern Mediterranean: insights into assemblage structure and population dynamics. *Estuar. Coast. Shelf Sci.* 82: 477–84. [3.3.6]

Barbrook, A. C., C. J. Howe, D. P. Kurniawan, and S. J. Tarr (2010). Organization and expression of organellar genomes. *Phil. Trans. R. Soc. B.* 365:785–97. [1.3.2]

Barile, P. J., and B. E. Lapointe (2005). Atmospheric nitrogen deposition from a remote source enriches macroalgae in coral reef ecosystems near Green Turtle Cay, Abacos, Bahamas. *Mar. Poll. Bull.* 50:1262–72. [9.6.3]

Barner, A. K., C. A. Pfister, and T. Wootton (2010). The mixed mating system of the sea palm kelp *Postelsia palmaeformis*: few costs to selfing. *P. Roy. Soc. B. Bio.* 278:1347–55. [1.4.2]

Barnes, D. J., and B. E. Chalker (1990). Calcification and photosynthesis in reef-building corals and algae. In Z. Dubinsky (ed.), *Ecosystems of the World* 25 (pp. 109–31). Amsterdam: Elsevier. [6.5.3]

Barnett, B. E., and C. R. Ashcroft (1985). Heavy metals in *Fucus vesiculosus* in the Humber Estuary. *Environ. Pollut. B-9*:193–201. [9.3.2]

Barr, N. G., and T. A. V. Rees (2003) Nitrogen status and metabolism in the green seaweed *Enteromorpha intestinalis*: an examination of three natural populations. *Mar. Ecol. Prog. Ser.* 249:133–44. [6.5.1]

Barr, N. G., A. Kloeppel, T. A. V. Rees, *et al.* (2008). Wave surge increases rates of growth and nutrient uptake in the green seaweed *Ulva pertusa* maintained at low bulk flow velocities. *Aquat. Biol.* 3:179–86. [8.2.1]

Barrington, K., T. Chopin, and S. Robinson (2009). Integrated multi-trophic aquaculture (IMTA) in marine temperate waters. In D. Soto (ed.) *Integrated Mariculture: A Global Review*. FAO Fisheries and Aquaculture Technical Paper No. 529:7–49. [10.9]

Bartsch, I., C. Wiencke, K. Bischof, *et al.* (2008). The genus *Laminaria sensu lato*: recent insights and developments. *Eur. J. Phycol.* 43: 1–86. [1.4.2, 2.3, 2.3.1, 2.3.2, 2.3.3, 3.2, 3.3.2, 3.5.2, 4.5.2, 10.3.2]

Basu, S., H. G. Sun, L. Brian, R. L. Quatrano, and G. K. Muday (2002). Early embryo development in *Fucus distichus* is auxin sensitive. *Plant Physiol.* 130:292–302. [2.6.3]

Bates, C. R., G. W. Saunders, and T. Chopin (2009). Historical versus contemporary measures of seaweed biodiversity in the Bay of Fundy. *Botany* 87:1066–76. [3.4]

Baumann, H. A., L. Morrison, and D. B. Stengel (2009). Metal accumulation and toxicity measured by PAM chlorophyll fluorescence in seven species of marine macroalgae. *Ecotox. Environ. Safety* 72:1063–75. [9.3.4]

Baweja, P., D. Sahoo, P. Garcia-Jiménez, and R. R. Robaina (2009). Seaweed tissue culture as applied to biotechnology: problems, achievements and prospects. *Phycol. Res.* 57:45–58. [10.11]

Bayer-Giraldi, M., C. Uhlig, U. John, T. Mock, and K. Valentin (2010). Antifreeze proteins in polar sea ice diatoms: diversity and gene expression in the genus *Fragilariopsis*. *Environ Microbiol* 12:1041–52. [7.5.2]

Bayne, B. L. (1989). The biological effects of marine pollutants. In J. Albaigés (ed.), *Marine Pollution* (pp. 131–51). New York: Hemisphere. [9.2]

Beck, C. B. (2010). *An Introduction to Plant Structure and Development. Plant Anatomy for the the Twenty-First Century*. Cambridge: Cambridge University Press, 464 pp. [1.3]

Becker, S., B. Walter, and K. Bischof (2009). Freezing tolerance and photosynthetic performance of polar seaweeds at low temperatures. *Bot. Mar.* 52: 609–16. [7.3.4, 7.3.5, 7.3.6]

Becker, S., M. Graeve, and K. Bischof (2010). Photosynthesis and lipid composition of the Antarctic endemic rhodophyte *Palmaria decipiens*: effects of changing light and temperature levels. *Polar Biol.* 33:945–55. [7.3.4, 7.3.5, 7.3.6]

Beer, S., and L. Axelsson (2004). Limitations in the use of PAM fluorometry for measuring photosynthetic rates of macroalgae at high irradiances. *Eur. J. Phycol.* 39:1–7. [5.7.1]

Beer, S., and A. Eshel (1983). Photosynthesis of *Ulva* sp. I. Effects of desiccation when exposed to air. *J. Exp. Mar. Biol. Ecol.* 70:91–7. [7.5.1, 7.9]

Bégin, C., and R. E. Scheibling (2003). Growth and survival of the invasive green alga *Codium fragile* ssp. *tomentosoides* in tide pools on a rocky shore in Nova Scotia. *Bot. Mar.* 46:404–12. [3.4]

Bekheet, I. A., K. M. Kandil, and N. Z. Shaban (1984). Studies on urease extracted from *Ulva lactuca*. *Hydrobiologia* 116/117:580–3. [6.5.1]

Bell, E. C. (1995). Environmental and morphological influences on thallus temperature and desiccation of the intertidal alga *Mastocarpus papillatus* Kützing. *J. Exp. Mar. Biol. Ecol.* 191:29–55. [2.6.2]

Bell, E. C. (1999). Applying flow tank measurements to the surf zone: predicting dislodgment of the Gigartinaceae. *Phycol. Res.* 47:159–66. [8.1.4]

Bell, E. C., and M. W. Denny (1994). Quantifying wave exposure: a simple device for recording maximum velocity and results of its use at several field sites. *J. Exp. Mar. Biol. Ecol.* 181:9–29. [8.1.2, 8.1.4, 8.3.1]

Bell, G. (1997). The evolution of the life cycle of brown seaweeds. *Biol. J. Linn. Soc.* 60:21–38. [2.2]

Bellgrove, A., H. Kihara, A. Iwata, M. N. Aoki, and P. Heraud (2009). Fourier transform infrared microspectroscopy as a tool to identify macroalgal propagules. *J. Phycol.* 45:560–70. [2.2]

Belliveau, D. J., D. J. Garbary, and J. L. McLachlan (1990). Effects of fluorescent brighteners on growth and morphology of the red alga *Antithamnion kylinii*. *Stain Technol.* 65:303–11. [1.3.4]

Bellwood, D. R., T. P. Hughes, C. Folke, and M. Nyström (2004). Confronting the coral reef crisis. *Nature* 429:827–33. [Essay 6]

Bender, E. A., T. J. Case, and M. E. Gilpin (1984). Perturbation experiments in community ecology: theory and practice. *Ecology* 65: 1–13. [8.3.2]

Benedetti-Cecchi, L. (2001). Variability in abundance of algae and invertebrates at different spatial scales on rocky sea shores. *Mar. Ecol. Prog. Ser.* 215:79–92. [3.1.2, 3.1.3, 3.5.1]

Benedetti-Cecchi, L. (2005). Unanticipated impacts of spatial variance of biodiversity on plant productivity. *Ecol. Lett.* 8:791–9. [3.5.1]

Benedetti-Cecchi, L., I. Bertocci, S. Vaselli, and E. Maggi (2006). Morphological plasticity and variable spatial patterns in different populations of the red alga *Rissoella verrucosa*. *Mar. Ecol. Prog. Ser.* 315:87–98. [2.6.2]

Benson, E. E., J. C. Rutter, and A. H. Cobb (1983). Seasonal variation in frond morphology and chloroplast physiology of the intertidal alga *Codium fragile* (Suringar) Hariot. *New Phytol.* 95:569–80. [2.6.2]

Berges, J. A. (1997). Algal nitrate reductases. *Eur. J. Phycol.* 32:3–8. [6.4.2, 6.5, 6.5.1]

Berges, J. A., and P. J. Harrison (1995). Nitrate reductase activity quantitatively predicts the rate of nitrate incorporation under steady state light limitation: a revised assay and characterization of the enzyme in three species of marine phytoplankton. *Limnol. Oceanogr.* 40:82–93. [6.5.1]

Berges, J. A., and M. Mulholland (2008). Enzymes and cellular N cycling. In D. G. Capone, D. A. Bronk, M. R. Mulholland, and E. J. Carpenter (eds), *Nitrogen in the Marine Environment* (pp. 1361–420). New York: Academic Press. [6.5.1]

Bergström, L., R. Berger, and L. Kautsky (2003). Negative direct effects on nutrient enrichment on the establishment of *Fucus vesiculosus* in the Baltic Sea. *Eur. J. Phycol.* 38:41–6. [9.6.1]

Berndt, M-L., J. A. Callow, and S. H. Brawley (2002). Gamete concentrations and timing and success of fertilization in a rocky shore seaweed. *Mar. Ecol. Prog. Ser.* 226:273–85. [2.4]

Bernstein, B. B., and N. Jung (1979). Selective pressures and coevolution in a kelp canopy community in southern California. *Ecol. Monogr.* 49:335–55. [4.2.2]

Bertness, M. D., and R. Callaway (1994). Positive interactions in communities. *Trends Ecol. Evol.* 9:191–3. [4.1]

Bertness, M. D., G. H. Leonard, J. M. Levine, P. R. Schmidt, and A. O. Ingraham (1999). Testing the relative contribution of positive and negative interactions in rocky intertidal communities. *Ecology* 80:2711–26. [4.1]

Bertness, M. D., S. D. Gaines, and M. E. Hay (eds) (2001). *Marine Community Ecology.* Sunderland, MA: Sinauer Associates, Inc., 550 pp. [3.1.2, 4.0]

Bertness, M. D., G. C. Trussell, P. J. Ewanschuk, and B. R. Silliman (2002). Do alternate stable community states exist in the Gulf of Maine rocky intertidal zone? *Ecology* 83:3434–48. [4.3.1]

Besada, V., J. M. Andrade, F. Schultze, and J. J. González (2009). Heavy metals in edible seaweeds commercialized for human consumption. *J. Mar. Systems* 75:305–15. [9.3.6]

Bessho, K., and Y. Iwasa (2009). Heteromorphic and isomorphic alternations of generations in macroalgae as adaptations to a seasonal environment. *Evol. Ecol. Res.* 11:691–711. [2.2]

Bessho, K., and Y. Iwasa (2010). Optimal seasonal schedules and the relative dominance of heteromorphic and isomorphic life cycles in macroalgae. *J. Theor. Biol.* 267:201–12. [2.2]

Beutlich, A., B. Borstelmann, R. Reddemann, K. Speckenbach, and R. Schnetter (1990). Notes on the life histories of *Boergesenia* and *Valonia* (Siphonocladales, Chlorophyta). *Hydrobiologia* 204/205:425–34. [2.2]

Bhattacharya, D., and L. D. Druehl (1988). Phylogenetic comparison of the small-subunit ribosomal DNA sequence of *Costaria costata* (Phaeophyta) with those of other algae, vascular plants and oomycetes. *J. Phycol.* 24:539–43. [1.4.1]

Bidwell, J. R., K. W. Wheeler, and T. R. Burridge (1998). Toxicant effects on the zoospore stage of the marine macroalga *Ecklonia radiata* (Phaeophyta: Laminariales). *Mar. Ecol. Prog. Ser.* 163:259–65. [9.3.4]

Bidwell, R. G. S. (1979). *Plant Physiology*, 2nd edn. New York: Macmillan. [7.3.4]

Bidwell, R. G. S., and J. McLachlan (1985). Carbon nutrition of seaweeds: photosynthesis, photorespiration and respiration. *J. Exp. Mar. Biol. Ecol.* 86:15–46. [5.4.1, 5.7.1]

Biebl, R. (1962). Seaweeds. In R. A. Lewin (ed.), *Physiology and Biochemistry of the Algae* (pp. 799–815). New York: Academic Press. [7.4.1, 7.4.4]

Biebl, R. (1970). Vergleichende Untersuchungen zur Temperaturresistenz von Meeresalgen entlang der pazifischen Küste Nordamerikas. *Protoplasma* 69: 61–83. [7.2.3, 7.3.8]

Biedka, R. F., J. M. Gosline, and R. E. DeWreede (1987). Biomechanical analysis of wave-induced mortality in the marine alga *Pterygophora californica*. *Mar. Ecol. Prog. Ser.* 36: 163–70. [8.3.1]

Biggins, J., and D. Bruce (1989). Regulation of excitation energy transfer in organisms containing phycobilins. *Photosynth. Res.* 20: 1-34. [5.3.1]

Billard, E., E. Serrão, G. Pearson, C. Destombe, and M. Valero (2010). *Fucus vesiculosus* and *spiralis* species complex: a nested model of local adaptation at the shore level. *Mar. Ecol. Prog. Ser.* 405:163-74. [1.4.2]

Billoud, B., A. Le Bail, and B. Charrier (2008). A stochastic 1D nearest-neighbour automaton models early development of the brown alga *Ectocarpus siliculosus. Funct. Plant Biol.* 35:1014-24. [2.6]

Bingham, S., and J. A. Schiff (1979). Conditions for attachment and development of single cells released from mechanically disrupted thalli of *Prasiola stipitata* Suhr. *Biol. Bull.* 156:257-71. [2.5.2, 2.6.1]

Bird, K. T., C. Habig, and T. DeBusk (1982). Nitrogen allocation and storage patterns in *Gracilaria tikvahiae* (Rhodophyta). *J. Phycol.* 18:344-8. [6.5.1]

Birrell, C. L., L. J. McCook, B. L. Willis, and G. A. Diaz-Pulido (2008). Effects of benthic algae on the replenishment of corals and the implications for the resilience of coral reefs. *Oceanogr. Mar. Biol.* 46:25-63. [3.3.1]

Bischof, K., and F. Steinhoff (2012). Impacts of stratospheric ozone depletion and solar UVB radiation on seaweeds. In C. Wiencke, and K. Bischof (eds), *Seaweed Biology: Novel Insights into Ecophysiology, Ecology and Utilization, Ecological Studies*, Vol. 219 (pp. 433-8). New York: Springer. [7.1]

Bischof, K., D. Hanelt, and C. Wiencke (1998). UV-radiation can affect depth-zonation of Antarctic macroalgae. *Mar Biol* 131: 597-605. [5.2.2, 7.9]

Bischof, K., D. Hanelt, and C. Wiencke (2000). Effects of ultraviolet radiation on photosynthesis and related enzyme reactions of marine macroalgae. *Planta* 211:555-62. [5.2.2, 5.4.2]

Bischof, K., G. Kräbs, C. Wiencke, and D. Hanelt (2002a). Solar ultraviolet radiation affects the activity of ribulose-1,5-bisphosphate carboxylase/oxygenase and the composition of photosynthetic and xanthophyll cycle pigments in the intertidal green alga *Ulva lactuca* L. *Planta* 215: 502-9. [5.3.3, 5.4.2]

Bischof, K., G. Peralta, G. Kräbs *et al.* (2002b). Effects of solar UV-B radiation on canopy formation of natural *Ulva* communities from Southern Spain. *J. Exp. Bot.* 53:2411-21. [5.7.1, 5.7.2, 7.3.5, 7.9]

Bischof, K., P. J. Janknegt, A. G. J. Buma, *et al.* (2003). Oxidative stress and enzymatic scavenging of superoxide radicals induced by solar UV-B radiation in *Ulva* canopies from southern Spain. *Sci. Mar.* 67:353-9. [7.9]

Bischof, K., I. Gómez, M. Molis *et al.* (2006a). Ultraviolet radiation shapes seaweed communities. *Rev. Environ. Sci. Bio-Technol.* 5:141-66. [5.2.2, 5.3.3, 7.3.7, 7.6]

Bischof, K., R. Rautenberger L. Brey, and J. L. Perez-Llorens (2006b). Physiological acclimation along gradients of solar irradiance within mats of the filamentous green macroalga *Chaetomorpha linum* from southern Spain. *Mar. Ecol. Prog. Ser.* 306:165-75. [5.7.1]

Bischoff, B., and C. Wiencke (1995). Temperature adaptation in strains of the amphi-equatorial green alga *Urospora penicilliformis* (Acrosiphoniales): biogeographical implications. *Mar. Biol.* 122:681-8. [7.3.3]

Bischoff, K., and R. Rautenberger (2012). Seaweed responses to environmental stress: reactive oxygen and antioxidative strategies In C. Wienke and K. Bischof (eds) *Seaweed Biology: Novel Insights into Ecophysiology, Ecology and Utilization* (pp. 109-33). New York: Springer. [9.3.3, 9.3.4]

Bischoff-Bäsmann, B. (1997). Temperature requirements and biogeography of marine macroalgae: adaptation of marine macroalgae to low temperatures. *Berichte zur Polarforschung* 245, 134 pp. [7.3.6]

Bischoff-Bäsmann, B., and C. Wiencke (1996). Temperature requirements for growth and survival of Antarctic Rhodophyta. *J. Phycol.* 32:525-35. [7.3.6]

Bisgrove, S. R. (2007). Cytoskeleton and early development in fucoid algae. *J. Integr. Plant Biol.* 49:1192-8. [1.3.3, 2.5.3]

Bisgrove, S. R., and D. L. Kropf (2001). Cell wall deposition during morphogenesis in fucoid algae. *Planta* 212:648-58. [1.3.1]

Bisgrove, S. R., and D. L. Kropf (2007). Asymmetric cell divisions: zygotes of fucoid algae as a model system. *Cell Division Control in Plants; Series Plant Cell Monogr.* 9:323-41. [2.5.3, 2.6]

Bixler H. J., and H. Porse (2011). A decade of change in the seaweed hydrocolloids industry. *J. Appl. Phycol.* 23:321-35. [10.1, 10.3.2, 10.5, 10.6.1]

Björk, M., L. Axelsson, and S. Beer (2004). Why is *Ulva intestinalis* the only macroalga inhabiting isolated rockpools along the Swedish Atlantic coast? *Mar. Ecol. Prog. Ser.* 284:109-16. [4.2.2]

Björk, M., S. M. Mohammed, M. Björklund, and A. Semesi (1995). Coralline algae, important coral reef builders threatened by pollution. *Ambio. Stockholm* 24:502-5. [Essay 6]

Björn, L. O., T. V. Callaghan, C. Gehrke, U. Johanson, and M. Sonesson (1999). Ozone depletion, ultraviolet radiation and plant life. *Chemosphere* 1:449-54. [5.2.2, 7.6]

Björnsater, B. R., and P. A. Wheeler (1990). Effect of nitrogen and phosphorus supply on growth and tissue composition of *Ulva fenestrata* and *Enteromorpha intestinalis* (Ulvales, Chlorophyta). *J. Phycol.* 26:603-11. [6.7.2, 6.8.2]

Black, W. A. P. (1949). Seasonal variation in chemical composition of some of the littoral seaweeds common to Scotland. Part II. *Fucus serratus, Fucus vesiculosus, Fucus spiralis* and *Pelvetia canaliculata. J. Soc. Chem. Ind.* 68:183-9. [5.5.1]

Black, W. A. P. (1950). The seasonal variation in weight and chemical composition of the common British Laminariaceae. *J. Mar. Biol. Assoc. UK* 29:45-72. [5.5.1]

Blanchette, C. A. (1996). Seasonal patterns of disturbance influence recruitment of the sea palm, *Postelsia palmaeformis. J. Exp. Mar. Biol. Ecol.* 197:1-14. [8.3.2]

Blanchette, C. A. (1997). Size and survival of intertidal plants in response to wave action: a case study with *Fucus gardneri. Ecology* 78:1563-78. [8.3.1]

Blanchette, C. A., C. M. Miner, P. T. Raimondi, *et al.* (2008). Biogeographical patterns of rocky intertidal communities along the Pacific coast of North America. *J. Biogeogr.* 35:1593-607. [3.5.1]

Blouin, N., F. Xiugeng, J. Peng, C. Yarish, and S. H. Brawley (2007). Seeding nets with neutral spores of the red alga *Porphyra umbilicals* (L.) Kützing for use in integrated multitrophic aquaculture (IMTA). *Aquaculture* 27:77-91. [10.12]

Blunden, G., and S. M. Gordon (1986). Betaines and their sulphonio analogues in marine algae. *Prog. Phycol. Res.* 4:39-80. [7.4.2]

Blunt, J. W., B. R. Copp, W.-P. Hu, *et al.* (2007). Marine natural products. *Nat. Prod. Rep.* 23:31-86. [4.2.2, 4.4.1]

Boaventura, D., M. Alexander, P. Della Santina, *et al.* (2002). The effects of grazing on the distribution and composition of low-shore algal communities on the central coast of Portugal and on the southern coast of Britain. *J. Exp. Mar. Biol. Ecol.* 267:185-206. [4.3.1]

Bogorad, L. (1975). Phycobiliproteins and complementary chromatic adaptation. *Annu. Rev. Plant Physiol.* 26: 369-401. [5.3.4]

Bokn, T. (1987). Effects of diesel oil and subsequent recovery of commercial benthic algae. *Hydrobiologia* 151/152:277-84. [9.4.2]

Bokn, T.L., F. E. Moy, and S. N. Murray (1993). Long-term effects of the water-accommodated fraction (WAF) of diesel oil on rocky shore populations maintained in experimental mesocosms. *Bot. Mar.* 36:313-19. [9.4.3]

Bokn, T. L., C. M. Duarte, M. F. Pedersen, *et al.* (2003). The response of experimental rocky shore communities to nutrient additions. *Ecosystems* 6:577-94. [6.8.5]

Bold, H. C., and M. J. Wynne (1985). *Introduction to the Algae: Structure and Reproduction*, 2nd edn. Englewood Cliffs, NJ: Prentice Hall, 706 pp. [2.2]

Boller, M. L., and E. Carrington (2006a). *In situ* measurements of hydrodynamic forces imposed on *Chondrus crispus* Stackhouse. *J. Exp. Mar. Biol. Ecol.* 337:159-70. [8.1.4, 8.3.1]

Boller, M. L., and E. Carrington (2006b). The hydrodynamic effects of shape and size during reconfiguration of a flexible macroalga. *J. Exp. Biol.* 209:1895-903. [8.1.4, 8.3.1]

Boller, M. L., and E. Carrington (2007). Interspecific comparison of hydrodynamic performance and structural properties among intertidal macroalgae. *J. Exp. Biol.* 210:1874-84. [8.3.1]

Bolton, J. J. (1983). Ecoclinal variation in *Ectocarpus siliculosus* (Phaeophyceae) with respect to temperature growth optima and survival limits. *Mar. Biol.* 73:131-8. [7.3.3, 7.3.8]

Bolwell, G. P, J. A. Callow, M. E. Callow, and L. V. Evans (1979). Fertilization in brown algae. II. Evidence for lectin-sensitive complementary receptors involved in gamete recognition in *Fucus serratus. J. Cell Sci.* 36:19-30. [2.4]

Bolwell, G.P, J. A. Callow, and L. V. Evans (1980). Fertilization in brown algae. III. Preliminary characterization of putative gamete receptors from eggs and sperm of *Fucus serratus. J. Cell Sci.* 43:209-24. [2.4]

Bond, P. R., M. T. Brown, R. M. Moate, *et al.* (1999). Arrested development in *Fucus spiralis* (Phaeophyceae) germlings exposed to copper. *Eur. J. Phycol.* 34:513-21. [9.3.2, 9.3.4]

Booth, D., J. Provan, and C. A. Maggs (2007). Molecular approaches to the study of invasive seaweeds. *Bot. Mar.* 50:385-96. [3.4]

Borowitzka, L. M. (1986). Osmoregulation in blue-green algae. *Prog. Phycol. Res.* 4:243-56. [7.4.2]

Borowitzka, M. A. (1977). Algal calcification. *Oceanogr. Mar. Biol. Annu. Rev.* 15:189-223. [6.5.3]

Borowitzka, M. A. (1979). Calcium exchange and the measurement of calcification rates in the calcareous coralline red alga *Amphiroa foliacea. Mar. Biol.* 50:339-47. [6.5.3]

Borowitzka, M. A. (1982). Mechanisms in algal calcification. *Prog. Phycol. Res.* 1:137-77. [5.4.1]

Borowitzka, M. A. (1987). Calcification in algae: mechanisms and the role of metabolism. *CRC Crit. Rev. Plant Sci.* 6:1-45. [6.5.3]

Borowitzka, M. A. (1989). Carbonate calcification in algae initiation and control. In S. Mann, J. Webb, and R. J. P. Williams (eds), *Biomineralization: Chemical and Biochemical Perspectives* (pp. 63-94). Weinheim, Germany: VCH Publications. [6.5.3]

Bothwell, J. H., D. Marie, A. F. Peters, J. M. Cock, and S. M. Coelho (2010). Role of endoreduplication in the model brown alga *Ectocarpus. New Phytol.* 188:111-21. [2.2, Fig. 2.4]

Bouarab, K., P. Potin, J. Correa, and B. Kloareg (1999). Sulfated oligosaccharides mediate the interaction between a marine red alga and its green algal pathogenic endophyte. *Plant Cell* 11:1635-50. [4.2.2]

Bouarab, K., B. Kloareg, P. Potin, and J. A. Correa (2001). Ecological and biochemical aspects in algal infectious diseases. *Cah. Biol. Mar.* 42:91-100. [4.2.2]

Bouarab, K., F. Adas, E. Gaquerel, B, *et al.* (2004). The innate immunity of a marine red alga involves oxylipins from both the eicosanoid and octadecanoid pathways. *Plant Physiol.* 135:1838-48. [4.2.2]

Bouzon, Z. L., and L. C. Ouriques (2007). Characterization of *Laurencia arbuscula* spore mucilage and cell walls with stains and FITC-labelled lectins. *Aquat. Bot.* 86:301-8. [2.5.1, 2.5.2, 2.5.3]

Bouzon, Z. L., L. C. Ouriques, and E. C. Oliveira (2005). Ultrastructure of tetraspore germination in the agar-producing seaweed *Gelidium floridanum* (Gelidiales, Rhodophyta). *Phycologia* 44:409-15. [2.5.3]

Bouzon, Z. L., L. C. Ouriques, and E. C. Oliveira (2006). Spore adhesion and cell wall formation in *Gelidium floridanum* (Rhodophyta, Gelidiales). *J. Appl. Phycol.* 18:287-94. [2.5.2, 2.5.3]

Boyd, P. W., and M. J. Elwood (2010). The biogeochemical cycle of iron in the ocean. *Nature Geosci.* 3:675-82. [6.5.5]

Boyen, C., B. Kloareg, and V. Vreeland (1988). Comparison of protoplast wall regeneration and native wall deposition in zygotes of *Fucus distichus* by cell wall labelling with monoclonal antibodies. *Plant Physiol. Biochem.* 26:653-9. [1.3.1, 2.5.2]

Boyer, K. E., P. Fong, A. R. Armitage, and R. A. Cohen (2004). Elevated nutrient content of tropical macroalgae increases rates of herbivory in coral, seagrass, and mangrove habitats. *Coral Reefs* 23:530-8. [Essay 6]

Boyer, K. E., J. S. Kertesz, and J. F. Bruno (2009). Biodiversity effects on productivity and stability of marine macroalgal communities: the role of environmental context. *Oikos* 118:1062-72. [3.5.1]

Boyle, T. P. (1985). *Validation and Predictability of Laboratory Methods for Assessing the Fate and Effects of Contaminants in Aquatic Ecosystems.* ASTM STP 865. Philadelphia: American Society for Testing and Materials. [9.2]

Bracken, M. E. S., and K. J. Nielsen (2004). Diversity of intertidal macroalgae increases with nitrogen loading by invertebrates. *Ecology* 85:2828-36. [6.8.4, 6.8.5]

Bracken, M. E. S., and J. J. Stachowicz (2006). Seaweed diversity enhances nitrogen uptake via complementary use of nitrate and ammonium. *Ecology* 87:2397-403. [6.8.4]

Bracken, M. E. S., and J. J. Stachowicz (2007). Top-down modification of bottom-up processes: selective grazing reduces macroalgal nitrogen uptake. *Mar. Ecol. Prog. Ser.* 330:75-82. [6.8.4]

Bracken, M. E. S., C. A. Gonzalez-Dorantes, and J. J. Stachowicz (2007). Whole-community mutualism: associated invertebrates facilitate a dominant habitat-forming seaweed. *Ecology* 88:2211-19. [4.5.1]

Bracken, M. E. S., E. Jones, and S. L. Williams (2011). Herbivores, tidal elevation, and species richness simultaneously mediate nitrate uptake by seaweed assemblages. *Ecology* 92:1083-93. [6.8.4]

Bradley, P M. (1991). Plant hormones do have a role in controlling growth and development of algae. *J. Phycol.* 27:317-21. [2.4, 2.6.3]

Branch, G. M. (1975). Mechanisms reducing intraspecific competition in *Patella* spp.: migration, differentiation and territorial behaviour. *J. Animal Ecol.* 44:575-600. [4.2.1]

Branch, G. M., and C. L. Griffiths (1988). The Benguela ecosystem. Part V. The coastal zone. *Oceanogr. Mar. Bio. Annu. Rev.* 26:395-486. [Fig. 5.30]

Brawley, S. H. (1987). A sodium-dependent, fast block to polyspermy occurs in eggs of fucoid algae. *Devel. Biol.* 124:390-7. [2.4]

Brawley, S. H. (1991). The fast block against polyspermy in fucoid algae is an electrical block. *Devel. Biol.* 144: 94-106. [2.4]

Brawley, S. H. (1992a). Fertilization in natural populations of the dioecious brown alga *Fucus ceranoides* and the importance of the polyspermy block. *Mar. Biol.* 113:145-57. [2.4]

Brawley, S. H. (1992b). Mesoherbivores. In D. M. John, S. J. Hawkins, and J. H. Price (eds), *Plant-Animal Interactions in the Marine Benthos* (pp. 235-63). Oxford: Oxford University Press. [4.3.1]

Brawley, S. H., and W. H. Adey (1977). Territorial behavior of three-spot damselfish (*Eupomacentrus planifrons*) increases reef algal biomass and productivity. *Env. Biol. Fish.* 2:45-51. [4.3.2]

Brawley, S. H., and W. H. Adey (1981). The effect of micrograzers on algal community structure in a coral reef microcosm. *Mar. Biol.* 61:167-77. [4.3.1]

Brawley, S. H., and X. G. Fei (1987). Studies of mesoherbivory in aquaria and in an unbarricaded mariculture farm on the Chinese coast. *J. Phycol.* 23:614-23. [4.3.1]

Brawley, S. H., and L. E. Johnson (1992). Gametogenesis, gametes, and zygotes - an ecological perspective on sexual reproduction in the algae. *Brit. Phycol. J.* 27:233-52. [2.4]

Breeman, A. M. (1988). Relative importance of temperature and other factors in determining geographic boundaries of seaweeds: experimental and phenological evidence. *Helgol Meeresunters* 42: 199-241. [7.3.8]

Breeman, A. M. (1990). Expected effects of changing seawater temperatures on the geographic distribution of seaweed species. In J. J. Beukema, W. J. Wolff, and J. W. M. Brouns (eds) *Expected Effects of Climate Change on Coastal Ecosystems* (pp. 69–76). Amsterdam: Kluwer Academic Publishers. [7.3.8]

Breeman, A. M., and M. D. Guiry (1989). Tidal influences on the photoperiodic induction of tetrasporogenesis in *Bonnemaisonia hamifera* (Rhodophyta). *Mar. Biol.* 102:5–14. [2.3.4]

Breeman, A. M., S. Bos, S. van Essen, and L. L. van Mulekom (1984). Light-dark regimes in the intertidal zone and tetrasporangial periodicity in the red alga *Rhodochorton purpureum* (Lightf.) Rosenv. *Helgol. Meeresunters.* 37:365–87. [2.3.4]

Breeman, A. M., E. J. S. Meulenhoff, and M. D. Guiry (1988). Life history regulation and phenology of the red alga *Bonnemaisonia hamifera. Helgol. Meeresunters.* 42:535–51. [2.3.4]

Breen, P. A., and K. H. Mann (1976). Changing lobster abundance and the destruction of kelp beds by sea urchins. *Mar. Biol.* 34:137–42. [4.3.1]

Brekke, C., and A. H. S. Solberg (2005). Oil detection by satellite remote sensing. *Rem. Sens. Environ.* 95:1–13. [9.4]

Brenchley, J. L., J. A. Raven, and A. M. Johnston (1996). A comparison of reproductive allocation and reproductive effort between semelparous and iteroparous fucoids (Fucales, Phaeophyta). *Hydrobiologia* 326/327:185–90. [3.5.2]

Brinkhuis, B. H. (1985). Growth patterns and rates. In M. M Littler and D. S. Littler (eds), *Handbook of Phycological Methods: Ecological Field Methods: Macroalgae* (pp. 461–77). Cambridge: Cambridge University Press. [6.7.1]

Brinkhuis, B. H., H. G. Levine, C. G. Schlenk, and S. Tobin (1987). Laminaria cultivation in the Far East and North America. In K. T. Bird, and P. H. Benson (eds), *Seaweed Cultivation for Renewable Resources* (pp. 107–46). Amsterdam: Elsevier. [10.3.2, 10.8]

Brinkhuis, B. H., R. Li, C. Wu, and X. Jiang (1989). Nitrite uptake transients and consequences for in vivo algal nitrate reductase assays. *J. Phycol.* 25:539–45. [6.5.1]

Brinza, L., C. A. Nygård, M. J. Dring, M. Gavrilescu, and L. G. Benning (2009). Cadmium tolerance and adsorption by the marine brown alga *Fucus vesiculosus* from the Irish Sea and Bothnian Sea. *Bioresource Technol.* 100:1727–33. [9.3.6]

Britz, S. J., and W. R. Briggs (1976). Circadian rhythms of chloroplast orientation and photosynthetic capacity in *Ulva. Plant Physiol.* 58:22–7. [5.7.2]

Broadwater, S. T., and J. Scott (1982). Ultrastructure of early development in the female reproductive system of *Polysiphonia harveyi* Bailey (Ceramiales, Rhodophyta). *J. Phycol.* 18:427–41. [Fig. 2.16]

Brodie, J. A., and J. Lewis (2007). *Unravelling the Algae: the Past, Present and Future of Algae Systematics.* Boca Raton, FL: CRC Press Inc., 376 pp. [1.1.1]

Brodie, J., R. A. Andersen, M. Kawachi, and A. J. K. Millar. (2009). Endangered algal species and how to protect them. *Phycologia* 48:423–38. [3.5.1]

Brodie, J. E., F. J. Kroon, B. Schaffelke, *et al.* (2012). Terrestrial pollutant runoff to the Great Barrier Reef: an update of issues, priorities and management responses. *Mar. Poll. Bull.* 65:81–100. [Essay 6]

Brokovich, E., I. Ayalon, S. Einbinder, *et al.* (2010). Grazing pressure on coral reefs across a wide depth gradient in the Gulf of Aqaba, Red Sea. *Mar. Ecol. Prog. Ser.* 399:69–80. [3.3.6]

Brown, J. H. (1995). *Macroecology.* Chicago, IL: University of Chicago Press. [Essay 2]

Brown, M. T., and J. E. Newman (2003). Physiological responses of *Gracilariopsis longissima* (S. G. Gmelin) Steentoft L. M. Irvine and Farnham (Rhodophyceae) to sublethal copper concentrations. *Aquat. Toxic.* 64:201–13. [9.3.4]

Brown, M. T., J. E. Newman, and T. Han (2012). Inter-population comparisons of copper resistance and accumulation in the red seaweed, *Gracilariopsis longissima. Ecotox.* 21:591–600. [9.3.2]

Bruhn, J., and V. A. Gerard (1996). Photoinhibition and recovery of the kelp *Laminaria saccharina* at optimal and superoptimal temperatures. *Mar. Biol.* 125: 639–48. [7.3.1]

Bruland, K. W. (1983). Trace elements in sea-water. In J. P. Riley (ed.), *Chemical Oceanography* 8 (pp. 157–219). London: Academic Press. [Table 6.2]

Bruland, K. W., and M. C. Lohan (2004). Controls of trace metals in seawater. *Treatise on Geochemistry* 6:23–47. [9.3.1]

Bruno, E., and B. Eklund (2003). Two new growth inhibition tests with the filamentous algae *Ceramium strictum* and *C. tenuicorne* (Rhodophyta). *Envir. Poll.* 125:287–93. [9.2, 9.3.4]

Bruno, J. F., and M. D. Bertness (2001). Habitat modification and facilitation in benthic marine communities. In M. D. Bertness, S. D. Gaines, and M. E. Hay (eds), *Marine Community Ecology* (pp. 201–18). Sunderland, MA: Sinauer Associates. [3.5.1, 4.1]

Bruno, J. F., H. Sweatman, W. F. Precht, E. R. Selig, and V. G. W. Schutte (2009). Assessing evidence of phase shifts from coral to macroalgal dominance on coral reefs. *Ecology* 90:1478–84. [3.3.1]

Brzezinski, M. A., D. C. Reed, and C. D. Amsler (1993). Neutral lipids as major storage products in zoospores of the giant-kelp *Macrocystis pyrifera* (Phaeophyceae). *J. Phycol.* 29:16–23. [2.5.1]

Buchanan, B. B., W. Gruissem, and R. L. Jones (eds) (2000). *Biochemistry and Molecular Biology of Plants*. Rockville, MD: ASPP, 1367 pp. [1.3, 1.4.3, 2.3.3, 2.6.3, 5, 5.3.1, 6.3.1, 7.4.1]

Büchel, C., and C. Wilhelm (1993). *In vivo* analysis of slow chlorophyll fluorescence induction kinetics in algae: progress, problems and perspectives. *Photochem. Photobiol.* 58:137–48. [7.9]

Buchholz, C. M., G. Krause, and B. H. Buck (2012). Seaweed and man. In C. Wiencke and K. Bischof (eds) *Seaweed Biology: Novel Insights into Ecophysiology, Ecology and Utilization* (pp. 471–93). New York: Springer. [10.1, 10.4, 10.5, 10.6.1]

Buck, B. H., and C. M. Buchholz (2004). The offshore ring: a new system design for the open ocean aquaculture of macroalgae. *J. Appl. Phycol.* 16:355–68. [10.8]

Buck, B. H., and C. M. Buchholz (2005). Response of offshore cultivated *Laminaria saccharina* to hydrodynamic forcing in the North Sea. *Aquaculture* 250:674–91. [10.8]

Buesseler, K., M. Aoyama, and M. Fukasawa (2011). Impacts of the Fukushima nuclear power plants on marine radioactivity. *Envir. Sci. Technol.* 45:9931–5. [9.7]

Buesseler, K., S. R. Jayne, N. S. Fisher, *et al.* (2012). Fukushima-derived radionuclides in the ocean and biota off Japan. *PNAS* 109:5984–8. [9.7]

Buggeln, R. G. (1974). Negative phototropism of the haptera of *Alaria esculenta* (Laminariales). *J. Phycol.* 10:80–2. [2.6.2]

Buggeln, R. G. (1981). Morphogenesis and growth substances. In C. S. Lobban and M. J. Wynne (eds), *The Biology of Seaweeds* (pp. 627–60). Oxford: Blackwell Scientific. [2.6.2]

Buggeln, R. G. (1983). Photoassimilate translocation in brown algae. *Prog Phycol Res* 2:283–332. [5.6]

Buggeln, R. G., D. S. Fensom, and C. J. Emerson (1985). Translocation of ^{11}C-photoassimilate in the blade of *Macrocystis pyrifera* (Phaeophyceae). *J Phycol.* 21:35–40. [5.6]

Bulleri, F. (2006). Duration of overgrowth affects survival of encrusting coralline algae. *Mar. Ecol. Prog. Ser.* 321:79–85. [4.2.1]

Bulleri, F. (2009). Facilitation research in marine systems: state of the art, emerging patterns and insights for future developments. *J. Ecol.* 97:1121–30. [4.1]

Bulleri, F., and L. Benedetti-Cecchi (2008). Facilitation of the introduced green alga *Caulerpa racemosa* by resident algal turfs: experimental evaluation of underlying mechanisms. *Mar. Ecol. Prog. Ser.* 364:77–86 [3.4, 4.1]

Bulleri, F., D. Balata, I. Bertocci, L. Tamburello, and L. Benedetti-Cecchi (2010). The seaweed *Caulerpa racemosa*

on Mediterranean rocky reefs: from passenger to driver of ecological change. *Ecology* 91:2205–12. [3.4]

Burgess, R. M., M. C. Pelletier, K. T. Ho, *et al.* (2003). Removal of ammonia toxicity in marine sediment TIEs: a comparison of *Ulva lactuca*, zeolite and aeration methods. *Mar. Poll. Bull.* 46:607–18. [9.6.2]

Burgman, M. A., and V. A. Gerard (1990). A stage-structured, stochastic population model for the giant kelp, *Macrocystis pyrifera*. *Mar. Biol.* 105: 15–23. [5.7.4]

Burke, L., K. Reytar, M. Spalding, and A. Perry (2011). *Reefs at Risk Revisited*. Washington, DC: World Resources Institute. [Essay 6]

Burkhardt, E., and A. F. Peters (1998). Molecular evidence from nrDNA its sequences that *Laminariocolax* (Phaeophyceae, Ectocarpales *sensu lato*) is a worldwide clade of closely related kelp endophytes. *J. Phycol.* 34:682–91. [4.5.1]

Burki, F., N. Okamoto, J.-F. Pombert, and P. J. Keeling (2012). The evolutionary history of haptophytes and cryptophytes: phylogenomic evidence for separate origins. *Proc. R. Soc. B.* 179:2246–54. [1.1.1]

Burridge, T. R., and J. Bidwell (2002). Review of the potential use of brown algal ecotoxicological assays in monitoring effluent discharge and pollution in southern Australia. *Mar. Poll. Bull.* 45:140–7. [9.3.2]

Burridge, T. R., and M. A. Shir (1995). The comparative effects of oil dispersants and oil/dispersant conjugates on germination of the marine macroalga *Phyllospora comosa* (Fucales: Phaeophyta). *Mar. Poll. Bull.* 31:446–52. [9.4.3]

Burritt, D. J., J. Larkindale, and C. L. Hurd (2002). Antioxidant metabolism in the intertidal red seaweed *Stictosiphonia arbuscula* following desiccation. *Planta* 215: 829–38. [7.5.1]

Burrows, M. T., R. Harvey, and L. Robb (2008). Wave exposure indices from digital coastlines and the prediction of rocky shore community structure. *Mar. Ecol. Prog. Ser.* 353:1–12. [8.1.4]

Burrows, M. T., R. Harvey, L. Robb, *et al.* (2009). Spatial scales of variance in abundance of intertidal species: effects of region, dispersal mode, and trophic level. *Ecology* 90:1242–54. [3.5.1]

Burtin, P. (2003). Nutritional value of seaweeds. *J. Environmental Agric. Food Chem.* 2:1–9. [10.2, 10.10]

Buschmann, A. H., and A. Bravo (1990). Intertidal amphipods as potential dispersal agents of carpospores of *Iridaea laminarioides* (Gigartinales, Rhodophyta). *J. Phycol.* 26:417–20. [4.3.1]

Buschmann, A. H., R. Westermeier, and C. A. Retamales (1995). Cultivation of *Gracilaria* on the sea-bottom in southern Chile: a review. *J. Appl. Phycol.* 7:291–301. [10.6.2]

Buschmann, A. H., J. A. Correa, R. Westermeier, M. C. Hernández-Gonález, and R. Norambuena (2001). Red algal farming in Chile: a review. *Aquaculture* 194:203–20. [10.6.2]

Buschmann, A. H., J. A. Vásquez, P. Osorio, *et al.* (2004). The effect of water movement, temperature and salinity on abundance and reproductive patterns of *Macrocystis* spp. (Phaeophyta) at different latitudes in Chile. *Mar. Biol.* 145:849–62. [3.5.2]

Buschmann, A. H., C. Moreno, J. A. Vásquez, and M. C. Hernández-González (2006). Reproduction strategies of *Macrocystis pyrifera* (Phaeophyta) in Southern Chile: the importance of population dynamics. *J. Appl. Phycol.* 18:575–82. [3.5.2]

Buschmann, A. H., M. C. Hernández-Gonález, and D. Varela (2008). Seaweed future cultivation in Chile: perspectives and challenges. *Internat. J. Envir. Poll.* 33:432–55. [10.3, 10.6.1, 10.6.2, 10.10]

Butler, D. M., K. Ostgaard, C. *et al.* (1989). Isolation conditions for high yields of protoplasts from *Laminaria saccharaina* and *L. digitata* (Phaeophyceae). *J. Exp. Bot.* 40: 1237–46. [2.6.1]

Butler, J. N., B. F. Morris and T. D. Sleeter (1976). The fate of petroleum in the open ocean. In *Sources, Effects, and Sinks of Hydrocarbons in the Aquatic Environment* (pp. 287–97). Washington DC: American Institute of Biological Sciences. [Table 9.4]

Butterworth, J., P. Lester, and G. Nickless (1972). Distribution of heavy metals in the Severn estuary. *Mar. Pollut. Bull.* 3:72–4. [8.3.5]

Cabello-Pasini, A., and R. S. Alberte (1997). Seasonal patterns of photosynthesis and light-independent carbon fixation in marine macrophytes. *J. Phycol.* 33: 321–9. [7.3.1]

Cabello-Pasinsi, A., and R. S. Alberte (2001). Enzymatic regulation of photosynthetic and light-independent carbon fixation in *Laminaria setchelli* (Phaeophyta), *Ulva lactuca* (Chlorophyta) and *Iridea cordata* (Rhodophyta). *Rev. Chil. Hist. Nat.* 74:226–36. [5.4.2]

Cabioch, J. (1988). Morphogenesis and generic concepts in coralline algae-a reappraisal. *Helgol. Meeresunters.* 42:493–509. [1.2.1]

Cabioch, J., and G. Giraud (1986). Structural aspects of biomineralization in the coralline algae (calcified Rhodophyceae). In B. S. C. Leadbeater and R. Riding (eds), *Biomineralization in Lower Plants and Animals* (pp. 141–56). Oxford: Clarendon Press. [6.5.3]

Cáceres, E. J., and E. R. Parodi (1998). Fine structure of zoosporogenesis, zoospore germination, and early gametophyte development in *Cladophora surera* (Cladophorales, Chlorophyta). *J. Phycol.* 34:825–34. [2.5.2]

Cairrão, E., M. Couderchet, A. M. Soares, and L. Guilhermino (2004). Glutathione-S-transferase activity of *Fucus* spp. as a biomarker of environmental contamination. *Aquat. Toxicol.* 70:277–86. [9.4.2]

Caldeira, K., and M. E. Wickett (2003). Anthropogenic carbon and ocean pH. *Nature* 425:365–7. [Fig. 7.26]

Callaway, R. M. (2007). *Positive Interactions and Interdependence in Plant Communities.* Dordrecht, The Netherlands: Springer, 415 pp. [4.1]

Callow, J. A., and M. E. Callow (2006). The *Ulva* spore adhesive system. In A. M. Smith, and J. A. Callow (eds), *Biological Adhesives* (pp. 63–78). Berlin and Heidelberg: Springer-Verlag. [2.5.1, 2.5.2]

Callow, J. A., M. E. Callow, and L. V. Evans (1979). Nutritional studies on the parasitic red alga *Choreocolax polysiponiae.* *New Phytol.* 83:451–62. [4.5.2]

Callow, M. E., L. V. Evans, G. P. Bolwell, and J. A. Callow (1978). Fertilization in brown algae. I. SEM and other observations on *Fucus serratus.* *J. Cell. Sci.* 32:45–54. [2.4]

Callow, M. E., J. A. Callow, J. D. Pickett-Heaps, and R. Wetherbee (1997). Primary adhesion of *Enteromorpha* (Chlorophyta, Ulvales) propagules: quantitative settlement studies and video microscopy. *J. Phycol.* 33:938–47. [2.5.1, 2.5.2]

Camargo, J. A., and A. Alonso (2006). Ecological and toxicological effects of inorganic nitrogen pollution in aquatic ecosystems: a global assessment. *Environ. Internat.* 32:832–49. [9.6.1, 9.6.2]

Cambridge, M., A. M. Breeman, and C. van den Hoek (1984). Temperature responses of some North Atlantic *Cladophora* species (Chlorophyceae) in relation to their geographic distribution. *Helgol Meeresunters* 38: 349–63. [7.3.8]

Camilli, R., C. M. Reddy, D. Yoerger, *et al.* (2010). Tracking hydrocarbon plume transport and biodegradation at Deepwater Horizon. *Science* 330:201–4. [9.4]

Campbell, J. W., and T. Aarup (1989). Photosynthetically available radiation at high latitudes. *Limnol. Oceanogr.* 34:1490–9. [5.2.2]

Cancino, J. M., J. Muñoz, M. Muñoz, and M. C. Orellana (1987). Effects of the Bryozoan *Membranipora tuberculata* (Bosc.) on the photosynthesis and growth of *Gelidium rex* Santelices et Abbott. *J. Exp. Mar. Biol. Ecol.* 113:105–12. [4.2.2]

Canuel, E. A., J. N. Cloern, D. B. Ringelberg, J. B. Guckert, and G. H. Rau (1995). Molecular and isotopic tracers used to examine source of organic matter and its incorporation into the food webs of San Francisco Bay. *Limnol. Oceanogr.* 40 (1):67–81 [Essay 3]

Capone, D. G., and M. F. Bautista (1985). A groundwater source of nitrate in nearshore marine sediments. *Nature* 313:214–16. [6.2]

Carballo, J. L., C. Olabarria, and T. G. Osuna (2002). Analysis of four macroalgal assemblages along the Pacific Mexican Coast during and after the 1997-1998 El Niño. *Ecosystems* 5:749-60. [7.3.7]

Carillo, S., E. López, M. M. Casas, *et al.* (2008). Potential use of seaweeds in the laying hen ration to improve the quality of n-3 fatty acids enriched eggs. *J. Appl. Phycol.* 20:721-8. [10.10]

Carney, L. T. (2011). A multispecies laboratory assessment of rapid sporophyte recruitment from delayed kelp gametophytes. *J. Phycol.* 47:244-51. [3.5.2]

Caron, L., J. P. Dubacq, C. Berkaloff, and H. Jupin (1985). Subchloroplast fractions from the brown alga *Fucus serratus*: phosphatidylglycerol contents. *Plant Cell Physiol.* 26:131-9. [5.3.1]

Caron, L., R. Remy, and C. Berkaloff (1988). Polypeptide composition of light harvesting complexes from some brown algae and diatoms. *FEBS Letters* 229:11-15. [5.3.1]

Carpenter, R. C. (1988). Mass mortality of a Caribbean sea urchin: immediate effects on community metabolism and other herbivores. *Proc. Natl. Acad. Sci. USA* 85:511-14. [4.3.1]

Carpenter, R. C. (1990). Competition among marine macroalgae: a physiological perspective. *J. Phycol.* 26:6-12. [4.0, 4.2, 4.3.1, 4.3.2, 6.8.1]

Carpenter, R. C., and P. J. Edmunds (2006). Local and regional-scale recovery of *Diadema* promotes recruitment of scleractinian corals. *Ecol. Lett.* 9:268-77. [4.3.1]

Carpenter, R. C., J. M. Hackney, and W. H. Adey (1991). Measurements of primary productivity and nitrogenase activity of coral reef algae in a chamber incorporating oscillatory flow. *Limnol. Oceanogr.* 36:40-9. [8.1.4, 8.2.1]

Carrington, E. (1990). Drag and dislodgement of an intertidal macroalga: consequences of morphological variation in *Mastocarpus papillatus* Kützing. *J. Exp. Mar. Biol. Ecol.* 139:185-200. [8.3.1]

Carrington, E., S. P. Grace, and T. Chopin (2001). Life history phases and the biomechanical properties of the red alga *Chondrus crispus* (Rhodophyta). *J. Phycol.* 37:699-704. [8.3.1]

Cashmore, A. R. (2005). Cryptochrome overview. In: M. Wada, K. Shimazaki, and M. Lino (eds), *Light Sensing in Plants* (pp.121-30). Tokyo: Springer-Verlag (Yamada Science Foundation and Springer-Verlag Tokyo). [2.3.3]

Cassab, G. I. (1998). Plant cell wall proteins. *Annu. Rev. Plant Phys.* 49:281-309. [1.3.1]

Castilla, J. C. (1988). Earthquake-caused coastal uplift and its effects on rocky intertidal kelp communities. *Science* 242:440-3. [8.3.2]

Castilla, J. C., P. H. Manriquez, and A. Camaño (2010). Effects of rocky shore coseismic uplift and the 2010 Chilean megaearthquake on intertidal biomarker species. *Mar. Ecol. Prog. Ser.* 418:17-23. [8.3.2]

Caut, S., E. Angulo, and F. Courchamp (2009). Variation in discrimination factors (Delta N-15 and Delta C-13): the effect of diet isotopic values and applications for diet reconstruction. *Journal of Applied Ecology* 46(2):443-53 [Essay 3]

Cavalier-Smith, T. (1999). Principles of protein and lipid targeting in secondary symbiogenesis: euglenoid, dinoflagellate, and sporozoan plastid origins and the eukaryote family tree. *J. Eukaryot. Microbiol.* 46:347-66. [1.1.1]

Cavanaugh, K. C., D. A. Siegel, B. P. Kinlan, and D. C. Reed (2010). Scaling giant kelp field measurements to regional scales using satellite observations. *Mar. Ecol. Prog. Ser.* 403:13-27. [3.5.1]

Ceccarelli, D. M., G. P. Jones, and L. J. McCook (2001). Territorial damselfishes as determinants of the structure of benthic communities on coral reefs. *Oceanogr. Mar. Biol.* 39:355-89. [4.3.2]

Cecere, E., A. Petrocelli, and M. Verlaque (2011). Vegetative reproduction by multicellular Rhodophyta: an overview. *Mar. Ecol.* 32:419-37. [2.1]

Cembella, A., N. J. Antia, and P. J. Harrison (1984). The utilization of inorganic and organic phosphorus compounds as nutrients by eukaryotic microalgae: a multidisciplinary perspective. *CRC Crit. Rev. Microbiol.* 10:317-91. [6.2, 6.5.2]

Cerda, O., I. A. Hinojosa, and M. Thiel (2010). Nest-building by the amphipod *Peramphithoe femorata* (Kroyer) on the kelp *Macrocystis pyrifera* (Linnaeus) C. Agardh from northern-central Chile. *Biol. Bull.* 218:248-58. [4.3.2]

Cetrulo, G. L., and M. E. Hay (2000). Activated chemical defenses in tropical versus temperate seaweeds. *Mar. Ecol. Prog. Ser.* 207:243-53. [4.4.2]

Chamberlain, Y. M. (1984). Spore size and germination in *Fosliella*, *Pneophyllum* and *Melobesia* (Rhodophyta, Corallinaceae). *Phycologia* 23:433-42. [2.5.3]

Chan, C. X., N. A. Blouin, Y. Zhuang, *et al.* (2012a). *Porphyra* (Bangiophyceae) transcriptomes provide insights into red algal development and metabolism. *J. Phycol.* 48:1328-42. [1.4, 2.2, 10.11]

Chan, C. X., S. Zauner, G. L. Wheeler, *et al.* (2012b). Analysis of *Porphyra* membrane transporters demonstrates gene transfer among photosynthetic eukaryotes and numerous sodium-coupled transport systems. *Plant Physiol.* 158:2001-12. [7.4.2]

Chan, K., P. K. Wong, and S. L. Ng (1982). Growth of *Enteromorpha linza* in sewage effluent and sewage effluent-seawater mixtures. *Hydrobiologia* 97:9-13 [9.6.2]

Chapman, A. R. O. (1973). Methods for macroscopic algae. In J. R. Stein (ed.), *Handbook of Phycological Methods: Culture Methods and Growth Measurements* (pp. 87-104). Cambridge: Cambridge University Press. [2.3.4]

Chapman, A. R. O. (1984). Reproduction, recruitment and mortality in two species of *Laminaria* in south-west Nova Scotia. *J. Exp. Mar. Biol. Ecol.* 78:99-108. [3.5.2]

Chapman, A. R. O. (1985). Demography. In M. M. Littler and D. S. Littler (eds), *Handbook of Phycological Methods: Ecological Field Methods: Macroalgae* (pp. 251-68). Cambridge: Cambridge University Press. [3.5.2]

Chapman, A. R. O. (1986). Population and community ecology of seaweeds. In J. H. S. Blaxter and A. J. Southward (eds), *Advances in Marine Biology* (vol. 23, pp. 1-161). London: Academic Press. [3.1.2, 3.5.2, 4.3.1, 7.5.1]

Chapman, A. R. O. (1987). The wild harvest and culture of *Laminaria longicruris* in eastern Canada. In M. S. Doty, J. F. Caddy, and B. Santelices (eds), *Case Studies of Seven Commercial Seaweed Resources* (pp. 193-237). FAO Fish. Tech. Pap. 281. [10.3]

Chapman, A. R. O. (1993). Hard data for matrix modeling of *Laminaria digitata* (Laminariales, Phaeophyta) populations. *Hydrobiologia* 261:263-7. [3.5.2]

Chapman, A. R. O. (1995). Functional ecology of fucoid algae: twenty-three years of progress. *Phycologia* 34:1-32. [3.1.3]

Chapman, A. R. O., and E. M. Burrows (1970). Experimental investigations into the controlling effects of light conditions on the development and growth of *Desmarestia aculeata* (L.) Lamour. *Phycologia* 9:103-8. [2.3, 2.3.3]

Chapman, A. R. O., and J. S. Craigie (1977). Seasonal growth in *Laminaria longicruris*: relations with dissolved inorganic nutrients and internal reserves of nitrogen. *Mar. Biol.* 40:197-205. [5.5.1, 6.5.1]

Chapman, A. R. O., and C. L. Goudey (1983). Demographic study of the macrothallus of *Leathesia difformis* (Phaeophyta) in Nova Scotia. *Can J. Bot.* 61:319-23. [3.5.2]

Chapman, A. R. O., J. W. Markham, and K. Lüning (1978). Effects of nitrate concentration on the growth and physiology of *Laminaria saccharina* (Phaeophyta) in culture. *J. Phycol.* 14:195-8. [6.7.2]

Chapman, A. S. (1999). From introduced species to invader: what determines variation in the success of *Codium fragile* ssp. *tomentosoides* (Chlorophyta) in the North Atlantic Ocean? *Helgolander. Meeresun.* 52:277-89. [3.4]

Chapman, M. G., and A. J. Underwood (1998). Inconsistency and variation in the development of rocky intertidal algal assemblages. *J. Exp. Mar. Biol. Ecol.* 224: 265-89. [3.5.1]

Chapman, M. S., S. W. Suh, P. M. G. Curmi, *et al.* (1988). Tertiary structure of plant RuBisCO: domains and their contacts. *Science* 241: 71-4. [5.4.2]

Charpy, L., B. E. Casareto, M. J. Langlade, and Y. Suzuki (2012). Cyanobacteria in coral reef ecosystems: a review. *J. Mar. Biol.* 2012 Article ID 259571. [1.1.1]

Charrier, B., S. M. Coelho, A. Le Bail, *et al.* (2008). Development and physiology of the brown alga *Ectocarpus siliculosus*: two centuries of research. *New Phytol.* 177:319-32. [1.3.1, 2.2, 2.3.2]

Charters, A. C., M. Neushul, and C. Barilotti (1969). The functional morphology of *Eisenia arborea*. *Proc. Intl. Seaweed Symp.* 6:89-105. [8.3.1]

Chavez, F. P., P. G. Strutton, G. E. Friederich, *et al.* (1999). Biological and chemical response of the Equatorial Pacific Ocean to the 1997-1998 El Niño. *Science* 286: 2126-31. [7.4.3]

Chen, L. C.-M. (1986). Cell development of *Porphyra miniata* (Rhodophyceae) under axenic culture. *Bot. Mar.* 29:435-9. [10.2.4]

Chen, L. C.-M., T. Edelstein, E. Ogata, and J. McLachlan (1970). The life history of *Porphyra miniata*. *Can. J. Bot.* 48:385-9. [2.3.2]

Cheshire, A. C., and N. D. Hallam (1985). The environmental role of alginates in *Durvillaea potatorum* (Fucales, Phaeophyta). *Phycologia* 24:147-53. [5.5.2]

Cheshire, A. C., and N. D. Hallam (1989). Methods for assessing the age composition of native stands of subtidal macroalgae: a case study on *Durvillaea potatorum*. *Bot. Mar.* 32:199-204. [3.5.2]

Chisolm, J. R. M. (2000). Calcification by crustose coralline algae on the northern Great Barrier Reef, Australia. *Limnol. Oceanogr.* 45:1476-84. [6.5.3]

Chisholm, J. R. M., M. Marchioretti, and J. M. Jaubert (2000). Effect of low water temperature on metabolism and growth of a subtropical strain of *Caulerpa taxifolia* (Chlorophyta). *Mar. Ecol. Prog. Ser.* 201:189-98. [3.4]

Choi, D., J. H. Kim, and Y. Lee (2008). Expansins in plant development. *Adv. Bot. Res.* 47:47-97. [1.3.4]

Choi, H.-G., and I. K. Lee (1996). Mixed-phase reproduction in *Dasysiphonia chejuensis* (Rhodophyta) from Korea. *Phycologia* 35:9-18. [1.4.2]

Choi, H. G., and T. A. Norton (2005a). Competition and facilitation between germlings of *Ascophyllum nodosum* and *Fucus vesiculosus*. *Mar. Biol.* 147:525-32. [4.1, 4.2.3]

Choi, H. G., and T. A. Norton (2005b). Competitive interactions between two fucoid algae with different growth forms, *Fucus serratus* and *Himanthalia elongata*. *Mar. Biol.* 146:283-91. [4.2.3]

Choo, K. S., J. Nilsson, M. Pedersen, and P. Snoeijs (2005). Photosynthesis, carbon uptake and antioxidant defense in two coexisting filamentous green algae under different stress conditions. *Mar. Ecol. Prog. Ser.* 292:127–38. [5.4.1]

Chopin, T. (2011) Progression of the Integrated Multi-Trophic Aquaculture (IMTA) concept and upscaling of IMTA systems towards commercialization. *Aquacult. Europe* 36:5–12. [10.9, 10.10, 10.11]

Chopin, T. (2012) Aquaculture, Integrated Multi-Trophic (IMTA) In R.A. Meyers (ed.), *Encyclopedia of Sustainability Science and Technology* (pp. 542–64). Dordrecht, The Netherlands: Springer. [10.10]

Chopin, T., and M. Sawhney (2009). Seaweeds and their mariculture. In J. H Steele, S. A. Thorpe, and K. K. Turekian (eds), *The Encyclopedia of Ocean Sciences* (pp. 4477–87). Oxford: Elsevier. [10.1, 10.3]

Chopin, T. and B. T. Wagey (1999). Factorial study of the effects of phosphorus and nitrogen enrichments on nutrient and carrageenan content in *Chondrus crispus* (Rhodophyceae) and on residual nutrient concentration in seawater. *Bot. Mar.* 42:23–31. [10.7]

Chopin, T., A. Hourmat, J.-Y. Floc'h, and M. Penot (1990). Seasonal variations in the red alga *Chondrus crispus* on the Atlantic French coast. II. Relations with phosphorus concentration in seawater and intertidal phosphorylated fractions. *Can. J. Bot.* 68:512–17. [6.5.2]

Chopin, T., H. Lehmal, and K. Halcrow (1997). Polyphosphates in the red macroalga *Chondrus crispus* (Rhodophyta). *New Phytol.* 135:587–94. [6.5.2]

Chopin, T., B. F. Kerin, and R. Mazerolle (1999) Phycocolloid chemistry as a taxonomic indicator of phylogeny in the Gigartinales, Rhodophyceae: a review and current developments using Fourier transform infrared diffuse reflectance spectroscopy. *Phycol. Res.* 47:167–88. [10.5.3]

Chopin, T., A. H. Buschmann, C. Halling, *et al.* (2001). Integrating seaweeds into marine aquaculture systems: a key towards sustainability. *J. Phycol.* 37:975–86. [10.1, 10.9]

Chopin, T., S. M. C. Robinson, M. Troell, *et al.* (2008). Multitrophic integration for sustainable marine aquaculture. In S. E. Jørgensen and B. D. Fath (eds), *The Encyclopedia of Ecology, Ecological Engineering* (Vol. 3, pp. 2463–75). Oxford: Elsevier. [10.3, 10.9]

Chopin, T., J. A. Cooper, G. Reid, S. Cross and C. Moore (2012) Open-water integrated multi-trophic aquaculture: environmental biomitigation and economic diversification of fed aquaculture by extractive aquaculture. *Rev. Aquaculture* 4:209–20. [10.1, 10.9]

Chow, F., and M. C. de Oliveira (2008). Rapid and slow modulation of nitrate reductase activity in the red macroalga *Gracilaria chilensis* (Gracilariales, Rhodophyta): influence of different nitrogen sources. *J. Appl. Phycol.* 20:775–82. [6.5.1]

Christensen, T. (1988). Salinity preferences of twenty species of *Vaucheria* (Tribophyceae). *J Mar Biol Assoc UK* 68:531–45. [7.4.3]

Christie, A. O., and M. Shaw (1968). Settlement experiments with zoospores of *Enteromorpha intestinalis* (L.) Link. *Br. Phycol. Bull.* 3:529–34. [2.5.2]

Christie, H., K. M. Norderhaug, and S. Fredriksen. (2009). Macrophytes as habitat for fauna. *Mar. Ecol. Prog. Ser.* 396:221–33. [3.3.2]

Chu, Z. X., and J. M. Anderson (1985). Isolation and characterization of a siphonoxanthin-chlorophyll a/b protein complex of photosystem I from a *Codium* species (Siphonales). *Biochim Biophys Acta* 806:154–60. [5.3.1]

Chung, I. K., and B. Brinkhuis (1986). Copper effects in early stages of the kelp, *Laminaria saccharina. Mar. Poll. Bull.* 17:213–18. [9.4.3, 10.10]

Chung, I. K., J. Beardall, S. Mehta, D. Sahoo, and S. Stojkovic (2011). Using marine macroalgae for carbon sequestration: a critical appraisal. *J. Appl. Phycol.* 23:877–86. [10.10]

Clark, M. S., A. Tanguy, D. Jollivet *et al.* (2010). Populations and pathways: genomic approaches to understanding population structure and environmental adaptation. In J. M. Cock, K. Tessmar-Raible, C. Boyen, and F. Viard (eds), *Introduction to Marine Genomics* (pp. 73–118). Dordrecht, The Netherlands: Springer. [1.4.1]

Clark, R.B. (2001). *Marine Pollution* (5th edn). Oxford: Oxford University Press. 237 pp. [9.4]

Claudet, J., C. W. Osenberg, L. Benedetti-Cecchi, *et al.* (2008). Marine reserves: size and age do matter. *Ecol. Lett.* 11:481–9. [3.5.1]

Clayton, M. N. (1981). Correlated studies on seasonal changes in the sexuality, growth rate and longevity of complanate *Scytosiphon* (Scytosiphonaceae, Phaeophyta) from southern Australia growing *in situ. J. Exp. Mar. Biol. Ecol.* 51:87–96. [2.1]

Clayton, M. N. (1984). An electron microscope study of gamete release and settling in the complanate form of *Scytosiphon* (Scytosiphonaceae, Phaeophyta). *J. Phycol.* 20:276–85. [2.5.2]

Clayton, M. N. (1988). Evolution and life histories of brown algae. *Bot. Mar.* 31:379–87. [2.1]

Clayton, M. N. (1992). Propagules of marine macroalgae: structure and development. *Brit. Phycol. J.* 27:219–32. [2.5.1]

Clayton, M. N., and C. M. Ashburner (1990). The anatomy and ultrastructure of "conducting channels" in *Ascoseira mirabilis* (Ascoseirales, Phaeophyceae). *Bot Mar* 33: 63–70. [5.6]

Clayton, M. N., K. Kevekordes, M. E. A. Schoenwaelder, C. E. Schmid, and C. M. Ashburner (1998). Parthenogenesis in *Hormosira banksii* (Fucales, Phaeophyceae). *Bot. Mar.* 41:23–30. [2.2]

Clements, K. D., D. Raubenheimer, and J. H. Choat (2009). Nutritional ecology of marine herbivorous fishes: ten years on. *Funct. Ecol.* 23:79–92. [4.3.2.]

• Clifton, K. E. (1997). Mass spawning by green algae on coral reefs. *Science* 275:1116–18. [2.3.3, 2.4]

Clifton, K. E., and L. M. Clifton (1999). The phenology of sexual reproduction by green algae (Bryopsidales) on Caribbean coral reefs. *J. Phycol.* 35:24–34. [2.3.3]

Cock, J. M., L. Sterck, P. Rouzé, *et al.* (2010a). The *Ectocarpus* genome and the independent evolution of multicellularity in brown algae. *Nature* 465:617–21. [1.1.1, 1.4.1, 1.4.2, 5.5.2]

Cock, J.M., K. Tessmar-Raible, C. Boyen, and F. Viard (eds) (2010b). *Introduction to Marine Genomics, Advances in Marine Genomics 1*. Dordrecht, The Netherlands: Springer. [1.4.1]

Cocquyt, E., H. Verbruggen, F. Leliaert, and O. De Clerck (2010). Evolution and cytological diversification of the green seaweeds (Ulvophyceae). *Mol. Biol. Evol.* 27:2052–61. [1.1.1]

Coelho, L., J. Prince, and T. G. Nolen (1998). Processing of defensive pigment in *Aplysia californica*: acquisition, modification and mobilization of the red algal pigment r-phycoerythrin by the digestive gland. *J. Exp. Biol.* 201:425–38. [4.4.2]

Coehlo, S. M., J. W. Rijstenbil, and M. J. Brown (2000). Impacts of anthropogenic stresses on the early development stages in seaweeds. *J. Aquat. Ecosyst. Stress Recovery* 7:317–33. [9.4.2]

Coelho, S. M., A. F. Peters, B. Charrier, *et al.* (2007). Complex life cycles of multicellular eukaryotes: new approaches based on the use of model organisms. *Gene* 406:152–70. [2.2]

Coelho, S. M., C. Brownlee, and J. H. F. Bothwell.(2008). A tip-high, Ca^{2+}-interdependent, reactive oxygen species gradient is associated with polarized growth in *Fucus serratus* zygotes. *Planta* 227:1037–46. [2.5.3]

Coelho, S. M., S. Heesch, N. Grimsley, H. Moreau, and M. J. Cock (2010). Genomics of marine algae. In J. M. Cock, K. Tessmar-Raible, C. Boyen, and F. Viard (eds), *Introduction to Marine Genomics, Advances in Marine Genomics 1* (pp. 179–212). The Netherlands: Springer. [3.5.1]

Coelho, S. M., O. Godfroy, A. Arun, G, *et al.* (2011). OUROBOROS is a master regulator of the gametophyte to

sporophyte life cycle transition in the brown alga *Ectocarpus*. *PNAS* 108:11, 518–23. [2.2]

Cohen, R. A., and P. Fong (2005). Experimental evidence supports the use of $\delta15N$ content of the opportunistic green macroalga *Enteromorpha intestinalis* (Chlorophyta) to determine nitrogen sources to estuaries. *J. Phycol.* 41:149–56. [9.6.3]

Cohen-Bazire, G., and D. A. Bryant (1982). Phycobilisomes: composition and structure. In N. G. Carr and B. A. Whitton (eds) *The Biology of Cyanobacteria* (pp. 143–190). Oxford: Blackwell Scientific. [5.3.1]

Cole, D. G. (2003). The intraflagellar transport machinery of *Chlamydomonas reinhardtii*. *Traffic* 4:435–42. [1.3.3]

Coleman, M. A. (2002). Small-scale spatial variability in inter-tidal and subtidal turfing algal assemblages and the tem-poral generality of these patterns. *J. Exp. Mar. Biol. Ecol.* 267:53–74. [3.5.1]

Coleman, M. A., and S. H. Brawley (2005). Are life history characteristics good predictors of genetic diversity and structure? A case study of the intertidal alga *Fucus spiralis* (Heterokontophyta; Phaeophyceae). *J. Phycol.* 41:753–62. [1.4.2]

Coleman, R. A., A. J. Underwood, L. Benedetti-Cecchi, *et al.* (2006). A continental scale evaluation of the role of limpet grazing on rocky shores. *Community Ecology* 147:556–64. [4.3.1]

Coleman, R. A., S. J. Ramchunder, A. J. Moody, and A. Foggo (2007). An enzyme in snail saliva induces herbivore-resistance in a marine alga. *Funct. Ecol.* 21:101–6. [4.4.2]

Collado-Vides, L. (2002a). Clonal architecture in marine macroalgae: ecological and evolutionary perspectives. *Evol. Ecol.* 15:531–45. [1.2.3]

Collado-Vides, L. (2002b). Morphological plasticity of *Caulerpa prolifera* (Caulerpales, Chlorophyta) in relation to growth form in a coral reef lagoon. *Bot. Mar.* 45:123–9. [2.6.2, 3.4]

Collado-Vides, L., L. M. Rutten, and J. W. Fourqurean (2005). Spatiotemporal variation of the abundance of calcareous green macroalgae in the Florida Keys: a study of synchrony within a macroalgal functional-form group. *J. Phycol.* 41:742–52. [1.2.2]

Collén, J. and I. R. Davison (1999a). Reactive oxygen metabol-ism in intertidal *Fucus* spp. (Phaeophyceae). *J Phycol* 35:62–9. [7.3.4, 7.3.5, 7.5.1, 7.5.2, 7.6, 7.7, Essay 5]

Collén, J., and I. R. Davison (1999b). Production and damage of reactive oxygen in intertidal *Fucus* spp. (Phaeophyceae). *J Phycol* 35: 54–61. [7.5.2, 7.6, 7.5.1, 7.7, Essay 5]

Collén, J., and I. R. Davison (1999c). Stress tolerance and reactive oxygen metabolism in the intertidal red seaweeds

Mastocarpus stellatus and *Chondrus crispus. Plant Cell. Environ.* 22:1143–51. [7.5.2, 7.6, 7.7, Essay 5]

Collén, J., and I. R. Davison (2001). Seasonality and thermal acclimation of reactive oxygen metabolism in *Fucus vesiculosus* (Phaeophyceae). *J. Phycol.* 37:474–81. [7.3.4, 7.8]

Collén, J., M. J. Del Río, G. García-Reina, and M. Pedersén (1995). Photosynthetic H_2O_2 production by *Ulva rigida. Planta* 196:225–30. [Essay 5]

Collén, J., V. Roeder, S. Rousvoal, *et al.* (2006). An expressed sequence tag analysis of thallus and regenerating protoplasts of *Chondrus crispus* (Gigarinales, Rhodophyceae). *J. Phycol.* 42:104–12. [1.4.1]

Collén, J., I. Guisle-Marsollier, J. Léger, and C. Boyen (2007). Response of the transcriptome of the intertidal red seaweed *Chondrus crispus* to controlled and natural stresses. *New Phytol.* 176:45–55. [7.3.5, 7.6, Essay 5]

Collén, J., B. Porcel, W. Carré, *et al.* (2013). Genome structure and metabolic features in the red seaweed *Chondrus crispus* shed light on evolution of the Archaeplastida. *PNAS* 110: 5247–52. [1.1.1, 1.4.1, 1.4.2, 10.11]

Collins, C. J., C. I. Fraser, A. Ashcroft, and J. M. Waters (2010). Asymmetric dispersal of southern bull-kelp (*Durvillaea antarctica*) adults in coastal New Zealand: testing an oceanographic hypothesis. *Mol. Ecol.* 19:4572–80. [3.3.7]

Collins, S., and G. Bell (2004). Phenotypic consequences of 1,000 generations of selection at high CO_2 in a green alga. *Nature* 431:566–69. [Essay 4]

Collos, Y., M. Y. Siddiqi, M. Y. Yang, A. D. M. Glass, and P. J. Harrison (1992). Nitrate uptake kinetics by two marine diatoms using the radioactive tracer ^{13}N. *J. Exp. Mar. Biol. Ecol.* 163:251–60. [6.4]

Colman, J. (1933). The nature of the intertidal zonation of plants and animals. *J. Mar. Biol. Assoc. UK* 18:435–76. [3.1.3]

Conklin, E. J., and J. E. Smith (2005). Abundance and spread of the invasive red algae, *Kappaphycus* spp., in Kane'ohe Bay, Hawai'i and an experimental assessment of management options. *Biological Invasions* 7:1029–39. [3.4, Essay 6]

Connan, S., and D. B. Stengel (2011). Impacts of ambient salinity and copper on brown algae: 1. Interactive effects on photosynthesis, growth and copper accumulation. *Aquat. Toxicol.* 104:94–107. [9.3.2]

Connell, J. H. (1961a). Effects of competition, predation by *Thais lapillus*, and other factors on natural populations of the barnacle *Balanus balanoides. Ecol. Monogr.* 31:61–104. [Essay 2]

Connell, J. H. (1961b). The influence of interspecific competition and other factors on the distribution of the barnacle *Chthamalus stellatus. Ecology* 42:710–23. [Essay 2]

Connell, J. H. (1978). Diversity in tropical rain forests and coral reefs. *Science* 199:1302–10. [Essay 2]

Connell, J. H., and R. O. Slatyer (1977). Mechanisms of succession in natural communities and their role in community stability and organization. *Am. Nat.* 111:1119–44. [3.5.1.]

Connell, J. H., T. P. Hughes, and C. C. Wallace (1997). A 30-year study of coral abundance, recruitment, and disturbance at several scales in space and time. *Ecol. Monogr.* 67:461–88. [8.3.2]

Connell, S. D. (2007a). Subtidal temperate rocky habitats: habitat heterogeneity at local to continental scales. In S. D. Connell and B. M. Gillanders (eds), *Marine Ecology* (pp. 378–401). Melbourne, Australia: Oxford University Press. [3.1.2, 3.3.2]

Connell, S. D. (2007b). Water quality and the loss of coral reefs and kelp forests: alternative states and the influence of fishing. In S. D. Connell and B. M. Gillanders (eds), *Marine Ecology* (pp. 556–68). Melbourne, Australia: Oxford University Press. [3.3.1]

Connell, S. D., and B. M. Gillanders (eds) (2007). *Marine Ecology.* Melbourne, Australia: Oxford University Press, 630 pp. [3.1.2, 3.1.3, 4.0]

Connell, S. D., and A. D. Irving (2009). The subtidal ecology of rocky coasts: local–regional biogeographic patterns and their experimental analysis. In J. D. Witman and K. Roy (eds), *Marine Macroecology* (pp. 392–417). Chicago, IL: University of Chicago Press. [1.1.3, 3.1.2, 3.2, 3.5.1]

Connell, S. D., and B. D. Russell (2010). The direct effects of increasing CO_2 and temperature on non-calcifying organisms: increasing the potential for phase shifts in kelp forest. *P. Roy. Soc. B Biol. Sci.* 227:1409–15. [7.3.8]

Connolly, R. M., M. A. Guest, A. J. Melville, and J. M. Oakes (2004). Sulfur stable isotopes separate producers in marine food-web analysis. *Oecologia* 138(2): 161–7. [Essay 3]

Connolly, S. R., and J. Roughgarden (1999). Theory of marine communities: competition, predation, and recruitment-dependent interaction strength. *Ecol. Monogr.* 69:277–96. [Essay 2]

Contreras, L., D. Mella, A. Moenne, and J. A. Correa (2009). Differential responses to copper-induced oxidative stress in the marine macroalgal *Lessonia migrescens* and *Scytosiphon lomentaria* (Phaeophyceae). *Aquat. Tox.* 94:94–102. [9.3.4]

Contreras-Porcia, L., G. Dennett, A. González, *et al.* (2011). Identification of copper-induced genes in the marine alga *Ulva compressa* (Chlorophyta). *Mar. Biotechnol.* 13:544–56. [9.3.4]

Cook, P. L. M., E. V. C. Butler, and B. D. Eyre (2004). Carbon and nitrogen cycling on intertidal mudflats of a temperate

Australian estuary. I. Benthic metabolism. *Mar. Ecol. Prog. Ser.* 280:25–38 [Essay 3]

Coomans, R. J., and M. H. Hommersand (1990). Vegetative growth and organization. In K. M. Cole and R. G. Sheath (eds), *Biology of the Red Algae* (pp. 275–304). Cambridge: Cambridge University Press. [2.6.2]

Coon, D. A., M. Neushul, and A. C. Charters (1972). The settling behavior of marine algal spores. *Proc. Intl. Seaweed Symp.* 7:237–42. [2.5.1]

Cooper, T. F., J. P. Gilmour, and K. E. Fabricus (2009). Bioindicators of changes in water quality on coral reefs: review and recommendations for monitoring programmes. *Coral Reefs* 28:589–606.[9.6.1]

Corbit, J. D., and D. J. Garbary (1993). Computer-simulation of the morphology and development of several species of seaweed using Lindenmayer-systems. *Comput. Graph.* 17:85–8. [2.6]

Cornwall, C. E., C. D. Hepburn, C. A. Pilditch, and C. L. Hurd (2013). Concentration boundary layers around complex assemblages of macroalgae: implications for the effects of ocean acidification on understory coralline algae. *Limnol. Oceanogr.* 58: 121–30. [8.1.3, 8.2.1]

Correa, J. A., and V. Flores (1995). Whitening, thallus decay and fragmentation in *Gracilaria chilensis* associated with an endophytic amoeba. *J. Appl. Phycol.* 7:421–5. [10.6.2]

Correa, J. A., and E. A. Martinez (1996). Factors associated with host specificity in *Sporocladopsis novae-zelandiae* (Chlorophyta). *J. Phycol.* 32:22–7. [4.2.2]

Correa, J. A., and J. McLachlan (1991). Endophytic algae of *Chondrus crispus* (Rhodophyta). III. Host specificity. *J. Phycol.* 27:448–59. [4.2.2, 4.5.2]

Correa, J. A., and J. L. Mclachlan (1994). Endophytic algae of *Chondrus crispus* (Rhodophyta). V. Fine structure of the infection by *Acrochaete operculata* (Chlorophyta). *Eur. J. Phycol.* 29:33–47. [4.5.2]

Correa, J., I. Novaczek, and J. McLachlan (1986). Effect of temperature and daylength on morphogenesis of *Scytosiphon lomentaria* (Scytosiphonales, Phaeophyta) from eastern Canada. *Phycologia* 25: 469–75. [2.3.2]

Correa, J.A., P. González, P. Sánchez, J. Muñoz, and M. C. Orellana (1996). Copper-algae interactions: inheritance or adaptation. *Environ. Monitor. Assess.* 40:41–54. [9.3.3]

Corzo, A., and F. X. Niell (1991). Determination of nitrate reductase activity in *Ulva rigida* C. Agardh by the *in situ* method. *J. Exp. Mar. Biol. Ecol.* 146:181–91. [6.5.1]

Corzo, A., and F. X. Niell (1994). Nitrate reductase activity and *in vivo* nitrate reduction rate in *Ulva rigida* illuminated by blue light. *Mar. Biol.* 120:17–23. [6.5.1]

Cosgrove, D. J. (1981). Analysis of the dynamic and steady-state responses of growth rate and turgor pressure to changes in cell parameters. *Plant Physiol.* 68:1439–46. [7.4.1]

Cosse, A., C. LeBlanc, and P. Potin (2008). Dynamic defense of marine macroalgae against pathogens: from early activated to gene-regulated responses. *Adv. Bot. Res.* 46:221–66. [4.2.2]

Costa, S., D. Crespo, B. M. G. Henriques. *et al.* (2011). Kinetics of mercury accumulation and its effects on *Ulva lactuca* growth rate at two salinities and exposure conditions. *Water Air Soil Poll.* 217:689–99. [9.3.1, 9.3.4]

Cotton, A. D. (1912). Marine algae. In R. L. Praeger (ed.), *A Biological Survey of Clare Island in the County of Mayo, Ireland and the Adjoining District.* Section 1, Part 15 (pp. 1–178). Dublin, Ireland: Hodges, Figgis and Co., Ltd. [3.3.5]

Coughlan, S. (1977). Sulphate uptake in *Fucus serratus.* 1. *Exp. Bot.* 28:1207–15. [6.5.4]

Coughlan, S., and L. V. Evans (1978). Isolation and characterization of Golgi bodies from vegetative tissue of the brown alga *Fucus serratus. J. Exp. Bot.* 29:55–68. [5.5.3]

Court, G. J. (1980). Photosynthesis and translocation studies of *Laurencia spectabilis* and its symbiont *Janczewskia gardneri* (Rhodophyceae). *J Phycol.* 16:270–9. [5.4.1]

Coutinho, R., and Y. Yoneshigue (1990). Diurnal variation in photosynthesis vs. irradiance curves from "sun" and "shade" plants of *Pterocladia capillacea* (Gmelin) Bornet et Thuret (Gelidiaceae [sic]: Rhodophyta) from Cabo Frio, Rio de Janeiro, Brazil. *J. Exp. Mar. Biol. Ecol.* 118:217–28. [5.3.2]

Coyer, J. A., A. F. Peters, G. Hoarau, W. T. Stam, and J. L. Olsen (2002). Hybridization of the marine seaweeds, *Fucus serratus* and *Fucus evanescens* (Heterokontophyta: Phaeophyceae) in a 100-year-old zone of secondary contact. *P. Roy. Soc. Lond. B. Biol. Sci.* 269:1829–34. [1.4.2]

Coyer, J. A., G. Hoarau, M.-P. Oudot-Le Secq, W. T. Stam and J. L. Olsen (2006a). A mtDNA-based phylogeny of the brown algal genus *Fucus* (Heterokontophyta; Phaeophyta). *Mol. Phylogenet. Evol.* 39:209–22. [1.4.1]

Coyer, J. A., G. Hoarau, G. A. Pearson, *et al.* (2006b). Convergent adaptation to a marginal habitat by homoploid hybrids and polyploidy ecads in the seaweed genus *Fucus. Biol. Lett.* 2:405–8. [1.4.2]

Craigie, J. S. (1974). Storage products. In W. D. P. Stewart (ed.) *Algal Physiology and Biochemistry* (pp. 206–35). Oxford: Blackwell Scientific. [5.4.2, 5.5.1]

Craigie, J. S. (1990). Cell walls. In K. M. Cole and R. G. Sheath (eds), *Biology of the Red Algae* (pp. 221–57). Cambridge: Cambridge University Pres. [5.5.2]

Craigie, J. S. (2011). Seaweed extract stimuli in plant science and agriculture. *J. Appl. Phycol.* 23:371–93. [10.10]

Craigie, J. S., and P. F. Shacklock (1995). Culture of Irish moss. In A. D. Boghen (ed.), *Cold-water Aquaculture in Atlantic Canada* (pp. 363–90). Canadian Institute for Research and Regional Development, Moncton, NB, Canada. [10.7]

Craigie, J. S., E. R. Morris, D. A. Rees, and D. Thorn (1984). Alginate block structure in Phaeophyceae from Nova Scotia: variation with species, environment and tissue-type. *Carbohydro Polymers* 4:237–52. [1.3.1, 8.3.1]

Craigie, J. S., J. A. Correa, and M. E. Gordon (1992). Cuticles from *Chondrus crispus* (Rhodophyta). *J. Phycol.* 28:777–86. [1.3.1]

Crawford, N. M., and A. D. M. Glass (1998). Molecular and physiological aspects of nitrate uptake in plants. *Trends Plant Sci.* 3:389–95. [6.4]

Crawley, K. R., and G. A. Hyndes (2007). The role of different types of detached macrophytes in the food and habitat choice of a surf-zone inhabiting amphipod. *Mar. Biol.* 151 (4):1433–43. [Essay 3]

Crawley, K. R., G. A. Hyndes, and S. G. Ayvazian (2006). Influence of different volumes and types of detached macrophytes on fish community structure in surf zones of sandy beaches. *Mar. Ecol. Prog. Ser.* 307:233–46. [Essay 3]

Crawley, K.R., G. A. Hyndes, M. A. Vanderklift, A. T. Revill, and P. D. Nichols (2009). Allochthonous brown algae are the primary food source for consumers in a temperate, coastal environment. *Mar. Ecol. Prog. Ser.* 376:33–44. [Essay 3]

Crayton, M. A., E. Wilson, and R. S. Quatrano (1974). Sulfation of fucoidan in *Fucus* embryos. II. Separation from initiation of polar growth. *Devel. Biol.* 39:134–7. [2.5.1]

Creed, J. C., T. A. Norton, and S. P. Harding (1996). The development of size structure in a young *Fucus serratus* population. *Eur. J. Phycol.* 31:203–9. [3.5.2]

Creed, J. C., T. A. Norton, and J. M. Kain (Jones) (1997). Intraspecific competition in *Fucus serratus* germlings: the interaction of light, nutrients and density. *J. Exp. Mar. Biol. Ecol.* 212:211–23. [4.2.3]

Creed, J. C., J. M. Kain (Jones), and T. Norton (1998). An experimental evaluation of density and plant size in two large brown seaweeds. *J. Phycol.* 34(1):39–52. [4.2.3]

Critchley, A. T., and M. Ohno (eds) (1998). *Seaweed Resources of the World*. Yokosuka, Japan: Japan International Cooperation Agency, 429 pp. [10.1, 10.6.1]

Critchley, A. T., M. Ohno, and D. B. Largo (2006). *World Seaweed Resources: An Authoritative Reference System*. DVD-ROM. Wokingham, UK: ETI Information Services. [10.1]

Cronin, G., and M. E. Hay (1996a). Susceptibility to herbivores depends on recent history of both the plant and the animal. *Ecology* 77:1531–43. [4.3.2, Essay 3]

Cronin, G., and M. E. Hay (1996b). Induction of seaweed chemical defenses by amphipod grazing. *Ecology* 77:2287–301. [4.4.2]

Crothers, J. H. (1983). Field experiments on the effects of crude oil and dispersant on the common animals and plants of rocky seashores. *Mar. Environ. Res.* 8:215–39. [9.4.3]

Crowe, T.P., R. C. Thompson, S. Bray, and S. J. Hawkins (2000). Impacts of anthropogenic stress on rocky intertidal communities. *J. Aquat. Ecosy. Stress Recov.* 7:273–97. [9.4.3]

Cruces, E., P. Huovinen, and I. Gómez (2012). Phlorotannin and antioxidant responses upon short-term exposure to UV-radiation and elevated temperature in three South Pacific kelps. *Photochem. Photobiol.* 88:58–66. [7.9]

Cruz-Rivera, E., and M. E. Hay (2000). Can quantity replace quality? Food choice, compensatory feeding, and fitness of marine mesograzers. *Ecology* 81:201–19. [4.3.2]

Cunningham, E. M., M. D. Guiry, and A. M. Breeman (1993). Environmental regulation of development, life history and biogeography of *Helminthora stackhousei* (Rhodophyta) by daylength and temperature. *J. Exp. Mar. Biol.* 171:1–21. [2.3.2]

Dalby, D. H. (1980). Monitoring and exposure scales. In J. H. Price, D. E. G. Irvine, and W. F. Farnham (eds), *The Shore Environment. Vol. 1: Methods* (pp. 117–36). New York: Academic Press. [8.1.4]

Daly, M. A., and A. C Mathieson (1977). The effects of sand movement on intertidal seaweeds and selected invertebrates at Bound Rock, New Hampshire, USA. *Mar. Biol.* 43:45–55. [8.3.2]

Davenport, A. C., and T. W. Anderson (2007). Positive indirect effects of reef fishes on kelp performance: the importance of mesograzers. *Ecology* 88:1548–61. [4.3.1]

Davis, T. A., B. Volesky, and A. Mucci (2003). A review of the biochemistry of heavy metal biosorption by brown algae. *Water Res.* 37:4311–30. [9.3.6]

Davison, I. R. (1991). Environmental effects on algal photosynthesis: temperature. *J. Phycol.* 27:2–8. [7.3.5]

Davison, I. R., and J. O. Davison (1987). The effect of growth temperature on enzyme activities in the brown alga *Laminaria saccharina*. *Br. Phycol. J.* 22:77–87. [7.3.1]

Davison, I. R., and G. A. Pearson (1996). Stress tolerance in intertidal seaweeds. *J. Phycol.* 32:197–222. [7.1, 7.5, 7.5.1]

Davison, I. R., and R. H. Reed (l985). Osmotic adjustment in *Laminaria digitata* (Phaeophyta) with particular reference to seasonal changes in internal solute concentrations. *J. Phycol.* 21:41–50. [7.4.2]

Davison, I. R., and W. D. P. Stewart (1983). Occurrence and significance of nitrogen transport in the brown alga *Laminaria digitata*. *Mar. Biol.* 77:107–12. [6.4.2]

Davison, I. R., and W. D. P. Stewart (1984). Studies on nitrate reductase activity in *Laminaria digitata* (Huds.) Lamour. I. Longitudinal and transverse profiles of nitrate reductase activity within the thallus. *J. Exp. Mar. Biol. Ecol.* 74:201–10. [6.5.1]

Davison, I. R., M. Andrews, and W. D. P. Stewart (1984). Regulation of growth in *Laminaria digitata*: use of in vivo nitrate reductase activities as an indicator of nitrogen limitation in field populations of *Laminaria* spp. *Mar. Biol.* 84:207–17. [6.5.1]

Davison, I. R., S. R. Dudgeon, and H-M. Ruan (1989). Effect of freezing on seaweed photosynthesis. *Mar Ecol Prog Ser* 58: 123–31. [7.3.4, 7.5.2]

Davison, I. R., R. M. Greene, and E. J. Podolak (1991). Temperature acclimation of respiration and photosynthesis in the brown alga *Laminaria saccharina*. *Mar Biol* 110: 449–54. [7.3.1, 7.3.2]

Dawes, C. J. (1979). Physiological and biochemical comparisons of *Eucheuma* spp. (Florideophyceae) yielding iotacarrageenan. *Proc. Intl. Seaweed Symp.* 9:188–207. [5.5.2]

Dawes, C. J. (1987). The biology of commercially important tropical marine algae. In K. Bird and P. H. Benson (eds), *Seaweed Cultivation for Renewable Resources* (pp. 155–90). Amsterdam: Elsevier. [10.5.3]

Dawes, C. J. (1989). Temperature acclimation in cultured *Eucheuma isiforme* from Florida and *E. alvarezii* from the Phillipines. *J. Appl. Phycol.* 1: 59–65. [7.3.2]

Dawes, C. J., and C. Barilotti (1969). Cytoplasmic organization and rhythmic streaming in growing blades of *Caulerpa prolifera*. *Am. J. Bot.* 56:8–15. [2.6.1]

Dawes, C. J., and R. P. McIntosh (1981). The effect of organic material and inorganic ions on the photosynthetic rate of the red alga *Bostrychia binderi* from a Florida estuary. *Mar Biol* 64:213–18. [7.4.3]

Dawes, C. J., J. M. Lawrence, D. P. Cheney, and A. C. Mathieson (1974). Ecological studies of Floridian *Eucheuma* (Rhodophyta, Gigartinales). Ill. Seasonal variation of carrageenan, total carbohydrate, protein and lipid. *Bull. Mar. Sci.* 24:286–99. [5.5.1]

Dawes, C. J., N. F. Stanley, and O. J. Stanicoff (1977) Seasonal and reproductive aspects of plant chemistry, and *iota-carrageenan* from Florida *Eucheuma* (Rhodophyta, Gigartinales). *Bot Mar* 20: 137–47. [5.5.1]

Dawson, E. Y. (1950). A giant new *Codium* from Pacific Baja California. *B. Torrey Bot. Club* 77:298–300. [1.2.1]

Dawson, E. Y. (1951). A further study of upwelling and associated vegetation along Pacific Baja California, Mexico. *J. Mar. Res.* 10:39–58. [7.3.8]

Dayton, P. K. (1971). Competition, disturbance, and community organization: the provision and subsequent utilization of space in a rocky intertidal community. *Ecolog. Monogr.* 41:351–89. [Essay 2]

Dayton, P K. (1973). Dispersion, dispersal, and persistence of the annual intertidal alga, *Postelsia palmaeformis* Ruprecht. *Ecology* 54:433–8. [2.5.1, 8.3.2]

Dayton, P. K. (1975). Experimental evaluation of ecological dominance in a rocky intertidal algal community. *Ecol. Monogr.* 45:137–59. [Essay 2]

Dayton, P. K. (1985). The structure and regulation of some South American kelp communities. *Ecol. Monogr.* 55:447–68. [7.2.2]

Dayton, P. K., and M. J. Tegner (1984a). The importance of scale in community ecology: a kelp forest example with terrestrial analogs. In P. W. Price, C. M. Slobodchikoff, and W. S. Gaud (eds), *A New Ecology: Novel Approaches to Interactive Systems* (pp. 457–81). New York: John Wiley and Sons. [Essay 2]

Dayton, P. K., and M. J. Tegner (1984b). Catastrophic storms, El Niño and patch stability in southern California kelp community. *Science* 224:283–5. [7.3.6, 7.3.7]

De Angelis, L., P. Rise, F. Giavarini, *et al.* (2005). Marine macroalgae analyzed by mass spectrometry are rich sources of polyunsaturated fatty acids. *J.f Mass Spect.* 40(12):1605–8 [Essay 3]

De Beer, D., and A. W. D. Larkum (2001). Photosynthesis and calcification in the calcifying algae *Halimeda discoidea* studied with microsensors. *Plant Cell Environ.* 24:1209–17. [5.7, 6.5.3, 8.2.1]

De Burgh, M. E., and P. V. Fankboner (1978). A nutritional association between the bull kelp *Nereocystis luetkeana* and its epizoic bryozoan *Membranipora membranacea*. *Oikos* 31:69–72. [4.2.2]

de Lestang-Bremond, G., and M. Quillet (1981). The turnover of sulphates on the lambda-carrageenan of the cell-walls of the red seaweed Gigartinale: *Catenella opuntia* (Grev.). *Proc. Intl Seaweed Symp.* 10:449–54. [5.5.3]

de los Santos, C. B., J. L. Pérez-Lloréns, and J. J. Vergara (2009). Photosynthesis and growth in macroalgae: linking functional-form and power-scaling approaches. *Mar. Ecol. Prog. Ser.* 377:113–22. [1.2.2]

De Martino, A., D. Douady, M. Quinet-Szely, *et al.* (2000). The light-harvesting antenna of brown algae. *Eur. J. Biochem.* 267:5540–9. [5.3.1]

de Nys, R., P. E. Jameson, N. Chin, *et al.* (1990). The cytokinins as endogenous growth-regulators in *Macrocystis pyrifera* (L) C. Agardh (Phaeophyceae). *Bot. Mar.* 33:467–75. [2.6.3]

de Nys, R., P. E. Jameson, and M. T. Brown, (1991). The influence of cytokinins on the growth of *Macrocystis pyrifera*. *Bot. Mar.* 34:465–7. [2.6.3] de Nys, R., P. D. Steinberg, P. Willemsen, *et al.* (1995). Broad-spectrum effects of secondary metabolites from the red alga *Delisea pulchra* in antifouling assays. *Biofouling* 8:259-71. [4.2.2]

de Ruyter van Steveninck, E. D., and R. P. M. Bak (1986). Changes in abundance of coral reef bottom components related to mass mortaility of the sea urchin *Diadema antillarum*. *Mar. Ecol. Prog. Ser.* 34:87-94. [4.3.1]

de Wit, C. T. (1960). On competition. *Versl. Landbouwkd Onderz (Agric. Res. Rep.)* 66:1-82. [4.2.3]

De Wreede, R. E. (1985). Destructive (harvest) sampling. In M. M. Littler and D. S. Littler (eds), *Handbook of Phycological Methods: Ecological Field Methods: Macroalgae* (pp. 147-60). Cambridge: Cambridge University Press. [3.5.1]

De Wreede, R. E., and T. Klinger (1988). Reproductive strategies in algae. In J. Lovett-Doust and L. Lovett-Doust (eds), *Plant Reproductive Ecology: Patterns and Strategies* (pp. 267-84). Oxford: Oxford University Press. [3.5.2]

De Wreede, R E., R. Scrosati, and E. Servière-Zaragoza (2000). Ramet dynamics for the clonal seaweed *Pterocladiella capillacea* (Rhodophyta): a comparison with *Chondrus crispus* and with *Mazzaella cornucopiae* (Gigartinales). *J. Phycol.* 36:1061-8 [4.2.3, 4.3.1]

Deal, M. S., M. E. Hay, D. Wilson, and W. Fenical (2003). Galactolipids rather than phlorotannins as herbivore deterrents in the brown seaweed *Fucus vesiculosus*. *Oecologia* 136:107-14. [4.2.3, 4.4.1]

Dean, T. A., and F. R. Jacobsen (1986). Nutrient-limited growth of juvenile kelp, *Macrocystis pyrifera*, during the 1982-1983 "El Niño" in southern California (USA). *Mar. Biol.* 90:597-602. [7.3.7]

DeBoer, J. A. (1981). Nutrients. In C. S. Lobban and M. J. Wynne (eds), *The Biology of Seaweeds* (pp. 356-91). Oxford: Blackwell Scientific. [2.3.4, 2.6.2, 6.1.1, 6.1.2, 6.4.2, 6.5.1, 6.5.2, 6.7.2, 6.8.1, 6.8.2]

DeBoer, J. A., and F. G. Whoriskey (1983). Production and role of hyaline hairs of *Ceramium rubrum*. *Mar. Biol.* 77:229-34. [6.4.2, 6.8.1]

DeBoer, J. A., H.J. Guigli, T. L. Israel, and C. F. D'Elia (1978). Nutritional studies of two red algae. I. Growth rate as a function of nitrogen source and concentration. *J. Phycol.* 14:261-6. [6.4, 6.4.2, 6.7.1, 6.7.2, 6.8.2]

DeCew, T. C., and J. A. West (1982). A sexual life history in *Rhodophysema* (Rhodophyceae): a re-interpretation. *Phycologia* 21:67-74. [Essay 1]

Deckert, R. J., and D. J. Garbary (2005). *Ascophyllum* and its symbionts. VI. Microscopic characterization of the *Ascophyllum nodosum* (Phaeophyceae), *Mycophycias ascophylli* (Ascomycetes) symbiotum. *Algae* 20:225-32. [4.5.1]

Deegan, L. A., A. Wright, S. G. Ayvazian, *et al.* (2002). Nitrogen loading alters seagrass ecosystem structure and support of higher trophic levels. *Aquat. Conserv.* 12(2):193-212 [Essay 3]

Delage, L., C. Leblanc, P. N. Collén, *et al.* (2011). *In silico* survey of the mitochondrial protein uptake and maturation systems in the brown alga *Ectocarpus siliculosus*. *PloS ONE* 6:e19540. doi:10.1371/journal.pone.0019540. [1.3.2]

de León-Chavira, F., M. A. Huerta-Diaz, and A. Chee-Barragán (2003). New methodology for extraction of total metal from macroalgae and its application to selected samples collected in pristine zones from Baja California, Mexico. *Bull. Environ. Contam. Toxicol.* 70:809-16. [Fig. 9.6]

Delgado, O., and B. E. Lapointe (1994) Nutrient-limited productivity of calcareous versus fleshy macroalgae in a eutrophic, carbonate-rich tropical marine environment. *Coral Reefs* 13:151-9. [6.5.2]

D'Elia, C. F., and J. A. DeBoer (1978). Nutritional studies of two red algae. II. Kinetics of ammonium and nitrate uptake. *J. Phycol.* 14:266-72. [6.4, 6.4.2, 6.8.2]

Demes, K. W., M. H. Graham, and T. S. Suskiewicz (2009a). Phenotypic plasticity reconciles incongruous molecular and morphological taxonomies: the giant kelp, *Macrocystis* (Laminariales, Phaeophyceae), is a monospecific genus. *J. Phycol.* 45:1266-9. [2.6.2]

Demes, K. W., S. S. Bell, and C. J. Dawes (2009b). The effects of phosphate on the biomineralization of the green alga, *Halimeda incrassata* (Ellis) Lam. *J. Exp. Mar. Biol. Ecol.* 374:123-7. [6.5.3]

Demes, K. W., E. Carrington, J. Gosline, and P. T. Martone (2011). Variation in anatomical and material properties explains differences in hydrodynamic performances of foliose red macroalgae (Rhodophyta). *J. Phycol.* 47:1360-7. [8.3.1]

Demmig-Adams, B., and W. W. Adams III (1992). Photoprotection and other responses of plants to high light stress. *Annu. Rev. Plant Physiol. Plant Mol. Biol.* 43:599-626. [5.3.3]

Demmig-Adams, B., W. W. Adams III, and A. K. Mattoo (2008). Photoprotection, photoinhibition, gene regulation, and environment. *Advances in Photosynthesis and Respiration*, Volume 21. Dordrecht, The Netherlands: Springer.

den Hartog, C. (1972). Substratum. Multicellular plants. In O. Kinne (ed.), *Marine Ecology* (vol. I pt. 3, pp. 1277-89). New York: Wiley. [4.2.2]

Deniaud, E., J. Fleurence, and M. Lahaye (2003). Preparation and chemical characterization of cell wall fractions enriched in structural proteins from *Palmaria palmate* (Rhodophyta). *Bot. Mar.* 46:366-77. [1.3.1]

DeNiro, M. J., and S. Epstein (1978). Influence of diet on the distribution of carbon isotopes in animals. *Geochimica Cosmochim. Ac.* 42:495–506 [Essay 3]

Denley, E. J., and P. K. Dayton (1985). Competition among macroalgae. In M. M. Littler and D. S. Littler (eds), *Handbook of Phycological Methods: Ecological Field Methods: Macroalgae* (pp. 511–30). Cambridge: Cambridge University Press. [4.2]

Denny, M. W. (1982). Forces on intertidal organisms due to breaking ocean waves: design and application of a telemetry system. *Limnol. Oceanogr.* 27:178–83. [8.1.2]

Denny, M. W. (1988). *Biology and the Mechanics of the Wave Swept Environment,* Princeton, NJ: Princeton University Press. [2.5.1, 3.2, 8.0, 8.1.2, 8.1.3, 8.3.1]

Denny, M. W. (1993). *Air and Water: The Biology and Physics of Life's Media.* Princeton, NJ: Princeton University Press, 341 pp. [8.0, 8.1.3]

Denny, M. W. (2006). Ocean waves, nearshore ecology, and natural selection. *Aquat. Ecol.* 40:439–61. [8.0, 8.1.2, 8.3.1]

Denny, M. W., and B. Gaylord (2010). Marine ecomechanics. *Annu. Rev. Mar. Sci.* 2:89–114. [8.0]

Denny, M., and B. Helmuth (2009). Confronting the physiological bottleneck: a challenge from ecomechanics. *Integr. Comp. Biol.* 49:197–201. [8.0]

Denny, M., and D. Wethey (2001). Physical processes that generate patterns in marine communities. In M. D. Bertness, S. D. Gaines, and M. E. Hay (eds), *Marine Community Ecology* (pp. 3–37). Sunderland, MA: Sinauer Assocs. [3.1.1, 8.0, 8.1.2, 8.1.3, 8.1.4, 8.3.1]

Denny, M. W., T. L. Daniel, and M. A. R. Koehl (1985). Mechanical limits to size in wave-swept organisms. *Ecol. Monogr.* 55:69–102. [8.1.2, 8.3.1]

Denny, M. [W.], V. Brown, E. Carrington, G. Kraemer, and A. Miller (1989). Fracture mechanics and the survival of wave-swept macroalgae. *J. Exp. Mar. Biol. Ecol.* 127:211–28. [8.3.1]

Denny, M. W., B. P. Gaylord, and E. A. Cowen (1997). Flow and flexibility. II. The roles of size and shape in determining wave forces on the bull kelp *Nereocystis luetkeana. J. Exp. Biol.* 200:3165–83. [8.3.1]

Derenbach, J. B., and M. V. Gerneck (1980). Interference of petroleum hydrocarbons with the sex pheromone reaction of *Fucus vesiculosus* (L.) *J. Exp. Mar. Biol. Ecol.* 44:61–5. [9.4.2]

Destombe, C., and L. V. Oppliger (2011). Male gametophyte fragmentation in *Laminaria digitata*: a life history strategy to enhance reproductive success. *Cah. Biol. Mar.* 52 (4SI):385–94. [2.1]

Dethier, M. N. (1982). Pattern and process in tidepool algae: factors influencing seasonality and distribution. *Bot. Mar.* 25:55–66. [3.3,4, 4.3.2]

Dethier, M. N. (1984). Disturbance and recovery in intertidal pools: maintenance of mosaic patterns. *Ecol. Monogr.* 54:99–118. [3.1.2]

Dethier, M. N., and R. S. Steneck (2001). Growth and persistence of diverse intertidal crusts: survival of the slow in a fast-paced world. *Mar. Ecol. Prog. Ser.* 223:89–100. [4.2.1]

Dethier, M. N., and S. L. Williams (2009). Seasonal stresses shift optimal intertidal algal habitats. *Mar. Biol.* 156:555–67. [3.1.2, 3.1.3]

Devinny, J. S., and L. A. Volse (1978). Effects of sediments on the development of *Macrocystis pyrifera* gametophytes. *Mar. Biol.* 48:343–8. [8.3.2]

Deysher, L., and T. A. Norton (1982). Dispersal and colonization in *Sargassum muticum* (Yendo) Fensholt. *J. Exp. Mar. Biol. Ecol.* 56:179–95. [2.5.1]

Diamond, J. (1986). Overview: laboratory experiments, field experiments, and natural experiments. In J. Diamond and T. J. Case, (ed.), *Community Ecology* (pp. 3–22). New York, NY: Harper and Row Publishers, Inc. [Essay 2]

Dias, P. F., J. M. Siqueira Jr, L. F. Vendruscolo, *et al.* (2005). Antiangiogenic and antitumoral properties of a polysaccharide isolated from the seaweed *Sargassum stenophyllum. Cancer Chemother. Pharmacol.* 56: 436–46. [5.5]

Diaz-Pulido, G., and L. J. McCook (2002). The fate of bleached corals: patterns and dynamics of algal recruitment. *Mar. Ecol. Prog. Ser.* 232:115–28. [3.3.1, Essay 6]

Diaz-Pulido, G., and L. J. McCook (2004). Effects of live coral, epilithic algal communities and substrate type on algal recruitment. *Coral Reefs* 23:225–33. [Essay 6]

Diaz-Pulido, G., L. Villamil, and V. Almanza (2007a). Herbivory effects on the morphology of the brown alga *Padina boergesenii* (Phaeophyta). *Phycologia* 46:131–6. [2.6.2]

Diaz-Pulido, G., L. J. McCook, A. W. D. Larkum, *et al.* (2007b). Chapter 7: Vulnerability of macroalgae of the Great Barrier Reef to climate change. In J. E. Johnson and P. A. Marshall, (eds), *Climate Change and the Great Barrier Reef* (pp. 153–92) Great Barrier Reef Marine Park Authority and Australian Greenhouse Office, Townsville. [Essay 6]

Diaz-Pulido, G., K. N. R. Anthony, D. L. Kline, S. Dove, and O. Hoegh-Guldberg (2012). Interactions between ocean acidification and warming on the mortality and dissolution of coralline algae. *J. Phycol.* 48:32–9. [6.5.3]

Dickson, L. G., and J. R. Waaland (1985). *Porphyra nerepcystis*: a dual-daylength seaweed. *Planta* 165:548–53. [2.3.3]

Dieckmann, G. S. (1980). Aspects of the ecology of *Laminaria pallida* (Grev.) 1. Ag. off the Cape Pennisula (South Africa). I. Seasonal growth. *Bot. Mar.* 23: 579–85. [5.7.4]

Dierssen, H. M., R. C. Zimmerman, L. A. Drake, and D. Burdige (2010). Benthic ecology from space: optics and net primary production in seagrass and benthic algae across the Great Bahama Bank. *Mar. Ecol. Prog. Ser.* 411:1–15. [3.5.1]

Digby, P. S. B. (1977). Photosynthesis and respiration in the coralline algae *Clathromorphum circumscriptum* and *Corallina officinalis* and the metabolic basis of calcification. *J. Mar. Biol. Assoc. UK* 57:1111–24. [Fig. 6.19]

Dillon, P S., J. S. Maki, and R. Mitchell (1989). Adhesion of *Enteromorpha* swarmers to microbial films. *Micro. Ecol.* 17:39–47. [2.5.2]

Ding, H., and J. Ma (2005). Simultaneous infection by red rot and chytrid diseases in *Porphyra yezoensis* Ueda. *J. Appl. Phycol.* 17:51–6. [10.2.3]

Dittami, S. M., C. Proux, S. Rousvoal, *et al.* (2011). Microarray estimation of genomic inter-strain variability in the genus *Ectocarpus* (Phaeophyceae). *BMC Mol. Biol.* 12:1–12. [1.4.1, 1.4.2]

Dixon, P S., and W. N. Richardson (1970). Growth and reproduction in red algae in relation to light and dark cycles. *Ann. N.Y. Acad. Sci.* 175:764–77. [2.5.3]

Doblin, M. S., I. Kurek, D. Jacob-Wilk, and D. P. Delmer (2002). Cellulose biosynthesis in plants: from genes to rosettes. *Plant Cell Physiol.* 43:1407–20. [1.3.1]

Done, T. J. (1992). Phase shifts in coral reef communities and their ecological significance. *Hydrobiologia* 247:121–32. [Essay 6]

Done, T. J., P. K. Dayton, A. E. Dayton, and R. Steger (1996). Regional and local variability in recovery of shallow coral communities: Moorea, French Polynesia and central Great Barrier Reef. *Coral Reefs* 9:183–92. [Fig. 4.11]

Doney, S.C., V. J. Fabry, R.A. Feely, and J. A. Kleypas (2009). Ocean acidification. The other CO_2 problem. *Ann. Rev. Mar. Sci.* 1:169–92. [Essay 4]

Doropoulos, C., G. A. Hyndes, P. S. Lavery, and F. Tuya (2009). Dietary preferences of two seagrass inhabiting gastropods: Allochthonous vs. autochthonous resources. *Estuar. Coast. Shelf S.* 83(1):13–18. [Essay 3]

Dortch, Q. (1982). Effect of growth conditions on accumulation of internal nitrate, ammonium, and protein in three marine diatoms. *J. Exp. Mar. Biol. Ecol.* 61:243–64. [6.5.1]

Dortch, Q. (1990). The interaction between ammonium and nitrate uptake in phytoplankton. *Mar. Ecol. Prog. Ser.* 61:183–201. [6.4, 6.4.2]

Doty, M. S. (1946). Critical tide factors that are correlated with the vertical distribution of marine algae and other organisms along the Pacific Coast. *Ecology* 27:315–28. [3.1.3]

Doty, M. S. (1971). Measurement of water movement in reference to benthic algal growth. *Bot. Mar.* 14:32–5. [8.1.4]

Drechsler, Z., R. Sharkia, Z. L. Cabantchik, and S. Beer (1994). The relationship of arginine groups to photosynthetic

HCO_3^- uptake in *Ulva* sp. mediated by a putative anion exchanger. *Planta* 194:250–5. [5.4.1]

Drew, E. A. (1977). The physiology of photosynthesis and respiration in some Antarctic marine algae. *Br. Antarct. Survey Bull.* 46:59–76. [7.3.2]

Drew, E. A. (1983). Light. In R. Earll and D. G. Erwin (eds), *Sublittoral Ecology. The Ecology of the Shallow Sublittoral Benthos* (pp. 10–57). Oxford: Clarendon Press. [Fig. 2.6]

Drew, E. A., and K. M. Abel (1990). Studies on *Halimeda*. III. A daily cycle of chloroplast migration within segments. *Bot. Mar.* 33:31–45. [1.3.2, 4.4.2]

Drew, E. A., and R. M. Hastings (1992). A year-round ecophysiological study of *Himantothallus grandifolius* (Desmarestiales, Phaeophyta) at Signy Island, Antarctica. *Phycologia* 31:262–77. [5.4.3]

Drew, K. M. (1949). Conchocelis phase in the life history of *Porphyra umbilicalis* (L.) *Kiitz. Nature* 164:748 [2.2, 10.21]

Dring, M. J. (1967). Phytochrome in red alga, *Porphyra tenera. Nature* 215:1411–12. [2.3.3]

Dring, M. J. (1974). Reproduction. In W. D. P. Stewart (ed.), *Algal Physiology and Biochemistry* (pp. 814–37). Oxford: Blackwell Scientific. [2.3.2]

Dring, M. J. (1981). Photosynthesis and development of marine macrophytes in natural light spectra. In H. Smith (ed), *Plants and the Daylight Spectrum* (pp. 297–314). London: Academic Press. [5.3.4]

Dring, M. J. (1982). *The Biology of Marine Plants.* London: Arnold. [2.2]

Dring, M. J. (1984a). Photoperiodism and phycology. *Prog. Phycol. Res.* 3:159–92. [2.3.3]

Dring, M. J. (1984b). Blue light effects in marine macroalgae. In H. Senger (ed.), *Blue Light Effects in Biological Systems* (pp. 509–16). Berlin: Springer-Verlag. [2.6.2]

Dring, M. J. (1987). Light climate in intertidal and subtidal zones in relation to photosynthesis and growth of benthic algae: a theoretical model. In R. M. M. Crawford (ed.), *Plant Life in Aquatic and Amphibious Habitats* (pp. 23–34). Oxford: Blackwell Scientific. [5.2.2]

Dring, M. J. (1988). Photocontrol of development in algae. *Annu. Rev. Plant Physiol. Plant Molec. Biol.* 39:157–74. [2.3.3]

Dring, M. J. (1989). Stimulation of light-saturated photosynthesis in *Laminaria* (Phaeophyta) by blue light. *J. Phycol.* 25:254–8. [5.3.3]

Dring, M. J. (1990). Light harvesting and pigment composition in marine phytoplankton and macroalgae. In P J. Herring, A. K. Campbell, M. Whitfield, and L. Maddock (eds), *Light and Life in the Sea* (pp. 89–103). Cambridge: Cambridge University Press. [1.3.2]

Dring, M. J. (2005). Stress resistance and disease resistance in seaweeds: the role of reactive oxygen metabolism. *Adv. Bot. Res.* 43:175-207. [7.8, 7.9, Essay 5]

Dring, M. J., and F. A. Brown (1982). Photosynthesis of intertidal brown algae during and after periods of emersion: a renewed search for physiological causes of zonation. *Mar. Ecol. Prog. Ser.* 8:301-8. [3.1.3, 7.5.1]

Dring, M. J., and K. Lüning (1975). A photoperiodic response mediated by blue light in the brown alga *Scytosiphon lomentaria. Planta* 125:25-32. [2.3.3]

Dring, M. J., and K. Lüning (1983). Photomorphogenesis in marine macroalgae. In W. Shropshire, Jr., and H. Mohr (eds), *Encyclopaedia of Plant Physiology, Vol. 16B: Photomorphogenesis* (pp. 545-68). Berlin: Springer-Verlag. [Fig. 2.7]

Dring, M. J., and J. A. West (1983). Photoperiodic control of tetrasporangium formation in the red alga *Rhodochorton purpureum. Planta* 159:143-50. [2.3.3]

Driskell, W. B., J. L. Ruesink, D. C. Lees, J. P. Houghton, and S. C. Lindstrom (2001). Long-term signal of disturbance: *Fucus gardneri* after the *Exxon Valdez* oil spill. *Ecol. Applic.* 11:815-27. [9.4.3]

Dromgoole, F. I. (1978). The effects of oxygen on dark respiration and apparent photosynthesis of marine macroalgae. *Aquat. Bot.* 4:281-97. [8.2.1]

Dromgoole, F. I. (1982). The buoyant properties of *Codium. Bot. Mar.* 25:391-7. [1.1.1]

Dromgoole, F. I. (1990). Gas-filled structures, buoyancy and support in marine macro-algae. *Prog. Phycol. Res.* 7: 169-211. [1.1.1]

Druehl, L. D. (1978) The distribution of *Macrocystis integrifolia* in British Columbia as related to environmental parameters. *Can. J. Bot.* 56: 69-79. [7.2.2]

Druehl, L. D. (1981). Geographic distribution. In C. S. Lobban and M. J. Wynne (eds), *The Biology of Seaweeds* (pp. 306-25). Oxford: Blackwell Scientific. [Fig. 7.2]

Druehl, L. D. (1988). Cultivated edible kelp. In C. A. Lembi and J. R. Waaland (eds), *Algae and Human Affairs* (pp. 119-34). Cambridge: Cambridge University Press. [10.3]

Druehl, L. D. (2001). *Pacific Seaweeds: A Guide to Common Seaweeds of the West Coast.* Madeira Park, Canada: Harbour Publishing. [3.3.2]

Druehl, L. D., and R. G. Footit (1985). Biogeographical analysis. In M. M. Littler and D. S. Littler (eds) *Handbook of Phycological Methods: Ecological Field Methods* (pp. 315-25). Cambridge: Cambridge University Press. [7.2.2]

Druehl, L. D., and J. M. Green (1982). Vertical distribution of intertidal seaweeds as related to patterns of submersion and emersion. *Mar. Ecol. Prog. Ser.* 9:163-70. [3.1.3]

Druehl, L. D., and G. W. Saunders (1992). Molecular explorations of kelp evolution. *Prog. Phycol. Res.* 8:47-83. [Essay 1]

Druehl, L. D., R. Baird, A. Lindwall, K. E. Lloyd, and S. Pakula (1988). Longline cultivation of some Laminariaceae in British Columbia. *Aquacult. Fish. Management* 19:253-63. [10.3]

Druehl, L. D., J. D. Collins, C. E. Lane, and G. W. Saunders (2005a). An evaluation of methods used to assess intergeneric hybridization in kelp using Pacific Laminariales (Phaeophyceae). *J. Phycol.* 41:250-62. [10.11]

Druehl, L.D., J. D. Collins, C. E. Lane, and G. W. Saunders (2005b). A critique of intergeneric kelp hybridization protocol, employing Pacific Laminariales (Phaeophyceae). *J. Phycol.* 41: 250-62. [Essay 1]

Duarte, C. M. (1992). Nutrient concentration of aquatic plants patterns across species. *Limnol. Oceanogr.* 37:882-9. [6, 6.4.2]

Duarte, P., and J. G. Ferreira (1993). A methodology for parameter estimation in seaweed productivity modelling. *Hydrobiologia* 260/262:183-9. [5.7.4]

Duarte, P., and J. G. Ferreira (1997). A model for the simulation of macroalgal population dynamics and productivity. *Ecol. Model.* 98:199-214. [5.7.4]

Dube, M. A., and E. Ball (1971). *Desmarestia* sp. associated with the seapen *Ptilosarcus gurneyi* (Gray). *J. Phycol.* 7:218-20. [3.3.8]

Ducreux, G. (1984). Experimental modification of the morphogenetic behavior of the isolated sub-apical cell of the apex of *Sphacelaria cirrosa* (Phaeophyceae). *J. Phycol.* 20:447-54. [2.6.1]

Ducreux, G., and B. Kloareg (1988). Plant regeneration from protoplasts of *Sphacelaria* (Phaeophyceae). *Planta* 174:259. [2.6.1]

Dudgeon, S., and P. S. Petraitis (2005). First year demography of the foundation species, *Ascophyllum nodosum*, and its community implications. *Oikos* 109:405-15. [3.5.2]

Dudgeon, S. R., I. R. Davison, and R. L. Vadas (1989). Effect of freezing on photosynthesis of intertidal macroalgae: relative tolerance of *Chondrus crispus* and *Mastocarpus stellatus* (Rhodophyta). *Mar. Biol.* 101:107-14. [7.3.4]

Dudgeon, S. R., I. R. Davison, and R. L. Vadas (1990). Freezing tolerance in the intertidal red algae *Chondrus crispus* and *Mastocarpus stellatus*: relative importance of acclimation and adaptation. *Mar Biol* 106:427-36. [3.3.2, 7.3.4]

Dudgeon, S. R., J. E. Kubler, R. L. Vadas, and I. R. Davison (1995). Physiological responses to environmental variation in intertidal red algae; does thallus morphology matter? *Mar. Ecol. Prog. Ser.* 117:193-206. [6.4.2]

Duffy, J. E. (1990). Amphipods on seaweeds: partners or pests? *Oecologia* 83:267-76. [4.3.1]

Duffy, J. E. (2009). Why biodiversity is important to the functioning of real-world ecosystems. *Front. Ecol. Environ.* 7:437-44. [3.5.1]

Duffy, J. E., and M. E. Hay (1990). Seaweed adaptations to herbivory. *BioScience* 40:368-75. [3.2.2]

Duffy, J. E., and M. E. Hay (2000). Strong impacts of grazing amphipods on the organization of a benthic community. *Ecological Monographs* 70(2): 237-63. [4.3.1, Essay 3]

Duffy, J. E., and M. E. Hay (2001). The ecology and evolution of marine consumer-prey interactions. In M. D. Bertness, S. D. Gaines, and M. E. Hay (eds), *Marine Community Ecology* (pp. 131-57). Sunderland, MA: Sinauer Assocs. [4.3.1, 4.3.2]

Duggins, D. O. (1980). Kelp beds and sea otters: an experimental approach. *Ecology* 61:447-53. [4.3.2]

Duggins, D. O., C. A. Simenstad, and J. A. Estes (1989). Magnification of secondary production by kelp detritus in coastal marine ecosystems. *Science* 245:170-3. [4.3.1, 5.7.5]

Duke, C. S., R. W. Litaker, and J. Ramus (1987). Seasonal variation in RuBPCase activity and N allocation in the chlorophyte seaweeds *Viva curvata* (Kiitz.) de Toni and *Codium decorticatum* (Woodw.) Howe. *J. Exp. Mar. Biol. Ecol.* 112:145-64. [6.8.1]

Duke, C. S., W. Litaker, and J. Ramus (1989). Effects of temperature, nitrogen supply, and tissue nitrogen on ammonium uptake rates of the chlorophyte seaweeds *Ulva curvata* and *Codium decorticatum*. *J. Phycol.* 25:113-20. [6.4, 6.8.2]

Duncan, M. J., and R. E. Foreman (1980). Phytochrome-mediated stipe elongation in the kelp *Nereocystis* (Phaeophyceae). *J. Phycol.* 16:138-42. [2.6.2]

Dunn, E. K., D. A. Shoue, X. Huang, *et al.* (2007). Spectroscopic and biochemical analysis of regions of the cell wall of the unicellular "Mannan Weed", *Acetabularia acetabulum*. *Plant Cell Physiol.* 48:122-33. [1.3.1]

Dunton, K. H., and D. M. Schell (1986). Seasonal carbon budget and growth of *Laminaria solidungula* in the Alaskan high Arctic. *Mar. Ecol. Prog. Ser.* 31:57-66. [5.6]

Dunton, K. H., and P. K. Dayton (1995). The biology of high latitude kelp. In H. R. Skjoldal, C. Hopkins, K. E. Erikstad, and H. P. Leinaas (eds) *Ecology of Fjords and Coastal Waters* (pp. 499-507). Amsterdam:Elsevier Science. [7.3.6]

Durako, M. J., and C. J. Dawes (1980). A comparative seasonal study of two populations of *Hypnea musciformis* from the east and west coasts of Florida. USA II Photosynthetic and respiratory rates. *Mar. Biol.* 59:157-62. [5.5.1]

Durante, K. M., and F. S. Chia (1991). Epiphytism on *Agarum fimbriatum*: can herbivore preferences explain distributions

of epiphytic bryozoans? *Mar. Ecol. Prog. Ser.* 77:279-87. [4.2.2]

Dworjanyn, S. A., R. de Nys, and P. D. Steinberg (1999). Localisation and surface quantification of secondary metabolites. *Mar. Biol.* 133:727-36. [4.2.2]

Dworjanyn, S. A., R. de Nys, and P. D. Steinberg. (2006a). Chemically mediated antifouling in the red alga *Delisea pulchra. Mar. Ecol. Prog. Ser.* 318: 153-63. [4.4.2]

Dworjanyn, S. A., J. T. Wright, N. A. Paul, R. de Nys, and P. D. Steinberg (2006b). Cost of chemical defence in the red alga *Delisea pulchra. Oikos* 113:13-22. [4.4.2]

Dyck, L. J., and R. E. DeWreede (2006a). Seasonal and spatial patterns of population density in the marine macroalga *Mazzaella splendens* (Gigartinales, Rhodophyta). *Phycol. Res.* 54:21-31. [3.5.2]

Dyck, L. J., and R. E. DeWreede (2006b). Reproduction and survival in *Mazzaella splendens* (Gigartinales, Rhodophyta). *Phycologia* 45:302-10. [3.5.2]

Eardley, D. D., C. W. Sutton, W. M. Hempel, D. C. Reed, and A. W. Ebeling (1990). Monoclonal-antibodies specific for sulfated polysaccharides on the surface of *Macrocystis pyrifera* (Phaeophyceae). *J. Phycol.* 26:54-62. [1.3.1]

Edelstein, T., and J. McLachlan (1975). Autecology of *Fucus distichus* ssp. *distichus* (Phaeophyceae: Fucales) in Nova Scotia, Canada. *Mar Biol* 30:305-24. [7.2.3]

Edwards, D. M., R. H. Reed, J. A. Chudek, R. M. Foster, and W. D. P. Stewart (1987). Organic solute accumulation in osmotically-stressed *Enteromorpha intestinalis*. *Mar. Biol.* 95: 583-92.

Edwards, M. S., and S. D. Connell (2012). Competition, a major factor structuring seaweed communities. In C. Wiencke, and K. Bischof (eds), *Seaweed Biology: Novel Insights into Ecophysiology, Ecology and Utilization, Ecological Studies* 219 (pp. 135-56). Berlin and Heidelberg: Springer-Verlag. [4.2, 4.2.1]

Edwards, M. S., and J. A. Estes (2006). Catastrophe, recovery and range limitation in NE Pacific kelp forests: a large-scale perspective. *Mar. Ecol. Prog. Ser.* 320:79-87. [7.3.7]

Edwards, M. S., and G. Hernández-Carmona (2005). Delayed recovery of giant kelp near its southern range limit in the North Pacific following El Niño. *Mar Biol* 147:273-9. [7.3.7]

Eggert, A., and U. Karsten (2010). Low molecular weight carbohydrates in red algae – an ecophysiological and biochemical perspective. In J. Seckbach, D. Chapman, and A. Weber (eds) *Cellular Origins, Life in Extreme Habitats and Astrobiology, Red Algae in the Genomic Age* (pp. 445-56). Berlin and Heidelberg: Springer-Verlag. [5.4.2]

Eggert, A., E. M. Burger, and A. M. Breeman (2003). Ecotypic differentiation in thermal traits in the tropical to warm-temperate green macrophyte *Valonia utricularis*. *Bot. Mar.* 46:69–81. [7.3.3]

Eggert, A., R. J. W. Visser, P. R. van Hasselt, and A. M. Breeman (2006). Differences in acclimation potential of photosynthesis in seven isolates of the tropical to warm temperate macrophyte *Valonia utricularis* (Chlorophyta). *Phycologia* 45:546–56. [7.9]

Eide, I., S. Myklestad, and S. Melson (1980). Long-term uptake and release of heavy metals by *Ascophyllum nodosum* (L.) Le Jol. (Phaeophyceae) *in situ. Environ. Pollut.* 23:19–28. [9.3.2]

Eklund, B. (2005). Development of a growth inhibition test with the marine and brackish water red alga *Ceramium tenuicorne*. *Mar. Poll. Bull.* 50:921–30. [9.2, 9.3.2, 9.3.4]

Eklund, B., and L. Kautsky (2003). Review on toxicity testing with marine macroalgae and the need for method standardization – exemplified with copper and phenol. *Mar. Poll. Bull.* 46:171–81. [9.3.2]

Eklund, B., M. Elfström, and H. Borg (2008). TBT originates from pleasure boats in Sweden in spite of firm restrictions. *Open Environ Sci.* 2:124–32. [9.5]

Eklund, B., M. Elfström, I. Gallego, B-E. Bengtsson, and M. Breitholtz (2010). Biological and chemical characterization of harbour sediments from the Stockholm area. *Soil Sed. Poll.* 10:127–41. [9.2]

Elner, R. W., and R. L. Vadas, Sr. (1990). Inference in ecology: the sea urchin phenomenon in the northwest Atlantic. *Am. Nat.* 136:108–25. [4.3.1]

Emerson, C. J., R. G. Buggeln, and A. K. Bal (1982). Translocation in *Saccorhiza dermatodea* (Laminariales, Phaeophyceae): anatomy and physiology. *Can. J. Bot.* 60: 2164–84. [5.6]

Engel, C. R., R. Wattier, C. Destombe, and M. Valero (1999). Performance of non-motile male gametes in the sea: analysis of paternity and fertilization success in a natural population of a red seaweed, *Gracilaria gracilis. Proc. Roy. Soc. Lond. B. Bio.* 266:1879–86. [2.4]

Engel, C. R., C. Daguin, and E. A. Serrão (2005). Genetic entities and mating system in hermaphroditic *Fucus spiralis* and its close dioecious relative *F. vesiculosus* (Fucaceae, Phaeophyceae). *Mol. Ecol.* 14:2033–46. [1.4.2]

Engelen, A., and R. Santos (2009). Which demographic traits determine population growth in the invasive brown seaweed *Sargassum muticum*? *J. Ecol.* 97:675–84. [3.5.2]

Engelmann, T. W. (1883). Farbe und Assimilation. *Bot Zeit* 41:1–13. [5.3.4]

Engelmann, T. W. (1884). Untersuchungen iiber die quantitativen Beziehungen zwischen Absorption des Lichtes und Assimilation in Pflanzenzellen. *Bot Zeit* 42:81–93. [5.3.4]

Enright, C. T. (1979). Competitive interaction between *Chondrus crispus* (Florideophyceae) and *Ulva lactuca* (Chlorophyceae) in *Chondrus* aquaculture. *Proc. Intl. Seaweed Symp.* 9:209–18. [4.2.3]

Enríquez, S., and A. Rodríquez-Román (2006). Effect of water flow on the photosynthesis of three marine macrophytes from a fringing-reef lagoon. *Mar. Ecol. Prog. Ser.* 323:119–32. [8.2.1]

Enríquez, S., C. M. Duarte, K. Sand-Jensen, and S. L. Nielsen (1996). Broad-scale comparison of photosynthetic rates across phototrophic organisms. *Oecologia* 108:197–206. [1.2.2]

Enríquez, S., E. Ávila, and J. L. Carballo (2009). Phenotypic plasticity induced in transplant experiments in a mutualistic association between the red alga *Jania adhaerens* (Rhodophyta, Corallinales) and the sponge *Haliclona caerulea* (Porifera: Haplosclerida): morphological responses of the alga. *J. Phycol.* 45:81–90. [4.5.1]

Estes, J. A., and P. D. Steinberg, (1988). Predation, herbivory, and kelp evolution. *Paleobiology* 14:19–36. [Essay 1]

Estes, J. A., E. M. Danner, D. F. Doak, *et al.* (2004). Complex trophic interactions in kelp forest ecosystems. *B. Mar. Sci.* 74:621–38. [4.3.1]

Estes, J. A., M. T. Tinker, T. M. Williams, *et al.* (1998). Killer whale predation on sea otters linking oceanic and nearshore ecosystems. *Science* 282:473–6. [4.3.1]

Estevez, J. M., and E. J. Cáceres (2003). Fine structural study of the red seaweed *Gymnogongrus torulosus* (Phyllophoraceae, Rhodophyta). *Biocell* 27:181–7. [1.3.1]

Evans, L. V., and D. M. Butler (1988). Seaweed biotechnology: current status and future prospects. In L. J. Rogers and J. R. Gallon (eds.) *Biochemistry of the Algae and Cyanobacteria* (pp. 335–50). Oxford: Clarendon Press. [10.11]

Evans, L. V., and A. O. Christie (1970). Studies on the shipfouling alga *Enteromorpha*. I. Aspects of the fine-structure and biochemistry of swimming and newly settled zoospores. *Ann. Bot.* 34:451–66. [2.5.2]

Evans, L. V., J. A. Callow, and M. E. Callow (1973). Structural and physiological studies on the parasitic red alga *Holmsella*. *New Phytol.* 72:393–402. [4.5.2]

Evans, L. V., J. A. Callow, and M. E. Callow (1982). The biology and biochemistry of reproduction and early development in *Fucus. Prog. Phycol. Res.* 1:67–110. [2.4]

Fabricius, K. E. (2011). Factors determining the resilience of coral reefs to eutrophication: a review and conceptual model. In Z. Dubinsky and N. Stambler (eds), *Coral Reefs: An Ecosystem in Transition* (pp. 493–506) Berlin and Heidelberg: Springer-Verlag. [Essay 6]

Fabricus, K., G. De'ath, L. McCook, E. Turak, and D. McB. Williams (2005). Changes in algal, coral and fish assemblages along water quality gradients on the inshore Great Barrier Reef. *Mar. Poll. Bull.* 51:384-98. [9.6.1]

Fabricius, K. E., C. Langdon, S. Uthicke, *et al.* (2011). Losers and winners in coral reefs acclimatized to elevated carbon dioxide concentrations. *Nature Clim. Change* 1:165-9. [Essay 6]

Faes, V. A., and R. M. Viejo (2003). Structure and dynamics of a population of *Palmaria palmata* (Rhodophyta) in northern Spain. *J. Phycol.* 39:1038-49. [3.5.2]

Fagerberg, W. R., and C. J. Dawes (1977). Studies on *Sargassum*. II. Quantitative ultrastructural changes in differentiated stipe cells during wound regeneration and regrowth. *Protoplasma* 92:211-27. [2.6.4]

Fagerberg, W. R., E. (Lavoie) Hodges, and C. J. Dawes (2010). The development and potential roles of cell wall trabeculae in *Caulerpa mexicana* (Chlorophyta). *J. Phycol.* 46:309-15. [2.6.1]

Fain, S. R., L. D. Druehl, and D. L. Baillie (1988). Repeat and single copy sequences are differentially conserved in the evolution of kelp chloroplast DNA. *J. Phycol.* 24:292-302. [1.4.1]

Fairhead, V. A., C. D. Amsler, J. B. McClintock, and B. J. Baker (2005). Within-thallus variation in chemical and physical defences in two species of ecologically dominant brown macroalgae from the Antarctic Peninsula. *J. Exp. Mar. Biol. Ecol.* 322:1-12.[4.4.2]

Falcão, V. R., M. C. Oliviera, and P. Colepicolo (2010). Molecular characterization of nitrate reductase gene and its expression in the marine red alga *Gracilaria tenuistipitata* (Rhodophyta). *J. Appl. Phycol.* 22:613-22.

Falkowski, P. G., and J. LaRoche (1991). Acclimation to spectral irradiance in algae. *J. Phycol.* 27:8-14. [5.3.3]

Falkowski, P. G., and J. A. Raven (2007). *Aquatic Photosynthesis*, 2nd edn. Princeton, NJ: Princeton University Press. 484 pp. [5]

Falkowski, P. G., L. McClosky, L. Muscatine, and Z. Dubinsky (1993). Population control in symbiotic corals. *BioSci.* 43:606-11. [9.6.1]

Fantle, M. S., A. I. Dittel, S. M. Schwalm, C. E. Epifanio, and M. L. Fogel (1999). A food web analysis of the juvenile blue crab, *Callinectes sapidus*, using stable isotopes in whole animals and individual amino acids. *Oecologia* 120(3): 416-26. [Essay 3]

FAO (2005). Cultured aquaculture species information program: *Porphyra* spp. Rome: FAO. Available at: www.fao.org/fishery/culturedspecies/*Porphyra* spp/en. [10.2, 10.2.2]

FAO (2006). Cultured aquaculture species information program: *Laminaria japonica*. Rome: FAO. Available at: www.fao.org/fishery/culturedspecies/Laminaria_japonica/en. [10.3.1]

FAO (2008). Cultured aquaculture species information program: *Euchema* spp. Rome: FAO. Available at: www.fao.org/fishery/culturedspecies/Euchema_spp./en. [10.5]

FAO (2009). *World Review of Fisheries and Aquaculture.* FAO Fisheries and Aquaculture Dept. Rome: FAO. 84 pp. [10.1, 10.2, 10.4].

FAO (2010). *The State of World Fisheries and Aquaculture 2010.* FAO Fisheries and Aquaculture Dept. Rome: FAO. [10.1, 10.3].

Fath, B. (eds), *The Encyclopedia of Ecology, Ecological Engineering*, Vol. 3 (pp. 2463-75). Oxford: Elsevier. [10.3, 10.9]

Fawley, M. W., C. A. Douglas, K. D. Stewart, and K. R. Mattox (1990). Light-harvesting pigment-protein complexes of the Ulvophyceae (Chlorophyta): characterization and phylogenetic significance. *J. Phycol.* 26:186-95. [5.3.1]

Fei, X. G. (2004). Solving the coastal eutrophication problem by large scale seaweed cultivation. *Hydrobiologia*, 512(1-3):145-51. [10.9]

Fernández, P. V., M. Ciancia, A. B. Miravalles, and J. M. Estevez (2010). Cell wall polymer mapping in the coenocytic macroalga *Codium vermilara* (Bryopsidales, Chlorophyta). *J. Phycol.* 46:456-65. [1.3.1]

Ferreira, J. G., and L. Ramos (1989). A model for the estimation of annual production rates of macrophyte algae. *Aquat. Bot.* 33:53-70. [5.7, 5.7.4]

Fetter, R., and M. Neushul (1981). Studies in developing and released spermatia in the red alga *Tiffaniella snyderae* (Rhodophyta). *J. Phycol.* 17:141-59. [2.4]

Fielding, A. H., and G. Russell (1976). The effect of copper on competition between marine algae. *J. Appl. Ecol.* 13:871-6. [9.3.4]

Fierst, J., C. terHorst, J. E. Kübler, and S. Dudgeon (2005). Fertilization success can drive patterns of phase dominance in complex life histories. *J. Phycol.* 41:238-49. [2.2]

Figueroa, F. L., J. Aguilera, and F. X. Niell (1994). End-of-day light control of growth and pigmentation in the red alga *Porphyra umbilicalis* (L.) Kützing. *Z. Naturforsch.* 49:593-600. [2.3.3]

Figurski, J. D., D. Malone, J. R. Lacy, and M. Denny (2011). An inexpensive instrument for measuring wave exposure and water velocity. *Limnol. Oceanogr.: Methods* 9:204-14. [8.1.4]

Filion-Myklebust, C., and T. A. Norton (1981). Epidermis shedding in the brown seaweed *Ascophyllum nodosum* (L.) Le Jolis, and its ecological significance. *Mar. Biol. Lett.* 2:45-51. [4.2.2]

Fischer, G., and C. Wiencke (1992). Stable carbon isotope composition, depth distribution and fate of macroalgae from the Antarctic Peninsula region. *Polar Biology* 12:341-8. [5.7.5]

Fjeld, A. (1972). Genetic control of cellular differentiation in *Ulva mutabilis*. Gene effects in early development. *Devel. Biol.* 28:326-43. [2.6.1]

Fjeld, A., and A. Løvlie (1976). Genetics of multicellular marine algae. In R. A. Lewin (ed.), *The Genetics of Algae* (pp. 219-35). Oxford: Blackwell Scientific. [2.6.1]

Fletcher, R. L., and M. E. Callow (1992). The settlement, attachment and establishment of marine algal spores. *Br. Phycol. J.* 27:303-29. [2.5, 2.5.1, 2.5.2, 2.5.3]

Fletcher, R. L., R. E. Baier, and M. S. Fornalik (1985). The effects of surface energy on germling development of some marine macroalgae (abstract). *Br. Phycol. J.* 20: 184-5. [2.5.2]

Floc'h, J.-Y. (1982). Uptake of inorganic ions and their long distance transport in Fucales and Laminariales. In L. M. Srivastava (ed.), *Synthetic and Degradative Processes in Marine Macrophytes* (pp. 139-65). Berlin: Walter de Gruyter. [6.6]

Floeter, S. R., M. D. Behrens, C. E. L. Ferreira, M. J. Paddack, and M. H. Horn (2005). Geographical gradients of marine herbivorous fishes: patterns and processes. *Mar. Biol.* 147:1435-47. [3.3.1, 4.3.1]

Florence, T. M., B. G. Lumsden, and J. J. Fardy (1984). Algae as indicators of copper speciation. In C. J. M. Kramer and J. C. Duinker (eds), *Complexation of Trace Metals in Natural Waters* (pp. 411-18). The Netherlands: Dr. W. Junk. [9.3.4]

Flores-Moya, A. (2012). Warm temperate seaweed communities: a case study of deep water kelp forests from the Alboran Sea (SW Mediterranean Sea) and the Strait of Gibraltar. In C. Wiencke and K. Bischof (eds), *Seaweed Biology: Novel Insights Into Ecophysiology, Ecology and Utilization* (pp. 315-27). Series: Ecological Studies, Volume 219. Berlin and Heidelberg: Springer. [3.3.2]

Flores-Moya, A., J. A. Fernández, and F. X. Niell (1997). Growth pattern, reproduction and self-thinning in seaweeds: a re-evaluation in reply to Scrosati. *J. Phycol.* 33:1080-1. [4.2.3]

Fong, P., and V. J. Paul (2011). Coral reef algae. In Z. Dubinsky and N. Stambler (eds), *Coral Reefs: An Ecosystem in Transition* (pp. 241-72). Dordrecht, The Netherlands: Springer. [3.3.1, 4.2.2]

Fong, P., R. M. Donohoe, and J. B. Zedler (1994). Nutrient concentration in the tissue of the macroalga *Enteromorpha* as a function of nutrient history: an experimental evaluation using field microcosms. *Mar. Ecol. Prog. Ser.* 106:273-81. [9.5]

Fork, D. C. (1963). Observations on the function of chlorophyll *a* and accessory pigments in photosynthesis. In *Photosynthetic Mechanisms in Green Plants* (pp. 352-61). Publ. no. 1145, NAS-NRC, Washington, DC. [5.3.4]

Forrest, B. M., S. N. Brown, M. D. Taylor, C. L. Hurd, and C. H. Hay (2000). The role of natural dispersal mechanisms in the spread of *Undaria pinnatifida* (Laminariales, Phaeophyta). *Phycologia* 39:547-53. [3.4]

Forsberg, A., S. Soderlund, A. Frank, L. R. Petersson, and M. Pedersen (1988). Studies on metal content in the brown seaweed, *Fucus vesiculosus*, from the Archipelago of Stockholm. *Environ. Pollut.* 49:245-63. [9.3.2]

Forsberg, J., and J. F. Alen (2001). Molecular recognition in thylakoid structure and function. *Trends in Plant Sciences* 6:317-26. [5.3.1]

Förster, T. (1948). Zwischenmolekulare Energiewanderung und Fluoreszenz. *Ann. Physik* 437:5.

Foster, M. S. (1972). The algal turf community in the nest of the ocean goldfish (*Hypsypops rubicunda*). *Proc. Intl. Seaweed Symp.* 7:55-60. [4.2.1]

Foster. M. S. (1975). Regulation of algal community development in a *Macrocystis pyrifera* forest. *Mar. Biol.* 32:331-42. [4.2.1]

Foster, M. S., and D. R. Schiel (2010). Loss of predators and the collapse of southern California kelp forests (?): alternatives, explanations and generalizations. *J. Exp. Mar. Biol. Ecol.* 393:59-70. [4.3.1]

Foster M. S., and W. P. Sousa (1985). Succession. In M. M. Littler and D. S. Littler (eds), *Handbook of Phycological Methods: Ecological Field Methods: Macroalgae* (pp. 269-90). Cambridge: Cambridge University Press. [3.5.1]

Foster, M. S., M. S. Edwards, D. C. Reed, *et al.* (2006). Top-down vs. bottom-up effects in kelp forests. *Science* 313:1737-8. [4.3.1]

Foster, N. L., S. J. Box, and P. J. Mumby (2008). Competitive effects of macroalgae on the fecundity of the reef-building coral *Montastraea annularis*. *Mar. Ecol. Prog. Ser.* 367:143-52. [4.2.2]

Foster, P. (1976). Concentrations and concentration factors of heavy metals in brown algae. *Environ. Pollut.* 10:45-54. [Table 9.2]

Fowler, J. E., and R. S. Quantrano (1997). Plant cell morphogenesis: plasma membrane interactions with the cytoskeleton and cell wall. *Annu. Rev. Cell Dev. Biol.* 13:697-743. [1.3.3, 2.5.3]

Fowler-Walker, M. J., and T. Wernberg (2006). Differences in kelp morphology between wave sheltered and exposed localities: morphologically plastic or fixed traits? *Mar. Biol.* 148:755-67. [2.6.2]

Fox, C.H., and A. K. Swanson (2007). Nested PCR detection of microscopic life-stages of laminarian macroalgae and comparison with adult forms along intertidal height gradients. *Mar. Ecol. Prog. Ser.* 332: 1-10. [Essay 1]

Fralick, R. A., and A. C. Mathieson (1975). Physiological ecology of four *Polysiphonia* species (Rhodophyta, Ceramiales). *Mar. Biol.* 29:29–36. [Fig. 7.14]

Franklin, L. A., and R. M. Forster (1997). The changing irradiance environment: consequences for marine macrophyte physiology, productivity and ecology. *Eur. J. Phycol.* 32:207–32. [7.6]

Franklin, L. A., B. Osmond, and A. W. D. Larkum (2003): Photoinhibition, UV-B and algal photosynthesis. In A. W. D. Larkum, S. E. Douglas, and J. A. Raven (eds) *Photosynthesis in Algae. Advances in Photosynthesis and Respiration* (Vol. 14, pp. 11–28). Dordrecht, The Netherlands: Kluwer Academic Publishers. [5.1]

Fraschetti, S., A. Terlizzi, and L. Benedetti-Cecchi (2005). Patterns of distribution of marine assemblages from rocky shores: evidence of relevant scales of variation. *Mar. Ecol. Prog. Ser.* 296:13–29. [3.5.1]

Fraser, C. I., C. H. Hay, H. G. Spenser, and J. M. Waters (2009). Genetic and morphological analyses of the southern bull kelp *Durvillaea antarctica* (Phaeophyceae: Durvillaeales) in New Zealand reveal cryptic species. *J. Phycol.* 45:436–43. [1.1.1, 1.4.1, 3.3.7]

Frederick, J. E., H. E. Snell, and E. K. Haywood (1989). Solar ultraviolet radiation at the earth´s surface. *Photochem. Photobiol.* 50:443–50. [7.6]

Fredersdorf, J., and K. Bischof (2007). Irradiance of photosynthetically active radiation determines UV-susceptibility of photosynthesis in *Ulva lactuca* L. (Chlorophyta). *Phycol. Res.* 55:295–301. [5.3.3]

Fredersdorf, J., R. Müller, S. Becker, C. Wiencke, and K. Bischof (2009). Interactive effects of radiation, temperature and salinity on different life history stages of the Arctic kelp *Alaria esculenta* (Phaeophyceae). *Oecologia* 160:483–92. [7.3.3, 7.3.6, 7.4.3]

Freidenburg, T. L. (2002). Macroscale to Local Scale Variation in Rocky Intertidal Community Structure and Dynamics in Relation to Coastal Upwelling. PhD Dissertation, Oregon State University. [Essay 2]

Fretwell, S. D. (1977). The regulation of plant communities by food chains exploiting them. *Perspect. Biol. Med.* 20:169–85. [Essay 2]

Fricke, A., M. Teichberg, S. Beilfuss, and K. Bischof (2011a). Succession patterns in algal turf vegetation on a Caribbean coral reef. *Bot. Mar.* 54:111–26. [3.3.1]

Fricke, A., T. V. Titlyanova, M. M. Nugues, and K. Bischof (2011b). Depth-related variation in epiphytic communities growing on the brown alga *Lobophora variegata* in a Caribbean coral reef. *Coral Reefs* 30:967–73. [3.3.1]

Friedlander, M. (2008). Advances in cultivation of Gelidiales. *J. Appl. Phycol.* 20:451–6. [10.7]

Friedlander, M., and C. J. Dawes (1985). *In situ* uptake kinetics of ammonium and phosphate and chemical composition of the red seaweed *Gracilaria tikvahiae*. *J. Phycol.* 21:448–53. [6.4, 6.5.2]

Friedlander. M., M. D. Krom, and A. Ben-Amotz (1991). The effect of light and ammonium on growth, epiphytes and chemical constituents of *Gracilaria conferta* in outdoor cultures. *Bot. Mar.* 34:161–6. [6.5.1]

Friedmann, E. I. (1961). Cinemicrography of spermatozoids and fertilization in Fucales. *Bull. Res. Counc. Israel* 10D:73–83. [2.4]

Friedrich, M. W. (2012). Bacterial communities on macroalgae. In C. Wiencke and K. Bischof (eds) (2012). *Seaweed Biology: Novel Insights into Ecophysiology, Ecology and Utilization, Ecological Studies* 219 (pp. 189–202). Berlin, Heidelberg: Springer-Verlag. [4.2.2]

Fries, L. (1966). Temperature optima of some red algae in axenic culture. *Bot. Mar.* 9:12–14. [7.3.3]

Fries, L. (1982). Selenium stimulates growth of marine macroalgae in axenic culture. *J. Phycol.* 18:328–31. [6.5.6]

Fries, N. (1979). Physiological characteristics of *Mycosphaerella ascophylli*, a fungal endophyte of the marine brown alga *Ascophyllum nodosum*. *Physiol. Plant.* 45:117–21. [4.5.1]

Fritsch, F. E. (1945). *The Structure and Reproduction of the Algae*, Vol. 2. Cambridge: Cambridge University Press. [Fig. 1.4]

Fry, B., R. S. Scalan, J. J. Winters, and P. L. Parker (1982). Sulphur uptake by salt grasses, mangroves and seagrasses in anaerobic sediments. *Geochim. Cosmochim. Acta* 43:1121–4. [Essay 3]

Fu, W., J. Yao, X. Wang, *et al.* (2009). Molecular cloning and expression analysis of a cytosolic HSP70 gene from *Laminaria japonica* (Laminariaceae, Phaeophyta). *Mar. Biotechnol.* 11:738–47. [7.3.5]

Fujimura, T., T. Kawai, M. Shiga, T. Kajiwara, and A. Hatanaka (1989). Regeneration of protoplasts into complete thalli in the marine green alga *Ulva pertusa*. *Nippon Suisan Gakkaishi* 55:1353–9. [2.6.1]

Fujita, R. M. (1985). The role of nitrogen status in regulating transient ammonium uptake and nitrogen storage by macroalgae. *J. Exp. Mar. Biol. Ecol.* 99:283–301. [6.5.1]

Fujita, R. M., and S. Migata (1986). Isolation of protoplasts from leaves of red algae *Porphyra yezoensis*. *Jpn. J. Phycol.* 34:63. [10.2.4]

Fujita, R. M., P. A. Wheeler, and R. L. Edwards (1988). Metabolic regulation of ammonium uptake by *Ulva rigida* (Chlorophyta): a compartmental analysis of the rate-limiting step for uptake. *J. Phycol.* 24:560–6. [6.4.1]

Fujita, R. M., P. A. Wheeler, and R. L. Edwards (1989). Assessment of macroalgal nutrient limitation in a seasonal upwelling region. *Mar. Ecol. Prog. Ser.* 53:293–303. [6.4]

Fujita, S., M. Iseki, S. Yoshikawa, Y. Makino, *et al.* (2005). Identification and characterization of a fluorescent flagellar protein from the brown alga *Scytosiphon lomentaria* (Scytosiphonales, Phaeophyceae): a flavoprotein homologous to old yellow enzyme. *Eur. J. Phycol.* 40:159-67. [1.3.3]

Fukuhara, Y., H. Mizuta, and H. Yasui (2002). Swimming activities of zoospores in *Laminaria japonica* (Phaeophyceae). *Fisheries Sci.* 68:1173-81. [2.5.1]

Fulcher, R. G., and M. E. McCully (1969). Histological studies on the genus *Fucus*. IV. Regeneration and adventive embryony. *Can. J. Bot.* 47:1643-9. [2.6.4]

Fulcher, R. G., and M. E. McCully (1971). Histological studies on the genus *Fucus*. V. An autoradiographic and electron microscopic study of the early stages of regeneration. *Can. J. Bot.* 49:161-5. [2.6.4]

Gacesa, P. (1988). Alginates. *Carbohydr. Polymers* 8: 161-82. [5.5.2]

Gachon, C. M. M., T. Sime-Ngando, M. Strittmatter, A. Chambouvet, and G.H. Kim (2010). Algal diseases: spotlight on a black box. *Trends Pl. Sci.* 15:633-40. [10.2.3]

Gacia, E., C. Rodriguez-Prieto, O. Delgado, and E. Ballesteros (1996). Seasonal light and temperature responses of *Caulerpa taxifolia* from the northwestern Mediterranean. *Aquat. Bot.* 53:215-25. [3.4]

Gagne, J. A., and K. H. Mann (1987). Evaluation of four models used to estimate kelp productivity from growth measurements. *Mar. Ecol. Prog. Ser.* 37: 35-44. [5.7.4]

Gagne, J. A., K. H. Mann, and A. R. O. Chapman (1982). Seasonal patterns of growth and storage in *Laminaria longicruris* in relation to differing patterns of availability of nitrogen in the water. *Mar. Biol.* 69:91-101. [5.5.1, 5.6, 6.8.4]

Gagnon, P., R. E. Scheibling, W. Jones, and D. Tully (2008). The role of digital bathmetry in mapping shallow marine vegetation from hyperspectral image data. *Int. J. Remote Sens.* 29:879-904. [3.5.1]

Gaillard, J., and M. T. L'Hardy-Halos (1990). Morphogenèse du *Dictyota dichotoma* (Dictyotales, Phaeophyta). III. Ontogenese et croissance des frondes adventives. *Phycologia* 29:39-53. [1.2.1, 2.6.4]

Gaines, S. D., and J. Lubchenco (1982). A unified approach to marine plant-herbivore interactions. II. Biogeography. *Annu. Rev. Ecol. Syst.* 13: 111-38. [4.4.2]

Gaines, S. D., and J. Roughgarden (1985). Larval settlement rate: a leading determinant of structure in an ecological community of the marine intertidal zone. *Proc. Nat. Acad. Sci., USA* 82:3707-11. [Essay 2]

Gaines, S. D., Lester, S. E., G. Eckert, *et al.* (2009). Dispersal and geographic ranges in the sea. In J. D. Witman, and K. Roy (eds), *Marine Macroecology* (pp. 227-49). Chicago: University of Chicago Press. [3.3, 3.5.2]

Gall, E. (Le Gall, Y.), A. Asensi, D. Marie, and B. Kloareg (1996). Parthenogenesis and apospory in the Laminariales: a flow cytometry analysis. *Eur. J. Phycol.* 31:369-80. [2.2]

Galloway, J. N., H. Levy II, and P.S. Kasibhatla (1994). Year 2020: Consequences of population growth and development of deposition of oxidized nitrogen. *Ambio* 23:120-3. [9.6.3]

Gansert, D., and S. Blossfeld (2008). The application of novel optical sensors (optodes) in experimental plant ecology. *Progr. Bot.* 69:333-58.

Gantt, E. (1975). Phycobilisomes: light harvesting pigment complexes. *BioScience* 25:781-8. [Fig. 5.4]

Gantt, E., G. M. Berg, D. Bhattacharya, *et al.* (2010). Porphyra: complex life histories in a harsh environment: *P. umbilicalis*, an intertidal red alga for genomic analysis. In J. Seckbach and D. J. Chapman (eds), *Red Algae in the Genomic Age (Cellular Origin, Life in Extreme Habitats and Astrobiology 13)* (pp. 129-48). Dordrecht, The Netherlands: Springer. [1.1.1, 1.4.1, 10.2.4, 10.11]

Gao, K., and I. Umezaki (1989a). Studies on diurnal photosynthetic performance of *Sargassum thunbergii*. I. Changes in photosynthesis under natural light. *Sorui (Jpn J. Phycol.)* 37:89-98.

Gao, K., and I. Umezaki (1989b). Studies on diurnal photosynthetic performance of *Sargassum thunbergii*. II. Explanation of diurnal photosynthesis patterns from examinations in the laboratory. *Sorui (Jpn J. Phycol.)* 37:99-104.

Garbary, D. J., and D. J. Belliveau (1990). Diffuse growth, a new pattern of cell wall deposition for the Rhodophyta. *Phycologia* 29:98-102. [1.3.4]

Garbary, D. J., and B. Clarke (2002). Intraplant variation in nuclear DNA content in *Laminaria saccharina* and *Alaria esculenta* (Phaeophyceae). *Bot. Mar.* 45:211-16. [2.2]

Garbary, D. J., and A. Gautam (1989). The *Ascophyllum, Polysiphonia, Mycosphaerella* symbiosis. I. Population ecology of *Mycosphaerella* from Nova Scotia. *Bot. Mar.* 32:181-6. [4.5.1]

Garbary, D. J., and J. F. London (1995). The *Ascophyllum / Polysiphonia / Mycosphaerella* symbiosis V. Fungal infection protects *A. nodosum* from desiccation. *Bot. Mar.* 38:529-33. [4.5.1]

Garbary, D. J., and K. A. MacDonald (1995). The *Ascophyllum/Polysiphonia/Mycosphaerella* symbiosi IV. Mutualism in the *Ascophyllum/Mycosphaerella* interaction. *Bot. Mar.* 38:221-5. [4.5.1]

Garbary, D. J., and A. R. McDonald (1996). Fluorescent labeling of the cytoskeleton in *Ceramium strictum* (Rhodophyta). *J. Phycol.* 32:85–93. [1.3.3]

Garbary, D., D. Belliveau, and R. Irwin (1988). Apical control of band elongation in *Antithamnion defectum* (Ceramiaceaae, Rhodophyta). *Can. J. Bot.* 66: 1308–15. [1.3.4]

Garbary, D. J., K. Y. Kim, T. Klinger, and D. Duggins (1999). Red algae as hosts for endophytic kelp gametophytes. *Mar. Biol.* 135:35–40. [3.3.8, 4.5.2]

Garbary, D. J., G. Lawson, K. Clement, and M. E. Galway (2009). Cell division in the absence of mitosis: the unusual case of the fucoid *Ascophyllum nodosum* (L.) Le Jolis (Phaeophyceae). *Algae.* 24:239–48. [1.3.5]

García-Jiménez, P., P. M. Just, A. M. Delgado, and R. R. Robaina (2007). Transglutaminase activity decrease during acclimation to hyposaline conditions in marine seaweed *Grateloupia doryphora* (Rhodophyta, Halymeniaceae). *J. Plant Physiol.* 164:367–70. [2.6.3]

García-Jiménez, P., F. García-Maroto, J. A. Garrido-Cárdenas, C. Ferrandiz, and R. R. Robaina (2009). Differential expression of the ornithine decarboxylase gene during carposporogenesis in the thallus of the red seaweed *Grateloupia imbricata* (Halymeniaceae). *J. Plant. Physiol.* 166:1745–54. [2.6.3]

Garreta, A. G., M. A. R. Siguan, N. S. Soler, J. R. Lluch, and D. F. Kapraun (2010). Fucales (Phaeophyceae) from Spain characterized by large-scale discontinuous nuclear DNA contents consistent with ancestral cryptopolyploidy. *Phycologia* 49:64–72. [1.4.2]

Garske, L. E. (2002). Macroalgas Marinas. In E. Danulat and G. J. Edgar (eds), *Reserva Marina de Gálapagos. Línea Base de la Biodiversidad* (pp. 419–431). Santa Cruz, Gálapagos, Ecuador: Estación Científica Charles Darwin/Servicio del Parque Nacional. [7.3.7]

Gartner, A., P. Lavery, and A. J. Smit (2002). Use of δ^{15}N signatures of different functional forms of macroalgae and filter-feeders to reveal temporal and spatial patterns in sewage dispersal. *Mar. Ecol. Prog. Ser.* 235:63–73. [9.6.3]

Gattuso, J. P., B. Gentili, C. M. Duarte, *et al.* (2006). Light availability in the coastal ocean: impact on the distribution of benthic photosynthetic organisms and contribution to primary production. *Biogeosci.* 3:489–513. [5.7.5]

Gaur, J. P., and L. C. Rai (2001). Heavy metal tolerance in algae. In L. C. Rai and J. P Gaur (eds), *Algal Adaptation to Environmental Stresses: Physical, Biochemical and Molecular Mechanisms* (pp. 363–8). New York: Springer. [9.3.1, 9.3.4, 9.3.5]

Gaylord, B. (2000). Biological implications of surf-zone flow complexity. *Limnol. Oceanogr.* 45:174–88. [8.3.1]

Gaylord, B., and M. W. Denny (1997). Flow and flexibility. I. Effects of size, shape and stiffness in determining wave forces on the stipitate kelps *Eisenia arborea* and *Pterygophora californica. J. Exp. Biol.* 200:3141–64. [8.3.1]

Gaylord, B., D. C. Reed, P. T. Raimondi, *et al.* (2002). A physically based model of macroalgal spore dispersal in the wave and current-dominated nearshore. *Ecology* 83:1239–51. [3.5.2]

Gaylord, B., J. H. Rosman, D. C. Reed, *et al.* (2007). Spatial patterns of flow and their modification within and around a giant kelp forest. *Limnol. Oceanogr.* 52:1838–52. [8.1.4]

Gaylord, B., M. W. Denny, and M. A. R. Koehl (2008). Flow forces on seaweeds: field evidence for roles of wave impingement and organism inertia. *Biol. Bull.* 215:295–308. [8.1.4, 8.3.1]

Gekeler, W., E. Grill, E.-L. Winnacker, and M. H. Zenk (1988). Algae sequester heavy metals via synthesis of phytochelatin complexes. *Arch. Microbiol.* 150:197–202. [9.3.3]

Genty, B., J. M. Briantais, and N. R. Baker (1989). The relationship bretween the quantum yield of photosynthetic electron transport and quenching of chlorophyll fluorescence. *Biochim. Biophys. Acta* 990:87–92. [5.7.1]

Gerard, V. A. (1982a). Growth and utilization of internal nitrogen reserves by the giant kelp *Macrocystis pyrifera* in a low-nitrogen environment. *Mar. Biol.* 66:27–35. [6.4.2]

Gerard, V. A. (l982b). *In situ* rates of nitrate uptake by giant kelp, *Macrocystis pyrifera* (L.) C. Agardh: tissue differences, environmental effects, and predictions of nitrogen limited growth. *J. Exp. Mar. Biol. Ecol.* 62:211–24. [7.3.7]

Gerard, V. A. (1987). Hydrodynamic stream lining of *Laminaria saccharina* Lamour, in response to mechanical stress. *J. Exp. Mar. Biol. Ecol.* 107:237–44. [8.2.1]

Gerard, V. (1997). The role of nitrogen nutrition in high temperature tolerance of the kelp, *Laminaria saccharina* (Chromophyta). *J. Phycol.* 33:800–10. [7.3.4, 7.3.5, 7.3.7]

Gerard, V. A., and K. R. Du Bois (1988). Temperature ecotypes near the southern boundary of the kelp *Laminaria saccharina. Mar. Biol.* 97:575–80. [7.3.3]

Gerard, V. A., and K. H. Mann (1979). Growth and production of *Laminaria longicruris* (Phaeophyta) populations exposed to different intensities of water movement. *J. Phycol.* 15:33–41. [8.1.4, 8.3.1]

Gerard, V. A., S. E. Dunham, and G. Rosenberg (1990). Nitrogen-fixation by cyanobacteria associated with *Codium fragile* (Chlorophyta): environmental effects and transfer of fixed nitrogen. *Mar. Biol.* 105:1–8. [6.2]

Gerlach, S. A. (1982). *Marine Pollution: Diagnoses and Therapy*. Berlin: Springer-Verlag, 218 pp. [9.4.3, 9.7]

Germann, I. (1988). Effects of the 1983-EI Niño on growth and carbon and nitrogen metabolism of *Pleurophycus gardneri* (Phaeophyceae: Laminariales) in the northeastern Pacific. *Mar. Biol.* 99:445-55. [7.3.7]

Gerwick, W. H., and N. J. Lang (1977). Structural, chemical and ecological studies on iridescence in *Iridaea* (Rhodophyta). *J. Phycol.* 13:121-7. [1.3.1]

Gessner, F. (1970). Temperature: plants. In O. Kinne (ed.), *Marine Ecology* (vol. I, pt. I, pp. 363-406). New York: Wiley. [7.3.8]

Gessner, F., and L. Hammer (1968). Exosmosis and "free space" in marine benthic algae. *Mar. Biol.* 2:88-91. [7.4.1]

Gessner, F., and W. Schramm (1971). Salinity: plants. In O. Kinne (ed.), *Marine Ecology* (Vol. I, pt. 2, pp. 705-820). New York: Wiley. [7.2.3, 7.4.3, 7.4.4, 7.5.1]

Gevaert, F., N. G. Barr, and T. A. V. Rees (2007). Diurnal cycle and kinetics of ammonium assimilation in the green alga *Ulva pertusa*. *Mar. Biol.* 151:1517-24. [6.4.2, 6.5.1]

Giddings, T. H., C. Wassman, and L.A. Staehelin (1983). Structure of the thylakoids and envelope membranes of the cyanelles of *Cyanophora paradoxa*. *Plant Physiol.* 71:409-19. [Fig. 5.10]

Gilman, S. E., C. D. G. Harley, D. C. Strickland, *et al.* (2006). Evaluation of effective shore level as a method of characterizing intertidal wave exposure regimes. *Limnol. Oceanogr.: Methods* 4:448-57. [3.1.3]

Ginger, M. L., N. Portman, and P. G. McKean (2008). Swimming with protests: perception, motility and flagellum assembly. *Nature* 6:838-50. [1.3.3]

Giordano, M., J. Beardall, and J. A. Raven (2005a). CO_2 concentrating mechanisms in algae: mechanisms, environmental modulation, and evolution. *Ann. Rev. Plant Biol.* 56:99-131. [5.4.3]

Giordano, M., A. Norici, and R. Hell (2005b). Sulfur and phytoplankton: acquisition, metabolism and impact on the environment. *New Phytol.* 166:371-82. [6.5.4]

Giordano, M., A. Norci, S. Ratti, and J. A. Raven (2008). Role of sulfur in algae: acquisition, metabolism, ecology and evolution. In R. Heel, C. Dahl, D. B. Knaff, and T. Leustek (eds), *Sulfur Metabolism in Phototrophic Organisms* (pp. 397-415). Dordrecht, The Netherlands: Springer. [6.5.4]

Givskov, M., R. de Nys, M. Manefield, *et al.* (1996). Eukaryotic interference with homoserine lactone-mediated prokaryotic signaling. *J. Bacteriol.* 178:6618-22. [4.2.2]

Glass, A. D. M. (1989). *Plant Nutrition: An Introduction to Current Concepts*. Boston, MA: Jones and Barttell, 234 pp. [6.3.1, 6.3.3, 6.3.4]

Glazer, A. N. (1985). Light harvesting by phycobilisomes. *Annu. Rev. Biochem.* 14:47-77. [5.3.1]

Gledhill, M., M. Nimmo, S. J. Hill, and M. T. Brown (1997). The toxicity of copper(II) species to marine algae, with particular reference to macroalgae. *J. Phycol.* 33:2-11. [9.3.2, 9.3.4, 9.3.5]

Gledhill, M., M. Nimmo, S. J. Hill, and M. T. Brown (1999). The release of copper-complexing ligands by the brown alga *Fucus vesiculosis* (Phaeophyceae) in response to increasing total copper levels. *J. Phycol.* 35:501-9. [9.3.3, 9.3.4, 9.3.5]

Gledhill, M., M. T. Brown, M. Nimmo, R. Moate, and S. J. Hill (1998). Comparison of techniques for the removal of particulate material from seaweed tissue. *Mar. Envir. Res.* 45:295-307. [9.3.2]

Glibert, P. M., and D. G. Capone (1993). Mineralization and assimilation in aquatic, sediment, and wetland systems. In R. Knowles and R. Blackburn (eds). *Nitrogen Isotope Techniques* (pp. 243-72). San Diego: Academic Press. [6.4.1]

Glibert, P. M., F. Lipshultz, J. J. McCarthy, and M. A. Altabet (1982). Isotope dilution models of uptake and remineralization of ammonium by marine plankton. *Limnol. Oceanogr.* 27:639-50. [6.4.1]

Glynn, P. W. (1965). Community composition, structure, and interrelationships in the marine intertidal *Endocladia muricata-Balanus glandula* association in Monterey Bay, California. *Beaufortia* 12(148):1-198. [7.2.3]

Glynn, P. W. (1988). El Niño-Southern Oscillation 1982-1983: nearshore population, community and ecosystem responses. *Annu. Rev. Ecol. Syst.* 19:309-45. [7.3.7]

Goecke, F., A. Labes, J. Weise, *et al.* (2010). Chemical interactions between marine macroalgae and bacteria. *Mar. Ecol. Prog. Ser.* 409:267-99. [4.2.2]

Goff, L. J. (1982). Biology of parasitic red algae. *Prog. Phycol. Res.* 1:289-369. [4.5.2]

Goff, L. J., and A. W. Coleman (l988). The use of plastid DNA restriction endonuclease patterns in delimiting red algal species and populations. *J. Phycol.* 24:357-68. [1.4.1, 1.4.2]

Goff, L. J., and A. W. Coleman (1990). Red algal plasmids. *Curr. Genet.* 18:557-65. [1.4.2]

Goff, L. J., and A. W. Coleman (1995). Fate of parasite and host organelle DNA cellular-transformation of red algae by their parasites. *Plant Cell* 7:1899-911. [4.5.2]

Goff, L. J., and G. Zuccarello (1994). The evolution of parasitism in red algae: cellular interactions of adelphoparasites and their hosts. *J. Phycol.* 30:695-720. [4.5.2]

Goff, L. J., D. A. Moon, P. Nyvall, *et al.* (1996). The evolution of parasitism in the red algae: molecular comparisons of adelphoparasites and their hosts. *J. Phycol.* 32:297-312. [4.5.2]

Goff, L. J., J. Ashen, and D. Moon (1997). The evolution of parasites from their hosts: a case study in the parasitic red algae. *Evolution* 51:1068–78. [4.5.2]

Gómez, I., and P. Huovinen (2011). Morpho-functional patterns and zonation of South Chilean seaweeds: the importance of photosynthetic and bio-optical traits. *Mar. Ecol. Progr. Ser.* 422:77–91. [5.3.2]

Gómez, I., and P. Huovinen (2012). Morpho-functionality of carbon metabolism in seaweeds. In C. Wiencke and K. Bischof (eds) *Seaweed Biology: Novel Insights into Ecophysiology, Ecology and Utilization, Ecological Studies* (Vol. 219, pp. 25–45). Berlin: Springer-Verlag.

Gómez, I., and C. Wiencke (1998). Seasonal changes in C, N and major organic compounds and their significance to morpho-functional processes in the endemic Antarctic brown alga *Ascoseira mirabilis*. *Polar Biol.* 19:115–24. [5.4.3, 5.6]

Gómez, I., F. L. Figueroa, I. Sousa-Pinto, *et al.* (2001). Effects of UV radiation and temperature on photosynthesis as measured by PAM fluorescence in the red alga *Gelidium pulchellum* (Turner) Kützing. *Bot. Mar.* 44:9–16. [7.3.1]

Gómez, I., N. Ulloa, and M. Orostegui (2005). Morpho-functional patterns of photosynthesis and UV sensitivity in the kelp *Lessonia nigrescens* (Laminariales, Phaeophyta). *Mar. Biol.* 148:231–40.

Gómez, I., M. Orostegui, and P. Huovinen (2007) Morpho-functional patterns of photosynthesis in the South Pacific kelp *Lessonia nigrescens*: effects of UV radiation on ^{14}C fixation and primary photochemical reactions. *J. Phycol.* 43:55–64. [5.4.3, 5.6]

Gómez, I., A. Wulff, M. Y. Roleda, *et al.* (2009). Light and temperature demands of marine benthic micro-algae and seaweeds in the polar regions. *Bot. Mar.* 52:593–608. [5.3.3]

Gonen, Y., E. Kimmel, and M. Friedlander (1995). Diffusion boundary layer transport in *Gracilaria conferta* (Rhodophyta). *J. Phycol.* 31:768–73. [8.2.1]

Gonen, Y., E. Kimmel, E. Tel-Or, and M. Friedlander (1996). Intercellular assimilate translocation in *Gracilaria cornea* (Gracilariaceae, Rhodophyta). *Hydrobiologia* 326/327:421–8. [5.6]

Goodwin, T. W., and E. I. Mercer (1983). *Introduction to Plant Biochemisty*, 2nd edn. Oxford: Pergamon. [5.3.1]

Gordillo F. J. L. (2012). Environment and algal nutrition. In C. Weincke and K. Bischoff (eds) *Seaweed Biology: Novel Insights into Ecophysiology and Utilization* (pp. 67–86). Ecology Study Series. Berlin: Springer. [6.4.1, 6.5]

Gordillo, F. J., J. Aguilera, and C. Jimenez (2006). The response of nutrient assimilation and biochemical composition of Arctic seaweeds to a nutrient input in summer. *J. Exp. Bot.* 57:2661–71. [5.4.1]

Gordon, D. M., P. B. Birch, and A. J. McComb (1980). The effects of light, temperature and salinity on photosynthetic rates of an estuarine *Cladophora. Bot. Mar.* 29:749–55. [7.2.2]

Gordon, D. M., P. B. Birch, and A. J. McComb (1981). Effects of inorganic phosphorus and nitrogen on the growth of an estuarine *Cladophora* in culture. *Bot. Mar.* 24:93–106. [6.7.2]

Gordon, R., and S. H. Brawley (2004). Effects of water motion on propagule release from algae with complex life histories. *Mar. Biol.* 145:21–9. [8.2.2]

Gorman, D., and S. D. Connell (2009). Recovering subtidal forests in human-dominated landscapes. *J. Appl. Ecol.* 46:1258–65. [3.5.1]

Gorospec, K. D., and S. A. Karl (2011). Small-scale spatial analysis of *in situ* sea temperature throughout a single coral patch reef. *J. Mar. Biol.* Article ID 719580. 12 pp. [1.1.3]

Graeve, M., W. Hagen, and G. Kattner (1994). Herbivorous or omnivorous? On the significance of lipid compositions as trophic markers in Antarctic copepods. *Deep Sea Res.* I 41:915–24. [Essay 3]

Graeve, M., G. Kattner, C. Wiencke, and U. Karsten (2002). Fatty acid composition of Arctic and Antarctic macroalgae: indicator of phylogenetic and trophic relationships. *Mar. Ecol. Prog. Ser.* 231:67–74. [5.7.5, 7.3.6, Essay 3]

Graham, L. E., J. M. Graham, and L. W. Wilcox (2009). *Algae*. London: Pearson College Division, 616 pp. [1.1.1, 1.1.3, 1.2.1, 1.3, 1.3.2, 1.3.5, 1.4.1, 2.2, 2.4, 2.6]

Graham, M. H. (1997). Factors determining the upper limit of giant kelp, *Macrocystis pyrifera* Agardh, along the Monterey Peninsula, central California, USA. *J. Exp. Mar. Biol. Ecol.* 218:127–49. [3.2]

Graham, M. H. (1999). Identification of kelp zoospores from *in situ* plankton samples. *Mar. Biol.* 135:709–20. [2.5.1]

Graham, M. H. (2002). Prolonged reproductive consequences of short-term biomass loss in seaweeds. *Mar. Biol.* 140:901–11. [3.5.2, 4.3.1]

Graham, M. H., J. A. Vásquez, and A. H. Buschmann (2007a). Global ecology of the giant kelp *Macrocystis* from ecotypes to ecosystems. *Oceanogr. Mar. Biol.* 45:39–88. [1.1.1, 1.1.3, 3.2, 3.3.2, 4.1, 4.2.1, 4.3.1]

Graham, M. H., B. P. Kinlan, L. D. Druehl, L. E. Garske, and S. Banks (2007b). Deep-water kelp refugia as potential hotspots of tropical marine diversity and productivity. *PNAS* 104:16,576–80. [5.2.2]

Graham, M. H., B. P. Kinlan, and R. K. Grosberg (2010). Post-glacial redistribution and shifts in productivity of giant kelp forests. *Proc. Royal Soc. B* 277:399–406. [5.7.4]

Grall, J., and L. Chauvaud (2002). Marine eutrophication and benthos: the need for new approaches and concepts. *Global Change Biol.* 8:813–30. [9.6.1]

Granbom, L. M., M. Pedersén, P. Kadel, and K. Lüning (2001). Circadian rhythm of photosynthetic oxygen evolution in *Kappaphycus alvarezii* (Rhodophyta): dependence on light quantity and quality. *J. Phycol.* 37:1020–5. [5.7.2]

Granhag, M., A. I. Larsson, and P. R. Jonsson (2007). Algal spore settlement and germling removal as a function of flow speed. *Mar. Ecol. Prog. Ser.* 344:63–9. [2.5.1]

Grant, A. J., D. A. Trautman, I. Menz, *et al.* (2006). Separation of two cell signaling molecules from a symbiotic sponge that modify algal carbon metabolism. *Biochem. Bioph. Res. Co.* 348:92–8. [4.5.1]

Grant, B. R., and M. A. Borowitzka (1984). The chloroplasts of giant-celled and coenocytic algae: biochemistry and structure. *Bot. Rev.* 50:267–307. [1.3.2]

Gravel, D., F. Guichard, M. Loreau, and N. Mouquet (2010a). Source and sink dynamics in meta-ecosystems. *Ecology* 91:2172–84. [Essay 2]

Gravel, D., N. Mouquet, M. Loreau, and F. Guichard (2010b). Patch dynamics, persistence, and species coexistence in metaecosystems. *Am. Nat.* 176:289–302. [Essay 2]

Gravot, A., S. M. Dittami, S. Rousvoal, *et al.* (2010). Diurnal oscillations of metabolite abundances and gene analysis provide new insights into central metabolic processes of the brown alga *Ectocarpus siliculosus*. *New Phytol.* 188:98–110. [1.4.1]

Gray, J. S. (2002). Biomagnification in marine systems: the perspective of an ecologist. *Mar. Poll. Bull.* 45:46–52. [9.3.2]

Green, B. R., and D. G. Durnford (1996). The chlorophyll-carotenoid proteins of oxygenic photosynthesis. *Ann. Rev. Plant Physiol. Plant Mol. Biol.* 47:685–714. [5.3.1]

Greene, R. M., and V. A. Gerard (1990). Effects of high frequency light fluctuations on growth and photoacclimation of the red alga *Chondrus crispus*. *Mar. Biol.* 105:337–44. [5.2.2]

Greene, R. W. (1970). Symbiosis in sacoglossan opisthobranchs: functional capacity of symbiotic chloroplasts. *Mar. Biol.* 72:138–42. [4.3.2, 4.5.3]

Greer, S. P., and C. D. Amsler (2004). Clonal variation in phototaxis and settlement behaviors of *Hincksia irregularis* (Phaephyceae) spores. *J. Phycol.* 40:44–53. [2.5.1]

Gregory, T. R. (2005). The C-value enigma in plants and animals: a review of parallels and an appeal for partnership. *Ann. Bot.-London* 95:133–46. [1.4.2]

Gressler, V., P. Colepicolo, and E. Pinto (2009). Useful strategies for algal volatile analysis. *Curr. Analyt. Chem.* 5:271–92. [4.2.2]

Grime, J.P (1979). *Plant Strategies and Vegetation Processes.* New York: Wiley. [3.5.2, 8.3.2]

Groen, P. (1980). Oceans and seas. I. Physical and chemical properties. In *The New Encyclopaedia Britannica* (vol. 13, pp. 484–97). London: Macropaedia [7.2.1]

Gross, M. G. (1996). *Oceanography.* New York: Prentice Hall, 236 pp. [3.1.1]

Gross, W. (1990). Occurrence of glycolate oxidase and hydro-pyruvate reductase in *Egreggia menziesii* (Phaeophyta). *J. Phycol.* 26:381–3. [5.4.2]

Grossman, A. (2005). Regeneration of a cell from protoplasm. *J. Phycol.* 42:1–5. [2.6.4]

Grossman, A. R., L. K. Anderson, P. B. Conley, and P. G. Lemaux (1989). Molecular analyses of complementary chromatic adaptation and the biosynthesis of a phycobilisome. In A. W. Coleman, L. J. Goff, and J. R. Stein-Taylor (eds) *Algae as Experimental Systems* (pp. 269–88). New York: Alan R. Liss. [5.3.4]

Grzymski, J., G. Johnsen, and E. Sakshaug (1997). The significance of intracellular self-shading on the bio-optical properties of brown, red and green macroalgae. *J. Phycol.* 33:408–14. [5.3.2]

Gualtieri, P., and K. R. Robinson (2002). A rhodopsin-like protein in the plasma membrane of *Silvetia compressa* eggs. *Photochem. Photobiol.* 75:76–8. [2.5.3]

Guerry, A. D. (2006). Grazing, nutrients, and marine benthic algae: insights into the drivers and protection of diversity. PhD Dissertation, Oregon State University. [Essay 2]

Guerry, A. D., B. A. Menge, and R. A. Dunmore (2009). Effects of consumers and enrichment on abundance and diversity of benthic algae in a rocky intertidal community. *J. Exp. Mar. Biol. Ecol.* 369:155–64. [4.3.1]

Guest, M. A., S. D. Frusher, P. D. Nichols, C. R. Johnson, and K. E. Wheatley (2009). Trophic effects of fishing southern rock lobster *Jasus edwardsii* shown by combined fatty acid and stable isotope analyses. *Mar. Ecol. Prog. Ser.* 388:169–84. [Essay 3]

Guillemin, M. L., S. Faugeron, C. Destombe, *et al.* (2008). Genetic variation in wild and cultivated populations of the haploid-diploid red alga *Gracilaria chilensis*: how farming practices favor asexual reproduction and heterozygosity. *Evolution* 62:1500–19. [10.6.2]

Guimarães, S. M. P. B., M. R. A. Broga, M. Cordeiro-Marino, and A. G. Pedrini (1986). Morphology and taxonomy of *Jolyna laminarioides*, a new member of the Scytosiphonales (Phaeophyceae) from Brazil. *Phycologia* 25:99–108. [5.6]

Guiry, M. (2011). Available online: www.seaweed/ie/aquaculture. [10.10]

Guiry, M.D. (1974). A preliminary consideration of the taxonomic position of *Palmaria palmata* (Linnaeus) Stackhouse = *Rhodymenia palmata* (Linnaeus) Greville. *J. Mar. Biol. Assoc. UK* 54:509-28. [Essay 1]

Guiry, M. D. (1978). The importance of sporangia in the classification of the Florideophycidae. In D. E. G. Irvine and J. H. Price (eds), *Modern Approaches to the Taxonomy of Red and Brown Algae* (pp. 111-44). London: Academic Press. [Essay 1]

Guiry, M. D. (1990). Sporangia and spores. In K. M. Cole and R. G. Sheath (eds), *Biology of the Red Algae* (pp. 43-71). Cambridge: Cambridge University Press. [2.2]

Gundlach, E., and M. Hayes (1978). Vulnerability of coastal environments to oil spill impacts. *Mar. Technol. Soc. J.* 12:1827. [9.4.3]

Gundlach, E. R., P. D. Boehm, M. Marchand, *et al.* (1983). The fate of Amoco. Cadiz oil. *Science* 221:122-9. [9.4, 9.4.1, 9.4.3]

Haavisto, F., T. Välikangas, and V. Jormalainen (2010). Induced resistance in a brown alga: phlorotannins, genotypic variation and fitness costs for the crustacean herbivore. *Oecologia* 162:685-95. [4.4.2]

Hable, W. E., and D. L. Kropf (2000). Sperm entry induces polarity in fucoid zygotes. *Development* 127:493-501. [2.5.3]

Hable, W. E., N. R. Miller, and D. L. Kropf (2003). Polarity establishment requires dynamic actin in fucoid zygotes. *Protoplasma* 221:193-204. [1.3.3, 2.5.3]

Hackney, J. M., and P. Sze (1988). Photorespiration and productivity rates of a coral reef algal turf assemblage. *Mar. Biol.* 98:483-92. [5.7.1]

Hackney, J. M., R. C. Carpenter, and W. H. Adey (1989). Characteristic adaptations to grazing among algal turfs on a Caribbean coral reef. *Phycologia* 28: 109-19. [1.1.1]

Häder, D. P., H. D. Kumar, R. C. Smith, and R. C. Worrest (1998). Effects on aquatic ecosystems. *J. Photochem. Photobiol. B: Biol.* 46:53-68. [5.2.2]

Hafting, J. F. (1999). A novel technique for propagation of *Porphyra yezoensis* Ueda blades in suspension cultures via monospores. *J. Appl. Phycol.* 11:361-7. [10.2.4]

Hagen, N. T. (1995). Recurrent destructive grazing of successionally immature kelp forests by green sea urchins in vestfjorden, Nothern Norway. *Mar. Ecol. Prog. Ser.* 123:95-106. [4.3.1]

Hagopian, J. C., M. Reis, J. P. Kitajima, *et al.* (2004). Comparative analysis of the complete plastid genome sequence of the red alga *Gracilaria tenuistipitata* var. *liui* provides insights into the evolution of rhodoplasts and their relationship to other plastids. *J. Mol. Evol.* 59:464-77. [1.3.2]

Haines, K. C., and P. A. Wheeler (1978). Ammonium and nitrate uptake by the marine macrophyte *Hypnea musciformis* (Rhodophyta) and *Macrocystis pyrifera* (Phaeophyta). *J. Phycol.* 14:319-24. [6.4, 6.4.2]

Hairston, N. G., F. E. Smith, and L. B. Slobodkin (1960). Community structure, population control, and competition. *Am. Nat.* 94:421-5. [Essay 2]

Hall, J. D., and S. N. Murray (1998). The life history of a Santa Catalina Island population of *Liagora californica* (Nemaliales, Rhodophyta) in the field and in laboratory culture. *Phycologia* 37:184-94. [2.3.2]

Hall, J. L. (2002). Cellular mechanisms for heavy metal detoxification and tolerance. *J. Exp. Bot.* 53(386):1-11. [9.3.3]

Hall, Jr. L.W., J. M. Giddings, K. R. Solomon, and R. Balcomb (1999). An ecological assessment for the use of Irgarol 1051 an an algacide for antifouling paints. *Crit. Rev. Toxicol.* 29:367-437. [9.5]

Hall, L. W., and A. E. Pinkney (1984). Acute and sublethal effects of organotin compounds on aquatic biota: an interpretive literature evaluation. *CRC Crit. Rev. Toxicol.* 14:159-209. [9.5]

Halldal, P. (1964). Ultraviolet action spectra of photosynthesis and photosynthetic inhibition in a green alga and a red alga. *Physiol. Plant* 17:414-21. [5.2.1]

Hall-Spencer, J. M., R. Rodolfo-Metalpa, S. Martin, *et al.* (2008). Volcanic carbon dioxide vents show ecosystem effects of ocean acidification. *Nature* 354:96-9. [6.5.3]

Halpern, B. S., S. Walbridge, K. A. Selkoe, *et al.* (2008) A global map of human impact on marine ecosystems. *Science* 319:948-52. [3.5.1]

Hamilton, J. G., A. R. Zangerl, E. H. DeLucia, and M. R. Berenbaum (2001). The carbon-nutrient balance hypothesis: its rise and fall. *Ecol. Lett.* 4:86-95. [4.4.2]

Han, T., and G-W. Choi (2005). A novel marine algal toxicity bioassay based on sporulation inhibition in the green macroalga *Ulva pertusa* (Chlorophyta). *Aquat. Toxicol.* 75:202-12. [9.2]

Han, T., Y.-S. Han, J. M. Kain, and D.-P. Häder (2003). Thallus differentiation of photosynthesis, growth, reproduction and UV-B sensitivity in the green alga *Ulva pertusa* (Chlorophyceae). *J. Phycol.* 39:712-21. [1.1.1]

Han, T., Y. S Han, C. Y Park, *et al.* (2008). Spore release by the green alga *Ulva*: A quantitative assay to evaluate aquatic toxicants. *Environ. Poll.* 153:699-705. [9.2, 9.3.4]

Han, T., J. A. Kong, and M. T. Brown (2009). Aquatic toxicity tests of *Ulva pertusa* Kjellman (Ulvales, Chlorophyta) using spore germination and gametophyte growth. *Eur. J. Phycol.* 44:357-63. [9.2]

Han, Y. S., M. T. Brown, G. S. Park, and T. Han (2007). Evaluating aquatic toxicity by visual inspection of thallus color in the green macroalga *Ulva*: Testing a novel bioassay. *Environ. Sci. Technol.* 47:3667-71. [9.2]

Hanelt, D. (1998). Capability of dynamic photoinhibition in Arctic macroalgae is related to their depth distribution. *Mar. Biol.* 131:361-9. [5.3.3, 7.9]

Hanelt, D., and W. Nultsch (1990). Daily changes of the phaeoplast arrangement in the brown alga *Dictyota dichotoma* as studied in field experiments. *Mar. Ecol. Prog. Ser.* 61:273-9. [1.3.2, 5.3.2]

Hanelt, D., and W. Nultsch (1991). The role of chromatophore arrangement in protecting the chromatophores of the brown alga *Dictyota dichotoma* against photodamage. *J. Plant Physiol.* 138:470-5. [1.3.2]

Hanelt, D., and W. Nultsch (1995). Field studies of photoinhibition show non-correlations between oxygen and fluorescence measurements in the Arctic red alga *Palmaria palmata*. *J. Plant Physiol.* 145:31-8. [5.7.1, 5.7.2]

Hanelt, D., and W. Nultsch (2003). Photoinhibition in seaweeds. In G. Heldmaier and D. Werner (eds) *Environmental Signal Processing and Adaptation* (pp. 1414-67). Heidelberg and Berlin: Springer Publishers.

Hanelt, D., and M. Y. Roleda (2009). UVB radiation may ameliorate photoinhibition in specific shallow-water tropical marine macrophytes. *Aquat. Bot.* 91:6-12. [5.2.2, 7.6]

Hanelt, D., H. Tüg, K. Bischof, *et al.* (2001) Light regime in an Arctic fjord: a study related to stratospheric ozone depletion as a basis for determination of UV effects on algal growth. *Mar. Biol.* 138:649-58. [5.2.1, 5.2.2]

Hanelt, D., C. Wiencke, and K. Bischof (2003). Photosynthesis in marine macroalgae. In A. W. D. Larkum, S. E. Douglas, and J. A. Raven (eds) *Photosynthesis in Algae. Advances in Photosynthesis and Respiration* (Vol. 14, pp. 413-35). Dordrecht, The Netherlands: Kluwer Academic Publishers. [7.9]

Hanelt, D., I. Hawes, and R. Rae (2006). Reduction of UV-B radiation causes an enhancement of photoinhibition in high light stressed aquatic plants from New Zealand lakes. *J. Photochem. Photobiol. B: Biol.* 84:89-102. [7.6]

Hanic, L. A., and J. S. Craigie (1969). Studies on the algal cuticle. *J. Phycol.* 5:89-109. [1.3.1]

Hanisak, M. D. (1979). Nitrogen limitation of *Codium fragile* ssp. *tomentosoides* as determined by tissue analysis. *Mar. Biol.* 50:333-7. [6.7.2]

Hanisak, M. D. (1983). The nitrogen relationships of marine macroalgae. In E. J. Carpenter and D. G. Capone (eds). *Nitrogen in the Marine Environment* (pp. 699-730). New York: Academic Press. [6.5]

Hanisak, M. D. (1990). The use of *Gracilaria tikvahiae* (Gracilariales, Rhodophyta) as a model system to understand the nitrogen limitation of cultured seaweeds. *Hydrobiologia* 204/205:79-87. [6.7.2]

Hanisak, M. D., and M. M. Harlin (1978). Uptake of inorganic nitrogen by *Codium fragile* subsp. *tomentosoides* (Chlorophyta). *J. Phycol.* 14:450-4. [6.5.1]

Hanisak, M. D., M. M. Littler, and D. S. Littler (1988). Significance of macroalgal polymorphism: intraspecific tests of the functional-form model. *Mar Biol* 99:157-65. [5.3.2]

Hansen, A. T., M. Hondzo, and C. L. Hurd (2011). Photosynthetic oxygen flux by *Macrocystis pyrifera*: a mass transfer model with experimental validation. *Mar. Ecol. Prog. Ser.* 434:45-55. [8.1.4]

Hanson, C. E., G. A. Hyndes, and S. F. Wang (2010). Differentiation of benthic marine primary producers using stable isotopes and fatty acids: implications to food web studies. *Aquat Bot* 93: 114-22. [5.7.5]

Harder, D. L., O. Speck, C. L. Hurd, and T. Speck (2004b). Reconfiguration as a pre-requisite for survival in highly unstable flow-dominated habitats. *J. Plant Growth Regul.* 23:98-107. [8.3.1]

Harder, T., S. Dobretsov, and P.-Y. Qian (2004a). Waterborne polar macromolecules act as algal antifoulants in the seaweed *Ulva reticulata*. *Mar. Ecol. Prog. Ser.* 274:133-41. [4.2.2, 5.7.3]

Hardy, F. G., and B. L. Moss (1979). Attachment and development of the zygotes of *Pelvetia canaliculata* (L.) Dcne. et Thur. (Phaeophyceae, Fucales). *Phycologia* 18:203-12. [2.5.2]

Harley, C. D. G. (2003). Abiotic stress and herbivory interact to set range limits across a two-dimensional stress gradient. *Ecology* 84:1477-88. [4.3.1]

Harley, C. D., and B. S. T. Helmuth (2003). Local- and regional-scale effects of wave exposure, thermal stress, and absolute versus effective shore level on patterns of intertidal zonation. *Limnol. Oceanogr.* 48:1498-508. [3.1.3]

Harlin, M. M. (1978). Nitrate uptake by *Enteromorpha* spp. (Chlorophyceae): applications to aquaculture systems. *Aquaculture* 15:373-6. [Table 6.3]

Harlin, M. M. (1987). Allelochemistry in marine macroalgae. *CRC Crit. Rev. Plant Sci.* 5:237-49. [4.2.2]

Harlin, M. M., and J. S. Craigie (1975). The distribution of photosynthate in *Ascophyllum nodosum* as it relates to epiphytic *Polysiphonia lanosa*. *J. Phycol.* 11:109-13. [6.4.2]

Harlin, M. M., and J. S. Craigie (1978). Nitrate uptake by *Laminaria longicruris* (Phaeophyceae). *J. Phycol.* 14:464-7. [Table 6.3]

Harlin, M. M., and P. A. Wheeler (1985). Nutrient uptake. In M. M. Littler and D. S. Littler (eds), *Handbook of Phycological Methods: Ecological Field Methods: Macroalgae* (pp. 493-508). Cambridge: Cambridge University Press. [6.4.1]

Harmer, S. L. (2009). The circadian system in higher plants. *Annu. Rev. Plant Biol.* 60:357-77. [1.3.5]

Harper, J. L. (1977). *Population Biology of Plants.* New York: Academic Press, 892 pp. [8.3.2]

Harpole, W. S., J. T. Ngai, E. E. Cleland, *et al.* (2011). Nutrient co-limitation of primary producers. *Ecol. Lett.* 14:852-62. [6.1.3]

Harrington, L., K. Fabricius, G. De'Ath, and A. Negri (2004). Recognition and selection of settlement substrata determine post-settlement survival in corals. *Ecology* 85:3428-37. [3.3.1]

Harrington, L., K. E. Fabricius, and A. Negri (2005). Synergistic effects of diuron and sedimentation on the photophysiology and survival of crustose coralline algae. *Marine Pollution Bulletin* 51:415-27. [Essay 6]

Harris, L. G., and A. C. Jones (2005). Temperature, herbivory and epibiont acquisition as factors controlling the distribution and ecological role of an invasive seaweed. *Biol. Invas.* 7:913-24. [3.4]

Harrison, P J., and L. D. Druehl (1982). Nutrient uptake and growth in the Laminariales and other macrophytes: a consideration of methods. In L. M. Srivastava (ed.), *Synthetic and Degradative Processes in Marine Macrophytes* (pp. 99-120). Berlin: Walter de Gruyter. [6.4.1, 8.2.1]

Harrison, P. J., and C. L. Hurd (2001). Nutrient physiology of seaweeds: application of concepts to aquaculture. *Cah. Biol. Mar.* 42:71-82. [10.7]

Harrison, P. J., L. D. Druehl, K. E. Lloyd, and P. A. Thompson (1986). Nitrogen uptake kinetics in three year-classes of *Laminaria groenlandica* (Laminariales: Phaeophyta). *Mar. Biol.* 93:29-35. [6.4.2, 6.5.1]

Harrison, P. J., J. S. Parslow, and H. L. Conway (1989). Determination of nutrient uptake kinetic parameters: a comparison of methods. *Mar. Ecol. Prog. Ser.* 52:301-12. [6.4, 6.4.1, 6.5.1]

Harrison, P. J., M. H. Hu, Y. P. Yang, and X. Lu (1990). Phosphate limitation in estuarine and coastal waters of China. *J. Exp. Mar. Biol. Ecol.* 140:79-87. [6.5.2]

Hartnoll, R. G., and Hawkins, S. J. (1982). The emersion curve in semidiurnal tidal regimes. *Estuar. Coast. Shelf S.* 15:365-71. [3.1.3]

Haslin, C., M. Lahaye, M. Pellegrini, and J. C. Chermann (2001). *In vitro* anti-HIV activity of sulfated cell wall polysaccharides from gametic, carposporic and tetrasporic stages of the Mediterranean red alga *Asparagopsis armata*. *Planta Med* 67: 301-5. [5.5]

Hata, H., and M. Kato (2004). Monoculture and mixed-species algal farms on a coral reef are maintained through intensive and extensive management by damselfishes. *J. Exp. Mar. Biol. Ecol.* 313:285-96. [4.3.2]

Hata, H., and M. Kato (2006). A novel obligate cultivation mutualism between damselfish and *Polysiphonia* algae. *Biol. Lett.* 2:593-6. [4.3.2]

Hatcher, B. G. (1977). An apparatus for measuring photosynthesis and respiration of intact large marine algae and comparison of results with those from experiments with tissue segments. *Mar. Biol.* 43:381-5. [5.7.1]

Hatcher, B. G., A. R. O. Chapman, and K. H. Mann (1977). An annual carbon budget for the kelp *Laminaria longicruris*. *Mar Biol* 44: 85-96. [5.7.1, 5.7.4, 6.8.4]

Haug, A. (1976). The influence of borate and calcium on the gel formation of a sulfated polysaccharide from *Ulva lactuca*. *Acta Chem. Scand.* B30:562-6. [5.5.2]

Haug, A., and B. Larsen (1974). Biosynthesis of algal polysaccharides. In J. B. Pridham (ed), *Plant Carbohydrate Biochemistry* (pp. 207-18). New York: Academic Press. [5.5.3]

Hauri, C., K. E. Fabricius, B. Schaffelke, and C. Humphrey (2010). Chemical and physical environmental conditions underneath mat- and canopy-forming macroalgae, and their effects on understorey corals. *PLoS ONE* 5:e12685.

Hawkes, M. W. (1990). Reproductive strategies. In K. M. Cole and R. G. Sheath (eds), *Biology of the Red Algae* (pp. 455-76). Cambridge: Cambridge University Press. [2.2]

Hawkins, S. J., and A. J. Southward (1992). The Torrey Canyon oil spill: recovery of rock shore communities. In G.W. Thayer (ed.). *Restoring the Nation's Environment* (pp. 583-631). College Park, MD: Maryland Sea Grant Coll. [9.4, 9.4.3]

Hawkins, S. J., H. E. Sugden, N. Mieszkowskz, *et al.* (2009). Consequences of climate-driven biodiversity changes for ecosystem functioning of North European rocky shores. *Mar. Ecol. Prog. Ser.* 396:245-59. [3.5.1]

Haxen, P.G., and O. A. M. Lewis (1981). Nitrate assimilation in the marine kelp, *Macrocystis angustifolia* (Phaeophyceae). *Bot. Mar.* 24:631-5. [6.5.1]

Haxo, F. T., and L. R. Blinks (1950). Photosynthetic action spectra of marine algae. *J. Gen. Physiol.* 33:389-425. [5.3.4]

Hay, M. E. (1981a). The functional morphology of turf-forming seaweeds: persistence in stressful marine habitats. *Ecology* 62:739-50. [4.3.2, 4.4.2]

Hay, M. E. (1981b). Spatial patterns of grazing intensity on a Caribbean barrier reef: herbivory and algal distribution. *Aquat. Bot.* 11: 97-109. [4.3.2]

Hay, M. E. (1986). Functional geometry of seaweeds: ecological consequences of thallus layering and shape in contrasting light environments. In: Givnish, T. J. (ed) *On the Economy of Plant Form and Function* (pp. 635-66). Cambridge: Cambridge University Press. [5.3.2, 5.3.3]

Hay, M. E. (1988). Associational plant defenses and the maintenance of species diversity: turning competitors into accomplices. *Am. Nat.* 128:617–41. [4.3.2]

Hay, M. E. (2009). Marine chemical ecology: chemical signals and cues structure marine populations, communities, and ecosystems. *Annu. Rev. Mar. Sci.* 1:193–212. [4.4]

Hay, M. E., J. D. Parker, D. E. Burkepile, *et al.* (2004). Mutualisms and aquatic community structure: the enemy of my enemy is my friend. *Annu. Rev. Ecol. Evol. Syst.* 35:175–97. [4.5.1]

Hay, M. E., J. J. Stachowicz, E. Cruz-Rivera, *et al.* (1998). Bioassays with marine and freshwater macroorganisms. In K. F. Haynes and J. G. Millar (eds), *Methods in Chemical Ecology, Vol. 2: Bioassay Methods* (pp. 39–141). New York: Chapman and Hall. [4.3.2]

Hay, M. E., Q. E. Kappel, and W. Fenical (1994). Synergisms in plant defenses against herbivores: interactions of chemistry, calcification, and plant quality. *Ecology* 75:1714–26. [4.3.2, 6.5.2]

Hay, M. E., V. J. Paul, S. M. Lewis, *et al.* (l988). Can tropical seaweeds reduce herbivory by growing at night? Diel patterns of growth, nitrogen content, herbivory, and chemical versus morphological defenses. *Oecologia* 75:233–45. [4.4.2]

Hays, C. G. (2007). Adaptive phenotypic differentiation across the intertidal gradient in the alga *Silvetia compressa*. *Ecology* 88:149–57. [3.1.3]

He, P., and C. Yarish (2006). The developmental regulation of mass cultures of free-living conchocelis for commercial net seeding of *Porphyra leucosticta* from North America. *Aquaculture* 257:373–81. [10.2.2, 10.2.4]

He, P., S. Xu, H. Zhang, *et al.* (2008). Bioremediation efficiency in the removal of dissolved inorganic nutrients by the red seaweed, *Porphyra yezoensis*, cultivated in the open sea. *Water Res.* 42:1281–9. [10.9]

Heaton, T. H. E. (1986). Isotopic studies of nitrogen pollution in the hydrosphere and atmosphere: a review. *Chem Geol.* 59:87–102. [9.6.3]

Hebert, P.D.N., A. Cywinska, S. L. Ball, and J. R. deWaard (2003). Biological identifications through DNA barcodes. *Proc. R. Soc. Lond. B* 270: 313–22. [Essay 1]

Heesch, S., A. F. Peters, J. E. Broom, and C. L. Hurd (2008). Affiliation of the parasite *Herpodiscus durvillaeae* (Phaeophyceae) with the Sphacelariales based on DNA sequence comparisons and morphological observations. *Eur. J. Phycol.* 43:283–95. [4.5.2]

Heesch, S., G. Y. Cho, A. F. Peters, *et al.* (2010). A sequence-tagged genetic map for the brown alga *Ectocarpus siliculosus* provides large-scale assembly of the genome sequence. *New Phytol.* 188:42–51. [1.4.1]

Hegemann, P. (2008). Algal sensory photoreceptors. *Annu. Rev. Plant Biol.* 59:167–89. [1.3.3, 2.3.3, 2.6.2]

Hein, M., M. F. Pedersen, and K. Sand-Jensen (1995). Size-dependent nitrogen uptake in micro- and macroalgae. *Mar. Ecol. Prog. Ser.* 118:247–53. [6.4.2, 6.5.1, 6.8.1]

Hellebust, J. A. (1976). Osmoregulation. *Annu. Rev. Plant Physiol.* 27: 485–505. [7.4.2]

Helmuth, B., and M. W. Denny (2003). Predicting wave exposure in the rocky intertidal zone: do bigger waves always lead to larger forces?. *Limnol. Oceanogr.* 48:1338–45. [8.1.4]

Helmuth, B., N. Mieszkowska, P. Moore, and S. J. Hawkins (2006). Living on the edge of two changing worlds: forecasting the responses of rocky intertidal ecosystems to climate change. *Annu. Rev. Ecol. Evol. Syst.* 37:373–404. [3.5.1]

Hemmi, A., T. Honkanen, and V. Jormalainen (2004). Inducible resistance to herbivory in *Fucus vesiculosus*: duration, spreading and variation with nutrient availability. *Mar. Ecol. Prog. Ser.* 273:109–20. [4.4.1]

Hemminga, M. A., and C. M. Duarte (eds) (2000). *Seagrass Ecology*. Cambridge: Cambridge University Press, 298 pp. [3.3.1]

Henry, E. C. (1988). Regulation of reproduction in brown algae by light and temperature. *Bot. Mar.* 31:353–7. [2.3.4]

Henry, E. C., and K. M. Cole (1982). Ultrastructure of swarmers in the Laminariales. I. Zoospores. *J. Phycol.* 18:550–69. [2.5.2]

Hepburn, C. D., and C. L. Hurd (2005). Conditional mutualism between the giant kelp *Macrocystis pyrifera* and colonial epifauna. *Mar. Ecol. Prog. Ser.* 302:37–48. [4.2.2, 4.5.1]

Hepburn, C. D., C. L. Hurd, and R. D. Frew (2006). Colony structure and seasonal differences in light and nitrogen modify the impact of sessile epifauna on the giant kelp *Macrocystis pyrifera* (L.) C Agardh. *Hydrobiologia* 560:373–84. [4.2.2]

Hepburn, C. D., J. D. Holborrow, S. R. Wing, R. D. Frew, and C. L. Hurd (2007). Exposure to waves enhances the growth rate and nitrogen status of the giant kelp *Macrocystis pyrifera*. *Mar. Ecol. Prog. Ser.* 339:99–108. [8.1.4, 8.3.1]

Hepburn, C. D., D. W. Pritchard, C. E. Cornwall, *et al.* (2011). Diversity of carbon use strategies in a kelp forest community: implications for a high CO_2 ocean. *Glob. Change Biol.* 17:2488–97. [Essay 4, 7.6]

Hepburn, C. D., R. D. Frew, and C. L. Hurd (2012). Uptake and transport of nitrogen derived from sessile epifauna in the giant kelp *Macrocystis pyrifera*. *Aquat. Biol.* 14:121–8. [6.6]

Herbert, R. A. (1999). Nitrogen cycling in coastal marine ecosystems. *FEMS Microbiol. Rev.* 23:563–90. [6.2]

Herbert, S. K. (1990). Photoinhibition resistance in the red alga *Porphyra perforata*. The role of photoinhibition repair. *Plant Physiol.* 92:514-19. [5.3.3]

Herbert, S. K., and J. R. Waaland (1988). Photoinhibition of photosynthesis in a sun and a shade species of the red algal genus *Porphyra*. *Mar. Biol.* 97:1-7. [5.3.3]

Hernández, I., M. Christmas, J. M. Yelloly, and B. A. Whitton (1997). Factors affecting surface alkaline phosphatase activity in the brown alga *Fucus spiralis* at a North Sea intertidal site (Tyne Sands, Scotland). *J. Phycol.* 33:569-75. [6.5.2]

Hernández, I., F. X. Niell, and B. A. Whitton (2002). Phosphatase activity of benthic marine algae. An overview. *J. Appl. Phycol.* 14:475-87. [6.5.2]

Heyward, A. J., and A. P. Negri (1999). Natural inducers for coral larval metamorphosis. *Coral Reefs* 18:273-9. [3.3.1]

Higgins, H. W., and D. J. Mackey (1987). Role of *Ecklonia radiata* (C. Ag.) J. Agardh in determining trace metal availability in coastal waters. I. Total trace metals. *Aust. J. Mar. Freshw. Res.* 38:307-15. [9.2, 9.3.2]

Hill, N. A., A. R. Pepper, M. L. Poutinen, *et al.* (2010). Quantifying wave exposure in shallow temperate reef systems: applicability of fetch models for predicting algal biodiversity. *Mar. Ecol. Prog. Ser.* 417:83-95. [8.1.4]

Hillebrand, H., D. S. Gruner, E. T. Borer, *et al.* (2007). Consumer versus resource control of producer diversity depends on ecosystem type and producer community structure. *P. Natl. Acad. Sci. USA* 104:10,904-9. [4.3.1]

Hillis, L. (1997). Coral reefs from a calcareous green alga perspective, and a first carbonate budget. *Proc. 8th Int. Coral Reef Symp.* 1:761-6. [3.3.1]

Hillis-Colinveaux, L. (1985). *Halimeda* and other deep forereef algae at Enewetak Atoll. In V. M. Harmelin and B. Salvat (eds), *Proceedings of the 5th International Coral Reef Congress, Tahiti* (vol. 5, pp. 9-14). Moorea, French Polynesia: Antenne Museum-EPHE. [3.2]

Hillman, W. S. (1976). Biological rhythms and physiological timing. *Annu. Rev. Plant Physiol.* 27:159-79. [5.7.2]

Hincha, D. K. (2002). Cryoprotectin: a plant lipid-transfer protein homolog that stabilizes membranes during freezing. *Phil. Trans. R. Soc. London B* 357:909-15. [7.3.4]

Hinds, P.A., and D. L. Ballentine (1987). Effects of the Caribbean threespot damselfish, *Stegastes planifrons* (Cuvier), on algal lawn composition. *Aquat. Bot.* 27:299-308. [4.3.2]

Hinojosa, I. A., M. Pizarro, M. Ramos, and M. Thiel (2010). Spatial and temporal distribution of floating kelp in the channels and fjords of southern Chile. *Estuar. Coast. Shelf S.* 87:367-77. [3.3.7]

Hiscock, K. (1983). Water movement. In R. Earll and D. G. Erwin (eds), *Sublittoral Ecology* (pp. 58-96). Oxford: Clarendon Press. [8.1.3]

Ho, Y. B. (1990). Metals in *Ulva lactuca* in Hong Kong intertidal waters. *Bull. Mar. Sci.* 47:79-85. [9.3.2]

Hoarau, G., J. A. Coyer, and J. L. Olsen (2009). Paternal leakage of mitochondrial DNA in a *Fucus* (Phaeophyceae) hybrid zone. *J. Phycol.* 45:621-4. [1.4.2]

Hoegh-Guldberg, O., P. J. Mumby, A. J. Hooten, *et al.* (2007). Coral reefs under rapid climate change and ocean acidification. *Science* 318:1737-42. [5.7.5]

Hoffmann, A. J., and P. Camus (1989). Sinking rates and viability of spores from benthic algae in central Chile. *J. Exp. Mar. Biol. Ecol.* 126:281-91. [2.5.1]

Hofmann, L. C., G. Yildiz, D. Hanelt, and K. Bischof (2012a). Physiological responses of the calcifying rhodophyte, *Corallina officinalis* (L.), to future CO_2 levels. *Mar. Biol.* 159:783-92. [7.7, 7.9]

Hofmann, L. C., S. Straub, and K. Bischof (2012b). Competitive interactions between calcifying and noncalcifying temperate marine macroalgae under elevated CO_2 levels: a mesocosm study. *Mar. Ecol. Prog. Ser.* 464:89-105. [7.7]

Holbrook, N. M., M. W. Denny, and M. A. R. Koehl (1991). Intertidal "trees": consequences of aggregation on the mechanical and photosynthetic properties of sea-palms *Postelsia palmaeformis* Ruprecht. *J. Exp. Mar. Biol. Ecol.* 146:39-67. [8.3.1]

Holmes, M. A., M. T. Brown, M. W. Loutit, and K. Ryan (1991). The involvement of epiphytic bacteria in zinc concentration by the red alga *Gracilaria sordida*. *Mar. Environ. Res.* 31:56-67. [9.3.2]

Hommersand, M. H., and S. Fredericq (1990). Sexual reproduction and cystocarp development. In K. M. Cole and R. G. Sheath (eds) *Biology of the Red Algae* (pp. 305-45). Cambridge: Cambridge University Press. [2.4]

Hong, Y., C. H. Sohn, M. Polne-Fuller, and A. Gibor (1995). Differential display of tissue-specific messenger RNAs in *Porphyra perforata* (Rhodophyta) thallus. *J. Phycol.* 31:640-3. [1.1.1]

Honkanen, T., and V. Jormalainen (2005). Genotypic variation in tolerance and resistance to fouling in the brown alga *Fucus vesiculosus*. *Oecologia* 144:196-205. [4.2.2]

Hood, D. W., A. Schoener, P. K. Park, and I. W. Duedall (1989). Evolution of at-sea scientific monitoring strategies. In D. W. Hood, A. Schoener, and P. K. Park (eds), *Oceanic Processes in Marine Pollution. Vol. 4: Scientific Monitoring Strategies for Ocean Waste Disposal* (pp. 4-28). Malabau, FL: E. W. Krieger Publishing. [Fig. 9.1]

Hooper, R. G., and G. R. South (1977). Distribution and ecology of *Papenfussiella callitricha* (Rosenv.) Kylin (Phaeophyceae, Chordariaceae). *Phycologia* 16:153-7. [7.2.1, 7.3.4]

Hooper, R. G., E. C. Henry, and R. Kuhlenkamp (1988). *Phaeosiphoniella cryophila* gen. et sp. nov., a third member of the Tilopteridales (Phaeophyceae). *Phycologia* 27:395-404 [7.3.4]

Hop, H., C. Wiencke, B. Vögele, and N. A. Kovaltchouk (2012). Species composition, zonation, and biomass of marine benthic macroalgae in Kongsfjorden, Svalbard. *Bot. Mar.* 55:399-414. [3.3.3]

Hopkin, R., and J. M. Kain (1978). The effects of some pollutants on the survival, growth and respiration of *Laminaria hyperborea*. *Estu. Cstl. Mar. Sci.* 7:531-53. [9.3.4, 9.4.2, 9.6.2]

Hou, X., and X. Yan (1998). Study on concentration and seasonal variation of inorganic elements in 35 species of marine algae. *Sci. Total Environ.* 222:141-56. [9.3.2]

Hou, X. L., H. Dahlgaard, and S. P. Nielsen (2000). Iodine-129 time series in Danish, Norwegian and northwest Greenland coast and the Baltic Sea by seaweed. *Est. Coast. Shelf Sci.* 52:571-84. [9.7]

Howard, B. M., and W. Fenical (1981). The scope and diversity of terpenoid biosynthesis by the marine alga *Laurencia*. *Prog. Phytochem.* 7:263-300. [4.4.1]

Howard, R. J., K. R. Gayler, and B. R. Grant (1975). Products of photosynthesis in *Caulerpa simpliciuscula*. *J. Phycol.* 11: 463-71. [5.4.2]

Howe, C. J., A. C. Barbrook, R. E. R. Nisbet, P. J. Lockhart, and A. W. D. Larkum (2008). The origin of plastids. *Philos. T. Roy. Soc. B.* 363:2675-85. [1.3.2]

Hoyer, K., U. Karsten, T. Sawall, and C. Wiencke (2001). Photoprotective substances in Antarctic macroalgae and their variation with respect to depth distribution, different tissues and developmental stages. *Mar. Ecol. Prog. Ser.* 211:117-29. [7.6]

Hsiao, S. I.-C. (1969). Life history and iodine nutrition of the marine brown alga *Petalonia fascia* (O. F. Mull.) Kuntze. *Can. J. Bot.* 47:1611-16. [2.6.2]

Huang, I., J. Rominger, and H. Nepf (2011). The motion of kelp blades and the surface renewal model. *Limnol. Oceanogr.* 56:1453-62. [8.2.1]

Hubbard, C. B., D. J. Garbary, K. Y. Kim, and D. M. Chiasson (2004). Host specificity and growth of kelp gametophytes symbiotic with filamentous red algae (Ceramiales, Rhodophyta). *Helgoland Mar. Res.* 58:18-25. [4.5.2]

Hughes, J. A., and S. P. Otto (1999). Ecology and evolution of biphasic life cycles. *Am. Nat.* 154:306-20. [2.2]

Hughes, T. P. (1994). Catastrophes, phase shifts, and large-scale degradation of a Caribbean coral reef. *Science* 265:1547-51. [4.3.1]

Hughes, T. P, D. C. Reed, and M.-J. Boyle (1987). Herbivory on coral reefs: community structure following mass mortality of sea urchins. *J. Exp. Mar. Biol. Ecol.* 113:39-60. [4.3.1]

Hughes, T., A. M. Szmant, R. Steneck, R. Carpenter, and S. Miller (1999). Algal blooms on coral reefs: what are the causes? *Limnol. Oceanogr.* 44:1583-6. [6.8.5]

Huisman, J.M. (2000). *Marine Plants of Australia.* Nedlands, Australia: University of Western Australia Press. [Essay 3]

Hulatt, C. J., D. N. Thomas, D. G. Bowers, L. Norman, and C. Zhang (2009). Exudation and decomposition of chromophoric dissolved organic matter (CDOM) from some temperate macroalgae. *Estuar. Coast Shelf Sci.* 84: 147-53. [5.7.2]

Hunter, K. A., and R. Strzepek (2008). Iron cycle. In S.E. Jorgensen and B.D. Fath (eds), *Global Ecology, Vol. 3 of Encyclopedia of Ecology* (pp. 2028-33).Oxford: Elsevier. [6.5.5]

Huovinen, P., and I. Gómez (2011). Spectral attenuation of solar radiation in Patagonian fjords and coastal waters and implications for algal photobiology. *Cont. Shelf Res.* 31:254-9. [5.2.1, 5.2.2, 5.3.2]

Huovinen, P., and I. Gómez (2012). Cold temperate seaweed communities of the southern hemisphere. In C. Wiencke, and K. Bischof (eds), *Seaweed Biology: Novel Insights Into Ecophysiology, Ecology and Utilization* (Series: Ecological Studies, Volume 219, pp. 293-313). Berlin and Heidelberg: Springer. [3.3, 3.3.2, 3.3.3, 7.3.6, 7.3.8]

Huovinen, P., I. Gómez, F. L. Figueroa, *et al.* (2004). UV absorbing mycosporine-like amino acids in red macroalgae from Chile. *Bot. Mar.* 47:21-9. [7.6]

Huovinen, P., I. Gómez, and M. Orostegui (2007). Patterns and UV sensitivity of carbonic anhydrase and nitrate reductase activities in south Pacific macroalgae. *Mar. Biol.* 151:1813-21. [5.4.1]

Huovinen, P., P. Leal, and I. Gómez (2010). Impact of interaction of copper, nitrogen and UV radiation on the physiology of three south Pacific kelps. *Mar. Freshw. Res.* 61:330-41. [7.7, 7.9]

Huppertz, K., D. Hanelt, and W. Nultsch (1990). Photoinhibition of photosynthesis in the marine brown alga *Fucus serratus* as studied in field experiments. *Mar. Ecol. Prog. Ser.* 66:175-82. [5.3.3, 7.5.1]

Hurd, C. L. (2000). Water motion, marine macroalgal physiology, and production. *J. Phycol.* 36:453-72. [6.4.2, 8.1.3, 8.2.1]

Hurd, C. L., and M. J. Dring (1990). Phosphate uptake by intertidal fucoid algae in relation to zonation and season. *Mar. Biol.* 107:281-9. [6.5.2]

Hurd, C. L., and M. J. Dring (1991). Desiccation and phosphate uptake by intertidal fucoid algae in relation to zonation. *Br. Phycol. J.* 26:327-33. [6.4.2, 7.5.1]

Hurd, C. L., and C. A. Pilditch (2011). Flow-induced morphological variations affect diffusion boundary-layer thickness of *Macrocystis pyrifera* (Heterokontophyta, Laminariales). *J. Phycol.* 47:341–51. [8.1.3, 8.1.4, 8.2.1]

Hurd, C. L., and C. L. Stevens (1997). Flow visualization around single- and multiple-bladed seaweeds with various morphologies. *J. Phycol.* 33:360–7. [8.1.3]

Hurd, C.L., R. S. Galvin, T. A. Norton, and M. J. Dring (1993). Production of hyline hairs by intertidal species of *Fucus* (Fucales) and their role in phosphate uptake. *J. Phycol.* 29:160–5. [2.6.2, 6.5.2, 6.8.1]

Hurd, C. L., K. M. Durante, F. S. Chia, and P. J. Harrison (1994a). Effect of bryozoan colonization on inorganic nitrogen acquisition by the kelps *Agarum fimbriatum* and *Macrocystis integrifolia*. *Mar. Biol.* 121:167–73. [4.2.2]

Hurd, C. L., M. Quick, C. L. Stevens, *et al.* (1994b). A low-volume flow tank for measuring nutrient uptake by large macrophytes. *J. Phycol.* 30:892–6. [8.1.4]

Hurd, C.L., J. A. Berges, J. Osbourne, and P. J. Harrison (1995). An *in vitro* nitrate reductase assay for marine macroalgae: optimization and characterization of the enzyme for *Fucus gardneri* (Phaeophyta). *J. Phycol.* 31:835–43. [6.5.1]

Hurd, C. L., Harrison, P. J., and L. D. Druehl (1996). Effect of seawater velocity on inorganic nitrogen uptake by morphologically distinct forms of *Macrocystis integrifolia* from wave-sheltered and exposed sites. *Mar. Biol.* 126:205–14. [8.2.1]

Hurd, C. L., C. L. Stevens, B. E. Laval, *et al.* (1997). Visualization of seawater flow around morphologically distinct forms of the giant kelp *Macrocystis integrifolia* from wave-sheltered and exposed sites. *Limnol. Oceanogr.* 42:156–63. [8.1.4, 8.2.1]

Hurd, C. L., C. D. Hepburn, K. I. Currie, J. A. Raven, and K. A. Hunter (2009). Testing the effects of ocean acidification on algal metabolism: considerations for experimental designs. *J. Phycol.* 45:1236–51. [7.7]

Hurd, C. L., C. E. Cornwall, K. Currie, *et al.* (2011). Metabolically-induced pH fluctuations by some coastal calcifiers exceed projected 22nd century ocean acidification: a mechanism for differential susceptibility? *Glob. Change Biol.* 17:3254–62. [7.7, 8.2.1]

Hurlbert, S. H. (1984). Pseudoreplication and the design of ecological field experiments. *Ecol. Monogr.* 54:187–211. [1.1.2]

Hutchins, L. W. (1947). The basis for temperature zonation in geographical distribution. *Ecol. Monogr.* 17:325–35. [7.3.8]

Hwang, E. K., and M. J. Dring (2002). Quantitative photoperiodic control of erect thallus production in *Sargassum muticum*. *Bot. Mar.* 45:471–5. [2.3.3]

Hyndes, G. A., and P. S. Lavery (2005). Does transported seagrass provide an important trophic link in unvegetated, nearshore areas? *Est. Coast. Shelf Sci.* 63 (4):633–43. [Essay 3]

Hyndes, G. A., P. S. Lavery, and C. Doropoulos (2012). Dual processes for cross-boundary subsidies: incorporation of nutrients from reef-derived kelp into a seagrass ecosystem. *Mar. Ecol. Prog. Ser.* 445:97–107. [Essay 3]

Ianora, A., M. Boersma, R. Casotti, *et al.* (2006). The H. T. Odum synthesis essay: new trends in marine chemical ecology. *Estuar. Coast.* 29:531–51. [4.4]

Ikawa, M., T Watanabe, and K. Nisizawa (1972). Enzymes involved in the last steps of the biosynthesis of mannitol in brown algae. *Plant Cell Physiol.* 3:1017–29. [5.4.2]

Iken, K. (2012). Grazers on benthic seaweeds. In C. Wiencke and K. Bischof (eds), *Seaweed Biology: Novel Insights Into Ecophysiology, Ecology and Utilization* (Series: Ecological Studies, Volume 219, pp. 157–75). Berlin and Heidelberg: Springer. [4.3.2, 4.4]

Ilvessalo, H., and J. Tuomi (1989). Nutrient availability and accumulation of phenolic compounds in the brown alga *Fucus vesiculosus*. *Mar. Biol.* 101:115–19. [6.5.1]

Inderjit, D. Chapman, M. Ranelletti, and S. Kaushik (2006). Invasive marine algae: an ecological perspective. *Bot. Rev.* 72:153–78. [3.4]

Innes, D. J. (1988). Genetic differentiation in the intertidal zone in populations of the alga *Enteromorpha linza* (Ulvales: Chlorophyta). *Mar. Biol.* 97:9–16. [7.3.3]

Inouye, R. S., and W. M. Schaffer (1981). On the ecological meaning of ratio (de Wit) diagrams in plant ecology. *Ecology* 62: 1679–81. [4.2.3]

IPCC (2007). Climate Change 2007: The physical science basis. Summary for policymakers. Contribution of working group I to the fourth assessment report. *The Intergovernmental Panel on Climate Change*, http://www.ipcc.ch/SPM2feb07.pdf. [7.3.6]

Ireland, H. E., S. J. Harding, G. A. Bonwick, *et al.* (2004). Evaluation of heat shock protein 70 as a biomarker of environmental stress in *Fucus serratus* and *Lemna minor*. *Biomarkers* 9:139–55. [7.3.5]

Irving, A. D., and S. D. Connell (2006a). Predicting understorey structure from the presence and composition of canopies: an assembly rule for marine algae. *Oecologia* 148:491–502. [3.1.2]

Irving, A. D., and S. D. Connell (2006b). Physical disturbance by kelp abrades erect algae from the understorey. *Mar. Ecol. Prog. Ser.* 324:127–37. [4.2.1]

Irving, A. D., D. Balata, F. Colosio, G. A. Ferrando, and L. Airoldi (2009). Light, sediment, temperature, and the early life-history of the habitat-forming alga *Cystoseira barbata*. *Mar. Biol.* 156:1223–31. [8.3.2]

Irwin, S., and J. Davenport (2002). Hyperoxic boundary layers inhabited by the epiphytic meiofauna of *Fucus serratus*. *Mar. Ecol. Prog. Ser.* 244:73-9. [8.2.1]

Ishikawa, M., F. Takahashi, and H. Nozaki (2009). Distribution and phylogeny of the blue light receptors aureochromes in eukaryotes. *Planta* 230:543-52. [2.3.3, 2.6.2]

ISO (2006). Water quality - Marine algal growth inhibition test with *Skeletonema costatum* and *Phaeodactylum tricornutum*. ISO 10253:2006. [9.2]

ISO (2010). Water quality — Growth inhibition test with the marine and brackish water macroalga *Ceramium tenuicorne*. ISO 10710:2010. [9.2]

Israel, A., I. Levy, and M. Friedlander (2006). Experimental tank cultivation of *Porphyra* in Israel. *J. Appl. Phycol.* 18:235-40. [10.7]

Iwamoto, K., and T. Ikawa (1997). Glycolate metabolism and subcellular distribution of glycolate oxidase in *Patoglossum pacificum* (Phaeophyceae, Chromophyta). *Phycol. Res.* 45:77-83. [5.4.2]

Jackson, G. A. (1984). The physical and chemical environment of a kelp community. In W. Bascom (ed.), *The Effects of Waste Disposal on Kelp Communities* (pp. 11-37). Long Beach, CA: So. Calif. Coastal Water Res Proj. [5.7.4]

Jackson, G. A. (1987). Modelling the growth and harvest yield of the giant kelp *Macrocystis pyrifera*. *Mar. Biol.* 95:611-24. [5.3.2]

Jackson, G. A., D. E. James, and W. J. North (1985). Morphological relationships among fronds of giant kelp *Macrocystis pyrifera* off La Jolla, California. *Mar. Ecol. Prog. Ser.* 26:261-70. [5.7.4]

Jacobs, J. P. (1993). A search for some angiosperm hormones and their metabolites in *Caulerpa paspaloides* (Chlorophyta). *J. Phycol.* 29:595-600. [2.6.3]

Jacobs, W. P. (1970). Develoment and regeneration of the algal giant coenocyte *Caulerpa*. *Ann. N. Y. Acad. Sci.* 175:732-48. [2.6.4]

Jacobs, W. P. (1994). *Caulerpa*. *Scientific American*, December 1994. [2.6.4]

Jacobs, W. P., and J. Olson (1980). Developmental changes in the algal coenocyte *Caulerpa prolifera* (Siphonales) after inversion with respect to gravity. *Am. J. Bot.* 67: 141-6. [2.6.4]

Janouškovec, J., S.-L. Liu, P. T. Martone, et al. (2013). Evolution of red algal plastid genomes: ancient architectures, introns, horizontal gene transfer, and taxonomic utility of plastid markers. *PLoS ONE* 8(3), e59001. [1.1.1]

Jaschinski, S., D. C. Brepohl, and U. Sommer (2008). Carbon sources and trophic structure in an eelgrass *Zostera marina* bed, based on stable isotope and fatty acid analyses. *Mar. Ecol. Prog. Ser.* 358:103-14. [Essay 3]

Jassby, A. D., and T. Platt (1976). Mathematical formulation of the relationship between photosynthesis and light for phytoplankton. *Limnol. Oceanogr.* 21:540-7. [5.3.3]

Jékely, G. (2009). Evolution of phototaxis. *Phil. Trans. R. Soc. B.* 364:2795-808. [1.3.3]

Jelinek, R., and S. Kolusheva (2004). Carbohydrate biosensors. *Chem. Rev.* 104:5987-6015. [1.3.1]

Jenkins, S. R., P. Moore, M. T. Burrows, et al. (2008). Comparative ecology of North Atlantic shores: do differences in players matter for process? *Ecology* 89:S3-S23. [4.3.1]

Jennings, J. G., and P. D. Steinberg (1997). Phlorotannins versus other factors affecting epiphyte abundance on the kelp *Ecklonia radiata*. *Oecologia* 109:461-73. [4.2.2]

Jensen, A. (1995). Production of alginate. In W. Wiessner, E. Schnepf, and R. C. Star (eds) *Algae, Environment and Human Affairs* (pp. 79-92). Bristol, UK: Biopress Ltd. [10.3.2]

Jensen, A., and A. Haug (1956). Geographical and seasonal variation in the chemical compositon of *Laminaria hyperborea* and *Laminaria digitata* from the Norwegian Coast. Norwegian Institute of Seaweed Research, Report 14. [5.5.1]

Jensen, R. G., and J. T. Bahr (1977). Ribulose 1,5-bisphosphate carboxylase-oxygenase. *Annu. Rev. Plant Physiol.* 28:379-400. [5.7.2]

Jerlov, N. G. (1970). Light: general introduction. In O. Kinne (ed), *Marine Ecology* (vol. I, pt. I, pp. 95-102). New York: Wiley. [5.2.2]

Jerlov, N. G. (1976). *Marine Optics*. Amsterdam: Elsevier, 231 pp. [5.2.2]

Jiménez, C., T. Berl, C. J. Rivard, C. L. Edelstein, and J. M. Capasso (2004). Phosphorylation of MAP kinase-like proteins mediate the response of the halotolerant alga *Dunaliella viridis* to hypertonic shock. *Biochim. Biophys. Acta: Mol. Cell Res.* 1644: 61-9. [7.8]

Jin, Q., and S. Dong (2003). Comparative studies on the allelopathic effects of two different strains of *Ulva pertusa* on *Heterosigma akashiwo* and *Alexandrium tamarense*. *J. Exp. Mar. Biol Ecol.* 293:41-55. [4.2.2]

Jin, Q., S. Dong, and C. Wang (2005). Allelopathic growth inhibition of *Prorocentrum micans* (Dinophyta) by *Ulva pertusa* and *Ulva linza* (Chlorophyta) in laboratory cultures. *Eur. J. Phycol.* 40:31-7. [4.2.2]

Johannesson, K. (1989). The bare zone of Swedish rocky shores: why is it there? *Oikos* 54:77-86. [3.1.1]

Johansson, G., and P. Snoeijs (2002). Macroalgal photosynthetic responses to light in relation to thallus morphology and depth zonation. *Mar. Ecol. Prog. Ser.* 244:63-72. [5.3.2]

Johansson, G., P. A. Sosa, and P. Snoeijs (2003). Genetic variability and level of differentiation in North Sea and

Baltic Sea populations of the green alga *Cladophora rupestris*. *Mar. Biol.* 142:1019-27. [7.2.3]

Johnson, A. S. (2001). Drag, drafting, and mechanical interactions in canopies of the red alga *Chondrus crispus*. *Biol. Bull.* 201:126-35. [8.3.1]

Johnson, A., and M. Koehl (1994). Maintenance of dynamic strain similarity and environmental stress factor in different flow habitats: thallus allometry and material properties of a giant kelp. *J. Exp. Biol.* 195:381-410. [8.3.1]

Johnson, C. H. (2010). Circadian clocks and cell division. *Cell Cycle* 9:3864-73. [1.3.5]

Johnson, C. R., and A.R.O. Chapman (2007). Seaweed invasions: introduction and scope. *Bot. Mar.* 50:321-5. [3.4]

Johnson, C. R., S. C. Banks, N. S. Barrett, *et al.* (2011). Climate change cascades: shifts in oceanography, species' ranges and subtidal marine community dynamics in eastern Tasmania. *J. Exp. Mar. Biol. Ecol.* 400:17-32. [4.3.1]

Johnson, L. E., and S. H. Brawley (1998). Dispersal and recruitment of a canopy-forming intertidal alga: the relative roles of propagule availability and post-settlement processes. *Oecologia* 117:517-26. [2.5.1]

Johnson, M. P., S. J. Hawkins, R. G. Hartnoll, and T. A. Norton (1998). The establishment of fucoid zonation on algal-dominated rocky shores: hypotheses derived from a simulation model. *Funct. Ecol.* 12:259-69. [3.1.3]

Johnson, W. S., A. Gigon, S. L. Gulmon, and H. A. Mooney (1974). Comparative photosynthetic capacities of intertidal algae under exposed and submerged conditions. *Ecology* 55:450-3. [7.5.1]

Johnston, A. M., and J. A. Raven (1986). Dark carbon fixation studies on the intertidal macroalga *Ascophyllum nodosum* (Phaeophyta). *J. Phycol.* 22:78-83. [5.4.3]

Johnston, A. M., and J. A. Raven (1990). Effects of culture in high CO_2 on the photosynthetic physiology of *Fucus serratus*. *Br. Phycol. J.* 25:75-82. [5.4.1]

Johnston, A. M., S. C. Maberly, and J. A. Raven (1992). The acquisition of inorganic carbon by four red macroalgae from different habitats. *Oecologia* 92:317-26. [5.4.1]

Joint, I., M. E. Callow, J. A. Callow, and K. R. Clarke (2000). The attachment of *Enteromorpha* zoospores to a bacterial biofilm assemblage. *Biofouling* 16:151-8. [2.5.1]

Joint, I., K. Tait, and M. E. Callow (2002). Cell-to-cell communication across the prokaryote–eukaryote boundary. *Science* 298:1207. [2.5.1]

Joly, A. B., and E. C. de Oliveira Filho (1967). Two Brazilian *Laminaria*. *Publ. Inst. Pesq. Mar.* 4:1-13. [7.3.8]

Jones, A. B., W. C. Dennison, and G. R. Stewart (1996). Macroalgal responses to nitrogen source and availability: amino

acid metabolic profiling as a bioindicator using *Gracilaria edulis* (Rhodophyta). *J. Phycol.* 32:757-66. [6.5.1]

Jones, G. P., L. Santana, L. J. McCook, and M. I. McCormick (2006). Resource use and impact of three herbivorous damselfishes on coral reef communities. *Mar. Ecol. Prog. Ser.* 328:215-24. [4.3.2]

Jones, J. L., J. A. Callow, and J. R. Green (1988). Monoclonal antibodies to sperm surface antigens of the brown alga *Fucus serratus* exhibit region-, gamete-, species- and genus preferential binding. *Planta* 176:298-306. [1.3.1, 2.4]

Jónsson, S., M.-H. Laur, and L. Pham-Quang (1985). Mise en evidence de differents types de glycoproteines dans un extrait inhibiteur de la gametogenese chez *Enteromorpha prolifera*, Chlorophyceae marine. *Crypt. Algol.* 6:253-64. [2.3.3]

Jormalainen, V., and T. Honkanen (2008). Macroalgal chemical defenses and their roles in structuring temperate marine communities. In C. D. Amsler (ed.), *Algal Chemical Ecology* (pp. 57-90). London: Springer (Limited). [4.4]

Jormalainen, V., T. Honkanen, and N. Heikkila (2001). Feeding preferences and performance of a marine isopod on seaweed hosts: cost of habitat specialization. *Mar. Ecol. Prog. Ser.* 220:219-30. [Essay 3]

Jormalainen, V., T. Honkanen, R. Koivikko, *et al.* (2003). Induction of phlorotannin production in a brown alga: defense or resource dynamics? *Oikos* 103:640-50. [4.4.2]

Joska, M. A. P., and J. J. Bolton (1987). *In situ* measurement of zoospore release and seasonality of reproduction in *Ecklonia maxima* (Alariaceae, Laminariales). *Br. Phycol. J.* 22:209-14. [5.7.3]

Josselyn, M. N., and A. C. Mathieson (1978). Contribution of receptacles from the fucoid *Ascophyllum nodosum* to the detrital pool of a north temperate estuary. *Estuaries* 1:258-61. [5.7.3]

Juanes, J. A., X. Guinda, A. Puente, and J. A. Revilla (2008). Macroalgae, a suitable indicator of the ecological status of coastal rocky communities in the NE Atlantic. *Ecol. Indicators* 8:351-9. [3.5.1]

Jung, V., and G. Pohnert (2001). Rapid wound-activated transformation of the green algal defensive metabolite caulerpenyne. *Tetrahedron* 57:7169-72. [2.6.4]

Kaczmarska, I., and L. L. Dowe (1997). Reproductive biology of the red alga *Polysiphonia lanosa* (Ceramiales) in the Bay of Fundy, Canada. *Mar. Biol.* 128:695-703. [2.4]

Kagami, Y., Y. Mogi, T. Arai, *et al.* (2008). Sexuality and uniparental inheritance of chloroplast DNA in the isogamous green alga *Ulva compressa* (Ulvophyceae). *J. Phycol.* 44:691-702. [1.3.2]

Kai, T., K. Nimura, H. Yasui, and H. Mizuta (2006). Regulation of sorus formation by auxin in Laminariales sporophyte. *J. Appl. Phycol.* 18:95-101. [2.6.3]

Kain, J. M. (1969). The biology of *Laminaria hyperborea*. V. Comparison with early stages of competitors. *J. Mar. Biol. Assoc. UK* 49:455–73. [7.3.3]

Kain, J. M. (1979). A view of the genus *Laminaria*. *Oceanogr. Mar. Biol. Annu. Rev.* 17:101–61. [5.7.4]

Kain, J. M. (1989). The seasons in the subtidal. *Brit. Phycol. J.* 24:203–15. [2.3.1]

Kain (Jones), J. M. (2006). Photoperiodism in *Delesseria sanguinea* (Ceramiales, Rhodophyta) 2. Daylengths are shorter underwater. *Phycologia* 45:624–31. [2.3.3]

Kain (Jones), J. M. (2008). Winter favours growth and survival of *Ralfsia verrucosa* (Phaeophyta) in high intertidal rockpools in southeast Australia. *Phycologia* 47:498–509. [3.3.4]

Kain (Jones), J. M., and C. Destombe (1995). A review of the life history, reproduction and phenology of *Gracilaria*. *J. Appl. Phycol.* 7:269–81. [1.4.2]

Kaiser, M. J., M. J. Attrill, S. Jennings, *et al.* (2011). *Marine Ecology: Processes, Systems and Impacts*. Oxford: Oxford University Press, 528 pp. [3.1.2, 3.3.5]

Kajiwara, T., A. Hatanaka, Y. Tanaka, *et al.* (1989). Volatile constituents from marine brown algae of Japanese *Dictyopteris*. *Phytochemistry* 28:636–8. [2.4]

Kalle, K. (1945). *Der Stoffhaushalt der Meere*. Leipzig, Germany: Becker & Erler, 263 pp. [Fig. 5.16]

Kalle, K. (1971). Salinity: general introduction. In O. Kinne (ed.), *Marine Ecology* (vol. I, pt. 2, pp. 683–8). New York: Wiley. [7.4]

Kalle, K. (1972). Dissolved gases: general introduction. In O. Kinne (ed.), *Marine Ecology* (vol. I, pt. 3, pp. 1451–7). New-York: Wiley. [5.4.1]

Kamiya, M., and J. A. West (2010). Investigations on reproductive affinities in red algae. In J. Seckbach and D. J. Chapman (eds), *Red Algae in the Genomic Age (Cellular Origin, Life in Extreme Habitats and Astrobiology 13)* (pp. 77–109). Dordrecht, The Netherlands: Springer. [1.4.1, 1.4.2, 2.2]

Kapraun, D. F. (2005). Nuclear DNA content estimates in multicellular green, red and brown algae: phylogenetic considerations. *Ann. Bot.-London.* 95:7–44. [1.4.2]

Kapraun, D. E., and P. W. Boone (1987). Karyological studies of three species of Scytosiphonaceae (Phaeophyta) from coastal North Carolina. *J. Phycol.* 23:318–22. [1.3.5, 2.2]

Kapraun, D. E., and D. J. Martin (1987). Karyological studies of three species of *Codium* (Codiales, Chlorophyta) from coastal North Carolina. *Phycologia* 26:228–34. [2.2]

Karez, R., S. Engelbert, and U. Sommer (2000). "Co-consumption" and "protective coating": two new proposed effects of epiphytes on their macroalgal hosts in mesograzer-epiphyte–host interactions. *Mar. Ecol. Prog. Ser.* 205:85–93. [4.2.2]

Karlsson, J., and B. Eklund (2004). New biocide-free antifouling paints are toxic. *Mar. Poll. Bull.* 49:456–64. [9.5]

Karlsson, J., M. Breitholtz, and B. Eklund (2006). A practical ranking system to compare toxicity of antifouling paints. *Mar. Poll. Bull.* 52:1661–7. [9.5]

Karlsson, J., E. Ytreberg, and B. Eklund (2010). Toxicity of antifouling paints for use on pleasure boats and vessels to non-target organisms representing three trophic levels. *Environ. Poll.* 158:681–7. [9.5]

Karsten, U. (2007). Salinity tolerance of Arctic kelps from Spitsbergen. *Phycol. Res.* 55:257–62. [7.4.4, 7.9]

Karsten, U., and G. O. Kirst (1989). Incomplete turgor pressure regulation in the "terrestrial" red alga, *Bostrychia scorpioides* (Huds.). *Mont. Plant Sci.* 61:29–36. [7.4.2]

Karsten, U., R. J. King, and G. O. Kirst (1990). The distribution of D-sorbitol and D-dulcitol in the red algal genera *Bostrychia* and *Stictosiphonia* (Rhodomelaceae, Rhodophyta) a re-evaluation. *Br. Phycol. J.* 25:363–6. [5.4.2]

Karsten, U., K. D. Barrow, O. Nixdorf, J. A. West, and R. J. King (1997). Characterization of mannitol metabolism in the mangrove red alga *Caloglossa leprieurii* (Montagne) J. Agardh. *Planta* 201:173–8. [5.4.2]

Karsten, U., D. Michalik, M. Michalik, and J. A. West (2005). A new unusual low molecular weight carbohydrate in the red algal genus *Hypoglossum* (Delesseriaceae, Ceramiales) and its possible function as osmolyte. *Planta* 222:319–26. [7.4.2]

Karsten, U., S. Görs, A. Eggert, and J. A. West (2007). Trehalose, digeneaside and floridoside in the Florideophyceae (Rhodophyta) – a re-evaluation of its chemotaxonomic value. *Phycologia* 46:143–50. [7.4.2, 7.4.4, 7.9]

Karyophyllis, D., C. Katsaros, and B. Galatis (2000). F-actin involvement in apical cell morphogenesis of *Sphacelaria rigidula* (Phaeophyceae): mutual alignment between cortical actin filaments and cellulose microfibrils. *Eur. J. Phycol.* 35:195–203. [2.6.1]

Katoh, T., and T. Ehara (1990). Supramolecular assembly of fucoxanthin-chlorophyll-protein complexes isolated from a brown alga, *Petalonia fascia*. Electron microscopic studies. *Plant Cell Physiol.* 31:439–47. [5.3.1]

Katsaros, C., and B. Galatis (1988). Thallus development in *Dictyopteris membranacea* (Phaeophyta, Dictyotales). *Br. Phycol. J.* 23:71–88. [2.6.2]

Katsaros, C., D. Karyophyllis, and B. Galatis (2006). Cytoskeleton and morphogenesis in brown algae. *Ann. Bot.-London* 97:679–93. [1.3, 1.3.3]

Katsaros, C., T. Motomura, C. Nagasato, and B. Galantis (2009). Diaphragm development in cytokinetic vegetative cells of brown algae. *Bot. Mar.* 52:150–61. [1.3.5]

Kavanaugh, M. T., K. J. Nielsen, F. T. Chan, *et al.* (2009). Experimental assessment of the effects of shade on an intertidal kelp: do phytoplankton blooms inhibit growth of open-coast macroalgae? *Limnol. Oceanogr.* 54:276-88. [4.2.2]

Kawai, H., D. G. Müller, E. Fölster, and D. P Häder (1990). Phototactic responses in the gametes of the brown alga, *Ectocarpus siliculosus. Planta* 182:292-7. [1.3.3]

Kawai, H., S. Nakamura, M. Mimuro, M. Furuya, and M. Watanabe (1996). Microspectrofluorometry of the auto-fluorescent flagellum in phototactic brown algal zoids. *Protoplasma* 191:172-7. [1.3.3]

Kawamata, S. (1998). Effect of wave-induced oscillatory flow on grazing by a subtidal sea urchin *Strongylocentrotus nudus* (A. Agassiz). *J. Exp. Mar. Biol. Ecol.* 224:31-48. [4.3.2]

Kawamata, S. (2001). Adaptive mechanical tolerance and dislodgement velocity of the kelp *Laminaria japonica* in wave-induced water motion. *Mar. Ecol. Prog. Ser.* 211:89-104. [8.3.1]

Kawamata, S. (2010). Inhibitory effects of wave action on destructive grazing by sea urchins: a review. *Bull. Fish. Res. Agen.* 32:95-102. [4.3.2]

Kawamata, S., S. Yoshimitsu, S. Tokunaga, S. Kubo, and T. Tanaka (2012). Sediment tolerance of *Sargassum* algae inhabiting sediment-covered rocky reefs. *Mar. Biol.* 159:723-33. [8.3.2]

Kawamitsu, Y., and J. S. Boyer (1999). Photosynthesis and carbon storage between tides in a brown alga, *Fucus vesiculosus. Mar. Biol.* 133:361-9. [5.4.1]

Kawashima, S. (1984). Kombu cultivation in Japan for human foodstuff. *Jpn. J. Phycol.* 32:379-94. [10.3, 10.3.1]

Keats, D. W., M. A. Knight, and C. M. Pueschel (1997). Antifouling effects of epithallial shedding in three crustose coralline algae (Rhodophyta, Coralinales) on a coral reef. *J. Exp. Mar. Biol. Ecol.* 213:281-93. [4.2.2]

Keeling, P. J. (2009). Chromalveolates and the evolution of plastids by secondary endosymbiosis. *J. Eukaryot. Microbiol.* 56:1-8. [1.1.1]

Keeling, P. J. (2010). The endosymbiotic origin, diversification and fate of plastids. *Phil. Trans. R. Soc. B.* 365:729-48. [1.1.1]

Keiji, I., and H. Kanji (1989). Seaweeds: chemical composition and potential food uses. *Food Rev. Internat.* 5:101-44. [10.1, 10.2, 10.10]

Keiter, S., T. Braunbeck, S. Heise, *et al.* (2009). A fuzzy logic-classification of sediments based on data from *in vitro* biotests. *J. Soils Sed.* 9:168-79. [9.2]

Kelly, G. J. (1989). A comparison of marine photosynthesis with terrestrial photosynthesis: a biochemical perspective. *Oceanogr. Mar. Biol. Annu. Rev.* 27:11-44. [5.7]

Kendrick, G. A. (1991). Recruitment of coralline crusts and filamentous turf algae in the Galapagos Archipelago: effect of simulated scour, erosion and accretion. *J. Exp. Mar. Biol. Ecol.* 147:47-63. [8.3.2]

Kennelly, S. J. (1987a). Inhibition of kelp recruitment by turfing algae and consequences for an Australian kelp community. *J. Exp. Mar. Biol. Ecol.* 112:49-60. [8.3.2]

Kennelly, S. J. (l987b). Physical disturbances in an Australian kelp community. I. Temporal effects. *Mar. Ecol. Prog. Ser.* 40: 145-53. [8.3.2]

Kenyon, K. W., and D. W. Rice (1959). Life history of the Hawaiian monk seal. *Pac. Sci.* 13:215-52. [3.3.8]

Keogh, S. M., A. Aldahan, and G. Possnert (2007). Trends in the spatial and temporal distribution of ^{129}I and ^{99}Tc in coastal waters surrounding Ireland using *Fucus vesiculosus* as a bio-indicator. *J. Environ. Radioact.* 95:23-38. [9.7]

Kerby, N. W., and J. A. Raven (1985). Transport and fixation of inorganic carbon by marine algae. *Adv. Bot. Res.* 11:71-123. [5.4.1, 5.4.2]

Kerby, N. W., and L. V. Evans (1983). Phosphoenolpyruvate carboxykinase activity in *Ascophyllum nodosum* (Phaeophyceae). *J Phycol* 19:1-3. [5.4.3]

Kerr, R. A. (2011). First detection of ozone hole recovery claimed. *Science* 332:160. [7.6]

Kershaw, P. J., D. McCubbin, and K. S. Leonard (1999). Continuing contamination of north Atlantic and Arctic waters by Sellafield radionuclides. *Sci. Total Envir.* 237/238:119-32. [9.7]

Kerswell, A. P. (2006). Global biodiversity patterns of benthic marine algae. *Ecology* 87:2479-88. [4.4.2]

Keser, M., J. T. Swenarton, and J. F. Foertch (2005). Effects of thermal input and climate change on growth of *Ascophyllum nodosum* (Fucales, Phaeophyceae) in eastern Long Island Sound. *J. Sea Res.* 54:211-20. [9.8]

Kessler, W. S. (2006). The circulation of the eastern tropical Pacific: a review. *Prog. Oceanogr.* 69:181-217. [7.3.7]

Khailov, K. M., and Z. P. Burlakova (1969). Release of dissolved organic matter by marine seaweeds and distribution of their total organic production to inshore communities. *Limnol. Oceanogr.* 14:521-7. [5.7.3]

Khailov, K. M., V. I. Kholodov, Y. K. Firsov, and A. V. Prazukin (1978). Thalli of *Fucus vesiculosus* in ontogenesis: changes in morpho-physiological parameters. *Bot. Mar.* 21:289-311. [5.7.2]

Kharlamenko, V. I., S. I. Kiyashko, A. B. Imbs, and D. I. Vyshkvartzev (2001). Identification of food sources of invertebrates from the seagrass *Zostera marina* community using carbon and sulfur stable isotope ratio and fatty acid analyses. *Mar. Ecol. Prog. Ser.* 220:103-17. [Essay 3]

Khotimchemko, S.V. (2003). The fatty acid composition of glycolipids of marine macrophytes. *Russian J. Mar. Biol.* 29(2):126–8. [Essay 3]

Kilar, J. A., M. M. Littler, and D. S. Littler (1989). Functional-morphological relationships in *Sargassum polyceratium* (Phaeophyta): phenotypic and ontogenetic variability in apparent photosynthesis and dark respiration. *J. Phycol.* 25:713–20. [5.7.2]

Kim, G. H., I. K. Lee, and L. Fritz (1996). Cell–cell recognition during fertilization in the red alga, *Aglaothamnion sparsum* (Ceramiaceae, Rhodophyta). *Plant Cell Physiol.* 37:621–8. [2.4]

Kim, G. H., T. A. Klotchkova, B.-C. Lee, and S. H. Kim (2001a). FITC-phalloidin staining of F-actin in *Aglaothamnion oosumiense* and *Griffithsia japonica* (Rhodophyta). *Bot. Mar.* 44:501–8. [1.3.3]

Kim, G. H., T. A. Klotchkova, and Y. M. Kang (2001b). Life without a cell membrane: regeneration of protoplasts from disintegrated cells of the marine green alga *Bryopsis plumosa. J. Cell Sci.* 114:2009–14. [2.6.4]

Kim, G. H., T. A. Klotchkova, and J. A. West (2002). From protoplasm to swarmer: regeneration of protoplasts from disintegrated cells of the multicellular marine green alga *Microdictyon umbilicatum* (Chlorophyta). *J. Phycol.* 38:174–83. [2.6.4]

Kim, G. H., T. A. Klochkova, K.-S. Yoon, Y.-S. Song, and K. P. Lee (2005). Purification and characterization of a lectin, bryohealin, involved in the protoplast formation of a marine green alga *Bryopsis plumosa* (Chlorophyta). *J. Phycol.* 42:86–95. [2.6.4]

Kim J. K., G. P. Kraemer, and C. Yarish (2013). Integrated multi-tropic aquaculture in the United States. In T. Chopin, A. Neori, S. Robinson, and M. Troell (eds), *Integrated Multi-Trophic Aquaculture (IMTA).* New York: Springer Science. [10.10]

Kim, S.-H., and G. H. Kim (1999). Cell–cell recognition during fertilization in the red alga, *Aglaothamnion oosumiense* (Ceramiaceae, Rhodophyta). *Hydrobiologia* 398/399:81–9. [2.4]

Kimura, K., C. Nagasto, K. Kogame, and T. Motomura (2010). Disappearance of male mitochondrial DNA after the four-cell stage in sporophytes of the isogamous brown alga *Scytosiphon lomentaria* (Scytosiphonaceae, Phaeophyceae). *J. Phycol.* 46:143–52. [1.3.2]

Kindig, A. C., and M. M. Littler (1980). Growth and primary productivity of marine macrophytes exposed to domestic sewage effluents. *Mar. Environ. Res.* 3:81–100. [9.6.1]

King, R. J. (1984). Oil pollution and marine plant systems. In M. H. Cheng and C. D. Field (eds). *Pollution and Plants* (pp. 127–42). Melbourne: Insearch Ltd. [9.4.1]

King, R. J. (1990). Macroalgae associated with the mangrove vegetation of Papua New Guinea. *Bot. Mar.* 33:55–62. [7.4.4]

Kingham, D. L., and L. V. Evans (1986). The *Pelvetia-Mycosphaerella* interrelationship. In S. T. Moss (ed.), *The Biology of Marine Fungi* (pp. 177–87). Cambridge: Cambridge University Press. [4.5.1]

Kingsford, M., and C. Battershill (eds) (1998). *Studying Temperate Marine Environments: A Handbook for Ecologists.* Christchurch, NZ: Canterbury University Press, 335 pp. [3.5.1]

Kinne, O. (1970). Temperature: general introduction. In O. Kinne (ed), *Marine Ecology* (vol. I, pt. 1, pp. 321–46). New York: Wiley.

Kirk, J. T. O. (2010). *Light and Photosynthesis in Aquatic Ecosystems*, 3rd edn. Cambridge: Cambridge University Press. 649 pp. [5.2.1, 5.2.2, 5.3.1, 5.3.3]

Kirkman, H. (1981). The first year in the life history and the survival of the juvenile marine macrophyte, *Ecklonia radiata* (Turn.) 1. Agardh. *J. Exp. Mar. Biol. Ecol.* 55:243–54. [8.3.2]

Kirst, G. O. (1988). Turgor pressure regulation in marine macroalgae. In C. S. Lobban, D. J. Chapman, and B. P. Kremer (eds), *Experimental Phycology: A Laboratory Manual* (pp. 203–9). Cambridge: Cambridge University Press. [7.4.1]

Kirst, G. O. (1990). Salinity tolerance of eukaryotic marine algae. *Annu. Rev. Plant Physiol. Plant Mol. Biol.* 41:21–53. [7.4.2]

Kirst, G. O., and C. Wiencke (1995). Ecophysiology of polar algae. *J. Phycol.* 31:181–99. [7.3.1]

Kitade, Y., M. Nakamura, T. Uji, *et al.* (2008). Structural features and gene-expression profiles of actin homologs in *Porphyra yezoensis* (Rhodophyta). *Gene* 423:79–84. [1.3.3, 2.2]

Kitagawa, D., I. Vakonakis, N. Olieric, *et al.* (2011). Structural basis of the 9-fold symmetry of centrioles. *Cell* 144:364–75. [1.3.3]

Klenell, M., P. Snoeijs, and M. Pedersen (2002) The involvement of a plasma membrane H+-ATPase in the blue-light enhancement of photosynthesis in *Laminaria digitata* (Phaeophyta). *J. Phycol.* 38:1143–9. [5.3.4]

Klerks, P. L., and J. S. Weis (1987). Genetic adaptation to heavy metals in aquatic organisms: a review. *Environ. Pollut.* 45:173–205. [9.3.3]

Kling, R., and M. Bodard (1986). La construction du thalle de *Gracilaria verrucosa* (Rhodophyceae, Gigartinales): edification de la fronde; essai d'interpretation phylogenetique. *Crypt. Algol.* 7:231–46. [1.2.1]

Klinger, T., and R. E. De Wreede (1988). Stipe rings, age, and size in populations of *Laminaria setchellii* Silva

(Laminariales, Phaeophyta) in British Columbia, Canada. *Phycologia* 27:234-40. [3.5.2]

Kloareg, B., and B. R. S. Quatrano (1988). Structure of the cell walls of marine algae and ecophysiological functions of the matrix polysaccharides. *Oceanogr. Mar. Biol. Annu. Rev.* 26:259-315. [1.3.1, 5.5.2]

Kloareg, B., M. Demarty, and S. Mabeau (1986). Polyanionic characteristics of purified sulphated homofucans from brown algae. *Int. J. Biol. Macromol.* 8:380-6. [Fig. 1.8]

Kloareg, B., M. Demarty, and S. Mabeau (1987). Ion-exchange properties of isolated cell walls of brown algae: the interstitial solution. *J. Exp. Bot.* 38:1652-62. [6.3.2]

Klotchkova, T. A., O.-K. Chah, J. A. West, and G. H. Kim (2003). Cytochemical and ultrastructural studies on protoplast formation from disintegrated cells of the marine alga *Chaetomorpha aerea* (Chlorophyta). *Eur. J. Phycol.* 38:205-16. [2.6.4]

Klumpp, D. W., and A. D. McKinnon (1989). Temporal and spatial patterns in primary productivity of a coral reef epilithic algal community. *J. Exp. Mar. Biol. Ecol.* 131:1-22. [3.3.1]

Klumpp, D. W., and N. V. C. Polunin (1989). Partitioning among grazers of food resources within damselfish territories on a coral reef. *J. Exp. Mar. Biol. Ecol.* 125:145-69. [4.3.2]

Klumpp, D. W., D. McKinnnon, and P. Daniel (1987). Damselfish territories: zones of high productivity on coral reefs. *Mar. Ecol. Prog. Ser.* 40:41-51. [4.3.2]

Knight, M., and M. Parke (1950). A biological study of *Fucus vesiculosus* L. and *F serratus* L. *J. Mar. Biol. Assoc. UK* 29:439-514. [5.7.3]

Knoop, W. T., and G. C. Bate (1990). A model for the description of photosynthesis-temperature responses by subtidal Rhodophyta. *Bot. Mar.* 33:165-71. [7.3.2]

Knoth, A., and C. Wiencke (1984). Dynamic changes of protoplasmic volume and of fine structure during osmotic adaptation in the intertidal red alga *Porphyra umbilicalis. Plant Cell Envir.* 7:113-19. [7.4.2]

Koch, E. W. (1994). Hydrodynamics, diffusion-boundary layers and photosynthesis of the seagrasses *Thalassia testudinum* and *Cymodocea nodosa. Mar. Biol.* 118:767-76. [8.1.4]

Koehl, M. A. R. (1982). The interaction of moving water and sessile organisms. *Sci. Am.* 247(6): 124-35. [8.3.1]

Koehl, M. A. R. (1986). Seaweeds in moving water: form and mechanical function. In T. J. Givnish (ed.), *On the Economy of Plant Form and Function* (pp. 603-34). Cambridge: Cambridge University Press. [8.1.3, 8.3.1]

Koehl, M. A. R., and R. S. Alberte (1988). Flow, flapping, and photosynthesis of *Nereocystis luetkeana*: a functional comparison of undulate and flat blade morphologies. *Mar. Biol.* 99:435-44. [8.1.4, 8.2.1, 8.3.1]

Koehl, M. A. R., and S. A. Wainwright (1977). Mechanical adaptations of a giant kelp. *Limnol. Oceanogr.* 22:1067-71. [8.3.1]

Koehl, M. A. R., and S. A. Wainwright (1985). Biomechanics. In M. M. Littler and D. S. Littler (eds) *Handbook of Phycological Methods: Ecological Field Methods: Macroalgae* (pp. 291-313). Cambridge: Cambridge University Press. [8.3.1]

Koehl, M. A. R., W. K. Silk, H. Liang, and L. Mahadevan (2008). How kelp produce blade shapes suited to different flow regimes: a new wrinkle. *Integr. Comp. Biol.* 48:834-51. [8.2.1, 8.3.1]

Kohlmeyer, J., and E. Kohlmeyer (1972). Is *Ascophyllum* lichenized? *Bot. Mar.* 15:109-12. [4.5.1]

Kohlmeyer, J., and M. W. Hawkes (1983). A suspected case of mycophycobiosis between *Mycosphaerella apophlaeae* (Ascomycetes) and *Apophlaea* spp. (Phodophyta). *J. Phycol.* 19:257-60. [4.5.1]

Kongelschatz, J., L. Solórzano, R. Barber, and P. Mendoza (1985). Oceanographic conditions in the Galapagos Islands during the 1982/1983 El Niño. In G. Robinson and E. M. del Pino (eds) *El Niño en las Islas Galapagos. El evento de 1982/1983*. Fund (pp. 91-123). Quito: Charles Darwin. [7.3.7]

Konstantinou, I. K., and T. A. Albanis (2004). Worldwide occurrence and effects of antifouling paint booster biocides in the aquatic environment: a review. *Environ. Internat.* 30:235-48. [9.5]

Kooistra, W. H. C. F., A. M. T. Joosten, and C. van den Hoek (1989). Zonation patterns in intertidal pools and their possible causes: a multivariate approach. *Bot. Mar.* 32:9-26. [3.3.4]

Kopczak, C. D., R. C. Zimmerman, and J. N. Kremer (1991). Variation in nitrogen physiology and growth among geographically isolated populations of the giant kelp, *Macrocystis pyrifera* (Phaeophyta). *J. Phycol.* 27:149-58. [1.1.3]

Kopp, D., Y. Bouchon-Navaro, S. Cordonnier, *et al.* (2010). Evaluation of algal regulation by herbivorous fishes on Caribbean coral reefs. *Helgoland Mar. Res.* 64:181-90. [3.3.1]

Korb, R. E., and V. A. Gerard (2000). Nitrogen assimilation characteristics of polar seaweeds from differing nutrient environments. *Mar. Ecol. Prog. Ser.* 198:83-92. [6.8.4, Fig. 6.25]

Kornmann, P. (1970). Advances in marine phycology on the basis of cultivation. *Helgol. Meeresunters.* 20:39-61. [1.3.5]

Kornmann, P., and P.-H. Sahling (1977). Marine algae of Helgoland: benthic green brown and red algae. *Helgol. Meeresunters.* 29:1-289. [2.5.1]

Korpinen, S., V. Jormalainen, and T. Honkanen (2007). Bottom-up and cascading top-down control of macroalgae along a depth gradient. *J. Exp. Mar. Biol. Ecol.* 343:52-63. [4.3.1]

Kottmeier, S. T., and C. W. Sullivan (1988). Sea ice microbial communities (SIMCO). *Polar Biol.* 8:293-304. [7.3.1]

Kraan, S., and M. D. Guiry (2000). Molecular and morphological character inheritance in hybrids of *Alaria esculenta* and *A. praelonga* (Alariaceae, Phaeophyceae). *Phycologia* 39:554-9. [1.3.2]

Kraberg, A. C., and T. A. Norton (2007). Effect of epiphytism on reproductive and vegetative lateral formation in the brown, intertidal seaweed *Ascophyllum nodosum* (Phaeophyceae). *Phycol. Res.* 55:17-24. [4.2.2]

Kraemer, G. P., and D. J. Chapman (1991a). Biomechanics and alginic acid composition during hydrodynamic adaptation by *Egregia menziesii* (Phaeophyceae) juveniles. *J. Phycol.* 27:47-53. [8.2.1, 8.3.1]

Kraemer, G. P., and D. J. Chapman (1991b). Effects of tensile force and nutrient availability on carbon uptake and cell wall synthesis in blades of juvenile *Egregia menziesii* (Turn.) Aresch. (Phaeophyceae). *J. Exp. Mar. Biol. Ecol.* 149:267-77. [8.2.1]

Kraufvelin, P., F. E. Moy, H. Christie, and T. L. Bokn (2006). Nutrient addition to experimental rocky shore communities revisited: delayed responses, rapid recovery. *Ecosystems* 9:1076-93. [6.8.5]

Kraufvelin, P., A. Lindholm, M. F. Pedersen, L. A. Kirkerud, and E. Bonsdorff (2010). Biomass, diversity and production of rocky shore macroalgae at two nutrient enrichment and wave action levels. *Mar. Biol.* 157:29-47. [6.8.5]

Krause, G. H., and E. Weis (1991). Chlorophyll fluorescence and photosynthesis: the basics. *Annu. Rev. Plant Physiol. Plant Mol. Biol.* 42: 313-49. [5.7.1]

Kregting, L. T., C. L. Hurd, C. A. Pilditch, and C. L. Stevens (2008a). The relative importance of water motion on nitrogen uptake by the subtidal macroalga *Adamsiella chauvinii* (Rhodophyta) in winter and summer. *J. Phycol.* 44:320-30. [6.4.2]

Kregting, L. T., C. D. Hepburn, C. L. Hurd, and C. A. Pilditch (2008b). Seasonal patterns of growth and nutrient status of the macroalga *Adamsiella chauvinii* (Rhodophyta) in soft sediment environments. *J. Exp. Mar. Biol. Ecol.* 360:94-102. [8.3.1]

Kregting, L. T., C. L. Stevens, C. D. Cornelisen, C. A. Pilditch, and C. L. Hurd (2011). Effects of a small-bladed macroalgal canopy on benthic boundary layer dynamics: implications for nutrient transport. *Aquat. Biol.* 14:41-56. [8.1.4]

Kreimer, G. (2001). Light perception and signal modulation during photoorientation of flagellate green algae. In D.-P.

Häder and M. Lebert (eds), *Photomovement; Comprehensive Series in Photosciences* (Vol. 1, pp. 193-227). Amsterdam: Elsevier. [1.3.3]

Kreimer, G., H. Kawai, D. G. Müller, and M. Melkonian (1991). Reflective properties of the stigma in male gametes of *Ectocarpus siliculosus* (Phaeophyceae) studied by confocal laser scanning microscopy. *J. Phycol.* 27:268-76. [1.3.3]

Kremer, B. P. (1977). Biosynthesis of polyols in *Pelvetia canaliculata*. *Z. Pflanzenphysiol.* 81:68-73. [5.4.2]

Kremer, B. P. (1981a). Carbon metabolism. In C. S. Lobban and M. J. Wynne (eds) *The Biology of Seaweeds* (pp. 493-533). Oxford: Blackwell Scientific. [5.4.1, 5.4.2, 7.3.2]

Kremer, B. P. (1981b). Metabolic implications of nonphotosynthetic carbon fixation in brown macroalgae. *Phycologia* 20:242-50. [5.4.3]

Kremer, B. P. (1983). Carbon economy and nutrition of the alloparasitic red alga *Harveyella mirabilis*. *Mar. Biol.* 76:231-9. [5.4.1]

Kremer, B. P. (1985). Aspects of cellular compartmentation in brown marine macroalgae. *J. Plant Physiol.* 120:401-7. [5.4.2]

Kremer, B. P., and J. W. Markham (1982). Primary metabolic effects of cadmium in the brown alga *Laminaria saccharina*. *Z. Pflanzenphysiol.* 1008:125-30. [9.3.4]

Kroeker, K. J., R. J. Kordas, R. N. Crim, and G. G. Singh (2010). Meta-analysis reveals negative yet variable effects of ocean acidification on marine organisms. *Ecol. Lett.* 13: 1419-34. [7.6]

Krom, M. D., S. Ellner, J. van Rijin, and A. Neori (1995). Nitrogen and phosphorus cycling and transformations in a prototype "non-polluting" integrated mariculture system, Eilat, Israel. *Mar. Ecol. Prog. Ser.* 118:25-36. [10.9]

Krumhansl, K. A., J. M. Lee and R. S. Scheibling (2011). Grazing damage and encrustation by an invasive bryozoan reduce the ability of kelps to withstand breakage by waves. *J. Exp. Mar. Biol. Ecol.* 407:12-18. [4.2.2]

Kubanek, J. K., S. E. Lester, W. Fenical, and M. E. Hay (2004). Ambiguous role of phlorotannins as chemical defenses in the brown alga *Fucus vesiculosus*. *Mar. Ecol. Prog. Ser.* 277:79-93. [Essay 3]

Kübler, J. E., and S. R. Dudgeon (1996). Temperature dependent change in the complexity of form of *Chondrus crispus* fronds. *J. Exp. Mar. Biol. Ecol.* 207:15-24. [2.6.2]

Kübler, J. E., and J. A. Raven (1994). Consequences of light limitation for carbon acquisition in three rhodophytes. *Mar. Ecol. Prog. Ser.* 110:203-9. [5.4.1]

Kübler, J. E., and J. A. Raven (1996). Nonequilibrium rates of photosynthesis and respiration under dynamic light supply. *J. Phycol.* 32:963-9. [8.3.1]

Kübler, J. E., A. M. Johnston, and J. A. Raven (1999). The effects of reduced and elevated CO_2 and O_2 on the seaweed *Lomentaria articulata*. *Plant Cell Environ.* 22:1303-10. [8.2.1]

Kucera, H., and G. W. Saunders (2008). Assigning morphological variants of *Fucus* (Fucales, Phaeophyceae) in Canadian waters to recognized species using DNA barcoding. *Botany* 86:1065-79. [Essay 1]

Kuffner, I. B., A. J. Andersson, P. L. Jokel, and K. S. Rogers (2008). Decreased abundance of crustose coralline algae due to ocean acidification. *Nature Geosci.* 1:114-17. [6.5.3]

Kumar, M., P. Kumari, V. Gupta, C. R. K. Reddy and B. Jha (2010a). Biochemical responses of red alga *Gracilaria corticata* (Gracilariales, Rhodophyta) to salinity induced oxidative stress. *J. Exp. Mar. Biol. Ecol.* 391:27-34. [7.8]

Kumar, M., P. Kumari, V. Gupta, P. A. Anisha, and C. R. K. Reddy (2010b). Differential responses to cadmium induced oxidative stress in marine macroalga *Ulva lactuca* (Ulvales, Chlorophyta). *Biometals* 23:315-25. [7.8, 9.3.4]

Kumar, M., V. Gupta, N. Trivedi, *et al.* (2011). Desiccation induced oxidative stress and its biochemical responses in intertidal red alga *Gracilaria corticata* (Gracilariales, Rhodophyta). *Envir. Exp. Bot.* 72:194-201.

Kumar, M., A. J. Bijo, R. S. Baghel, C. R. K. Reddy, and B. Jha (2012). Selenium and spermine alleviate cadmium induced toxicity in the red seaweed *Gracilaria dura* by regulating antioxidants and DNA methylation. *Plant Physiol. Biochem.* 51:129-38. [9.3.4]

Küpper, F. C., N. Schweigert, E. Ar Gall, *et al.* (1998). Iodine uptake in Laminariales involves extracellular, haloperoxidase-mediated oxidation of iodide. *Planta* 207:163-71. [6.5.6]

Küpper, F. C., D. G. Müller, A. F. Peters, B. Kloareg, and P. Potin (2002a). Oligoalginate recognition and oxidative burst play a key role in natural and induced resistance of sporophytes of Laminariales. *J. Chem. Ecol.* 28:2057-81. [4.2.2]

Küpper, H., I. Šetlik, M. Spiller, F. C. Küpper, and O. Prášil (2002b). Heavy metal-induced inhibition of photosynthesis: targets of in vivo heavy metal chlorophyll formation. *J. Phycol.* 38:429-41. [9.3.4]

Küpper, F. C., L. J. Carpenter, G. B. McFiggans, *et al.* (2008). Iodine accumulation provides kelp with inorganic antioxidant impacting atmospheric chemistry. *PNAS* 105:6954-8. [6.5.6]

Küppers, U., and B. P. Kremer (1978). Longitudinal profiles of carbon dioxide fixation capacities in marine macroalgae. *Plant Physiol.* 62:49-53. [5.7.2]

Küppers, U. and M. Weidner (1980). Seasonal variation of enzyme activities in *Laminaria hyperborea*. *Planta* 148:222-30. [7.3.1]

Kurihara, A., T. Abe, M. Tani, and A. R. Sherwood (2010). Molecular phylogeny and evolution of red algal parasites: a case study of *Benzaitenia*, *Janczewskia*, and *Ululania* (Ceramiales). *J. Phycol.* 46:580-90. [4.5.2]

Kurle, C. M., D. A. Croll, and B. R. Tershy (2008). Introduced rats indirectly change marine rocky intertidal communities from algae- to invertebrate-dominated. *PNAS* 105:3800-4. [4.3.1]

Kurogi, M., and K. Hirano (1956). Influences of water temperature on the growth, formation of monosporangia and monospore-liberation in the *Conchocelis*-phase of *Porphyra tenera* Kjellm. *Bull. Tohoku Reg. Fish. Res. Lab.* 8:45-61. [2.3.2]

Kutser, T., E. Vahtmäe, and G. Martin (2006). Assessing suitability of multispectral satellites for mapping benthic macroalgal cover in turbid coastal waters by means of model simulations. *Estuar. Coast. Shelf S.* 67:521-9. [3.5.1]

Kuwano, K., R. Sakurai, Y. Motozu, Y. Kitade, and N. Saga (2008). Diurnal cell division regulated by gating the G_1/S transition in *Enteromorpha compressa* (Chlorophyta). *J. Phycol.* 44:364-73. [1.3.5]

Kylin, H. (1956). *Die Gattungen der Rhodophyceen.* Lund, Sweden: Gleerups Förlag. [Essay 1]

La Claire, J. W., II (1982a). Cytomorphological aspects of wound healing in selected Siphonocladales (Chlorophyceae). *J. Phycol.* 18:379-84. [2.6.4]

La Claire, J. W., II (1982b). Wound-healing motility in the green alga *Ernodesmis*: calcium ions and metabolic energy are required. *Planta* 156:466-74. [2.6.4]

La Claire, J. W., II (1989). Actin cytoskeleton in intact and wounded coenocytic green algae. *Planta* 177:47-57. [2.6.4]

La Claire, J. W., II, and J. Wang (2000). Localization of plasmidlike DNA in giant-celled marine green algae. *Protoplasma* 213:157-64. [1.4.2]

Ladah, L. B., and J. A. Zertuche-González (2007). Survival of microscopic stages of a perennial kelp (*Macrocystis pyrifera*) from the center and the southern extreme of its range in the northern hemisphere after exposure to simulated El Niño stress. *Mar. Biol.* 152:677-86. [7.3.7]

Ladah, L. B., F. Feddersen, G. A. Pearson, and E. A. Serrão (2008). Egg release and settlement patterns of dioecious and hermaphroditic fucoid algae during the tidal cycle. *Mar. Biol.* 155:583-91. [2.3.2, 2.3.4]

Lago-Lestón, A., C. Mota, L. Kautsky, and P. A. Pearson (2010). Functional divergence in heat shock response following rapid speciation of *Fucus* spp. in the Baltic Sea. *Mar. Biol.* 157: 683-8. [7.3.5]

Lahaye, M., and A. Robic (2007). Structure and functional properties of ulvan, a polysaccharide from green seaweeds. *Biomacromol.* 8:1765-74. [1.3.1, 5.5.1, 5.5.3]

Lane, C. E., and G. W. Saunders (2005). Molecular investigation reveals epi/endophytic extrageneric kelp (Laminariales, Phaeophyceae) gametophytes colonizing *Lessoniopsis littoralis* thalli. *Bot. Mar.* 48:426-36. [Essay 1]

Lane, C. E., C. Mayes, L. D. Druehl, and G. W. Saunders (2006). A multi-gene molecular investigation of the kelp (Laminariales, Phaeophyceae) supports substantial taxonomic reorganization. *J. Phycol.* 42: 493-512. [Essay 1, 3.3.2]

Lane, C.E., S. Lindstrom, and G. W. Saunders (2007). A molecular assessment of northeast Pacific *Alaria* species (Laminariales, Phaeophyceae) with reference to the utility of DNA barcoding. *Mol. Phylogenet. Evol.* 44:634-48. [Essay 1]

Lang, J. C. (1974). Biological zonation at the base of a reef. *Amer. Sci.* 62: 271-81. [5.3.4]

Langer, G., G. Nehrke, I. Probert, J. Ly, and P. Ziveri (2009). Strain-specific responses of *Emiliania huxleyi* to changing seawater carbonate chemistry. *Biogeosci.* 6: 2637-46. [7.3.7]

Langford, T. E. L. (1990). *Ecological Effects of Thermal Discharges*. New York: Elsevier Applied Science, 468 pp. [9.8]

Langston, W. J. (1990). Toxic effects of metals and the incidence of metal pollution in marine ecosystems. In R. W. Furness and P.S. Rainbow (eds), *Heavy Metals in the Marine Environment* (pp. 101-22). Boca Raton, FL: CRC Press. [9.5, 9.8]

Lapointe, B. E. (1985). Strategies for pulsed nutrient supply to *Gracilaria* cultures in the Florida Keys: interactions between concentration and frequency of nutrient pulses. *J. Exp. Mar. Biol. Ecol.* 93:211-22. [5.5.2]

Lapointe, B. E. (1986). Phosphorus-limited photosynthesis and growth of *Sargassum natans* and *Sargassum fluitans* (Phaeophyceae) in the western North Atlantic. *Deep Sea Res.* 33:391-9. [6.5.2]

Lapointe, B. E. (1987). Phosphorus- and nitrogen-limited photosynthesis and growth of *Gracilaria tikvahiae* (Rhodophyceae) in the Florida Keys: an experimental field study. *Mar. Biol.* 93:561-8. [6.5.2]

Lapointe, B. E. (1997). Nutrient thresholds for bottom-up control of macroalgal blooms on coral reefs in Jamaica and southeast Florida. *Limnol. Oceanogr.* 42:1119-31. [6.2, 6.5.2, Essay 6]

Lapointe, B. E. (1999). Simultaneous top-down and bottom-up forces control macroalgal blooms on coral reefs (Reply to comment by Hughes *et al.*). *Limnol. Oceanogr.* 44:1586-92. [6.5.2, 6.8.5]

Lapointe, B. E., and B. J. Bedford (2010). Ecology of nutrition of invasive *Caulerpa brachypus f. parvifolia* blooms on coral reefs off southeast Florida, USA *Harmful Algae* 9:1-12. [6.8.5]

Lapointe, B. E., and C. S. Duke (1984). Biochemical strategies for growth of *Gracilaria tikvahiae* (Rhodophyta) in relation to light intensity and nitrogen availability. *J. Phycol.* 20:488-95. [6.7.2, 6.8.2]

Lapointe, B. E., and J. O'Connell (1989). Nutrient-enhanced growth of *Cladophora prolifera* in Harrington Sound, Bermuda: eutrophication of a confined, phosphorus-limited marine ecosystem. *Estu. Cstl. Shelf Sci.* 28:347-60. [9.6.1]

Lapointe, B. E., and J. H. Ryther (1979). The effects of nitrogen and seawater flow rate on the growth and biochemical composition of *Gracilaria foliifera* v. angustissima in mass outdoor cultures. *Bot. Mar.* 22:529-37. [6.5.1]

Lapointe, B. E., M. M. Littler, and D. S. Littler (1987). A comparison of nutrient-limited productivity in macroalgae from a Caribbean barrier reef and from a mangrove ecosystem. *Aquat. Bot.* 28:243-55. [6.8.2]

Lapointe, B. E., M. M. Littler, and D. S. Littler (1992). Nutrient availability to marine macroalgae in siliciclastic versus carbonate-rich coastal waters. *Estuaries* 15:75-82. [6.5.2, 6.8.2, 6.8.5, Essay 6]

Larkum, A. W. D., and J. Barrett (1983). Light-harvesting processes in algae. *Adv. Bot. Res.* 10:1-219. [1.3.2, 5.3.1, 5.3.4, 5.7.2]

Larkum, A. W. D., and M. Kühl (2005). Chlorophyll d: the puzzle resolved. *Trends Plant Sci.* 10:355-7. [5.3.1]

Larkum, A. W. D., and M. Vesk (2003). Algal plastids: their fine structure and properties. In A. W. D. Larkum, S. E. Douglas, and J. A. Raven (eds), *Photosynthesis in Algae. Advances in Photosynthesis and Respiration* (Vol. 14, pp. 11-28). Dordrecht, The Netherlands:Kluwer Academic Publishers. [1.3, 1.3.2, 5.3.1]

Larkum, A. W. D., E. A. Drew, and R. N. Crossett (1967). The vertical distribution of attached marine algae in Malta. *J. Ecol.* 55:361-71. [5.3.4]

Larkum, A. W. D., R. J. Orth, and C. M. Duarte (eds) (2006). *Seagrasses: Biology, Ecology and Conservation*. Dordrecht, The Netherlands: Springer, 691 pp. [1.1.1]

Larkum, A. W. D., A. Salih, and M. Kühl (2011). Rapid mass movement of chloroplasts during segment formation of the calcifying siphonalean green alga *Halimeda macroloba*. *PloS ONE* 6:e20841. [1.3.2, 4.4.2]

Larned, S. T., and M. J. Atkinson (1997). Effects of water velocity on NH_4 and PO_4 uptake and nutrient-limited growth in the macroalga *Dictyosphaeria cavernosa*. *Mar. Ecol. Prog. Ser.* 157:295-302. [8.1.4, 8.2.1]

Larsen, B., A. Haug, and T. J. Painter (1970). Sulphated polysaccharides in brown algae. III. The native state of fucoidan in *Ascophyllum nodosum* and *Fucus vesiculosus*. *Acta Chem. Scand.* 24:3339-52. [5.5.2]

Lartigue, J., and T. D. Sherman (2005). Response of *Enteromorpha* sp. (Chlorophyceae) to a nitrate pulse: nitrate uptake, inorganic nitrogen storage and nitrate reductase activity. *Mar. Ecol. Prog. Ser.* 292:147-57. [9.4.1, 9.5]

Lary, D. J. (1997). Catalytic destruction of stratospheric ozone. *J. Geophys. Res.* 102(21):515-26. [7.6]

Lasley-Rasher, R. S., D. B. Rasher, A. H. Marion, R. B. Taylor, and M. E. Hay (2011). Predation constrains host choice for a marine mesograzer. *Mar. Ecol. Prog. Ser.* 434:91-9. [4.3.2]

Lassuy, D. R. (1980). Effects of "farming" behavior by *Eupomacentrus lividus* and *Hemiglyphidodon plagiometopon* on algal community structure. *Bull. Mar. Sci.* 30:304-12. [4.3.2]

Lauzon-Guay, J.-S., and R. E. Scheibling (2007). Seasonal variation in movement, aggregation and destructive grazing of the green sea urchin (*Strongylocentrotus droebachiensis*) in relation to wave action and sea temperature. *Mar. Biol.* 151:2109-18. [4.3.1]

Lauzon-Guay, J.-S., and R. E. Scheibling (2010). Spatial dynamics, ecological thresholds and phase shifts: modeling grazer aggregation and gap formation in kelp beds. *Mar. Ecol. Prog. Ser.* 403:29-41. [4.3.1]

Lavery, P. S., and A. J. McComb (l991a). Macroalgal–sediment nutrient interactions and their importance to macroalgal nutrition in a eutrophic estuary. *Estu. Cstl. Shelf Sci.* 32:281-95. [6.2, 6.7.2]

Lavery, P. S., and A. J. McComb (l991b). The nutritional ecophysiology of *Chaetomorpha linum* and *Ulva rigida* in Peel Inlet, Western Australia. *Bot. Mar.* 34:251-60. [6.5.1, 6.5.2, 6.7.2, Table 6.3]

Lavery, P. S., and M. A. Vanderklift (2002). A comparison of spatial and temporal patterns in epiphytic macroalgal assemblages of the seagrasses *Amphibolis griffithii* and *Posidonia coriacea. Mar. Ecol. Prog. Ser.* 236:99-112. [Essay 3]

Lavery, P. S., R. J. Lukatelich, and A. J. McComb (1991). Changes in the biomass and species composition of macroalgae in a eutrophic estuary. *Estu. Cstl. Shelf Sci.* 33:1-22. [6.7.2]

Laws, E. A. (2000). *Aquatic Pollution: An Introductory Text.* New York: J. Wiley, 655 pp. [9.2, 9.3.1, 9.4, 9.4.3, 9.8]

Laycock, M. V., and J. S. Craigie (1977). The occurrence and seasonal variation of gigartinine and L-citrullinyl-L-arginine in *Chondrus crispus* Stackh. *Can. J. Biochem.* 55:27-30. [6.5.1]

Laycock, M. V., K. C. Morgan, and J. S. Craigie (1981). Physiological factors affecting the accumulation of L-citrullinyl-L-arginine in *Chondrus crispus. Can. J. Bot.* 59:522-7. [6.5.1, 6.8.2]

Le Bail, A., B. Billoud, C. Maisonneuve, *et al.* (2008). Early development pattern of the brown alga *Ectocarpus siliculosus* (Ectocarpales, Phaeophyceae) sporophyte. *J. Phycol.* 44:1269-81. [2.6, 2.6.1]

Le Bail, A., B. Billoud, N. Kowalczyk, *et al.* (2010). Auxin metabolism and function in the multicellular brown alga *Ectocarpus siliculosus. Plant Physiol.* 153:128-44. [2.6.3]

Le Bail, A., B. Billoud, S. Le Panse, S. Chenivesse, and B. Charrier (2011). ETOILE regulates developmental patterning in the filamentous brown alga *Ectocarpus siliculosus. The Plant Cell* 23:1666-78. [2.6.1]

Le Corguillé, G., G. Pearson, M. Valente, C. Viegas, *et al.* (2009). Plastid genomes of two brown algae, *Ectocarpus siliculosus* and *Fucus vesiculosus*: further insights on the evolution of red algal derived plastids. *BMC Evol. Biol.* 9:240-53. [1.3.2]

Le Gall, L., and G. W. Saunders (2007). A nuclear phylogeny of the Florideophyceae (Rhodophyta) inferred from combined EF2, small subunit and large subunit ribosomal DNA: establishing the new red algal subclass Corallinophycidae. *Mol. Phylogenet. Evol.* 43:1118-30. [Essay 1]

Le Gall, L., and G. W. Saunders (2010). DNA barcoding is a powerful tool to uncover algal diversity: a case study of the Phyllophoraceae (Gigartinales, Rhodophyta) in the Canadian flora. *J. Phycol.* 46:374-89. [Essay 1]

Le Gall, L., A.-M. Rusig, and J. Cosson (2004). Organisation of the microtubular cytoskeleton in protoplasts from *Palmaria palmate* (Palmariales, Rhodophyta). *Bot. Mar.* 47:231-7. [1.3.3]

Leal, M. F. C., M. T. Vasconcelos, I. Sousa-Pinto, and J. P. S. Cabral (1997). Biomonitoring with benthic macroalgae and direct assay of heavy metals in seawater of the Oporto coast (NW Portugal). *Mar. Poll. Bull.* 34:1006-15. [9.3.2]

Lechat, H., M. Amat, J. Mazoyer, *et al.* (2000). Structure and distribution of glucomannan and sulfated glucan in the cell walls of the red alga *Kappaphycus alvarezii* (Gigartinales, Rhodophyta). *J. Phycol.* 36:891-902. [1.3.1]

Ledlie, M., N. Graham, J. Bythell, *et al.* (2007). Phase shifts and the role of herbivory in the resilience of coral reefs. *Coral Reefs* 26:641-53. [Essay 6]

Lee, S.-H., T. Motomura, and T. Ichimura (2002). Light and electron microscopic observations of preferential destruction of chloroplast and mitochondrial DNA at early male gametogenesis of the anisogamous green alga *Derbesia tenuissima* (Chlorophyta). *J. Phycol.* 38:534-42. [1.3.2]

Lee, T. M. (1998). Investigations of some intertidal green macroalgae to hyposaline stress: detrimental role of putrescine under extreme hyposaline conditions. *Plant Sci.* 138:1-8. [2.6.3]

Lee, T. M., P.-F. Tsac, Y.T. Shyu, and F. Sheu (2005). The effects of phosphate on phosphate starvation responses of *Ulva lactuca* (Ulvales, Chlorophyta). *J. Phycol.* 41:975-82.

Lehninger, A. L. (1975). *Biochemistry*, 2nd edn. New York: Worth, 1104 pp. [Fig. 5.20]

Leichter, J. J., M. D. Stokes, and S. J. Genovese (2008). Deep water macroalgal communities adjacent to the Florida Keys reef tract. *Mar. Ecol. Prog. Ser.* 356:123-38. [3.3.6]

Leigh, E. G., Jr., R. T. Paine, 1. F. Quinn, and T. H. Suchanek (1987). Wave energy and intertidal productivity. *Proc. Natl. Acad. Sci. USA* 84:1314-18. [8.0, 8.3.1]

Leighton, D. L. (1971). Grazing activities of benthic invertebrates in southern California kelp beds. *Nova Hedwegia Beiheft* 32:421-53. [4.3.2]

Leighton, D. L., L. G. Jones, and W. J. North (1966). Ecological relationships between giant kelp and sea urchins in southern California. *Proc. Intl. Seaweed Symp.* 5:141-53. [4.3.1]

Lenanton, R. C. J., A. I. Robertson, and J. A. Hansen (1982). Nearshore accumulations of detached macrophytes as nursery areas for fish. *Mar. Ecol. Prog. Ser.* 9:51-7. [Essay 3]

Leonardi, P. I., and J. A. Vasquez (1999). Effects of copper on the ultrastructure of *Lessonia* spp. *Hydrobiol.* 398/399:375-83. [9.3.3]

Leonardi, P. I., A. B. Miravalles, S. Faugeron *et al.* (2006). Diversity, phenomenology and epidemiology of epiphytism in farmed *Gracilaria chilensis* (Rhodophyta) in northern Chile. *Eur. J. Phycol.* 41:247-57. [10.6.2]

Lepoint, G., F. Nyssen, S. Gobert, P. Dauby, and J. M. Bouquegneau (2000). Relative impact of a seagrass bed and its adjacent epilithic algal community in consumer diets. *Mar. Biol.* 136(3):513-18. [Essay 3]

Leroux, S. J., and M. Loreau (2008). Subsidy hypothesis and strength of trophic cascades across ecosystems. *Ecology Letters* 11:1147-56. [Essay 2]

Lesser, M. P. (2006). Oxidative stress in marine environments: biochemistry and physiological ecology. *Annu. Rev. Physiol.* 68:253-78. [7.8, 9.3.2]

Lessios, H. A. (1988). Mass mortality of *Diadema antillarum* in the Caribbean: what have we learned? *Annu. Rev. Ecol. Syst.* 19:371-93. [4.3.1]

Lessios, H. A. (2005). *Diadema antillarum* populations in Panama twenty years following mass mortality. *Coral Reefs* 24:125-7. [4.3.1]

Leustek, T., M. N. Martin, J-A. Bick and J. P. Davies (2000). Pathways and regulation of sulfur metabolism revealed through molecular and genetic studies. *Annu. Rev. Plant Phys.* 51:141-65. [6.5.4]

Levenbach, S. (2008). Behavioral mechanism for an associational refuge for macroalgae on temperate reefs. *Mar. Ecol. Prog. Ser.* 370:45-52. [4.3.2]

Levin, S. A. (1992). The problem of pattern and scale in ecology. *Ecology* 73:1943-67. [Essay 2]

Lewis, J. R. (1964). *The Ecology of Rocky Shores*. London: English Universities Press, 323 pp. [3.1.2, Essay 2, 4.3.1, 8.1.4]

Lewis, J. R. (1980). Objectives in littoral ecology: a personal viewpoint. In J. H. Price, D. E. G. Irvine, and W. F. Farnham (eds), *The Shore Environment. Vol. 1: Methods* (pp. 118). New York: Academic Press. [3.1.2]

Lewis, S. M., J. N. Norris, and R. B. Searles (1987). The regulation of morphological plasticity in tropical reef algae by herbivory. *Ecology* 68:636-41. [2.6.2]

Lewis, S. E., J. E. Brodie, Z. T. Bainbridge, *et al.* (2009). Herbicides: a new threat to the Great Barrier Reef. *Environ. Poll.* 157:2470-84. [9.1]

Li, N., and R. A. Cattolico (1987). Chloroplast genome characterization in the red alga *Griffithsia pacifica*. *Mol. Gen. Genet.* 209:343-51. [1.4.1]

Li, T., C. Wang, and J. Miao (2007). Identification and quantification of indole-3-acetic acid in the kelp *Laminaria japonica* Areschoug and its effect on growth of marine microalgae. *J. Appl. Phycol.* 19:479-84. [2.6.3]

Libes, S. M. (1992). *Introduction to Marine Biogeochemistry* (2nd edn). Amsterdam: Elsevier, 909 pp. [9.3.1]

Lignell, A., and M. Pedersen (1987). Nitrogen metabolism in *Gracilaria secundata*. *Hydrobiologia* 151/152:431-41. [6.5.1]

Lilley, S. A., and D. R. Schiel (2006). Community effects following the deletion of a habitat-forming alga from rocky marine shores. *Oecologia* 148:672-81. [8.3.2]

Lin, D. T., and P. Fong (2008). Macroalgal bioindicators (growth, tissue N, δ^{15}N) detect nutrient enrichment from shrimp farm effluent entering Opunohu Bay, Moorea, French Polynesia. *Mar. Poll. Bull.* 56:245-9. [9.6.3]

Lin, R., and M. S. Stekoll (2007). Effects of plant growth substances on the conchocelis phase of Alaskan *Porphyra* (Bangiales, Rhodophyta) species in conjunction with environmental variables. *J. Phycol.* 43:1094-103. [2.6.3]

Lin, T. Y., and W. Z. Hassid (1966). Pathway of alginic acid synthesis in the marine brown alga, *Fucus gardneri Silva*. *J. Biol Chem* 241: 5284-97. [5.5.3]

Lindstrom, S. C. (2008). Cryptic diversity and phylogenetic relationships within the *Mastocarpus papillatus* species complex (Rhodophyta, Phyllophoraceae). *J. Phycol.* 44:1300-8. [Essay 1]

Ling, S. D., and C. R. Johnson (2009). Population dynamics of an ecologically important range-extender: kelp beds versus sea urchin barrens. *Mar. Ecol. Prog. Ser.* 374:113-25. [4.3.1]

Littler, M. M. (1979). The effects of bottle volume, thallus weight, oxygen saturation levels, and water movement on apparent photosynthetic rates in marine algae. *Aquat. Bot.* 7:21-34. [5.7.1, 8.2]

Littler, M. M., and K. E. Arnold (1980). Sources of variability in macroalgal primary productivity: sampling and interpretative problems. *Aquat Bot* 8: 141-56. [5.7.2]

Littler, M. M., and K. E. Arnold (1982). Primary productivity of marine macroalgal functional-form groups from southwestern North America. *J Phycol* 18: 307-11. [5.3.2, 5.7.2]

Littler, M. M., and B. J. Kauker (1984). Heterotrichy and survival strategies in the red alga *Corallina officinalis* L. *Bot. Mar.* 27:37-44. [1.2.2]

Littler, M. M., and D. S. Littler (1980). The evolution of thallus form and survival strategies in benthic marine macroalgae: field and laboratory tests of a functional form model. *Am. Nat.* 116:25-44. [1.2.2, 1.2.3, 3.5.2, 5.7.2]

Littler, M. M., and D. S. Littler (1984). Models of tropical reef biogenesis: the contribution of algae. *Prog. Phycol. Res.* 3:323-64. [4.3.2, Essay 6]

Littler, M. M., and D. S. Littler (1987). Effects of stochastic processes on rocky intertidal biotas: an unusual flash flood near Corona del Mar, California. *Bull. So. Cal. Acad. Sci.* 86:95-106. [1.1.2, 7.2.3]

Littler, M. M., and D. S. Littler (1988). Structure and role of algae in tropical reef communities. In C. A. Lembi and J. R. Waaland (eds), *Algae and Human Affairs* (pp. 29-56). Cambridge: Cambridge University Press. [3.3.1, 4.3.2]

Littler, M. M., and D. S. Littler (1990). Productivity and nutrient relationships in psammophytic versus epilithic forms of Bryopsidales (Chlorophyta): comparisons based on a short-term physiological assay. *Hydrobiologia* 204/205:49-55. [6.5.2]

Littler, M. M., and D. S. Littler (1999). Blade abandonment/proliferation: a novel mechanism for rapid epiphyte control in marine macrophytes. *Ecology* 80:1736-46. [4.2.2]

Littler, M. M., and D. S. Littler (2007). Assessment of coral reefs using herbivory/nutrient assays and indicator groups of benthic primary producers: a critical synthesis, proposed protocols, and critique of management strategies. *Aq. Cons.: Mar. Fresh. Ecosyst.* 17:195-215. [Essay 6]

Littler, M. M., and D. S. Littler (2011a). Algae: macro. In D. Hopley (ed.) *Encyclopedia of Modern Coral Reefs: Structure, Form and Process* (pp. 30-8). Berlin: Springer-Verlag. [1.1.1, 3.3.1]

Littler, M. M., and D. S. Littler (2011b). Algae, turf. In D. Hopley (ed.) *Encyclopedia of Modern Coral Reefs: Structure, Form and Process* (pp. 38-9). Berlin: Springer-Verlag. [1.1.1, 3.3.1]

Littler, M. M., and D. S. Littler (2011c). Algae, coralline. In D. Hopley (ed.) *Encyclopedia of Modern Coral Reefs: Structure, Form and Process* (pp. 20-30). Berlin: Springer-Verlag. [3.3.1]

Littler, M. M., and D. S. Littler (2011d). Algae, blue-green boring. In D. Hopley (ed.) *Encyclopedia of Modern Coral Reefs: Structure, Form and Process* (pp. 18-20). Berlin: Springer-Verlag. [3.3.1, 3.3.8, 4.3.2]

Littler, M. M., and S. N. Murray (1975). Impact of sewage on the distribution, abundance and community structure of rocky intertidal macro-organisms. *Mar. Biol.* 30:277-91. [9.6.1]

Littler, M. M., D. S. Littler, and P R. Taylor (1983a). Evolutionary strategies in a tropical barrier reef system: functional-form groups of marine macroalgae. *J. Phycol.* 19:229-37. [1.2.2, 3.5.2, 5.7.2]

Littler, M. M., D. R. Martz, and D. S. Littler (1983b) Effects of recurrent sand deposition on rocky intertidal organisms: importance of substrate heterogeneity in a fluctuating environment. *Mar. Ecol. Prog. Ser.* 11: 129-39. [4.3.2, 8.3.2]

Littler, M. M., D. S. Littler, S. M. Blair, and J. N. Norris (1985). Deepest known plant life discovered on an uncharted seamount. *Science* 227:57-9. [3.3.1, 5.2.2, 5.3.4, 6.5.3]

Littler, M. M., D. S. Littler, S. M. Blair, and J. N. Norris (1986a). Deep-water plant communities from an uncharted seamount off San Salvador Island, Bahamas: distribution, abundance, and primary productivity. *Deep Sea Res.* 33:881-92. [3.2]

Littler, M. M., P. R. Taylor, and D. S. Littler (1986b). Plant defense associations in the marine environment. *Coral Reefs* 5:63-71. [4.3.2]

Littler, M. M., D. S. Littler, and P. R. Taylor (1987). Animal-plant defense associations: effects on the distribution and abundance of tropical reef macrophytes. *J. Exp. Mar. Biol. Ecol.* 105:107-21. [4.3.2]

Littler, M. M., D. S. Littler, and B. E. Lapointe (1988). A comparison of nutrient- and light-limited photosynthesis in psammophytic versus epilithic forms of *Halimeda* (Caulerpales, Halimediaceae) from the Bahamas. *Coral Reefs* 6:219-25. [1.1.1, 6.8.3]

Littler, M. M., D. S. Littler, and B. L. Brooks (2006). Harmful algae on tropical coral reefs: bottom-up eutrophication and top-down herbivory. *Harmful Algae* 5:565-85. [3.3.1]

Liu, D., J. K. Keesing, Q. Xing, and P. Shi (2009). World's largest macroalgal bloom caused by expansion of seaweed aquaculture in China. *Mar. Poll. Bull.* 58:888-95. [3.3.7, 10.10]

Liu, J., S. Dong, X. Liu, and S. Ma (2000). Responses of the macroalgae *Gracilaria tenuistipitata* var. *liui* (Rhodophyta) to iron stress. *J. Appl. Phycol.* 12:605-12. [6.5.5]

Lobban, C. S. (1978a). The growth and death of the *Macrocystis* sporophyte (Phaeophyceae, Laminariales). *Phycologia* 17: 196-212. [2.6.2, 5.3.2]

Lobban, C. S. (l978b). Translocation of ^{14}C in *Macrocystis pyrifera* (giant kelp). *Plant Physiol.* 61: 585–9. [5.6]

Lobban, C. S. (1978c). Translocation of 14C in *Macrocystis integrifolia* (Phaeophyceae). 1. *Phycol.* 14:178–82. [5.6]

Lobban, C. S. (1989). Environmental factors, plant responses, and colony growth in relation to tube-dwelling diatom blooms in the Bay of Fundy, Canada, with a review of the biology of tube-dwelling diatoms. *Diatom Res.* 4:89–109. [1.1.1]

Lobban, C. S., and D. M. Baxter (1983). Distribution of the red algal epiphyte *Polysiphonia lanosa* on its brown algal host *Ascophyllum nodosum* in the Bay of Fundy, Canada. *Bot. Mar.* 26:533–8. [4.2.2]

Lobban, C. S., and P. J. Harrison (1994). *Seaweed Ecology and Physiology.* Cambridge: Cambridge University Press. [10.7]

Lobban, C. S., and R. W. Jordan (2010). Diatoms on coral reefs and in tropical marine lakes. In J. P. Smol and E. F. Stoermer (eds), *The Diatoms: Applications for the Environmental and Earth Sciences* (2nd edn) (pp. 346–56). Cambridge: Cambridge University Press. [4.3.2]

Lodish, H., A. Berk, C. A. Kaiser, *et al.* (2008). *Molecular Cell Biology.* New York: W. H. Freeman, 973 pp. [1.3.3, 2.4]

Longstaff, B. J., T. Kildea, J. W. Runcie, *et al.* (2002). An *in situ* study of photosynthetic oxygen exchange and electron transport rate in the marine macroalga *Ulva lactuca* (Chlorophyta). *Photosynth. Res.* 74:281–93. [5.7]

Longtin, C. M., and R. A. Scrosati (2009). Role of surface wounds and brown algal epiphytes in the colonization of *Ascophyllum nodosum* (Phaeophyceae) fronds by *Vertebrata lanosa* (Rhodophyta). *J. Phycol.* 45:535–9. [4.2.2]

Lopez-Figueroa, F., and X. Niell (1990). Effects of light quality on chlorophyll and biliprotein accumulation in seaweeds. *Mar. Biol.* 104:321–7. [5.3.1]

López-Figueroa, F., P. Lindemann, K. Braslavsky, *et al.* (1989). Detection of a phytochrome-like protein in macroalgae. *Bot. Acta* 102:178–80. [2.3.3]

Lorb, R. E., and V. A. Gerard (2000). Nitrogen assimilation characteristics of polar seaweeds from differing nutrient environments. *Mar. Ecol Prog. Ser.* 198:83–92. [6.8.5]

Loreau, M., N. Mouquet, and R. D. Holt (2003). Meta-ecosystems: a theoretical framework for a spatial ecosystem ecology. *Ecol. Lett.* 6:673–9. [Essay 2]

Lotze, H., and W. Schramm (2000). Ecophysiological traits explain species dominance in macroalgal blooms. *J. Phycol.* 36:287–95. [6.8.5]

Lotze, H. K., B. Worm, and U. Sommer (2000). Propagule banks, herbivory and nutrient supply control population development and dominance patterns in macroalgal blooms. *Oikos* 89:46–58. [Essay 6]

Lotze, H. K., B. Worm and U. Sommer (2001). Strong bottom-up and top-down control of early life stages of macroalgae. *Limnol. Oceanogr.* 46:749–57. [4.3.1]

Løvstad Holdt, S., and S. Kraan (2011). Bioactive compounds in seaweed: Functional food applications and legislation. *J. Appl. Phycol.* 23:543–97. [10.10]

Lowe, R. J., J. R. Koseff, and S. G. Monismith (2005). Oscillatory flow through submerged canopies: 1. Velocity structure. *J. Geophys. Res.* 110, C10016, 17 pp. [8.1.4]

Loya, Y., and B. Rinkevich (1980). Effects of oil pollution on coral reef communities. *Mar. Ecol. Prog. Ser.* 3:176–80. [9.4.3]

Lü, F., W. Xü, C. Tian, *et al.* (2011). The *Bryopsis hypnoides* plastid genome: multimeric forms and complete nucleotide sequence. *PloS ONE.* 6:e14663. [1.3.2]

Lubchenco, J. (1978). Plant species diversity in a marine intertidal community: importance of herbivore food preference and algal competitive abilities. *Am. Nat.* 112:23–39. [4.3.1]

Lubchenco, J. (1983). *Littorina* and *Fucus*: effects of herbivores, substratum heterogeneity, and plant escapes during succession. *Ecology* 64:1116–23. [Essay 2, 4.3.2]

Lubchenco, J. (1986). Relative importance of competition and predation: early colonization by seaweeds in New England. In J. M. Diamond and T. Case (eds). *Community Ecology* (pp. 537–55). New York: Harper and Row. [Essay 2]

Lubchenco, J., and J. Cubit (1980). Heteromorphic life histories of certain marine algae as adaptations to variations in herbivory. *Ecology* 61:676–87. [4.3.2]

Lubchenco, J., and S. D. Gaines (1981). A unified approach to marine plant-herbivore interactions. I. Populations and communities. *Annu. Rev. Ecol. Syst.* 12:405–37. [4.3.2]

Lubchenco, J., and B. A. Menge (1978). Community development and persistence in a low rocky intertidal zone. *Ecol. Monogr.* 48:67–94. [Essay 4, 4.3.1]

Lubimenko, V., and Q. Tichovskaya (1928). *Recherches sur la Photosynthese et ['Adaptation Chromatique chez les Algues Marines.* Moscow: Academy of Sciences. [5.3.4]

Lück, J., H. B. Lück, M.-Th. L'Hardy-Halos, and C. Lambert (1999). Simulation of the thallus development of *Antithamnion plumula* (Ellis) Le Jolis, (Rhodophyceae, Ceramiales). *Acta Biotheor.* 47:329–51. [2.6]

Lüder, U. H., and M. N. Clayton (2004). Induction of phlorotannins in the brown macroalga *Ecklonia radiata* (Laminariales, Phaeophyta) in response to simulated herbivory - the first microscopic study. *Planta* 218:928–37. [2.6.4]

Lundberg, P., R. G. Weich, P. Jensen, and H. J. Vogel (1989). Phosphorus-31 and nitrogen-14 studies of the uptake of phosphorus and nitrogen compounds in the marine macroalga *Ulva lactuca. Plant Physiol.* 89:1380–7. [6.5.2]

Lundheim, R. (1997). Ice nucleation in seaweeds in relation to vertical zonation. *J. Phycol.* 33: 739-42. [7.3.4]

Lüning, K. (1969). Growth of amputated and dark-exposed individuals of the brown alga *Laminaria hyperborea*. *Mar. Biol.* 2: 218-23. [5.7.3]

Lüning, K. (1980). Control of algal life-history by daylength and temperature. In J. H. Price, D. E. G. Irvine, and W. F. Farnham (eds), *The Shore Environments. Vol. 2: Ecosystems* (pp. 915-45). New York: Academic Press. [2.3.3]

Lüning, K. (l981a). Light. In: Lobban, C. S., and M. J. Wynne (eds), *The Biology of Seaweeds* (pp. 326-55). Oxford: Blackwell Scientific. [4.2.2, 5.3.3]

Lüning, K. (1981b). Photomorphogenesis of reproduction in marine macroalgae. *Ber Deutsch. Bot. Ges.* 94:401-17. [2.6.2]

Lüning, K. (1984). Temperature tolerance and biogeography of seaweeds: the marine algal flora of Helgoland (North Sea) as an example. *Helgol. Meeresunters* 38: 305-17. [7.3.4, 7.3.6, 7.3.8]

Lüning, K. (1986). New frond formation in *Laminaria hyperborea* (Phaeophyta): a photoperiodic response. *Br. Phycol. J.* 21:269-73.[5.6]

Lüning, K. (1988). Photoperiodic control of sorus formation in the brown alga *Laminaria saccharina*. *Mar. Ecol. Prog. Ser.* 45:137-44. [2.3.2]

Lüning, K. (1990). *Seaweeds. Their Environment, Biogeography, and Ecophysiology* (trans. and ed. C. Yarish and H. Kirkman). New York: Wiley-Interscience. [2.3, 3.3, 3.3.2, 3.3.3, 7.3.8]

Lüning, K. (1991). Circannual growth rhythm in a brown alga, *Pterygophora californica*. *Bot. Acta* 104:157-62. [2.3.1]

Lüning, K. (1994). When do algae grow? The third Founders' Lecture. *Eur. J. Phycol.* 29:61-7. [1.3.5, 2.3.1, Essay 6]

Lüning, K., and I. tom Dieck (1989). Environmental triggers in algal seasonality. *Bot. Mar.* 32:389-97. [2.3]

Lüning, K., and I. tom Dieck (1990). The distribution and evolution of the Laminariales: North Pacific-Atlantic relationships. In D. J. Garbary, and G. R. South (eds), *Evolutionary Biogeography of the Marine Algae of the North Atlantic* (pp. 187-204). Berlin: Springer-Verlag. [Essay 1]

Lüning, K., and M. J. Dring (1972). Reproduction induced by blue light in female gametophytes of *Laminaria saccharina*. *Planta* 104:252-6. [10.3.1, 10.3.2]

Lüning, K., and M. J. Dring (1979). Continuous underwater light measurements near Helgoland (North Sea) and its significance for characteristic light limits in the sublittoral region. *Helgol. Meeresunters* 32:403-24. [3.0, 5.2.2]

Lüning, K., and W. Freshwater (1988). Temperature tolerance of northeast Pacific marine algae. *J. Phycol.* 24:310-15. [7.3.4, 7.3.8]

Lüning, K., and M. Neushul (1978). Light and temperature demands for growth and reproduction of laminarian gametophytes in southern and central California. *Mar. Biol.* 45:297-309. [2.3, 2.3.3]

Lüning, K., K. Schmitz, and J. Willenbrink (1973). CO_2 fixation and translocation in benthic marine algae. III. Rates and ecological significance of translocation in *Laminaria hyperborea* and *L. saccharina*. *Mar. Biol.* 23:275-81. [5.6, 5.7.3]

Lüning, K., E. A. Titlyanov, and T. Titlyanov (1997). Diurnal and circadian periodicity of mitosis and growth in marine macroalgae. III. The red alga *Porphyra umbilicalis*. *Eur. J. Phycol.* 32:167-73. [1.3.5]

Lüning, K., P. Kadel, and S. Pang (2008). Control of reproduction rhythmicity by environmental and endogenous signals in *Ulva pseudocurvata* (Chlorophyta). *J. Phycol.* 44:866-73. [2.3.1, 2.3.3]

Lyngby, J. E. (1990). Monitoring of nutrient availability and limitation using the marine macroalga *Ceramium rubrum* (Huds.) C. Ag. *Aquat. Bot.* 38:153-61. [6.7.2]

Lyons, D. A., and R. E. Scheibling (2009). Range expansion by invasive marine algae: rates and patterns of spread at a regional scale. *Diversity Distrib.* 15:762-75. [3.4]

Maberly, S. C. (1990). Exogenous sources of inorganic carbon for photosynthesis by marine macroalgae. *J. Phycol.* 26:439-49. [5.4.1]

Maberly, S. C., and T. V. Madsen (1990). Contribution of air and water to the carbon balance of *Fucus spiralis*. *Mar. Ecol. Prog. Ser.*62:175-83. [5.3.3, 5.4.1]

Maberly, S. C., J. A. Raven, and A. M. Johnston (1992). Discrimination of ^{12}C and ^{13}C by marine plants. *Oecologia* 91:481-92. [5.4.3]

MacArtain, P., C. I. R. Gill, M. Brooks, R. Campbell, and I. R. Rowland (2007). Nutritional value of edible seaweeds. *Nutr. Rev.* 65:535-43. [10.2, 10.10]

MacArthur, R. H. (1955). Fluctuations of animal populations and a measure of community stability. *Ecology* 36:533-6. [Essay 2]

MacArthur, R. H. (1965). Patterns of species diversity. *Biol. Rev.* 40:510-33. [Essay 2]

MacArthur, R. H. (1972). Strong, or weak, interactions? *Transactions of the Connecticut Academy of Arts and Sciences* 44:177-88. [Essay 2]

MacArthur, R. H., and R. Levins (1967). The limiting similarity, convergence, and divergence of coexisting species. *Am. Nat.* 101:377-85. [Essay 2]

MacArthur, R. H., and E. O. Wilson (1963). An equilibrium theory of insular zoogeography. *Evolution* 17:373-87. [Essay 2]

Macaya, E. C., and G. C. Zuccarello (2010). Genetic structure of the giant kelp *Macrocystis pyrifera* along the southeastern Pacific. *Mar. Ecol. Prog. Ser.* 420:103-12. [3.3.7]

Mach, K. J. (2009). Mechanical and biological consequences of repetitive loading: crack initiation and fatigue failure in the red macroalga *Mazzaella*. *J. Exp. Biol.* 212:961-76. [8.3.1]

Mach, K. J., D. V. Nelson, and M. W. Denny (2007). Techniques for predicting the lifetimes of wave-swept macroalgae: a primer on fracture mechanics and crack growth. *J. Exp. Biol.* 210:2213-30. [8.1.2, 8.3.1]

Machalek, K. M., I. R. Davison, and P. G. Falkowski (1996). Thermal acclimation and photoacclimation of photosynthesis in the brown alga *Laminaria saccarina*. *Plant Cell. Envir.* 19: 1005-16. [7.3.1, 7.3.2]

Mackerness, S. A.-H., and B. R. Jordan (1999). Changes in gene expression in response to UV-B induced stress. In M. Pessarakli (ed.), *Handbook of Plant and Crop Stress* (pp. 749-68). New York: Marcel Dekker Inc.

Mackie, W., and R. D. Preston (1974). Cell wall and intercellular region polysaccharides. In W. D. P Stewart (ed.), *Algal Physiology and Biochemistry* (pp. 40-85). Oxford: Blackwell Scientific. [1.3.1]

Madronich, S., R. L. McKenzie, L. O. Björn, and M. M. Caldwell (1998). Changes in biologically active ultraviolet radiation reaching the earth´s surface. *J. Photochem. Photobiol. B: Biol.* 46:5-19. [5.2.2]

Madsen, T. V., and S. C. Maberly (1990). A comparison of air and water as environments for photosynthesis by the intertidal alga *Fucus spiralis* L. *J. Phycol.* 26:24-30. [5.3.3, 5.4.1]

Maggs, C. A. (1988). Intraspecific life history variability in the Florideophycidae (Rhodophyta). *Bot. Mar.* 31:465-90. [2.1, 2.2]

Maggs, C. A. (1989). *Erythrodermis allenii* Batters in the life history of *Phyllophora traillii* Holmes ex Batters (Phyllophoraceae, Rhodophyta). *Phycologia* 28:305-17. [2.2]

Maggs, C. A. (1998). Life history variation in *Dasya ocellata* (Dasyaceae, Rhodophyta). *Phycologia* 37:100-5. [2.2]

Maggs, C. A., and D. P. Cheney (1990). Competition studies of marine macroalgae in laboratory culture. *J. Phycol.* 26:17-24. [4.0]

Maggs, C. A., and M. D. Guiry (1987). Environmental control of macroalgal phenology. In R. M. M. Crawford (ed.), *Plant Life in Aquatic and Amphibious Habitats* (pp. 359-73). Oxford: Blackwell Scientific. [2.3.2]

Maggs, C. A., and C. M. Pueschel (1989). Morphology and development of *Ahnfeltia plicata* (Rhodophyta): proposal of Ahnfeltiales ord. novo *J. Phycol.* 25:333-51. [1.4.2]

Maggs, C. A., H. L. Fletcher, and D. Fewer (2011). Speciation in red algae: members of the Ceramiales as model organisms. *Integr. Comp. Biol.* 51:492-504. [2.2, 2.4]

Magruder, W. H. (1984). Specialized appendages on spermatia from the red alga *Aglaothamnion neglectum*

(Ceramiales, Ceramiaceae) specifically bind with trichogynes. *J. Phycol.* 20:436-40. [2.4]

Maier, C. M., and A. M. Pregnall (1990). Increased macrophyte nitrate reductase activity as a consequence of groundwater input of nitrate through sandy beaches. *Mar. Biol.* 107:263-71. [6.2]

Maier, I. (1997). Fertilization, early embryogenesis and parthenogenesis in *Durvillaea potatorum* (Durvillaeales, Phaeophyceae). *Nova Hedwigia* 64:41-50. [2.2]

Maier, I., and D. G. Müller (1986). Sexual pheromones in algae. *Biol. Bull.* 170:145-75. [9.3.4]

Maier, I., C. Hertweck, and W. Boland (2001). Stereochemical specificity of lamoxirene, the sperm-releasing pheromone in kelp (Laminariales, Phaeophyceae). *Biol. Bull.* 201:121-5. [2.4]

Makarov, V. N., E. V. Schoschina, and K. Lüning (1995). Diurnal and circadian periodicity of mitosis and growth in marine macroalgae. I. Juvenile sporophytes of Laminariales (Phaeophyta). *Eur. J. Phycol.* 30:261-6. [1.3.5]

Malta, E. J., D. G. Ferreira, J. J. Vergara, and J. L. Perez-Llorens (2005). Nitrogen load and irradiance affect morphology, photosynthesis and growth of *Caulerpa prolifera* (Bryopsidales: Chlorophyta). *Mar. Ecol. Prog. Ser.* 298:101-14. [3.4]

Mance, G. (1987). *Pollution Threat of Heavy Metals in Aquatic Environments*. Amsterdam: Elsevier. [9.3.1]

Mandal, P., C. G. Mateu, K. Chattopadhay, *et al.* (2007). Structural features and antiviral activity of sulphated fucans from the brown seaweed *Cystoseira indica*. *Antivir. Chem. Chemother.* 18:153-62. [5.5]

Mandoli, D. F. (1998a). What ever happened to *Acetabularia*? Bringing a once-classic model system into the age of molecular genetics. *Int. Rev. Cytol.* 182:1-67. [1.3.2, 1.4.1, 1.4.3]

Mandoli, D. F. (1998b). Elaboration of body plan and phase change during development of *Acetabularia*: how is the complex architecture of a giant unicell built? *Annu. Rev. Plant Physiol. Plant Mol. Biol.* 49:173-98. [1.3.3, 1.4.3]

Maneveldt, G. W., and D. W. Keats (2008). Effects of herbivore grazing on the physiognomy of the coralline alga *Spongites yendoi* and on associated competitive interactions. *Afr. J. Mar. Sci.* 30:581-93. [4.3.2]

Maneveldt, G. W., R. C. Eager, and A. Bassier (2009). Effects of long-term exclusion of the limpet *Cymbula oculus* (Born) on the distribution of intertidal organisms on a rocky shore. *Afr. J. Mar. Sci.* 31:171-9. [4.3.1]

Manley, S. L. (1983). Composition of sieve tube sap from *Macrocystis pyrifera* (Phaeophyta) with emphasis on the inorganic constituents. *J. Phycol.* 19:118-21. [5.6, 6.6]

Manley, S. L., and W.J. North (1984). Phosphorus and the growth of juvenile *Macrocystis pyrifera* (Phaeophyta) sporophytes. *J. Phycol.* 20:389-93. [6.7.2]

Mann, K. H. (1972). Ecological energetics of the seaweed zone in a marine bay on the Atlantic coast of Canada. II. Productivity of the seaweeds. *Mar. Biol.* 14:199–209. [5.7.3, 5.7.4]

Mann, K. H., N. Jarman, and G. Dieckmann (1979). Development of a method for measuring the productivity of the kelp *Ecklonia maxima* (Osbeck) Papenf. *Trans. R. Soc. S. Afr.* 44:27–41. [5.7.4]

Manney, G. L. , M. L. Santee, M. Rex, *et al.* (2011). Unprecedented Arctic ozone loss in 2011. *Nature* 478:469–75. [7.6]

Manning, W.M., and H. H. Strain (1943). Chlorophyll d, a green pigment of red algae. *J. Biol. Chem.* 151:1–19. [5.3.1]

Mantyka, C. S., and D. R. Belllwood (2007). Macroalgal grazing selectivity among herbivorous coral reef fishes. *Mar. Ecol. Prog. Ser.* 352:177–85. [5.7.5]

Marande, W., and L. Kohl (2011). Flagellar kinesins in protests. *Future Microbiol.* 6:231–46. [1.3.3]

Marian, F. D., P. Garcia-Jiménez, and R. R. Robaina (2000). Polyamines in marine macroalgae: levels of putrescine, spermidine and spermine in the thalli and changes in their concentration during glycerol-induced cell growth *in vitro*. *Physiol. Plantarum* 110:530–4. [2.6.3]

Markager, S., and K. Sand-Jensen (1990). Heterotrophic growth of *Ulva lactuca* (Chlorophyceae). *J. Phycol.* 26:670–3. [5.4.1]

Markham, J. W. (1973). Observations on the ecology of *Laminaria sinclairii* on three northern Oregon beaches. *J. Phycol.* 9:336–41. [8.3.2]

Markham, J. W., and P. R. Newroth (1972). Observations on the ecology of *Gymnogongrus linearis* and related species. *Proc. Intl. Seaweed Symp.* 7:127–30. [8.3.2]

Markham, J. W., B. P. Kremer, and K. R. Sperling (1980). Effect of cadmium on *Laminaria saccharina* in culture. *Mar. Ecol. Prog. Ser.* 3:31–9. [9.3.4]

Marquardt, R., H. Schubert, D. A. Varela *et al.* (2010). Light acclimation strategies of three commercially important red algal species. *Aquaculture* 299:140–8. [5.3.3]

Marsden, A. D., and R. E. DeWreede (2000). Marine macroalgal community structure, metal content and reproductive function near an acid mine drainage outflow. *Environ. Poll.* 110:431–40. [9.3.3]

Marshall, K., I. Joint, M. E. Callow, and J. A. Callow (2006). Effect of marine bacterial isolates on the growth and morphology of axenic plantlets of the green alga *Ulva linza*. *Microbial Ecol.* 52:302–10. [2.6.2]

Martinez, B., and J. M. Rico (2004). Inorganic nitrogen and phosphorus uptake kinetics in *Palmaria palmata* (Rhodophyta). *J. Phycol.* 40:642–50. [6.5.1, 6.5.2]

Martinez, E. A., C. Destombe, M. C. Quillet, and M. Valero (1999). Identification of random amplified polymorphic

DNA (RAPD) markers highly linked to sex determination in the red alga *Gracilaria gracilis*. *Mol. Ecol.* 8:1533–8. [1.4.2]

Martins, G. M., S. J. Hawkins, R. C. Thompson, and S. R. Jenkins (2007). Community structure and functioning in intertidal rock pools: effects of pool size and shore height at different successional stages. *Mar. Ecol. Prog. Ser.* 329:43–55. [3.3.4]

Martone, P. T. (2006). Size, strength and allometry of joints in the articulated coralline *Calliarthron*. *J. Exp. Biol.* 209:1678–89. [8.3.1]

Martone, P. T. (2007). Kelp versus coralline: cellular basis for mechanical strength in the wave-swept seaweed *Calliarthron* (Corallinaceae, Rhodophyta). *J. Phycol.* 43:882–91. [8.3.1]

Martone, P. T., and M. W. Denny (2008a). To break a coralline: mechanical constraints on the size and survival of a wave-swept seaweed. *J. Exp. Biol.* 211: 3433–41. [8.1.4, 8.3.1]

Martone, P. T., and M. W. Denny (2008b). To bend a coralline: effect of joint morphology on flexibility and stress amplification in an articulated calcified seaweed. *J. Exp. Biol.* 211:3421–32. [8.3.1]

Martone, P. T., J. M. Estevez, F. Lu, *et al.* (2009). Discovery of lignin in seaweed reveals convergent evolution of cell wall architecture. *Curr. Biol.* 19:169–75. [1.3.1]

Martone, P. T., M. Boller, I. Burgert, *et al.* (2010a). Mechanics without muscle: biomechanical inspiration from the plant world. *Integr. Comp. Biol.* 50:888–907. [8.3.1]

Martone, P. T., D. A. Navarro, C. A. Stortz, and J. M. Estevez (2010b). Differences in polysaccharide structure between calcified and uncalcified segments in the coralline *Calliarthron cheilosporioides*. *J. Phycol.* 46:507–15. [6.5.3]

Martone, P. T., L. Kost, and M. Boller (2012). Drag reduction in wave-swept macroalgae: alternative strategies and new predictions. *Am. J. Bot.* 99:806–15. [8.1.4, 8.3.1]

Maschek, J. A., and B. J. Baker (2008). The chemistry of algal secondary metabolites. In C. D. Amsler (ed.), *Algal Chemical Ecology* (pp. 1–24). London: Springer (Limited). [4.4.1]

Mass, T., A. Genin, U. Shavit, M. Grinstein, and D. Tchernov (2010). Flow enhances photosynthesis in marine benthic autotrophs by increasing the efflux of oxygen from the organism to the water. *P. Natl. Acad. Sci. USA* 107:2527–31. [8.2.1]

Masterson, P., F. A. Arenas, R. C. Thompson, and S. R. Jenkins (2008). Interaction of top down and bottom up factors in intertidal rockpools: effects on early successional macroalgal community composition, abundance and productivity. *J. Exp. Mar. Biol. Ecol.* 363:12–20. [3.3.4]

Mathews, S. (2006). Phytochrome-mediated development in land plants: red light sensing evolves to meet the challenges of changing light environments. *Mol. Ecol.* 15:3483–503. [2.3.3]

Mathieson, A. C., and T. L. Norall (l975). Physiological studies of subtidal red algae. *J. Exp. Mar. Biol. Ecol.* 20:237–47. [7.3.2]

Mathieson, A. C., and C. A. Penniman (1991). Floristic patterns and numerical classification of New England estuarine and open coastal seaweed populations. *Nova Hedwigia* 52:453–85. [3.3.5]

Mathieson, A. C., C. J. Dawes, M. L. Anderson, and E. J. Hehre (2001). Seaweeds of the Brave Boat Harbor salt marsh and adjacent open coast of southern Maine. *Rhodora* 103:1–46. [3.3.5]

Mathieson, A. C., C. J. Dawes, A. L. Wallace, and A. S. Klein (2006). Distribution, morphology, and genetic affinities of dwarf embedded *Fucus* populations from the Northwest Atlantic Ocean. *Bot. Mar.* 49:283–303. [3.3.5]

Mathieson, A. C., E. J. Hehre, C. J. Dawes, and C. D. Neefus (2008). An historical comparison of seaweed populations from Casco Bay, Maine. *Rhodora* 110:1–102. [3.4]

Matilsky M. B., and W. Jacobs (1983). Accumulation of amyloplasts on the bottom of normal and inverted rhizome tips of *Caulerpa prolifera* (Forsskal) Lamouroux. *Planta* 159:189–92. [2.6.2]

Matson, P. G., and M. S. Edwards (2007). Effects of ocean temperature on the southern range limits of two understory kelps, *Pterygophora californica* and *Eisenia arborea*, at multiple life-stages. *Mar. Biol.* 151:1941–9. [7.3.8]

Matsunaga, S., H. Uchida, M. Iseki, M. Watanabe, and A. Murakami (2010). Flagellar motions in phototactic steering in a brown algal swarmer. *Photochem. Photobiol.* 86:374–81. [2.5.1]

Matsuo, Y., H. Imagawa, M. Nishizawa, and Y. Shizuri (2005). Isolation of an algal morphogenesis inducer from a marine bacterium. *Science* 307:1598. [1.2.1, 2.6.2]

Matz, C. (2011). Competition, communication, cooperation: molecular crosstalk in multi-species biofilms. In H.-C. Flemming, J. Wingender, and U. Szewzyk (eds), *Biofilm Highlights* (pp. 29–40). Springer Series on Biofilms 5. Berlin and Heidelberg: Springer-Verlag. [4.2.2]

Maumus, F., P. Rabinowicz, C. Bowler, and M. Rivarola (2011). Stemming epigenetics in marine stramenopiles. *Curr. Genomics* 12:357–70. [1.4.2]

Maxell, B. A., and K. A. Miller (1996). Demographic studies of the annual kelps *Nereocystis luetkeana* and *Costaria costata* (Laminariales, Phaeophyta) in Puget Sound, Washington. *Bot. Mar.* 39:479–89. [3.5.2]

Maximilien, R., R. de Nys, C. Holmström, *et al.* (1998). Chemical mediation of bacterial surface colonization by secondary metabolites from the red alga *Delisea pulchra*. *Aquat. Microb. Ecol.* 15:233–46. [4.2.2]

Maximova, O. V., and A. F. Sazhin (2010). The role of gametes of the macroalgae *Ascophyllum nodosum* (L.) Le Jolis and *Fucus vesiculosus* L. (Fucales, Phaeophyceae) in summer nanoplankton of the White Sea coastal waters. *Mar. Biol.* 50:218–29. [1.1.1]

Mayhoub H., P. Gayral, and R. Jacques (1976). Action de la composition spectrale de la lumière sur la croissance et la reproduction de *Calosiphonia vermicularis* (J. Agardh) Schmitz (Rhodophycées Gigartinales). *C. R. Acad. Sci. Paris* 283(D):1041–4. [2.6.2]

Mayr, E. (1982). *The Growth of Biological Thought: Diversity, Evolution, and Inheritance.* Cambridge, MA: Belknap/ Harvard Press, 992 pp. [1.1.2]

McArthur D. M., and B. L. Moss (1977). The ultrastructure of cell walls in *Enteromorpha intestinalis* (L.) Link. *Br. Phycol. J.* 12:359–68. [4.2.2]

McCandless, E. L. (1981). Polysaccharides of the seaweeds. In C. S. Lobban and M. J. Wynne (eds) *The Biology of Seaweeds* (pp. 559–88). Oxford: Blackwell Scientific. [5.5.1]

McCandless, E. L. J., and J. S. Craigie (1979). Sulfated polysaccharides in red and brown algae. *Annu. Rev. Plant Physiol.* 30:41–53. [5.5.2]

McClanahan, T. R., E. Sala, P. Stickels, *et al.* (2003). Interaction between nutrients and herbivory in controlling algal communities and coral condition on Glover's Reef, Belize. *Mar. Ecol Prog. Ser.* 261:135–47. [9.6.1]

McClanahan, T. R., E. Sala, P. J. Mumby, and S. Jones (2004). Phosphorus and nitrogen enrichment do not enhance brown frondose "macroalgae". *Mar. Poll. Bull.* 48:196–9. [9.6.1]

McClanahan, T. R., R. S. Steneck, D. Pietri, B. Cokos, and S. Jones (2005). Interaction between inorganic nutrients and organic matter in controlling coral reef communities in Glovers Reef Belize. *Mar. Poll. Bull.* 50:566–75. [Essay 6]

McClintock, J. B., and B. J. Baker (2010). *Marine Chemical Ecology.* Taylor and Francis, 624 pp. [4.4]

McClintock, M., N. Higinbotham, E. G. Uribe, and R. E. Cleland (1982). Active, irreversible accumulation of extreme levels of H_2SO_4 in the brown alga, *Desmarestia*. *Plant Physiol.* 70:771–4. [4.4.2]

McClintock, J. B., C. D. Amsler, and B. J. Baker (2010). Overview of the chemical ecology of benthic marine invertebrates along the western Antarctic Peninsula. *Integr. Compar. Biol. (Advanced Access)* 50:967–80. [4.4]

McConnaughey, T. (1991). Calcification in *Chara corallina*: CO_2 hydroxylation generates protons for bicarbonate assimilation. *Limnol. Oceanogr.* 36:619–28. [6.5.3]

McConnaughey, T. (1998). Acid secretion, calcification, and photosynthetic carbon concentrating mechanisms. *Can. J. Bot.* 76:1119-26. [6.5.3]

McConnaughey, T., and J. F. Whelan (1997). Calcification generates protons for nutrient and bicarbonate uptake. *Earth Sci. Rev.* 42:95-117. [6.5.3]

McConnico, L. A., and M. S. Foster (2005). Population biology of the intertidal kelp, *Alaria marginata* Postels and Ruprecht: a non-fugitive annual. *J. Exp. Mar. Biol. Ecol.* 324:61-75. [3.5.2]

McCook, L. J., and A. R. O. Chapman (1992). Vegetative regeneration of *Fucus* rockweed canopy as a mechanism of secondary succession on an exposed rocky shore. *Bot. Mar.* 35:35-46. [2.6.4]

McCook, L. J., J. Jompa, and G. Diaz-Pulido (2001). Competition between corals and algae on coral reefs: a review of evidence and mechanisms. *Coral Reefs* 19:400-17. [Essay 6]

McCord, J. M., and I. Fridovich (1969). Superoxide dismutase —an enzymatic function for erythrocuprein (hemocuprein). *J. Biol. Chem.* 22:6049-55. [7.9]

McCracken, D. A., and J. R. Cain (1981). Amylose in floridean starch. *New Phytol.* 88:67-71. [5.5.1]

McDevit, D.C., and G. W. Saunders (2009). On the utility of DNA barcoding for species differentiation among brown macroalgae (Phaeophyceae) including a novel extraction protocol. *Phycol. Res.* 57:131-41. [Essay 1]

McDevit, D.C., and G. W. Saunders (2010). A DNA barcode examination of the Laminariaceae (Phaeophyceae) in Canada reveals novel biogeographical and evolutionary insights. *Phycologia* 49: 235-48. [Essay 1]

McGlathery, K. J., R. Marino, and R. W. Howarth (1994). Variable rates of phosphorus uptake by shallow marine carbonate sediments: Mechanisms and ecological significance. *Biogeochem.* 25:127-46. [6.5.2]

McGlathery, K. J., M. F. Pedersen, and J. Borum (1996). Changes in intracellular nitrogen pools and feedback controls on nitrogen uptake in *Chaetomorpha linum* (Chlorophyta). *J. Phycol.* 32:393-401. [6.5.1]

McGlathery, K. J., K. Sundback, and I. C. Anderson (2007). Eutrophication in shallow coastal bays and lagoons: the role of plants in the coastal filter. *Mar. Ecol. Prog. Ser.* 348:1-18. [9.6.1, 9.6.2]

McHugh, D. J. (2004). *A Guide to the Sewaweed Industry.* FAO Fosheries Technical Paper. No. 441, Rome: FAO, 105 pp. [10.1, 10.3, 10.3.1, 10.3.2, 10.4.1, 10.5, 10.5.3, 10.6.1, 10.6.2, 10.10, 10.11]

McKay, R. M. L., and S. P. Gibbs (1990). Phycoerythrin is absent from the pyrenoid of *Porphyridium cruentum*: photosynthetic implications. *Planta* 180:249-56. [5.4.3]

McKenzie, G. H., A. L. Ch'ng, and K. R. Gayler (1979). Glutamine synthetase/glutamine:α-ketoglutarate aminotransferase in chloroplasts from the marine alga *Caulerpa simpliciuscula*. *Plant Physiol.* 63:578-82. [6.5.1]

McKenzie, P. F., and A. Bellgrove (2009). Dislodgement and attachment strength of the intertidal macroalga *Hormosira banksii* (Fucales, Phaeophyceae). *Phycologia* 48:335-43. [8.3.1]

McLachlan, P. J., and R. G. S. Bidwell (1978). Photosynthesis of eggs, sperm, zygotes, and embryos of *Fucus serratus*. *Can. J. Bot.* 56:371-3. [5.7.2]

Meinesz, A. (1980). Connaissances actuelles et contribution a l'etude de la reproduction et du cycle des Udoteacees (Caulerpales,Chlorophytes). *Phycologia* 19:110-38. [Fig. 2.3]

Meinesz, A. (2007). Methods for identifying and tracking seaweed invasions. *Bot. Mar.* 50:373-84. [3.4]

Mejia, A. Y., G. N. Puncher, and A. H. Engelen (2012). Macroalgae in tropical marine coastal systems. In C. Wiencke and K. Bischof (eds), *Seaweed Biology: Novel Insights Into Ecophysiology, Ecology and Utilization* (Series: Ecological Studies, Volume 219, pp. 329-57). Berlin and Heidelberg: Springer. [3.3.1]

Melkonian, M., and H. Robenek (1984). The eyespot apparatus of flagellated green algae. *Prog. Phycol. Res.* 3:193-268. [1.3.3]

Meneses, I., and B. Santelices (1999). Strain selection and genetic variation in *Gracilaria chilensis* (Gracilariales, Rhodophyta). *J. Appl. Phycol.* 11:241-6. [10.6.2]

Menge, B. A. (1972a). Competition for food between two intertidal starfish species and its effect on body size and feeding. *Ecology* 53:635-44. [Essay 2]

Menge, B. A. (1972b). Foraging strategy of a starfish in relation to actual prey availability and environmental predictability. *Ecol. Monogr.* 42:25-50. [Essay 2]

Menge, B. A. (1976). Organization of the New England rocky intertidal community: role of predation, competition and environmental heterogeneity. *Ecol. Monogr.* 46:355-93. [Essay 2]

Menge, B. A. (1978a). Predation intensity in a rocky intertidal community. Effect of an algal canopy, wave action and desiccation on predator feeding rates. *Oecologia Berlin* 34:17-35. [Essay 2]

Menge, B. A. (1978b). Predation intensity in a rocky intertidal community. Relation between predator foraging activity and environmental harshness. *Oecologia Berlin* 34:1-16. [Essay 2]

Menge, B. A. (1983). Components of predation intensity in the low zone of the New England rocky intertidal region. *Oecologia* 58:141-55. [Essay 2]

Menge, B. A. (1991). Generalizing from experiments: is predation strong or weak in the New England rocky intertidal? *Oecologia* 88:1-8. [Essay 2]

Menge, B. A. (1992). Community regulation: under what conditions are bottom-up factors important on rocky shores? *Ecology* 73:755-65. [Essay 2]

Menge, B. A., and G. M. Branch (2001). Rocky intertidal communities. In M. D. Bertness, S. D. Gaines, and M. E. Hay (eds), *Marine Community Ecology* (pp. 221-51). Sunderland, MA: Sinauer Associates. [3.1.3, 4.2.1]

Menge, B. A., and J. P. Sutherland (1976). Species diversity gradients: synthesis of the roles of predation, competition, and temporal heterogeneity. *Am. Nat.* 110:351-69. [Essay 2]

Menge, B. A., and J. P. Sutherland (1987). Community regulation: variation in disturbance, competition, and predation in relation to environmental stress and recruitment. *Am. Nat.* 130:730-57. [Essay 2]

Menge, B. A., J. Lubchenco, S. D. Gaines, and L. R. Ashkenas (1986). A test of the Menge-Sutherland model of community organization in a tropical rocky intertidal food web. *Oecologia* (Berlin) 71:75-89. [Essay 2]

Menge, B. A., E. L. Berlow, C. A. Blanchette, S. A. Navarrete, and S. B. Yamada (1994). The keystone species concept: variation in interaction strength in a rocky intertidal habitat. *Ecol. Monogr.* 64:249-86. [Essay 2, 4.3.1]

Menge, B. A., J. Lubchenco, M. E. S Bracken, *et al.* (2003). Coastal oceanography sets the pace of rocky intertidal community dynamics. *Proc. Nat. Acad. Science USA* 100: 12,229-34. [Essay 2, 7.3.1]

Menge, J. L., and B. A. Menge (1974). Role of resource allocation, aggression and spatial heterogeneity in coexistence of two competing intertidal starfish. *Ecol. Monogr.* 44:189-209. [Essay 2]

Menzel, D. (1988). How do giant plant cells cope with injury? The wound response in siphonous green algae. *Protoplasma* 144:73-91. [2.6.4]

Menzel, D. (1994). Cell differentiation and the cytoskeleton in *Acetabularia*. *New Phytol.* 128:369-93. [1.3.3, 1.4.3]

Menzel, D., and C. Elsner-Menzel (1989). Actin-based chloroplast rearrangements in the cortex of the giant coenocytic green alga *Caulerpa*. *Protoplasma* 150: 1-8. [1.3.2]

Mercado, J., C. Jimenez, F. X. Niell, and F. L. Figueroa (1996). Comparison of methods for measuring light absorption by algae and their application to the estimation of the package effect. *Sci. Mar.* 60:39-45. [5.3.2]

Mercado, J. M., R. Carmona, and F. X. Niell (1998). Bryozoans increase available CO_2 for photosynthesis in *Gelidium sesquipedale* (Rhodophyceae). *J. Phycol.* 34:925-7. [4.2.2]

Mercado, J. M., C. B. de los Santos, J. L. Perez-Llorens, and J. J. Vergara (2009). Carbon isotopic fractionation in macroalgae from Cadiz Bay (Southern Spain): comparision with other bio-geographic regions. *Estuar. Coast Shelf Sci.* 85:449-58. [5.4.1, 5.4.3]

Metaxas, A., and R. E. Scheibling (1994). Spatial and temporal variability of tidepool hyperbenthos on a rocky shore in Nova Scotia, Canada. *Mar. Ecol. Prog. Ser.* 108:175-84. [3.3.4]

Michael, T. S. (2009). Glycoconjugate organization of *Enteromorpha* (=*Ulva*) *flexuosa* and *Ulva fasciata* (Chlorophyta) zoospores. *J. Phycol.* 45:660-77. [2.5.1]

Michel, G., T. Tonon, D. Scorner, J. M. Cock, and G. Kloareg (2010a). The cell wall polysaccharide metabolism of the brown alga *Ectocarpus siliculosus*. Insights into the evolution of extracellular matrix polysaccharides in eukaryotes. *New Phytol.* 188:82-97. [1.1.1, 1.3.1, 1.3.4]

Michel, G., T. Tonon, D. Scornet, J. M. Cock, and B. Kloareg (2010b). Central and storage carbon metabolism of the brown alga *Ectocarpus siliculosus*: insights into the origin and evolution of storage carbohydrates in eukaryotes. *New Phytol.* 188:67-81. [1.1.1, 5.5.2]

Michener, R. H., and D. M. Schell (1994). Stable isotope ratios as tracers in marine aquatic food webs. In K. Lajtha, and R. H. Michener (eds), *Stable Isotopes in Ecology and Environmental Science* (pp. 138-57). Oxford: Blackwell Scientific Publications. [Essay 3]

Miller III, H. L., and K. H. Dunton (2007). Stable isotope (^{13}C) and O_2 micro-optode alternatives for measuring photosynthesis in seaweeds. *Mar. Ecol. Prog. Ser.* 329:85-97. [5.7.1]

Miller III, H. L., P. J. Neale, and K. H. Dunton (2009). Biological weighting functions for UV inhibition of photosynthesis in the kelp *Laminaria hyperborea* (Phaeophyceae). *J. Phycol.* 45:571-84. [5.7.1]

Miller, K. A., J. L. Olsen, and W. T. Stam (2000). Genetic divergence correlates with morphological and ecological subdivision in the deep-water elk kelp, *Pelagophycus porra* (Phaeophyceae). *J. Phycol.* 36:862-70. [2.6.2]

Miller, R. J., H. S. Lenihan, E. B. Muller, *et al.* (2010). Impacts of metal oxide nanoparticles on marine phytoplankton. *Environ. Sci. Technol.* 44:7329-34. [9.1]

Miller, R. J., S. Bennett, A. A. Keller, S. Pease, and H. S. Lenihan (2012). TiO_2 nanoparticles are phototoxic to marine phytoplankton. *PLos ONE* 7(1): doi:10.1371/journal.pone.0030321. [9.1]

Miller, S. M., S. R. Wing, and C. L. Hurd (2006). Photoacclimation of *Ecklonia radiata* (Laminariales, Heterokontophyta) in Doubtful Sound, Fjordland, Southern New Zealand. *Phycologia* 45:44-52. [8.2.1]

Milligan, A. J., and P. J. Harrison (2000). Effects of non-steady-state iron limitation on nitrogen assimilatory enzymes in the marine diatom *Thalassiosira weissfloggi* (Bacillario-phyta). *J. Phycol.* 36:78–86. [6.5.1]

Milligan, K. L. D., and R. E. De Wreede (2000). Variations in holdfast attachment mechanics with developmental stage, substratum-type, season, and wave-exposure for the inter-tidal kelp species *Hedophyllum sessile* (C. Agardh) Setchell. *J. Exp. Mar. Biol. Ecol.* 254:189–209. [8.3.1]

Millner, A., and L. V. Evans (1980). The effects of triphenyltin chloride on respiration and phytosynthesis in the green algae *Enteromorpha intestinalis* and *Ulothrix pseudoflacca*. *Plant Cell Environ.* 3:339–48. [9.5]

Millner, A., and L. V. Evans (1981). Uptake of triphenyltin chloride by *Enteromorpha intestinalis* and *Ulothrix pseudo-flacca*. *Plant Cell Environ.* 4:383–9. [9.5]

Mimura, T., R. J. Reid, and F. A. Smith (1998). Control of phosphate transport across the plasmalemma of *Chara corallina*. *J. Exp. Bot.* 49:13–19.

Mimuro, M., and S. Akimoto (2003). Carotenoids of light har-vesting systems: energy transfer processes from fucoxanthin and peridinin to chlorophyll. In A. W. D. Larkum, S. E. Doug-las, and J. A. Raven (eds) *Photosynthesis in Algae. Advances in Photosynthesis and Respiration*, (Vol. 14 pp. 335–49). Dor-drecht, The Netherlands: Kluwer Academic Publishers. [5.3.1]

Minagawa, M., and E. Wada (1984). Stepwise enrichment of ^{15}N along food chains: further evidence and the relation-ship between σ ^{15}N and animal age. *Geochim. Cosmochim. Acta* 48:1135–40.[Essay 3]

Mine, I., Y. Anota, D. Menzel, and K. Okuda (2005). Poly (A)$^+$RNA and cytoskeleton during cyst formation in the cap ray of *Acetabularia peniculus*. *Protoplasma* 226:199–206. [1.4.3]

Mine, I., D. Menzel, and K. Okuda (2008). Morphogenesis in giant-celled algae. *Int. Rev. Cell Mol. Biol.* 266:37–83. [1.3.3, 1.4.3]

Miner, B. G., S. E. Sultan, S. G. Morgan, D. K. Padilla, and R. A. Relyea (2005). Ecological consequences of phenotypic plas-ticity. *Trends Ecol. Evol.* 20:685–92. [2.6.2]

Mishkind, M., D. Mauzerall, and S. I. Beale (1979). Diurnal variation *in situ* of photosynthetic capacity in *Ulva* caused by a dark reaction. *Plant Physiol.* 64: 896–9. [5.7.2]

Mitman, G. G., and J. P. van der Meer (1994). Meiosis, blade development, and sex determination in *Porphyra purpurea* (Rhodophyta). *J. Phycol.* 30:147–59. [1.4.2]

Miura, A. (1975). Porphyra cultivation in Japan. In J. Tokida and H. Hirose (eds), *Advance of Phycology in Japan* (pp. 273–304). The Hague: Dr. W. Junk. [7.3.4]

Miyagishima, S.-Y., and H. Nakanishi (2010). The chloroplast division machinery: origin and evolution. In J. Seckbach and D. J. Chapman (eds), *Red Algae in the Genomic Age (Cellular Origin, Life in Extreme Habitats and Astrobiology 13)* (pp. 3–23). Dordrecht, The Netherlands: Springer. [1.3.2]

Miyamura, S. (2010). Cytoplasmic inheritance in green algae: patterns, mechanisms and relation to sex type. *J. Plant Res.* 123:171–84. [1.3.2]

Miyamura, S., S. Sakaushi, T. Hori, and T. Nagumo (2010). Behavior of flagella and flagellar root systems in the plano-zygotes and settled zygotes of the green alga *Bryopsis maxima* Okamura (Ulvophyceae, Chlorophyta) with refer-ence to spatial arrangement of eyespot and cell fusion site. *Phycol. Res.* 58:258–69. [1.3.3, 2.5.1]

Miyashita, H., H. Ikemoto, N. Kurano, and S. Miyachi (2003). *Acaryochloris marina* ge. Et. Sp. nov (Cyabobacteria), an oxygenic photosynthetic prokaryote containing chlorophyll *d* as a major pigment. *J. Phycol.* 39:1247–53. [5.3.1]

Mizuno, M. (1984). Environment at the front shore of the Institute of Algological Research of Hokkaido University. *Sci. Pap. Inst. Algol. Res., Fac. Sci. Hokkaido* U 7:263–92. [7.5.1]

Mizuta, H., T. Kai, K. Tabuchi, and H. Yasui (2007). Effects of light quality on the reproduction and morphology of sporo-phytes of *Laminaria japonica* (Phaeophyceae). *Aquac. Res.* 38:1323–9. [2.6.2]

Mobley, C. D. (1989). A numerical model for the computation of radiance distributions in natural waters with windrough-ened surfaces. *Limnol. Oceanogr.* 34: 1473–83. [5.2.2]

Moe, R. L., and P. C. Silva (1981). Morphology and taxonomy of *Himanthothallus* (including *Phaeoglossum* and *Phyllogi-gas*), an Antarctic member of the Desmarestiales (Phaeo-phyceae). *J. Phycol.* 17: 15–29. [5.6]

Molis, M., J. Körner, Y. W. Ko, J. H. Kim, and M. Wahl (2006). Inducible responses in the brown seaweed *Ecklonia cava*: the role of grazer identity and season. *J. Ecol.* 94:243–9. [4.4.2]

Molis, M., H. Wessels, W. Hagen, *et al.* (2009). Do sulphuric acid and the brown alga *Desmarestia viridis* support com-munity structure in Arctic kelp patches by altering grazing impact, distribution patterns, and behaviour of sea urchins? *Polar Biol.* 32:71–82. [4.4.2]

Moncreiff, C. A., and M. J. Sullivan (2001). Trophic importance of epiphytic algae in subtropical seagrass beds: evidence from multiple stable isotope analyses. *Mar. Ecol. Prog. Ser.* 215:93–106. [Essay 3]

Monro, K., and A. G. B. Poore (2009a). The potential for evolutionary responses to cell-lineage selection on growth form and its plasticity in a red seaweed. *Am. Nat.* 173:151–63. [1.4.2, 2.6.2]

Monro, K., and A. G. B. Poore (2009b). Performance benefits of growth-form plasticity in a clonal red seaweed. *Biol. J. Linn. Soc.* 97:80-9. [2.6.2]

Monro, K., A. G. B. Poore, and R. Brooks (2007). Multivariate selection shapes environment-dependent variation in the clonal morphology of a red seaweed. *Evol. Ecol.* 21:765-82. [2.6.2]

Monteiro, C. A., A. H. Engelen, and R. Santos (2009). Macro- and mesoherbivores prefer native seaweeds over the invasive brown seaweed *Sargassum muticum*: a potential regulating role on invasions. *Mar. Biol.* 156:2505-15. [3.4]

Moon, D. A., and L. J. Goff (1997). Molecular characterization of two large DNA plasmids in the red alga *Porphyra pulchra*. *Curr. Genet.* 32:132-8. [1.4.2]

Morel, F. M. M., J. G. Rueter, D. M. Anderson, and R. R. L. Guillard (1979). Aquil: a chemically defined phytoplankton culture medium for trace metal studies. *J. Phycol.* 15:135-41. [9.3.4]

Moreno, C. A., and E. Jaramillo (1983). The role of grazers in the zonation of intertidal macroalgae on the Chilean coast. *Oikos* 41:73-6. [4.3.1]

Morris, C.A., B. Nicolaus, V. Sampson, J. L. Harwood, and P. Kille (1999). Identification and characterization of a recombinant metallothionein protein from a marine alga, *Fucus vesiculosus. Biochem. J.* 338:553-60. [9.3.3]

Moss, B. (1964). Wound healing and regeneration in *Fucus vesiculosus* L. *Proc. Intl. Seaweed Symp.* 4: 117-22. [2.6.4]

Moss, B. (1974). Attachment and germination of the zygotes of *Pelvetia canaliculata* (L.) Dcne. et Thur. (Phaeophyceae, Fucales). *Phycologia* 13:317-22. [2.5.2]

Moss, B. L. (1982). The control of epiphytes by *Halidrys siliquosa* (L.) Lyngb. (Phaeophyta, Cystoseiraceae). *Phycologia* 21:185-91. [4.2.2]

Moss, B. L. (1983). Sieve elements in the Fucales. *New Phytol* 93:433-7. [5.6]

Motomura, T. (1990). Ultrastructure of fertilization in *Laminaria angustata* (Phaeophyta, Laminariales) with emphasis on the behavior of centrioles, mitochondria and chloroplasts of the sperm. *J. Phycol.* 26:80-9. [1.3.2]

Motomura, T., and C. Nagasato (2004). The first spindle formation in brown algal zygotes. *Hydrobiologia* 512:171-6. [1.3.5]

Motomura, T., and Y. Sakai (1988). The occurrence of flagellated eggs in *Laminaria angustata* (Phaeophyta, Laminariales). *J. Phycol.* 24:282-5. [2.4]

Motomura, T., C. Nagasato, and K. Kimura (2010). Cytoplasmic inheritance of organelles in brown algae. *J. Plant Res.* 123:185-92. [1.3.2, 1.3.5]

Muhlin, J. F., C. R. Engel, R. Stessel, R. A. Weatherbee, and S. H. Brawley (2008). The influence of coastal topography, circulation patterns, and rafting in structuring populations of an intertidal alga. *Mol. Ecol.* 17:1198-210. [3.3.7]

Mulholland, M. R., and M. W. Lomas (2008). Nitrogen uptake and assimilation. In D.G. Capone, D.A. Bronk, M.R. Mulholland, and E.J. Carpenter (eds), *Nitrogen in the Marine Environment* (pp. 303-84). New York: Academic Press. [6.4.1, 6.5]

Müller, D. G. (1963). Die Temperaturabhangigkeit der Sporangienbildung bei *Ectocarpus siliculosus* von verschiedenen Standorten. *Publ. Staz. Zool. Napoli* 33:310-14. [2.3.1, 2.3.2]

Müller, D. G. (1981). Sexuality and sex attraction. In C. S. Lobban and M. J. Wynne (eds), *The Biology of Seaweeds* (pp. 661-74). Oxford: Blackwell Scientific. [2.4]

Müller, D. G. (1989). The role of pheromones in sexual reproduction of brown algae. In A. W. Coleman, L. 1. Goff, and 1. R. Stein-Taylor (eds), *Algae as Experimental Systems* (pp. 201-13). New York: Alan R. Liss. [2.4]

Müller, D. G., I. Maier, and G. Gassmann (1985). Survey on sexual pheromone specificity in Laminariales (Phaeophyceae). *Phycologia* 24:475-7. [2.4]

Müller, R., C. Wiencke, and K. Bischof (2008). Interactive effects of UV radiation and temperature on microstages of Laminariales (Phaeophyceae) from the Arctic and North Sea. *Clim. Res.* 37: 203-13. [7.3.3, 7.3.4, 7.3.6]

Müller, R., C. Wiencke, K. Bischof, and B. Krock (2009a). Zoospores from three Arctic Laminariales under different UV radiation and temperature conditions: exceptional spectral absorbance properties and lack of phlorotannin induction. *Photochem. Photobiol.* 85: 970-7. [5.7.3]

Müller, R., T. Laepple, I. Bartsch, and C. Wiencke (2009b). Impact of oceanic warming on the distribution of seaweeds in polar and cold-temperate waters *Bot. Mar.* 52 617-38. [7.3.6, 7.3.8, 7.6, 7.9]

Mumby, P. J., and R. S. Steneck (2011). The resilience of coral reefs and its implications for reef management. In Z. Dubinsky, and N. Stambler (eds), *Coral Reefs: An Ecosystem in Transition* (pp. 509-19). Dordrecht, The Netherlands: Springer. [Essay 6]

Mumford, T. E., Jr. (1990). Nori cultivation in North America: growth of the industry. *Hydrobiol.* 204/205:89-98. [10.2]

Mumford, T. E., Jr., and A. Miura (1988). Porphyra as food: cultivation and economics. In C. A. Lembi and J. R. Waaland (eds), *Algae and Human Affairs* (pp. 87-117). Cambridge: Cambridge University Press. [10.2.3]

Munda, I. M. (1984). Salinity dependent accumulation of Zn, Co and Mn in *Scytosiphon lomentaria* (Lyngb.) Link and *Enteromorpha intestinalis* (L.) from the Adriatic Sea. *Bot. Mar.* 27:371-6. [9.3.5]

Munda, I. M., and V. Hudnik (1986). Growth response of *Fucus vesiculosus* to heavy metals, singly and in dual combinations, as related to accumulation. *Bot. Mar.* 29:401–12. [9.3.5]

Munda, I. M., and V. Hudnik (1988). The effects of Zn, Mn, and Co accumulation on growth and chemical composition of *Fucus vesiculosus* L. under different temperature and salinity conditions. *Mar. Ecol.* 9:213–25. [9.3.5]

Munns, R., H. Greenway, and G. O. Kirst (1983). Halotolerant eukaryotes. In A. Pirson, and M. H. Zimmerman (eds) *Encyclopedia of Plant Physiology. Vol. 12: Physiological Plant Ecology III* (pp. 59–135). Berlin: Springer-Verlag. [7.4.3]

Muñoz, J., J. M. Cancino, and M. X. Molina (1991). Effect of encrusting bryozoans on the physiology of their algal substratum. *J. Mar. Biol. Assoc. UK.* 71:877–82. [4.2.2]

Murray, S. N., and P. S. Dixon (1992). The Rhodophyta: some aspects of their biology. III. *Oceanogr. Mar. Biol.* 30:1–148. [2.2, 2.5.3, 2.6.2]

Murray, S. N., and M. M. Littler (1978). Patterns of algal succession in a perturbated marine intertidal community. *J. Phycol.* 14:506–12. [9.6.1]

Murray, S. N., R. F. Ambrose, and M. N. Dethier (2006). *Monitoring Rocky Shores*. Princeton, NJ: University of California Press, 240 pp. [3.5.1, 3.5.2]

Murthy, M. S., T. Ramakrishna, G. V. Sarat Babu, and Y. N. Rao (1986). Estimation of net primary productivity of intertidal seaweeds: limitations and latent problems. *Aquat. Bot.* 23:383–7. [5.7]

Mutchler, T., M. J. Sullivan, and B. Fry (2005). Potential of N-14 isotope enrichment to resolve ambiguities in coastal trophic relationships. *Mar. Ecol. Prog. Ser.* 266:27–33. [Essay 3]

Myers, J. H., L. Gunthorpe, G. Allinson, and S. Duda (2006). Effects of antifouling biocides to the germination and growth of the marine macroalga *Hormosira banksii* (Turner) Desicaine. *Mar. Poll. Bull.* 52:1048–55. [9.5]

Myers, J. H., S. Duda, L. Gunthorpe, and G. Allinson (2007). Evaluation of the *Hormosira banksii* (Turner) Desicaine germination and growth inhibition bioassay for use as a regulatory assay. *Chemosphere* 69:955–60. [9.2]

Nagasato, C. (2005). Behavior and function of paternally inherited centrioles in brown algal zygotes. *J. Plant Res.* 118:361–9. [1.3.2]

Nagasato, C., A. Inoue, M. Mizuno, *et al.* (2010). Membrane fusion process and assembly of cell wall during cytokinesis in the brown alga, *Silvetia babingtonii* (Fucales, Phaeophyceae). *Planta.* 232:287–98. [1.3.5]

Nagashima, H., S. Nakamura, K. Nisizawa, and T. Hori (1971). Enzymic synthesis of floridean starch in a red alga, *Serraticardia maxima*. *Plant Cell Physiol.* 12:243–53. [5.5.3]

Nakahara, H., and Y. Nakamura (1973). Parthenogenesis, apogamy and apospory in *Alaria crassifolia* (Laminariales). *Mar. Biol.* 18:327–32. [2.2, 2.6.1]

Nakajima, Y., Y. Endo, Y. Ionue, *et al.* (2006). Ingestion of *Hijiki* seaweed and risk of arsenic poisoning. *Appl. Organomet. Chem.* 20:557–64. [9.3.6]

Naldi, M., and P. A. Wheeler (1999). Changes in nitrogen pools in *Ulva fenestrata* (Chlorophyta) and *Gracilaria pacifica* (Rhodophyta) under nitrate and ammonium enrichment. *J. Phycol.* 35:70–7. [6.5.1]

Naldi, M., and P. A. Wheeler (2002). ^{15}N measurements of ammonium and nitrate uptake by *Ulva fenestrata* (Chlorophyta) and *Gracilaria pacifica* (Rhodophyta): comparison of net nutrient disappearance, release of ammonium and nitrate and ^{15}N accumulation in algal tissue. *J. Phycol.* 38:135–44. [6.4.1]

Nanba, N., R. Kado, and H. Ogawa (2005). Effects of irradiance and water flow on formation and growth of spongy and filamentous thalli of *Codium fragile*. *Aquat. Bot.* 81:315–25. [2.6.2]

Navarrete, S. A., B. R. Broitman, E. A. Wieters, and J. C. Castilla (2005). Scales of benthic-pelagic coupling and the intensity of species interactions: from recruitment limitation to top down control. *Proc. Nat. Acad. Sci. USA* 102:18,046–51. [Essay 2]

Neefus, C., A. C. Mathieson, T. L. Bray, and C. Yarish (2008). The occurrence of three introduced Asiatic species of *Porphyra* (Bangiales, Rhodophyta) in the northwestern Atlantic. *J. Phycol.* 44:1399–414. [10.2]

Neill, K., S. Heesch, and W. Nelson (2008). *Diseases, Pathogens and Parasites of Undaria pinnatifida*. Ministry of Agriculture and Forestry, Biosecurity New Zealand, Tech. Paper No. 2009/44. Available at www.maf.govt.nz/publications. [10.2.3, 10.3.1, 10.4.1]

Neish, A. C., P. F. Shacklock, C. H. Fox, and F. J. Simpson (1977). The cultivation of *Chondrus crispus*. Factors affecting growth under greenhouse conditions. *Can. J. Bot.* 55:2263–71. [5.5.1]

Neiva, J., G. A. Pearson, M. Valero, and E. A. Serrão (2010). Surfing the wave on a borrowed board: range expansion and spread of introgressed organellar genomes in the seaweed *Fucus ceranoides* L. *Mol. Ecol.* 19:4812–22. [1.4.2]

Nelson, S. G., and A. W. Siegrist (1987). Comparison of mathematical expressions describing light-saturation curves for photosynthesis by tropical marine macroalgae. *Bull. Mar. Sci.* 41:617–22. [5.3.3]

Nelson, T. A., D. J. Lee, and B. C. Smith (2003). Are "green tides" harmful algal blooms? Toxic properties of water-soluble extracts from two bloom-forming macroalgae, *Ulva fenestrata* and *Ulvaria obscura* (Ulvophyceae). *J. Phycol.* 39:874-9. [4.2.2]

Nelson, W. A. (2005). Life history and growth in culture of the endemic New Zealand kelp *Lessonia variegata* J. Agardh in response to differing regimes of temperature, photoperiod and light. *J. Appl. Phycol.* 17:23-8. [2.3.2]

Nelson, W.A. (2009). Calcified macroalgae – critical to coastal ecosystems and vulnerable to climate change: a review. *Mar. Freshwat. Res.* 60:787-801. [2.6.2, 3.3.2, 6.5.3]

Nelson, W. A., J. Brodie, and M. D. Guiry (1999). Terminology used to describe reproduction and life history stages in the genus *Porphyra* (Bangiales, Rhodophyta). *J. Appl. Phycol.* 11:407-10. [2.2]

Nelson, W. G. (1982). Experimental studies of oil pollution on the rocky intertidal community of a Norwegian fjord. *J. Exp. Mar. Biol. Ecol.* 65:121-38. [9.4.3]

Nelson-Smith, A. (1972). *Oil Pollution and Marine Ecology.* London: Elek Science Press, 260 pp. [9.4.2, 9.4.3]

Neori, A., T. Chopin, M. Troell, *et al.* (2004). Integrated aquaculture: rationale, evolution and state of the art emphasizing seaweed biofiltration in modern mariculture. *Aquaculture* 321:361-91. [10.9, 10.12]

Neori, A., M. Troell, T. Chopin, *et al.* (2007) The need for a balanced ecosystem approach to blue revolution aquaculture. *Environment* 49:36-43. [10.9]

Neushul, M. (1972). Functional interpretation of benthic marine algal morphology. In I. A. Abbott and M. Kurogi (eds), *Contributions to the Systematics of Benthic Marine Algae of the North Pacific* (pp. 47-73). Tokyo: Japan Society for Phycology. [2.5.1, 8.3.1]

Neushul, M. (1981). The ocean as a culture dish: experimental studies of marine algal ecology. *Proc. Intl. Seaweed Symp.* 8:19-35. [1.1.3]

Neville, A. C. (1988). The helicoidal arrangement of microfibrils in some algal cell walls. *Prog. Phycol. Res.* 6:1-21. [Fig. 1.8]

Newcombe, E. M., and R. B. Taylor (2010) Trophic cascade in a seaweed-epifauna-fish food chain. *Mar. Ecol. Prog. Ser.* 408:161-7. [4.3.1]

Nezlin, N. P., K. Kamer, and E. D. Stein (2007). Application of color infrared aerial photography to assess macroalgal distribution in an eutrophic estuary, Upper Newport Bay, California. *Estuar. Coasts* 30:855-68. [3.5.1]

Nicotri, M. E. (1980). Factors involved in herbivore food preference. *J. Exp. Mar. Biol. Ecol.* 42:13-26. [4.3.2]

Niell, F. X. (1976). C:N ratio in some marine macrophytes and its possible ecological significance. *Bot. Mar.* 14:347-50. [6.8.2]

Nielsen, H. D., and S. L. Nielsen (2008). Evaluation of imaging and conventional PAM as a measure of photosynthesis in thin- and thick-leaved marine macroalgae. *Aquat. Biol.* 3:121-31. [5.7.1]

Nielsen, H. D., and S. L. Nielsen (2010). Adaptation to high light irradiances enhances the photosynthetic Cu^{2+} resistance in Cu^{2+} tolerant and non-tolerant populations of the brown macroalga *Fucus serratus. Mar. Poll. Bull.* 60:710-17. [9.3.4]

Nielsen, H. D., C. Brownlee, S. M. Coelho, and M. T. Brown (2003a). Inter-population differences in inherited copper tolerance involve photosynthetic adaptation and exclusion mechanisms in *Fucus serratus. New Phytol.* 160:157-65. [7.9, 9.3.3]

Nielsen, H. D., M. T. Brown, and C. Brownlee (2003b). Cellular responses of developing *Fucus serratus* embryos exposed to elevated concentrations of Cu. *Pl. Cell Environ.* 26:1737-44. [9.3.3]

Nielsen, H. D., T. R. Burridge, C. Brownlee, and M. T. Brown (2005). Prior exposure to Cu contamination influences the outcome of toxicological testing of *Fucus serratus* embryos. *Mar. Poll. Bull.* 50:1675-80. [9.3.4]

Nielsen, K. J. (2001). Bottom-up and top-down forces in tide pools: test of a food chain model in an intertidal community. *Ecol. Monogr.* 71:187-217. [4.3.1]

Nielsen, K. J., and S. A. Navarrete (2004). Mesoscale regulation comes from the bottom-up: intertidal interactions between consumers and upwelling. *Ecol. Lett.* 7:31-41. [4.3.1]

Nienhuis, P. H. (1987). Ecology of salt-marsh algae in the Netherlands. In A. H. L. Huiskes, C. W. P. M. Blom, and J. Rozema (eds), *Vegetation Between Land and Sea* (pp. 66-83). Dordrecht, The Netherlands: Dr. W. Junk. [7.4.4]

Niklas, K. J. (2009). Functional adaptation and phenotypic plasticity at the cellular and whole plant level. *J. Biosci.* 34:613-20. [1.3, 2.3, 8.2.1]

Nilsen, G., and Ø. Nordby (1975). A sporulation-inhibiting substance from vegetative thalli of the green alga, *Ulva mutabilis* Føyn. *Planta* 125:127-39. [2.3.3]

Nishihara, G. N., and J. D. Ackerman (2006). The effect of hydrodynamics on the mass transfer of dissolved inorganic carbon to the freshwater macrophyte *Vallisneria americana. Limnol. Oceanogr.* 51:2734-45. [8.1.3]

Nishihara, G. N., and J. D. Ackerman (2007). On the determination of mass transfer in a concentration boundary layer. *Limnol. Oceanogr. Methods* 5:88-96. [8.1.3]

Nishihara, G. N., and R. Terada (2010). Species richness of marine macrophytes is correlated to a wave exposure gradient. *Phycol. Res.* 58:280-92. [8.1.4, 8.3.2]

Nishikawa, T., and M. Yamaguchi (2008). Effect of temperature on light-limited growth of the harmful diatom *Coscinodiscus wailesii*, a causative organism in the bleaching of aquacultured *Porphyra* thalli. *Harmful Algae* 7:561-6. [10.2.3]

Nishimura, N. J., and D. F. Mandoli (1992). Population analysis of reproductive cell structures of *Acetabularia acetabulum* (Chlorophyta). *Phycologia* 31(3/4):351-8. [Fig. 1.3]

Nisizawa, K., H. Noda, R. Kikuchi, and T. Watanabe (1987). The main seaweed foods in Japan. *Hydrobiologia* 151/152:5-29. [10.3, 10.4, 10.10]

Niwa, K., and T. Sakamoto (2010). Allopolyploidy in natural and cultivated populations of *Porphyra* (Bangiales, Rhodophyta). *J. Phycol.* 46:1097-105. [10.2.3]

Niwa, K., H. Furuita, and T. Yamamoto (2008). Changes of growth characteristics and free amino acid content of cultivated *Porphyra yezoensis* Ueda (Bangiales Rhodophyta) blades with the progression of the number of harvests in a nori farm. *J. Appl. Phycol.* 20:687-93. [10.2.3]

Niwa, K., S. Iida, A. Kato, *et al.* (2009a). Genetic diversity and introgression in two cultivated species (*Porphyra yezoensis* and *Porphyra tenera*) and closely related wild species of *Porphyra* (Bangiales, Rhodophyta). *J. Phycol.* 45:493-502. [10.2]

Niwa, K., Y. Hayashi, T. Abe, and Y. Aruga (2009b). Induction and isolation of pigmentation mutants of *Porphyra yezoensis* (Bangiales, Rhodophyta) by heavy-ion beam irradiation. *Phycol. Res.* 57:194-202. [1.4.2]

Niwa, K., T. Yamamoto, H. Furuita, and T. Abe (2011). Mutation breeding in the marine crop *Porphyra yezoensis* (Bangiales, Rhodophyta): cultivation experiment of the artificial red mutant isolated by heavy ion beam mutagenesis. *Aquaculture* 314:182-7. [10.2.3]

Nöel, L. M.-L. J., S. J. Hawkins, S. R. Stuart, *et al.* (2009). Grazing dynamics in intertidal rockpools: connectivity of microhabitats. *J. Exp. Mar. Biol. Ecol.* 370:9-17. [4.3.1]

North, W. J. (1987) Biology of the *Macrocystis* resource in North America. In M. S. Doty, 1. F. Caddy, and B. Santelices (eds.), *Case Studies of Seven Commerical Seaweed Resources* (pp. 265-311). FAO Fish. Tech. Pap. 281. [10.8]

Norton, T. A. (1977). Ecological experiments with *Sargassum muticum*. *J. Mar. Biol. Assoc. UK* 57:33-43. [7.3.3, 7.3.4]

Norton, T. A. (1991). Conflicting constraints on the form of intertidal algae. *Brit. Phycol. J.* 26:203-18. [1.2.2, 3.1.2]

Norton, T. A. (1992). Dispersal by algae. *Br. Phycol. J.* 27:293-301. [2.5.1]

Norton, T. A., and R. Fetter (1981) The settlement of *Sargassum muticum* propagules in stationary and flowing water. *J. Mar. Biol. Assoc. UK* 61:929-40. [1.1.3, 2.5.1]

Norton, T. A., and A. C. Mathieson (1983). The biology of unattached seaweeds. *Prog. Phycol. Res.* 2:333-86. [2.6.2, 3.3.5]

Norton, T. A., A. C. Mathieson, and M. Neushul (1981). Morphology and environment. In C. S. Lobban and M. J. Wynne (eds), *The Biology of Seaweeds* (pp. 421-51). Oxford: Blackwell Scientific. [2.6.2]

Nott, A., H.-S. Jung, S. Koussevitzky, and J. Chory (2006). Plastid-to-nucleus retrograde signaling. *Annu. Rev. Plant Biol.* 57:739-59. [1.3.2, 1.4.3]

Nugues, M. M., and R. P. M. Bak (2006). Differential competitive abilities between Caribbean coral species and a brown alga: a year of experiments and a long-term perspective. *Mar. Ecol. Prog. Ser.* 315:75-86. [Essay 6]

Nultsch, W., J. Pfau, and U. Rüffer (1981). Do correlations exist between chromatophore arrangement and photosynthetic activity in seaweeds? *Mar Biol* 62: 111-17. [5.7.2]

Nyberg, C. D., and I. Wallentinus (2005). Can species traits be used to predict marine macroalgal introductions? *Biol. Invasions* 7:265-79. [3.4]

Nylund, G. M., and H. Pavia (2005). Chemical versus mechanical inhibition of fouling in the red alga *Dilsea carnosa*. *Mar. Ecol. Prog. Ser.* 299:111-21. [4.2.2]

Nylund, G. M., G. Cervin, M. Hermansson, and H. Pavia (2005). Chemical inhibition of bacterial colonization by the red alga *Bonnemaisonia hamifera*. *Mar. Ecol. Prog. Ser.* 302:27-36. [4.2.2]

Oates, B. R. (1988). Water relations of the intertidal saccate alga *Colpomenia peregrina* (Phaeophyta, Scytosiphonales). *Bot. Mar.* 31:57-63. [7.5.1]

Oates, B. R. (1989). Articulated coralline algae as a refuge for the intertidal saccate species, *Colpomenia peregrina* and *Leathesia difformis* in southern California. *Bot. Mar.* 32:475-8. [8.3.2]

Oates, B. R., and K. M. Cole (1994). Comparative studies on hair cells of two agarophyte algae: *Gelidium vagum* (Gelidiales, Rhodophyta) and *Gracilaria pacifica* (Gracilariales, Rhodophyta). *Phycologia* 33:420-33. [6.4.2]

O'Brien, M. C., and P. A. Wheeler (1987). Short-term uptake of nutrients by *Enteromorpha prolifera* (Chlorophyceae). *J. Phycol.* 23:547-56. [6.4.1, 6.5.2]

O'Brien, P. Y., and P. S. Dixon (1976). The effects of oils and oil components on algae: a review. *Br. Phycol. J.* 11: 115-42. [9.4.2, 9.4.3]

Ogata, E. (1971). Growth of conchocelis in artificial medium in relation to carbon dioxide and calcium metabolism. *J. Shimonoseki U. Fish.* 19:123-9. [5.4.1]

Ogawa, H. (1984). Effects of treated municipal wastewater on the early development of sargassaceous plants. *Hydrobiologia* 116/117:389-92. [9.6.2]

Ogden, J. C., R. A. Brown, and N. Salesky (1973). Grazing by the echinoid *Diadema antillarum* Philippi: formation of halos around West Indian patch reefs. *Science* 182:715-17. [4.3.2]

Okabe, Y., and M. Okada (1990). Nitrate reductase activity and nitrate in native pyrenoids purified from the green alga *Bryopsis maxima*. *Plant Cell Physiol.* 31:429-32. [1.3.2]

O'Kelly, C. J., and B. J. Baca (1984). Time course of carpogonial branch and carposporophyte development in *Callithamnion cordatum* (Rhodophyta) Ceramiales. *Phycologia* 23:407-17. [2.4]

Okuda, T., T. Noda, T. Yamamoto, M. Hori, and M. Nakaoka (2010). Contribution of environmental and spatial processes to rocky intertidal metacommunity structure. *Acta Oecol.* 36:413-22. [3.1.2]

Olson, A. M., and J. Lubchenco (1990). Competition in seaweeds: linking plant traits to competitive outcomes. *J. Phycol.* 26:1-6. [4.0]

Oltmanns, F. (1892). Ueber die Cultur-und Lebensbedingungen der Meeresalgen. *Jahr. Wissensch Bot.* 23:349-440. [5.3.4]

Oohusa, T. (1980). Diurnal rhythm in the rates of cell division, growth and photosynthesis of *Porphyra yezoensis* (Rhodophyceae) cultured in the laboratory. *Bot. Mar.* 23:1-5. [5.7.2]

Oppliger, L. V., J. A. Correa, and A. Peters (2007). Parthenogenesis in the brown alga *Lessonia nigrescens* (Laminariales, Phaeophyceae) from Central Chile. *J. Phycol.* 43:1295-301. [2.2]

Orduña-Rojas, J., and D. Robledo (1999). Effects of irradiance and temperature on the release and growth of carpospores from *Gracilaria cornea* J. Agardh (Gracilariales, Rhodophyta). *Bot. Mar.* 42:315-19. [2.3.2]

Osborne, B. A., and J. A. Raven (1986). Light absorption by plants and its implications for photosynthesis. *Biol. Rev.* 61:1-61. [1.3.2, 5.2.2, 5.3.2]

Osmond, C. B. (1994). What is photoinhibition? Some insights from comparisons of shade and sun plants. In N. R. Baker and N. R. Bowyer (eds), *Photoinhibition of Photosynthesis, From the Molecular Mechanisms to the Field* (pp. 1-24). Oxford: BIOS Scientific Publ. [5.3.3, 7.3.5]

Ouriques, L. C., and Z. L. Bouzon (2003). Ultrastructure of germinating tetraspores of *Hypnea musciformis* (Gigartinales, Rhodophyta). *Plant Biosys.* 137:193-202. [2.5.3]

Ouriques, L. C., É. C. Schmidt, and Z. L. Bouzon (2012). The mechanism of adhesion and germination in the carospores of *Porphyra spiralis* var. *amplifolia* (Rhodophyta, Bangiales). *Micron* 43:269-77. [2.5.2, 2.5.3]

Owens, N. J. P. (1987). Natural variations in ^{15}N in the marine environment. *Adv. Mar. Biol.* 24:389-451. [Essay 3]

Padilla, D. K., and B. J. Allen (2000). Paradigm lost: reconsidering functional form and group hypotheses in marine ecology. *J. Exp. Mar. Biol. Ecol.* 250:207-21. [1.2.2, 4.3.2]

Paerl, H. W., J. Rudek, and M. A. Mallin (1990). Stimulation of phytoplankton in coastal waters by natural rainfall inputs: nutritional and trophic implications. *Mar. Biol.* 107:247-54. [6.2]

Paine, R. T. (1966). Food web complexity and species diversity. *Am. Nat.* 100:65-75. [Essay 2]

Paine, R.T. (1979). Disaster, catastrophe, and local persistence of the sea palm *Postelsia palmaeformis*. *Science* 205:685-7. [8.3.2]

Paine, R. T. (1986). Benthic community-water column coupling during the 1982-1983 El Nino. Are community changes at high latitudes attributable to cause or coincidence? *Limnol. Oceanogr.* 31:351-60. [1.1.2, 7.3.7]

Paine, R. T. (1988). Habitat suitability and local population persistence of the sea palm *Postelsia palmaeformis*. *Ecology* 69:1787-94. [4.2.1, 8.3.2]

Paine, R. T. (1990). Benthic macroalgal competition: complications and consequences. *J. Phycol.* 26:12-17. [4.0, 4.2.1]

Paine, R. T. (1994). *Marine Rocky Shores and Community Ecology: An Experimentalist's Perspective, Excellence in Ecology 4.* Oldendorf/Luhe, Germany: Ecology Institute, 152 pp. [3.1.2, 3.1.3, 3.5.2, 4.2]

Paine, R. T. (2010). Macroecology: does it ignore or can it encourage further ecological synthesis based on spatially local experimental manipulations? *Am. Nat.* 176:385-93. [Essay 4,, 3.5.1]

Paine, R. T., and S. A. Levin (1981). Intertidal landscapes: disturbance and the dynamics of pattern. *Ecol. Monogr.* 51:145-78. [Essay 4,, 4.3.2, 8.3.2]

Paine, R. T., J. L. Ruesink, A. Sun, *et al.* (1996). Trouble in oiled waters: lessons from the *Exxon Valdez* oil spill. *Ann. Rev. Ecol. Syst.* 27:197-235. [9.4.3]

Pak, J. Y., C. Solorzano, M. Arai, and T. Nitta (1991). Two distinct steps for spontaneous generation of subprotoplasts from a disintegrated *Bryopsis* cell. *Plant Physiol.* 96:819-25. [2.6.4]

Palmieri, M., and J. Z. Kiss (2007). The role of plastids in gravitropism. In R. R. Wise and J. K. Hoober (eds), *The Structure and Function of Plastids* (pp. 507-25). Dordrecht, The Netherlands: Springer. [1.3.2]

Palumbi, S. R. (1984). Measuring intertidal wave forces. *J. Exp. Mar. Biol. Ecol.* 81:171-9. [8.1.2]

Palumbi, S. R. (2001). The ecology of marine protected areas. In M. D. Bertness, S. D. Gaines, and M. E. Hay (eds), *Marine Community Ecology* (pp. 509-30). Sunderland, MA: Sinauer Associates, Inc. [3.5.1]

Pandolfi, J. M., R. H. Bradbury, E. Sala, *et al.* (2003). Global trajectories of the long-term decline of coral reef ecosystems. *Science* 301:955-8. [Essay 6]

Pang, S.-J., and K. Lüning (2004). Photoperiodic long-day control of sporophyll and hair formation in the brown alga *Undaria pinnatifida. J. Appl. Phycol.* 16:83-92. [2.3.3, 2.6.2]

Papenfuss, G. F. (1958). Die Gattungen der Rhodophyceen. By Harold Kylin. *Bull. Torrey Botan. Club* 85: 142-3. [Essay 1]

Pareek, M., A. Mishra, and B. Jha (2010). Molecular phylogeny of *Gracilaria* species inferred from molecular markers belonging to three different genomes. *J. Phycol.* 46:1322-8. [1.4.1]

Park, C. S., K. Y. Park, J. M. Back, and E. K. Hwang (2008). The occurrence of pinhole disease in relation to developmental stage in cultivated *Undaria pinnatifida* (Harvey) Suringar (Phaeophyta) in Korea. *J. Appl. Phycol.* 20:485-90. [10.4.1]

Park, H. S., W. J. Jeong, E. C. Kim, *et al.* (2011). Heat shock protein gene family of the *Porphyra seriata* and enhancement of heat stress tolerance by PsHSP70 in *Chlamydomonas. Mar. Biotechnol.* 14:332-42. [7.3.5]

Parke, M. W. (1948). Studies of the British Laminariaceae. I. Growth in *Laminaria saccharina* (L.) Lamour. *J. Mar. Biol. Assoc. UK* 27:651-709. [5.7.4]

Parsons, T. R., M. Takahashi, and B. Hargrave (1977). *Biological Oceanographic Processes* (2nd edn). New York: Pergamon Press, 332 pp. [6.8.2]

Parsons, T. R., Y. Maita, and C. M. Lalli (1984). *A Manual of Chemical and Biological Methods of Seawater Analysis.* New York: Pergamon Press, 173 pp. [6.2]

Pastorok, R. A., and G. R. Bilyard (1985). Effects of sewage pollution on coral reef communities. *Mar. Ecol. Prog. Ser.* 21:175-89. [9.6.1]

Patterson, D. J. (1989a). Stramenopiles: chromophytes from a protistan perspective. In J. C. Green, B. S. C. Leadbeater, and W. L. Diver (eds), *The Chromophyte Algae: Problems and Perspectives* (Systematics Association Special Volume 38, pp. 357-79). Oxford: Clarendon Press. [1.1.1]

Patterson, M. R. (1989b). Nearshore biomechanics [review of Denny's book]. *Science* 243:1374. [8.0]

Patwary, M. U., and J. P. van der Meer (1994). Application of RAPD markers in an examination of heterosis in *Geldium vagum* (Rhodophyta). *J. Phycol.* 30:91-7. [1.4.2]

Paul, N. A., L. Cole, R. de Nys, and P. D. Steinberg (2006a). Ultrastructure of the gland cells of the red alga *Asparagopsis armata* (Bonnemaisoniaceae). *J. Phycol.* 42:637-45. [1.3, 4.2.2]

Paul N.A., R. de Nys R, and P. D. Steinberg (2006b). Chemical defense against bacteria in the red alga *Asparagopsis armata*: linking structure with function. *Mar. Ecol. Prog. Ser.* 306:87-101. [4.2.2]

Paul, V. J., and M. E. Hay (1986). Seaweed susceptibility to herbivory: chemical and morphological correlates. *Mar. Ecol. Prog. Ser.* 33:255-64. [4.4.2]

Paul, V. J., and K. L. Van Alstyne (1988a). Chemical defense and chemical variation in some tropical Pacific species of *Halimeda* (Halimedaceae; Chlorophyta). *Coral Reefs* 6:263-9. [4.4.2]

Paul, V. J., and K. L. Van Alstyne (1988b). Use of ingested algal diterpenoids by *Elysia halimedae* Macnae (Opisthobranchia: Ascoglossa) as antipredator defenses. *J. Exp. Mar. Biol. Ecol.* 119:15-29. [4.4.2]

Paul, V. J., and K. L. Van Alstyne (1992). Activation of chemical defenses in the tropical green algae *Halimeda* spp. *J. Exp. Mar. Biol. Ecol.* 160:191-203. [4.4.2]

Paul, V. J., M. P. Puglisi, and R. Ritson-Williams (2006c). Marine chemical ecology. *Nat. Prod. Rep.* 23:153-80. [4.4]

Paul, V. J., R. Ritson-Williams, and K. Sharp (2011). Marine chemical ecology in benthic environments. *Nat. Prod. Rep.* 28:345-87. [4.4]

Pavia, H., and G. B. Toth (2000). Inducible chemical resistance to herbivory in the brown seaweed *Ascophyllum nodosum. Ecology* 81:3212-25. [4.2.2]

Pavia, H., and G. B. Toth (2008). Macroalgal models in testing and extending defense theories. In C. D. Amsler (ed.), *Algal Chemical Ecology* (pp. 147-72). London: Springer (Limited). [4.4.2]

Pavia, H., G. Cervin, A. Lindgren, and P. Åberg (1997). Effects of UV-B radiation and simulated herbivory on phlorotannins in the brown alga *Ascophyllum nodosum. Mar. Ecol. Prog. Ser.* 157:139-46. [5.7.3, 7.8]

Pavia, H., G. Toth, and P. Åberg (1999). Trade-offs between phlorotannin production and annual growth in natural populations of the brown seaweed *Ascophyllum nodosum. J. Ecol.* 87:761-71. [4.4.2]

Pawlik-Skowrońska, B., J. Pirszel, and M. T. Brown (2007). Concentrations of phytochelatins and glutathione found in natural assemblages of seaweeds depend on species and metal concentrations of the habitat. *Aquat. Toxicol.* 83:190-9. [7.9]

Pearson, G. A., and I. R. Davison (1993). Freezing rate and duration determine the physiological response of intertidal fucoids to freezing. *Mar Biol* 115: 353-62. [7.3.4]

Pearson, G. A., and I. R. Davison (1994). Freezing stress and osmotic dehydration in *Fucus distichus* (Phaeophyta): evidence for physiological similarity. *J. Phycol.* 30:257-67. [7.3.4]

Pearson, G. A., and L. V. Evans (1990). Settlement and survival of *Polysiphonia lanosa* (Ceramiales) spores on *Ascophyllum nodosum* and *Fucus vesiculosus* (Fucales). *J. Phycol.* 26:597-603. [4.2.2]

Pearson, G. A., and E. A. Serrão (2006). Revisiting synchronous gamete release by fucoid algae in the intertidal zone: fertilization success and beyond? *Integr. Comp. Biol.* 46:587–97. [2.3.1, 2.3.4, 2.4, 8.2.2]

Pearson, G. A., E. A. Serrão, and S. H. Brawley (1998). Control of gamete release in fucoid algae: sensing hydrodynamic conditions via carbon acquisition. *Ecology* 79:1725–39. [8.2.2]

Pearson, G., L. Kautsky, and E. Serrão (2000). Recent evolution in Baltic *Fucus vesiculosus*: reduced tolerance to emersion stresses compared to intertidal (North Sea) populations. *Mar. Ecol. Prog. Ser.* 202:67–79. [7.5.1]

Pearson, G. A., E. A. Serrão, and M. Dring (2004). Blue- and green-light signals for gamete release in the brown alga, *Silvetia compressa*. *Oecologia* 133:193–201. [8.2.2]

Pearson, G. A., A. Lago-Leston, and C. Mota (2009). Frayed at the edges: selective pressure and adaptive response to abiotic stressors are mismatched in low diversity edge populations. *J. Ecol.* 97:450–62. [3.1.2, 3.1.3]

Pearson, G. A., G. Hoarau, A. Lago-Leston, *et al.* (2010). An expressed sequence tag analysis of the intertidal brown seaweeds *Fucus serratus* (L.) and *F. vesiculosus* (L.) (Heterokontophyta, Phaeophyceae) in response to abiotic stressors. *Mar. Biotechnol.* 12:195–213. [1.4.1, 7.8]

Peckol, P., and J. Ramus (1988). Abundances and physiological properties of deep-water seaweeds from Carolina outer continental shelf. *J. Exp. Mar. Biol. Ecol.* 115:25–39. [5.3.2]

Peckol, P., J. M. Krane, and J. L. Yates (1996). Interactive effects of inducible defense and resource availability on phlorotannins in the North Atlantic brown alga *Fucus vesiculosus*. *Mar. Ecol. Prog. Ser.* 138:209–17. [4.4.1]

Peddigari, S., W. Zhang, K. Takechi, H. Takano, and S. Takio (2008). Two different clades of *copia*-like retrotransposons in the red alga, *Porphyra yezoensis*. *Gene* 424:153–8. [1.4.2]

Pedersen, M. F. (1994). Transient ammonium uptake in the macroalga *Ulva lactuca* (Chlorophyta): nature, regulation, and the consequences for the choice of measuring technique. *J. Phycol.* 30:980–6. [6.4.1, 6.4.2, 6.5.1]

Pedersen, M. F., and J. Borum (1996). Nutrient control of algal growth in estuarine waters. Nutrient limitation and the importance of nitrogen requirements and nitrogen storage among phytoplankton and species of macroalgae. *Mar. Ecol. Prog. Ser.* 142:261–72. [6.5.1, 6.8.3]

Pedersen, M. F., and J. Borum (1997). Nutrient control of estuarine macroalgae: growth strategy and the balance between nitrogen requirements and uptake. *Mar. Ecol. Prog. Ser.* 161:155–63. [6.5.1, 6.8.3]

Pedersen, M. F., and P. Snoeijs (2001). Patterns of macroalgal diversity, community composition and long-term changes along the Swedish west coast. *Hydrobiologia* 459:83–102. [5.3.2]

Pedersen, L. B., S. Geimer, and J. L. Rosenbaum (2006). Dissecting the molecular mechanisms of intraflagellar transport in *Chlamydomonas*. *Curr. Biol.* 16:450–9. [Fig. 1.15]

Pedersen, M. F., J. Borum, and F. L. Fotel (2010). Phosphorus dynamics and limitation of fast- and slow-growing temperate seaweeds in Oslofjord, Norway. *Mar. Ecol. Prog. Ser.* 399:103–15. [6.5.2, 6.7.2, 6.8.3]

Pedersen, P. M. (1981). Phaeophyta: life histories. In C. S. Lobban and M. J. Wynne (eds), *The Biology of Seaweeds* (pp. 194–217). Oxford: Blackwell Scientific. [2.5.3]

Pelevin, V. N., and V. A. Rutkovskaya (1977). On the optical classification of ocean waters from the spectral attenuation of solar radiation. *Oceanology* 17:28–32. [5.2.2]

Pellegrini, L. (1980). Cytological studies on physodes in the vegetative cells of *Cystoseira stricta* Sauvageau (Phaeophyta, Fucales). *J. Cell Sci.* 41:209–31.[Fig. 1.7]

Pelletreau, K. N., and G. Muller-Parker (2002). Sulfuric acid in the phaeophyte alga *Desmarestia munda* deters feeding by the sea urchin *Strongylocentrotus droebachiensis*. *Mar. Biol.* 141:1–9. [4.4.2]

Pelletreau, K. N., and N. M. Targett (2008). New perspectives for addressing patterns of secondary metabolites in marine algae. In C. D. Amsler (ed.), *Algal Chemical Ecology* (pp. 121–46). London: Springer (Limited). [4.4, 4.4.2]

Penela-Arenaz, M., J. Bellas, and E. Vázquez (2009). Effects of the *Prestige* oil spill on the biota in NW Spain: 5 years of learning. *Adv. Mar. Biol.* 56:365–94. [9.4]

Pennings, S. C. (1990). Size-related shifts in herbivory: specialization in the sea hare *Aplysia californica* Cooper. *J. Exp. Mar. Biol. Ecol.* 142:43–61. [4.3.2]

Pennings, S. C., and M. D. Bertness (2001). Salt-marsh communities. In M. D. Bertness, S. D. Gaines, and M. E. Hay (eds), *Marine Community Ecology* (pp. 289–337). Sunderland, MA: Sinauer Associates, Inc. [3.3.5]

Pennings, S. C., T. H. Carefoot, M. Zimmer, J. P. Danko, and A. Ziegler (2000). Feeding preferences of supralittoral isopods and amphipods. *Canadian Journal of Zoology-Revue Canadienne De Zoologie* 78(11): 1918–29. [Essay 3]

Pentecost, A. (1985). Photosynthetic plants as intermediary agents between environmental HCO3- and carbonate deposition. In W. J. Lucas and J. A. Berry (eds), *Inorganic Carbon Uptake by Aquatic Photosynthetic Organisms* (pp. 459–80). Bethesda, MD: American Society of Plant Physiologists. [5.4.1, 6.5.3]

Percival, E. (1979). The polysaccharides of green, red and brown seaweeds: their basic structure, biosynthesis and function. *Br. Phycol. J.* 14:103–17. [5.5.1, 5.5.2, 5.5.3]

Percival, E., and R. H. McDowell (1967). *Chemistry and Enzymology of Marine Algal Polysaccharides.* New York: Academic Press, 219 pp. [5.5.2]

Pereira, L., A. T. Critchley, A. M. Amado, and P. J. A. Ribeiro-Claro (2009). A comparative analysis of phycolloids produced by underutilized versus industrially utilized carrageenophytes (Gigartinales, Rhodophyta). *J. Appl. Phycol.* 21:599-605. [10.5]

Pereira, R. C., and B. A. P. da Gama (2008). Macroalgal chemical defenses and their roles in structuring tropical marine communities. In C. D. Amsler (ed.), *Algal Chemical Ecology* (pp. 25-56). London: Springer (Limited). [4.4, 4.4.2]

Pereira, R. C., E. M. Bianco, L. B. Bueno, *et al.* (2010). Associational defense against herbivory between brown seaweeds. *Phycologia* 49:424-8. [4.3.2]

Pereira, R., and C. Yarish (2010). The role of *Porphyra* in sustainable culture systems: physiology and applications. In A. Israel and R. Einav (eds), *Role of Seaweeds in a Globally Changing Environment* (pp. 339-54). New York: Springer. [10.2, 10.2.2, 10.2.4]

Pereira, R., G. Kramer, C. Yarish, and I. Sousa-Pinto (2008). Nitrogen uptake by gametophytes of *Porphyra dioica* (Bangiales, Rhodophyta) under controlled-culture conditions. *Eur. J. Phycol.* 43:107-18. [6.4.1]

Pereira, R., C. Yarish, and A. Critchley (2012). Seaweed aquaculture for human foods, in land-based and IMTA systems. In R. A. Meyers (ed.). *Encyclopedia of Sustainability Science and Technology* (pp. 9109-28). New York: Springer Science. [10.5, 10.7, 10.10]

Pereira, T. R., A. H. Engelen, G. A. Pearson, *et al.* (2011). Temperature effects on the microscopic haploid stage development of *Laminaria ochroleuca* and *Saccorhiza polyschides*, kelps with contrasting life histories. *Cah. Biol. Mar.* 52:395-403. [7.3.3]

Perez, R., P. Durand, R. Kaas, *et al.* (1988). *Undaria pinnatifida* on the French coasts: cultivation method, biochemical composition of the sporophyte and the gametophyte. In T. Stadler (ed.), *Algal Biotechnology* (pp. 315-27). Amsterdam: Elsevier. [10.4.1]

Pérez-Rodríguez, E., J. Aguilera, I. Gómez, and F. L. Figueroa (2001). Excretion of coumarins by the Mediterranean green alga *Dasycladus vermicularis* in response to environmental stress. *Mar. Biol.* 139:633-9. [7.8]

Perrin, C., C. Daguin, M. van de Vliet, *et al.* (2007). Implications of mating system for genetic diversity of sister algal species: *Fucus spiralis* and *Fucus vesiculosus* (Heterokontophyta, Phaeophyceae). *Eur. J. Phycol.* 42:219-30. [1.4.2]

Perrone, C., and G. P. Felicini (1976). Les bourgeons adventifs de *Gigartina acicularis* (Wulf.) Lamour. (Rhodophyta, Gigartinales) en culture. *Phycologia* 15:45-50. [2.6.4]

Perrot-Rechenmann, C. (2010). Cellular responses to auxin: division versus expansion. *Cold Spring Harb. Perspect. Biol.* 2:a001446. [1.3.4]

Peters, A. F., and M. N. Clayton (1998). Molecular and morphological investigations of three brown algal genera with stellate plastids: evidence for Scytothamnales ord. Nov. (Phaeophyceae). *Phycologia* 37:106-13. [1.3.2]

Peters, A. F., and B. Schaffelke (1996). *Streblonema* (Ectocarpales, Phaeophyceae) infection in the kelp *Laminaria saccharina* (Laminariales, Phaeophyceae) in the western Baltic. *Hydrobiologia* 326/327:111-16. [4.5.2]

Peters, A. F., van Oppen, M. J. H., C. Wiencke, *et al.* (1997). Phylogeny and historical ecology of the Desmarestiaceae (Phaeophyceae) support a southern hemisphere origin. *J. Phycol.* 33:294-309. [3.3.3]

Peters, A. F., D. Scornet, D. G. Müller, B. Kloareg, and J. M. Cock (2004a). Inheritance of organelles in artificial hybrids of the isogamous multicellular chromist alga *Ectocarpus siliculosus* (Phaeophyceae). *Eur. J. Phycol.* 39:235-42. [1.3.2]

Peters, A. F., D. Marie, D. Scornet, B. Kloareg, and J. M. Cock (2004b). Proposal of *Ectocarpus siliculosus* (Ectocarpales, Phaeophyceae) as a model organism for brown algal genetics and genomics. *J. Phycol.* 40:1079-88. [1.4.1, 2.3.2]

Peters, A. F., D. Scornet, M. Ratin, *et al.* (2008). Life-cycle-generation-specific developmental processes are modified in the immediate upright mutant of the brown alga *Ectocarpus siliculosus*. *Development* 135:1503-12. [2.2]

Peterson, B. J., and B. Fry (1987). Stable isotopes in ecosystem studies. *Ann. Rev. Ecol. Syst.* 18:293-320. [Essay 3]

Peterson, D. H., M. J. Perry, K. E. Bencala, and M. C. Talbot (1987). Phytoplankton productivity in relation to light intensity: a simple equation. *Estuar. Coast Shelf Sci.* 24:813-32. [5.3.3]

Peterson, R. D. (1972). Effects of light intensity on the morphology and productivity of *Caulerpa racemosa* (Forsskal) 1. Agardh. *Micronesica* 8:63-86. [2.6.2]

Petraitis, P. S., E. T. Methratta, E. C. Rhile, N. A. Vidargas, and S. R. Dudgeon (2009). Experimental confirmation of multiple community states in a marine ecosystem. *Oecologia* 161:139-48. [4.3.1]

Petrell, R. J., K. Mazhari Tabrizi, P. J. Harrison, and L. D. Druehl (1993). Mathematical model of *Laminaria* production near a British Columbian salmon sea cage farm. *J. Appl. Phycol.* 5:1-4. [10.3.1]

Pfister, C. A. (2007). Intertidal invertebrates locally enhance primary production. *Ecology* 88:1647-53. [4.5.1]

Pfister, C. A., and M. E. Hay (1988). Associational plant refuges: convergent patterns in marine and terrestrial communities result from differing mechanisms. *Oecologia* 77:118–29. [4.3.2]

Philips, J. C., and C. L. Hurd (2003). Nitrogen ecophysiology of intertidal seaweeds from New Zealand: N uptake, storage and utilization in relation to shore position and season. *Mar. Ecol. Prog. Ser.* 264:31–40. [6.4.2]

Philips, J. C., and C. L. Hurd (2004). Kinetics of nitrate, ammonium and urea uptake by four intertidal seaweeds from New Zealand. *J. Phycol.* 40:534–45. [6.4.2, 6.5.1]

Phillips, D. J. H. (1990). Use of macroalgae and invertebrates as monitors of metal levels in estuaries and coastal waters. In R. W. Furness and P. S. Rainbow (eds), *Heavy Metals in the Marine Environment* (pp. 82–99). Boca Raton, FL: CRC Press. [9.2, 9.3.2]

Phillips, D. J. H. (1991). Selected trace elements and the use of biomonitors in subtropical and tropical marine ecosystems. *Rev. Environ. Contamin. Toxicol.* 120:105–29. [9.3.2]

Phillips, J. A., M. N. Clayton, I. Maier, W. Boland, and D. G. Müller (1990). Sexual reproduction in *Dictyota diemensis* (Dictyotales, Phaeophyta). *Phycologia* 29:367–79. [2.3.4]

Phillips, J. C., G. A. Kendrick, and P. S. Lavery (1997). A test of a functional group approach to detecting shifts in macroalgal communities along a disturbance gradient. *Mar. Ecol. Prog. Ser.* 153:125–38. [1.2.2]

Phlips, E. J., and C. Zeman (1990). Photosynthesis, growth and nitrogen fixation by epiphytic forms of filamentous cyanobacteria from pelagic *Sargassum*. *Bull. Mar. Sci.* 47:613–21. [6.2]

Pianka, E. R. (1966). Latitudinal gradients in species diversity: a review of concepts. *Am. Nat.* 100:33–46. [Essay 2]

Pickett-Heaps, J. D., and J. West (1998). Time-lapse video observations on sexual plasmogamy in the red alga *Bostrychia*. *Eur. J. Phycol.* 33:43–56. [2.4]

Pickett-Heaps, J. D., J. A. West, S. M. Wilson, and D. L. McBride (2001). Time-lapse videomicroscopy of cell (spore) movement in red algae. *Eur. J. Phycol.* 36:9–22. [1.3.3]

Pils, B., and A. Heyl (2009). Unraveling the evolution of cytokinin signaling. *Plant Physiol.* 151:782–91. [2.6.3]

Pinto, E., T. C. S. Sigaud-Kutner, M. A. S. Leitão, O. K., *et al.* (2003). Heavy metal-induced oxidative stress in algae. *J. Phycol.* 39:1008–18. [9.3.3, 9.3.4]

Plastino, E., S. Ursi, and M. T. Fujii (2003). Color inheritance, pigment characterization, and growth of a rare light green strain of *Gracilaria birdiae* (Gracilariales, Rhodophyta). *Phycol. Res.* 51:45–52. [1.4.2]

Plettner, I., M. Steinke, and G. Malin (2005). Ethene (ethylene) production in the marine macroalga *Ulva* (*Enteromorpha*) *intestinalis* L. (Chlorophyta, Ulvophyceae): effect of light-stress and co-production with dimethyl sulphide. *Plant Cell Environ.* 28:1136–45. [2.6.3]

Pohnert, G., and W. Boland (2002). The oxylipin chemistry of attraction and defense in brown algae and diatoms. *Nat. Prod. Rep.* 19:108–22. [2.4, 2.4]

Polis, G. A., W. B. Anderson, and R. D. Holt (1997). Toward an integration of landscape and food web ecology: the dynamics of spatially subsidized food webs. *Ann. Rev. Ecol. Syst.* 28:289–316. [Essay 3]

Polle, A. (1996). Mehler reaction: friend or foe in photosynthesis. *Bot. Acta* 109:84–9. [7.3.5]

Polne-Fuller, M., and A. Gibor (1984). Development studies in *Porphyra*. I. Blade differentiation in *Porphyra perforata* as expressed by morphology, enzymatic digestion, and protoplast regeneration. *J. Phycol.* 20:609–16. [2.6.1, 10.2.4]

Polne-Fuller, M., and A. Gibor (1987). Tissue culture of seaweeds. In K. T. Bird and P. H. Benson (eds), *Seaweed Cultivation for Renewable Resources* (pp. 219–40). Amsterdam: Elsevier. [10.11]

Polne-Fuller, M., and A. Gibor (1990). Development studies in *Porphyra* (Rhodophyceae). III. Effect of culture conditions on wall regeneration and differentiation of protoplasts. *J. Phycol.* 26:674–82. [10.2.4]

Polunin, N. V. C. (1988). Efficient uptake of algal production by a single resident herbivorous fish on the reef. *J. Exp. Mar. Biol. Ecol.* 123:61–76. [4.3.2]

Pomin, V. H. (2010). Structural and functional insights into sulfated galactans: a systematic review. *Glycoconjugate Journal* 27:1–12. [5.5.2]

Pomin, V. H., and P. A. S. Mourao (2008). Structure, biology, evolution, and medical importance of sulfated fucans and galactans. *Glycobiology* 18:1016–27. [5.5.2]

Poore, A. G. B., A. H. Campbell, and P. D. Steinberg (2009). Natural densities of mesograzers fail to limit growth of macroalgae or their epiphytes in a temperate algal bed. *J. Ecol.* 97:164–75. [4.3.1]

Poore, A. G., A. H. Campbell, R. A. Coleman, *et al.* (2012). Global patterns in the impact of marine herbivores on benthic primary producers. *Ecol. Lett.* 15:912–22. [4.3.1, 4.3.2, 4.4.2]

Porter, E. T., P. S. Lawrence, and S. E. Suttles (2000). Gypsum dissolution is not a universal integrator of "water motion". *Limnol. Oceanogr.* 45:145–58. [8.1.4]

Potin, P. (2008). Oxidative burst and related responses in biotic interactions of algae. In C. D. Amsler (ed.) *Algal*

Chemical Ecology (pp. 245-72). Berlin and Heidelberg: Springer-Verlag. [Essay 5, 4.4.2. 7.8]

Potin, P. (2012). Intimate associations between epiphytes, endophytes, and parasites of seaweeds. In C. Wiencke, and K. Bischof (eds), *Seaweed Biology: Novel Insights into Ecophysiology, Ecology and Utilization, Ecological Studies* 219 (pp. 203-34). Berlin and Heidelberg: Springer-Verlag. [4.2.2, 4.5.2]

Preston, C. H. (2001). The *Exxon Valdez* oil spill in Alaska: acute, indirect and chronic effects on the ecosystem. *Adv. Mar. Biol.* 39:1-101. [9.4, 9.4.3]

Preston, M. R. (1988). Marine pollution. In J. P. Riley (ed.), *Chemical Oceanography* (pp. 53-196). Orlando, FL: Academic Press. [9.4.1, 9.4.3]

Price, I. R. (1989). Seaweed phenology in a tropical Australian locality (Townsville, North Queensland). *Bot. Mar.* 32:399-406. [2.3.3]

Price, I. R., R. L. Fricker, and C. R. Wilkinson (1984) *Ceratodictyon spongiosum* (Rhodophyta), the macroalgal partner in an alga-sponge symbiosis, grown in unialgal culture. *J. Phycol.* 20:156-8. [3.3.8, 4.5.1]

Price, N. M., and P. J. Harrison (1988). Specific selenium-containing macromolecules in the marine diatom *Thalassiosira pseudonana*. *Plant Physiol.* 86:192-9. [6.5.6]

Prince, E. K., and G. Pohnert (2010). Searching for signals in the noise: metabolomics in chemical ecology. *Anal. Bioanal. Chem.* 396:193-7. [4.4]

Prince, J. S., and C. D. Trowbridge (2004). Reproduction in the green macroalga *Codium* (Chlorophyta): characterization of gametes. *Bot. Mar.* 47:461-70. [2.2]

Probyn, T. A. (1984). Nitrate uptake by *Chordaria flagelliformis* (Phaeophyta). *Bot. Mar.* 17:271-5. [Table 6.3]

Probyn, T. A., and A. R. O. Chapman (1982). Nitrogen uptake characteristics of *Chordaria flagelliformis* (Phaeophyta) in batch mode and continuous mode experiments. *Mar. Biol.* 71:129-33. [6.4.1, 6.5.1, 6.8.4]

Probyn, T. A., and A. R. O. Chapman (1983). Summer growth of *Chordaria flagelliformis* (0. F. Muell.) C. Ag.: physiological strategies in a nutrient stressed environment. *J. Exp. Mar. Biol. Ecol.* 73:243-71. [6.7.2, 6.8.4]

Probyn, T. A., and C. D. McQuaid (1985). *In situ* measurements of nitrogenous nutrient uptake by kelp (*Ecklonia maxima*) and phytoplankton in a nitrate-rich upwelling environment. *Mar. Biol.* 88:149-54. [6.2]

Provasoli, L., and I. J. Pintner (1980). Bacteria induced polymorphism in an axenic laboratory strain of *Ulva lactuca* (Chlorophyceae). *J. Phycol.* 16:196-201. [2.6.2]

Pu, R., and K. R. Robinson (2003). The involvement of Ca^{2+} gradients, Ca^{2+} fluxes, and CaM kinase II in polarization

and germination of *Silvetia compressa* zygotes. *Planta* 217:407-16. [2.5.3]

Pueschel, C. M. (1989). An expanded survey of the ultrastructure of red algal pit plugs. *J. Phycol.* 25:625-36. [1.2.1]

Pueschel, C. M. (1990). Cell structure. In K. M. Cole and R. G. Sheath (eds), *Biology of the Red Algae* (pp. 7-41). Cambridge: Cambridge University Press. [1.3, 5.6]

Pueschel, C. M., and K. M. Cole (1982). Rhodophycean pit plugs: an ultrastructural survey with taxonomic implications. *Am. J. Bot.* 69:703-20. [Essay 1]

Pueschel, C. M., and R. E. Korb (2001). Storage of nitrogen in the form of protein bodies in the kelp *Laminaria solidungula*. *Mar. Ecol. Prog Ser.* 218:107-14. [6.5.1]

Purton, S. (2002). Algal chloroplasts. In: *eLS* (pp. 1-9). Chichester: John Wiley and Sons Ltd. [1.3.2]

Quillet, M., and G. de Lestang-Bremond (1981). The MeCDPS, a carrying sulphate's nucleotide of the red seaweed *Catenella opuntia* (Grev.) *Proc. Intl Seaweed Symp.* 10:503-7. [5.5.3]

Raffaelli, D. G., and S. J. Hawkins (1996). *Intertidal Ecology*. Dordrecht, The Netherlands: Kluwer Academic Pubs, 356 pp. [3, 3.1.1, 3.1.2, 3.1.3, 3.3.5, 4.3.1, 8.3.2]

Ragan, M. A., and K.-W. Glombitza (1986). Phlorotannins, brown algal polyphenols. *Prog. Phycol. Res.* 4:129-241. [4.4.1]

Raghothama, K. G. (1999). Phosphate acquisition. *Ann. Rev. Plant Physiol. Plant Mol. Biol.* 50:655-93. [6.5.2]

Raikar, V., and M. Wafar (2006). Surge ammonium uptake in macroalgae from a coral atoll. *J. Exp. Mar. Biol. Ecol.* 339:236-40. [6.4.2]

Raimondi, P. T., D. C. Reed, B. Gaylord, and L. Washburn (2004). Effects of self-fertilization in the giant kelp *Macrocystis pyrifera*. *Ecology* 85:3267-76. [1.4.2]

Raimonet, M., G. Guillon, F. Mornet, and P. Richard (2013). Macroalgal $\delta^{15}N$ values in well-mixed estuaries: indicator of anthropogenic nitrogen input or macroalgal metabolism? *Estu. Coastal Shelf Sci.* 119:126-38. [9.6.3]

Rainbow, P. S. (1995). Biomonitoring of heavy metal availability in the marine environment. *Mar. Poll. Bull.* 31:183-92. [9.3.2]

Rainbow, P. S., and D. J. H. Phillips (1993). Cosmopolitan biomonitors of trace metals. *Mar. Poll. Bull.* 26:593-601. [9.3.2]

Ramirez, M. E., D. G. Müller, and A. F. Peters (1986). Life history and taxonomy of two populations of ligulate *Desmarestia* (Phaeophyceae) from Chile. *Can. J. Bot.* 64:2948-54. [2.2]

Ramon, E. (1973). Germination and attachment of zygotes of *Himanthalia elongata* (L.) S. F. Gray. *J. Phycol.* 9:445-9. [2.5.3]

Ramus, J. (1978). Seaweed anatomy and photosynthetic performance: the ecological significance of light guides, heterogeneous absorption and multiple scatter. *J. Phycol.* 14: 352-62. [5.3.2, 5.3.4]

Ramus, J. (1981). The capture and transduction of light energy. In C. S. Lobban and M. J. Wynne (eds) *The Biology of Seaweeds* (pp. 458-92). Oxford: Blackwell Scientific. [5.3.3, 5.3.4, 5.7.1, 5.7.2]

Ramus, J. (1982). Engelmann's theory: the compelling logic. In L. M. Srivastava (ed.), *Synthetic and Degradative Processes in Marine Macrophytes* (pp. 29-46). Berlin: Walter de Gruyter. [5.3.4]

Ramus, J. S. (1990). A form-function analysis of photon capture for seaweeds. *Proc. Intl Seaweed Symp.* 13:65-71. [5.3.2, 5.3.3]

Ramus, J., and G. Rosenberg (1980). Diurnal photosynthetic performance of seaweeds measured under natural conditions. *Mar. Biol.* 56:21-8. [5.7.2]

Ramus, J., and M. Venable (1987). Temporal ammonium patchiness and growth rate in *Codium* and *Ulva* (Ulvophyceae). *J. Phycol.* 23:518-23. [6.8.1]

Rasher, D. B., and M. E. Hay (2010). Chemically rich seaweeds poison corals when not controlled by herbivores. *PNAS* 107:9683-8. [4.2.2]

Rasher, D., S. Engel, V. Bonito, *et al.* (2012). Effects of herbivory, nutrients, and reef protection on algal proliferation and coral growth on a tropical reef. *Oecologia* 169:187-98. [Essay 6]

Rausch, C., and M. Bucher (2002). Molecular mechanisms of phosphate transport in plants. *Planta* 216:23-37. [6.5.2]

Rautenberger, R., and K. Bischof (2006). Impact of temperature on UV susceptibility of two species of *Ulva* (Chlorophyta) from Antarctic and Subantarctic regions. *Polar Biol.* 29:988-96. [7.8, 7.9]

Raven, J. A. (1984). *Energetics and Transport in Aquatic Plants (MBL Lectures in Biology*, Vol. 4). New York: Alan R. Liss. [6.6]

Raven, J. A. (1991). Implications of inorganic carbon utilization: ecology, evolution, and geochemistry. *Can. J. Bot.* 69:908-24. [8.2.1]

Raven, J. A. (1996). Into the voids: the distribution, function, development and maintenance of gas spaces in plants. *Ann. Bot.-London.* 78:137-42. [1.1.1]

Raven, J. A. (1997a). Miniview: multiple origins of plasmodesmata. *Eur. J. Phycol.* 32:95-101. [1.2.1]

Raven, J. A. (1997b). Putting the C in phycology. *Eur. J. Phycol.* 32:319-33. [5.4.2]

Raven, J. A. (2003). Long-distance transport in non-vascular plants. *Plant Cell Envir.* 26:73-85. [5.2.2, 5.6, 6.6]

Raven, J. A. (2010). Inorganic carbon acquisition by eukaryotic algae: four current questions. *Photosynth. Res.* 106:123-34. [5.4.1, 5.4.2, 5.4.3]

Raven, J. A. (2011). The cost of photoinhibition. *Physiol. Plant* 142:87-104. [5.3.3]

Raven, J. A., and J. Beardall (1981). Respiration and photorespiration. *Can. Bull. Fish Aquat. Sci.* 210:55-82. [5.4.1, 5.7.1]

Raven, J. A., and R. J. Geider (1988). Temperature and algal growth. *New Phytol.* 110:441-61. [6.4.2, 7.3.1, 7.3.2]

Raven, J. A., and R. J. Geider (2003). Adaptation, acclimation and regulation in algal photosynthesis. In A. W. D. Larkum, S. E. Douglas, and J. A. Raven (eds), *Photosynthesis in Algae. Advances in Photosynthesis and Respiration* (Volume 14, pp. 385-412). Dordrecht, The Netherlands: Kluwer Academic Publishers. [1.1.3]

Raven, J. A., and C. L. Hurd (2012). Ecophysiology of photosynthesis in macroalgae. *Photosynth. Res.* 113:105-25. [8.1.3, 8.2.1]

Raven, J. A., and W. J. Lucas (1985). The energetics of carbon acquisition. In W. J. Lucas and J. A. Berry (eds) *Inorganic Carbon Uptake by Aquatic Photosynthetic Organisms* (pp 305-24). Rockville MD: The American Society of Plant Physiologists. [5.4.1, 5.4.2]

Raven, J. A., and R. Taylor (2003). Macroalgal growth in nutrient-enriched estuaries: a biochemical and evolutionary perspective. *Water, Air Soil Poll.* 3:7-26. [6.8.1]

Raven, J. A., A. M. Johnston, and J. J. MacFarlane (1990). Carbon metabolism. In K. M. Cole and R. G. Sheath (eds) *Biology of the Red Algae* (pp. 171-202). Cambridge: Cambridge University Press. [5.1]

Raven, J. A., J. E. Kübler, and J. Beardall (2000). Put out the light, and then put out the light. *J. Mar. Ass. UK* 80:1-25. [5.2.2, 5.3.3]

Raven, J. A., A. M. Johnston, J. E. Kübler, *et al.* (2002a). Mechanistic interpretation of carbon isotope discrimination by marine macroalgae and seagrass. *Funct. Plant Biol.* 29:355-78. [5.4.3]

Raven, J. A., A. M. Johnston, J. E. Kübler, *et al.* (2002b). Seaweeds in cold seas: evolution and carbon acquisition. *Ann. Bot.* 90:525-36. [5.4.3, 7.3.1]

Raven, J. A., M. Giordano, and J. Beardall (2008). Insights into the evolution of CCMs from comparisons with other resource acquisition and assimilation processes. *Physiol. Plant.* 133:4-14. [6.3.4]

Raven, J.A., J. Beardall, M. Giordano, and S. C. Maberly (2012). Algal evolution in relation to atmospheric CO_2: carboxylases, carbon concentrating mechanisms and carbon oxidation cycles. *Phil. Trans. Roy. Soc. London B* 367:493-507. [Essay 4]

Raven, P. H., R. F. Evert, and S. E. Eichhorn (2005). *Biology of Plants* (7th edn) Madison, NY: Freeman and Co. Publ., 686 pp. [5.4.2, 6.3.1, 6.5.3]

Rayko, E., F. Maumus, U. Maheswari, K. Jabbari, and C. Bowler (2010). Transcription factor families inferred from genome sequences of photosynthetic stramenopiles. *New Phytol.* 188:52-66. [1.4.1]

Reddy, C. R. K., B. Jha, Y. Fujita, and M. Ohno (2008a). Seaweed protoplasts and their potentials: an overview. *J. Appl. Phycol.* 20:609-17. [2.6.1, 10.2.4, 10.11]

Reddy, C. R. K., M. K. Gupta, V. A. Mantri, and B. Jha (2008b). Seaweed protoplasts: status, biotechnological perspectives and needs. *J. Appl. Phycol.* 20:619-52. [10.11]

Reed, D. C. (1990a). The effects of variable settlement and early competition on patterns of kelp recruitment. *Ecology* 71:776-87. [2.4, 3.5.2, 4.2.3]

Reed, D. C. (1990b). An experimental evaluation of density dependence in a subtidal algal population. *Ecology* 71:2286-96. [3.5.2, 4.2.3]

Reed, D. C., D. R. Laur, and A. W. Ebeling (1988). Variation in algal dispersal and recruitment: the importance of episodic events. *Ecol. Monogr.* 58:321-35. [3.5.2]

Reed, D. C., C. D. Amsler, and A. W. Ebeling (1992). Dispersal in kelps: factors affecting spore swimming and competency. *Ecology* 73:1577-85. [2.5.1]

Reed, D. C., T. W. Anderson, A. W. Ebeling, and M. Anghera (1997). The role of reproductive synchrony in the colonization potential of kelp. *Ecology* 78:2443-57. [2.3.1]

Reed, D. C., M. A. Brzezinski, D. A. Coury, W. M. Graham, and R. L. Petty (1999). Neutral lipids in macroalgal spores and their role in swimming. *Mar. Biol.* 133:737-44. [2.5.1]

Reed, D. C., B. P. Kinlan, P. T. Raimondi, *et al.* (2006). A metapopulation perspective on the patch dynamics of giant kelp in Southern California. In J. Kritzer, and P. Sale (eds), *Marine Metapopulations* (pp. 353-86) Burlington, MA:Elsevier Academic Press. [3.3.2, 3.5.2]

Reed, R. H. (1990c). Solute accumulation and osmotic adjustment. In K. M. Cole and R. G. Sheath (eds), *Biology of Red Algae* (pp. 147-70). Cambridge: Cambridge University Press. [6.3.3, 6.3.5, 7.4.1]

Reed, R. H., and J. C. Collins (1980). The ionic relations of *Porphyra purpurea* (Roth) C. Ag. (Rhodophyta, Bangiales). *Plant Cell Environ.* 3:399-407. [6.4.1]

Reed, R. H., and L. Moffat (1983). Copper toxicity and copper tolerance in *Enteromorpha compressa* (L.) Grev. *J. Exp. Mar. Biol. Ecol.* 63:85-103. [9.3.3]

Reed, R. H., J. C. Collins, and G. Russell (l980). The effects of salinity upon galactosyl-glycerol content and concentration

of the marine red alga *Porphyra purpurea* (Roth) C. Ag. *J. Exp. Bot.* 31:1539-54. [7.4.2]

Reed, R. H., I. R. Davison, J. A. Chudek, and R. Foster (1985). The osmotic role of mannitol in the Phaeophyta: an appraisal. *Phycologia* 24:35-47. [7.4.2]

Rees, D. A. (1975). Stereochemistry and binding behaviour of carbohydrate chains. In W. J. Whelan (ed.) *Biochemistry of Carbohydrates* (pp. 1-42). London: Butterworth. [5.5.2]

Rees, T. A. V. (2003). Safety factors and nutrient uptake by seaweeds. *Mar. Ecol. Prog. Ser.* 263:29-40. [6.5, 6.5.1]

Rees, T. A. V. (2007). Metabolic and ecological constraints imposed by similar rates of ammonium and nitrate uptake per unit surface area at low substrate concentrations in marine phytoplankton and macroalgae. *J. Phycol.* 43:197-207. [6.5.1]

Rees, T. A. V., C. M. Grant, H. E. Harmens, and R. B. Taylor (1998). Measuring rates of ammonium assimilation in marine algae: use of the protonophore carbonyl cyanide *m*-chlorophenylhydrazone to distinguish between uptake and assimilation. *J. Phycol.* 34:264-72. [6.5.3]

Rees, T. A. V., B. C. Dobson, M. Bijl, and B. Morelissen (2007). Kinetics of nitrate uptake by New Zealand marine macroalgae and evidence for two nitrate transporters in *Ulva intestinalis* L. *Hydrobiol.* 586:135-41. [6.4.1, 6.5.1]

Reise, K. (1983). Sewage, green algal mats anchored by lugworms, and the effects on *Turbellaria* and small Polychaeta. *Helgol. Meeresunters.* 36:151-62. [9.6.1]

Reiskind, J. B., P. T. Seamon, and G. Bowes (1988). Alternative methods of photosynthetic carbon assimilation in marine macroalgae. *Plant Physiol.* 87: 686-92. [5.4.3]

Reiskind, J. B., S. Beer, and G. Bowes (1989). Photosynthesis, photorespiration and ecophysiological interaction in marine macroalgae. *Aquat. Bot.* 34:131-52. [5.3.3, 5.4.3]

Reith, M., and J. Munholland (1995). Complete nucleotide sequence of the *Porphyra purpurea* choloroplast genome. *Plant Molec. Biol. Reporter* 13:333-5. [1.3.2]

Ren, G.-Z., J.-C. Wang, and M.-Q. Chen (1984). Cultivation of *Gracilaria* by means of low rafts. *Hydrobiologia* 116/117:72-6. [10.6.2]

Rensing, L., and P. Ruoff (2002). Temperature effect on entrainment, phase shifting, and amplitude of circadian clocks and its molecular bases. *Chronobiol. Int.* 19:807-64. [2.3.1]

Revsbech, N. P. (1989). An oxygen microelectrode with a guard cathode. *Limnol. Oceanogr.* 55:1907-10. [5.7.1]

Reyes-Prieto, A., A. P. M. Weber, and D. Bhattacharya (2007). The origin and establishment of the plastid in algae and plants. *Annu. Rev. Genet.* 41:147-68. [1.3.2]

Rhoades, D. F. (1979). Evolution of plant chemical defense against herbivores. In G. A. Rosenthal and D. H. Janzen (eds), *Herbivores: Their Interaction with Secondary Plant Metabolites* (pp. 3-54). New York: Academic Press. [4.2.2]

Richoux, N. B., and P. W. Froneman (2008). Trophic ecology of dominant zooplankton and macrofauna in a temperate, oligotrophic South African estuary: a fatty acid approach. *Mar. Ecol. Prog. Ser.* 357:121-37. [Essay 3]

Rico, J. M., and M. D. Guiry (1996). Phototropism in seaweeds: a review. *Sci. Mar.* 60:273-81. [2.6.2]

Ridler, N., M. Wowchuk, B. Robinson, *et al.* (2007) Integrated multi-trophic aquaculture (IMTA): a potential strategic choice for farmers. *Aquacult. Econ. Management* 11:99-110. [10.9]

Ries, J. B. (2010). Review: geological and experimental evidence for secular variation in seawater Mg/Ca (calcite-aragonite seas) and its effects on marine biological calcification. *Biogeosciences* 7:2795-848. [6.5.3]

Rietema, H. (1982). Effects of photoperiod and temperature on macrothallus initiation in *Dumontia contorta* (Rhodophyta). *Mar. Ecol. Prog. Ser.* 8:187-96. [2.3.2, 2.6.2]

Rietema, H. (1984). Development of erect thalli from basal crusts in *Dumontia contorta* (Gmel.) Rupr. (Rhodophyta, Cryptonemiales). *Bot. Mar.* 27:29-36. [2.6.2]

Rietema, H., and A. M. Breeman (1982). The regulation of the life history of *Dumontia contorta* in comparison to that of several other Dumontiaceae (Rhodophyta). *Bot. Mar.* 25:569-76. [2.3.3, 2.6.2]

Riquelme, C., A. Rojas, V. Flores, and J. A. Correa (1997). Epiphytic bacteria in a copper-enriched environment in northern Chile. *Mar. Poll. Bull.* 34:816-20. [9.3.3]

Risk, M. J., B. E. Lapointe, O. A. Sherwood, and B. J. Bedford (2009). The use of $\delta^{15}N$ in assessing sewage stress on corals. *Mar. Poll. Bull.* 58:793-802. [9.6.3]

Ritchie, R. J., and A. W. D. Larkum (1987). The ionic relations of small-celled marine algae. *Prog. Phycol. Res.* 5:179-222. [7.4.2]

Ritter, A., S. Goulitquer, J. P. Salaün, *et al.* (2008). Copper stress induces biosynthesis of octadecanoid and eicosanoid oxygenated derivatives in the brown algal kelp *Laminaria digitata. New Phytol.* 180:809-21. [7.7]

Rivera, M., and R. Scrosati (2006). Population dynamics of *Sargassum lapazeanum* (Fucales, Phaeophyta) from the Gulf of California, Mexico. *Phycologia* 45:178-89. [3.5.2]

Rivera, M., and R. Scrosati (2008). Self-thinning and size inequality dynamics in a clonal seaweed (*Sargassum lapazeanum*, Phaeophyceae). *J. Phycol.* 44:45-9 [4.2.3]

Roberson, L. M., and J. A. Coyer (2004). Variation in blade morphology of the kelp *Eisenia arborea*: incipient speciation due to local water motion? *Mar. Ecol. Prog. Ser.* 282:115-28. [2.6.2, 8.1.3, 8.2.1]

Roberts, D. A., A. G. B. Poore, and E. L. Johnston (2006). Ecological consequences of copper contamination in macroalgae: effects on epifauna and associated herbivores. *Envir. Tox. Chem.* 25:2470-9. [9.3.6]

Roberts, E., and A. W. Roberts (2009). A cellulose synthase (CESA) gene from the red alga *Porphyra yezoensis* (Rhodophyta). *J. Phycol.* 45:203-12. [1.3.1]

Roberts, M., and F. M. Ring (1972). Preliminary investigations into conditions affecting the growth of the microscopic phase of *Scytosiphon lomentarius* (Lyngbye) Link. *Mem. Soc. Bot. Fr.* 1972:117-28. [2.6.2]

Roberts, S. K., I. Gillot, and C. Brownlee (1994). Cytoplasmic calcium and *Fucus* egg activation. *Development* 120:155-63. [2.5.2]

Robertson, A. I., and J. J. Lucas (1983). Food choice, feeding rates, and the turnover of macrophyte biomass by a surf-zone inhabiting amphipod. *J. Exp. Mar. Biol. Ecol.* 72:99-124. [Essay 3]

Robles, C., and R. Desharnais (2002). History and current development of a paradigm of predation in rocky intertidal communities. *Ecology* 83:1521-36. [Essay 2]

Roenneberg, T., and M. Mittag (1996). The circadian program of algae. *Cell Dev. Biol.* 7:753-63. [2.3.1]

Rogers, K. M. (2003). Stable carbon and nitrogen isotope signatures indicate recovery of marine biota from pollution at Moa Point, New Zealand. *Mar. Poll. Bull.* 46:821-7. [9.6.3]

Rohde, S., and M. Wahl (2008). Antifeeding defense in Baltic macroalgae: induction by direct grazing versus waterborne cues. *J. Phycol.* 44:85-90. [4.4.2]

Rohde, S., M. Molis, and M. Wahl (2004). Regulation of antiherbivore defence by *Fucus vesiculosus* in response to various cues. *J. Ecol.* 92:1011-18. [4.4.2]

Rohde, S., C. Hebenthal, M. Wahl, R. Karez, and K. Bischoff (2008). Decreased depth distribution of *Fucus vesiculosus* (Phaeophyceae) in the Western Baltic: effects of light deficiency and epibionts on growth and photosynthesis. *Eur. J. Phycol.* 43:143-50. [4.4.2]

Roleda, M. Y., and D. Dethleff (2011). Storm-generated sediment deposition on rocky shores: simulating burial effects on the physiology and morphology of *Saccharina latissima* sporophytes. *Mar. Biol. Res.* 7:213-23. [8.3.2]

Roleda, M., C. Wiencke, D. Hanelt, W. Van de Poll, and A. Gruber (2005). Sensitivity of Laminariales zoospores from Helgoland (North Sea) to ultraviolet and photosynthetically active radiation: implications for depth distribution and seasonal reproduction. *Plant Cell Envir.* 28: 466-79. [7.6]

Rosell, K.-G., and L. M. Strivastava (1985). Seasonal variations in total nitrogen, carbon and amino acids in *Macrocystis integrifola* and *Nereocystis luetkeana* (Phaeophyta). *J. Phycol.* 21:304-9. [6.8.2]

Rosenberg, G., and H. W. Paerl (1981). Nitrogen fixation by blue-green algae associated with the siphonous green seaweed *Codium decorticatum*: effects on ammonium uptake. *Mar. Biol.* 61:151-8. [6.2]

Rosenberg, G., and J. Ramus (1982). Ecological growth strategies in the seaweed *Gracilaria foliifera* (Rhodophyceae) and *Ulva* sp. (Chlorophyceae): soluble nitrogen and reserve carbohydrates. *Mar. Biol.* 66:251-9. [6.5.2, 6.8.1, 6.8.2]

Rosenberg, G., and J. Ramus (1984). Uptake of inorganic nitrogen and seaweed surface area: volume ratios. *Aquat. Bot.* 19:65-72. [6.4.1, 6.8.1]

Rosenberg, G., T. A. Probyn, and K. H. Mann (1984). Nutrient uptake and growth kinetics in brown seaweeds: response to continuous and single additions of ammonium. *J. Exp. Mar. Biol. Ecol.* 80:125-46. [6.4.2]

Rosenberg, G., D. S. Littler, M. M. Littler, and E. C. Oliveira (1995). Primary production and photosynthetic quotients of seaweeds from Sao Paulo State, Brazil. *Bot. Mar.* 38:369-77. [5.7.1]

Rosman, J. H., S. G. Monismith, M. W. Denny, and J. R. Koseff (2010). Currents and turbulence within a kelp forest (*Macrocystis pyrifera*): insights from a dynamically scaled laboratory model. *Limnol. Oceanogr.* 55:1145-58. [8.1.4]

Ross, C., and K. L. Van Alstyne (2007). Intraspecific variation in stress-induced hydrogen peroxide scavenging by the ulvoid macroalga *Ulva lactuca*. *J. Phycol.* 43:466-74. [Essay 5]

Ross, C., V. Vreeland, J. H. Waite, and R. S. Jacobs (2005a). Rapid assembly of a wound plug: stage one of a two-stage wound repair mechanism in the giant unicellular chlorophyte *Dasycladus vermicularis* (Chlorophyceae). *J. Phycol.* 41:46-54. [2.6.4]

Ross, C., F. C. Küpper, V. Vreeland, J. H. Waite, and R. S. Jacobs (2005b). Evidence of a latent oxidative burst in relation to wound repair in the giant unicellular chlorophyte *Dasycladus vermicularis*. *J. Phycol.* 41:531-41. [2.6.4, 7.8]

Ross, C., F. C. Küpper, and R. S. Jacobs (2006). Involvement of reactive oxygen species and reactive nitrogen species in the wound response of *Dasycladus vermicularis*. *Chem. Biol.* 13:353-64. [2.6.4]

Rossi, F., C. Olabarria, M. Incera, and J. Garido (2010). The trophic significance of the invasive seaweed *Sargassum muticum* in sandy beaches. *J. Sea Res.* 63:52-61.

Rothäusler, E., I. Gómez, I. A. Hinojosa, *et al.* (2009). Effect of temperature and grazing on growth and reproduction of floating *Macrocystis* spp. (Phaeophyceae) along a latitudinal gradient. *J. Phycol.* 45:547-59. [3.3.7]

Rothäusler, E., L. Gutow, and M. Thiel (2012). Floating seaweeds and their communities. In C. Wiencke, and K. Bischof (eds), *Seaweed Biology: Novel Insights Into Ecophysiology, Ecology and Utilization* (Ecological Studies, Volume 219, pp. 359-80). Berlin and Heidelberg: Springer. [3.3.7]

Röttgers, R. (2007). Comparison of different variable chlorophyll a fluorescence techniques to determine photosynthetic parameters of natural phytoplankton. *Deep Sea Res.* I 54:437-51. [5.7.1]

Roughgarden, J., S. D. Gaines, and H. Possingham (1988). Recruitment dynamics in complex life cycles. *Science* 241:1460-6. [Essay 2]

Rovilla, M., J. Luhan, and H. Sollesta (2010). Growing the reproductive cells (carpospores) of the seaweed, *Kappaphycus striatum*, in the laboratory until outplanting in the field and maturation to tetrasporophyte. *J. Appl. Phycol.* 22:579-85. [10.5.3]

Rowan, K. S. (1989). *Photosynthetic Pigments of Algae*. Cambridge: Cambridge University Press. [5.3.1]

Rüdiger, W., and F. López-Figueroa (1992). Photoreceptors in algae. *Photochem. Photobiol.* 55:949-54. [2.3.3]

Rueness, J. (1973). Pollution effects on littoral algal communities in the inner Oslofjord, with special reference to *Ascophyllum nodosum*. *Helgol. Meeresunters.* 24:446-54. [9.6.1]

Ruesink, J. L. (1998). Diatom epiphytes on *Odonthalia floccosa*: the importance of extent and timing. *J. Phycol.* 34:29-38. [4.2.2]

Rugg, D. A., and T. A. Norton (1987). *Pelvetia canaliculata*, a high shore seaweed that shuns the sea. In R. M. M. Crawford (ed.), *Plant Life in Aquatic and Amphibious Habitats* (pp. 347-58). Oxford: Blackwell Scientific. [4.5.1]

Rui, F., and W. Boland (2010). Algal pheromone biosynthesis: stereochemical analysis and mechanistic implications in gametes of *Ectocarpus siliculosus*. *J. Org. Chem.* 75:3958-64. [2.4]

Ruiz, D. J., and M. Wolff (2011). The Bolivar Channel Ecosystem of the Galapagos Marine Reserve: Energy flow structure and role of keystone groups. *J. Sea Res.* 66: 123-34. [5.7.5]

Rumpho, M. E., F. P. Dastoor, J. R. Manhart, and J. Lee (2006). The kleptoplast. In R. R. Wise, and J. K. Hoober (eds), *The Structure and Function of Plastids* (pp. 451-73).Series: Advances in Photosynthesis and Respiration, Vol. 23. Springer. [1.3.2]

Rumpho, M. E., J. M. Worful, J. Lee, *et al.* (2008). Horizontal gene transfer of the algal nuclear gene *psbO* to the photosynthetic sea slug *Elysia chlorotica*. *PNAS* 105:17,867-71. [4.5.3]

Rumpho, M. E., K. N. Pelletreau, A. Moustafa, and D. Bhatta-charya (2011). The making of a photosynthetic animal. *J. Exp. Biol.* 214:303-11. [4.5.3]

Runcie, J. W., and A. W. Larkum (2001). Estimating internal phosphorus pools in macroalgae using radioactive phosphorus and trichloroacetic acid extracts. *Anal. Biochem.* 297:191-2. [6.5.2]

Runcie, J. W., R. J. Ritchie, and A. D. W. Larkum (2004). Uptake kinetics and assimilation of phosphorus by *Catenella nipae* and *Ulva lactuca* can be used to indicate ambient phosphate availability. *J. Appl. Phycol.* 16:181-94. [6.5.2]

Runcie, J. W., C. F. D. Gurgel, and K. J. McDermid (2008) *In situ* photosynthetic rates of tropical marine macroalgae at their lower depth limit. *Eur. J. Phycol.* 43:377-88. [5.2.2, 5.3.4]

Ruperez, P., O. Ahrazem, and J. A. Leal (2002). Potential antioxidant capacity of sulfated polysaccharides from the edible marine brown seaweed *Fucus vesiculosus*. *J. Agric. Food Chem.* 50:840-5. [5.5]

Rusig, A.-M., H. Le Guyader, and G. Ducreux (1994). Dedifferentiation and microtubule reorganization in the apical cell protoplast of *Sphacelaria* (Phaeophyceae). *Protoplasma* 179:83-94. [2.6.1]

Russell, B. D., and S. D. Connell (2007). Response of grazers to sudden nutrient pulses in oligotrophic versus eutrophic conditions. *Mar. Ecol Prog. Ser.* 349:73-80. [6.8.5]

Russell, G. (1983). Formation of an ectocarpoid epiflora on blades of *Laminaria digitata*. *Mar. Ecol. Prog. Ser.* 11:1817. [4.2.2]

Russell, G. (1986). Variation and natural selection in marine macroalgae. *Oceanogr. Mar. Biol. Annu. Rev.* 24:309-77. [1.2.2, 2.1]

Russell, G. (1987). Salinity and seaweed vegetation. In R. M. M. Crawford (ed.), *Plant Life in Aquatic and Amphibious Habitats* (pp. 35-52). Oxford: Blackwell Scientific. [7.4.3]

Russell, G. (1988). The seaweed flora of a young semienclosed sea: the Baltic. Salinity as a possible agent of flora divergence. *Helgol Meeresunters* 42: 243-50. [7.4.2, 7.4.3]

Russell, G. (1991). Vertical distribution. In A. C. Mathieson, and P. H. Nienhuis (eds), *Ecosystems of the World. 24. Intertidal and Littoral Ecosystems* (pp. 43-65). New York: Elsevier. [3.1.3]

Russell, G., and J. J. Bolton (1975). Euryhaline ecotypes of *Ectocarpus siliculosus* (Dillw.) *Lyngb. Estu. Cstl. Mar. Sci.* 3:91-4. [7.3.3]

Russell, G., and A. H. Fielding (1974). The competitive properties of marine algae in culture. *J. Ecol.* 62:689-98. [4.2.3]

Russell, G., and A. H. Fielding (1981). Individuals, populations and communities. In C. S. Lobban and M. J. Wynne (eds),

The Biology of Seaweeds (pp. 393-420). Oxford: Blackwell Scientific. [3.5.1, 3.5.2]

Russell, G., and C. J. Veltkamp (1984). Epiphyte survival on skin-shedding macrophytes. *Mar. Ecol. Prog. Ser.* 18:149-53. [4.2.2]

Russell, L. K., C.D. Hepburn, C.L. Hurd, and M.D. Stuart (2008). The expanding range of *Undaria pinnatifida* in southern New Zealand: distribution, dispersal mechanisms and the invasion of wave-exposed environments. *Biol. Inv.* 10:103-15. [3.4]

Russell-Hunter, W. D. (1970). *Aquatic Productivity. An Introduction to Some Basic Aspects of Biological Oceanography and Limnology.* New York: Macmillan. [5.2.2]

Ryther, J. H., J. C. Goldman, C. E. Gifford, *et al.* (1979). Physical models of integrated waste recycling marine polyculture systems. *Aquaculture* 5:163-77. [10.8]

Sacramento, A. T., P. Garcia-Jiménez, R. Alcázar, A. F. Tiburcio, and R. R. Robaina (2004). Influence of polymines on the sporulation of *Grateloupia* (Halymeniaceae, Rhodophyta). *J. Phycol.* 40:887-94. [2.6.3]

Sacramento, A. T., P. Garcia-Jiménez, and R. R. Robaina (2007). The polyamine spermine induces cystocarp development in the seaweed *Grateloupia* (Rhodophyta). *Plant Growth Regul.* 53:147-54. [2.6.3]

Saffo, M. B. (1987). New light on seaweeds. *BioSci.* 37:654-64. [5.3.4]

Saga, N., T. Uchida, and Y. Sakai (1978). Clone *Laminaria* from single isolated cell. *Bull. Jpn. Soc. Sci. Fish.* 44:87. [2.6.1]

Sagert, S., and H. Schubert (1995). Acclimation of the photosynthetic aparatus of *Palmaria palmata* (Rhodophyta) to light qualities that preferentially excite photosystem I or PS II. *J. Phycol.* 31:547-54. [5.3.1]

Sahoo, D., and C. Yarish (2005). Mariculture of seaweeds. In R. Andersen (ed.), *Algal Culturing Techniques* (pp. 219-37). New York: Academic Press. [10.2.2, 10.3.1, 10.4.1]

Saito, A., H. Mizuta, H. Yasui, and N. Saga (2008). Artificial production of regenerable free cells in the gametophyte of *Porphyra pseudolinearis* (Bangiales, Rhodophyceae). *Aquaculture* 281:138-44. [10.2.4]

Salgado, L. T., N. B. Viana, L. R. Andrade, *et al.* (2008). Intracellular storage, transport and exocytosis of halogenated compounds in marine red alga *Laurencia obtusa*. *J. Struct. Biol.* 162:345-55. [1.3, 4.2.2]

Salisbury, J. L. (2007). A mechanistic view on the evolutionary origin for centrin-based control of centriole duplication. *J. Cell. Physiol.* 213:420-8. [1.3.3]

Salles, S., J. Aguilera, and F. L. Figueroa (1996). Light fields in algal canopies: Changes in spectral light ratios and growth of *Porphyra leucosticta* Thur. *Le Jol. Sci. Mar.* 60:29–38. [5.3.3]

Salvucci, M. E. (1989). Regulation of Rubisco activity in vivo. *Physiol. Plant* 77:164–71. [5.4.2]

Sánchez, P. C., J. A. Correa, and G. Garcia-Reina (1996). Host-specificity of *Endophyton ramosum* (Chlorophyta), the causative agent of green patch disease in *Mazzaella laminarioides* (Rhodophyta). *Eur. J. Phycol.* 31:173–9. [4.5.2]

Sanders, H. L., J. F. Grassle, G. R. Hampson *et al.* (1980). Anatomy of an oil spill: long-term effects from the grounding of the barge *Florida* off West Falmouth, Massachusetts. *J. Mar. Res.* 38:265–380. [9.4.3]

Sand-Jensen, K. (1987). Environmental control of bicarbonate use among freshwater and marine macrophytes. In R. M. M. Crawford (ed.) *Plant Life in Aquatic and Amphibious Habitats* (pp. 99–112). Oxford: Blackwell Scientific. [5.3.4]

Sanina, N. M., S. N. Goncharova, and E. Y. Kostetsky (2004). Fatty acid composition of individual polar lipid classes from marine macrophytes. *Phytochemistry* 65:721–30. [Essay 3]

Santelices, B. (1990). Patterns of reproduction, dispersal and recruitment in seaweeds. *Oceanogr. Mar. Biol. Annu. Rev.* 28:177–276. [2.1, 2.3, 2.3.1, 3.5.2]

Santelices, B. (1991). Production ecology of *Gelidium*. *Hydrobiol.* 221:31–44. [10.6.1]

Santelices, B. (1999). How many kinds of individual are there? *Tree* 14:152–5. [1.2.3]

Santelices, B. (2002). Recent advances in fertilization ecology of macroalgae. *J. Phycol.* 38:4–10. [2.4]

Santelices, B. (2004a). A comparison of ecological responses among aclonal (unitary), clonal and coalescing macroalgae. *J. Exp. Mar. Biol. Ecol.* 300:31–64. [1.2.3]

Santelices, B. (2004b). Mosaicism and chimerism as components of intraorganismal genetic heterogeneity. *J. Evol. Biol.* 17:1187–8. [1.2.3]

Santelices, B., and M. S. Doty (1989). A review of *Gracilaria* farming. *Aquaculture* 78:95–133. [10.6.2]

Santelices, B., and E. Martinez (1988). Effects of filter feeders and grazers on algal settlement and growth in mussel beds. *J. Exp. Mar. Biol. Ecol.* 118:281–306. [4.2.1, 4.3.2]

Santelices, B., and F. P. Ojeda (1984). Recruitment, growth and survival of *Lessonia nigrescens* (Phaeophyta) at various tidal levels in exposed habitats of central Chile. *Mar. Ecol. Prog. Ser.* 19:73–82. [4.2.1]

Santelices, B., and I. Paya (1989). Digestion survival of algae: some ecological comparisons between free spores and propagules in fecal pellets. *J. Phycol.* 25:693–9. [2.5.1]

Santelices, B., S. Montalva, and P. Oliger (1981). Competitive algal community organization in exposed intertidal habitats from central Chile. *Mar. Ecol. Prog. Ser.* 6:267–76. [4.2.1]

Santelices, B., J. Correa, and M. Avila (1983). Benthic algal spores surviving digestion by sea urchins. *J. Exp. Mar. Biol. Ecol.* 70:263–9. [2.5.1, 4.3.2]

Santelices, B., A. J. Hoffmann, D. Aedo, M. Bobadilla, and R. Otaiza (1995). A bank of microscopic forms on disturbed boulders and stones in tide pools. *Mar. Ecol. Prog. Ser.* 129:215–28. [Essay 6]

Santelices, B., J. A. Correa, D. Aedo, M. Hormazábal, and P. Sànchez (1999). Convergent biological processes in coalescing Rhodophyta. *J. Phycol.* 35:1127–49. [1.2.1, 1.2.3]

Santelices, B., J. A. Correa, I. Meneses, D. Aedo, and D. Varela (1996). Sporeling coalescence and intraclonal variation in *Gracilaria chilensis* (Gracilariales, Rhodophyta). *J. Phycol.* 32:313–22. [1.2.3, 1.4.2]

Santelices, B., J. J. Bolton, and I. Meneses (2009). Marine algal communities. In J. D. Witman, and K. Roy (eds), *Marine Macroecology* (pp. 153–92). Chicago: University of Chicago Press. [3.1.2]

Santelices, B., J. L. Alvarado, and V. Flores (2010). Size increments due to interindividual fusions: how much and for how long? *J. Phycol.* 46:685–92. [1.2.3]

Santos, R. (1995). Size structure and inequality in a commercial stand of the seaweed *Gelidium sesquipedale*. *Mar. Ecol. Prog. Ser.* 119:253–63. [4.2.3]

Saroussi, S., and S. Beer (2007). Alpha and quantum yield of aquatic plants derived from PAM fluorometry: uses and misuses. *Aquat. Bot.* 86: 89–92. [5.7.1, 7.9]

Saunders, G. W. (2005). Applying DNA barcoding to red macroalgae: a preliminary appraisal holds promise for future applications. *Phil. Trans. R. Soc. London B* 360:1879–88. [Essay 1]

Saunders, G. W. (2008). A DNA barcode examination of the red algal family Dumontiaceae in Canadian waters reveals substantial cryptic species diversity. 1. The foliose *Dilsea-Neodilsea* complex and *Weeksia*. *Botany* 86:773–89. [Essay 1]

Saunders, G. W. (2009). Routine DNA barcoding of Canadian Gracilariales (Rhodophyta) reveals the invasive species *Gracilaria vermiculophylla* in British Columbia. *Molec. Ecol. Res.* 9(s1):140–50. [Essay 1]

Saunders, G. W., and L. D. Druehl (1992). Nucleotide sequences of the small-subunit ribosomal RNA genes from selected Laminariales (Phaeophyta): implications for kelp evolution. *J. Phycol.* 28:544–9. [Essay 1]

Saunders, G.W., and L. D. Druehl (1993). Revision of the kelp family Alariaceae and the taxonomic affinities of *Lessoniopsis* Reinke (Laminariales, Phaeophyta). *Hydrobiologia* 260/261:689–97. [Essay 1]

Saunders, G.W., and M. Hommersand (2004). Assessing red algal supraordinal diversity and taxonomy in the context of contemporary systematic data. *Amer. J. Bot.* 91:1494–507. [Essay 1]

Saunders, G. W., and G. T. Kraft (1997). A molecular perspective on red algal evolution: focus on the Florideophycidae. *Plant Systematics and Evolution (Supplement)* 11:115–38. [Essay 1]

Saunders, G.W., and K. V. Lehmkuhl (2005). Molecular divergence and morphological diversity among four cryptic species of *Plocamium* (Plocamiales, Florideophyceae) in northern Europe. *Eur. J. Phycol.* 40: 293–312. [Essay 1]

Saunders, G.W., C. J. Bird, M. A. Ragan, and E. L. Rice (1995). Phylogenetic relationships of species of uncertain taxonomic position within the Acrochaetiales/Palmariales complex (Rhodophyta): inferences from phenotypic and 18S rDNA sequence data. *J. Phycol.* 31:601–11. [Essay 1]

Saxena, I. M., and M. Brown Jr. (2005). Cellulose biosynthesis: current views and evolving concepts. *Ann. Bot.-London* 96:9–21. [1.3.1]

Scanlan, C. M., and M. Wilkinson (1987). The use of seaweeds in biocide toxicity testing. Part I. The sensitivity of different stages in the life-history of *Fucus*, and of other algae, to certain biocides. *Mar. Environ. Res.* 21:11–29. [9.3.4]

Schaffelke, B. (1995a). Storage carbohydrates and abscisic acid contents in *Laminaria hyperborea* are entrained by experimental daylengths. *Eur. J. Phycol.* 30:313–17. [2.6.3]

Schaffelke, B. (1995b). Abscisic acid in sporophytes of 3 *Laminaria* species (Phaeophyta). *J. Plant Physiol.* 146:453–8. [2.6.3]

Schaffelke, B. (1999a). Short-term nutrient pulses as tools to assess responses of coral reef macroalgae to enhanced nutrient availability. *Mar. Ecol. Prog. Ser.* 182:305–10. [Essay 6]

Schaffelke, B. (1999b). Particulate organic matter as an alternative nutrient source for tropical *Sargassum* species (Fucales, Phaeophyceae). *J. Phycol.* 35:1150–7. [8.2.1, Essay 6]

Schaffelke, B. (2001). Surface alkaline phosphatase activities of macroalgae on coral reefs of the central Great Barrier Reef, Australia. *Coral Reefs* 19:310–17. [Essay 6]

Schaffelke, B., and D. Deane (2005). Desiccation tolerance of the introduced marine green alga *Codium fragile* ssp. *tomentosoides*: clues for likely transport vectors? *Biol. Invas.* 7:557–65. [3.4]

Schaffelke, B., and C. L. Hewitt (2007). Impacts of introduced seaweeds. *Bot. Mar.* 50:397–417. [Essay 6]

Schaffelke, B., and D. W. Klumpp (1998a). Nutrient-limited growth of the coral reef macroalga *Sargassum baccularia* and experimental growth enhancement by nutrient addition in continuous-flow culture. *Mar. Ecol. Progr. Ser.* 164:199–211. [Essay 6]

Schaffelke, B., and D. W. Klumpp (1998b). Short-term nutrient pulses enhance growth and photosynthesis of the coral reef macroalga *Sargassum baccularia*. *Mar. Ecol. Progr. Ser.* 170:95–105. [Essay 6]

Schaffelke, B., and K. Lüning (1994). A circannual rhythm controls seasonal growth in the kelps *Laminaria hyperborea* and *L. digitata* from Helgoland (North Sea). *Eur. J. Phycol.* 29:49–56. [2.3.1, Essay 6]

Schaffelke, B., A. F. Peters, and T. B. H. Reusch (1996). Factors influencing depth distribution of soft bottom inhabiting *Laminaria saccharina* (L.) Lamour. in Kiel Bay, Western Baltic. *Hydrobiologia* 326/327:117–23. [Essay 6]

Schaffelke, B., J. E. Smith, and C. L. Hewitt (2006). Introduced macroalgae: a growing concern. *J. Appl. Phycol.* 18:529–41. [3.4]

Schagerl, M., and M. Möstl (2011). Drought stress, rain and recovery of the intertidal seaweed *Fucus spiralis*. *Mar. Biol.* 158:2471–9. [7.5.1]

Schatz, S. (1980). Degradation of *Laminaria saccharina* by higher fungi: a preliminary report. *Bot. Mar.* 23: 617–22. [5.7.1]

Scheibling, R. E., and P. Gagnon (2006). Competitive interactions between the invasive alga *Codium fragile* ssp. *tomentosoides* and native canopy-forming seaweeds in Nova Scotia (Canada). *Mar. Ecol. Prog. Ser.* 325:1–14. [4.2.1]

Scheibling, R. E., and B. G. Hatcher (2007). Ecology of *Strongylocentrotus droebachiensis*. In J. M. Lawrence (ed.), *Edible Sea Urchins: Biology and Ecology* (pp. 353–92). New York: Elsevier. [4.3.1]

Schiel, D. R. (2006). Rivets or bolts? When single species count in the function of temperate rocky reef communities. *J. Exp. Mar. Biol. Ecol.* 338:233–52. [4.1]

Schiel, D. R., and J. H. Choat (1980). Effects of density on monospecific stands of marine algae. *Nature* 285:324–6. [4.2.3]

Schiel, D. R., and M. S. Foster (1986). The structure of subtidal algal stands in temperate waters. *Oceanogr. Mar. Biol. Annu. Rev.* 24:265–307. [1.1.2, 1.1.3, 3.3, 3.5.1, 3.5.2, 4.1, 4.3.2]

Schiel, D. R., and M. S. Foster (2006). The population biology of large brown seaweeds: ecological consequences of multiphase life histories in dynamic coastal environments. *Annu. Rev. Ecol. Evol. Syst.* 37:343–72. [3.5.2, 4.2.3]

Schiel, D. R., and S. A. Lilley (2011). Impacts and negative feedbacks in community recovery over eight years following removal of habitat-forming macroalgae. *J. Exp. Mar. Biol. Ecol.* 407:108–15. [4.2.1]

Schiel, D. R., J. R. Steinbeck, and M. S. Foster (2004). Ten years of induced ocean warming causes comprehensive changes in marine benthic communities. *Ecol.* 85:1833–9. [3.5.1, 9.8]

Schiel, D. R., S. A. Wood, R. A. Dunmore, and D. I. Taylor (2006). Sediment on rocky intertidal reefs: effects on early post-settlement stages of habitat-forming seaweeds. *J. Exp. Mar. Biol. Ecol.* 331:158–72. [8.3.2]

Schiff, J. A. (1983). Reduction and other metabolic reactions of sulfate. In A. Uiuchli and R. L. Bieleski (eds), *Encyclopaedia of Plant Physiology* (Vol. 15, pp. 382–99). Berlin:Springer-Verlag. [6.5.4]

Schils, T., and S. C. Wilson (2006). Temperature threshold as a biogeographic barrier in northern Indian Ocean macroalgae. *J. Phycol.* 42:749–56. [7.3.8]

Schmid, C. E. (1993). Cell-cell-recognition during fertilization in *Ectocarpus siliculosus* (Phaeophyceae). *Hydrobiologia* 260/261:437–43. [2.4]

Schmid, C. E., N. Schroer, and D. G. Müller (1994). Female gamete membrane glycoproteins potentially involved in gamete recognition in *Ectocarpus siliculosus* (Phaeophyceae). *Plant Sci.* 102:61–7. [2.4]

Schmid, R. (1984). Blue light effects on morphogenesis and metabolism in *Acetabularia*. In H. Senger (ed.), *Blue Light Effects in Biological Systems* (pp. 419–32). Berlin: Springer-Verlag. [2.6.2]

Schmid, R., and M. J. Dring (1996). Blue light and carbon acquisition in brown algae: an overview and recent developments. *Sci. Mar.* 60:115–24. [5.3.3]

Schmid, R., M. Tunnermann, and E.-M. Idziak (1990). Role of red light in hair-formation induced by blue light in *Acetabularia mediterranea*. *Planta* 181:144–7. [2.6.2]

Schmitt, T. M., N. Lindquist, and M. E. Hay (1998). Seaweed secondary metabolites as antifoulants: effects of *Dictyota* spp. diterpenes on survivorship, settlement, and development of marine invertebrate larvae. *Chemoecology* 8:125–31. [4.2.2]

Schmitz, K. (1981). Translocation. In C. S. Lobban and M. J. Wynne (eds), *The Biology of Seaweeds* (pp. 534–58) Oxford: Blackwell Scientific. [5.6]

Schmitz, K., and W. Riffarth (1980). Carrier-mediated uptake of L-leucine by the brown alga *Giffordia mitchellliae*. *Z. Pflanzenphysiol.* 67:311–24. [5.4.1, 6.5.1]

Schmitz, K., and L. M. Srivastava (1974). Fine structure and development of sieve tubes in *Laminaria groenlandica* Rosenv. *Cytobiologie* 10:66–87. [Fig. 5.23]

Schneider, C. W. (1976). Spatial and temporal distributions of benthic marine algae on the continental shelf of the Carolinas. *Bull. Mar. Sci.* 26:133–51. [5.3.4]

Schoch, G. C., B. A. Menge, G. Allison, *et al.* (2006). Fifteen degrees of separation: latitudinal gradients of rocky intertidal biota along the California Current. *Limnol. Oceanogr.* 51:2564–85. [3.5.1]

Schoenwaelder, M. E. A. (2002). The occurrence and cellular significance of physodes in brown algae. *Phycologia* 41:125–39. [1.3]

Schoenwaelder, M. E. A., and M. N. Clayton (1999). The role of the cytoskeleton in brown algal physode movement. *Eur. J. Phycol.* 34:223–9. [1.3.3]

Schofield, O., T. J. Evens, and D. F. Millie (1998). Photosystem II quantum yields and xanthophyll-cycle pigments of the macroalga *Sargassum natans* (Phaeophyceae): responses under natural sunlight. *J. Phycol.* 34: 104–12. [7.3.1]

Schonbeck, M., and T. A. Norton (1978). Factors controlling the upper limits of fucoid algae on the shore. *J. Exp. Mar. Biol. Ecol.* 31:303–13. [3.1.3, 7.3.4, 7.5.1]

Schonbeck, M. W., and T. A. Norton (1979a). An investigation of drought avoidance in intertidal fucoid algae. *Bot. Mar.* 22:133–44. [3.1.3]

Schonbeck, M. W., and T. A. Norton (1979b). Drought hardening in the upper shore seaweeds *Fucus spiralis* and *Pelvetia canaliculata*. *J. Ecol.* 67:687–96. [3.1.3]

Schonbeck, M. W., and T. A. Norton (1979c). The effects of brief periodic submergence on intertidal algae. *Estu. Cstl. Mar. Sci.* 8:205–11. [3.1.3]

Schonbeck, M. W., and T. A. Norton (1980a). Factors controlling the lower limits of fucoid algae on the shore. *J. Exp. Mar. Biol. Ecol.* 43:131–50. [3.1.3, 7.3.4]

Schonbeck, M. W., and T. A. Norton (1980b). The effects on intertidal fucoids of exposure to air under various conditions. *Bot. Mar.* 23:141–7. [3.1.3]

Schramm, W. (1999). Factors influencing seaweed responses to eutrophication: some results from EU-project EUMAC. *J. Appl. Phycol.* 11:69–78. [9.6.1, 9.6.2, Essay 6]

Schreiber, U., W. Bilger, and C. Neubauer (1994). Chlorophyll fluorescence as a non-intrusive indicator for rapid assessment of in vivo photosynthesis. In E. D. Schulze and M. M. Caldwell (eds) *Ecophysiology of Photosynthesis* (pp. 49–70). Berlin: Springer-Verlag. [5.7.1, 7.9]

Schubert, H., S. Sagert, and R. M. Forster (2001) Evaluation of the different levels of variability in the underwater light field of a shallow estuary. *Helgoland Mar. Res.* 55:12–22. [5.2.2]

Schuenhoff, A., M. Shpigel, I. Lupatsch, *et al.* (2003). A semi-recirculating integrated system for the culture of fish and seaweed. *Aquaculture* 221:167–81. [10.9]

Schulze, T., K. Prager, H. Dathe, *et al.* (2010). How the green alga *Chlamydomonas reinhardtii* keeps time. *Protoplasma* 244:3–14. [1.3.5]

Schumacher, J. F., M. L. Carman, T. G. Estes, *et al.* (2007). Engineered antifouling microtopographies – effect of feature size, geometry, and roughness on settlement of zoospores of the green alga *Ulva. Biofouling* 23:55–62. [2.5.2]

Schweikert, K., J. E. S. Sutherland, C. L. Hurd, and D. J. Burritt (2011). UV-B radiation induces changes in polyamine metabolism in the red seaweed *Porphyra cinnamomea. Plant Growth Regul.* 65:389–99. [2.6.3]

Schwenk, K., D. K. Padilla, G. S. Bakken, and R. J. Full (2009). Grand challenges in organismal biology. *Integr. Comp. Biol.* 49:7–14. [2.6.2]

Scott, F. J., R. Wetherbee, and G. T. Kraft (1984). The morphology and development of some prominently stalked southern Australian Halymeniaceae (Cryptonemiales, Rhodophyta). II. The sponge-associated genera *Thamnoclonium* Kuetzing and *Codiophyllum* Gray. *J. Phycol.* 20:286–95. [4.5.1]

Scrosati, R. (2002a). An updated definition of genet applicable to clonal seaweeds, bryophytes, and vascular plants. *Basic Appl. Ecol.* 3:97–9. [1.2.3]

Scrosati, R. (2002b). Morphological plasticity and apparent loss of apical dominance following the natural loss of the main apex in *Pterocladiella capillacea* (Rhodophyta, Gelidiales) fronds. *Phycologia* 41:96–8. [2.6.2]

Scrosati, R. (2005). Review of studies on biomass-density relationships (including self-thinning lines) in seaweeds: main contributions and persisting misconceptions. *Phycol. Res.* 53:224–33. [1.2.3, 4.2.3]

Scrosati, R., and C. Heaven (2008). Trends in abundance of rocky intertidal seaweeds and filter feeders across gradients of elevation, wave exposure, and ice scour in eastern Canada. *Hydrobiologia* 603:1–14. [3.5.1]

Scrosati, R., and B. Mudge (2004). Persistence of gametophyte predominance in *Chondrus crispus* (Rhodophyta, Gigartinaceae) from Nova Scotia after 12 years. *Hydrobiologia* 519:215–18. [2.2]

Scrosati, R., and R. E. De Wreede (1997). Dynamics of the biomass-density relationship and frond biomass inequality for *Mazzaella cornucopiae* (Gigartinaceae, Rhodophyta): implications for the understanding of frond interactions. *Phycologia* 36:506–16. [4.2.3]

Searles, R. B. (1980). Strategy of the red algal life-history. *Amer. Nat.* 115:113–20. [2.4]

Sears, J. R., and R. T. Wilce (1975). Sublittoral, benthic marine algae of southern Cape Cod and adjacent islands: seasonal periodicity, associations, diversity, and floristic composition. *Ecol. Monogr.* 45:337–65. [8.3.2]

Seery, C. R., L. Gunthorpe, and P. J. Ralph (2006). Herbicide impact on *Hormosira banksii* gametes measured by fluorescence and germination bioassays. *Environ Poll.* 140:43–51. [9.2]

Semesi, I., S. Beer, and M. Björk (2009). Seagrass photosynthesis controls rates of calcification and photosynthesis of calcareous macroalgae in a tropical seagrass meadow. *Mar. Ecol. Prog. Ser.* 382:41–7. [3.3.1, 6.5.3]

Serikawa, K. A., and D. F. Mandoli (1999). *Aaknox1*, a *kn1*-like homeobox gene in *Acetabularia acetabulum*, undergoes developmentally regulated subcellular localization. *Plant Mol. Biol.* 41:785–93. [1.4.3]

Serikawa, K. A., D. M. Porterfield, and D. F. Mandoli (2001). Asymmetric subcellular mRNA distribution correlates with carbonic anhydrase activity in *Acetabularia acetabulum*. *Plant Physiol.* 125:900–11. [1.4.3]

Serrão, E. A., G. Pearson, L. Kautsky, and S. H. Brawley (1996). Successful external fertilization in turbulent environments. *P. Natl. Acad. Sci. USA.* 93:5286–90. [2.4, 8.2.2]

Serrão, E. A., L. A. Alice, and S. H. Brawley (1999). Evolution of the Fucaceae (Phaeophyceae) inferred from nrDNA-ITS. *J. Phycol.* 35:382–94. [1.4.2]

Setchell, W. A., and N. L. Gardner (1919). The marine algae of the Pacific Coast of North America. Part I. Myxophyceae. *Univ. Calif. Publ. Bot.* 8:1–138. [1.1.1]

Setchell, W. A., and N. L. Gardner (1925). *III. Melanophyceae. The Marine Algae of the Pacific Coast of North America* (pp. 383–898). Berkley, CA: Univ. of California Press. [Essay 1]

Seymour, R. J., M. J. Tegner, P. K. Dayton, and P. E. Parnell (1989). Storm wave induced mortality of giant kelp, *Macrocystis pyrifera*, in southern California. *Estu. Cstl. Shelf Sci.* 28:277–92. [8.1.2, 8.1.4, 8.3.2]

Sfriso, A., A. Marcomini, and B. Pavoni (1987). Relationships between macroalgal biomass and nutrient concentrations in a hypertrophic area of the Venice Lagoon. *Mar. Environ. Res.* 22:297–312. [6.2, 9.6.1]

Sharp, G. (1987). *Ascophyllum nodosum* and its harvesting in eastern Canada. In M. S. Doty, J. F. Caddy, and B. Santelices (eds), *Case Studies of Seven Commercial Seaweed Resources* (pp. 3–48). FAO Fish. Tech. Pap. 281. [10.3]

Sharrock, R. A. (2008). The phytochrome red/far-red photoreceptor superfamily. *Genome Biol.* 9:230. [2.3.3]

Shaughnessy, F. J., R. E. De Wreede, and E. C. Bell (1996). Consequences of morphology and tissue strength to blade survivorship of two closely related Rhodophyta species. *Mar. Ecol. Prog. Ser.* 136:257–66. [8.3.1]

Shears, N. T., and R. C. Babcock (2003). Continuing trophic cascade effects after 25 years of no-take marine reserve protection. *Mar. Ecol. Prog. Ser.* 246:1–16. [4.3.1]

Shepherd, V. A., M. J. Beilby, and M. A. Bisson (2004). When is a cell not a cell? A theory relating coenocytic structure to the unusual electrophysiology of *Ventricaria ventricosa* (*Valonia ventricosa*). *Protoplasma* 223:79-91. [2.6.4]

Shimshock, N., G. Sennefelder, M. Dueker, F. Thurberg, and C. Yarish (1992). Patterns of metal accumulation in *Laminaria longicruris* from Long Island Sound (Connecticut). *Arch. Environ. Contam. Toxicol.* 22:305-12. [9.3.2]

Shivji, M. S. (1985). Interactive effects of light and nitrogen on growth and chemical composition of juvenile *Macrocystis pyrifera* (L.) C. Ag. (Phaeophyta) sporophytes. *J. Exp. Mar. Biol. Ecol.* 89:81-96. [6.7.2]

Shivji, M. S. (1991). Organization of the chloroplast genome in the red alga *Porphyra yezoensis. Curr. Genet.* 19:49-54. [1.4.1]

Sieburth, J. M. (1969). Studies on algal substances in the sea. III. The production of extracellular organic matter by littoral marine algae. *J. Exp. Mar. Biol. Ecol.* 3: 290-309. [5.7.3]

Sieburth, J. M., and J. T. Tootle (1981). Seasonality of microbial fouling on *Ascophyllum nodosum* (L.) Lejol., *Fucus vesiculosus* L., *Polysiphonia lanosa* (L.) Tandy and *Chondrus crispus* Stackh. *J. Phycol.* 17:57-64. [Fig. 4.3]

Silberfeld, T., J. W. Leigh, H. Verbruggen, *et al.* (2010). A multi-locus time-calibrated phylogeny of the brown algae (Heterokonta, Ochrophyta, Phaeophyceae): investigating the evolutionary nature of the "brown algal crown radiation". *Mol. Phylogenet. Evol.* 56:659-74. [Essay 1]

Simkiss, K., and K. M. Wilbur (1989). *Biomineralization: Cell Biology and Mineral Deposition.* Orlando, FL: Academic Press, 340 pp. [6.5.3]

Simpson, C. L., and D. B. Stern (2002). The treasure trove of algal chloroplast genomes. Surprises in architecture and gene content, and their functional implications. *Plant Physiol.* 129:957-66. [1.3.2, 1.4.3]

Singh, P, P. A. Kumar, Y. P. Abroi, and M. S. Naik (1985). Photorespiratory nitrogen cycle - a critical evaluation. *Physiol. Plant.* 66:169-76. [6.5.1]

Sjøtun, K. S. Fredriksen, and J. Rueness (1998). Effect of canopy biomass and wave exposure on growth in *Laminaria hyperborea* (Laminariaceae, Phaeophya). *Eur. J. Phycol.* 33:337-43. [8.3.1]

Sjøtun, K., H. Christie, and J. H. Fosså (2006). The combined effect of canopy shading and sea urchin grazing on recruitment in kelp forests (*Laminaria hyperborea*). *Mar. Biol. Res.* 2:24-32. [5.7.5]

Skene, K. H. (2004). Key differences in photosynthetic characteristics of nine species of intertidal macroalgae are related to their position on the shore. *Can. J. Bot.* 82:177-84. [7.5.1]

Slocum, C. J. (1980). Differential susceptibility to grazers in two phases of an intertidal alga: advantages of heteromorphic generations. *J. Exp. Mar. Biol. Ecol.* 46:99-110. [4.3.2]

Smetacek, V., B. von Bodungen, K. von Brödsel, and B. Zeitzschel (1976). The plankton tower.II. Release of nutrients from sediments due to changes in the density of bottom water. *Mar. Biol.* 34:373-8. [6.2]

Smit, A.J. (2004). Medicinal and pharmaceutical uses of seaweed natural products: A review. *J. Appl. Phycol.* 16:245-62. [10.10]

Smit, A. J., A. Brearley, G. A. Hyndes, P. S. Lavery, and D. I. Walker (2005). Carbon and nitrogen stable isotope analysis of an *Amphibolis griffithii* seagrass bed. *Estu. Cstl. Shelf Sci.* 65(3):545-56. [Essay 3]

Smit, A. J., A. Brearley, G. A. Hyndes, P. S. Lavery, and D. I. Walker (2006). delta N-15 and delta C-13 analysis of a *Posidonia sinuosa* seagrass bed. *Aquat. Bot.* 84(3):277-82. [Essay 3]

Smith, A. H., K. Nichols, and J. McLachlan (1984). Cultivation of seamoss (*Gracilaria*) in St. Lucia, West Indies. *Hydrobiologia* 116/117:249-51. [10.6.2]

Smith, A. M., J. E. Sutherland, L. Kregting, T. J. Farr, and D. J. Winter (2012). Phylominerology of the Coralline red algae: Correlation of skeletal mineralogy with molecular phylogeny. *Phytochem.* 81:97-108. [6.5.3]

Smith, C. M., and L. J. Walters (1999). Fragmentation as a strategy for *Caulerpa* species: Fates of fragments and implications for management of an invasive weed. *P. S. Z. N.; Mar. Ecol.* 20: 307-19. [3.4]

Smith, D. R., J. Hua, R. W. Lee, and P. J. Keeling (2012). Relative rates of evolution among the three genetic compartments of the red alga *Porphyra* differ from those of green plants and do not correlate with genome architecture. *Mol. Phylogen. Evol.* 65:339-44. [1.3.2]

Smith, G. M. (1947). On the reproduction of some Pacific coast species of *Ulva. Am. J. Bot.* 34:80-7. [2.3.3]

Smith, J. E., C. L. Hunter, E. J. Conklin, *et al.* (2004). Ecology of the invasive red alga *Gracilaria salicornia* (Rhodophyta) on O'ahu, Hawai'i. *Pacific Science* 58:325-43. [Essay 6]

Smith, J. E., E. J. Conklin, C. M. Smith, and C. L. Hunter (2008). Fighting algae in Kaneohe Bay (response). *Science* 319:157-8. [3.4]

Smith, J., C. Hunter, and C. Smith (2010). The effects of top-down versus bottom-up control on benthic coral reef community structure. *Oecologia* 163:497-507. [Essay 6]

Smith, R. C., and J. E. Tyler (1974). In N. G. Jerlov and E Steemann-Nielsen (eds), *Optical Aspects of Oceanography.* Orlando, FL: Academic Press. [5.2.1]

Smith, R. C., and J. E. Tyler (1976). Transmission of solar radiation into natural waters. *Photochem. Photobiol. Rev.* 1:117-55. [5.2.2]

Smith, R. G., W. N. Wheeler, and L. M. Srivastava (1983). Seasonal photosynthetic performance of *Macrocystis integrifolia* (Phaeophyceae). *J. Phycol.* 19:352-9. [6.8.2]

Smith, S. D. A. (2002). Kelp rafts in the Southern Ocean. *Global Ecol. Biogeogr.* 11:67-9. [3.3.7]

Smith, S. V., W. Kimmerer, E. Laws, R. Brock, and T. Walsh (1981). Kaneohe Bay sewage diversion experiment: perspectives on ecosystem responses to nutritional perturbation. *Pacific Sci.* 35:270-395. [Essay 6]

Solis, M. J. L., S. Draeger, and T. E. dela Cruz (2010). Marine-derived fungi from *Kappaphycus alvarezii* and *K. striatum* as potential causative agents of ice-ice disease in farmed seaweeds. *Bot. Mar.* 53:587-94. [10.5.2]

Sorte, C. J. B., S. L. Williams, and J. T. Carlton (2010). Marine range shifts and species introductions: comparative spread rates and community impacts. *Global Ecol. Biogeogr.* 19:303-16. [3.4]

Sotka, E. E. (2005). Local adaptation in host use among marine invertebrates. *Ecol. Lett.* 8:448-59. [4.3.2]

Sotka, E. E., and P. L. Reynolds (2011). Rapid experimental shift in host use traits of a polyphagous marine herbivore reveals fitness costs on alternative hosts. *Evol. Ecol.* 25:1335-55. [4.3.2]

Sotka, E. E., and K. E. Whalen (2008). Herbivore offense in the sea: the detoxification and transport of secondary metabolites. In C. D. Amsler (ed.), *Algal Chemical Ecology* (pp. 203-28). London: Springer (Limited). [4.4.1, 4.4.2]

Sotka, E. E., M. E. Hay, and J. D. Thomas (1999). Host-plant specialization by a non-herbivorous amphipod: advantages for the amphipod and costs for the seaweed. *Oecologia* 118:471-82. [4.3.2]

Sotka, E. E., R. B. Taylor, and M. E. Hay (2002). Tissue-specific induction of resistance to herbivores in a brown seaweed: the importance of direct grazing versus waterborne signals from grazed neighbors. *J. Exp. Mar. Biol. Ecol.* 277:1-12. [4.2.2]

Sotka, E. E., J. P. Wares, and M. E. Hays (2003). Geographic and genetic variation in feeding preference for chemically defended seaweeds. *Evolution.* 57:2262-767. [4.3.2]

Sotka, E. E., J. Forbey, M. Horn, *et al.* (2009). The emerging role of pharmacology in understanding consumer-prey interactions in marine and freshwater systems. *Integr. Comp. Biol.* 49:291-313. [4.4.1]

Soulsby, P. G., D. Lowthion, M. Houston, and H. A. C. Montgomery (1985). The role of sewage effluent in the accumulation of macroalgal mats on intertidal mudflats in two basins in southern England. *Neth. J. Sea Res.* 19:257-63. [9.6.1]

Sousa, W. P. (1979). Experimental investigation of disturbance and ecological succession in a rocky intertidal algal community. *Ecol. Monogr.* 49:227-54. [4.2.1, 8.3.2]

Sousa, W. P. (2001). Natural disturbance and the dynamics of marine benthic communities. In M. D. Bertness, S. D. Gaines, and M. E. Hay (eds), *Marine Community Ecology* (pp. 85-130). Sunderland, MA: Sinauer Assocs. [3.1.2, 3.5.1, 8.3.2]

Sousa, W. P., S. C. Schroeter, and S. D. Gaines (1981). Latitudinal variation in intertidal algal community structure: the influence of grazing and vegetative propagation. *Oecologia* 48:297-307. [4.2.1, 8.3.2]

Spaargaren, D. H. (1984). On ice formation in sea water and marine animals at subzero temperatures. *Mar. Biol. Lett.* 5: 203-16. [7.3.4]

Spalding, H., M. S. Foster, and J. N. Heine (2003). Composition, distribution, and abundance of deep-water (>30 m) macroalgae in central California. *J. Phycol.* 39:273-84. [3.3.6]

Speransky, S. R., S. H. Brawley, and W. A. Halteman (2000). Gamete release is increased by calm conditions in the coenocytic green alga *Bryopsis* (Chlorophyta). *J. Phycol.* 36:730-9. [2.4, 8.2.2]

Speransky, V. V., S. H. Brawley, and M. E. McCully (2001). Ion fluxes and modification of the extracellular matrix during gamete release in fucoid algae. *J. Phycol.* 37:555-73. [8.2.2]

Spilling, K., J. Titelman, T. M. Greve, and M. Kühl (2010). Microsensor measurements of the external and internal microenvironment of *Fucus vesiculosus* (Phaeophyceae). *J. Phycol.* 46:1350-5. [8.1.3, 8.2.1]

Springer, Y. P., C. G. Hays, M. H. Carr, and M. R. Mackey (2010). Toward ecosytem-based management of marine macroalgae – the bull kelp, *Nereocystis luetkeana*. *Oceanogr. Mar. Biol.* 48:1-42. [2.5.1, 3.3.2]

Stachowicz, J. J., and M. E. Hay (1999). Reducing predation through chemically mediated camouflage: indirect effects of plant defenses on herbivores. *Ecology* 80:495-509. [4.4.2]

Stachowicz, J. J., and R. B. Whitlatch (2005). Multiple mutualists provide complementary benefits to their seaweed host. *Ecology* 86:2418-27. [4.5.1]

Stachowicz, J. J., J. F. Bruno, and J. E. Duffy (2007). Understanding the effects of marine biodiversity on communities and ecosystems. *Annu. Rev. Ecol. Evol. Syst.* 38:739-66. [3.5.1]

Stanley, S. M., J. B. Ries, and L. A. Hardie (2010). Increased production of calcite and slower growth for the major sediment-producing alga *Halimeda* as the Mg/Ca ratio of seawater is lowered to a "calcite sea" level. *J. Sedimentary Res.* 80:6-16. [6.5.3]

Stauber, J. L., and T. M. Florence (1985). Interactions of copper and manganese: a mechanism by which manganese alleviates copper toxicity to the marine diatom, *Nitzschia closterium* (Ehrenberg) W. Smith. *Aquat. Toxic.* 7:241-54. [9.3.3]

Stauber, J. L., and T. M. Florence (1987). Mechanisms of toxicity of ionic copper and copper complexes to algae. *Mar. Biol.* 94:511-19. [9.3.3, 9.3.4]

Steele, R. L., and M. D. Hanisak (1979). Sensitivity of some brown algal reproductive stages to oil pollution. *Proc. Intl. Seaweed Symp.* 9:181-91. [9.4.2]

Steele, R. L., G. B. Thursby, and J. P. van der Meer (1986). Genetics of *Champia parvula* (Rhodymeniales, Rhodophyta): Mendelian inheritance of spontaneous mutants. *J. Phycol.* 22:538A-42. [1.4.2]

Steemann-Nielsen, E. (1974). Light and primary production. In N. G. Jerlov (ed.) *Optical Aspects of Oceanography* (pp. 331-88). New York: Academic Press. [5.2.2]

Steen, H. (2004). Interspecific competition between *Enteromorpha* (Ulvales: Chlorophyceae) and *Fucus* (Fucales: Phaeophyceae) germlings: effects of nutrient concentration, temperature, and settlement density. *Mar. Ecol. Prog. Ser.* 278:89-101. [4.2.3]

Steen, H., and R. Scrosati (2004). Intraspecific competition in *Fucus serratus* and *F. evanescens* (Phaeophyceae: Fucales) germlings: effects of settlement density, nutrient concentration, and temperature. *Mar. Biol.* 14:61-70. [4.2.3]

Steinbeck, J. R., D. R. Schiel, and M. S. Foster (2005). Detecting long-term change in complex communities: a case study from the rocky intertidal zone. *Ecol. Appl.* 15:1813-32. [3.5.1]

Steinberg, P. D., and I. Altena (1992). Tolerance of marine invertebrate herbivores to brown algal phlorotannins in temperate Australasia. *Ecol. Monogr.* 62(2):189-222. [Essay 3]

Steinberg, P. D., and R. de Nys (2002). Chemical mediation of colonization of seaweed surfaces. *J. Phycol.* 38:621-9. [4.2.2]

Steinberg, P. D., R. de Nys, and S. Kjelleberg (2001). Chemical mediation of surface colonization. In J. B. McClintock and B. J. Baker (eds), *Marine Chemical Ecology* (pp. 355-87). Boca Raton, FL: CRC Press. [4.2.2]

Steinberg, P. D., R. de Nys, and S. Kjelleberg (2002). Chemical cues for surface colonization. *J. Chem. Ecol.* 28:1935-51. [4.2.2]

Steinhoff, F. S., C. Wiencke, R. Müller, and K. Bischof (2008). Effects of ultraviolet radiation and temperature on the ultrastructure of zoospores of the brown macroalga *Laminaria hyperborea*. *Plant Biol.* 10: 388-97. [7.6]

Stekoll, M. S., and L. Deysher (2000). Response of the dominant alga *Fucus garneri* (Silva) (Phaeophyceae) to the *Exxon Valdez* oil spill and clean-up. *Mar. Poll. Bull.* 40:1028-41. [9.4.3]

Steneck, R. S. (1982). A limpet-coralline alga association: adaptations and defenses between a selective herbivore and its prey. *Ecology* 63:507-22. [4.3.2]

Steneck, R. S. (1992). Plant-herbivore coevolution: a reappraisal from the marine realm and its fossil record. In D. M. John, S. J. Hawkins, and J. H. Price (eds), *Plant-Animal Interactions in the Marine Benthos*, Systematics Association Special Volume 46 (pp. 477-91). Oxford: Claredon Press. [4.3.2]

Steneck, R. S., and M. N. Dethier (1994). A functional group approach to the structure of algal-dominated communities. *Oikos* 69:476-98. [1.2.2]

Steneck, R. S., and P. T. Martone (2007). Calcified algae. In M.W. Denny and S.D. Gaines (eds) *Encyclopedia of Tidepools* (pp. 21-4) Berkeley, CA: University of California Press. [6.5.3]

Steneck, R. S., and L. Watling (1982). Feeding capabilities and limitation of herbivorous molluscs: a functional approach. *Mar. Biol.* 68:299-312. [4.3.2]

Steneck, R. S., S. D. Hacker, and M. N. Dethier (1991). Mechanisms of competitive dominance between crustose coralline algae: an herbivore-mediated competitive reversal. *Ecology* 72:938-50. [4.3.1]

Steneck, R. S., M. H. Graham, B. J. Bourque, *et al.* (2002). Kelp forest ecosystems: Biodiversity, stability, resilience and future. *Envir. Conserv.* 29:436-59. [1.1.3, 3.3.2, 7.3.7]

Stengel, D. B., and M. J. Dring (1997). Morphology and *in situ* growth rates of plants of *Ascophyllum nodosum* (Phaeophyta) from different shore levels and responses of plants to vertical transplantation. *Eur. J. Phycol.* 32:193-202. [2.6.2]

Stengel, D. B., and M. J. Dring (2000). Copper and iron concentrations in *Ascophyllum nodosum* (Fucales, Phaeophyta) from different sites in Ireland and after culture experiments in relation to thallus age and epiphytism. *J. Exp. Mar. Biol. Ecol.* 246:145-61. [9.3.2]

Stengel, D. B., A. Macken, L. Morrison, and N. Morely (2004). Zinc concentrations in marine macroalgae and a lichen from western Ireland in relation to phylogenetic grouping, habitat and morphology. *Mar. Poll. Bull.* 48:902-9. [9.3.2]

Stepanyan, O. V., and G. M. Voskoboinikov (2006). Effect of oil and oil products on morphofunctional parameters of marine macrophytes. *Russian J. Mar. Biol.* 32:S32-9. [9.4.2]

Stephenson, T. A., and A. Stephenson (1949). The universal features of zonation between tide-marks on rocky coasts. *J. Ecol.* 38:289-305. [3.1.2]

Stephenson, T. A., and A. Stephenson (1972). *Life Between Tidemarks on Rocky Shores*. San Francisco, CA: Freeman, 425 pp. [Essay 2, 3.1.2, 7.3.8]

Stevens, C. L., and C. L. Hurd (1997). Boundary-layers around bladed aquatic macrophytes. *Hydrobiologia* 346:119-28. [8.2.1]

Stevens, C. L., C. L. Hurd, and M. J. Smith (2001). Water motion relative to subtidal kelp fronds. *Limnol. Oceanogr.* 46:669–78. [8.3.1]

Stevens, C. L., C. L. Hurd, and M. J. Smith (2002). Field measurement of the dynamics of the bull kelp *Durvillaea antarctica* (Chamisso) Heriot. *J. Exp. Mar. Biol. Ecol.* 269:147–71. [8.1.4, 8.3.1]

Stevens, C. L., C. L. Hurd, and P. E. Isachsen (2003). Modelling of diffusion boundary-layers in subtidal macroalgal canopies: the response to waves and currents. *Aquat. Sci.* 65:81–91. [8.1.4]

Stevens, C. L., D. I. Taylor, S. Delaux, M. J. Smith, and D. R. Schiel (2008). Characterisation of wave-influenced macroalgal propagule settlement. *J. Marine Syst.* 74:96–107. [2.5.1]

Stewart, H. L. (2004). Hydrodynamic consequences of maintaining an upright posture by different magnitudes of stiffness and buoyancy in the tropical alga *Turbinaria ornata*. *J. Marine Syst.* 49:157–67. [8.3.1]

Stewart, H. L. (2006). Morphological variation and phenotypic plasticity of buoyancy in the macroalga *Turbinaria ornata* across a barrier reef. *Mar. Biol.* 149:721–30. [8.3.1]

Stewart, H. L., and R. C. Carpenter (2003). The effects of morphology and water flow on photosynthesis of marine macroalgae. *Ecology* 84:2999–3012. [5.7.2, 8.2.1]

Stewart, J. G. (1982). Anchor species and epiphytes in intertidal algal turf. *Pac. Sci.* 36:45–59. [4.2.2, 8.3.2]

Stewart, J. G. (1983). Fluctuations in the quantity of sediments trapped among algal thalli on intertidal rock platforms in southern California. *J. Exp. Mar. Biol. Ecol.* 73:205–11. [8.3.2]

Stewart, J. G. (1989). Establishment, persistence and dominance of *Corallina* (Rhodophyta) in algal turf. *J. Phycol.* 25:436–46. [4.2.1, 8.3.2]

Stimson, J., S. T. Larned, and E. Conklin (2001). Effects of herbivory, nutrient levels, and introduced algae on the distribution and abundance of the invasive macroalga *Dictyosphaeria cavernosa* in Kaneohe Bay, Hawaii. *Coral Reefs* 19:343–57. [Essay 6]

Stirk, W. A., O. Novák, M. Strnad, and J. van Staden (2003). Cytokinins in macroalgae. *Plant Growth Regul.* 41:13–24. [2.6.3]

Stirk, W. A., O. Novák, V. Hradecká, *et al.* (2009). Endogenous cytokinins, auxins and abscisic acid in *Ulva fasciata* (Chlorophyta) and *Dictyota humifusa* (Phaeophyta): towards understanding their biosynthesis and homeostasis. *Eur. J. Phycol.* 44:231–40. [2.6.3]

Storz, H., K. Müller, F. Ehrhart, *et al.* (2009) Physicochemical features of ultra-high viscosity alginates. *Carbohydrate Res.* 344:985–95. [5.5.2]

Stratmann, J., G. Paputsoglu, and W. Oertel (1996). Differentiation of *Ulva mutabilis* (Chlorophyta) gametangia and gamete release are controlled by extracellular inhibitors. *J. Phycol.* 32:1009–21. [2.2, 2.3.3]

Strickland, J. D. H., and T. R. Parsons (1972). *A Practical Handbook of Seawater Analysis*, 2nd edn. *Fish Res. Bd. Canada Bull.* 167, 310 pp. [5.7.1]

Strömgren, T. (1979). The effect of zinc on the increase in length of five species of intertidal Fucales. *J. Exp. Mar. Biol. Ecol.* 40:95–102. [9.3.4.]

Strömgren, T. (1980a). The effect of dissolved copper on the increase in length of four species of intertidal fucoid algae. *Mar. Environ. Res.* 3:5–13. [9.3.4]

Strömgren, T. (1980b). The effect of lead, cadmium, and mercury on the increase in length of five intertidal Fucales. *J. Exp. Mar. Biol. Ecol.* 43:107–19. [9.3.4]

Strömgren, T. (1980c). Combined effects of Cu, Zn, and Hg on the increase in length of *Ascophyllum nodosum* (L.) Le Jolis. *J. Exp. Mar. Biol. Ecol.* 48:225–31. [9.3.5]

Su, H.-N., B.-B. Xie, X.-Y. Zhang, B.-C. Zhou, and Y-Z. Zhang (2010). The supramolecular architecture, function, and regulation of thylakoid membranes in red algae: an overview. *Photosynth. Res.* 106:73–87. [1.3.2]

Suetsugu, N., F. Mittmann, G. Wagner, J. Hughes, and M. Wada (2005). A chimeric photoreceptor gene, N EOCHROME, has arisen twice during plant evolution. *P. Natl. Acad. Sci. USA.* 102:13,705–9. [2.3.3]

Sunda, W. G. (2009). Trace element nutrients. In J. H. Steele (ed.), *Encyclopedia of Ocean Science* (2nd edn), (pp. 75–86). New York: Springer. [9.3.1]

Suple, C. (1999). El Niño/La Niña. *Nat. Geog.* 4 (3):74–95. [7.3.7]

Surif, M. B., and J. A. Raven (1989). Exogenous inorganic carbon sources for photosynthesis in seawater by members of the Fucales and the Laminariales (Phaeophyta): ecological and taxonomic implications. *Oecologia* 78:97–105. [5.4.1]

Surif, M. B., and J. A. Raven (1990). Photosynthetic gas exchange under emersed conditions in eulittoral and normally submersed members of the Fucales and Laminariales: interpretation in relation to C isotope ratio and N and water use efficiency. *Oecologia* 82:68–80. [5.4.1, 5.7.1]

Sussmann, A. V., and R. E. De Wreede (2002). Host specificity of the endophytic sporophyte phase of *Acrosiphonia* (Codiolales, Chlorophyta) in southern British Columbia, Canada. *Phycologia* 41:169–77. [4.5.2]

Sutherland, J. E., S. C. Lindstrom, W. A. Nelson, *et al.* (2011). A new look at an ancient order: generic revision of the Bangiales (Rhodophyta). *J. Phycol.* 47:1131–51. [1.1.1, 1.4.1, 10.2]

Suto, S. (1950). Studies on shedding, swimming and fixing of the spores of seaweeds. *Bull. Jpn. Soc. Sci. Fish.* 16:1–9. [2.5.1]

Suttle, C.A. (2007). Marine viruses–major players in the global ecosystem. *Nature Reviews* 5: 801–12. [4.5.2]

Suzuki, K., K. Iwamoto, S. Yokoyama, and T. Ikawa (1991). Glycolate-oxidizing enzymes in algae. *J. Phycol.* 27:492–8. [5.4.1]

Suzuki, Y., K. Kuma, I. Kudo, and K. Matsunaga (1995). Iron requirement of the brown macroalgae, *Laminaria japonica*, *Undaria pinnatifida* (Phaeophyta) and the crustose coralline alga *Lithophyllum yessoense* (Rhodophyta), and their competition in the northern Japan Sea. *Phycologia* 34:201–5. [6.5.5]

Svendsen, H., A. Beszczynska-Møller, J. O. Hagen, *et al.* (2002). The physical environment of Kongsfjorden-Krossfjorden, an Arctic fjord system in Svalbard. *Polar Res.* 21:133–66. [7.3.6]

Svensson, J. R., M. Lindegarth, M. Siccha, *et al.* (2007). Maximum species richness at intermediate frequencies of disturbance: consistency among levels of productivity. *Ecology* 88:830–8. [8.3.2]

Svensson, J. R., M. Lindegarth, and H. Pavia (2010). Physical and biological disturbances interact differently with productivity: effects on floral and faunal richness. *Ecology* 91:3069–80. [8.3.2]

Svensson, J. R., M. Lindegarth, P. R. Jonsson, and H. Pavia (2012). Disturbance-diversity models: what do they really predict and how are they tested? *P. Roy. Soc. Lond. B. Bio.* 279:2163–70. [3.5.1, 8.3.2]

Swanson, A. K., and L. D. Druehl (2002). Induction, exudation and the UV protective role of kelp phlorotannins. *Aquat. Bot.* 73:241–53. [5.7.3]

Sweeney, B. M. (1974). A physiological model for circadian rhythms derived from the *Acetabularia* rhythm paradoxes. *Int. J. Chronobiol.* 2: 25–33. [5.7.2]

Sweeney, B. M., and B. B. Prezelin (1978). Circadian rhythms. *Photochem. Photobiol.* 27: 841–7.

Swinbanks, D. D. (1982). Intertidal exposure zones: a way to subdivide the shore. *J. Exp. Mar. Biol. Ecol.* 62:69–86. [3.1.3]

Syrett, P. J. (1981). Nitrogen metabolism of microalgae. *Can. Bull. Fish. Aquat. Sci.* 210:182–210. [6.3.5, 6.5.1]

Szymanski, D. B., and D. J. Cosgrove (2009). Dynamic coordination of cytoskeletal and cell wall systems during plant cell morphogenesis. *Curr. Biol.* 19:R800–11. [1.3.1, 1.3.4]

Tabita, F. R., S. Satagopan, T. E. Hanson, N. E. Kreel, and S. S. Scott (2008) Distinct form I, II, III, and IV Rubisco proteins from the three kingdoms of life provide clues about Rubisco evolution and structure/function relationships. *J. Exp. Bot.* 59:1515–24. [5.4.2]

Taiz, L., and E. Zeiger (eds) (2010). *Plant Physiology* (5th edn). Sunderland, MA: Sinauer Assoc., Inc., 782 pp. [1.3.3, 2.3.3, 2.6.3]

Takahashi, F., D. Yamagata, M. Ishikawa, *et al.* (2007). AUREOCHROME, a photoreceptor required for photomorphogenesis in stramenopiles. *PNAS* 104:19,625–30. [2.3.3, 2.6.2]

Talarico, L. (1990). R-phycoerythrin from *Audouinella saviana* (Nemaliales, Rhodophyta). Ultrastructural and biochemical analysis of aggregates and subunits. *Phycologia* 29:292–302. [5.3.1]

Talarico, L., and G. Maranzana (2000). Light and adaptive responses in red macroalgae: an overview. *J. Photochem. Photobiol. B: Biol.* 56: 1–11. [5.3.4]

Tanaka, A., C. Nagasato, S. Uwai, T. Motomura, and H. Kawai (2007). Re-examination of ultrastructures of the stellate chloroplast organization in brown algae: structure and development of pyrenoids. *Phycol. Res.* 55:203–13. [1.3.2]

Tang, Y. Z., and C. J. Gobler (2011). The green macroalga, *Ulva lactuca*, inhibits the growth of seven common harmful algal bloom species via allelopathy. *Harmful Algae* 10:480–8. [4.2.2]

Tanner, C. E. (1986). Investigations of the taxonomy and morphological variation of *Ulva* (Chlorophyta): *Ulva californica* Wille. *Phycologia* 25:510–20. [7.5.1]

Tarakhovskaya, E. R., Yu. I. Maslov, and M. F. Shishova (2007). Phytohormones in algae. *Russ. J. Plant Physl.* 54:163–70. [2.6.3]

Targett, N. M., and T. M. Arnold (2001). Effects of secondary metabolites on digestion in marine herbivores. In J. B. McClintock, and B. J. Baker, (eds), *Marine Chemical Ecology* (pp. 391–411). Boca Raton, FL: CRC Press. [4.4.1]

Tatewaki, M. (1970). Culture studies on the life history of some species of the genus *Monostroma*. *Sci. Pap. Inst. Algol. Res., Fac. Sci., Hokkaido U.* 6(1):1–56. [2.6.2]

Tatewaki, M., L. Provasoli, and I. J. Pintner (1983). Morphogenesis of *Monostroma oxyspermum* (Kiitz.) Doty (Chlorophyceae) in axenic culture, especially in bialgal culture. *J. Phycol.* 19:409–16. [2.6.2]

Taylor, D. I., and D. R. Schiel (2010). Algal populations controlled by fish herbivory across a wave exposure gradient on southern temperate shores. *Ecology* 91:201–11. [4.3.1]

Taylor, D., S. Delaux, C. Stevens, R. Nokes, and D. Schiel (2010). Settlement rates of macroalgal propagules: cross-species comparisons in a turbulent environment. *Limnol. Oceanogr.* 55:66–76. [2.5.1]

Taylor, M. W. R. B. Taylor, and T.A. V. Rees (1999). Allometric evidence for the dominant role of surface cells in ammonium metabolism and photosynthesis in northeastern New

Zealand seaweeds. *Mar. Ecol. Prog. Ser.* 184:73-81. [1.2.2, 6.4.2, 6.5.1]

Taylor, M. W., and T. A. V. Rees (1999). Kinetics of ammonium assimilation in two seaweeds, *Enteromorpha* sp. (Chlorophyta) and *Osmundaria colensoi* (Rhodophyceae). *J. Phycol.* 35:740-6. [6.5.1]

Taylor, M. W., N. G. Barr, C. M. Grant, and T. A. V. Rees (2006). Changes in amino acid composition of *Ulva intestinalis* (Chlorophyceae) following addition of ammonium or nitrate. *Phycologia* 45:270-6. [6.5.1]

Taylor, P. R., and M. M. Littler (1982). The roles of compensatory mortality, physical disturbance, and substrate retention in the development and organization of a sand-influenced rocky intertidal community. *Ecology* 63: 135-46. [7.5.1]

Taylor, R. B., and T. A. V. Rees (1998). Excretory products of mobile epifauna as a nitrogen source for seaweeds. *Limnol. Oceanogr.* 43:600-6. [4.5.1]

Taylor, R. B., J. T. A. Peek, and T. A. V. Rees (1998). Scaling of ammonium uptake by seaweeds to surface area:volume ratio: geographical variation and the role of uptake by passive diffusion. *Mar. Ecol. Prog. Ser.* 169:143-8. [6.4, 6.4.2, 6.5.1]

Taylor, R. B., E. Sotka, and M. E. Hay (2002). Tissue-specific induction of herbivore resistance: seaweed response to amphipod grazing. *Oecologia* 132: 68-76. [4.4.2]

Taylor, W. R. (1957). *Marine Algae of the Northeastern Coast of North America*. Ann Arbor, MI: University of Michigan Press. [Fig. 1.5]

Taylor, W. R. (1960). *Marine Algae of the Eastern Tropical and Subtropical Coasts of the Americas*. Ann Arbor, MI: University of Michigan Press. [Fig. 1.5]

Tegner, M. J., and P. K. Dayton (1987). El Niño effects on Southern California kelp forest communities. *Adv. Ecol. Res.* 17:243-79. [4.3.1]

Teichberg, M., S. Fox, C. Aguila, Y. Olsen, and I. Valiela (2008). Macroalgal responses to experimental nutrient enrichment in shallow coastal waters: growth, internal nutrient pools, and isotopic signatures. *Mar. Ecol. Prog. Ser.* 368:117-26. [9.6.1]

Teichberg, M., S. Fox, C. Aguila, *et al.* (2010). Eutrophication and macroalgal blooms in temperate and tropical coastal waters: nutrient enrichment experiments with *Ulva* spp. *Global Change Biol.* 16:2624-37. [9.6.1]

Telfer, A. (2002). What is β-carotene doing in the photosystem II reaction centre? *Phil. Trans. Roy. Soc. London B Biol. Sci.* 357:1431-40. [5.3.1]

Terumoto, I. (1964). Frost resistance in some marine algae from the winter intertidal zone. *Low Temp. Sci. (Ser. B)* 22:19-28. [7.3.4]

Tewari, A., and H. V. Joshi (1988). Effect of domestic sewage and industrial effluents on biomass and species diversity of seaweeds. *Bot. Mar.* 31:389-97. [9.6.1]

The Royal Society (2005). Ocean acidification due to increasing atmospheric CO2. Policy Document 12/05. *The Royal Society*, London. [Essay 4]

Thiel, M., and L. Gutow (2004). The ecology of rafting in the marine environment. I. The floating substrata. *Oceanogr. Mar. Biol.: Ann. Rev.* 42:181-264. [3.3.7]

Thiel, M., and L. Gutow (2005). The ecology of rafting in the marine environment. II. The rafting organisms and community. *Oceanogr. Mar. Biol.* 43:279-418. [3.3.7]

Thiel, M., E. C. Macaya, E. Acuña, *et al.* (2007). The Humboldt current system of northern-central Chile: oceanographic processes, ecological interactions and socio-economic feedback. *Oceanogr. Mar. Biol. Ann. Rev.* 45:195-345. [7.3.7]

Thomas, D. N., and C. Wiencke (1991). Photosynthesis, dark respiration and light independent carbon fixation of endemic Antarctic macroalgae. *Polar Biol.* 11:329-37. [5.4.3]

Thomas, D. N., G. E. Fogg, P. Convey, *et al.* (2008). *The Biology of Polar Regions*. Oxford: Oxford University Press, 416 pp. [3.3.3]

Thomas, F., A. Cosse, S. Goulitquer, *et al.* (2011). Waterborne signaling primes the expression of elicitor-induced genes and buffers the oxidative responses in the brown alga *Laminaria digitata*. *PloS ONE* 6:e21475. [4.4.2]

Thomas, M. L. H. (1986). A physically derived exposure index for marine shorelines. *Ophelia* 25:1-13. [8.14]

Thomas, T. E., and P. J. Harrison (1985). Effects of nitrogen supply on nitrogen uptake, accumulation and assimilation in *Porphyra perforata* (Rhodophyta). *Mar. Biol.* 85:269-78. [6.4.1, 6.5.1]

Thomas, T. E., and P. J. Harrison (1987). Rapid ammonium uptake and field conditions. *J. Exp. Mar. Biol. Ecol.* 107:1-8. [6.4.1]

Thomas, T. E., and P. J. Harrison (1988). A comparison of *in vitro* and in vivo nitrate reductase assays in three intertidal seaweeds. *Bot Mar.* 31:101-7. [6.5.1]

Thomas, T. E., and D. H. Turpin (1980). Desiccation enhanced nutrient uptake rates in the intertidal alga *Fucus distichus*. *Bot. Mar.* 23:479-81. [6.4.2, 7.5.1]

Thomas, T. E., P. J. Harrison, and E. B. Taylor (1985). Nitrogen uptake and growth of the germlings and mature thalli of *Fucus distichus*. *Mar. Biol.* 84:267-74. [6.4.2, 6.5.1]

Thomas, T. E., P. J. Harrison, and D. H. Turpin (1987a). Adaptations of *Gracilaria pacifica* (Rhodophyta) to nitrogen procurement at different intertidal locations. *Mar. Biol.* 93:569-80. [6.4.1, 6.5.1]

Thomas, T. E., D. H. Turpin, and P. J. Harrison (l987b). Desiccation enhanced nitrogen uptake rates in intertidal seaweeds. *Mar. Biol.* 94:293-8. [6.4.2]

Thompson, S. E. M., J. A. Callow, M. E. Callow, *et al.* (2007). Membrane recycling and calcium dynamics during settlement and adhesion of zoospores of the green alga *Ulva linza. Plant Cell Environ.* 30:733-44. [2.5.2]

Thompson, S. E. M., M. E. Callow, and J. A. Callow (2010). The effects of nitric oxide in settlement and adhesion of zoospores of the green alga *Ulva. Biofouling* 26:167-78. [2.5.2]

Thomsen, M. S., T. Wemberg, and G. A. Kendrick (2004). The effect of thallus size, life stage, aggregation, wave exposure and substratum conditions on the forces required to break or dislodge the small kelp *Ecklonia radiata. Bot. Mar.* 47:454-60. [8.3.1]

Thomsen, M. S., T. Wernberg, A. Altieri, *et al.* (2010). Habitat cascades: the conceptual context and global relevance of facilitation cascades via habitat formation and modification. *Integr. Compar. Biol.* 50:158-75. [4.1]

Thornber, C. S. (2006). Functional properties of the isomorphic biphasic algal life cycle. *Integr. Comp. Biol.* 46:605-14. [2.2]

Thornber, C. S., and S. D. Gaines (2004). Population demographics in species with biphasic life cycles. *Ecology* 85:1661-74. [2.2]

Thornber, C., J. J. Stachowicz, and S. Gaines. (2006). Tissue type matters: selective herbivory on different life history stages of an isomorphic alga. *Ecology* 87:2255-63. [2.2]

Thurman, H. V., and A. P. Trujillo (eds) (2004). *Introductory Oceanography* (10th edition). 608 pp. Upper Saddle River, NJ: Prentice Hall. [8.0]

Thursby, G. (1984). Development of toxicity test procedures for the marine alga *Champia parvula.* USEPA 60019844. [9.2]

Titlyanov, E. A. (1976). Adaptation of benthic plants to light. Role of light in distribution of attached marine algae. *Biol. Morya* 1:3-12. [5.3.4]

Titlyanov, E. A., and T. V. Titlyanova (2010). Seaweed cultivation: methods and problems. *Russian J. Mar. Biol.* 36:227-42. [10.7]

Togashi, T., and P. A. Cox (2001). Tidal-linked synchrony of gamete release in the marine green alga, *Monostroma angicava* Kjellman. *J. Exp. Mar. Biol. Ecol.* 264:117-31. [2.3.4]

Togashi, T., M. Nagisa, T. Miyazaki, *et al.* (2006). Gamete behaviours and the evolution of "marked anisogamy": reproductive strategies and sexual dimorphism in Bryopsidales marine green algae. *Evol. Ecol. Res.* 8:617-28. [2.5.1]

tom Dieck (Bartsch), I. (1991). Circannual growth rhythm and photoperiodic sorus induction in the kelp *Laminaria setchellii* (Phaeophyta). *J. Phycol.* 27:341-50. [2.3.1]

Toohey, B. D., and G. A. Kendrick (2007). Survival of juvenile *Ecklonia radiata* sporophytes after canopy loss. *J. Exp. Mar. Biol. Ecol.* 349:170-82. [8.3.2]

Toohey, B. D., G. A. Kendrick, and E. S. Harvey (2007). Disturbance and reef topography maintain high local diversity in *Ecklonia radiata* kelp forests. *Oikos* 116:1618-30. [8.3.2]

Toole, C. M., and F. C. T. Allnutt (2003). Red, cryptomonade and glaucocystophyte algal phycobiliproteins. In A. W. D. Larkum, S. E. Douglas, and J. A. Raven (eds) *Photosynthesis in Algae. Advances in Photosynthesis and Respiration,* Vol. 14. (pp. 305-34) Dordrecht, The Netherlands: Kluwer Academic Publishers. [5.3.1]

Topinka, J. A. (1978). Nitrogen uptake by *Fucus spiralis* (Phaeophyceae). *J. Phycol.* 14:241-7. [6.4.2]

Torres, J., A. Rivera, G. Clark, and S. J. Roux (2008). Participation of extracellular nucleotides in the wound response of *Dasycladus vermicularis* and *Acetabularia acetabulum* (Dasycladales, Chlorophyta). *J. Phycol.* 44:1504-11. [2.6.4]

Toth, G. B., and H. Pavia (2000a). Water-borne cues induce chemical defense in a marine alga (*Ascophyllum nodosum*). *P. Nat. Acad. Sci. USA* 97:14,418-20. [4.4.2]

Toth, G., and H. Pavia (2000b). Lack of phlorotannin induction in the brown seaweed *Ascophyllum nodosum* in response to increased copper concentrations. *Mar. Ecol. Prog. Ser.* 192:119-26. [9.3.2, 9.3.3]

Toth, G. B., and H. Pavia (2007). Induced herbivore resistance in seaweeds: a meta-analysis. *J. Ecol.* 95:425-34. [4.4.2]

Toth, G. B., O. Langhamer, and H. Pavia (2005). Inducible and constitutive defenses of valuable seaweed tissues: consequences for herbivore fitness. *Ecology* 86:612-18. [4.4.2]

Trautman, D. A., R. Hinde, and M. A. Borowitzka (2000). Population dynamics of an association between a coral reef sponge and a red macroalga. *J. Exp. Mar. Biol. Ecol.* 244:87-105. [4.5.1]

Troell, M. (2009). Integrated marine and brackish water aquaculture in tropical regions: research implementation and prospects. In D. Soto (ed.) *Integrated Mariculture: A Global Perspective.* FAO Fisheries and Aquaculture Technical Paper No. 529:47-131. [10.9]

Troell, M., P. Rönnbäck, C. Halling, N. Kautsky, and A. Buschmann (1999). Ecological engineering in aquaculture: use of seaweeds for removing nutrients from intensive mariculture. *J. Appl. Phycol.* 11:89-97. [10.9]

Troell, M., C. Halling, A. Neori, *et al.* (2003). Integrated mariculture: asking the right questions. *Aquaculture* 226:69-90. [10.9]

Troell, M., A. Joyce, T. Chopin, *et al.* (2009). Ecological engineering in aquaculture: potential for integrated multi-trophic aquaculture (IMTA) in marine offshore systems. *Aquaculture* 297:1-9. [10.1]

Trowbridge, C. D., and C. D. Todd (2001). Host-plant change in marine specialist herbivores: *Ascoglossan* sea slugs on introduced macroalgae. *Ecol. Monogr.* 71:219-43. [4.3.2]

Trujillo, A. P., and H. V. Thurman (2010). *Essentials of Oceanography*. Upper Saddle River, NJ: Prentice Hall, 551 pp. [3.1.1]

Tsekos, I. (1999). The sites of cellulose synthesis in algae: diversity and evolution of cellulose-synthesizing enzyme complexes. *J. Phycol.* 35:635-55. [1.3.1, 1.3.4]

Tsekos, I., and H.-D. Reiss (1994). Tip cell growth and the frequency and distribution of cellulose microfibril-synthesizing complexes in the plasma membrane of apical shoot cells of the red alga *Porphyra yezoensis*. *J. Phycol.* 30:300-10. [1.3.1]

Tseng, C. K. (1981). Commercial cultivation. In C. S. Lobban and M. J. Wynne (eds), *The Biology of Seaweeds* (pp. 680-725). Oxford: Blackwell Scientific. [7.3.4, 7.3.7, 10.1, 10.2.2, 10.2.3, 10.3, 10.3.2, 10.4.1]

Tseng, C. K. (l987). *Laminaria* mariculture in China. In M. S. Doty, J. F. Caddy, and B. Santelices (eds), *Case Studies of Seven Commercial Seaweed Resources* (pp. 239-63). FAO Fish. Tech. Pap. 281. [10.3, 10.3.1, 10.3.2]

Tsuda, R. T. (1965). Marine algae from Laysan Island with additional notes on the vascular flora. *Atoll Res. Bull.* 110, 31 pp. [3.3.8]

Tsuda, R. T., H. K. Larson, and R. J. Lujan (1972). Algal growth on beaks of live parrotfishes. *Pac. Sci.* 26:20-3. [3.3.8]

Tucker, J., N. Sheats, A. E. Giblin, C. S. Hopkinson, and J. P Montoya (1999). Using stable isotopes to trace sewage-derived material through Boston Harbor and Massachusetts Bay. *Mar. Environ. Res.* 48:353-75. [9.6.3]

Turner, A. (2010). Marine pollution from antifouling paint particles. *Mar. Poll. Bull.* 60:159-71. [9.5]

Turner, A., H. Pollock, and M. T. Brown (2009). Accumulation of Cu and Zn from antifouling paint particles by the marine macroalga, *Ulva lactuca*. *Environ. Poll.* 157:2314-19. [9.5]

Turner, D. R., M. Whitfield, and G. A. Dickson (1981). The equilibrium speciation of dissolved components in freshwater and seawater at 25ºC and I atm pressure. *Geochim. Cosmochim. Acta* 45:855-81. [6.2]

Turpin, D. H. (1980). Processes in nutrient based phytoplankton ecology. Ph.D. dissertation, University of British Columbia, Vancouver. [Fig. 6.1]

Turpin, D. H., and H. C. Huppe (1994) Integration of carbon and nitrogen metabolism in plant and algal cells. *Annu. Rev. Plant Physiol. Plant Mol. Biol.* 45:577-607. [6.4.1, 6.4.2, 6.5.1]

Turvey, J. R. (1978). Biochemistry of algal polysaccharides. *Int. Rev. Biochem.* 16: 151-77. [5.5.3]

Tyler, A. C., and K. J. McGlathery (2006). Uptake and release of nitrogen in the macroalgae *Gracilaria vermiculophylla* (Rhodophyta). *J. Phycol.* 42:515-25. [6.4.1]

Tyrrell, T. (2011). Anthropogenic modification of the oceans. *Phil. Trans. Roy. Soc. A* 309:887-908. [Essay 4]

Ueki, C., C. Nagasato, T. Motomura, and N. Saga (2008). Reexamination of the pit plugs and the characteristic membranous structures in *Porphyra yezoensis* (Bangiales, Rhodophyta). *Phycologia* 47:5-11. [1.2.1]

Ueki, C., C. Nagasato, T. Motomura, and N. Saga (2009). Ultrastructure of mitosis and cytokinesis during spermatogenesis in *Porphyra yezoensis* (Bangiales, Rhodophyta). *Bot. Mar.* 52:129-39. [1.3.5]

Uhrmacher, S., D. Hanelt, and W. Nultsch (1995). Zeaxanthin content and the degree of photoinhibition are linearly correlated in the brown alga *Dictyota dichotoma*. *Mar. Biol.* 123:159-65.

Umar, M. J., L. J. McCook, and I. R. Price (1998). Effects of sediment deposition on the seaweed *Sargassum* on a fringing coral reef. *Coral Reefs* 17:169-77. [8.3.2]

Underwood, A. J. (1978). A refutation of critical tidal levels as determinants of the structure of intertidal communities on British shores. *J. Exp. Mar. Biol. Ecol.* 33:261-76. [3.1.3]

Underwood, A. J. (1980). The effects of grazing by gastropods and physical factors on the upper limits of distribution of intertidal macroalgae. *Oecologia* 46:201-13. [1.1.2, 4.3.1]

Underwood, A. J. (1986). The analysis of competition by field experiments. In: J. Kikkawa and D. J. Anderson (eds) *Community Ecology: Pattern and Processes* (pp. 240-68). Oxford: Blackwell Scientific Publications. [4.2.3]

Underwood, A.J. (1992). Beyond BACI: The detection of environmental impacts on population in the real, but variable world. *J. Exp. Mar. Biol. Ecol.* 161:145-78. [9.2]

Underwood, A.J. (1994). On beyond BACI: Sampling designs that might reliably detect environmental disturbances. *Ecol. Appl.* 4:3-15. [9.2]

Underwood, A. J. (1997). *Experiments in Ecology: Their Logical Design and Interpretation Using Analysis of Variance*. Cambridge: Cambridge University Press, 504 pp. [3.5.1]

Underwood, A. J. (2006). Why overgrowth of intertidal encrusting algae does not always cause competitive exclusion. *J. Exp. Mar. Biol. Ecol.* 330:448-54. [4.2.1]

Underwood, A. J., and E. J. Denley (1984). Paradigms, explanations, and generalizations in models for the structure of intertidal communities on rocky shores. In J. D.R. Strong, D. Simberloff, L. G. Abele, and A. B. Thistle, (eds). *Ecological*

Communities: Conceptual Issues and the Evidence (pp. 151–80). Princeton, NJ: Princeton University Press. [Essay 2]

Underwood, A. J., and M. J. Keough (2001). Supply-side ecology: the nature and consequences of variations in recruitment of intertidal organisms. In M. D. Bertness, S. D. Gaines, and M. E. Hay (eds), *Marine Community Ecology* (pp. 183–200). Sunderland, MA: Sinauer Assocs. [3.1.2]

Underwood, A. J., and P. Jernakoff (1981). Effects of interactions between algae and grazing gastropods on the structure of a low-shore intertidal algal community. *Oecologia* 48:221–33. [4.3.1]

Uppalapati, S. R., and Y. Fujita (2001). The relative resistance of *Porphyra* species (Bangiales, Rhodophyta) to infection by *Pythium porphyrae* (Peronosporales, Oomycota). *Bot. Mar.* 44:1–7. [10.2.3]

Uthicke, S., B. Schaffelke, and M. Byrne (2009). A boom-bust phylum? Ecological and evolutionary consequences of density variations in echinoderms. *Ecol. Monogr.* 79:3–24. [4.3.1]

Vadas, R. L. (1977). Preferential feeding: an optimization strategy in sea urchins. *Ecol. Monogr.* 47:337–71. [4.3.2]

Vadas, R. L. (1985). Herbivory. In M. M. Littler and D. S. Littler (eds), *Handbook of Phycological Methods: Ecological Field Methods: Macroalgae* (pp. 531–72). Cambridge: Cambridge University Press. [4.3.2]

Vadas, R. L., and R. S. Steneck (1988). Zonation of deep-water benthic algae in the Gulf of Maine. *J. Phycol.* 24:338–46. [3.2]

Vadas, R. L., M. Keser, and B. Larson (1978). Effects of reduced temperatures on previously stressed populations of an intertidal alga. In J. H. Thorp and J. W. Gibbons (eds), (pp. 431–51). DOE Symposium Series 48 (CONF-721114). Washington DC: U.S. Government Printing Office. [9.8]

Vadas, R. L., W. A. Wright, and S. L. Miller (1990). Recruitment of *Ascophyllum nodosum*: wave action as a source of mortality. *Mar. Ecol. Prog. Ser.* 61:263–72. [2.5.2]

Vadas, R. L. Sr., S. Johnson, and T. A. Norton (1992). Recruitment and mortality of early post-settlement stages of benthic algae. *Brit. Phycol. J.* 27:331–51. [2.5.1]

Valdivia, N., R. A. Scrosati, M. Molis, and A. S. Knox (2011). Variation in community structure across vertical intertidal stress gradients: how does it compare with horizontal variation at different scales? *PLoS ONE* 6: e24062, doi:10.1371/journal.pone.0024062. [7.1]

Valentine, J. P., and C. R. Johnson (2004). Establishment of the introduced kelp *Undaria pinnatifida* following dieback of the native macroalga *Phyllospora comosa* in Tasmania, Australia. *Mar. Freshwater Res.* 55:223–30. [3.4]

Valiela, I., and M. L. Cole (2002). Comparative evidence that salt marshes and mangroves may protect seagrass meadows from land-derived nitrogen loads. *Ecosystems* 5:92–102. [3.3.5]

Valiella, I., J. McClelland, J. Hauxwell, *et al.* (1997). Macroalgal blooms in shallow estuaries: Controls and ecophysiological and ecosystem consequences. *Limnol. Oceanogr.* 42:1105–18. [6.8.5]

van Alystyne, K. L., J. M. Ehlig, and S. L. Whitman (1999). Feeding preferences for juvenile and adult algae depend on algal stage and herbivore species. *Mar. Ecol. Prog. Ser.* 180:179–85. [4.3.2]

van Alstyne, K. L., S. L. Whitman, and J. M. Ehlig (2001). Differences in herbivore preferences, phlorotannin production, and nutritional quality between juvenile and adult tissues from marine brown algae. *Mar. Biol.* 139:201–10. [4.3.2, 7.4.2]

Van Assche, F., and H. Clijsters (1990). Effects of metals on enzyme activity in plants. *Plant Cell Environ.* 13:195–206. [9.3.4]

van de Poll, W. H., A. Eggert, A. G. J. Buma, and A. M. Breeman (2001). Effects of UV-B-induced DNA damage and photoinhibition on growth of temperate marine red macrophytes: Habitat-related differences in UV-B tolerance. *J. Phycol.* 37:30–7. [7.6]

van den Hoek, C. (1982). Phytogeographic distribution groups of benthic marine algae in the North Atlantic Ocean. A review of experimental evidence from life history studies. *Helgol. Meeresunters* 35:153–214. [7.3.8]

van den Hoek, C., W. T, Stam, and J. L. Olsen (1988). The emergence of a new chlorophytan system, and Dr. Kornmann's contribution thereto. *Helgol. Meeresunters* 42:339–83. [1.3.5,1.4.1]

van den Hoek, C., D. G. Mann, and H. M. Jahns (1995). *Algae: An Introduction to Phycology.* Cambridge: Cambridge University Press, 627 pp. [1.1.3, 1.2.1, 1.3, 1.3.2, 1.3.3, 1.3.5, 2.6, 2.6.2]

van der Meer, J. P. (1978). Genetics of *Gracilaria* sp. (Rhodophyceae, Gigartinales). III. Non-Mendelian gene transmission. *Phycologia* 17:314–18. [1.4.2]

van der Meer, J. P. (1986a). Genetic contributions to research on seaweeds. *Prog. Phycol. Res.* 4:1–38. [1.4.2]

van der Meer, J. P. (1986b). Genetics of *Gracilaria tikvahiae* (Rhodophyceae). XI. Further characterization of a bisexual mutant. *J. Phycol.* 22:151–8. [1.4.2]

van der Meer, J. P. (1990). Genetics. In K. M. Cole and R. G. Sheath (eds), *Biology of the Red Algae* (pp. 103–21). Cambridge: Cambridge University Press. [1.4.2]

van der Meer, J. P., and N. L. Bird (1977). Genetics of *Gracilaria* sp. (Rhodophyceae, Gigartinales). I. Mendelian inheritance of two spontaneous green variants. *Phycologia* 16:159–61. [1.4.2] van der Meer, J. P., and E. R. Todd (1977). Genetics of *Gracilaria* sp. (Rhodophyceae, Gigartinales). IV. Mitotic recombination and its relationship to mixed phases in the life history. *Can. J. Bot.* 55:2810-17. [1.4.2]

van der Meer, J. P., and E. R. Todd (1980). The life history of *Palmaria palmata* in culture. A new type for the Rhodophyta. *Can. J. Bot.* 58:1250-6. [1.4.2, Essay 1, 2.2]

van der Meer, J. P., and X. Zhang (1988). Similar unstable mutations in three species of *Gracilaria* (Rhodophyta). *J. Phycol.* 24:198-202. [1.4.2]

van der Meer, J. P, M. U. Patwary, and C. J. Bird (1984). Genetics of *Gracilaria tikvahiae* (Rhodophyceae). X. Studies on a bisexual clone. *J. Phycol.* 20:42-6. [1.4.2]

van der Strate, H., S. Boele-Bos, J. Olsen, L. van de Zande, and W. Stam (2002a). Phylogeographic studies in the tropical seaweed *Chladophoropsis membranacea* (Chlorophyta, Ulvophyceae) reveal a cryptic species complex. *J. Phycol.* 38:572-82. [7.3.3]

van der Strate, H. J., L. van de Zande, W. T. Stam, and J. L. Olsen (2002b). The contribution of haploids, diploids and clones to fine-scale population structure in the seaweed *Cladophoropsis membranacea* (Chlorophyta). *Mol. Ecol.* 11:329-45. [2.2]

van Oppen, M. J. H., J. L. Olsen, W. T. Stam, C. Van der Hoek, and C. Wiencke (1993). Arctic–Antarctic disjunction in the benthic seaweeds *Acrosiphonia arcta* (Chlorophyta) and *Desmarestia viridis* (Phaeophyta) are of recent origin. *Mar. Biol.* 115:381-6. [7.3.6]

van Oppen, M. J. H., O. E. Diekmann, C. Wiencke, W. T. Stam, and J. L. Olsen (1994). Tracking dispersal routes: phylogeography of the Arctic–Antarctic disjunct seaweed *Acrosiphonia arcta* (Chlorophyta). *J. Phycol.* 30:67-80. [7.3.6]

van Tamelen, P. G. (1996). Algal zonation in tidepools: experimental evaluation of the roles of physical disturbance, herbivory and competition. *J. Exp. Mar. Biol. Ecol.* 201:197-231. [3.3.4]

van Tussenbroek, B. I., and M.G. Barba Santos (2011). Demography of *Halimeda incrassata* (Bryopsidales, Chlorophyta) in a Caribbean reef lagoon. *Mar. Biol.* 158:1461-71. [3.5.2]

Vanderklift, M. A., and S. Ponsard (2003). Sources of variation in consumer-diet delta N-15 enrichment: a meta-analysis. *Oecologia* 136(2):169-82. [Essay 3]

Vanderklift, M. A., and T. Wernberg (2008). Detached kelps from distant sources are a food subsidy for sea urchins. *Oecologia* 157(2): 327-35. [Essay 3]

Varvarigos, V., C. Katsaros, and B. Galatis (2004). Radial F-actin configurations are involved in polarization during protoplast germination and thallus branching of *Macrocystis pyrifera* (Phaeophyceae, Laminariales). *Phycologia* 43:693-702. [2.6.1]

Vasconcelos, M. T., and M. F. C. Leal (2001). Seasonal variability in the kinetics of Cu, Pb, Cd, and Hg accumulation by macroalgae. *Mar. Chem.* 74:65-85. [9.3.2]

Vásquez, J. A. (2008). Production, use and fate of Chilean brown seaweeds: resources for a sustainable fishery. *J. Appl. Phycol.* 20:457-67. [10.3]

Vásquez, J. A., D. Véliz, and L. M. Pardo (2001). Vida bajo las grandes algas pardas. In K. Alveal and T. Antezana (eds), *Sustentabilidad de la Biodiversidad. Un Problema Actual* (pp. 615-34). Concepción, Chile: Base Científico de Concepdión. [7.3.7]

Vass, I. (1997). Adverse effects of UV-B light on the structure and function of the photosynthetic apparatus. In M. Pessarakli (ed.) *Handbook of Photosynthesis* (pp. 931-49). New York: Marcel Dekker Inc. [5.2.2, 7.6]

Vayda, M. E., and M. L. Yuan (1994) The heat shock response of an Antarctic alga is evident at 5°C. *Plant Mol. Biol.* 24:229-33. [7.3.5, 7.3.6]

Venegas, M., B. Matsuhiro, and M. E. Edding (1993). Alginate composition of *Lessonia trabeculata* (Phaeophyta: Laminariales) growing in exposed and sheltered habitats. *Bot. Mar.* 36:47-51. [8.3.1]

Verges, A., N. A. Paul, and P. D. Steinberg (2008). Sex and life-history stage alter herbivore responses to a chemically defended red alga. *Ecology* 89:1334-43. [2.2]

Verhaeghe, E. F., A. Fraysse, J-L. Guerquin-Kern, *et al.* (2008). Microchemical imaging of iodine distribution in the brown alga *Laminaria digitata* suggests a new mechanism for its accumulation. *J. Biol Inorg. Chem.* 13:257-69. [6.5.6]

Vermeij, M. J. A., M. L. Dailer, and C. M. Smith (2011). Crustose coralline algae can suppress macroalgal growth and recruitment on Hawaiian coral reefs. *Mar. Ecol. Prog. Ser.* 422:1-7. [3.3.1]

Viano, Y., D. Bonhomme, M. Camps, *et al.* (2009). Diterpenoids from the Mediterranean brown alga *Dictyota* sp. evaluated as antifouling substances against a marine bacterial biofilm. *J. Nat. Prod.* 72:1299-1304. [4.2.2]

Vidondo, B., and C. M. Duarte (1998). Population structure, dynamics, and production of the Mediterranean macroalga *Codium bursa* (Chlorophyceae). *J. Phycol.* 34:918-24. [3.5.2]

Villares, R., X. Puente, and A. Carballeira (2001). *Ulva* and *Enteromorpha* as indicators of heavy metal pollution. *Hydrobiol.* 462:221-32. [9.3.2]

Villares, R., E. Carral, X. Puente, and A. Carballeira (2005). Metal levels in estuarine macrophytes: differences among species. *Estuaries* 28:948-56. [9.3.2]

Vincensini, L., T. Blisnick, and P. Bastin (2011). 1001 model organisms to study cilia and flagella. *Biol. Cell* 103:109-30. [1.3.3]

Vogel, H., G. E. Grieninger, and K. H. Zetsche (2002). Differential messenger RNA gradients in the unicellular alga *Acetabularia acetabulum*. *Plant Physiol.* 129:1407-16. [1.4.3]

Vogel, S. (1994). *Life in Moving Fluids: The Physical Biology of Flow* (2nd edn) Princeton, NJ: Princeton University Press, 467 pp. [8.0, 8.1.3, 8.1.4]

Vogel, S., and C. Loudon (1985). Fluid mechanics of the thallus of an intertidal red alga, *Halosaccion glandiforme*. *Biol. Bull.* 168:161-74. [7.5.1]

Vreeland, V., and B. Kloareg (2000). Cell wall biology in red algae: divide and conquer. *J. Phycol.* 36:793-7. [1.3.1]

Vreeland, V., and W. M. Laetsch (1989). Identification of associating carbohydrate sequences with labeled oligosaccharides. Localization of alginate-gelling subunits in walls of a brown alga. *Planta* 177:423-34. [1.3.1]

Vreeland, V., E. Zablackis, and W. M. Laetsch (1992). Monoclonal-antibodies as molecular markers for the intracellular and cell wall distribution of carrageenan epitopes in *Kappaphycus* (Rhodophyta) during tissue-development. *J. Phycol.* 28:328-42. [1.3.1]

Vreeland, V., J. H. Waite, and L. Epstein (1998). Polyphenols and oxidases in substratum adhesion by marine algae and mussels. *J. Phycol.* 34:1-8. [1.3.1, 2.5.2]

Vreeland, Y., E. Zablackis, B. Doboszewski, and W. M. Laetsch (1987). Molecular markers for marine algal polysaccharides. *Hydrobiologia* 151/152:155-60. [1.3.1]

Waaland, J. R., L. G. Dickson, and J. E. Carrier (1987). Conchocelis growth and photoperiodic control of conchospore release in *Porphyra torta* (Rhodophyta). *J. Phycol.* 23:399-406. [2.3.3]

Waaland, J. R., J. W. Stiller, and D. P. Cheney (2004). Macroalgal candidates for genomics. *J. Phycol.* 40:26-33. [1.4.1, 10.11]

Waaland, S. D. (1980). Development in red algae: elongation and cell fusion. In E. Gantt (ed.), *Handbook of Phycological Methods. Developmental and Cytological Methods* (pp. 85-93). Cambridge: Cambridge University Press. [1.3.4, 2.6.4]

Waaland, S. D. (1989). Cellular morphogenesis in the filaments of the red alga *Griffithsia*. In A. W. Coleman, L. J. Goff, and J. R. Stein-Taylor (eds), *Algae as Experimental Systems* (pp. 121-34). New York: Alan R. Liss. [2.6.4]

Waaland, S. D. (1990). Development. In K. M. Cole and R. G. Sheath (eds), *Biology of the Red Algae* (pp. 259-73). Cambridge: Cambridge University Press. [2.6.2, 2.6.4]

Waaland, S. D., and R. Cleland (1972). Development of the red alga *Griffithsia pacifica*: control by internal and external factors. *Planta* 105:196-204. [2.3, 2.6.4]

Waaland, S. D., and R. Cleland (1974). Cell repair through cell fusion in the red alga *Griffithsia pacifica*. *Protoplasma* 79:185-96. [2.6.3, 2.6.4]

Waaland, S. D., and J. R. Waaland (1975). Analysis of cell elongation in red algae by fluorescent labeling. *Planta* 126:127-38. [1.3.4]

Wada, S., M. N. Aoki, Y. Tsuchiya, *et al.* (2007) Quantitative and qualitative analyses of dissolved organic matter released from *Ecklonia cava* Kjellman, in Oura Bay, Shimoda, Izu Peninsula, Japan. *J. Exp. Mar. Biol. Ecol.* 349:344-58. [5.7.3]

Wada, S., M. N. Aoki, A. Mikami, T *et al.* (2008) Bioavailability of macroalgal dissolved organic matter in seawater. *Mar. Ecol. Prog. Ser.* 370:33-44. [5.7.3]

Wagner, F., and G. Falkner (2001). Phosphate limitation. In L.C. Rai and J.P. Gaur (eds). *Algal Adaptation to Environmental Stresses* (pp. 65-110). New York: Springer. [6.5.2]

Wahl, M. (2008). Ecological lever and interface ecology: epibiosis modulates the interactions between host and environment. *Biofouling* 24:427-38. [4.2.2]

Wai, T.-C., and G. A. Williams (2005). The relative importance of herbivore-induced effects on productivity of crustose coralline algae: sea urchin grazing and nitrogen excretion. *J. Exp. Mar. Biol. Ecol.* 324:141-56. [4.5.1]

Walker, C. H., S. P. Hopkin, R. M. Sibly, and D. B. Peakall (2006). *Principles of Ecotoxicology* (3rd edn). New York: Taylor and Francis, 315 pp. [9.2, 9.3.2]

Walker, G., R. G. Dorrell, A. Schlacht, and J. B. Dacks (2011). Eukaryotic systematics: a user's guide for cell biologists and parasitologists. *Parasitology* 138:1638-63. [1.1.1]

Wallentinus, I. (1984). Comparisons of nutrient uptake rates for Baltic macroalgae with different thallus morphologies. *Mar. Biol.* 80:215-25. [6.4.2, 6.8.1]

Walsh, R. S., and K. A. Hunter (1992). Influence of phosphorus storage on the uptake of cadmium by the marine alga *Macrocystis pyrifera*. *Linmol. Oceanogr.* 37:1361-9. [9.3.3]

Walters, L. J., and C. M. Smith (1994). Rapid rhizoid production in *Halimeda discoidea* decaisne (Chlorophyta, Caulerpales) fragments: a mechanism for survival after separation from adult thalli. *J. Exp. Mar. Biol. Ecol.* 175:105-20. [2.6.4]

Walters, L. J., C. M. Smith, J. A. Coyer, *et al.* (2002). Asexual propagation in the coral reef macroalga *Halimeda* (Chlorophyta, Bryopsidales): production, dispersal and attachment of small fragments. *J. Exp. Mar. Biol. Ecol.* 278:47-65. [2.1]

Warwick, R. M., K. R. Clarke, and Suharsono (1990). A statistic-analysis of coral community response to the El Niño in the Thousand Islands, Indonesia. *Coral Reefs* 8:171-9. [7.3.7]

Watanabe, S., A. Metaxas, and R. E. Scheibling (2009). Dispersal potential of the invasive green alga *Codium fragile* ssp. *fragile. J. Exp. Mar. Biol. Ecol.* 381:114-25. [3.4]

Watson, B. A., and S. D. Waaland (1983). Partial purification and characterization of a glycoprotein cell fusion hormone from *Griffithsia pacifica*, a red alga. *Plant Physiol.* 71:327-32. [2.6.4]

Watson, B. A., and S. D. Waaland (1986). Further biochemical characterization of a cell fusion hormone from the red alga *Griffithsia pacifica. Plant Cell Physiol.* 27:1043-50. [2.6.4]

Webb, W. L., M. Newton, and D. Starr (1974). Carbon dioxide exchange of *Alnus rubra. Oecologia* 17:281-91. [5.3.3]

Weber, A. P. M., and K. W. Osteryoung (2010). From endosymbiosis to synthetic photosynthetic life. *Plant Physiol.* 154:593-7. [1.3.2]

Wehr, J. D., and R. G. Sheath (eds) (2003). *Freshwater Algae of North America: Ecology and Classification*. New York: Academic Press. 918 pp. [1.1.1]

Weich, R. G., and E. Graneli (1989). Extracellular alkaline phosphatase activity in *Ulva lactuca* L. *J. Exp. Mar. Biol. Ecol.* 129:33-44. [6.5.2]

Weinberger, F., and M. Friedlander (2000). Response of *Gracilaria conferta* (Rhodophyta) to oligoagars results in defense against agar-degrading epiphytes. *J. Phycol.* 36:1079-86. [4.2.2]

Weinberger, F., B. Coquempot, S. Forner, *et al.* (2007). Different regulation of haloperoxidation during agar oligosaccharide-activated defence mechanisms in two related red algae, *Gracilaria* sp. and *Gracilaria chilensis. J. Exper. Bot.* 58:4365-72. [4.2.2]

Weinberger, F., M.-L. Guillemin, C. Destombe, *et al.* (2010). Defense evolution in the Gracilariaceae (Rhodophyta): substrate-regulated oxidation of agar oligosaccharides is more ancient than the oligoagar-activated oxidative burst. *J. Phycol.* 46:958-68. [4.2.2]

Weiner, S. (1986). Organization of extracellularly mineralized tissues: a comparative study of biological crystal growth. *Crit. Rev. Biochem.* 20:365-408. [Fig. 6.16]

Weissflog, J., S. Adolph, T. Wiesemeier, and G. Pohnert (2008). Reduction of herbivory through wound-activated protein cross-linking by the invasive macroalga *Caulerpa taxifolia. ChemBioChem.* 9:29-32. [2.6.4]

Welling, M., G. Pohnert, F. C. Küpper, and C. Ross (2009). Rapid biopolymerisation during wound plug formation in green algae. *J. Adhesion* 85:825-38. [2.6.4]

Wernberg, T. (2005). Holdfast aggregation in relation to morphology, age, attachment and drag for the kelp *Ecklonia radiata. Aquat. Bot.* 82:168-80. [8.3.1]

Wernberg, T., and M. A. Vanderklift (2010). Contribution of temporal and spatial components to morphological variation in the kelp *Ecklonia* (Laminariales). *J. Phycol.* 46:153-61. [2.6.2]

Wernberg, T., G. A. Kendrick, and J. C. Phillips (2003). Regional differences in kelp-associated algal assemblages on temperate limestone reefs in south-western Australia. *Diversity and Distributions* 9:427-41. [Essay 3]

Wernberg, T., M. A. Vanderklift, J. How, and P. S. Lavery (2006). Export of detached macroalgae from reefs to adjacent seagrass beds. *Oecologia* 147(4):692-701. [Essay 3]

Wernberg, T., M. S. Thomsen, F. Tuya, G. A., *et al.* (2010). Decreasing resilience of kelp beds along a latitudinal temperature gradient: potential implications for a warmer future. *Ecol. Lett.* 13:685-94. [7.3.8]

Wernberg, T., B. D. Russell, M. S. Thomsen, *et al.* (2011a). Seaweed communities in retreat from ocean warming. *Curr. Biol.* 21:1828-32. [7.3.8]

Wernberg, T., M. S. Thomsen, F. Tuya, and G. A. Kendrick (2011b). Biogenic habitat structure of seaweeds change along a latitudinal gradient in ocean temperature. *J. Exp. Mar. Biol. Ecol.* 400:264-71. [7.3.8]

West, J. A. (1972). Environmental regulation of reproduction in *Rhodochorton purpureum*. In I. A. Abbott and M. Kurogi (eds), *Contributions to the Systematics of the Benthic Marine Algae of the North Pacific* (pp. 213-30). Kobe: Japan Soc. Phycol. [7.3.3]

Westermeier, R., and I. Gómez (1996). Biomass, energy contents and major organic compounds in the brown alga *Lessonia nigrescens* (Laminariales, Phaeophyceae) from Mehuin, South Chile. *Bot. Mar.* 39:553-9. [5.6, 5.7.2]

Weykam, G., I. Gómez, C. Wiencke, K. Iken, and H. Klöser (1996). Photosynthetic characteristics and C:N ratios of macroalgae from King George Island (Antarctica). *J. Exp. Mar. Biol. Ecol.* 204:1-22. [5.3.3]

Weykam, G., D. N. Thomas, and C. Wiencke (1997). Growth and photosynthesis of the Antarctic red algae *Palmaria decipiens* (Palmariales) and *Iridaea cordata* (Gigartinales) during and following extended periods of darkness. *Phycologia* 36:395-405. [5.4.3]

Wheeler, P. A. (1979). Uptake of methylamine (an ammonium analogue) by *Macrocystis pyrifera* (Phaeophyta). *J. Phycol.* 15:12-17. [Table 6.3]

Wheeler, P. A. (1985). Nutrients. In M. M. Littler and D. S. Littler (eds), *Handbook of Phycological Methods: Ecological Field Methods: Macroalgae* (pp. 493-508). Cambridge: Cambridge University Press. [6.2]

Wheeler, P. A., and B. R. Björnsater (1992). Seasonal fluctuations in tissue nitrogen, phosphorus, and N: P for five macroalgal species common to the Pacific northwest coast. *J. Phycol.* 28:1-6. [6.7.2, 6.8.2]

Wheeler, P. A., and W. J. North (1980). Effect of nitrogen supply on nitrogen content and growth rate of juvenile *Macrocystis pyrifera* (Phaeophyta) sporophytes. *J. Phycol.* 16:577-82. [6.5.1, 6.8.2]

Wheeler, P. A., and W. J. North (1981). Nitrogen supply, tissue composition and frond growth rates for *Macrocystis pyrifera* off the coast of southern California. *Mar. Biol.* 64:59-69. [5.6, 6.8.2]

Wheeler, W. N. (1980). Effect of boundary layer transport on the fixation of carbon by the giant kelp *Macrocystis pyrifera*. *Mar. Biol.* 56:103-10. [6.4.2, 8.1.4, 8.2.1]

Wheeler, W. N. (1982). Nitrogen nutrition of *Macrocystis*. In L. M. Srivastava (ed.), *Synthetic and Degradative Processes in Marine Macrophytes* (pp. 121-37). Berlin:Walter de Gruyter. [Fig. 6.7]

Wheeler, W. N. (1988). Algal productivity and hydrodynamics-a synthesis. *Prog. Phycol. Res.* 6:23-58. [8.1.3, 8.2.1]

Wheeler, W. N., and L. M. Srivastava (1984). Seasonal nitrate physiology of *Macrocystis integrifolia* Bory. *J. Exp. Mar. Biol. Ecol.* 76:35-50. [6.4.2, 6.8.2]

Wheeler, W. N., and M. Weidner (1983). Effects of external inorganic nitrogen concentration on metabolism, growth and activities of key carbon and nitrogen assimilating enzymes of *Laminaria saccharina* (Phaeophyceae) in culture. *J. Phycol.* 19:92-6. [6.7.2, 7.3.1]

White, H. H. (1984). *Concepts in Marine Pollution Measurements.* College Park, MD: Maryland Sea Grant College Program, University of Maryland, 743 pp. [9.2]

White, P. J., and M. R. Broadley (2003). Calcium in plants. *Annals Bot.* 92:487-511. [6.5.3]

Whitton, B. A., and M. Potts (1982). Marine littoral. In N. G. Carr and B. A. Whitton (eds), *The Biology of Cyanobacteria* (pp. 515-42). Oxford: Blackwell Scientific. [1.1.1]

Wichard, T., and W. Oertel (2010). Gametogenesis and gamete release of *Ulva mutabilis* and *Ulva lactuca* (Chlorophyta): regulatory effects and chemical characterization of the "swarming inhibitor". *J. Phycol.* 46:248-59. [2.3.3]

Wiencke, C. (1990a). Seasonality of brown macroalgae from Antarctica - a long-term culture study under fluctuating Antarctic daylengths. *Polar Biol.* 10:589-600. [2.3.1, 2.3.3]

Wiencke, C. (1990b). Seasonality of red and green macroalgae from Antarctica - a long-term culture study under fluctuating Antarctic daylengths. *Polar Biol.* 10:60-7. [2.3.2]

Wiencke, C. (1996). Recent advances in the investigation of Antarctic macroalgae. *Polar Biol.* 16:231-40. [7.3.6]

Wiencke, C., and M. N. Clayton (1990). Sexual reproduction, life history, and early development in culture of the

Antarctic brown alga *Himantothallus grandifolius* (Desmarestiales, Phaeophyceae). *Phycologia* 29:9-18. [5.6]

Wiencke, C., and M. Clayton (2002). *Antarctic Seaweeds. Synopses of the Antarctic Benthos.* Lichtenstein: A.R.G. Gantner Verlag, 239 pp. [7.3.3, 7.3.6]

Wiencke, C., and M. N. Clayton (2009). Biology of polar benthic algae. *Bot. Mar.* 52:479-81. [3.3.3]

Wiencke, C., and J. Davenport (1987). Respiration and photosynthesis in the intertidal alga *Cladophora rupestris* (L.) Kütz. under fluctuating salinity regimes. *J. Exp. Mar. Biol. Ecol.* 114:183-97. [4.2.2, 7.4.3]

Wiencke, C., and A. Läuchli (1981). Inorganic ions and florido-side as osmostic solutes in *Porphyra umbilicalis*. *Z. Pflanzenphysiol.* 103:247-58. [7.4.2]

Wiencke, C., and I. tom Dieck (1989). Temperature requirements for growth and temperature tolerance of macroalgae endemic to the Antarctic region. *Mar. Ecol. Prog. Ser.* 54:189-97. [7.3.3, 7.6, 7.3.8]

Wiencke, C., J. Rahmel, U. Karsten, G. Weykam, and G. Kirst (1993). Photosynthesis of marine macroalgae from Antarctica: light and temperature requirements. *Bot. Acta* 106:78-87. [7.3.1, 7.3.2]

Wiencke, C., I. Bartsch, B. Bischoff, A. F. Peters, and A. M. Breeman (1994). Temperature requirements and biogeography of Antarctic, Arctic and amphiequatorial seaweeds. *Bot. Mar.* 37:247-59. [7.3.6]

Wiencke, C., I. Gómez, H. Pakker, *et al.* (2000). Impact of UV-radiation on viability, photosynthetic characteristics and DNA on brown algal zoospores: Implications for depth zonation. *Mar. Ecol. Progr. Ser.* 197:217-29. [5.2.2]

Wiencke, C., M. Roleda, A. Gruber, M. Clayton, and K. Bischof (2006). Susceptibility of zoospores to UV radiation determines upper depth distribution limit of Arctic kelps: evidence through field experiments. *J Ecol* 94:455-63. [5.2.2, 7.6, 7.9]

Wiencke, C., I. Gómez, and K. Dunton (2009). Phenology and seasonal physiological performance of polar seaweeds. *Bot. Mar.* 52:585-92. [2.3.1, 5.4.3]

Wiens, J. A. (1989). Spatial scaling in ecology. *Func. Ecol.* 3:385-97. [Essay 2]

Wiesemeier, T., M. E. Hay, and G. Pohnert (2007). The potential role of wound-activated volatile release in the chemical defence of the brown alga *Dictyota dichotoma*: blend recognition by marine herbivores. *Aquat. Sci.* 69:403-12. [4.4.1]

Wiesemeier, T., K. Jahn, and G. Pohnert (2008). No evidence for the induction of brown algal chemical defense by the phytohormones jasmonic acid and methyl jasmonate. *J. Chem. Ecol.* 34:1523-31. [2.6.3]

Wilce, R. T. (1990). Role of the Arctic Ocean as a bridge between the Atlantic and the Pacific Ocean: fact and hypothesis. In D. J. Garbary and G. R. South (eds), *Evolutionary Biogeography of the Marine Algae of the North Atlantic. NATO ASI. Ser. G. Ecol. Sci.* (vol. 22, pp. 323–47). Berlin: Springer-Verlag. [3.3.3]

Wilce, R. T., and A. N. Davis (1984). Development of *Dumontia contorta* (Dumontiaceae, Cryptonemiales) compared with that of other higher red algae. *J. Phycol.* 20:336–51. [1.2.1]

Wilkinson, C. (2008). *Status of Coral Reefs of the World: 2008.* Townsville, Australia: Global Coral Reef Monitoring Network and Reef and Rainforest Research Centre. [Essay 6]

Williams, I. D., V. C. Polunin, and V. J. Hendrick (2001). Limits to grazing by herbivorous fishes and the impact of low coral cover on macroalgal abundance on a coral reef in Belize. *Mar. Ecol. Prog. Ser.* 222:187–96. [Essay 6]

Williams, S. L. (1984). Uptake of sediment ammonium and translocation in a marine green macroalga *Caulerpa cupressoides. Limnol. Oceanogr.* 29:374–9. [6.4.2, 6.5.2]

Williams, S. L. (2007). Introduced species in seagrass ecosystems: status and concerns. *J. Exp. Mar. Biol. Ecol.* 350:89–110. [3.4]

Williams, S. L., and R. C. Carpenter (1990). Photosynthesis/photon flux density relationships among components of coral reef algal turfs. *J. Phycol.* 26:36–40. [1.2.2, 3.3.1, 4.3.2, 5.3.2]

Williams, S. L., and M. N. Dethier (2005). High and dry: variation in net photosynthesis of the intertidal seaweed *Fucus gardneri. Ecology* 86:2373–9. [3.1.3]

Williams, S. L., and T. R. Fisher (1985). Kinetics of nitrogen-15 labeled ammonium uptake by *Caulerpa cupressoides* (Chlorophyta). *J. Phycol.* 21:287–96. [6.4.1, 6.4.2]

Williams, S. L., and K. L. Heck Jr. (2001). Seagrass community ecology. In M. D. Bertness, S. D. Gaines, and M. E. Hay (eds), *Marine Community Ecology* (pp. 317–37). Sunderland, MA: Sinauer Associates, Inc. [3.3.1]

Williams, S. L., and S. K. Herbert (1989). Transient photosynthetic responses of nitrogen-deprived *Petalonia fascia* and *Laminaria saccharina* (Phaeophyta) to ammonium resupply. *J. Phycol.* 25:515–22. [6.4.1]

Williams, S. L., and S. L. Schroeder (2004). Eradication of the invasive seaweed *Caulerpa taxifolia* by chlorine bleach. *Mar. Ecol. Prog. Ser.* 272:69–76. [3.4]

Williams, S. L., and J. E. Smith (2007). A global review of the distribution, taxonomy, and impacts of introduced seaweeds. *Annu. Rev. Ecol. Evol. Syst.* 38:327–59. [3.4]

Williams, S. L., V. A. Breda, T. W. Anderson, and B. B Nyden (1985). Growth and sediment disturbances of *Caulerpa* spp.

(Chlorophyta) in a submarine canyon. *Mar. Ecol. Prog. Ser.* 21:275–81. [8.3.2]

Wilmotte, A., A. Goffart, and V. Demoulin (1988). Studies of marine epiphytic algae, *Calvi*, Corsica. I. Determination of minimal sampling areas for microscopic algal epiphytes. *Br. Phycol. J.* 23:251–8. [Fig. 3.13]

Wilson, J. B. (1999). Assembly rules in plant communities. In E. Weiher and P. Keddy (eds), *Ecological Assembly Rules: Perspectives, Advances and Retreats* (pp. 130–64). Cambridge: Cambridge University Press. [3.1.2]

Wilson, S. M., J. A. West, and J. D. Pickett-Heaps (2003). Time-lapse videomicroscopy of fertilization and the actin cytoskeleton in *Murrayella periclados* (Rhodomelaceae, Rhodophyta). *Phycologia* 42:638–45. [2.4]

Wiltens, J., U. Schreiber, and W. Vidaver (1978). Chlorophyll fluorescence induction: an indicator of photosynthetic activity in marine algae undergoing desiccation. *Can. J. Bot.* 56: 2787–94. [7.5.1]

Wing, S. R., and M. R. Patterson (1993). Effects of wave-induced light flecks in the intertidal zone on photosynthesis in the macroalgae *Postelsia palmaeformis* and *Hedophyllum sessile* (Phaeophyceae). *Mar. Biol.* 116:519–25. [3.3.2, 5.2.2, 8.1.4, 8.2.1, 8.3.1]

Wing, S. R., J.J. Leichter, C. Perrin *et al.* (2007). Topographic shading and wave exposure influence morphology and ecophysiology of *Ecklonia radiata* (C. Agardh 1817) in Fiordland, New Zealand. *Limnol. Oceanogr.* 52:1853–64. [2.6.2, 8.2.1]

Wise, R. R. (2007). The diversity of plastid form and function. In R. R. Wise and J. K. Hoober (eds), *The Structure and Function of Plastids* (pp. 3–26). Dordrecht, The Netherlands: Springer. [1.3.2]

Witman, J. D., and P. K. Dayton (2001). Rocky subtidal communities. In M. D. Bertness, S. D. Gaines, and M. E. Hay (eds), *Marine Community Ecology* (pp. 339–66). Sunderland, MA: Sinauer Assocs. [3.2, 3.3.2, 4.2.1]

Witman, J. D., and K. Roy (2009). Experimental marine macroecology: progress and prospects. In J. D. Witman, and K. Roy (eds), *Marine Macroecology* (pp. 341–56). Chicago: University of Chicago Press. [3.1.2]

Witman, J. D., M. Brandt, and F. Smith (2010). Coupling between subtidal prey and consumers along a mesoscale upwelling gradient in the Galapagos Islands. *Ecol. Monogr.* 80:153–77. [Essay 2]

Wolanski, E., and W. M. Hammer (1988). Topographically controlled fronts in the ocean and their biological influence. *Science* 241:177–81. [8.1.1]

Wolcott, B. D. (2007). Mechanical size limitation and life-history strategy of an intertidal seaweed. *Mar. Ecol. Prog. Ser.* 338:1–10. [8.3.1]

Wolfe, J. M., and M. M. Harlin (l988a). Tidepools in southern Rhode Island, USA. I. Distribution and seasonality of macroalgae. *Bot. Mar.* 31:525-36. [3.3.4]

Wolfe, J. M., and M. M. Harlin (l988b). Tidepools in southern Rhode Island, USA. II. Species diversity and similarity analysis of macroalgal communities. *Bot. Mar.* 31:537-46. [3.3.4]

Wolff, M. (1987). Population dynamics of the Peruvian scallop *Argopecten purpuratus* during the El Nino Phenomenon 1983. *Can. J. Fish. Aquat. Sci.* 44:1684-91. [7.3.7]

Womersley, H. B. S. (1971). *Palmoclathrus*, a new deep-water genus of Chlorophyta. *Phycologia* 10:229-33. [1.1.1]

Wong, K. F., and J. S. Craigie (1978). Sulfohydrolase activity and carrageenan biosynthesis in *Chondrus crispus* (Rhodophyceae). *Plant Physiol.* 61:663-6. [5.5.3]

Wong, P.-F., L.-J. Tan, H. Nawi, and S. AbuBakar (2006). Proteomics of the red alga, *Gracilaria changii* (Gracilariales, Rhodophyta). *J. Phycol.* 42:113-20. [1.4.1]

Wong, T. K.-M., C.-L. Ho, W.-W. Lee, R. A. Rahim, and S.-M. Phang (2007). Analyses of expressed sequence tags from *Sargassum binderi* (Phaeophyta). *J. Phycol.* 43:528-34. [1.4.1]

Wood, S. A., S. A. Lilley, D. R. Schiel, and J. B. Shurin (2010). Organismal traits are more important than environment for species interactions in the intertidal zone. *Ecol. Lett.* 13:1160-71. [4.1]

Woodley, J. D., E. A. Chornesky, P. A. Clifford, *et al.* (1981). Hurricane Allen's impact on Jamaican coral reefs. *Science* 214:749-55. [8.3.2]

Worm, B., and H. K. Lotze (2006). Effects of eutrophication, grazing and algal blooms on rocky shores. *Limnol. Oceanogr.* 51:569-79. [6.8.5]

Worm, B., H. K. Lotze, and U. Sommer (2001). Algal propagule banks modify competition, consumer and resource control on Baltic rocky shores. *Oecologia* 128:281-93. [3.5.2]

Worm, B., E. B. Barbier, N. Beaumont, *et al.* (2006). Impacts of biodiversity loss on ocean ecosystem services. *Science* 314:787-90. [3.5.1]

Wright, D. G., R. Pawlowicz, T. J. McDougall, R. Feistel, and G. M. Marion (2010). Absolute salinity, "density salinity" and the reference-composition salinity scale: present and future use in the seawater standard TEOS-10. *Ocean Sci. Discuss.* 7:1559-625. [7.2.1, 7.4]

Wright, J. T., S. A. Dworjanyn, C. N. Steinberg, J. E. Williamson, and A. G. B. Poore (2005). Density-dependent sea urchin grazing: differential removal of species, changes in community composition and alternative community states. *Mar. Ecol. Prog. Ser.* 298:143-56. [4.3.1]

Wright, P. J., J. A. Chudek, R. Foster, I. R. Davison, and R. H. Reed (1985). The occurrence of altritol in the brown alga *Himanthalia elongata. Br. Phycol. J.* 20:191-2. [5.4.2]

Wright, P. J., J. R. Green, and J. A. Callow (1995a). The *Fucus* (Phaeophyceae) sperm receptor for eggs. 1. Development and characteristics of a binding assay. *J. Phycol.* 31:584-91. [2.4]

Wright, P. J., J. A. Callow, and J. R. Green (1995b). The *Fucus* (Phaeophyceae) sperm receptor for eggs. 2. Isolation of a binding-protein which partially activates eggs. *J. Phycol.* 31:592-600. [2.4]

Wulff, A., K. Iken, M. L. Quartino, *et al.* (2009). Biodiversity, biogeography and zonation of marine benthic micro- and macroalgae in the Arctic and Antarctic. *Bot. Mar.* 52:491-507. [3.3.3]

Xu, B., Q. S. Zhang, S. C. Qu, Y. Z. Cong, and X. X. Tang (2009). Introduction of a seedling production method using vegetative gametophytes to the commercial farming of *Laminaria* in China. *J. Appl. Phycol.* 21:171-8. [10.3.2]

Xu, H., R. J. Deckert, and D. J. Garbary (2008). *Ascophyllum* and its symbionts. X. Ultrastructure of the interaction between *A. nodosum* (Phaeophyceae) and *Mycophycias ascophylli* (Ascomycetes). *Botany* 86:185-93. [4.5.1]

Yamada, T., K. Ikawa, and K. Nisizawa (1979). Circadian rhythm of the enzymes participating in the CO_2 photoassimilation of a brown alga, *Spatoglossum pacificum. Bot. Mar.* 22:203-9. [5.7.2]

Yan, X.-H., and M. Huang (2010). Identification of *Porphyra haitanensis* (Bangiales, Rhodophyta) meiosis by simple sequence repeat markers. *J. Phycol.* 46:982-6. [1.4.2, 2.2]

Yarish, C., and R. Pereira (2008). Mass production of marine macroalgae. In S. E. Jørgensen and B. D. Fath (eds), *Ecological Engineering. Vol. 3 of Encyclopedia of Ecology*, (pp. 2236-47). Oxford: Elsevier. [10.2, 10.3.1, 10.4.1, 10.4.2, 10.6.2]

Yarish, C., P. Edwards, and S. Casey (1979). Acclimation responses to salinity of three estuarine red algae from New Jersey. *Mar. Biol.* 51:289-94.

Yarish, C., P. Edwards, and S. Casey (1980). The effects of salinity, and calcium and potassium variation on the growth of two estuarine red algae. *J. Exp. Mar. Biol. Ecol.* 47:235-49. [7.4.3]

Yarish, C., A. M. Breeman, and C. van den Hoek (1984). Temperature, light, and photoperiod responses of some northeast American and west European endemic rhodophytes in relation to their geographical distribution. *Helgol. Meeresunters* 38:273-304. [7.3.8]

Yarish, C., H. Kirkman, and K. Lüning (1987). Lethal exposure times and preconditioning to upper temperature limits of some temperate North Atlantic red algae. *Helgol. Meeresunters* 41: 323-7. [7.3.8]

Yarish, C., B. H. Brinkhuis, B. Egan, and Z. Garcia-Ezquivel (1990). Morphological and physiological bases for

Laminaria selection protocols in Long Island Sound. In C. Yarish, C. A. Penniman, and P. Van Patten (eds), *Economically Important Plants of the Atlantic: Their Biology and Cultivation* (pp. 53-94). Groton, CT: Connecticut Sea Grant College Program. [10.3]

Yellowlees, D., T. A. V. Rees, and W. Leggat (2008). Metabolic interactions between algal symbionts and invertebrate hosts. *Plant Cell Environ.* 31:679-94. [4.5.1]

Yiou, F., G. M. Raisbeck, G. C. Christensen, and E. Holm (2002). $^{129}I/^{127}I$, $^{129}I/^{137}Cs$ and $^{129}I/^{99}Tc$ in the Norwegian coastal current from 1980 to 1998. *J. Environ. Radioact.* 60:61-1. [9.7]

Yokohama, Y., and T. Misonou (1980). Chlorophyll a:b ratios in marine benthic algae. *Jap. J. Phycol.* 28: 219-23. [5.3.3]

Yoon, H. S., J. D. Hackett, C. Ciniglia, *et al.* (2004). A molecular timeline for the origin of photosynthetic eukaryotes. *Mol. Biol. Evol.* 21:809-18. [1.1.1]

Yoon, H. S., G. C. Zuccarello, and D. Battacharya (2010). Evolutionary history and taxonomy of red algae. In J. Seckbach, and D. J. Chapman (eds) *Red Algae in the Genomic Age* (pp. 25-44). Dordrecht, The Netherlands: Springer. [1.1.1]

Yoon, K. S., K. P. Lee, T. A. Klochkova, and G. H. Kim (2008). Molecular characterization of the lectin, bryohealin, involved in protoplast regeneration of the marine alga *Bryopsis plumosa* (Chlorophyta). *J. Phycol.* 44:103-12. [2.6.4]

Young, E. B., P. S. Lavery, B. van Elven, M. J. Dring, and J. A. Berges (2005). Nitrate reductase activity in macroalgae and its vertical distribution in macroalgal epiphytes of seagrasses. *Mar. Ecol. Prog. Ser.* 288:103-14. [6.5.1]

Young, E. B., M. J. Dring, G. Savidge, D. A. Birkett, and J. A. Berges (2007). Seasonal variations in nitrate reductase activity and internal N pools in intertidal brown algae are correlated with ambient nitrate concentrations. *Plant Cell Envir.* 30:764-74. [6.5.1]

Young, E. B., J. A. Berges, and M. J. Dring (2009). Physiological responses of intertidal marine brown algae to nitrogen deprivation and resupply of nitrate and ammonium. *Physiol. Plant.* 135:400-11. [6.5.1]

Young-Sook, O., D. S. Sim, and S. J. Kim (2001). Effects of nutrients on crude oil biodegradation in the upper intertidal zone. *Mar. Poll. Bull.* 42:1367-72. [9.4.1]

Ytreberg, E., J. Karlsson, S. Hoppe, B. Eklund, and K. Ndungu (2011a). Effects of organic complexation on copper accumulation and toxicity to the estuarine red macroalga *Ceramium tenuicorne*: A test of the free ion activity model. *Environ. Sci. Technol.* 45:3145-53. [9.2, 9.3.5]

Ytreberg, E., J. Karlsson, K. Ndungu, *et al.* (2011b). Influence of salinity and organic matter on the toxicity of Cu to a brackish water and marine clone of the red macroalga *Ceramium tenuicorne*. *Ecotox. Environ. Safety* 74:636-42. [9.2, 9.3.5]

Zacher, K., A. Wulff, M. Molis, D. Hanelt, and C. Wiencke (2007). Ultraviolet radiation and consumer effects on a field-grown intertidal macroalgal assemblage in Antarctica. *Global Change Biol.* 13:1201-15. [7.6]

Zacher, K., R. Rautenberger, D. Hanelt, A. Wulff, and C. Wiencke (2009). The abiotic environment of polar marine benthic algae. *Bot. Mar.* 52:483-90. [3.3.3]

Zbikowski, R., P. Szefer, and A. Latala (2007). Comparison of green algae *Cladophora* sp. and *Enteromorpha* sp. as potential biomonitors of chemical elements in the southern Baltic. *Sci. Total Environ.* 387:320-32.

Zechman, F. W., H. Verbruggen, F. Leliaert, *et al.* (2010). An unrecognized ancient lineage of green plants persists in deep marine waters. *J. Phycol.* 46:1288-95. [1.1.1]

Zhang, Q. S., X. X. Tang, Y. Z. Cong, S. C. Qu, and G. P. Yang (2007). Breeding of an elite *Laminaria* variety 90-1 through inter-specific gametophyte crossing. *J. Appl. Phycol.* 19:303-11. [10.3.2]

Zhang, A. Q., K. M. Y. Leung, K. W. H. Kwok, V. W. W. Bao, and M. H. W. Lam (2008a). Toxicities of antifouling biocide Irgarol 1051 and its major degraded product to marine primary producers. *Mar. Poll. Bull.* 57:575-86. [9.5]

Zhang, Q. S., S. C. Qu, Y. Z. Cong, S. J Luo, and X. X. Tang (2008b). High throughput culture and gametogenesis induction of *Laminaria japonica* gametophyte clones. *J. Appl. Phycol.* 20:205-11. [10.3.2]

Zhang, X., and J. P. van der Meer (1988). Polyploid gametophytes of *Gracilaria tikvahiae* (Gigartinales, Rhodophyta). *Phycologia* 27:312-18. [2.2]

Zhuang, S. H. (2006). Species richness, biomass and diversity of macroalgal assemblages in tidepools of different sizes. *Mar. Ecol. Prog. Ser.* 309:67-73. [3.3.4]

Zimmerman, R. C., and D. L. Robertson (1985). Effects of El Niño on local hydrography and growth of the giant kelp, *Macrocystis pyrifera*, at Santa Catalina Island, California. *Limnol. Oceanogr.* 30: 1298-1302. [7.3.7]

Zuccarello, G. C., D. Moon, and L. J. Goff (2004). A phylogenetic study of parasitic genera placed in the family Choreocolacaceae (Rhodophyta). *J. Phycol.* 40:937-45. [4.5.2]

Zuccaro, A., and J. I. Mitchell (2005). Fungal communities of seaweeds. In J. Dighton, J. F. White, and P. Oudemans (eds), *The Fungal Community: Its Organization and Role in the Ecosystem* (3rd edn) (pp. 533-79). Boca Raton, FL: CRC Press. [4.5.1]

Subject Index

Page numbers in **boldface** indicate sections on the topic. Page numbers in *italics* indicate a table or figure. Index terms under genera include only places with substantial description of the subject and not just brief reference or mention of the genus.

abscisic acid (ABA), 92
absorption spectra. *See* light
accessory pigments. *See* carotenoids, etc.
Acetabularia
    cap morphology, *6*
    cyst development, 25
    cyst morphogenesis, *28*
    homeobox genes, 46
    life history, *27*
    morphogenetic substances, 46
    nucleo cytoplasmic interactions, 45, *46*
Acrochaete
    endophytic pathogen of *Chondrus*, 172
    endophytism in *Chondrus*, 145
adelphoparasites, 173
agar
    Gelidium production and products, **428**
    Gracilaria production and products, **429**
    structure, *214*
agars, 213
algae
    evolutionary origin, *2*, *3*
algal turf
    pre-emptive competition, 139
algal–herbivore interactions, **164**
    bioactive chemicals, **164**
alginate
    commercial use of *Saccharina*, 423
    composition, 212
    in propagule adhesion, 79

properties, 424
structure, *213*
synthesis, 41
alkaline phosphatase, 266
allelopathy, **140**, 142
    criteria, 143
    defense against pathogens, 145
alternation of generations, *49*, 53
amino acids
    uptake, 258
ammonium
    depletion and uptake, *248*
    incorporation, 262
    local excretion and seaweed diversity, 288
    uptake, *248*, 256
    uptake kinetics, *255*
    uptake vs. ammonium concentration, *257*
amyloplasts, 23, 25
    in rhizoid orientation, 90
amylose, 210
Antarctic
    Circumpolar Current, 118
    Desmarestia temperature optima, 306
    endophytes, 172
    environment, 117
    heat shock protein induction, 314
    Himantothallus ammonium uptake, 285
    Iridaea life history, 57
    Palmaria fatty acid composition, 314
    Q_{10} values, 301
    seaweed communities vs. global warming, 319
    seaweeds, temperature tolerance, 315
    Ulva susceptibility to UV, 343
antifouling compounds, 399
    booster biocides, 399
    volatile halogenated organic compounds, 143
antioxidant metabolism, **310**
    water-water cycle, *311*
Antithamnion
    cell growth, 32, *33*
apical development
    Gracilariopsis, *13*
    Sphacelaria, *86*
apparent free space, 243
aragonite, 268–9
Arctic
    Laminaria nitrogen uptake, 285
    seaweed communities vs. global warming, 319
    seaweeds, temperature tolerance, 315

Arctic seaweeds, 118
Ascophyllum, 124
    canopy facilitation of invertebrate communities, *137*
    competition with mussels, 154, *155*
    epiphytism by *Vertebrata*, 142
    facilitation along stress gradient, 137
    ice disturbance, 154
    inducible defense against grazers, 167
    phlorotannin content vs. grazing, 167
    phlorotannin induction, *169*
    pollution impact on photosynthesis and productivity, 410
    symbiosis with fungus *Mycophycias*, 171
assimilation
    kinetics, 261
attachment, **78**
    adhesive substances, 79
    in morphogenesis, 90
    ultrastructural changes, 78
auxins, 92
    role in *Ectcarpus* development, *93*

bacteria
    as morphogenetic factor, 91
basal bodies, 27
bicarbonate utilization, 203
bioaccumulation
    effects of environmental factors, 382
    of metals, 380
biomarkers, **230**
biomass budget, *235*
biotic interactions, **136**
    facilitation, 137
blue light, 89
    in morphogenesis, 59
    non-photosynthetic effects, *89*
Bonnemaisonia
    predicted tetrasporogenesis, *67*
    reproductive timing, 65
    tetrasporangium formation, *67*
boring and endolithic algae
    vs. grazers, 157
brackish waters
    seaweed flora, 330
branching, 88
    effects of light, 88
Bryopsis
    morphology, *4*
    regeneration, *96*
    wound healing, *95*

cadmium
    concentrations along estuary, *389*
    toxicity, 387
calcification, 268
    and ocean acidification, 273
    and pH stress, 341
    antagonism by orthophosphate, 272
    aragonite deposition, 269
    calcite deposition, 269
    Corallinales, 271
    disadvantages, 273
    factors, 271
    ion fluxes in *Halimeda*, *270*
    mechanisms, 271
    models, Corallinaceae, *272*
    sites and forms, *270*
    vs. CO_2 concentration, *340*
calcite, 268-9
    vs. CO_2 concentration, *340*
calcium
    cellular functions, 273
calcium carbonate
    crystalline forms, 268
callus
    in mariculture, 436
calmodulin, 273
Calvin cycle, 179, *180*, 206
carbon budgets
    autecological models, **227**
carbon concentrating mechanisms, 204
    $\delta^{13}C$, 209
carbon dioxide uptake, 204
carbon fixation, **202**
carbonate equilibrium, *202*
Carbon-Nutrient Balance Model, 166
carotenoids, 189
    structure, *190*
carrageenans, 213
    commercial production, **426**
    gel strengths, **427**
    production, uses, future prospects, **427**
    structure, *214*
Caulerpa
    morphology, *4*
    pre-emptive competition, 123
    regeneration, 97
    wound healing, 95, *98*
cell differentiation, **85**
cell division, **32**. *See also* cytokinesis.
    asymmetric, fucoids, *84*

cell fusion, *97*
cell growth, **32**, *33*
cell repair, *97*
cell structure, **17**
    basal bodies, 27
    eyespots, 30
    secondary metabolite bodies, 17
    walls, **17**, *20*
cell volume
    and osmotic control, **326**
    changes with salinity, 328
cell wall
    localized specialization, 21
cellulose microfibrils, 19, *20*
centrioles, 27, *29*, 35
Chaetomorpha
    nutrient competition with *Ulva*, 279
chemical composition and nutrient limitation, **280**
chemical defenses
    activated, 167
    algal compounds used by grazers, 170
    constitutive, 167
    costs, 169
    evolution/strategies, **166**
    geographic variation, 169
    inducible, 167
chemical ecology, **164**
chlorophylls, 188
    reaction centers, 191
    structure, *190*
Chondrus
    defense against *Acrochaete*, 145
    genome, 42
    sloughing epiphytes, *142*
    temperature effects on morphology, 90
    versus *Ulva* and grazers, 152
chromatic adaptation, **200**, 202
circadian (diel) rhythms
    in photosynthetic rate, 222
circadian clock, 35
circadian gating, 35
coastal waters
    light spectra with depth, 184
    subtidal light, 182
    temperatures and salinities, **295**
Codium
    dispersal as invasive species, 123
    filamentous vs. spongy thalli, 88
    growth vs. internal nitrogen, 277
    morphology, *4*

pre-emptive competition, 139
communities, 103
community analysis, **124**
community food web interactions, *286*
community interactions
  with nutrient enrichment, **286**
community structure
  impact of grazing, 148
  "top-down" versus "bottom-up" control, 155
  under nutrient enrichment, 287
compatible solutes, 327
competition, **138**
  between corals and seaweeds, 145
  mechanisms, *138*
concentration boundary layer, 243, 353–4, *354*
  and kelp blade morphology, 357
  and nutrient fluxes, 353
  synchronization of propagule release, 359
construction. *See* thallus construction.
copper
  cellular effects, 385
  effect on physiology, *387*
  effect on relative growth response, *377*
  effects on photosynthesis, 385
  effects on reproduction and growth, 386
  in antifouling paints, 399
  in seaweed food chains, 389
  seawater concentrations, 386
  toxicity, 385
coral reefs, *115*
  algal turf communities, 114
  calcifying seaweeds, 114
  eutrophication and macroalgal abundance, **404**
  eutrophication, 403
  foliose seaweeds, 115
  recovery after sewage diversion, 402
  seaweed communities, 113
  seaweed growth after urchin die-off, 149
Corallinales
  calcification, 269, 271
corps en cerise, 17, 143
critical tide levels, *101*, 104
cultivation mutualism, 159
currents, **349**
  effects of coastal topography, *350*
cuticle, 20
cyanelles, *193*
cytokinesis, *34*, 85
  brown seaweeds, 33
cytokinins, 92

cytoplasmic organelles, **21**
cytoskeleton, **25**

Dasycladus
  wound healing, 95
day length, 65. *See also* photoperiodism.
  and irradiance, latitudinal variation, *58*
  interacting factors, 65
  Scytosiphon erect thallus formation, *66*
Delisea
  allelopathic compounds, 143
demography, 128, 146
  frond dynamics in clonal seaweeds, 147
  Gini-coefficient (G'), 146
depth refugia
  from warm water, 186
desiccation
  amino acid leakage, *335*
  effect on photochemical yield, *336*
  effect on quantum yield, *336*
  inorganic carbon supply, 334
  recovery, 334
  stress, **333**
Desmarestia
  Antarctic, temperature requirements, *317*
  sulfuric acid content, 170
development of adult form, **88**
diatoms
  epiphytism, 141
  in farmer-fish territories, 159
Dictyosphaeria
  bloom in sewage eutrophication, 402
diffusion boundary layer, 353. *See also* concentration.
      boundary layer
dispersants, oil
  toxicity, 398
disturbance
  and community changes, 367
  by water motion, 366
  categories, **366**
  Dynamic Equilibrium Model, 368
  ice scour, 154
  Intermediate Disturbance Hypothesis, 368
  sand scour, 370
  seasonal burial, 370
  sediment, 369
  vs. stress, 132
DNA Barcoding, 37, *38*, 39
drag
  forces, 360

drag (cont.)
   forces vs. morphology, 362
   reduction in aggregates, 361

Ecklonia
   and disturbance, 367
   geographic variation, 9
   productivity, contribution to coastal ecosystems,
      231
ecosystem models, 235
ecotypic variation, 9
Ectocarpus
   and multicellularity, 5
   as model seaweeds, 41
   auxins in sporophyte development, *93*
   genome, 41
   life history, *52*
   model life history, 53
   morphological development, 85
effective shore level, 104
El Niño/Southern Oscillation (ENSO), **319**
   kelp canopy area, *228*
emersion
   component factors, 7
endophytes, **172**
   grazing defense, 173
endosymbiosis
   origin of plastids, 3
   plastids and mitochondria, 21
environmental stressors, **294**
   interactions, 343
   intertidal zone, **297**
   temperature, **300**
environmental-factor interactions, **5**
   between physico-chemical and biological factors, 7
   sequential effects, 7
epibionts. See also epiphytes.
   and allelopathy, **140**
epiphytes
   impact on basiphytes, 140–1
   obligate, 141
   removal by sloughing, *142*
essential elements, **238**
   concentrations in seawater and seaweeds, *240*
   functions and compounds, *239*
establishment and germination, **81**
estuaries and bays
   temperatures and salinities, **296**
ethylene, 94

Eucheuma/Kappaphycus
   cultivation, **426**
   use for carrageenans, **426**
eutrophication, **400**
   changes in communities, *401*
   effects on communities, **286**, 400
   non-sewage nutrient sources, **407**
exploitative competition, 138, 145
   laboratory studies, 147
exudation, 225
eyespot (stigma), 30, *31*

facilitation, **136**, 136–7, *137*
farmerfish, *159*, 159
fatty acids
   as biomarkers in food webs, 232
feeding deterrents
   calcified algae, 162
fertilization, **68**
   post-fertilization events, red algae, 72
   red algal mechanisms, 72
   spermatium ultrastructure, *74*
   strategies, 68
flagella, **27**
   in attachment process, 79
   in gamete recognition, 71
   roles, 30
   synthesis, 27, *31*
floridean starch, 210
fluorescence
   ratio of variable to maximal, as stress indicator,
      346
foundation species, **136**
freezing, **336**
   cryoprotectants, 337
   tolerance vs. vertical zonation, *337*
fucoidan
   in propagule adhesion, 79
fucoids
   asymmetric cell division, 84
   phosphorus uptake, zonation, season, 265
Fucus
   competition in nutrient-rich mesocosms, 287
   demography, 146
   desiccation, 336
   desiccation and recovery, 336
   desiccation tolerance, 335
   development, *10*
   effect of DIC and water flow on gamete release, *360*

fertilization, *71*
freezing tolerance, 309
growth vs. salinity, 331
hybridization, 44
in salt marshes, 120
intertidal zonation factors, 106
net production in water vs. air, *205*
nitrogen uptake during desiccation, *251*
nitrogen uptake, germlings, 253
plastid ultrastructure, *24*
recovery after oil spill cleanup, 397
recovery from desiccation stress, 334
sperm surface carbohydrates, 21
stress tolerance, 311
survival vs. density, *147*
thallus construction, 15
wound healing, 98
functional form
    absorptance spectra, *195*
    and opportunistic vs. late-successional seaweeds,
        132
    grazing susceptibility, 160
    groups, *14*, *114*
    growth rate–light relationships, *200*
    in light trapping, **194**
    light-harvesting ability, 194
    models, **11**
    monolayered vs. multilayered, 195
    photosynthesis, *223*
    resource acquisition, water motion, **356**
    SA : V ratios, micro- vs. macroalgae, *254*

gamete
    approach, brown algae, *70*
    recognition, 70–1
    release, 68
    release, effect of DIC and water flow, *360*
gametogenesis, *63*
Gelidium
    agar production and products, **428**
    productivity model, 227
    tetraspore development, *82*
    thallus construction, 15
genetics
    breeding experiments, 42
genomics, 41
geographic distribution
    amphi-equatorial seaweeds, 316, *318*
    and temperature, **321**

effect of currents, 321
poleward shifts in Australia, *323*
germination
    after UV irradiation, *329*
    cell division patterns, 83
global warming
    poleward shifts in Australia, *323*
glycolysis, *211*
Gracilaria
    agar production and products, 429
    C : N ratios, 281
    clonal thalli, *16*
    genetic inheritance, 42
    growth rate vs. tissue C : N ratio, *281*
    mariculture, 429
    mitotic recombination, *44*
    nitrogen storage, 263
    nitrogen uptake and desiccation, 251
    non-Mendelian inheritance, *43*
    oxidative burst against pathogens, 145
    spore coalescence, *17*
    unstable mutants, 43
Gracilariopsis
    apical development, *13*
    developmental control by parasitic red algae, 173
    effect of copper on growth, *377*
    effect of copper on physiology, *387*
Grateloupia
    role of polyamines, 93
grazing, **148**. *See also* mesograzers.
    amphipods, *160*, 160
    as morphogenetic factor, 91
    associational escapes, **158**, 170
    constitutive defenses, 167
    feeding choice assays, 161
    feeding preferences, 160
    generalists vs. specialists, 163
    impact on community structure and zonation, **148**
    impact on seaweed individuals, **156**
    impact on zonation, 153
    influence on seaweed competition, 152
    interactions with competition, wave action, 154
    limpets, latitudinal changes, 153
    refuges, 157, *158*
    seaweed defenses, 157
    structural and chemical deterrents, **160**
    susceptibility vs. functional group, *162*
    temporal and spatial escapes, **157**
    urchin barrens, 148

growth
    relation to nutrients, 276
    vs. internal nitrogen, *277*
growth kinetics, **276**
    and tissue nutrients, **276**
    measurement, **276**
growth rate
    nitrogen and light, **284**
    vs. nitrogen, *277*
    vs. tissue C : N ratio, *281*
growth substances, **91**

hairs
    and concentration boundary layer, 354, 357
    in nutrient acquisition, 90
    in nutrient uptake, 280
Halimeda
    activated chemical defense, 167
    aragonite crystals, *269*
    calcification, 268-9
    calcification schematic, *270*
    carbonate production, 114
    diel plastid migration, 25
    herbivore defenses, 170
    life history, *51*
    morphology, *4*
    plastid migration, *26*
Harveyella
    evolution and range extension, 173
heat shock proteins, 314, 345
herbivores. *See also* grazing.
    and nutrient impacts on seaweeds, 288
    community interations with nutrient enrichment, 287
    effect on kelp nitrate uptake, *288*
Hijikia
    arsenic content in hijiki, 390
Hincksia
    amino acid uptake, 258
horizontal gene transfer, 43
    alginate biosynthesis, 212
hybridization, *37*, 44
hydrodynamic force, 349
hydrodynamic forces
    component factors, 6
    destructiveness, 352
    field measurements, *361*
    on *Nereocystis* blades, *365*
    seaweed reconfiguration, 361
hyperparasitism, 173

inorganic carbon
    forms in seawater vs. pH and salinity, *203*
    sources and uptake, **202**
    transport and CO_2 accumulation, *208*
Integrated Multi-Trophic Aquaculture, **431**
    concept map, *433*
interference competition, **138**, 138. *See also* farmerfish.
    "gardening", 140
    whiplash, 139
intergeneric hybrids, 37
intertidal seaweed community, *106*
intertidal zonation, **100**, **101**, *102*
    biological factors, 105
    controlling factors, **104**
    effects of grazers, 153
    effects of grazing and competition, 153
    modeling communities, 111
    submersion-emersion data, *105*
    year-to-year changes, *103*
intertidal zone
    salinities, 298
    temperatures and salinities, 297
intraspecific competition, 146
invasive seaweeds, **121**, *122*
    biotic characteristics, 122
    ecological impacts, 123
iodine
    accumulation by kelp, 275
iridescence, 20
irradiance, **179**, 181
    compensation levels, 198
    measurement, **179**
    saturation levels, 198
    spectra vs. depth, *187*
    transmission, water types, *185*
    underwater heterogeneity, *183*

Kappaphycus. *See also Eucheuma/Kappaphycus*.
kelp
    Arctic, vs. UVB radiation, 339
    blade development and water motion, 359
    blade morphology and light harvesting, 358
    canopy vs. ENSO, *228*
    community recovery after ENSO, 320
    deep-water refuge model, *187*
    gametophytes endophytism, 172
    impact of global warming, 322, *324*
    productivity flow to other ecosystems, *232*
    productivity in near-shore, *234*

productivity, geologic past, *228*
sporophyte growth vs. temperature, *307*
subtidal, and surge, 365
temperature optima for growth, 306
tropical deep-water populations, 186
urchin grazing, 148
urchins and sea otters, 148
kleptoplasty, **173**
kombu mariculture. *See Saccharina.*

laboratory culture versus field experiments, **8**
laminaran, 210
Laminaria. See also Saccharina
   geographic shifts under global warming, 322
   life-history control, *61*
   nitrogen depletion, 285
   photoperiodism, 62
   seasonal changes in enzymes, 300
   thallus construction, 15
   translocation, 218
Laminariales
   carbon translocation, 215
   phlorotannin protection of spores, 226
laminarian life history, *55*
Lessonia
   whiplash interference, 139
Lessoniopsis
   phylogeny, 37
Liebig's law of the minimum, 240
life histories, **48**
   and grazing, 158
   environmental factors, **54**, *61*
   growth vs. reproduction, 54
   latitudinal variation, 65
   light photoperiod and wavelength, **58**
   lunar rhythms, 62
   ploidy levels, 50
   seasonal anticipators and responders, **55**
   short-day algae, *60*
   sporulation inhibitors, 62
   temperature, **56**
   temperature effects, 308
   types, 48
   variations, 52
light. *See also* irradiance.
   absorption spectra, *195*, *196*
   absorption spectra vs. solar spectrum, *181*
   and photosynthesis, **176**
   annual total irradiance, 186

changing radiation climate, **186**
component factors, 6
growth limitation, 284
intertidal and in estuaries, 182
light harvesting, **188**
light-harvesting complexes, 192
light-saturation curves, *197*
limits to growth, **185**
penetration into ocean water, **182**
photon flux density (PFD), 181
photosynthetically active radiation (PAR), 179
photosystems I, II, 191
physical properties, absorption, 176
P-I curves, *197*, *224*
spectral changes with depth, **184**
underwater heterogeneity, **182**
water surface effects, **184**
water types, 184
light-independent carbon fixation, 179, **207**, *211*
long-distance transport. *See* translocation.
low molecular weight carbohydrates, 206, *207*
   in osmoregulation, 328
lunar and tidal cycles, 66

Macrocystis
   as foundation species, 136
   blade morphology and boundary layers, 357
   carbon translocation, 216, 275
   forces from wind waves, 352
   geographic variation, 9
   kelp forest, *116*
   nitrogen composition, 282
   nitrogen-uptake rates, *250*
   ontogenetic gradients in photosynthesis, 222
   productivity models, 229
   recovery from ENSO, 320
   urchin impacts, 149, *150*
   variation in photosynthesis, *225*
magnesium
   cellular functions, 273
mannitol
   in osmoregulation, 327
mariculture, **413**
   Eucheuma/Kappaphycus, **426**
   for biomitigation, **431**
   global production, *414*
   Gracilaria, 429
   Integrated Multi-Trophic Aquaculture, **431**, *433*
   kelp cultivation, *423*

mariculture (cont.)

offshore/open ocean, **431**

Pyropia methods, *418*

Saccharina, **422**

tank cultivation, **430**

Undaria, **425**

mercury

toxicity, 385

mesochiton, 68

role in attachment, 79

mesograzers, 164. *See also* grazing.

control by predators, 152

feeding behavior with dictyol E, *168*

impact on seaweeds, 152

meta-ecosystem

dynamics, **107**

ecology, 110

food webs, *109*

metals, **378**

accumulation by seaweeds, 380

adsorption, uptake, accumulation, and biomonitors, **380**

associated with particulates, 382

detoxification, 384

ecological aspects, **389**

effects on algal metabolism, **384**

essential elements and performance, *378*

exclusion of ions, 383

factors affecting toxicity, **388**

relative toxicity, 385

sources and forms, **378**

tissue concentration factors, fucoids, *381*

tolerance mechanisms, *383*

toxicity tolerance mechanisms, **383**

trace elements baseline concentrations, *381*

trace metal concentrations and species in open ocean, *379*

Michaelis–Menten equation, 244

mitochondria

genome size, 21

inheritance, *22*, 22

mitosis

diurnal rhythm, 35

mitotic recombination, 42, *44*

molecular biology of seaweeds, **35, 36**

morphogenesis, **48**

blue light, 59, 89

effects of nutrients on *Petalonia*, 90

grazing effect on *Padina*, 91

red light, 59, 89

rhodomorphins, 97

temperature and daylength, *57*

thallus, **83**

morphological plasticity

Padina, 92

mucilage

in attachment, 79

in spermatia strands, 72

multicellularity

and algal evolution, 5

and alternatives for use of water column, 4

and tissue differentiation, 5

cell-cell interactions, 5

decreased SA : V ratio and nutrient uptake, 280

mycosporine-like amino acids

Antarctic red algal vs. photosynthesis, *338*

in UVB protection, 338

Nereocystis

and water motion, 364

effect of water motion on blade morphology, *358*

hydrodynamic forces, *365*

life history, *64*

nitrate

storage in vacuole, 261

uptake, 256

uptake kinetics, *255*

nitrate reductase, 258, 260

nitric oxide

in wound healing, 95

nitrite

uptake, 256

nitrogen. *See also* ammonium, nitrate, nitrite.

assimilation, 258

cycle, marine, *242*

incorporation, **263**

isotope ratios detect pollution sources, 407

limitation, 281

sources, 241

storage, 263

tissue concentrations, *278*

uptake, **256**

uptake and assimilation pathways, *262*

uptake by Antarctic kelp, *285*

uptake kinetic parameters, 258, *259*

uptake rate measurements, 246

non-Mendelian inheritance

somatic mutations, 42, *43*

nori. *See Pyropia*.

nuclear genome, 45

nucleocytoplasmic interactions, **45**

nutrient enrichment. *See* eutrophication.
nutrient requirements
    essential elements, **238**
    limiting nutrients, **240**
    vitamins, **239**
nutrient uptake
    active transport, **244**
    biological factors, **252**
    facilitated diffusion, **243**
    ion entry, **242**
    kinetics, **244**
    methods, *247*
    passive transport, **243**
    rate measurements, **246**
    variation over time, 247
nutrient-uptake rates, *245*
    chemical factors, **252**
    intrathallus variation, 253
    phytoplankton vs. macroalgae, 254
    "safety factor", 254
    vs. age classes, 252
    vs. desiccation, 251
    vs. light, **249**
    vs. surface-area
      volume ratio, 252–3
    vs. temperature, 250
    vs. water motion, 251
nutrients, **238**. *See also* specific elements, nutrient uptake,
    nutrient uptake rates, nutrient requirements.
    availability in seawater, **240**
    calcium and magnesium, **267**
    component factors, 6
    critical tissue concentrations, 277
    effect of supply, **279**
    functional-form effects, **279**
    in morphogenesis, 90
    iron, **274**
    movement to and through the membrane, **243**
    phosphorus, **264**
    relation to growth, 276
    storage and availability, **282**
    sulfur, **274**
    trace elements, **274**
    uptake, assimilation, incorporation, and metabolic roles,
      255

ocean acidification, **301**
    and calcification, 273
    community impacts, **340**
    inorganic carbon and calcification, *301*

    taxonomic variation of effects, *344*
oil, **390**
    crude oil components, 390
    diesel, effect on seaweed growth, 395
    dispersants in spill cleanup, 398
    disruption of cellular metabolism, 395
    effects on metabolism, life histories, communities,
      394
    effects on sexual reproduction, 395
    environmental variables in situ, **396**
    factors affecting spill, *393*
    fate of spills in the ocean, **392**
    hydrocarbon groups, *391*
    microbial degradation, 393
    pathways for environmental fate, *393*
    slick weathering, *392*
    spill and cleanup impacts on habitats, *396*
opportunistic vs. late-successional seaweeds
    characteristics, *133*
    costs and benefits, *134*
Optimal Defense Theory, 166
osmoregulation, 328
    cell volume, **326**
osmotic potential
    component factors, 6
oxidative burst, 145
oxidative stress, 295, **343**
ozone depletion, 337

Padina
    grazing and morphogenesis, 91, 92
parasitic red algae, 172–3
parthenogenesis, 53
patchiness, 125, *126*
pesticides, **398**
Petalonia
    nutrients in morphogenesis, 90
pH. *See also* ocean acidification.
    Desmarestia sulfuric acid content, 170
    fluctuations at thallus surface, *342*
phenotypic plasticity, 5, 88
    temperature tolerance, 306
pheromones, 68
    brown algal, *69*
phlorotannins, 143, 165, *169*
    UV protection, 226, 345
phosphorus
    assimilation in tropical seaweeds, 281
    availability, 264
    critical and maximum tissue concentrations, *283*

phosphorus (cont.)
  limitation, 266
  limitation in temperate seaweeds, 282
  phosphate organic sources, 266
  sources, 242
  tissue concentrations, *278*
  translocation in kelps and fucoids, 275
  uptake and assimilation, *264*
  uptake kinetics, 265
  uptake variation with shore position and season, 265
  uptake, assimilation and storage, 265
photoinhibition, 199
  by UV radiation, 199
  dynamic, mechanisms, 199
  from desiccation, 334
  temperature dependent, 313
photon growth yield (PGY), 200
photoperiodism, 59, 61, 64, 88
  life histories, *60*
  predicted tetrasporogenesis, *67*
photoreceptor proteins, 30
photorespiration, 207, 209
photosynthesis. *See also* photoinhibition
  and functional form, 222
  and salinity change, 330
  Antarctic red algae vs MAA concentrations, *338*
  Calvin cycle, 179
  circadian rhythms, 222
  effects of spectral composition, 199
  electron transport rate, 220
  intrinsic variation, **222**
  light reactions, 176
  measurement, **218**
  overview, **176**
  pathways, **206**
  photosynthetic quotient (PQ), 219
  Photosystem II energy flow, *221*
  photosystems I, II, 178
  quantum yield, 220
  rates, 218
  reaction centers, 177
  rhythms, *226*
  vs. irradiance, **197**
  vs. temperature, light, *305*
  vs. temperature, salinity, *328*
  vs. water content, *335*
  Z scheme, *177*, 177
phototaxis, 76
phototropism, 89–90

phycobilins
  structure, *190*
phycobiliproteins, 191
phycobilisomes, 191, 193
physodes, 17, *18*
phytochromes, 59
phytohormones. *See* growth substances. *See also* abscisic acid, auxins, cytokinins, ethylene, polyamines.
pigments. *See also* carotenoids, chlorophylls, phycobilins.
  antenna size vs. irradiance, 199
  structures, *190*
Pilinia
  obligate epiphytism, 141
pit connections, 11
  secondary, in parasitism, 173
plasmids, 45
plasmolysis, 326
plastids, 23, *24*
  autonomy and kleptoplasty, 25
  diel migration, 224
  evolutionary origin, *2*, *3*
  genome maps, *39*
  inheritance, 22
  migration, 25, *26*, 170
  sequenced genomes, 21
  types, *189*
plastids, pigments, and pigment-protein complexes, **188**
polar seaweeds
  and global warming, 319
  seasonal changes in enzyme activities, 303
  temperature tolerance, **315**
polarity
  of zygotes, 81
pollution, **374**. *See also* metals, oil.
  biogeochemical processes, *375*
  community effects, 374
  component factors, 6
  effects at various levels of organization, *375*
  impact on individuals, 376
  mixing of freshwater inputs, *380*
  oil, **390**
  radioactivity, **408**
  seaweeds in standard tests, 376
  synthetic organic chemicals, **398**
  test organisms, 377
polyamines, 93
polyphosphates, 267

polysaccharides, **210**
  gel strength, 214
  storage, *213*
  storage polymers, **210**
  synthesis, **215**
  wall matrix, **211**, *213*
  wall matrix, red algae, *214*
polyspermy, 71
population dynamics, **128**, 146
  dispersal, 130
  mortality, 131
  propagule bank, 131
  reproductive effort, 129
  reproductive success, 130
  survivorship curves, *131*
Porphyra, See also Pyropia
Porphyra mariculture. *See Pyropia*, nori.
Postelsia
  and disturbance, 367
  interference competition with mussels, 140
productivity, **218**
  autecological models, **227**
  carbon losses, **225**
  ecological impact, **229**
  tissue loss, 227
  tracking through food webs, **230**
proteomics, 41
protoplasts
  fusion, 436
  regeneration, 436
Pterygophora
  stipe flexibility, 365
  washed ashore, *366*
pyrenoids, 23
Pyropia
  cell growth, 32
  cultivation, **417**
  cultivation methods, *418*
  culture problems, **419**
  diseases in cultivation, 419
  genetic selection for mariculture, 420
  life history, *64*, 416
  mariculture, 415
  mariculture production, 415
  mariculture, future trends, **420**
  nori production flow chart, *417*
  vegetative propagation, 420
Pyropia nereocystis
  life history, *64*

radioactivity, **408**
reactive oxygen species, 295, 310, 312, 345–6
  in cell polarity, 83
  in wound healing, 95
red algal parasites
  transfer into host cells, *174*
red light
  in morphogenesis, 59, 89
Redfield ratio, 280, 282
regeneration, *86*, 87, **94**, *97*, 98
  from sub-protoplasts, *96*
  polarity, 97
Relative Dominance Model, *114*
relative growth rate
  vs. nitrogen, *289*
reproduction, **48**, *See also* life histories, fertilization
  asexual vs. sexual, 48
  synchronicity, 65
respiration
  measurement, **218**
Reynolds number, 353
rhodomorphins, 97
RuBisCO (ribulose 1,5-bisphosphate carboxylase-oxygenase)
  properties, 206

Saccharina
  breeding, 424
  commercial cultivation, **422**
  commercial cultivation of gametophytes, 424
  commercial harvest for alginate, 423
  commercial utilization and future prospects, **423**
  copper effects on reproduction and growth, 386
  for food and alginates, 421
  growth and chemistry vs. light, nitrogen, *284*
  mitochondrial inheritance, *22*
  offshore cultivation test, 431
  photosynthesis vs. temperature, *305*
  seasonal changes in enzymes, 300
  seasonal growth vs. nitrogen, light, 284
  temperature dependence of primary photosynthetic reactions, 302
  temperature, photosynthesis acclimation, 304
salinity. *See also* osmoregulation.
  adaptations, Baltic Sea, 330
  biochemical and physiological effects, **325**
  component factors, 6, 325
  effect on quantum yield, *332*

salinity. (cont.)
    effects on photosynthesis and growth, **328**
    estuaries, 296
    growth of *Fucus*, *331*
    in tide pools, *299*
    measurement, 295
    natural ranges, **295**
    short-term responses to change, *327*
    tolerance and acclimation, **330**
    tolerance, calcium and bicarbonate, 330
    water potential, 325
salt marshes
    seaweed communities, 119
sampling, *125*
sampling units
    area sampled vs. number of taxa, *125*
sand
    scour and burial, *371*
Sargassum
    propagule attachment, 78
    settlement, *78*
Scytosiphon
    day length and life history, 65
    erect thallus formation, *66*
    life history and anatomy, *50*
    mitochondrial inheritance, *22*
    morphogenesis cues, *57*
sea otters and kelp productivity, 148
sea urchin grazing, 148
    and otters, Alaska, 148
    Macrocystis beds, 149
    tropics, 149
seagrass, 3
    tropical, 115
seawater pH. *See* ocean acidification.
seaweed communities, **113**
    deep water, **120**
    estuaries, **119**
    floating seaweeds, **120**
    kelp beds, 116
    polar regions, **117**
    salt marshes, 119
    temperate floras, **116**
    tide pools, **118**
    tropical habitats, **113**
    vs. CO_2 concentration, *343*
seaweeds
    allelopathy in competition with corals, 145
    as animal feeds, 434
    as biomonitors, 380
    as soil conditioners, fertilizer, 434
    as term, 1
    biomechanical properties, **360**
    biotechnology, **435**
    commercial production, *414*
    consumption history and production, 413
    cultivated, 415
    edible, production, 414
    genetics, **42**
    heavy metal accumulation, 389
    herbivore interactions, **156**
    in cosmetics, 435
    in standard pollution tests, 376
    interference competition with sessile animals, 140
    morphology and anatomy, **9**
    morphology vs. wave exposure, 363
    nutrient removal from wastes, 435
    phycocolloid production, *415*
    productivity in near-shore ecosystems, *234*
    tank cultivation, **430**
    tensile strength, 363
    thalli and cells, **1**
    uses, 414
    wall composition and tensile strength, 364
secondary metabolites, 164
sediment
    as a substratum, 372
    effects on rocky coast communities, *369*
    impact on settlement, 369
self-fertilization, 45
self-thinning, 146
settlement, **74**
    and boundary layer, 74
    motility, 75
    on bacterial biofilm, *77*
    propagules in fecal pellets, 76
    selection, 77
    sinking rates, 75
    strategies, 75
    substratum roughness, *78*
    surface properties, 77–8
    Ulva spores, *81*
    zoospore ultrastructure, *80*
sewage
    ammonium toxicity, **403**
    coral reef recovery, Hawaii, 402
    effect on communities, **400**
    effect on succession, 402
    nutrient uptake by seaweeds, 435
    outfall impacts, 401

shell-boring algae
  bicarbonate source, 204
sieve tubes
  ultrastructure, *217*
signaling pathways
  in cell repair, 95
Siphonocladales
  wound healing, 97
siphonous green seaweeds
  morphology, *4*, 4
somatic cell hybridization
  in mariculture, 436
somatic embryogenesis, 44
spatial scale and methodology, *127*
sperm limitation, 72
Sphacelaria
  apical development, *86*
  development, *10*
  morphological development, 85
sponges
  symbiosis with seaweeds, 171
Spongites
  response to limpet grazing, *163*, 163
sporulation inhibitors, 62, *63*
stable isotopes
  as biomarkers in food webs, 232
starch, 210
stratification in coastal waters, 296
stress, **294**
  breaking, red and brown seaweeds, *363*
  disruptive, 294
  factor interactions, *295*
  limitation, 294
  physiological indicators, **346**
  types, 294
  vs. disturbance, 132
stress response
  heat shock proteins, molecular chaparones,
    314
subtidal zonation, **111**, 113
  cold temperate, *112*
  functional-form groups, 113
succession, 128
  following disturbance, 367
  investigational approaches, *128*
sulfation, 83
  in attachment, 79
surface-area : volume ratio and morphology, **279**
surge
  and growth rates, *357*

symbiosis, **170**
  Agarum and bryozoans, 141
  fungi in seaweeds, 171
  mutualism, **171**
  red seaweeds and sponges, 171
  Spongites and limpets, 163
  whole community mutualism, 171
synchronization, gamete and spore release, **359**

temperature
  adaptation, heat shock proteins and molecular chaparones,
    314
  adaptation, membrane composition, 314-15
  adaptation, physiological mechanisms, **312**
  and geographic distribution, **321**
  and reproductive timing, 308
  as environmental stressor, **300**
  chemical reaction rates, 300
  community responses, 322
  critical, 321
  effects on metabolic rates, **303**
  freezing stress, 308, *313*
  global warming, 322
  growth optima, **304**
  high, thallus damage, 309
  in morphogenesis, 90
  in tide pools, *299*
  microclimates, *298*
  natural ranges, **295**
  optima for kelp growth, *307*
  requirements, Antarctic *Desmarestia*, *317*
  requirements, Antarctic red algae, *316*
  tolerance, **308**
  tolerance, and global warming, 319
  tolerance, phenotypic variation, 306
  tolerance, polar seaweeds, **315**
temperature-photosynthesis acclimation, 304
temperature-salinity diagram, 296, *297*
terpenoids, 165
tetrasporangium formation
  temperature and daylength, *67*
thallus construction, **9**
  clonal, 15, *16*
  filamentous, 11, *12*
  modular, 15
  parenchymatous, 9, *10*, *18*
  spore coalescence, 16
  unitary, 15
thermal pollution, **409**. *See also* temperature.
  impact on photosynthesis and productivity, *410*

thermal pollution (cont.)
  physical disorders, 409
thigmotropism, 90
thylakoid
  structure, cyanelles, *193*
thylakoids
  distribution of pigments and electron transport, *178*
tide pools
  seaweed communities, **118**
tides, **100**
  predicted submersion/emersion curves, *104*
  sea-level changes, *101*
tin
  antifouling compounds, 399
tissue nitrogen
  in competition and succession, 278
totipotency, 87
transcriptomics, 41
translocation, **275**
  long distance, carbon, **215**
  mechanism, 216
transposons, 43
trophic cascade, 149, *151*
turbidity
  and downward irradiance, 182
turgor pressure, 325
  regulation, 326

ultraviolet radiation, **337**
  effect on spore germination, *339*
  impact on organisms, 338
  tolerance by Arctic seaweeds, 315
  UVB, 186
  UVB, ozone hole, 337
Ulva
  allelopathic compounds, 144
  ammonium uptake, *248*
  bubble mutant, 87
  circadian gating in cell cycle, 35
  diel plastid migration, 224
  gametogenesis and gamete release, *63*
  in toxicant bioassays, 377
  life-cycle regulation, 62
  nitrogen starvation, *252*
  nitrogen uptake, *249*
  nutrient competition with *Chaetomorpha*, 279
  phosphorus ecology, 265
  photoperiodism, 62
  settlement, *77*, 77

  settlement and adhesion, *81*
  ulvan, 212
  versus *Chondrus* and grazers, 152
  zoospore ultrastructure, *80*
ulvan, 212
Undaria
  cultivation, **425**
  dispersal as invasive species, 123
  food products and future trends, **425**
  photoperiodism, 62
  use as food, 424
uptake kinetics
  phosphorus, 264
urchins
  intertidal and flash flood, 7
urea
  assimilation, 263
  uptake, 258

vacuole
  nitrate storage, 261
vegetation analysis, **124**
  quantitative sampling, 124
  remote sensing, 128
  species diversity, 127
velocity boundary layer, 352, *354*
vitamins
  requirements, **239**

wakame mariculture. *See Undaria.*
wall polysaccharides
  as cation exchangers, 243
water motion, **349**
  acceleration, lift and impingement, 362
  and seaweed tensile strength, 363
  boundary layers, *354*
  drag, 360
  effect on morphology, *358*
  flow around *Macrocystis* blade, *356*
  flow over surfaces, **352**
  in giant kelp bed, *351*
  laboratory flumes, 356
  laminar flow and eddies, *353*
  laminar vs. turbulent flow, 352
  measuring, **355**
  resource acquisition, functional form, **356**
  tenacity, 362
  thallus reconfiguration, 361
  waves, *352*

water potential, **325**
    desiccation and freezing stress, **333**
    regulation, 326
water–water cycle, *310*
wave action
    and productivity, 365
    disturbance, **366**
        water velocity and acceleration, *352*
wave-exposure indices, 355
waves
    energies, 351

physical nature, **350**
    surge, 350
wound healing, **94**
    Bryopsis, 95
    in multicellular algae, 97
wound plug assembly
    biochemistry, 94

zonation. *See* intertidal zonation, subtidal zonation.
zygotes
    asymmetric cell division mechanism, *84*

Printed in the United States
by Baker & Taylor Publisher Services